# Observation of the Earth and its Environment

H.J. Kramer

## Survey of Missions and Sensors

Second Edition
with 170 Figures and 244 Tables

Springer-Verlag
Berlin Heidelberg New York
London Paris Tokyo
Hong Kong Barcelona Budapest

Dr. Herbert J. Kramer
DLR-DFD
German Aerospace Research Establishment
German Remote Sensing Data Center
82230 Oberpfaffenhofen
Germany

The first edition was published under the title "Earth Observation Remote Sensing".

ISBN 3-540-57858-7 2nd ed. Springer-Verlag Berlin Heidelberg New York
ISBN 0-387-57858-7 2nd ed. Springer-Verlag New York Berlin Heidelberg

ISBN 3-540-55579-X 1st ed. Springer-Verlag Berlin Heidelberg New York
ISBN 0-387-55579-X 1st ed. Springer-Verlag New York Berlin Heidelberg



Typesetting: Camera-ready by author
SPIN:10466606          61/3020-5 4 3 2 1 0 - Printed on acid -free paper

# Table of Contents

# List of Tables

# Introduction

The following listing represents a survey and a short description of 'Earth Observing Missions' in alphabetical order. The listing in Part A considers completed-, operational- as well as planned missions on an **international scale** (Earth observations from space know no national boundaries). A look into past activities is important for reasons of heritage, context and of perspective. The document is intended for all who want to keep track of missions and sensors in the fast-growing field of Earth observations. There cannot be any claim to completeness, although a considerable effort was made to collect and integrate all known missions and sensors into this book.

Earth observation by remote sensing changes our view and perception of the world. We begin to realize the global character of remote sensing, its multidimensional and complementary nature, its vast potential to many disciplines, its importance to mankind as a whole. Remote sensing permits for the first time in history a total system view of the Earth. The view from space toward Earth has brought about sweeping revisions in the Earth sciences, in particular in such fields as meteorology, oceanology, hydrology,  geology, geography, forestry, agriculture, geodynamics, solar-terrestrial interactions, and many others.

The field of Earth observation by remote sensing has different meanings and associations to different people as to what it is comprised of. Some consider for instance the measurement of solar energy fluxes as part of Solar physics while others regard the energy flux interaction with the Earth and its environment as a portion of the Earth sciences, hence, as part of remote sensing. The scope of Earth observation missions included in this survey considers all those missions and applications, which are related to one of the following discipline areas in the Earth sciences[1]. However, the survey itself does not classify the missions or sensors into any particular applications category, the potential uses of the data are ever increasing.

- Solar-Terrestrial Interactions
  [physical mechanisms controlling the transport of mass, momentum and energy in geospace]

- Dynamic Atmosphere
  [Atmospheric Chemistry, Atmospheric Physics, Meteorology, Hydrology, etc. ]

- Dynamic Oceans and Coastal Regions
  [ocean winds, circulation, wave heights, ocean color (productivity of the sunlit upper layer of the ocean), photosynthesis (conversion of inorganic matter into organic matter), phytoplankton, chlorophyll (marine food chain), fish populations, ocean pollution,]

- Solid Earth (lithosphere)
  [Geology, Geography (cartography,), Geodynamics (Earth's rotation, plate tectonics, crustal motions, glaciology, accurate point positioning, altimetry), Topography (structures, mapping), Resources, Geopotential fields (gravity field, geoid,), bathymetry, soil moisture, etc.]

- Biosphere
  [Atmosphere-Biosphere interactions and exchange processes, biomass, global plant cover, vegetation conditions, agriculture, forestry, snow cover, pollution, sediments, hydrology, flood observation,  water runoff, erosion, etc.]

- Earth Climate
  [Long-term climate effects, climate-related parameters, radiation budget, global energy balance, trace gases of the atmosphere, global warming effects, etc.]

---

[1]  J.H. McElroy, "Earthview-Remote Sensing of the Earth from Space", in 'Monitoring Earth's Ocean, Land, and Atmosphere from Space', Volume 97 Progress in Astronautics and Aeronautics, AIAA, 1985, pp. 3-44

While most sensors are inward looking, i.e. from the satellite orbit toward the Earth, there are also sensor applications that are outward-looking, in particular with regard to Solar-terrestrial interactions (energy transport, etc.). Some series of meteorological satellites carry such outward-pointing sensors, hence they are included in this survey (polar-orbiting and geostationary). The benefits of weather satellites do not end with observing clouds. The GPS (Global Positioning System) and Glonass Navigation- Satellite-Systems are listed for other reasons; they are regarded as general service providers by supporting directly or indirectly precision orbit determination (measurement) for many applications, they are in addition basic research tools in many fields, in particular in the area of Solid Earth research. The knowledge of accurate navigation parameters (position, velocity, time) is in fact an implicit essential in all Earth observation endeavors. The breadth of emerging GPS applications in the commercial and research fields demonstrate the universal nature of navigation for mankind.

A typical mission description of this survey has generally the following elements (if available) for layout: Mission definition, objectives, planned launch, areas of application, orbit parameters, sensor short descriptions (spectral ranges, resolutions, swath width, etc.), and data (rates, types, availability, onboard storage, etc.). In addition a good many drawings and graphs are presented (whenever available) featuring satellite or sensor models or illustrating particular observation configurations/concepts. There are also summaries of major programs and summary charts of all missions.

The occasional reader will be confronted with a lot of new acronyms and terminology in addition to all the concepts and methods. Many sensors turn up in several missions on a truly international satellite scenario (sometimes sensors may have different names (in their planning phases) by different agencies, or sensors are renamed to reflect better their observation goals). A glimpse of the complexity of sensor observation terminology may be demonstrated simply by the example of different possible measurement arrangements-configurations-geometries-techniques, etc. such as:

- Measurement frequencies (spectral ranges, single channel or multichannel device, etc.)
- Measurement coverage (repeat coverage, temporal scales, permanent coverage, )
- Measurement resolution (spatial resolution, radiometric resolution, spectral resolution)
- Measurement target (Earth surface, atmospheric layer(s), local, remote, global, column, etc.)
- Measurement pointing (nadir−looking, limb−looking, side−looking, forward/backward−looking, view angles, )
- Measurement calibration (standard, wavelength, stability, device calibration, auxiliary calibration station, data calibration, etc. )
- Measurement technique (active device, passive device, )
- Measurement (source) data rate
- Measurement of events (volcano eruption, floods, etc.)
- etc.

One service objective of Earth observation is the provision, utilization and exploitation of all the data products from the various missions. It turns out, that a lot of user data of flown and operational missions are still available in many archives around the world. In general, archives outlive missions by a substantial amount of time.

It is very difficult to present the right information in an overview due to the great variety of reader interests on specific topics and the variety of reader backgrounds. Researchers may be more interested in sensors and their measurement capabilities or their uses on particular missions, or in the availability and interpretation of science data, while data (service) people view everything from the perspective of data handling or the problems of communications, storage and access. An overview cannot satisfy all requirements for detail, but it can give to the reader a perspective and an awareness of the major issues involved in Earth observation. Portions of the overview may be totally irrelevant to some, while informative and relevant on the whole for others.

The survey demonstrates also the multitude of opportunities given through cooperation and participation in programs or activities or developments on a world-wide scale. The sharing of the data for analysis should be a prime objective. The data access and pricing policies of many organizations are vital issues for the general user community.

The status of sensor development or of entire missions or programs may range from 'intentional' to 'definitive'. Everything depends on long-range and continuous Government funding policies, everywhere. Change is the essence of life, this proverb does also apply for Earth observation.

This document contains an **index table (Appendix C)** listing all sensors and their (multiple) placements in the document. This should enable anyone to find information as a quick reference. Part B is intended for those who are wrestling with definitions or who simply want to get a better understanding of some basic principles.

Throughout the writing period of the book, there occurred a marked shift in the Earth Observation policies of practically all governments and space agencies of the world, namely the trend away from large observation platforms, with virtually all-round capabilities, to smaller spacecraft with very specific objectives. This can definitely be regarded as a sensible and pragmatic approach for all parties concerned (governments, science communities, service communities, etc.) from a standpoint of project manageability in all its multi-faceted aspects. The benefits of smaller projects lie in the far greater flexibility, accountability, and responsiveness of a program to serve specific needs in time periods, that can focus and hold human dedication and expectations on the problem to be solved (much shorter preparation times, as well as sensible observation times to account for technology advances; in addition the degrees of freedom on all levels increase by some orders of magnitude).

The scope of the survey is focussed on the space segment of Earth observation missions. The reader should realize, however, that there is also a ground segment, mostly of considerable size and expense, to every mission. For the purpose of context, there are many short descriptions or identifications of portions of ground segments included, whenever they benefit the general understanding, or if they are of such general utility to the user community, as for instance the NOAA meteorological missions.

The scope of the text is further limited to spaceborne missions and sensors as opposed to airborne remote sensing by aircraft, also referred to as 'aerial surveys'. Both fields of spaceborne and airborne surveys have their valid applications. For instance, high-quality topographic imaging (for reliable small-scale cartographic maps, 1:50 000 or even smaller) is still the domain of airial surveys. While the nature of observation capability is rather local, resolutions on airborne sensors are at least an order of magnitude better than those of spaceborne data due to the proximity of the sensor to the target (usually the Earth surface). Technology will eventually come to grips with these problems to handle immense data rates (along with onboard processing and data compression techniques) of wide-swath-very-high-resolution imaging data so that spaceborne surveys of the future offer data of almost the same spatial resolution qualities that are nowadays achieved in airborne applications.


**Acknowledgements:**

I wish to express my thanks to all the individuals from many organizations (in particular space agencies), who have directly or indirectly contributed to this volume through discussions and by providing papers or viewgraphs, or by simply reviewing previous updates with regard to correctness and/or sufficient degree of completeness. These individuals are:

B. H. Needham and H. Kroehl (NOAA), R. G. Kirk (GSFC), K. R. S. Murthi (ISRO), G. Esenwein, H. L. Theis and G. Wilson (NASA HQ), Ch. Elachi and L. L. Fu (JPL), E. W. Bergamini (INPE), Y. V. Trifonov (VNIIEM), B. S. Zhukov and V. Tarnopolsky (IKI), N. A. Armand and B. Kutuza (IRE), A. S. Selivanov and M. V. Novikov (Russian Institute of Space Device Engineering), V. Panchenko and W. N. Voronkov (NPO Energia), K. Maeda (NASDA), C. J. Readings (ESA HQ), B. Battrick and K. Pseiner (ESA-ESTEC), N. Denyer (CSA), M. L. Chanin (CNRS), D. P. Murtagh (University of Stockholm).

Comments on the subject matter or remarks on obvious errors or on program changes are always welcome (Phone: 49-8153-282604; Fax: 49-8153-281843). All writing and illustration work was done with INTERLEAF on a Sun workstation at DLR-DFD in Oberpfaffenhofen, Germany. Special thanks are due to Friedrich L. Porsch, who drew expertly virtually all of the spacecraft illustrations with his computer. I would also like to acknowledge the help I received from R. Bamler, F. Heel, F. Jochim, W. Kirchhof, F. Lanzl, W. Markwitz, G. Mayer, D. Oertel, H. Öttl, H. van der Piepen, W. Schneider, G. Schreier, P. Seige, Ch. Werner, R. Winter, G. Zimmermann (all of DLR), F. Ackermann (University of Stuttgart), K. Schmidt (DARA), and Ch. Reigber (DGFI) for their discussions, text reviews and their valued suggestions. It is planned, that the document will be updated in the future to serve as a reference for many people with vested interests in the field of Earth observation.


Oberpfaffenhofen, June 1992



Herbert J. Kramer

## Foreword to the 2. Edition

Since the publication of "Earth Observation Remote Sensing - Survey of Missions and Sensors" in August of 1992, there have been many changes and additions in the Earth observation programs of virtually every space agency. This by itself warrants a new addition of the book to present the readership with valid and up-to-date information. In addition there have been many improvements with regard to the scope of the book.

1.  Many new and old missions have been added to the text in the field of Solar-Terrestrial interactions and energy transport. Although this seems to be the prime domain of solar physicists, it is felt, that the energy transport from the Sun to the Earth is of prime importance to our Earthly environment. The Earth cannot be considered as an isolated system - Sun and Earth must be monitored together as a system.

2.  The 1. edition of the book considered only **spaceborne** missions and sensors. Now the scope has been extended to include also a survey of **airborne** sensors. I was approached by some readers who suggested to me, to include also airborne sensors, to broaden the spectrum of applications. Well, the new dimension in scope meant also a lot of additional work for me just to keep up with the flood of new material.

I am quite aware of the limitations of my 'survey of airborne sensors', in particular in view of the untold numbers of airborne sensors, along with their ever-changing configurations, that have been flown, and are planned to be to be flown, by so many different institutions around the world. However, after I added the first two dozen airborne sensors to the text, I realized how valuable and complementary the information is for a better overall understanding of the subject matter. Space programs with very sophisticated sensor technology require simply complementary airborne programs for experimentation and validation of the new technology. It really makes sense to be aware of the activities and programs that take place on all levels.

With so many dates compiled in this survey, a word should be said with regard to datum conventions. Most of the world is stating a datum in the sequence: "Day, Month, Year", the USA convention is "Month, Day, Year". Whenever the latter convention was used (such as in NASA or NOAA programs/tables), I have normally letter- abbreviated the month (i.e. Jan. 15, 1994, instead of 1. 15. 1994) in order to come up with the right answer.

The upkeep of Part C overview/summary tables of various sensor types is increasingly becoming a problematic chore; this applies for the great differences in instrument capabilities and applications that are difficult to classify, but it is also due to the very limited information that is normally available to me to specify, for instance, a particular antenna type for each microwave radiometer, or to come up with the peak power parameter of each SAR instrument.

The title of the book has been changed from: "Earth Observation Remote Sensing - Survey of Missions and Sensors" to: "Observation of the Earth and its Environment - Survey of Missions and Sensors " to reflect the scope and the major changes. The book layout contains three major parts along with 3 appendices.

- Part A    Survey of Spaceborne Missions and Sensors
- Part B    Survey of Airborne Sensors
- Part C    Reference Data and Definitions


**Acknowledgements:**

The very nature of this data collection, compilation and presentation function literally depends on hundreds of contact people across the world, willing to share information with their counterparts in the remote sensing community for the benefit of all. In addition to requesting/receiving source information, I have been living on hints and tips and comments, in particular on open minds, throughout the writing period of this book. The vast majority of my contacts are correspondence acquaintances. I would like to thank all of the individuals, who are listed under my first acknowledgements, virtually all participated again for the second book.

In addition I have to thank the following people for their help or contributions: J. S. Langford (Aurora),  S. M. Till (CCRS), J. P. Gastellu-Etchegorry (CNRS),  J. Vitko (Sandia N. L.), R. Beer, S. Durden, B. L. Gary, R. Green, D. Hagan, P. Li, Ch. Webster, G. C. Toon, and H. Zebker (NASA/JPL), P. Abel, M. Acuna, D. N. Baker, J. L. Bufton, B. J. Choudhury, J. Gass, G. M. Heymsfield, F. E. Hoge, M. D. King, H. G. McCain, R. Pfaff, V. Salomonson, J. Spinhirne, D. Vandemark, and J. R. Wang (NASA/GSFC), D. Dokken (NASA/HQ), M. E. Zolensky (NASA/JSC), R. Blakeslee, R. E. Hood, W. Koshak (NASA/MSFC), V. Connors (NASA/LaRC), M. C. Dudzik, B. Butchko, D. C. Carmer, and P. Wagner (ERIM), E. Weinstock and S. C. Wofsy (Harvard), T. W. Lawrence (LLNL), J. Hornstein, D. Pope, and J. Finkelstein (NRL), R. S. Vickers (SRI), D. H. Staelin, T. Murphy (MIT), T. H. Achtor (U. of Wisconsin), B. A. Spiering (NASA/SSC), E. S. Putnam (Hughes, SBRC), O. Fäst, S. Grahn, and S. Zenker (SSC, Sweden), A. Gustavsson (FOA, Sweden), E. A. Reshetov (Priroda Center, Moscow), B. Herrick (TRW Inc.), B. Speer (SAIC), C. Cattell (UCB), P. Lucey (U. of Hawaii), R. Lunetta (US/EPA), Ch. Oliver (DRA/UK), E. Mesquita (GER Corp.), R. Reuter (U. of Oldenburg), D. Cobb and W. Priedhorsky (LANL), P. Hoogeboom (TNO), B. M. Spee (NLR), N. Balteas (Kayser Threde), J. Vojta (GFU, Prague), Y. Krilov (Vega, Moscow), P. A. Shirokov (NPO Machinostroyenia), Yu. M. Mikhailov (IZMIRAN), S. N. Voyakin (NPO Planeta), A. Lotov (IRE), N. Skou and F. Primdahl (TUD, Lyngby, Denmark), M. Le Coz (ONERA), D. Clark, R. D. Zwickl, J. Wydick, and M. J. Nestlebush (NOAA), F. P. Kelly (DOD), P. Hildebrand, L. Radke, and P. Spyers-Duran (NCAR), M. Kikuchi (NASDA), K. R. Chan, M. Loewenstein, R. F. Pueschel, Ph. Russell, and F. Valero (NASA/ARC), L. Lalonde (Intera, Canada), J. Morgan (EUMETSAT), E. Attema, M. Schuyer, and K. P. Wenzel, (ESA/ESTEC), W. Flury  (ESA/ESOC), Zhimin Zhang (CAS/SITP, Shanghai), Tong Qingxi (CAS/IRSA, Beijing), J. Hyyppä and I. Panula−Ontto (HUT, Helsinki), F. M. Bréon (LMCE/CE, France), A. Rohrbach (Leica), G. England (Daedalus Inc.), G. Haerendel, J. Büchner (MPE), K. Pflug (AIP), L. Bayer (CLS Service ARGOS, Toulouse), A. Schönenberg (OHB), C. E. Santana (INPE), T. Miyake (ISAS, Tokyo), H. Shinohara (NEC Corp. Tokyo), G. Moody (OSC), W. P. Rudolf (Lockheed), P. Röber (Dornier), J. Radbone (U. of Surrey, UK).

I was very fortunate to enjoy the backing of some key contacts in several agencies, who opened doors that led to further contacts and/or contributed substantially with advice or reviews. These individuals are: J. R. Huning  (NASA/HQ), K. Staenz (CCRS), S. H. Melfi (NASA/GSFC), Y. H. Kerr (LERTS), Y. S. Sedunov (NPO Planeta), B. Zhukov (IKI), K. Schmidt (DARA), J. Myers (NASA/ARC), B. H. Needham (NOAA), and P. M. Fagundes (U. of Rio de Janeiro).

Whenever possible I tried to tap the advice and the services of my colleagues at DLR. In addition to those, listed under the acknowledgement of the first edition, I would like to thank: V. Harbers, D. Hounam, A. Pietraß, P. Meischner, T. Popp, H. Süß, F. Witt, M. Novak, J. Nithack, H. Schlager, M. Schröder, F. Köpp, A. Noelle, H. Wilhelms, P. Hausknecht, F. Pätzold, G. Lämmel, and M. Gottwald. Special thanks are again due to Friedrich L. Porsch who drew all spacecraft/sensor illustrations. At last, I would like to thank my institution, DFD of DLR, in particular W. Markwitz and G. V. Mayer, for their continued support in this undertaking.

Oberpfaffenhofen, February 2, 1994


Herbert J. Kramer

# Part A    Survey of Spaceborne Missions and Sensors

## A.1    ACE (Advanced Composition Explorer)

NASA Solar-Terrestrial mission in the explorer program with the objectives to determine: the elemental and isotopic composition of matter, the origin of the elements, the formation of the solar corona and acceleration of the solar wind. S/C builder: APL.
The mission is planned for a launch in 1997 with a Delta II launch vehicle. S/C weight = 650 kg, nominal life of mission = 1-3 years.

Orbit: Halo orbit about the Lagrangian (or libration) point $L_1$ (250 Earth radii toward the Sun, about 1.5 million km from Earth).

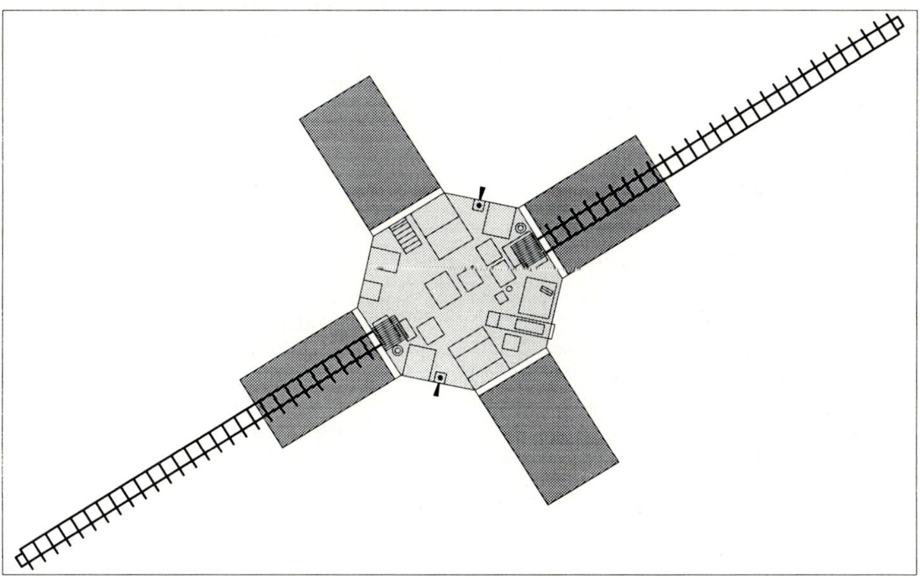

**Figure 1:**    ACE S/C Model (Top View of Orbital Configuration)

Sensors: (PI for mission: E. Stone, Caltech).

- **SWIMS** = Solar Wind Ion Mass Spectrometer (PI: G. Gloeckler, U. of Maryland). Objective: measurement of solar wind composition data over a wide range of solar wind bulk speeds and for all solar wind conditions. Abundances of most of the elements and several isotopes in the mass range from 4 - 60 amu every few minutes. SWIMS uses a time-of-flight (TOF) measurement technique to determine the mass of a solar wind ion with high accuracy. SWIMS consists of the Wide-Angle, Variable Energy/charge (WAVE) passband deflection system, the time-of flight High-Mass Resolution Spectrometer (HMRS), high-voltage supplies, and analog and digital electronics.

- **SWICS** = Solar Wind Ion Composition Spectrometer (PI: G. Gloeckler, U. of Maryland). Objective: measurement of the elemental and ionic-charge composition and the temperature and mean speeds of all major solar wind ions from H through Fe at solar wind speeds ranging from 145 km/s (for protons) to 1532 km/s (for $Fe^{+8}$). The instrument, which covers an energy per charge range from 110 eV/Q - 66.7 keV/Q in about 13 minutes, combines an electrostatic analyzer with post-acceleration, followed by a time-of-flight (TOF) and energy measurement.

- **ULEIS** = Ultra-low Energy Isotope Spectrometer (PI: G. Mason, U. of Maryland). Objective: measurement of ion fluxes over the charge range from He through Ni from $\sim 20$ keV/n to 10 MeV/n (superthermal and energetic particle ranges). ULEIS is a time-of-flight (TOF) mass spectrometer which identifies incident ion mass and energy by simultaneously measuring the time-of-flight, $\tau$, and residual kinetic energy, E, of particles which enter the telescope cone and stop in one of the six detectors in the telescope.

- **SEPICA** = Solar Energetic Particle Ionic Charge Analyzer (PI: E. Möbius, U. of New Hampshire and MPE Garching). Objective: measurement of the ionic charge state, Q, the energy, E, and the nuclear charge, Z, above 0.2 MeV/n.
  Energetic particles entering the multi-slit collimator will be electrostatically deflected between the six sets of electrode plates which are supplied with variable high voltages up to 30 kV. The deflection, which is inversely proportional to energy per charge, E/Q, is determined in the back portion of the instrument (dE/dX device and a position-sensitive silicon solid-state detector). The residual energy of the particle, $E_{res}$, and the amount of electrostatic deflection is directly determined in the detector, thus yielding the energy per ionic charge, E/Q, of the incoming particle, and its energy, E.

- **SIS** = Solar Isotope Spectrometer (PI: R. Mewaldt, California Institute of Technology). Objective: measurement of elemental and isotopic composition of solar energetic particles, anomalous cosmic rays, and interplanetary particles from He to Zn over the energy range from 8 - 150 MeV/nucleon. Measurements by a particle's energy loss technique $\Delta E$ in a detector (multiple $\Delta E$ versus residual energy E').

- **CRIS** = Cosmic Ray Isotope Spectrometer (PI: R. Mewaldt, California Institute of Technology). Objective: measurements of all stable and long-lived isotopes of galactic cosmic ray nuclei from He to Zn over the energy range from $\sim 100$ to 600 MeV/nucleon. CRIS provides also limited measurements of low energy H isotopes and exploratory studies of the isotopes of "ultra-heavy" (UH) nuclei. Measurements by a particle's energy loss technique $\Delta E$ in a detector ( multiple $\Delta E$ versus residual energy E'). CRIS is of CRRES, ISEE-3 and SAMPEX heritage.

- **EPAM** = Electron, Proton, and Alpha-particle Monitor (PI: T. Krimigis, APL). Objective: measurement of solar and interplanetary particle fluxes with a wide dynamic range and a directional coverage of nearly a full unit sphere.
  EPAM consists of five apertures in two telescope assemblies. It measures ions ($E_i \geq 50$ keV) and electrons ($E_e > 30$ keV) with essentially complete pitch angle coverage from the spinning ACE spacecraft. It also has an ion elemental abundance aperture using $\Delta E$ versus E technique in a three-element telescope.

| Instrument | Mass (kg) | Power (W) | Data rate (bit/s) | Measurement Technique | Type. Energy (MeV/nucleon) |
|---|---|---|---|---|---|
| CRIS | 21 | 11 | 462 | dE/dX x E | $\sim 300$ |
| SIS | 20 | 19 | 2000 | dE/dX x E | $\sim 50$ |
| ULEIS | 17 | 25 | 1000 | TOF x E | $\sim 5$ |
| SEPICA | 16 | 6 | 600 | $\Delta$E x E x E/Q | $\sim 1$ |
| SWICS | 5 | 5 | 500 | TOF x E x E/Q | $\sim 0.001$ |
| SWIMS | 8 | 3.5 | 505 | TOF thru special E-field | $\sim 0.001$ |
| SWEPAM | 6.7 | 5.4 | 1000 | Electrostatic Analyzer | $\sim 0.001$ |
| EPAM | 2.9 | 3 | 160 | dE/dX x E | $\sim 0.3$ |
| MAG | 4.7 | 3 | 300 | Triaxial Fluxgate | |

**Table 1:**     **ACE Instrument Summary**

- **SWEPAM** = Solar Wind Electron, Proton, and Alpha Monitor (PI: D. McComas, LANL). Objective: high quality measurements of electron and ion fluxes in the low energy solar wind range (electrons: 1 - 900 eV; ions: 0.26 - 35 keV). SWEPAM is of Ulysses mission heritage.

SWEPAM makes simultaneous and independent electron and ion measurements with two separate sensors. Both sensors make use of curved-plate electrostatic analyzers which are spherical sections cut off in the form of a sector.

- **MAG** = Magnetic Field Monitor (triaxial fluxgate). Objective: measurement of the three components of the magnetic field. MAG is boom-mounted.


## A.2    ACTIVE (AKTIVNY-IK)

Russian (IKI) Solar-Terrestrial mission within the Intercosmos program (International Space Plasma-Waves Laboratory, Intercosmos 24). Objectives: comprehensive study of VLF-wave (Very Low Frequency) propagation phenomena in the Earth's magnetosphere and wave interaction with the energetic particles of the radiation belts.[2],[3]

Mother-daughter pair of spacecraft. The mother craft is also designated as Intercosmos 24, the daughter spacecraft is provided by Czechoslovakia with designations of C2-AK as well as Magion-2.

Mother craft: 3-axis spin-stabilized; S/C diameter = 2 m; S/C height = 3 m; mass = 270 kg; power = 270 W;

Launch on Sept. 28, 1989 with a Cyclone launch vehicle from Plesetsk, releasing the Magion-2 subsatellite. Magion-2 orbits around the mother in a controlled mode. Magion-2 separation from mother satellite on October 3, 1989. Magion-2 began its science program in January 1991. ACTIVE is an operational mission as of 1993. Scientific program leader (since 1993): G. L. Gdalevich, (IKI, Moscow).

Orbit: apogee = 2500 km, perigee = 500 km, inclination = 82.5°, period = 116 minutes,

**Sensors of mother S/C:**
The mother craft carried a plasma generator for a modulated plasma environment around the antenna. A neutral xenon injector injected xenon around the S/C for monitoring of ionization of neutral gas flow and its propagation effects.

**VLF** = Very-Low Frequency Generator [PI: O. A. Molchanov, IFZ RAN (Institute of Earth Physics) and Yu. N. Agafonov, IFZ RAN]. Objectives: generation of VLF waves in the range 9-11 kHz via a 20 m diameter ORA-20 loop antenna (current about 10A). Note: ORA = O-formed loop antenna.

ORA-20 (transforms oscillations from the generator to electromagnetic emission). The tube from soft ductile aluminum alloy, with a wall thickness of about 1 mm, is used for the build-up of the antenna loop. The tube is tightly rolled up at launch and slowly unfolds in orbit (by its spring-load) on command into a loop antenna of 20 m diameter.

**PVP** = Generator of Electric Oscillations (PI: Yu. M. Mikhailov, IZMIRAN). Objectives: generation of VLF waves in the range 1.5 - 20.5 kHz. The PVP sensor consists of a self-opening band-type cylindrical dipole antenna of 15 m length and 25 mm diameter.
-   Output voltage:      50 - 300 V
-   Time of impulse:     0.25 s
-   Time of pause:       0.75 s

**NVK-ONCH** = VLF Analyzer (PI: Yu. M. Mikhailov, IZMIRAN). Objective: measurement of magnetic and electric field components in the range from 8 Hz - 20 kHz, three magnetic and two electric components.

---


2)   "The ACTIVE International Space Plasma-Wave Laboratory", The Solar-Terrestrial Science Project of the Inter-Agency Consultative Group for Space Science, esa SP-1107, November 1990,, pp. 45-49
3)   Aktivny-IK, interavia Space Directory, 1992-93, pp. 149

- Sensitivity on electric components: $10^{-7}$ V m$^{-1}$ Hz$^{-1/2}$
- Sensitivity on magnetic components: $10^{-5}$ nT Hz$^{-1/2}$
- Dynamic range: 80 dB
- Sensors for magnetic components: coil-type magnetic antenna with ferrite.
- Sensors for electric components: double spherical sondes with a diameter of 100 mm and with distances of 2.5 m between the sondes.

**SHASH** = VLF Spectroanalyzer (PI: J. Tarchai, Eötvös Univ., Budapest). Objective: digital analysis of VLF signals and noise levels from the NVK-ONCH instruments in the range of 20 Hz - 20 kHz. Independent transmission of the analyzed information is provided by a transmitter (frequency = 460 MHz, transmission rate = 100 kbit/s).

**VLF-2** = Very-Low Frequency Generator 2. (PI: Z. Klos, Center of Cosmic Research, Polish Academy of Sciences, Warsaw). Objective: provision of a parallel analysis of VLF spectra (one electric component with the use of 12 filters); measurement range = 20 Hz - 20 kHz. The VLF-2 sensor uses the same equipment as NVK-ONCH.

**PRS** = Plasma Wave Spectrometer (PI: Z. Klos, Center of Cosmic Research, Polish Academy of Sciences, Warsaw); Objective: measurements of HF-noise spectra in the range from 0.1 - 10 Hz; data rate = 10 kbit/s; mass = 5.1 kg. The sensor consists of a self-opening band-type antenna of 15 m length.

**ZL-A** = Langmuir Probe Experiment (PI: J. Rustenbach, MPE, Berlin). Objective: measurement of plasma electron/ion temperature and densities. Electron temperature range = $10^3$ - $10^4$ K, electron density range = $10^2$ - $10^8$ cm$^{-3}$. The sensor is a cylindrical sonde.

**KM-6** = Cold Plasma Analyzer (PI: J. Shmilauer, Geophysical Institute, Prague). Objective: measurement of plasma density and temperature, electron distribution and drift velocity; data rate = 23 kbit/s; mass = 5.7 kg; electron temperature range = $10^3$ - $10^4$ K; ion density $10^2$ - 5 x $10^6$ cm$^{-3}$. The sensor is a plane sonde that is made up of 4 electrodes.

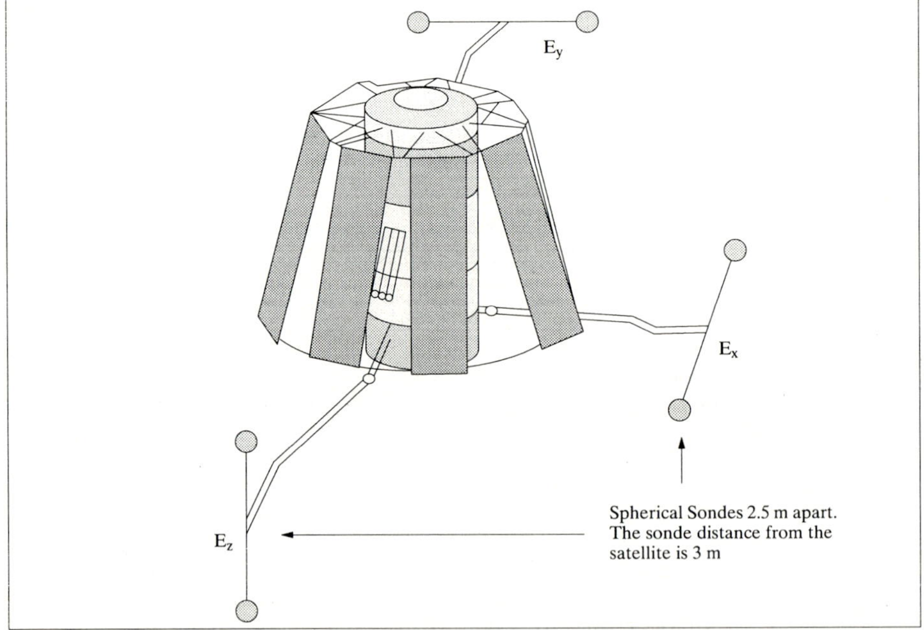


**Figure 2:** The ACTIVE S/C Model

**NAM-5** = Radiofrequency Mass-Spectrometer (PI: J. Shmilauer, Geophysical Institute, Prague). Objective: measurement of ions and neutral plasma composition. Range of mass = 1-60 a.m.u. (atomic mass unit); data rate = 29 kbit/s; mass = 6.1 kg

**DME** = Soft Electron Detector (PI: A. Melentiev, IKI). Objective: flux measurement of soft electrons in the energy spectrum of 0.01 - 10 keV; data rate = 6 kbit/s; mass = 6.1 kg.

**ANAPURNA** = Energy and Pitch Distribution Experiment (PI: A. Melentiev, IKI). Objective: measurement of energy and angular distributions of electrons in the range: 0.2 - 40 keV; data rate = 4 kbit/s; mass = 11 kg

**SPE-1** = High Energy Particles Experiment (PI: K. Kudela, Institute of Exp. Physics, Kosice, CSFR. Objective: measurement of spectra and anisotropy of electrons and protons; $E_e$ = 30-600 keV; $E_p$ = 15-600 keV; data rate = 14 kbit/s; mass = 4.5 kg.

### A.2.1    Subsatellite Magion-2 (C2-AK)

Magion = **Mag**netospheric and **Ion**ospheric research satellite (built by the Academy of Sciences of the Czech Republic, Geophysical Institute, Prague) with spring separation on command from mother craft. Objectives: monitoring of propagation VLF waves from the mother spacecraft. Full deployment of booms and 4 solar panels was confirmed on Dec. 24. 1989. The science program began on January 21, 1990. Magion-2 was operational until November 20, 1990. Magion-2 mass = 51.7 kg (11.4 kg of scientific instruments); 26 face polyhydron; 560 mm diagonal (see Figure 3 for Magion S/C). Data: Downlink transmission in VHF-Band (frequencies of 137 and 400 MHz). Selectable digital data rates of 5, 10, 20, and 41 kbit/s.

Orbit: perigee = 500 km, apogee = 2500 km, inclination = 83°

Sensors for the measurement of the electric and magnetic fields; VLF waves, plasma and energetic particles.

- **SGR-7** = 3-axis Fluxgate Magnetometer (PI: M. Ciobanu, IKI, Romania). Measurement range of ± 50048 nT or 6256 nT, resolution = 16 nT or 2 nT, frequency range = 0 - 20 Hz. Measurement of magnetic field vector and to serve as attitude sensor.
  - SGR6 = 1-component Fluxgate Variometer; dynamic range = ± 156 nT; resolution = 50 nT; frequency range = 0.1 - 20 Hz.
- **KEM-1** = Magnetic and Electric Field Analyzer (PI: P. Triska, GFU, Czech Republic);
  - ULF = Electric Field Experiment (3 components of the quasistatic electric field); dynamic range = 0.005 - 8000 mV, frequency range = 0.1 - 20 Hz.
  - VLF Wave Experiment (2-axis electric and 1-component magnet ELF-VLF field measurements, broadband waveforms, spectrum analyzer, filter bank;
    $E_{x,y,z}$ : 0.1 Hz - 120 kHz, sensitivity of $10^{-7}$ V/mHz$^{1/2}$ , 120 dB dynamic range.
    $B_x$ : 10Hz - 40 kHz, sensitivity of 5 x $10^{-6}$ nT/Hz$^{1/2}$ at 2 kHz and $10^{-4}$ nT/Hz$^{1/2}$ at 100 Hz, dynamic range 120 dB.
  - Data transmission modes selectable: broadband analog data 10 Hz - 60 kHz one channel, 10 Hz - 20 kHz broadband 3 channels, subcarriers of 1 kHz bandwidth 4 channels.
  - Filter bank; 17 Hz - 15 kHz, 8 filters, four independent sets
  - Frequency analyzer: range of 1 - 220 kHz, 32 frequency steps, full spectrum/2s, selection of any frequency.
- **KM-12** = Cold Plasma Analyzer (PI: J. Shmilauer, GFU, Czech Republic); HF probe measures 2 components of the electron temperature in the range: Te = $10^3$ - $10^5$ K,, spacecraft potential from -2 to +2V; Spherical ion-trap measures the ion density ($N_i$) in the range: $10^8$ - $10^{13}$ /m$^3$. $\Delta N_i/N_i$ fluctuations $f_{max}$ = 50 Hz.

- **ZL-A-S** = Langmuir Probe (PI: K. Sauer, MPE, Germany); measurement of electron and ion density in the range: $500 - 10^8$ cm$^{-3}$; electron temperature: 0.05 - 3 eV; current range: $10^{-10} - 10^{-3}$ A; in the current-mode density fluctuations f= < 200 Hz, $\Delta N_e/N_e$ resolution $10^{-3}$.

- **PRS-2-C** = Radiowave Spectrometer (PI: Z. Kloss, CBK PAN, Poland). Measurement of the HF wave spectra. Frequency range = 0.1 - 10 MHz, dynamic range = 1 μV - 10 mV, 0.2 s/spectrum using $\Delta f$=50 kHz or 2s/spectrum using $\Delta f$= 15 kHz; field fluctuations at fixed frequency (selectable), $\Delta t$= 1ms.

- **DOK-A-S** =Silicon Detector Spectrometer (PI: K. Kudela, UEF SAV, Slovakia); two sensors (parallel and perpendicular to the magnetic field vector), measurement of electrons and ions (20 keV - 1 MeV, 8 energy levels, geometric factor $10^{-2}$ cm$^2$ sr.

- **MPS SEA**= Energetic Particle Spectrometer (PI: Z. Nemecek, Prague University, Czech Republic); electrostatic analyzers, measurement of electrons and positive ions: 0.2 - 20 keV in 16 energy levels, pitch angle resolution 30°, geometric factor is $\sim 10^{-3}$ cm$^2$ sr.

- **FDS** = Photometer (PI: N. Petkov, IKI, Bulgaria); measurement of the optical ionospheric plasma emissions (630 and 577.7 nm).

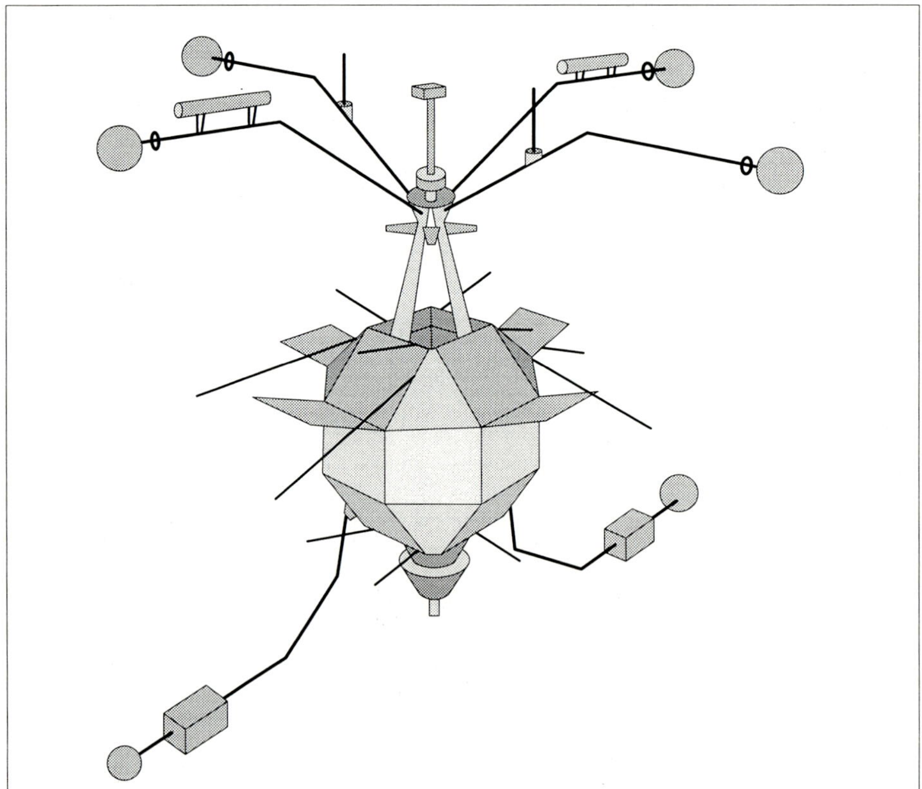

**Figure 3:**    **The Magion-2 S/C Model**

## A.3    ADEOS (Advanced Earth Observing Satellite)

ADEOS = Advanced Earth Observing Satellite. Japanese (NASDA) satellite mission.[4] Objective: Global observation of land, ocean and atmospheric processes (ocean color and sea surface temperature). In addition, communication experiments are planned for the study (feasibility) of inter orbit links, called IOCS (Inter-Orbital Communication Subsystem). Planned Launch (with H-II rocket): Feb. 1996[5] (3 year mission). Satellite mass = 3500 kg, payload mass = 1300 kg, power = 4.5 kW, 3-axis stabilized (attitude error $<0.3^\circ$, attitude stability $<0.003^\circ/s$).[6]

Orbit: Sun-synchronous sub-recurrent polar orbit; Altitude = 797 km; Orbital period = 101 Min, 10:30 AM local sun orbit (descending node). Ground repeat cycle=41 days (sub-cycle = 3 days), Inclination = $98.6^\circ$.

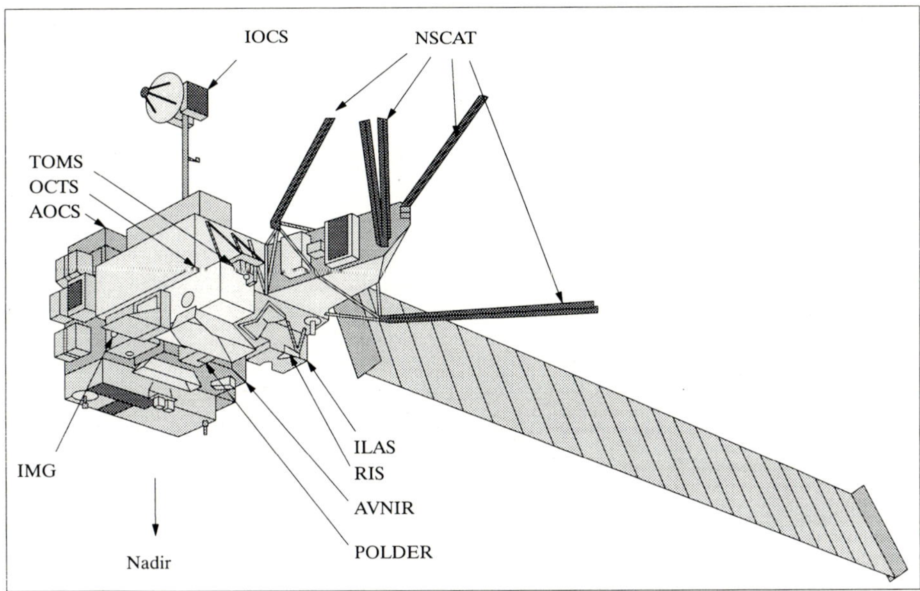


**Figure 4:**    **The Adeos S/C Model**

**Sensors:**

**OCTS** = Ocean Color and Temperature Scanner (mechanical scanning radiometer, NASDA sensor). Objectives: Ocean color and sea surface temperature measurements (ocean primary productivity, interaction between the ocean and the atmosphere and environmental studies). OCTS offers 12 measurement bands from 0.402 - 12.5 µm. Swath width = 1400 km. Spatial resolution: approx. 700 m. Operation requirements: global observation of the earth during daytime (TIR channel during night if required).

The OCTS sensor consists of a scanning radiometer with optical system, detector module and electrical unit. OCTS employs catoptric optical system and a mechanical rotating scanning method with mirror (due to wide spectral coverage). OCTS can be tilted about the along-track axis to prevent sea surface sun glitter. The IR detectors are cooled to 100 K by a large radiant cooler facing deep space.

---

4)    NASDA handout at the CEOS WGD-10 Meeting in Annapolis Md., April 16-19 1991
5)    'Japan Delays Remote-Sensing Missions', Space News, March 30/April5, 1992, p. 9
6)    'ADEOS',NASDA brochure, 1993

| Band Number | Spectral Band (μm) | Bandwidth (μm) | Radiance (W/m/sr/μm) | SNR |
|---|---|---|---|---|
| 1 | 0.402 - 0.422 | 0.020 | 145 | 450 |
| 2 | 0.433 - 0.453 | 0.020 | 150 | 500 |
| 3 | 0.480 - 0.500 | 0.020 | 130 | 500 |
| 4 | 0.510 - 0.530 | 0.020 | 120 | 500 |
| 5 | 0.555 - 0.575 | 0.020 | 90 | 500 |
| 6 | 0.655 - 0.675 | 0.020 | 60 | 500 |
| 7 | 0.745 - 0.785 | 0.040 | 40 | 500 |
| 8 | 0.845 - 0.885 | 0.040 | 20 | 450 |
| 9 | 3.55 - 3.88 | 0.33 | 0.15K | Target Temperature @ 300 K |
| 10 | 8.25 - 8.80 | 0.55 | 0.15K | |
| 11 | 10.3 - 11.4 | 1.1 | 0.15K | |
| 12 | 11.4 - 12.5 | 1.1 | 0.20K | |
| | IFOV | 0.85 mrad (~700 m) | Quantization | 10 bit/pixel |
| | Scanning angle | ~ ±40° | Tilting angle | -20°, 0°, +20° |
| | Polarization sensitivity | Band 1 ≤5% Band 2-8 ≤2% | Calibration VIS/NIR IR | Solar, internal light s. deep space,blackbody |

**Table 2:** **Definition of OCTS Parameters**

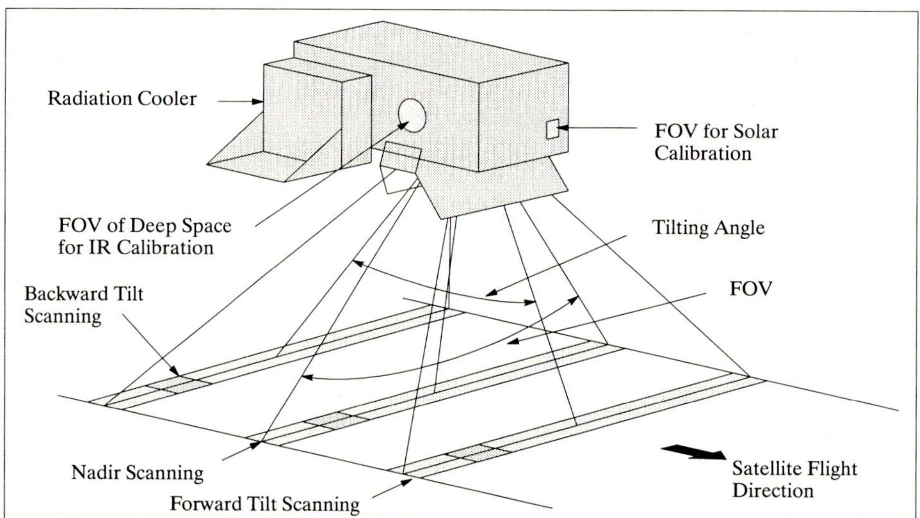


**Figure 5:** **The Observation Concept of the OCTS Instrument**

**AVNIR** = Advanced Visible and Near-Infrared Radiometer (NASDA sensor). Electronic scanning radiometer. Objective: Land and coastal zone observations, measurement of reflected sun light from the Earth's surface. Scanning method: electronic (CCD). Wavelength: 5 bands from 0.42 - 0.89 μm (visible: 3 bands 0.42-0.50, 0.52-0.60, 0.61-0.69 μm, near-infrared: 1 band 0.76-0.89 μm, panchromatic band (visible): 1 band 0.52-0.69 μm). Spatial resolution: multiband: ~16 m (IFOV = 20μrad), panchromatic band: ~8 m (IFOV = 10 μrad). Swath width = 80 km (FOV=5.7°). Observation requirements: regional observation according to user's requests; simultaneous operation of multispectral and panchromatic.

The AVNIR instrument has the capability to tilt the observation field by ±40° about the along-track axis. The 0.42-0.50 μm band is useful for coastal zones and lakes. Calibration of sensor using solar light and gray lamps. The radiometric absolute accuracy is ±10%, the on-board calibration accuracy = ± 5%. The CCD offers 5000 and 10000 detector elements for high spatial resolution. Instrument mass = 250 kg, power = 300W.

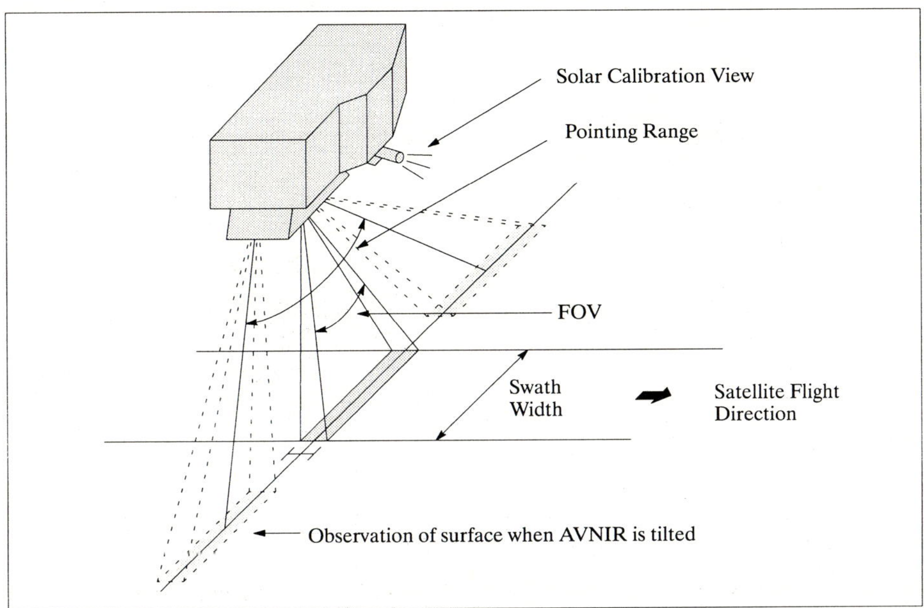


**Figure 6:**     **The Observation Geometries of the AVNIR Instrument**

**NSCAT** = NASA Scatterometer (NASA/JPL sensor). Objective: Measurement of surface wind speed and direction over the global oceans, coverage every 2 days under all weather and cloud conditions. NSCAT is a microwave radar instrument, using an array of antennas that radiate microwave pulses at a frequency of 13.995 GHz across broad regions of the Earth's surface. An array of six, 3 m long antennas scans two swaths of 600 km width each one band to each side of the flight path, separated by a gap of 300 km at nadir. Wind speed accuracy = 2 m/s, direction accuracy = 20°, resolution = 50 km. Operation requirements: continuous operation for global observation of the oceans (about 190,000 wind measurements/day). Note: NSCAT is an upgraded version of the Radar Scatterometer (SASS) on Seasat.

NSCAT transmits microwave pulse and receives a backscattered echo from the ocean surface. Changes in wind velocity cause changes in ocean surface roughness, modifying the radar cross section of the ocean and the magnitude of the backscattered power. Multiple collated measurements acquired from several directions can thus be used to solve wind speed and direction simultaneously.

**TOMS** = Total Ozone Mapping Spectrometer (NASA/GSFC sensor). Objective: Observation of total ozone changes, evaluation of changes in UV radiation and the observation of sulfur dioxide. Measurement wavelength: 308.6, 312.5, 317.5, 322.3, 331.3 and 360 nm with 1 nm bandpass. Swath width: 2795 km. IFOV = 50 km at nadir; cross-track scan = 105° (35 3° steps). Operation requirements: global observation of illuminated part.

TOMS measures the albedo of the Earth's atmosphere at six narrow spectral bands in the near-ultraviolet region. The albedo is measured by comparing the reflectivity of the Earth with the reflectivity of a calibrated on-band diffuser plate. Total ozone is derived from the differential albedo in three pairs of the spectral bands, which are selected to function at all latitudes and solar illumination conditions.

| Wind velocity | 2 m/s<br>10% | 3 - 20 m/s<br>20 - 30 m/s |
|---|---|---|
| Wind direction | 20° (rms) | 3 - 30 m/s |
| Spatial resolution | 25 km<br>50 km | 0°<br>wind cells |
| Location accuracy | 25 km (rms)<br>10 km (rms) | absolute<br>relative |
| Coverage | 90% of ocean surface every two days | |
| Instrument mass | 300 kg | |
| Instrument power | 275 W | |

**Table 3: Definition of NSCAT Parameters**

**POLDER** = Polarization and Directionality of the Earth's Reflectances (passive optical imaging radiometer, CNES sensor as passenger instrument on ADEOS)[7].
Objectives: Observation of bidirectionality and polarization of the solar radiation reflected by the atmosphere: tropospheric aerosols (inversion of the physical properties); ocean color (accurate determination of sea surface reflectances); land surfaces (determination of surface BRDF and improvement in the correction of the surface bidirectional and atmospheric effects on vegetation indices); Earth radiation budget (determination of cloud BRDF and classification of clouds according to their bidirectional properties).

a) measurement of polarized reflectance in VIS/NIR
b) observation of the Earth's target reflectance from 12 directions during a single S/C pass
c) operation in two dynamic modes for high SNR and wide dynamic range

Six of POLDER's eight frequencies are optimized for observing atmospheric aerosols, clouds, ocean color, and land surfaces. The other two frequencies are centered on the $H_2O$ and $O_2$ absorption bands for retrieving atmospheric water vapor amount and cloud top height, respectively.

| Wavelength<br>nm<br>(FWHM) | Bandwidth<br>(nm) | Polarization | Dynamic Range<br>(Normalized Radiance)<br>High | Low | Main Measurement Objective |
|---|---|---|---|---|---|
| 443 | 20 | no | NA | 0.05-0.22 | Ocean color |
| 443 | 20 | yes | 0.05-1.1 | NA | Aerosols, ERB |
| 490 | 20 | no | NA | 0.034-0.17 | Ocean color |
| 565 | 20 | no | NA | 0.019-0.11 | Ocean color |
| 670 | 20 | yes | 0.013-1.1 | 0.013-0.27 | Vegetation, aerosols, ERB |
| 763 | 10 | no | 0.007-1.1 | 0.007-0.27 | Cloud top temperature |
| 765 | 40 | no | 0.007-1.1 | 0.007-0.27 | Aerosols, CTP |
| 865 | 40 | yes | 0.007-1.1 | 0.007-0.27 | Vegetation, aerosols, ERB |
| 910 | 20 | no | 0.007-1.1 | 0.007-0.27 | Water vapor content |

**Table 4: Spectral Characteristics of POLDER**

Measurement channels: 15 channels (3 channels for each polarized band) Swath width = 1440 km x 2200 (across-track) km, ground spatial resolution of 7 km x 6 km at nadir. Data rate: 0.882 Mbit/s, 12 bit quantization. Operation requirements: global observation of the earth with more than 15° of Solar elevation, simultaneous operation with OCTS.

The detection unit of the POLDER instrument consists of a telecentric lens, a rotating wheel supporting filters and polarizers, and a matrix CCD sensor (242 x 548 photoelements,

7) P. Y. Deschamps, M. Herman, A. Podaire, M. Leroy, M Laporte, P. Vermande, "A Spatial Instrument for the Observation of Polarization and Directionality of Earth Reflectances: POLDER", IGARSS '90 Conference Proceedings, Washington D. C.

the pixels are binned one by two, resulting in 242 x 274 sensitive areas). FOV = $\pm 43°$ along-track, FOV = $\pm 51°$ across-track, FOV = $\pm 57°$ diagonal. Instrument mass = 33 kg, power = 42 W.

**IMG** = Interferometric Monitor for Greenhouse Gases, (MITI sensor developed by JA-ROS, this is a nadir-looking Fourier transform infrared spectrometer); Objectives: Mapping of greenhouse gases on a global scale ($CO_2$, $CH_4$, $N_2O$). Continuous measurements of the infrared spectrum in the range from 3.3 - 14.0 µm with a very fine spectral resolution of 0.1 cm$^{-1}$, absolute accuracy of measurement = < 1 K, stability of measurement < 0.1K, IFOV = 10mrad ($\sim 10$ km$^2$ footprint). vertical resolution $\sim$2-6 km depending on species; interferogram scan time $\leq$ 10s. Data rate = 0.9 Mbit/s. Operation requirements: full time operation for 3 days out of 13 days. Instrument mass = 115 kg, power $\leq$ 150 W.

IMG is a Michelson-type Fourier Transform Spectrometer (FTS) with two mirrors and a beam splitter. The incident radiation is divided by the beam splitter into two paths. One mirror is moved so that the two paths produce an interference pattern when recombined. The signal measured by the detector, the interferogram, can be Fourier transformed to obtain the incident spectrum. The diameter of the entrance aperture is 10 cm. The moving mirror is suspended on magnetic bearings and scans a 10 cm long path in 10 seconds.

**ILAS** = Improved Limb Atmospheric Spectrometer (Sensor of the Environment Agency of Japan, JEA). Objectives: Measurement of the limb atmospheric micro-ingredient over high-latitude regions. Spectral bands: 0.753-0.784 µm, 6.21-11.77 µm, 5.99-6.78 µm. Observation altitude = 10 - 60 km; Resolution = 13 km horizontal x 2 km vertical. Data rate = 0.5 Mbit/s. Operation requirements: twice 5 min per orbit around sunrise and sunset times.

ILAS is a solar occultation sensor which measures high latitude stratospheric constituents in both hemispheres. Measurement species: $O_3$, $H_2O$, $CO_2$, $CH_4$, $NO_2$, $N_2O$, $HNO_3$, aerosol, CFC11, atmospheric density and temperature.[8] ILAS measures the sequence of the atmospheric absorption spectrum which pass the various tangent heights, in 12 Hz (spectral sampling rate). Instrument: grating spectrometer with Lonear Linear Array Detector. Mass = 140 kg, power < 110 W.

**RIS** = Retroreflector in Space (Sensor of the Environment Agency of Japan).[9] RIS is a reflector for an Earth-satellite-Earth laser used in long-path absorption experiments. Instrument mass = 44 kg.
Objective: Measurement of ozone, fluorocarbon, carbon dioxide, etc. by laser beam absorption technique. Wave length: 0.4 - 1.4 µm. Corner-cube retroreflector: 50 cm diameter to derive column density of ozone and trace species from laser absorption measurements. Operation requirements: above ground tracking stations.

Data rates:

ADEOS provides onboard recording (MDR = Mission Data Recorder (3 instruments) and LMDR = Low Speed Mission Data Recorder).
Observation data rates: AVNIR (M): 60 Mbit/s, AVNIR (P): 60 Mbit/s, OCTS: 3.0 Mbit/s. Polder: 0.882 Mbit/s. IMG: 0.9 Mbit/s. ILAS: 0.5 Mbit/s. NSCAT: 2.9 kbit/s. TOMS: 0.7 kbit/s.

Data links: Uplink frequency = 2.0 GHz (CMD and ranging), downlink frequency = 2.2 GHz (CMD and ranging), command bit rate = 500 bit/s.
Science data transmission: 3 X-Band links (8.15, 8.25, 8.35 GHz) with QPSK modulation.
IOCS frequencies = S-Band (low rate mission data), Ka-Band (120 Mbit/s max).

8)   'Design of Improved Limb Atmospheric Spectrometer (ILAS) aboard ADEOS', Optical Remote Sensing of the Atmosphere, 1990 Technical Digest Series of the Optical Society of America, Volume 4, pp. 88-91

9)   'Retroreflector-In-Space for ADEOS: Earth-Space-Earth Laser Long-Path Absorption Measurements of Atmospheric Trace Species', Optical Remote Sensing of the Atmosphere, 1990 Technical Digest Series of the Optical Society of America, Volume 4, pp. 488-490

## A.4     ADEOS II (Advanced Earth Observing Satellite-II)

A planned Japanese (NASDA) mission for a launch in 1999/2000 with a mission life of 3 years minimum (5 years goal). Initially this mission/series was called **JPOP** (Japanese Polar Orbiting Platform) to be operational in the late nineties[10]. As of Fall 1991 the JPOP program has been formally renamed to ADEOS-II[11]. ADEOS-II is regarded the Japanese contribution in the framework of the International Earth Observation System (IEOS). Other parts of IEOS are EOS (USA), the POEM (ENVISAT and METOP) program of ESA (Europe). The overall mission objectives of ADEOS-II are:

- global change observation, dedicated to the following programs
- WCRP/GEWEX & CLIVER, IGBP and GCOS

S/C mass = 3500 kg, payload mass = 1200 kg, power = 5 kW, launch vehicle = H-II rocket, launch site = Tanegashima Space Center.

Orbit: Sun-synchronous subrecurrent orbit, altitude = 800 km, inclination = 99°, period = 101, recurrent period = 3 or 4 days, local time = 10:30 AM.

**Sensors**:

- **AMSR** = Advanced Microwave Scanning Radiometer (passive NASDA core sensor). Objectives: Ocean vapor profiles, precipitation, sea surface temperature, wind speed, ice. Microwave emission from the atmosphere, ocean, sea ice, and land are measured at multiple frequencies: 6.6, 10.65, 18.7 23.8, 36.5, (55), and 89 GHz. Polarization: H/V; temperature resolution: 0.3-1K; temperature accuracy: 1 K; spatial resolution: 5-60 km; swath width: 1700 km; scan rate: 40 rpm; instrument mass: 250 kg; power: 250 W; data rate: 100 kBit/s. From this information a number of geophysical data related to the Earth environment, such as water vapor content, water content of clouds, water equivalent of the snow cover, etc. are measured.

- **GLI** = Global Imager (NASDA core sensor)
  Spectral range: UV - TIR
  Objectives: Biological and physical processes, stratospheric ozone. GLI is for the study on and monitoring of the carbon cycle in the ocean, principally related to biological processes. Multispectral observations from the near UV to the near IR. Determination of chlorophyll pigment, phycobilin and dissolved organic matter (DOM) in the ocean; classification of phytoplankton according to their pigment.

  GLI instrument: spectral bands: 22 bands in VNIR, 5 bands in SWIR, 7 bands in TIR; spectral bandwidth: 10-15 nm (VNIR); SNR: 800; swath width: 2000 km; spatial resolution: 1 km; tilt angle: ± 20°; instrument mass: 350 kg; power: 400 W; data quantization: 12 bit; data rate: 5.2 Mbit/s.

- **DCS** = Data Collection System (NASDA system)

- **ILAS-II** = Improved Limb Atmospheric Spectrometer (JEA sensor)

- **SeaWinds** = NASA Scatterometer II (PI: M. Freilich, NASA/JPL). Scatterometer with six slotted waveguide Ku-Band "Stick" fan-beam antennas, previous designation was Stikscat. Objective: acquire accurate, high-resolution, continuous, all-weather measurements of near-surface vector winds over the ice-free global oceans. Application: studies of tropospheric dynamics and air-sea momentum fluxes.

- **TOMS** = Total Ozone Mapping Spectrometer (NASA/GSFC sensor). Objective: Observation of total ozone changes, evaluation of changes in UV radiation and the observation of sulfur dioxide (see description under ADEOS).

---

10)    "Monitoring the Earth Environment from Space", NASDA bulletin
11)    see CEOS WGD-11 Meeting Report, Nov. 5-7, 1991, Toulouse, Attachment 28a

- **POLDER-2** = Polarization and Directionality of the Earth's Reflectances (passive optical imaging radiometer, CNES/LERTS sensor, see description under ADEOS)

- **HiRDLS**  = High-Resolution Dynamics Limb Sounder (NASA and UK sensor, PIs: J. Barnett, Oxford University; J. Gille, NCAR). Option, see sensor description A.21.

The following sensors are candidates for the ADEOS-II follow-up missions (ADEOS-III, ).

- GLI-C = Global Imager (Continuous spectrum)
  This sensor performs the same type of observations as GLI. However, it does not observe in multispectral bands, but rather in a continuous spectrum.

- TERSE = Tunable Etalon Remote Sounder of Earth
  Spectral range: 2541 - 7941 $cm^{-1}$. Nadir scan. Swath width = 500 km
  Objective: Atmospheric chemistry, Tropospheric gases with high spectral resolution. Solar NIR radiation is measured by TERSE to determine the vertical distribution of tropospheric trace gases.

- TOMUIS = Total Ozone Mapping with UV Imaging Spectrometer
  Spectral range: 260 - 320 nm. Nadir scan, swath width = 1500 km.
  Objective: Atmospheric chemistry, ozone

- SLICES = Stratospheric Limb Infrared Emission Spectrometer
  Spectral range: 5 - 15 $\mu$. Nadir and limb scan
  Objectives: Atmospheric chemistry; stratospheric and tropospheric gases

- IMB = Investigator of the Micro-Biosphere
  Spectral range: 460 - 1200 nm. Nadir scan. Swath width of 100 and 100 km
  Objective: Ecological environment with high spatial resolution

- ADALT = Advanced Radar Altimeter
  Detailed measurements of the sea surface are used to determine the oceanic general circulation, ocean surface winds, wave height and wave spectrum In addition, extensions of sea ice and continental ice can also be observed.

- PR = Precipitation Radar (see also PR on TRMM)

- DPR = Dual-frequency Precipitation Radar
  Monitoring of hydrological processes (precipitation, evaporation and transpiration, snow cover, runoff, etc.). DPR provides measurements of 3-dimensional precipitation. It also permits the monitoring of rainfall and the estimation of of the sea surface wind velocity.


## A.5    AEM-2 (Applications Explorer Mission-2)

NASA/LaRC mission with the objective to monitor stratospheric aerosol and its influence on the climate (study of aerosol sources and sinks, aerosol transport, aerosol radiative and climatological implications, etc.).[12] Launch: Feb 18, 1979 by Scout vehicle from Wallops. The mission ended after 33 months of operation on November 11, 1981 due to a failure in the S/C power system.[13]

Application: stratospheric chemistry related to aerosol, ozone, and nitrogen dioxide. Provision of almost global profiles by solar occultation techniques.

Orbit: perigee = 549 km, apogee = 661 km, inclination = 55°, period = 96.8 min.

12)   M. P. McCormick, P. Hamill, T. J. Pepin, W. P. Chu, T. J. Swissler, L. R. McMaster, " Satellite Studies of the Stratospheric Aerosol", Bulletin of the American Meteorological Society, Vol. 60, No. 9, September 1979, pp. 1038-1046

13)   L. R. McMaster, M. W. Rowland, "SAGE-I Data User's Guide", NASA Reference Publication 1275, Aug. 1992

Sensor:

**SAGE-I** = Stratospheric Aerosol and Gas Experiment. The instrument is a sun photometer[14] [of SAM (on Apollo-Soyuz) and SAM-II (on Nimbus-7) heritage].

Objective: measurement of solar intensity profiles during each sunrise and sunset event (about 30 sampling opportunities per day, sunrise + sunset).
The instrument is a four-channel sun photometer. Spectral discrimination is achieved by using a holographic diffraction grating that disperses the different wavelengths in different directions. By placing four sensors at appropriate positions with respect to the grating, it is possible to measure light centered at wavelengths of 0.385 µm, 0.45 µm, 0.60 µm, and 1.0 µm.

The aerosol profiles can be interpreted to give concentrations of ozone, nitrogen, dioxide, and total molecular density. A number of ground truth measurements in the US, Japan, Europe, and with Nimbus-7 data are an integral part of the SAGE mission.

Data:
The instrument data include vertical profiles of stratospheric aerosol, ozone, and nitrogen dioxide. The SAGE-I aerosol data were validated by comparison with correlative lidar and dust-sonde in situ measurements, the ozone data were validated by comparison with balloon electrochemical cell ozone-sonde and sounding rocket measurements, the nitrogen dioxide measurements were compared with climatology data. All data are archived at NSSDC (National Space Science Data Center) at GSFC.


## A.6 ALEXIS (Array of Low Energy X−Ray Imaging Sensors)

ALEXIS is a sophisticated 'small satellite mission' of Los Alamos National Laboratory, Los Alamos, New Mexico. The spacecraft was launched by a Pegasus air-launched booster (OSC) on April 25, 1993. During powered flight a solar paddle was damaged, initial attempts to contact ALEXIS were unsuccessful. The satellite responded in June 1993 and soon was brought under control.[15]

The ALEXIS payload consists of an ultrasoft x-ray telescope array and a high-speed VHF receiver/digitizer with the name of Blackbeard. The S/C is spin-stabilized and uses solar pointing for orientation, S/C mass = 113 kg.

Orbit: Apogee = 844 km, Perigee = 749 km, inclination = 70°

Sensors:

**ALEXIS** = Array of Low Energy X−Ray Imaging Sensors (instrument has the same name as the S/C). Alexis is an outward-looking instrument for the detection of astrophysical signals. The instrument is an ultrasoft x-ray monitor, consisting of 6 compact normal-incidence telescopes tuned to narrow bands centered on 66, 71 and 93 eV. The 66 and 71 eV bandpasses are centered on a cluster of emission lines from Fe IX-XII. The 93 eV band, although designed as a continuum channel, includes Fe XXIII line characteristics of $10^7$ K plasma. [16]

The six ALEXIS telescopes are arranged in pairs covering three overlapping 33° FOVs. During each 45-second spin of the S/C, ALEXIS monitors the entire anti-solar hemisphere.

14) Note: A photometer is usually a broadband instrument capable of measuring thermal continuum radiation (i.e. flux) thereby permitting the study of energy balance and surface composition (also detection of infrared roughness of surface features)

15) W. Priedhorsky, B. W. Smith, J. J. Bloch, D. H. Holden, D. C. A. Roussel-Dupré, R. Dingler, R. Warner, G. Huffman, R. Miller, B. Dill, R. Fleeter, "The ALEXIS Small Satellite Project: Initial Flight Results", AIAA Space Programs and Technologies Conference, Sept. 21-23, 1993/ Huntsville, Al.; and Proc. SPIE Vol. 2006, 1993, pp.114-126

16) W. C. Priedhorsky, J. J. Bloch, S. P. Wallin, W. T. Armstrong, O. H. W. Siegmund, J. Griffee, R. Fleeter, "The ALEXIS Small Satellite Project: Better, Faster, Cheaper Faces Reality", IEEE Transactions on Nuclear Science, Vol. 40, Nr. 4, 4 August, 1993, pp. 863-873

The resolution of each telescope is limited by spherical aberration to about 0.5° diameter. The mass of the ALEXIS sensor is 30 kg, power = 30 W. Data: 10 kbit/s of event data. Measurement objectives: mapping of the diffuse background in three bands, performing a narrow-band survey of point sources, searching for transient phenomena, and monitoring variable ultrasoft x-ray sources.[17]

**Blackbeard**. The instrument looks at signals emitted near the Earth. Bleackbeard is a radio frequency experiment with the objective to study distortion and interference effects on transient trans-ionospheric VHF signals, such as lighting and artificial pulses. The instrument senses perturbations to the ionosphere, it can make a distinction between multi-path distortions resulting from large-scale coherent perturbations and from small-scale random perturbations to the ionosphere. The specific experiments of Bleackbeard include:

- Broad-band VHF measurements of transient signals originating from a controlled pulsed ground beacon, to characterize broad-band ionospheric distortion.
- Narrow-band VHF measurements of cw signals from a multichord interferometry ground beacon array, to characterize the ionospheric structure contributing to transmission distortion.
- Surveying of power envelopes of lightning and man-made interference in selectable VHF bands, for background rejection purposes.

Bleackbeard operation consists of an on-board 150 MHz digitization for 0.1 s in a broad-band reception mode; or 50 kHz digitization for 320 s in a narrow-band reception mode; or 120 kHz effective digitization for 130 s in a power-envelope survey mode. - The broad-band mode has selectable bandwidths up to 65 MHz within the ranges 25 - 100 MHz and 100 - 175 MHz, with a maximum of 30 dB SNR. The narrow-band mode has eight selectable 4 kHz bands between 32 and 36 MHz, with a maximum 40 dB SNR and 0.1 Hz Doppler resolution. A mixed-mode operation is available in which broad-band and narrow-band data are collected to allow correlated RF distortion and ionospheric structure analysis.

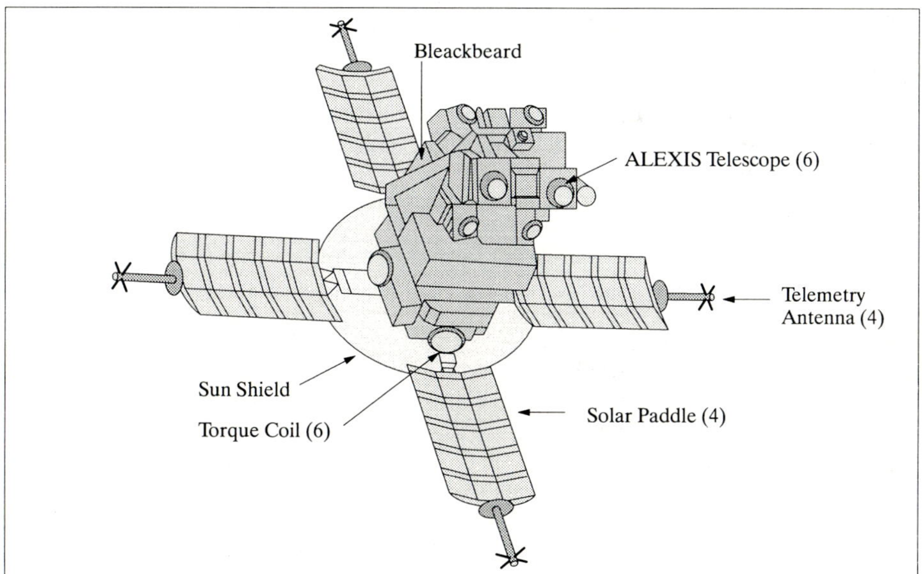


**Figure 7:** **The ALEXIS S/C Model**

17)   J. J. Bloch, et al., "Design, Performance and Calibration of the ALEXIS Ultrasoft X-Ray Telescopes", SPIE, Vol. 1344, 1990 , pp. 154-165

## A.7    ALMAZ Program

Background: The ALMAZ (="diamond") Earth Resources Satellite program of the former Soviet Union did not start with the launch of ALMAZ-1 in March 1991. At NPO Machinostroyenia in Reutov near Moscow, S/C of the ALMAZ series have been developed for more than 25 years. The program is based on the technical experience gained at NPO Machinostroyenia when developing manned- and automatic space stations - the Salyut Space Stations of the 60's and 70's were a very essential part of the ALMAZ program. The former USSR needed an all-weather remote sensing capability, based on SAR technology, due to the cloud coverage conditions predominant in the northern latitudes of its land mass. Even today (1993), ALMAZ automatic (i.e. unmanned) spacecraft are actually of the type 'Salyut space station' - a proven technology - such a S/C can even be visited by a crew in case of serious problems.

Some of the ALMAZ program objectives:

- Development of a series of automatic space stations on the basis of a universal ALMAZ platform
- Development of a sensor complement with an upgrading capability in the sensor performance characteristics with regard to spatial resolution, coverage, spectral characteristics, etc. Increase of the number of instruments and the number of solvable tasks with every mission.
- Development of a data system on-board and on-ground and a corresponding infrastructure to cope with the vast volumes of data.
- Provision of an all-weather remote sensing capability for the support of the following tasks:
    - Prospecting for natural resources and applying them to many fields of the national economy
    - Collection of remote sensing data for ecological analysis and use
    - Observation capability of natural catastrophes/disasters of actual interest.

## A.7.1    COSMOS-1870

This mission is regarded as the first USSR radar mission (ALMAZ prototype mission). The S/C was launched on 25. July, 1987 with a Proton vehicle from the Baikonur Cosmodrome launch facility, the COSMOS-1870 mission ended 30. July, 1989.

Orbit: altitude = 275 km, inclination = 73°, orbital period = 92 minutes

Spacecraft mass = 18550 kg, payload mass = 1950 kg, design life = 2 years, attitude precision = 15-20', stabilization precision = 4-6'.

Sensor:

- **SAR** = Synthetic Aperture Radar (S-Band; freq. = 3.125 GHz, instrument built by NPO Vega, Moscow). Wavelength = 9.6 cm, spatial resolution = 20 - 25 m, swath width = 20 km.

The S/C carried two SAR instruments, one on each side for surveying. The S/C had the capability to roll about its axis thereby extending the pointing range of the SAR antennas in the cross-track direction to 250 km swath range (swath width is nearly constant at 20 km). The S/C had 3 tons of fuel aboard permitting a lot of roll maneuvers for this type of operational support.

Data: On-board recording capability and subsequent stored data dump during passes over ground stations or of real-time data. Data transmission rate: 90 Mbit/s.

## A.7.2    ALMAZ-1

ALMAZ-1 [18),19),20),21),22),23)] was launched on March 31, 1991 (Proton booster from the Baikonur Cosmodrome launch facility, ALMAZ S/C designer and builder: NPO Machinostroyenia, Reutov, Moscow Region), the operational phase started in May 1991. End of mission: Oct. 17, 1992 (controlled descent into the Pacific ocean due to lack of fuel).

The ALMAZ-1 S/C has a total mass of 18550 kg, and a payload mass of 3420 kg. Attitude precision = 15-20', stabilization precision = 4-6'. The stabilization precision during the SAR operation is 1'.

**ALMAZ-1** is considered an important operational satellite (after SEASAT in 1978, SIR-A in 1981, SIR-B in 1984 and Cosmos-1870) with a radar sensor for earth observation. The Soviet space agency Glavkosmos has contractual agreements (Western marketing rights of Almaz data) with Almaz Corp., a subsidiary of the Houston-based Space Commerce Corp. (SCC). The Radar data of ALMAZ are complementary to the Spot and Landsat data. ALMAZ-1 data are also considered to be in direct competition with ERS-1 and JERS-1 data.

Application: Oceanology (study of the distribution and dynamics of currents and hydrospheric fronts, the spatial structure of wave formations and turbulence, the evaluation of surface winds and hurricanes, the topography of the ocean floor and its effects on wave patterns at the surface, the identification of oil spills and other forms of pollution, the state of the ice cover and its seasonal variations, boundaries of water exchange of rivers with the ocean, etc. );  Geology (structure of geological formations (such as folds, valleys and fractures) and the nature of volcanic activity, survey of mineral deposits, etc. ); Cartography and Geophysics (topographical maps, climatic changes (such as ice thawing and desertification)); Agriculture and Forestry (large scale evaluation of agricultural lands and crops, overall volume of the biomass in a region, soil moisture conditions, etc.); Statistic, Ecology, etc.

Orbit: Inclination = 72.7°; altitude = 270-380 km, Orbital period = 92 Minutes;

**Sensors:**

- **SAR** = Synthetic Aperture Radar (S-Band; freq. = 3.125 GHz, wavelength = 9.6 cm, built by NPO Vega).
    - Resolution = 10 - 15 m (depending on range and azimuth).
    - SAR images can be taken from each side of the satellite the swath width of each SAR is 40 km  within a swath range of 350 km (obtained by rolling the S/C)
    - Observation (incidence) angles: 30-60°
    - Radiometric resolution: 2-3 dB
- **UHF Radiometer**. Wavelengths of 0.8 cm, 5 cm, 11 μm, 12 μm, and 13.7 μm (or 37.5 GHz, 6 GHz, 2.72 THz, 2.5 THz, and 2.19 THz respectively); swath width = 10 - 30 km, swath range = 500 km; temperature resolution = 0.1 - 0.3 K, spatial resolution = 5km. Objective: compilation of an Earth surface temperature map.

**Data:**

Ground station in Moscow Region. The ALMAZ-1 schedule calls for a transmission of 60 images (scenes/day). A scene = 40 km x 300 km. Data rate: 10 Mbit/s. Operational modes: onboard data recording and subsequent transmission via relay satellite to a DRP (Data Reception Point).


18)    "Soviets Launch Largest Earth Resources Satellite on Modified Salyut Platform", Aviation Week & Space Technology/April 8 1991, pp. 21-22
19)    "Almaz to add Dimension to Earth Study", Space News, March 18-24 1991, pp. 1
20)    "ALMAS - Sowjetischer Erdsatellit mit Synthetic Aperture Radar zur Erderkundung", IKF Berlin, 1990, aus der Reihe: Informationen aus der internationalen Zusammenarbeit.
21)    "Almaz to add Dimension to Earth Study", Space News, March 18-24 1991, pp. 1
22)    "Sowjetisches Weltraumauge sammelt Ströme digitaler Daten", VDI Nachrichten, 21 Dez. 1990, Seite 20
23)    'Almaz Falls from Orbit', Space News, Oct 26-Nov. 1, 1992, p. 1

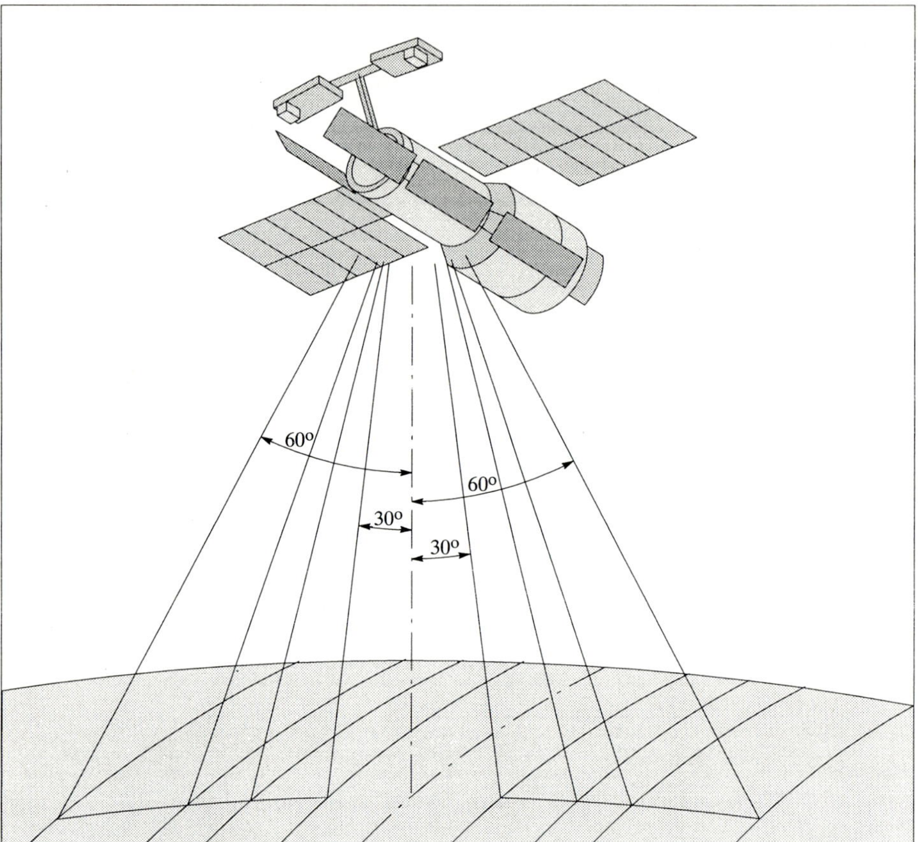


**Figure 8:** The Almaz-1 S/C Model and its SAR Sensor's Observation Geometry

## A.7.3 ALMAZ-1B

ALMAZ-1B is being developed on the basis of the universal ALMAZ space platform and is an upgrading of the flown ALMAZ-1 spacecraft. Major objectives: collection of complex data in the UHF, optical, infrared, and microwave ranges for a diverse variety of applications such as: cartography, land monitoring, geology, prospecting, ecological monitoring, oceanography, fishery and ship navigation, and a monitoring service of emergency events on Earth. S/C builder, integrator, operator and data archiver/distributor is NPO Machinostroyenia of Reutov, Moscow region.[24]

A launch is planned for the end of 1996 with a Proton launch vehicle from Baikonur, nominal life = 3 years (S/C operations for user community from 1996-98). Mass of station = 18550 kg, total payload mass = 4500 kg; orientation accuracy = 10 arc minutes, stabilization = 4 arc minutes, orientation control accuracy = 1 arc minute. On-board fuel >3 tons.

Data transmission modes:
- Direct transmission to existing ground stations in Russia and abroad
- On-board storage and data dumping capability to ground stations (up to 12 hours of storage capacity)

---

24) Information provided by NPO Machinostroyenia (P. A. Shirokov, L. A. Tararin, et. al.)

- On-board storage and downlink transmission via a relay satellite to a central ground station

Data transmission rates: (direct link to ground stations), some SAR data preprocessing capability onboard.

- Real-time transmission = 245.6 Mbit/s
- Store-and-forward transmission via a relay satellite at 10 Mbit/s
- Direct transmission to mobile and small user receiving points = 3 Mbit/s and 665 kBit/s.

RF link characteristics:

- S-Band (3.13 GHz) transmission at 2 Mbit/s (operative S/C data, TT&C)
- X-Band (8.6 GHz) downlink transmission of science data at 122.8 Mbit/s (2 links)
- P-Band (0.43 GHz) downlink transmission of science data at at 665 kbit/s (APT).

Orbit: Non-sun-synchronous circular orbit, altitude = 400 km, inclination = 73°, period = 90 minutes (16 orbits/day). The inclination of 73° and the swath range capability of the SAR instruments provide for an effective latitudinal coverage of ± 78°.

**Sensors:**

A three-frequency radar complex, consisting of an X-Band, S-Band, and P-Band SAR, as well as an X-Band SLR. The operation of the SAR instruments is featuring a survey coverage extension in the cross-range direction if requested by the data users (or if of general interest). Although the radar antennas are fixed to each side of the S/C, pointing at an angle of 35° from nadir, there is a capability of turning the S/C in the transverse direction (roll), thereby obtaining an incidence angle coverage from 25 - 51° for the SARs, and 38 - 60° for the SLR sensor. This roll capability of the S/C provides a potential swath range of 330 km to each side of the S/C in which the operational SAR instrument(s) can survey the terrain in the selected swath width configuration (in the range of 30-170 km depending on resolution requirements). - All ALMAZ-1B radar instruments are built by NPO Vega, Moscow.

The duty cycle for all radar instruments combined is 20 min/orbit (22.2%), i.e. one radar may operate for 20 minutes per orbit, or the 20 minutes may be divided up between all instruments to suit the observation requirements.

| Parameter | SLR-3 | SAR-3 | SAR-10 | SAR-70 |
|---|---|---|---|---|
| Wavelength Frequency | 3.5 cm; 8.6 GHz (incoherent mode) | 3.5 cm; 8.6 GHz coherent mode | 9.6 cm; 3.13 GHz | 70 cm; 0.430 GHz |
| Frequency band | X-Band | X-Band | S-Band | P-Band |
| Range of incidence angles | 38-60° | 25-51° | 21-51° | 25-51° |
| Swath range | 450 km | 330 km | 330 km | 330 km |
| Swath width | 450 km | 20-35 km | 120-170 km wide 60-70 km interm 30-55 km narrow | 120-170 km |
| Survey direction | left side | left side | left and right side | right side |
| Length of a scene | | | up to 2 min | up to 2 min |
| Resolution | 190-250 m, range 1.2-2 km, azimuth | 5-7 m | 15-40 m, wide 15 m, intermediate 5-7 m, narrow | 22-40 m |
| Polarization (transmit/receive) | VV | VV | VV, wide swath VHV or HHV, int. HH, narrow | VVH or HVH |
| SAR operation | SAR-3 + SAR-10 may operate simultaneously SAR-10 and SAR-70 may operate simultaneously SLR-3 and SAR-3 are one instrument operating either in incoherent mode (as SLR) or in coherent mode (as SAR) | | | |

**Table 5:**     **Main Characteristics of the ALMAZ-1B SAR Sensors**

- **SLR-3** = X-Band Side-looking Real-Aperture Radar ( wavelength = 3.5 cm, frequency = 8.6 GHz) **incoherent mode**. This instrument is functionally on another board, i.e. separated from the SAR-3 board. However, SLR-3 and SAR-3 use the same antenna; hence, they are operationally mutually exclusive.

- **SAR-3** = X-Band SAR (wavelength = 3.5 cm, frequency = 8.6 GHz) **coherent mode**.

- **SAR-10** = S-Band SAR (wavelength = 9.6 cm, frequency = 3.13 GHz). The instrument offers three different swath widths with corresponding resolutions. SAR-10 provides an observation coverage to each side of the ground track (left and right looking antenna).
    - Swath width = 120 - 170 km (wide mode) with a resolution of 15 - 40 m
    - Swath width = 60 - 70 km (intermediate mode), resolution = 15 m
    - Swath width = 30 - 55 km (narrow mode), resolution = 5 - 7 m

- **SAR-70** = P-Band SAR  (wavelength = 70 cm, frequency = 0.43 GHz). The longer wavelength of 70 cm has the ability of surface penetration (in the order of a few centimeters) this depends, however, very much on the soil moisture content.

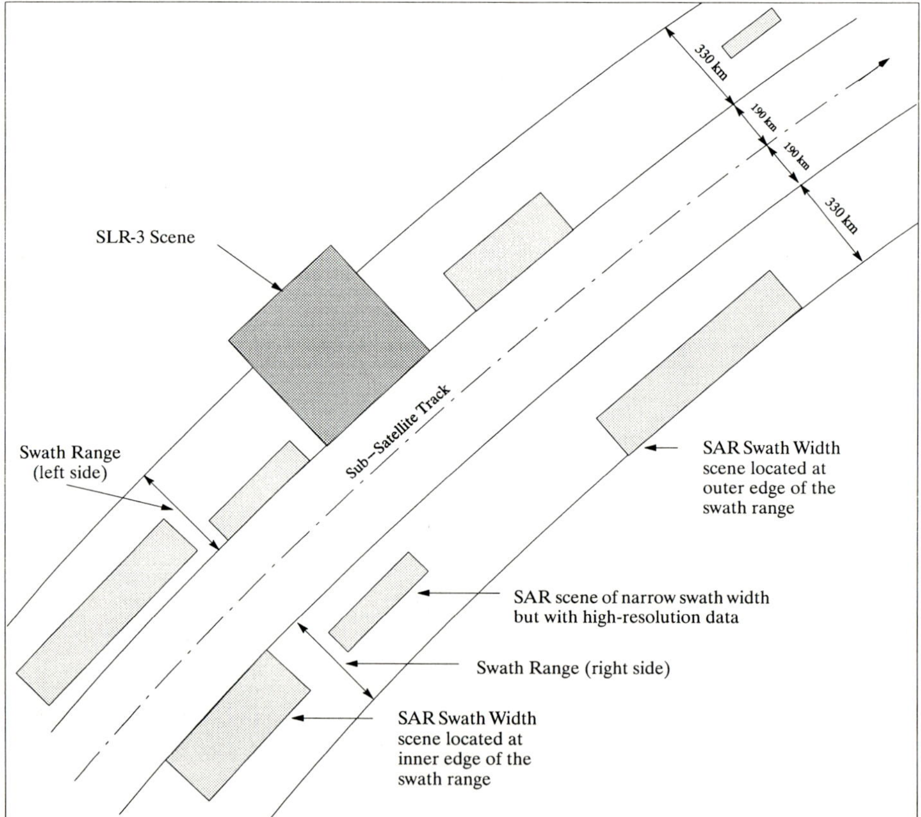


**Figure 9:**    **Observational Coverage Geometries of the ALMAZ-1B Radar Sensors**

- **MSU-E** = Optical Multispectral Scanner (electronic scanning, built by ISDE). Measures in three visible channels (0.5 to 0.6 µm, 0.6 to 0.7 µm, and 0.8 to 0.9 µm); the resolution is 10 m; the swath width is 24 km. There are 2 MSU-E devices on ALMAZ-1B, one on each side of the S/C, pointing 25° off nadir in the cross-range direction. Each MSU-E is looking into the same scene into which the operating radar is looking, allowing superposition of imagery. MSU-E capable to change its viewing angles in the cross-range direction from -5° to 55° due to a turn of a scan mirror. This feature provides a potential swath range of 400 km to each side of the S/C. The duty cycle of MSU-E is 100%, both instruments operate simultaneously.

- **MSU-SK** = Optical Multispectral Scanner - moderate resolution. (heritage: from Meteor and Cosmos 1639, the sensor is being built by the 'Institute of Space Device Engineering', ISDE). MSU-SK measures in 4 channels of the visible spectrum: (0.5 to 0.6 µm, 0.6 to 0.7 µm, 0.7 to 0.8 µm, and 0.8 to 1.1 µm) with a resolution of 80 m in the visual range; MSU-SK measures also in the infrared spectrum (10.4 to 12.6 µm) with a resolution of about 300 m. The swath width is 300 km. MSU-SK is nadir-looking; the duty cycle is 100%.

- **SROM** = Spectroradiometer for Ocean Monitoring (built by NPO Geophysica, Moscow; it is expected that the sensor will be ready and tested prior to launch time of AL-MAZ-1B). The SROM sensor is analogous to SeaWiFS in spectral range coverage offering additional ranges in the infrared region. SROM is nadir-pointing with a swath width of 2200 km (1100 km to either side of the ground track); the spatial resolution at nadir is 0.6 km; the temperature resolution = 0.1 K.
Objective: observation of chlorophyll content, evaluating the bio-productivity of the world ocean regions.
  - Band 1    0.405 - 0.422 µm
  - Band 2    0.433 - 0.453 µm
  - Band 3    0.480 - 0.500 µm
  - Band 4    0.510 - 0.530 µm
  - Band 5    0.555 - 0.575 µm
  - Band 6    0.655 - 0.675 µm
  - Band 7    0.745 - 0.785 µm
  - Band 8    0.843 - 0.884 µm
  - Band 9    3.6 - 3.9 µm (MWIR)
  - Band 10   10.5 - 11.5 µm (TIR)
  - Band 11   11.5 - 12.5 µm (TIR)

- **OSSI** = Optronic Sensor for Stereo Imagery (built by NPO ELAS, Moscow). Objective: cartographic applications. There are two instruments pointing at nadir with a look angle of ±25° (forward and backward of nadir). OSSI employs the CCD detector technology, it provides one telescope for each camera. The swath range is ±300 km, this is accomplished by rolling the S/C about its longitudinal axis (the S/C rolling is only done for special observation requirements). The swath width of OSSI is 80 km. The duty cycle of OSSI is 12 min/orbit (13.3%), the small duty cycle is due to on-board data recording limitations (400 Gbit of digital data is the recorder capacity for all data streams). The data rate of OSSI is 560 Mbit/s. Spatial resolution: 2.5 - 4 m.
Spectral ranges: 0.5 - 0.6 µm, 0.6 - 0.7 µm, 0.7 - 0.8 µm, and 0.58 - 0.8 µm

- **Balkan-2** = Lidar (built by the Institute of Atmospheric Optics, Tomsk; the instrument has a heritage of Balkan-1 on MIR-1). The lidar comprises a scanner capable of changing the sensing direction transverse to the flight path on either side within 10°.
  - Lidar type (solid state laser)           Nd: YAG
  - Wavelength                               532 nm
  - Sensing frequency                        1 Hz
  - Pulse duration                           10 ns
  - Altitude resolution in lidar mode        3 - 10 m
  - Alt. resolution in the range-finder mode 0.5 - 1 m
  - Angle of radiation                       40 seconds of arc

- Look angle from nadir (flight direction)     $\pm 10^\circ$
- Swath width / coverage                        90 m / 140 km
- Data volume per session                       64 kByte

NPO Machinostroyenia in Reutov is not only the S/C builder and integrator, but also the S/C operator, as well as the data archiver and distributor to the user community. The development of improved processing algorithms for radar imagery is a continuous effort that is supported by many organizations.

| Measured Parameter | | Measurement Error | Resolution Vertical/horizontal |
|---|---|---|---|
| Cloudiness | Upper boundary of clouds<br>Vertical profile of scattering coefficient<br>(water content) | 50 m<br>20% | 5 / 16000 m |
| Aerosol formations | Altitude of upper and lower aerosol layers<br>Vert. profile of scattering coefficient<br>(aerosol weight concentration) | 50 m<br>30% | 5 / 16000 m |
| Depth of shallow areas of shelf | | $\pm 0.5$ m | 0.5 / 16000 m |
| Average wave heights | | $\pm 0.5$ m | 0.5 / 16000 m |
| Muddiness of ocean upper layers | Secchi depth (optical depth)<br>Surface albedo | $\pm 2$ m<br>15% | - / 16000 m<br>- / 16000 m |
| Sand hills height and albedo | Height<br>Albedo | $\pm 0.5$ m<br>15% | 0.5 / 16000 m<br>- / 16000 m |
| Height of tree tops | | $\pm 0.5$ m | 0.5 / 16000 m |
| Detection of bioproductive zones in the ocean, shoals | | | |
| Detection of oil films on the water surface | | | |

**Table 6:     Potential Applications of Balkan-2 Measurements**

| Cartography | Land Monitoring | Geology and Prospecting | Ecological Monitoring | Oceanology, Fishery, Navigation aids | Emergency Information Provision |
|---|---|---|---|---|---|
| Major applications categories of ALMAZ-1B instruments | | | | | |
| Updating of topographic maps<br>Scale:<br>1:100.000;<br>1:50.000;<br>1:25.000 | Data for Earth cadential survey; lands, forests, water resources, etc. Compiling soil maps, determination of humus content | Areas for mineral prospecting; Geologic mapping scale:<br>1:50.000;<br>1:100.000<br>1:250.000 | Measuring of chemical compound content in the soil Monitoring technogenic actions[25] Soil salinity and desertification | Evaluation of sea ice conditions for ship pilotage Observation of chlorophyll content; evaluation of the bio-productivity | Observation of dangerous regions (Earthquakes, mudslides, volcanoes, etc.) Detection of forest and peat fires |
| Data for transport network planning | Harvest inventory; Determining the soil temperature and the moisture content | Survey area: 5-6 million km$^2$ yearly for geologic mapping | Soil erosion; Pollution monitoring of water surfaces | Compiling ocean temp. maps; studying fronts of currents; winds, etc. | Observation of large accidents (fires, explosions, et. |
| Mapping of coastal shelves | Detection of underground water sources | Monitoring of large open mining areas | Detection of underground waste deposits; Vertical profiles of scattering coefficient | Survey area of 0.5-1 million km$^2$ | Observation of transport accidents, including underground pipelines |
| Instruments for Observation Task Category | | | | | |
| SAR-3, SAR-10, SAR-70, OSSI, MSU-E, Balkan-2 | SAR-10, SAR-70, MSU-E, -SK, OSSI, SAR-3 | OSSI, SAR-10, SAR-70, MSU-E, -SK, SAR-3 | SLR-3, SAR-10, MSU-E, -SK, Balkan-2 SAR-3 | SLR-3, SAR-10, SAR-70, MSU-E, -SK, Balkan-2 SAR-3 | SAR-10, SAR-70, MSU-E, OSSI, Balkan-2 SAR-3 |

**Table 7:     Overview of potential Almaz-1B Data Applications**

---

25)   Note: the term 'technogenic actions' (in Table 7) refers to the variations in the natural environment caused by human actions or activities (e.g. constructions, operation of hydro-, thermal- and atomic power stations, etc.)

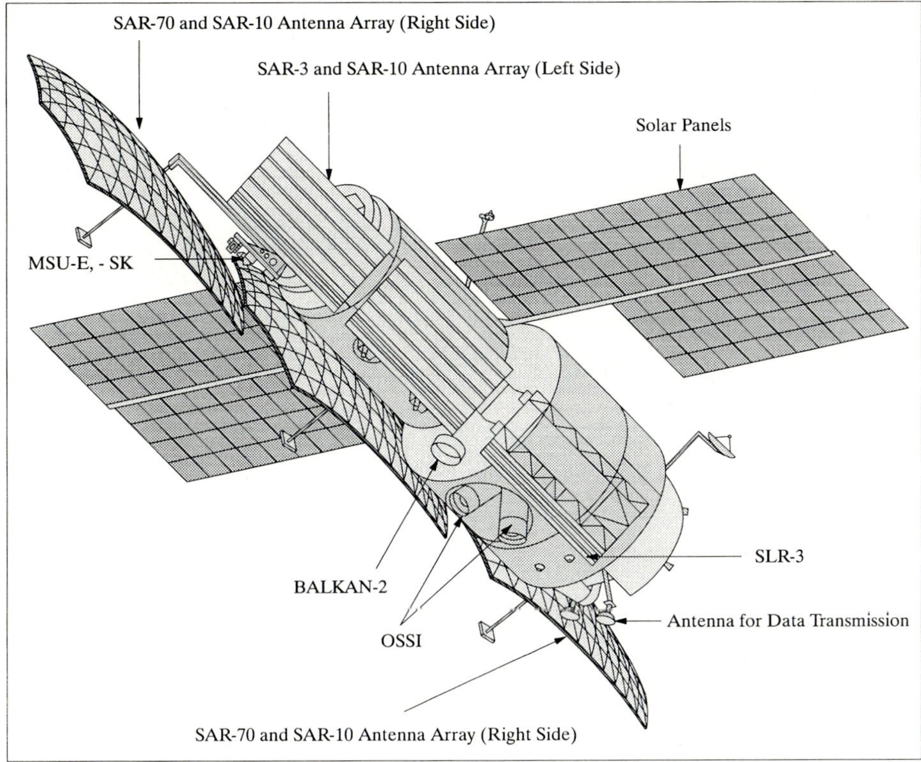


**Figure 10:**    **The ALMAZ-1B S/C Model**


## A.8    AMPTE (Active Magnetosphere Tracer Explorers)

AMPTE is a three satellite cooperative mission of the United States, Germany and the United Kingdom. Objectives: studies of solar-terrestrial interactions, in particular the interaction processes between two cosmic plasmas. Further aims of this mission are a systematic exploration of the highly variable mass and charge composition of the natural plasma population and a detailed investigation of the magnetospheric boundaries.[26),27),28)]

AMPTE uses artificial injection of rare ionic species for long-range tracing of mass transport into and through the magnetosphere system. Injection of lithium, barium and europium ions into the solar wind and magnetotail.

The AMPTE program consists of three spacecraft - the Ion Release Module (IRM) provided by Germany, the United Kingdom Subsatellite (UKS), and the US/NASA Charge Composition Explorer (CCE). All S/C were launched in a stack on a single Delta vehicle on Aug. 16, 1984 from Cape Canaveral.


26)    ampte brochure of MPE Garching
27)    Special Issue on the Active Magnetosphere Particle Tracer Explorer (AMPTE), in IEEE Trans. on Geoscience and Remote Sensing, May 1985, Volume GE-23, Nr. 3, pp. 175-314
28)    A. Valenzuela, G. Haerendel, H. Föppl, F. Melzner, H. Neuss, E. Riegler, J. Stöcker, O. Bauer, H. Höfner, J. Loidl, "The AMPTE artificial comet experiments", reprinted from Nature Vol. 320, Nr. 6064 pp. 700-723, 24 April 1986

| Final Orbits | CCE | IRM | UKS |
|---|---|---|---|
| Apogee (geocentric) | 8.8 $R_E$ | 18.7 $R_E$ | 18.7 $R_E$ |
| Perigee (height) | 1113 km | 552 km | 552 km |
| Period | 15.6 hrs | 44.3 hrs | 44.3 hrs |
| Inclination | 4.82º | 28.68º | 28.68º |
| Releases (active injections) | | 7 | |
| 1. releases (Sept. 1984) of lithium tracer ions into the solar wind near the 'nose' of the magnetosphere and close to the Earth-Sun line. | | | |
| 2. release (Dec. 27, 1984) called 'artificial comet' with IRM located in the dawn magnetosheath | | | |
| 3. and 4. releases (March and May 1985) with IRM located in the Earth's magnetotail region | | | |

**Table 8:** **Orbit Parameters and major Release Periods of the AMPTE Spacecraft**

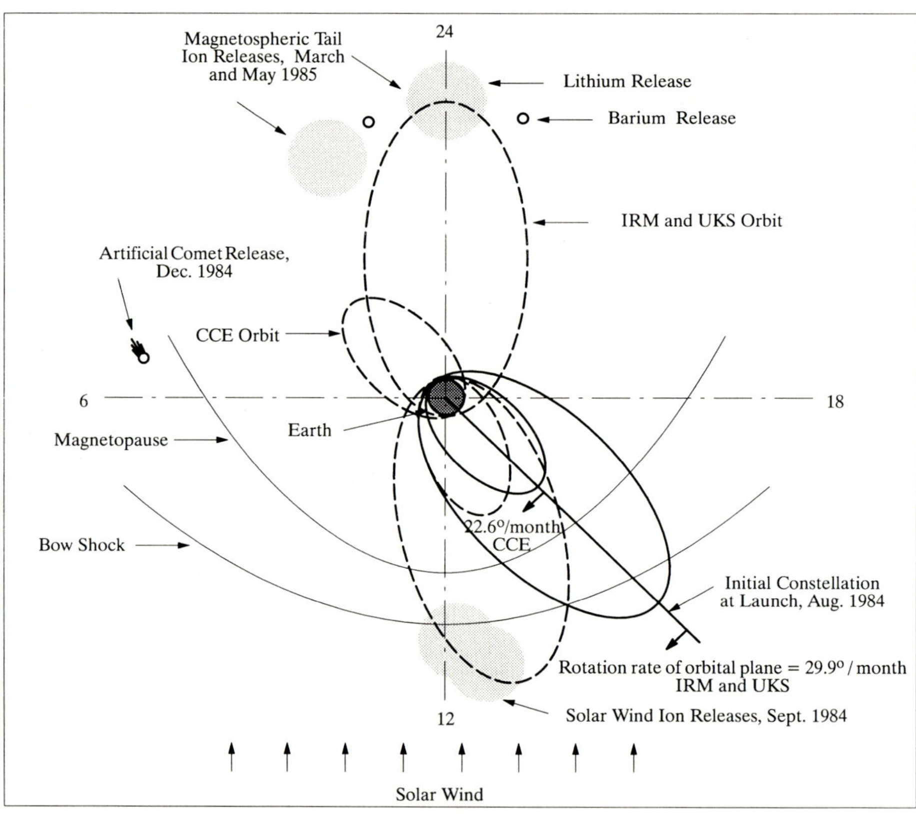


**Figure 11:** **Orbital Plane Constellations at major Events in the AMPTE Mission**

The IRM S/C has a mass of 705 kg. It carries scientific instruments and 16 canisters, 8 of which are filled with 5.8 kg of a copper-oxide-lithium mixture, the remaining eight are filled with a copper-oxide-barium mixture. The canisters are released in pairs by ground command, and the copper-oxide thermite reactions, which vaporize the tracer elements, are initiated by internal timers triggered during the release. A total of seven releases were planned over a period of 8 months which could be monitored by the maneuverable UKS following 3 minutes behind IRM and by CCE in a much lower orbit.

The IRM S/C was developed and built by MPE Garching and operated by GSOC (of DFVLR) at Oberpfaffenhofen. IRM failed in eclipse on Aug. 12, 1986.

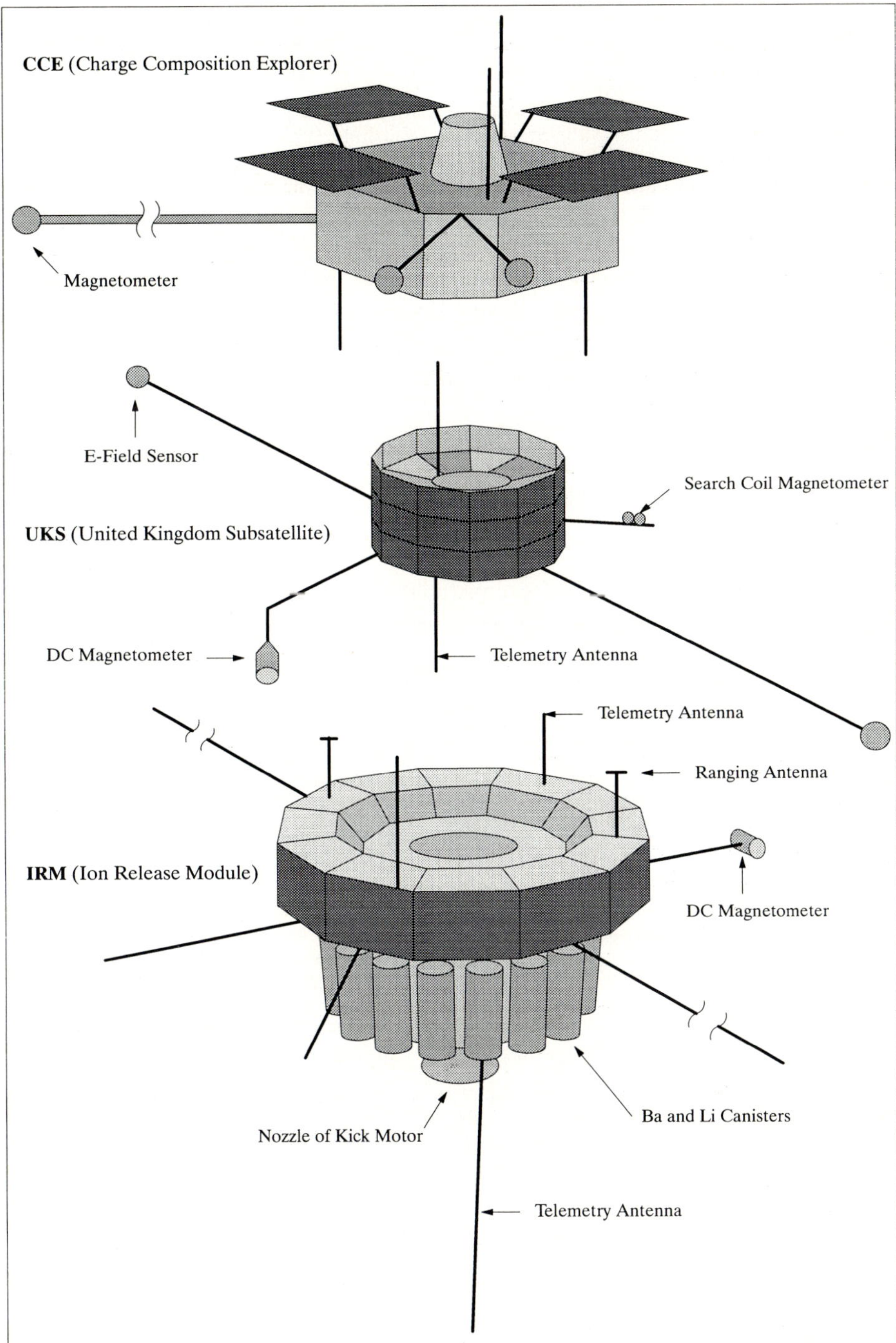


**Figure 12:    The AMPTE Mission S/C Models (Launch Stack Configuration)**

The UKS spacecraft has a mass of 78 kg. It is intended to detect and measure the ion releases in the vicinity of the IRM S/C. The UKS spacecraft was developed and built by RAL and operated by SERC. UKS failed unexpectedly on Jan. 16, 1985.

The CCE spacecraft weighs 240 kg. It's function is to detect and measure trace quantities of the injected ion releases at far distances. APL of Johns Hopkins developed and built the CCE spacecraft under NASA/GSFC contract, JPL provided tracking and S/C operations support. CCE operated fully until Jan. 10, 1989 and then intermittently until July 17, 1989 when the mission was terminated.

In September 1984, lithium was released into the solar wind by IRM, preliminary results indicated that less than 1% of the solar wind gained access to the Earth's magnetosphere. On December 27 1984, IRM released two barium canisters; they were observed to explode 10 minutes later by UKS, CCE, two high-flying aircraft and ground stations in the Pacific. The expanding cloud of the barium mixture (in a position of about 103000 km above the Pacific ocean) ionized within seconds by solar UV radiation, fluorescing in the dawn sky as seen from the western US. After about 12 minutes an 'artificial comet' of about 225 km diameter had formed, with a tail developing longer than 9000 km.[29]

## A.8.1    IRM Instrumentation (Sensors)

**IPIP** = IRM Plasma Instrument Package (3-D Plasma Analyzer). The instrument package consists of three sensors. Two of them measure the 3-D velocity distribution of ions and electrons. The third sensor is a retarding potential analyzer (RPA) for low-energy electron measurements; symmetrical quadrispheres. Energy ranges: $\sim 0$ eV - 25 eV, and 15 eV - 30 keV.

**MSIS** = Mass-Separating Ion Spectrometer. Objective: good mass-imaging characteristics and a wide fan-shaped angle of acceptance such that all ion species of interest can be measured simultaneously. Measurement technique: quadrispherical E/q analysis , magnetic analysis. Coverage: 0.01 - 12 keV/q.

**SULEICA** = Suprathermal Energy Ionic Charge Analyzer. Measurement of the ionic charge state and mass composition of all major ions from H through Fe for energies of the suprathermal plasma ($\sim 5$ - 270 keV/q) by the use of electrostatic deflection, time-of-flight (TOF) measurement, and energy analysis in solid-state detectors. Measurement technique: electrostatic analyzer, time-of-flight and total E. Coverage: 10 - 300 keV/q. The data contain counting rate information for the evaluation of absolute particle fluxes, live pulse-height events, and matrix rates for selected ions.

**Magnetometer**. Measurement technique: vector fluxgate (triaxial). Objective: measurement of the magnetic variations during the artificial plasma-cloud injections and the study of magnetospheric boundary layers and the ring-current region. Coverage: DC - 50 Hz. Dynamic range = 0.1 - 60000 nT; resolution = 16 bit analog to digital conversion.

**PWI** = Plasma Wave Instrumentation (spectrometer). Measurement of plasma wave activity in the magnetosphere/solar wind environment and inside the artificial plasma clouds. Measurement technique: 42 m tip-to-tip dipole antenna, boom-mounted search coils. Coverage: E-field: DC - 5 Hz; B-field: 30 Hz - 1 MHz.

**Lithium/Barium Experiments**. Measurement technique: Copper-oxide thermite reaction. The injection of lithium and barium plasmas into the solar wind and geomagnetic tail constitute the active element of the AMPTE mission. The objective of the releases is ion tracing and the study of the interaction of two vastly different plasmas in space.

-    8 Lithium release canisters (52 kg)
-    8 Barium release canisters (108 kg)

---

29)   'AMPTE', interavia Space Directory 1992-93, p. 149

## A.8.2    UKS Instrumentation (Sensors)

**3D-Ion Analyzer**. Objective: measurement of the 3-D distribution function of positive ions. Measurement technique: electrostatic analyzer. Coverage: 10 eV/q - 20 keV/q.

**3D-Electron Analyzer**. Objective: measurement of electrons (count rates). Measurement technique: electrostatic analyzer. Coverage: 6 eV - 25 keV.

**SPACE** = S/C Particle Correlator Experiment. Objective: Measurement of particle modulations (ions and electrons) resulting from local wave-particle interactions. Measurement technique: electron and ion signal processing. Coverage: ~1 Hz - ~1 MHz.

**Magnetometer** (of ISEE1/2 heritage). Measurement technique: vector fluxgate (triaxial), boom-mounted about 1 m from the S/C body. Coverage: DC - 10 Hz.

**Plasma Wave Spectrometer**. Objective: Measurement of the electric and magnetic field components. Measurement technique: 7 m tip-to-tip probe. Coverage: E-field: 30 Hz - 132 kHz, 4 spot frequencies to 2 MHz; B-Field: 30 Hz - 50 kHz (boom-mounted search coil).

## A.8.3    CCE Instrumentation (Sensors)

**HPCE** = Hot Plasma Composition Spectrometer (energetic ion-mass spectrometer and an electron background-environment monitor). Heritage of GEOS-1/2, ISEE-1 and DE-1 missions. Measurement technique: retarding potential electrostatic analyzer, ExB analyzer. Coverage of ion composition: 0 eV/q - 17 keV/q. Measurement technique of electrons: magnetic analyzers. Coverage: 50 eV - 25 keV.

**CHEM** = Charge Energy Mass Spectrometer. Measurement of energy spectra, pitch angle distributions, and ionization states. Measurement technique: electrostatic analyzer, time-of-flight and total E. Coverage: ion composition in the range from ~1 keV/q - 300 keV/q.

**MEPA** = Medium Energy Particle Analyzer. Measurement technique: time-of-flight and total E. Coverage: ion composition in the range of 10 keV/nucleon - > 1.0 MeV/nucleon.

**Magnetometer**. Measurement technique: vector fluxgate (triaxial). Coverage: DC - 50 Hz.

**Plasma Wave Spectrometer**. Objective: provision of first-order correlative information for studies of strong wave-particle interactions that develop close to the magnetic equator or have maximum effectiveness there. Measurement technique: electric dipole. Coverage: AC E-fields. 5 Hz - 178 kHz.

## A.9    APEX (Active Plasma Experiment)

APEX is a cooperative Solar-Terrestrial mission within the Intercosmos program (Russia/CSFR/Hungary/Germany/Poland/Bulgaria/Romania).[30],[31]
Objectives: study of terrestrial magnetospheric and auroral ionospheric relationships by the injection and monitoring of electron and ion beams by a parent/sub-satellite pair. The injection of electron and ion beams into the magnetosphere is provided by the parent craft.

- Simulation and initiation of aurora and radio frequency radiation in an auroral region.
- Study of the dynamics of modulated beams and plasmoids in the near-Earth plasma.
- Study of the nature of electrodynamic relationships of electromagnetic waves in the magnetosphere and ionosphere.
- Determination of radio emission characteristics of modulated beams of charged particles and plasmoids.

30)   "The Active Plasma Experiments in the Earth's Magnetometers", The Solar-Terrestrial Science Project of the Inter-Agency Consultative Group, esa SP-1107, November, 1990, pp. 55-60
31)   Information provided by Yu. M. Mikhailov, IZMIRAN

- Search for non-linear wave structures of the electromagnetic soliton type in a disturbed environment.

Launch: December 18, 1991 by SL-14 from Plesetsk. APEX mass = 1350 kg, payload mass = 340 kg (parent satellite), Magion-3 (sub-satellite) = 52 kg. Science program leader: V. N. Oraevski (IZMIRAN). APEX is an operational mission as of 1993.

Orbit: apogee = 3071 km, perigee = 438 km, inclination = 82.56°,

**Sensors:**

**UEM-2** = Electron Accelerator (PI: V. Dokukin, IZMIRAN). Objective: Injection of a modulated electron beam with an energy of 10 keV. The beam current is in the range from $10^{-2}$ A to 15 A. The frequency modulation is in the range from 10 Hz - $2.5 \times 10^5$ Hz.

**UPM** = Neutral Plasma Accelerator (PI: V. Dokukin, IZMIRAN). Objective: injection of xenon ions. The beam has an energy of 300 eV, the beam current is 2A, the frequency modulation range is from 60 - 1000 Hz.

**PEAS** = Electron and Ion Analyzer (PI: N. M. Shutte, IKI). Objective: measurement of the energetic and pitch-angle distributions of electrons and ions with energies in the range from 30 eV to 30 keV. The sensor is a toroidal electrostatic analysator.

**DEP-2E** = Electric Field Instrument (V. Chmirev, IZMIRAN, G. Stanev, IKI, Bulgaria). Objective: measurement of quasi-static electric fields. Sensor: three-component double-spheric sonde.

**DEP-2R** = AC Field Analyzer (V. Chmirev, IZMIRAN, G. Stanev, IKI, Bulgaria). Objective: measurement of the quasi-static electric field. The signals of DEP-2R are are analyzed by DEP-2E.

**DANI** = Potential and Soft Particle Analyzer (PI: Tc. Dachev, IKI, Bulgaria; V. Temnyi, IZ-MIRAN). Objective: measurements of energetic and pitch-angle distributions of electron and ions in the range from 0.1 - 30 keV as well as the potential of the satellite and the electron density. Sensors: toroidal analysator, parabolic analysator (Rogoski belt).

**KM-10** = Cold Plasma Measurements (PI: V. Afonin, IKI; J. Shmilauer, GFI CSAN, former CSFR now CR). Objective: measurement of the temperature distribution of electrons, the anisotropy of the temperature, the energetic distribution is in a range from o.1 - 10 eV, the density of ions, and the potential of the satellite. The sensor is a plane sonde with 4 electrodes.

**NVK-ONCH** = VLF Analyzer (PI: Ya. Sobolev, IZMIRAN). Same instrument as in AC-TIVE mission.

**UF-3K** = Photometer (PI: Yu. Ruzhin, IZMIRAN). Objective: observation of optical effects at the injection time of the electron and ion beams. Working wavelengths: 3914 Å, 5577 Å, and 6300 Å. The dynamical range is 80 dB and the look angle is 8° (from nadir).

**FS** = Photometer (PI: Yu. Ruzhin, IZMIRAN). Objective: measurement of the illumination intensity (at the injection of the electron and ion beams) in the range from 3914 - 6563 Å.

**SGR-5** = Fluxgate Magnetometer (PI: L. Zhusgov, IZMIRAN). Objective: measurement of the three components of the magnetic field in the range of ± 64.000 nT with an accuracy of 1 nT.

**MNCH** = Search-Coil Magnetometer (PI: V. Chmirev, IZMIRAN). Objective: measurement of the magnetic field variations in the frequency range from 0.1 - 10 Hz and an amplitude range of 40 - 400 nT. The information transfer is provided on the sensor DEP-2E.

**NAM-5** = Radiofrequency Mass-Spectrometer (PI: V. Istomin, IKI, J. Shmilauer, GFI CSAN, CSFR). Same instrument and same objectives as on ACTIVE mission.

**AVCH-2T** = HF-Field Analyzer (PI: Z. Klos, Center of Cosmic Research, Polish Academy of Sciences, Warsaw; S. A. Pulinets, IZMIRAN). Objective: Measurements of electrons in the energy range from 30-600 keV The sensor is a self-opening band-type antenna of 15 m length.

## A.9.1   APEX Subsatellite (Magion-3) Scientific Payload

Objective: monitoring of the propagation of electron beams and plasma injected by the parent spacecraft as well as plasma monitoring of natural origin. Magion-3 was separated from the parent on December 28, 1991, the experiments are almost identical to those of Magion-2 (ACTIVE mission).

Orbit: perigee = 500 km, apogee = 3200 km, inclination = 83°

**SGR-7** = 3-axis Fluxgate Magnetometer (PI: M. Ciobanu, IKI, Romania). Measurement range of $\pm 50048$ nT or 6256 nT, resolution = 16 nT or 2 nT, frequency range = 0 - 20 Hz. Measurement of magnetic field vector and to serve as attitude sensor.
-   SGR6 = 1-component Fluxgate Variometer; dynamic range = $\pm 156$ nT; resolution = 50 nT; frequency range = 0.1 - 20 Hz.

**KEM-1** = Magnetic and Electric Field Analyzer (PI: P. Triska, GFU, Czech Republic);
-   ULF = Electric Field Experiment (3 components of the quasistatic electric field), dynamic range = 0.005 - 8000 mV, frequency range = 0.1 - 20 Hz.
-   VLF Wave Experiment (3-axis electric and 1-component magnet ELF-VLF field measurements, broadband waveforms, spectrum analyzer, filter bank;
    $E_{x,y,z}$ : 0.1 Hz - 120 kHz, sensitivity of $10^{-7}$V/mHz$^{1/2}$ , 120 dB dynamic range.
    $B_x$ : 10Hz - 40 kHz, sensitivity of $5 \times 10^{-6}$ nT/Hz$^{1/2}$ at 2 kHz and $10^{-4}$ nT/Hz$^{1/2}$ at 100 Hz, dynamic range 120 dB.
-   Data transmission modes selectable: broadband analog data 10 Hz - 60 kHz one channel, 10 Hz - 20 kHz broadband 3 channels, subcarriers of 1 kHz bandwidth 4 channels.
-   Filter bank; 17 Hz - 15 kHz, 8 filters, four independent sets
-   Frequency analyzer: range of 1 - 220 kHz, 32 frequency steps, full spectrum/2s, selection of any frequency.

**KM-12** = Cold Plasma Analyzer (PI: J. Shmilauer, GFU, Czech Republic); HF probe measures 2 components of the electron temperature in the range: Te = $10^3$ - $10^5$ K,, spacecraft potential from -2 to +2V; Spherical ion-trap measures the ion density ($N_i$) in the range: $10^8$ - $10^{13}$ /m$^3$. $\Delta N_i/N_i$ fluctuations $f_{max}$ = 50 Hz.

**ZL-A-S** = Langmuir Probe (PI: K. Sauer, MPE, Germany); measurement of electron and ion density, range: 500 - $10^8$ cm$^{-3}$; electron temperature: 0.05 - 3 eV; current range: $10^{-10}$ - $10^{-3}$ A; in the current-mode density fluctuations f= < 200 Hz, $\Delta N_e/N_e$ resolution 10-3.

**PRS-2-C** = Radiowave Spectrometer (PI: Z. Kloss, CBK PAN, Poland). Measurement of the HF wave spectra. Frequency range = 0.1 - 10 MHz, dynamic range = 1 $\mu$V - 10 mV, 0.2 s/spectrum using $\Delta$f=50 kHz or 2s/spectrum using $\Delta$f= 15 kHz; field fluctuations at fixed frequency (selectable), $\Delta$t= 1ms.

**DOK-A-S** =Silicon Detector Spectrometer (PI: K. Kudela, UEF SAV, Slovakia); two sensors (parallel and perpendicular to the magnetic field vector), measurement of electrons and ions (20 keV - 1 MeV, 8 energy levels, geometric factor $10^{-2}$ cm$^2$ sr.

**MPS SEA**= Energetic Particle Spectrometer (PI: Z. Nemecek, Prague University, Czech Republic); electrostatic analyzers, measurement of electrons and positive ions: 0.2 - 20 keV in 16 energy levels, pitch angle resolution 30°, geometric factor is $\sim 10^{-3}$ cm$^2$ sr.

## A.10   ARGOS (Data Collection System)

ARGOS is a joint program of CNES, NASA and NOAA, started in 1974, for the purpose of longterm continued global satellite data collection services (in particular environmental data) from fixed and mobile platforms located anywhere in the world. The ARGOS system package is flown on all TIROS-N family satellite since 1978 (see chapter A.74.2). The space segment comprises two NOAA satellites in simultaneous orbit.

The ARGOS system exploitation, i.e. the data collection and distribution function and user interface, is a commercially provided service by CLS (Collecte Localisation Satellites), a CNES daughter in Toulouse, France, and by Service Argos of Landover MD, USA (a CLS subsidiary). Service to the user community has been continuously provided since the Fall of 1978.

Orbit: the ARGOS payload on NOAA satellites is in sun-synchronous polar orbit, altitude = 830-870 km, inclination = 98-99°, period = 102 minutes (approximately 14 orbits/day). The circle of visibility (or the footprint) is 5000 km in diameter at 5° elevation, which is identical with the swath width.

The ARGOS Space Segment:

Each ARGOS payload is equipped with a DCLS (Data Collection and Location System), also referred to simply as DCS, which receives all transmissions from the platforms in view during a pass. Functionally a DCLS is comprised of the following subsystems:

- housekeeping equipment, power supply and DCLS command interface
- receive assembly (receiver and search unit, both with full redundancy)
- signal processing assembly (four identical Data Recovery Units (DRUs, eight DRUs are planned for the next ARGOS series), and command unit, telemetry encoder, and buffer memory). All data are tape recorded on board the spacecraft.

The ARGOS Ground Segment:

A set of user platforms, fixed or mobile, deployed at sea, on land, or in the air. All platforms reporting to the ARGOS system must carry a certified PTT (Platform Transmitter Terminal) package for the satellite uplink communication. Each PTT outputs a short message (of duration from 0.36 to 0.92 seconds, or of length 32 bits to 256 bits max) modulating a carrier frequency. Message transmission intervals range from 90 s to 300 s, dependent on the application.

The ground segment of the service provider consists of two NOAA/NESDIS CDA (Command and Data Acquisition) stations, one at Wallops island Va., the other at Gilmore Creek, Alaska. In addition there is a downlink station at CMS (Centre de Météorologie Spatiale) Lannion, France. All these stations provide also real-time data during the pass. ARGOS provides two GPCs (Global Processing Centers), one in Landover Md., the other in Toulouse France. Each GPC receives the data from all the platforms but processes only the data that belong to "its" own users. Both centers will, however, immediately process all data in case of necessity, thereby ensuring a full redundancy.

**Communication Concept:**

Collection Uplink: ARGOS provides a total of 4 (8 in next series) parallel receiving channels for data collection, each at a rate of 400 bit/s. Each PTT in the ground segment transmits encoded messages at regular intervals (fixed platforms at 200-300 seconds, drifting or mobile platforms in the order of 90 to 150 seconds).

Note: the search unit is a spectrum analyzer that scans a 24 kHz band centered at 401.650 MHz. The next series of DCLS will have 80 kHz of bandwidth (100 kHz allocated, 2 safeguard bands of 10 kHz at each end). Time tagging and frequency measurements are made by the DRUs and processed on the ground for location determination.

Downlink: The data received by the ARGOS DCLS is multiplexed onboard by the TIP processor and transmitted to the ground via three paths:

- Real-time: the TIP output (8.32 kbit/s, see Figure 67) directly modulates a VHF beacon which transmits continuously.
- Real-time: the TIP output is multiplexed onboard the satellite with HRPT data and transmitted in S-Band
- Delayed Transfer: the TIP output is also recorded by a tape recorder and, each time the satellite passes over one of the ground stations, the recorded data is dumped via S-Band telemetry.

The ARGOS communication capability is limited to the function of data collection from the PTTs. The concept does not offer a remote configuration control capability of the data collection platforms in the ground segment.

**Access Method:**
The onboard DCLS receiver picks up messages from the transmitting platforms in its area of visibility. The receiving system can discriminate between message arrival times and between frequency shift due to the Doppler effect. Up to four (8 in next series) messages may be processed simultaneously.

The ARGOS access scheme employs 'pure (i.e. unslotted) ALOHA'. Messages from the PTTs are received onboard on a random access basis. The ARGOS Doppler system provides a position fix for drifter (or mobile) platforms. This setup requires between 3 and 5 successful transmissions, which must occur within a pass (footprint).

Within an average footprint of 10 minute duration each platform in the ground segment has usually a number of attempts to make contact with the DCLS in the space segment.

- Fixed platforms: the number of transmission attempts of fixed platforms is 3 at a repetition rate of 200 seconds (average = 3).
- Drifting (mobile) platforms: the repetition rate is 90-150 seconds, hence the maximum number of transmission attempts possible within a footprint is 5-6 (average = 5). [About 80% of the possible position fixes are actually achieved by the system, 20% are rejected during ground processing for various reasons, mainly geometrical configuration: number of messages, pass duration, distance to the track, etc.., according to CLS ARGOS].

The nature of random access degrades very much the data collection performance by the space segment. The scheme of pure ALOHA permits under normalized offered channel traffic a maximum channel throughput rate of 18%. Any two signals overlapping in time and frequency, may interfere, with the loss of both. The principal parameter that affects the performance of the ARGOS data relay system is "interference": it occurs when the demand for service exceeds the system's capability. The result is loss of data from system 'blockage'. The maximum number of platforms that a single ARGOS DCLS can actually service within a footprint, is in the order of 650. In this number, there is a certain mix of fixed (collection service only) platforms and drifting (collection and location services) platforms, a further assumption is a certain message length.[32),33)]

The probability of good message reception is 67% with a traffic density of 2.6 Erlang, and 8.3 Erlang for the next improved DCLS series which is scheduled to be launched starting in 1996 with NOAA-K.

The total number of platforms actually registered as active in the ARGOS system globally is around 4000 as of 6/1993, out of which around 2300 are transmitting every day. The remaining platforms are transmitting once every two or three days, or less. This information was provided by CLS ARGOS (6/1993), the service provider of the system.

---

32)    Note: the figure of 650 serviceable platforms in a footprint was provided by 'CLS ARGOS' of Toulouse
33)    "A Definition Study of an Advanced Data Collection and Location System (ADCLS)", prepared for GSFC by ECOSYSTEMS International Inc., January 1986

## A.11   ARISTOTELES

**ARISTOTELES**[34),35),36),37),38),39)] = **A**pplications and **R**esearch **I**nvolving **S**pace **T**echniques **O**bserving **T**he **E**arth's **F**ield from **L**ow **E**arth **O**rbiting **S**atellite. Planned ESA/NASA Gravity and Magnetic Field Mission. NASA Delta-2 Launch in 1998, nominal life = 4 years. Mission control from ESOC with a dedicated ground station at Kiruna. S/C mass = 2300 kg at launch; fuel mass = 960 kg, solar array power = 1.1 kW.

ARISTOTELES is viewed by the science community as the first of a series of space missions in the Solid Earth Program, with STEP and Gravity Probe-B following, and the planned Superconducting Gravity Gradiometer Mission (SGGM) of NASA thereafter.

Objective: High-precision measurements of the Earth's gravity field on a global and regional scale (geopotential and geodynamics for better geoid models).

Application: Geophysics, Geography, Geodesy, Oceanography, Model improvement for satellite orbits, earthquake research, etc.

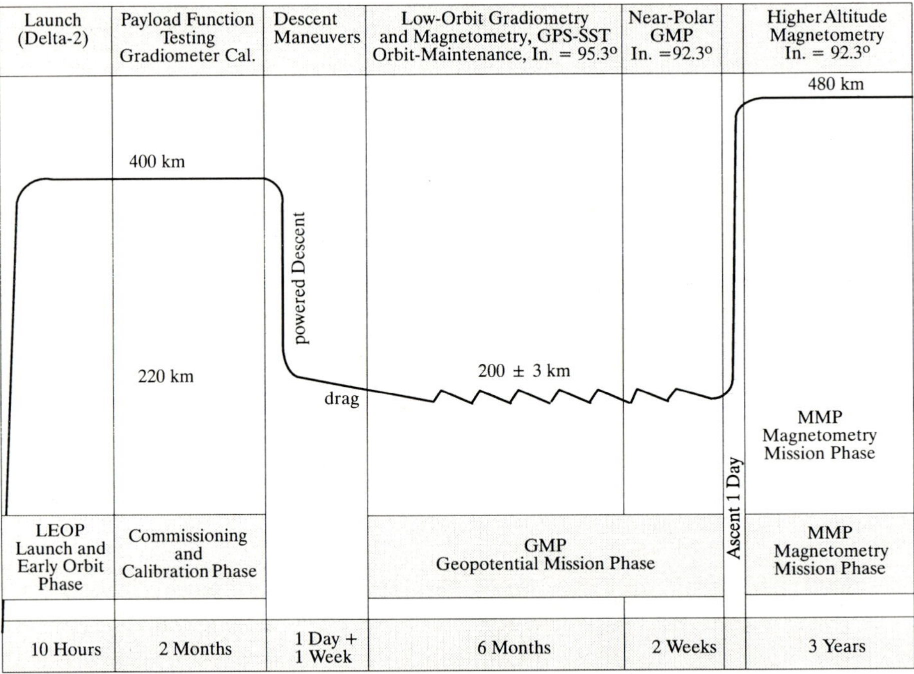


**Figure 13:**     **Concept of the Aristoteles Mission Orbit Profile**

The ARISTOTELES mission scenario consists of the following segments or phases: LEOP (Launch and Early Orbit Phase), CCP (Commissioning and Calibration Phase), GMP (Geopotential Mission Phase), and MMP (Magnetometry Mission Phase).


34)   "ARISTOTELES", Dornier booklet
35)   R. Rummel "Solid Earth Programme Considerations - a Contribution to the ESA Users Consulting Meeting", Paper presented at the ESA 'Earth Observation  User Consultation Meeting', at ESTEC, May 1991
36)   'Aristoteles data reductions: How to reconstitute the Earth's gravity field', earth observation quarterly, no. 27 Oct. 1989, ESA periodical
37)   'Aristoteles Programme Proposal', ESA/PB-EO (91) 1, Rev.2 of 1 June 1992
38)   M. Schuyer, P. Silvestrin, M. Aguirre, "Probing the Earth from Space - The Aristoteles Mission", esa bulletin, Nov. 1992, pp. 67-75
39)   **Note: The ARISTOTELES mission was not  selected for further funding by the European ministers conference at Granada in Nov. 1992. Some technical activities of the program are not terminated by ESA. The mission definition is kept in this text as a source of further reference.**

1. Geopotential Mission Phase (GMP). Objectives: Measurement of gravitational and magnetic anomalies (at an altitude of 200 km) originating in the Earth's lithosphere with good sensitivity. GMP will last for 6.5 months. The last two weeks of this phase will be preceded by a change in inclination to about 92°, so as to improve the coverage over the poles, and to achieve during the subsequent phase the rapid sampling of all solar times needed to separate diurnal and seasonal effects in geomagnetic studies.

2. Magnetometry Mission Phase (MMP) Objective: Measurement of long-term changes in the geomagnetic field (secular field variation). Detailed and greatly improved global determination of the gravity field with:

   - an accuracy of $< 2$ nT
   - a spatial resolution of 100 km x 100 km (arrangement of scalar and/or vector magnetometers)
   - for baselines up to 100 km:      5 cm
   - for baselines up to 1000 km:    10 cm

Orbit: Nearly-circular polar and quasi sun-synchronous orbit. Inclination $= 95.3°$ for GMP, and 92.3° for MMP.

**Sensors:**

- **GRADIO** (instrument provided by ESA). Gradiometer consisting of a two-dimensional arrangement of 4 accelerometers (two measurement axis are perpendicular to the satellite flight direction, while the third axis is in flight direction). Gradiometry = the measurement of the second derivatives of the gravity field potential.
  Instrument accuracy requirements:
  Gravity gradient:                      $= 0.01$ Eötvös ( 1 Eötvös $= 10^{-9}$ s$^{-2}$ ; i.e. difference of $10^{-9}$ ms$^{-2}$ acceleration per meter). This requires a sensitivity of $< 1$ pico-g from each accelerometer. An onboard calibration mechanism will be installed.
  Spatial resolution:                    100 km x 100 km
  Measurement Bandwidth:          0.005 Hz to 0.25 Hz

- **GPS Receiver** (instrument provided by NASA) for orbit determination. GPS provides continuous tracking of Aristoteles. The data from the onboard GPS receiver will be supplemented by data gathered from a worldwide ground network of GPS receivers, already deployed for other geodetic research activities. The tracking data from this network allows the use of a differential technique for precise orbit determination. An orbit accuracy of a few centimeters is required.
  Analysis of the tracking data provides information about the components of the gravitational field at long and medium wavelengths which, combined with the components obtained from GRADIO, will enable the reconstitution of the field to the expected accuracy of about 1 milliGal over the complete range of wavelengths (note: 1 milliGal $= 10^{-5}$ ms$^{-2}$ $\simeq 1$ μg).

- **Magnetometer** (instrument provided by NASA) requirements:
  Accuracy:                          $< 2$ nT
  Spatial resolution:            100 km x 100 km
  Arrangement of Scalar and Vector Magnetometers (fluxgate technique). Both magnetometers are boom-mounted (4.5 m boom). The vector magnetometer will be calibrated during the mission by the scalar magnetometer. An optical Attitude Transfer System (ATS) relates the orientation of the of the vector magnetometer with that of the star trackers.

Data: onboard storage provided (300 Mbit). Telemetry data rate = 1 Mbit/s.

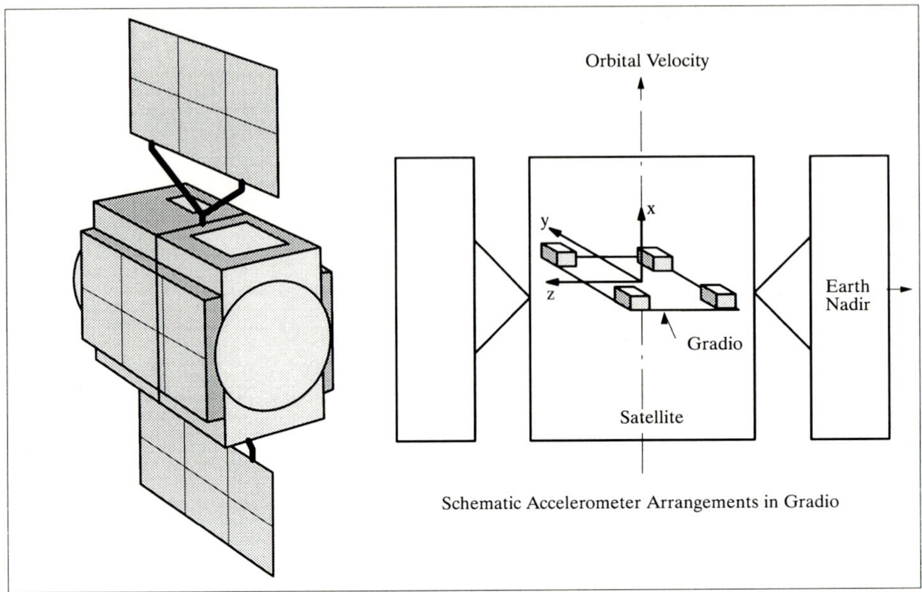


**Figure 14:     The ARISTOTELES S/C Model**

## A.12     ATLAS

ATLAS = Atmospheric Laboratory for Application and Science. The ATLAS program is part of NASA's 'Mission to Planet Earth', a unified study from the deep interior to the outermost regions of Earthspace. The ATLAS missions (on shuttle) investigate specifically how Earth's atmosphere and climate are affected by the Sun and by the products of industrial complexes and agricultural activities. Experiments flown on the ten ATLAS missions presently planned will gather data throughout the Sun's 11-year activity cycle (12 - 18 month launch centers)[40],[41].
There is a "core" of instruments which will be the same on the ATLAS missions, and there are additional instruments unique to several of the flights. ATLAS belongs in the 'Early Earth Observing Program', it is also coordinated with the UARS program for cross-support.

ATLAS-1 consists of 12 international experiments and supporting hardware mounted on a two-Spacelab pallet train in the orbiter payload bay, and one experiment contained in two canisters mounted on one adapter beam assembly (SSBUV/A). These experiments study the chemical makeup of the atmosphere between approximately 15 and 600 km above the Earth's surface, measure the total energy contained in sunlight and how that energy varies, investigate how Earth's electric and magnetic fields and atmosphere influence one another, and examine sources of ultraviolet light in the Universe. Many of the experiments are also scheduled for later ATLAS missions so that the data gathered during ATLAS-1 is the first collected in a series of long-range studies that measure changes in the atmosphere and the Sun.

ATLAS-1 had periods of solar pointing, Earth-limb pointing, and additional special attitudes throughout the mission. The orbiter is the primary experiment pointing system; how-

40)   Jack Kaye, 'Summary of ATLAS Shuttle Missions', Paper presented at the EOS-B Atmospheric Payload Panel Meeting Washington DC, Feb. 26-27 1991
41)   Information provided by the Earth Science Application Division (ESAD Office) at NASA HQ, Washington

ever, some of the experiments have a limited fine pointing capability for enhanced pointing accuracy.

Orbit: ATLAS-1 was launched by STS-45 (Atlantis) to a 296 km orbit, inclination = 57°, 8 day mission.

**Sensors (Core Instruments):**

**ATMOS** = Atmospheric Trace Molecule Spectroscopy (JPL Sensor). This is an infrared absorption instrument which works at occultations (sunrise, sunset) and measures a wide variety of species with good vertical resolution. The ATMOS experiment was already very successful on Spacelab 2 (flown in 1985). It employs a high-resolution FT (Fourier Transform) interferometer and limb viewing in the Bands of 2.2 µm to 16 µm.

**MAS** = Millimeter-Wave Atmospheric Sounder (G. Hartmann, MPAE). Measurement of millimeterwave emission from the atmosphere. Determination of the stratospheric amounts of temperature and ozone, and also stratospheric amounts of chlorine monoxide (ClO). As a limb emission instrument, it gets nearly global coverage.
MAS is the predecessor of AMAS, a Heterodyne Limb Sounder (like MLS), which measures in the channels: 61 GHz, 62 GHz, 63 GHz, 183 GHz, 184 GHz and 204 GHz. MAS determines pressure, temperature, ozone, water- and ClO profile.

The MAS experiment is a cooperation of: University of Bremen (K. Künzi, scientific investigation), University of Bern (Hardware), MPI Lindau (data processing and scientific investigation) and the Naval Research Lab (Washington).

**SUSIM** =Stratosphere Ultraviolet Spectral Irradiance Monitor (Naval Research Laboratory, US)[42]. Measures the solar UV flux as a function of wavelength from 110 to 410 nm (compatible with SUSIM on UARS).

**SOLCON** = (IRMB, Brussels). Measurement of the SOLar CONstant. The SOLCON instrument is a cooperation of IRMB, Space Science Dep. of ESA, and LaRC.

**ACR** = Active Cavity Radiometer (JPL). Measures the solar constant similar in technique to SOLCON.

**SOLSPEC** = SOLar SPECtrum Measurement (CNES). Measures the solar irradiance from 180 - 3200 nm using three double spectrometers and an onboard calibration device.

**SSBUV**[43] = Shuttle Solar Backscatter Ultraviolet Spectrometer (GSFC). Uses UV backscatter in nadir to measure vertical profiles of ozone in the stratosphere and in the lower mesosphere from 180 to 450 nm.
Note: SSBUV is not a core instrument; it is a separate ATLAS Shuttle payload (co-manifested with ATLAS and integrated into the ATLAS science plan)

SSBUV is a sensor of SBUV/2 heritage (from Nimbus-7 and NOAA-9 satellites onward). Objective: Calibration of Long-term satellite ozone data sets with complementary Shuttle flights.

SSBUV is scheduled to fly 10 times over the next 6 years on Shuttle missions. The first demonstration flight with SSBUV instrumentation occurred on October 19, 1989 on the Shuttle Atlantis. Throughout the Shuttle flight period coincident observations were taken with the SBUV on Nimbus-7 and the SBUV/2 on NOAA-9 and NOAA-11 satellites.

The SSBUV instrument is the SBUV/2 engineering model now flying on NOAA satellites. See A.74 (NOAA-POES) for a description of SBUV/2.

---

42)   SUSIM brochure of Naval Research Lab, available at NASA HQ's Document Resource Facility
43)   'Calibration of Long Term Satellite Ozone Data Sets Using the Space Shuttle', E. Hilsenrath, in Optical Remote
      Sensing of the Atmosphere, 1990 Technical Digest Series of the Optical Society of America, Vol. 4, pp. 409-412

The ATLAS Program on Shuttle ($\sim$ 1 week each time) has the following time schedule:
- ATLAS-1 Mission: March 24 - April 2, 1992 with Atlantis (correlative measurements to UARS). Flown mission on STS-45
- ATLAS-2 Mission: April 8 - 17, 1993 (correlative measurements to UARS). ATLAS-2 payloads gathered data on the relationship between the Sun's energy output and the Earth's middle-atmosphere chemical makeup. Study of how these factors affect the Earth's ozone level. Flown mission on STS-56.
- ATLAS-3 Launch: Oct. 1994 (with SPAS-01 payload CRISTA and MAHRSI, see Figure 15).
- ATLAS-4 Launch: August 1995
- ATLAS-5 Launch: Fall 1997
- ATLAS-6 Launch: 1998

Additional ATLAS sensors are:

**ALAE** = Atmospheric Lyman-Alpha Emissions (CNRS, France). Uses on-board hydrogen and deuterium cells to measure thermospheric/exospheric H and D concentrations, as well as Lyman alpha amounts in the interplanetary medium. ALAE will fly on ATLAS-1.

**GRILLE** = Infrared spectrometer (Belgian sensor). GRILLE was flown already on Spacelab-1 (1983). GRILLE has a high resolution in the infrared region (2 μm), it measures vertical profiles of: CO, $CO_2$, NO, $H_2O$, $CH_4$, $H_2O$ and HCl. GRILLE will fly on ATLAS-1.

**ISO** = Imaging Spectrometric Observatory (MSFC). The spectrometer measures 'low light observations' in the daylight and night-side of the Earth. There are 5 spectral bands from 30 - 1300 nm. ISO will fly on ATLAS-1.

ENAP = Energetic Neutral Atom Precipitation (University of Texas, Dallas). ENAP is an investigation which will use data from ISO.

**AEPI** = Atmospheric Emissions Photometric Imaging  (Lockheed, Palo Alto Research Lab). Optical emissions from the upper atmosphere/ionosphere and from the Space Shuttle environment. Images of natural and induced aurora and airglow. AEPI will fly on ATLAS-1.

**SEPAC** = Space Experiments with Particle Accelerators (Southwest Research Institute, US). Uses electron beam accelerator and other instruments to carry out active and interactive experiments on and in the Earth's ionosphere. SEPAC was flown on ATLAS-1.

**FAUST** = Far Ultraviolet Space Telescope (University of California, Berkeley). Far-ultraviolet images of large-scale phenomena. FAUST will fly on ATLAS-1.

**CRISTA** = Cryogenic Infrared Spectrometers and Telescopes for the Atmosphere (Offermann, U. Wuppertal). This is an infrared instrument with three independent telescopes pointed at the limb. Two telescopes are laterally directed (pointing angles = +/- 18° from center). Objective: analysis of dynamical processes (waves and turbulence) in the middle atmosphere with the use trace gases. The spectrometers provide 7 spectral bands from 4.6 - 14.1 μm, and 5 bands from 15.2 - 71 μm.[44]

**MAHRSI** = Middle Atmospheric High Resolution Spectrograph Investigation (Naval Research Lab). Measurement of dayglow in the 1900 -3200 Å region. Study of OH and NO in the mesosphere and thermosphere (30 - 150 km region in total). Vertical resolution will be 2 km.

**AMIE** = Airglow Measurement of Infrared Emitters (Utah State University/Stewart Radiance Lab).

Note: CRISTA, MAHRSI and AMIE are instruments on the ASTRO-SPAS III free-flyer platform, which is co-manifested with ATLAS-3. They are integrated in the ATLAS-3 science plan. MAHRSI will also be an ATLAS sensor for ATLAS-4 and ATLAS-5.

---

44)   P. Barthol, K. U. Grossmann, D. Offermann, "Telescope design of the CRISTA/SPAS experiment aboard the Space Shuttle", SPIE, Vol 1331, Stray Radiation in Optical Systems, 1990, pp. 54-63

| ATLAS-1 Investigative Topic | Instrument/Sensor | Agency/Country | Principal Investigator (PI) |
|---|---|---|---|
| Earth Science | Active Cavity Radiometer (ACR) | JPL | Willson |
| | Atmospheric Trace Molecule Spectroscopy (ATMOS) | JPL | Gunson |
| | Atmospheric Lyman-Alpha Emission (ALAE)* | CNRS, France | Bertaux |
| | Grille Spectrometer (GRILLE)* | Belgium (with CNES) | Ackerman |
| | Millimeter-wave Atmospheric Sounder (MAS) | MPAE, Germany | Hartmann |
| | Solar Constant (SOLCON) | IRMB,Belgium | Crommelynck |
| | Solar Spectrum (SOLSPEC) | CNES, France | Thuillier |
| | Shuttle Solar Backscatter Ultraviolet (SSBUV/A) | GSFC | Hilsenrath |
| | Solar Ultraviolet Spectral Irradiance Monitor (SUSIM) | Naval Research Laboratory | Brueckner |
| Space Physics | Atmospheric Emissions Photometric Imaging (AEPI)* | Lockheed | Mende |
| | Imaging Spectrometric Observatory (ISO)* | MSFC | Torr |
| | Space Experiments with Particle Accelerators (SEPAC)* | Southwest Research | Burch |
| Astrophysics | Far Ultraviolet Space Telescope (FAUST)* | University of California, Berkeley | Bowyer |

**Table 9:     ATLAS-1 Overview of Instrument Complement**[45]

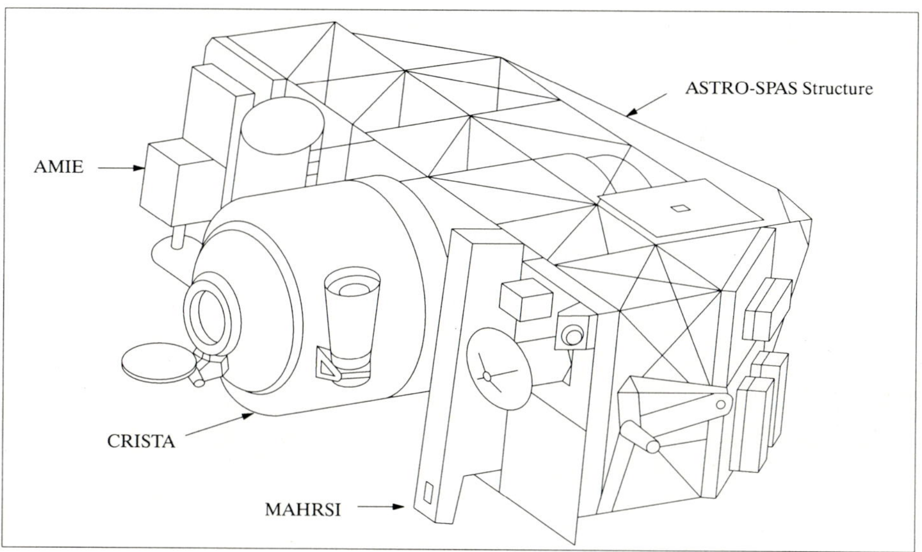


**Figure 15:     Configuration of ASTRO-SPAS III Platform with its Sensors**

---

45)    Sensors which are designated with a * are Spacelab-1 reflights

## A.13  ATMOS

ATMOS (Atmospheric and Oceanic Sensors) is a program initiated by the German ministry of science and technology (BMFT) to support scientists with environmental remote sensing data. The program has been designed according to user requirements defined by a German-French working group of scientists and refined on a number of international workshops. The program focuses on global long-term measurements of the chemistry and dynamics of the lower and middle atmosphere, on the interactions between the biosphere and the atmosphere and on the energy balance of the atmosphere.

From the above mentioned user requirements a sensor package consisting of AMAS, MI-PAS, SCIAMACHY, a radiation budget sensor (CERES or SCARAB), and an imaging spectrometer (ROSIS or MERIS for ocean and land applications) has been defined.

Initial plans included the realization of a dedicated ATMOS-satellite which would have allowed the launch of the ATMOS sensor package well before the launch of the European polar platform (POEM-ENVISAT). The dedicated ATMOS satellite would have been placed into a sun-synchronous polar orbit at 780 km altitude with a 3 day repeat cycle and an equatorial crossing of 11 AM (descending). As the sensor package is also envisaged for the European ENVISAT-1 and possible follow on missions, continuity of the data could have been guaranteed. Due to financial reasons (German reunification), the project to build a dedicated satellite has been cancelled after finalizing a phase-B2 study in February 1992. For the same reasons, the national development of the AMAS microwave instrument has been halted in February 1993.

The elements of the revised ATMOS program, which now mainly relies on ESA missions, are:

- Oct-1992:
  Start of implementation of an "ATMOS User Support Center" at the German Remote Sensing Data Center at DLR for processing the data of the ATMOS sensor package. As one of the first elements, ground processing for the GOME Sensor on ERS-2 is realized in close cooperation with ESA, RAL in England and Italian partners. The ATMOS User Support Center maintains continuous long-term archives for level 1 (calibrated geolocated radiances), level 2 (retrieved geophysical quantities such as ozone total column) and higher-level data. It will support a large user community with data from different sensors via an electronic user interface. The ATMOS User Support Center is closely coordinated with the CEO working groups and will be part of the ESA ground segment for ENVISAT as well as it will be a node in the Global Environmental Data Network.

- Late 1994:
  Launch of the GOME instrument on ERS-2; first continuous flow of global high quality data of stratospheric and tropospheric ozone.

- 1996:
  Possible launch of the Indian-German joint mission IRS-P2 with the Indian LISS-camera for moderate resolution land applications and the German MOS multispectral sensor for oceanic applications; scientists could be supported with data of biospheric relevance to span the time period until the launch of ENVISAT.

- 1998:
  Launch of the ENVISAT-1 satellite which includes, among others, the ATMOS sensor package; first continuous flow of the data which comprises to the "full-blown" scientific ATMOS user requirements. MERIS monitors of marine and terrestrial biosphere and synergistically supports cloud measurements for climatic studies. SCIAMACHY monitors tropospheric and stratospheric trace gases. MIPAS monitors a large number of

trace gases in the upper troposphere, the stratosphere and the mesosphere. SCARAB measures the radiation balance and supports climate studies.

- 2002:
  Launch of the ENVISAT-2 satellite which, in addition to the sensors flown on ENVISAT-1, might include the AMAS sensor for global measurements of ClO and HCl.

**Sensors:**

**AMAS** = Advanced Millimeter-Wave Atmospheric Sounder. This is a passive limb emission sounder operating in the spectral region between 62 and 206 GHz for day and night measurements of temperatures, ClO, $O_3$, $H_2O$ and CO in the middle atmosphere.

The objectives of the AMAS observations are:
- High-precision measurements of ozone and global ClO.
- Stratosphere: $O_3$ (trend analysis), global ClO and $H_2O$.
- Mesosphere: $O_3$, $H_2O$, and CO
- Temperature profiles in Stratosphere and Mesosphere

| | |
|---|---|
| Scan Range: | up to 150 km |
| Vertical Resolution (Stratosphere): | < 5 km |
| Vertical Resolution (Mesosphere): | 5 - 10 km |
| Horizontal Resolution (along track): | 300 km |
| Measurement Frequencies: | 63 - 204 GHz (8 channels) |
| Spectral Resolution (center): | 160 kHz |
| Spectral Resolution (wings): | 2 MHz |
| Date rate: | 26 kbit/s |

**MIPAS**[46] (Michelson Interferometer for Passive Atmospheric Sounding). This is a passive limb emission sounding instrument operating in the spectral region between 4.5 and 15 μm, measuring during day and night of the orbit a large variety of relevant trace gases, especially in the stratosphere and in cloud-free regions in the upper troposphere.

Major MIPAS Objectives:
- Chemistry of the stratosphere (global and polar Ozone): e.g. $O_3$, NO, $NO_2$, $HNO_3$, $HNO_3$, $HNO_4$, $N_2O_5$, $ClONO_2$, $COF_2$, HOCl.
- Climate Research: (global distribution of relevant parameters and clouds) e.g. $O_3$, $CH_4$, $H_2O$, $N_2O$, $CFC_s$, CO, OCS, aerosol and clouds.
- Transport Dynamics: $O_3$, $CH_4$, $N_2O$, $C_2H_2$, $C_2H_6$, $SF_6$.
- Tropospheric chemistry: NO, CO, $CH_4$, $O_3$, $HNO_3$.

| | |
|---|---|
| Vertical Scan Range: | 5 - 100 km |
| Vertical Resolution: | 3 km |
| Horizontal Resolution: | 30 km x 300 km |

| | |
|---|---|
| Horizontal Scan Geometry: | Backwards: 160 - 196°; Sidewards: 90 - 120° |
| Spectral Range: | 4.15 - 14.6 μm |
| Spectral Resolution: | 0.025 1/cm (unapodized) |
| Radiometric precision: | 1% |
| Interferometer Scan Time: | 4 sec |
| Data rate: | 800 kbit/s (for total measured spectrum) |

**SCIAMACHY** = Scanning Imaging Absorption Spectrometer for Atmospheric Cartography. This is a trace gas measurement device for the troposphere and stratosphere, operating by differential absorption spectroscopy of solar and lunar radiation in the spectral region between 240 nm and 2400 nm, viewing in occultation, nadir and limb scattering modes. Data rate: 400/100 kbit/s; 3 Mbit/s selective.

---

46) W. Posselt, "Michelson Interferometer for Passive Atmospheric Sounding", Proceedings of the Twenty-fourth International Symposium on Remote Sensing of Environment, 27-31 May 1991, Rio de Janeiro, Volume II, pp. 737-748, ERIM, Ann Arbor Mi.

SCIAMACHY Objectives:

- Global measurement of trace gases in the troposphere and stratosphere.
- Measurement of aerosol, cloud altitudes and spectral reflection of the surface.
- Troposphere: $O_3$, $NO_2$, $N_2O$, CO, $CO_2$, $CH_4$ and $H_2O$. Under special conditions also: HCHO, $SO_2$ and $NO_3$
- Stratosphere: $O_3$, $NO_2$, $N_2O$, CO, $CO_2$, $CH_4$, $H_2O$, HF and BrO. Observation of the NO column above 40 km. Measurement of OClO and ClO (Ozone Hole).

| | |
|---|---|
| IFOV (Instantaneous Field of View): | 2.3 x 0.023° |
| Pixel size (Nadir): | 32 km x 320 km |
| Nadir Resolution: | 32 km x 70 km |
| Vertical Resolution (Limb) | 3 km |
| Spectral Ranges: | 240 - 1700 nm |
| | 1980 - 2020 nm |
| | 2265 - 2380 nm |

## A.14    CBERS (China/Brazil - Earth Resources Satellite)

CBERS = China/Brazil - Earth Resources Satellite[47),48)]. The cooperative program between the People's Republic of China, PRC (CAST) and Brazil (INPE) was started in 1988 to establish a complete remote sensing system which is compatible to the 90's international scenario. The first CBERS satellite is projected to be launched in 1996 by a Chinese launch vehicle (CZ-4 series) from the Shanxi Launch Site in PRC. The second satellite is scheduled for launch one or two years after the first one. Each satellite has a design life of two years. Satellite mass = 1400 kg; Power = 1100 W. S/C is three-axis stabilized. The nominal design life is 2 years.

Orbit: Sun-synchronous polar orbit; altitude = 778 km; inclination = 98.5°; nodal period = 100.26 minutes; mean local solar time at descending node = 10:30 AM; revisit period = 26 days.

Objectives: Observation and monitoring of the Earth's resources and environment with a multi-sensor imaging payload providing different spatial resolutions.

**Sensors:**
CBERS is a multisensor payload with different spatial resolutions and collection frequencies.

- **WFI** = Wide-Field Imager
  Spectral bands: 0.63-0.69 µm, 0.77-0.89 µm; spatial resolution = 260 m; temporal resolution = 3-5 days; swath width = 856 km (FOV = 60°); data quantization = 8 bit; data rate = 1.35 Mbit/s; objective: acquisition of low-resolution wide-swath imaging.

- **CCD** (Charge-Coupled Device) Camera - prime sensor
  Spectral bands: 0.51-0.73 µm (panchromatic mode), 0.45-0.52 µm, 0.52-0.59 µm, 0.63-0.69 µm, 0.77-0.89 µm; spatial resolution = 19.5 m (at nadir); FOV = 8.32°; temporal resolution = nadir view 26 days, off nadir view (pointable side-looking capability of ±32°, providing an increased observation frequency for a given region) minimum time lag of 3 days; swath width = 120 km; objective: high resolution imaging.

- **IR-MSS** = Infrared Multispectral Scanner
  Spectral bands: 0.50-1.1 µm (panchromatic mode), 1.55-1.75 µm, 2.08-2.35 µm,

47)  'The China-Brazil Earth Resources Satellite Program', paper proved by G. Santana of INPE
48)  CBERS Spacecraft: Conception and Design", paper presented by E. A. Parada Tude of INPE and by C. Quinnan of CAST at the 1. Brazilian Symposium of Aerospace Technology, Sao Jose dos Campos, 27-31 Aug. 1990

10.40-12.50 µm; spatial resolution: = 78 m for (0.5 - 2.35 µm ranges), 156 m for the thermal band (10.4-12.5); temporal resolution = 26 days; swath width = 120 km; FOV = 8.78°; objective: medium resolution imaging (complementary information to CCD).

In addition to the imaging payload, the satellite carries a Data Collection System (DCS) for environmental monitoring; a Space Environment Monitor (SEM) for detecting high energy radiation, and an Experimental High Density Tape Recorder (HDTR) to record imagery onboard.
Each DCP transmits a short digital data message at a fixed rate (in UHF band).

## A.15    Cluster (Four S/C Mission in Concert with SOHO)

ESA/NASA collaborative mission within the ESA's 'Solar Terrestrial Science Programme' (STSP), part of ISTP (International Solar Terrestrial Physics Programme). Study of key interaction processes between two cosmic plasmas [study of small-scale structures (from a few to a few tens of ion Larmor radii) in the Earth's plasma environment].[49]

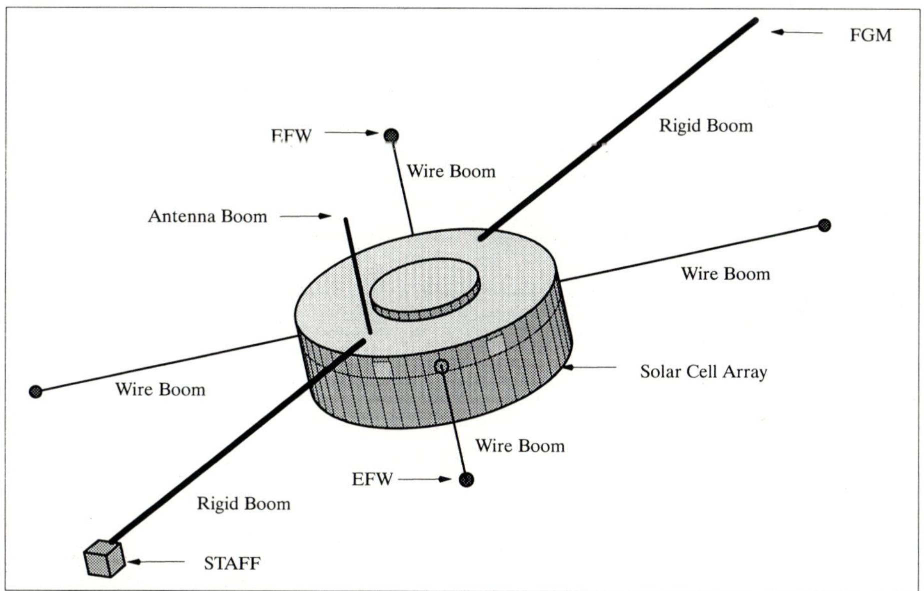


**Figure 16:    The Cluster S/C Model**

The four Cluster S/C are spin-stabilized (15 rpm, stringent requirements on electromagnetic cleanliness). Conductive surfaces and an extremely low S/C-generated electromagnetic background noise are mandatory for accurate electric field and cold plasma measurements.

The S/C are cylindrical in shape with a diameter of 2.9 m and a height of 0.9 m (mass = 354 kg for each S/C). The platform accommodates on one side the instruments. Each satellite carries two high capacity redundant tape recorders (data return of 50% per orbit). Two rigid booms, each 5 m, carry the magnetometers. Two pairs of wire booms, each with a tip-to-tip length of 100 m, permit electric field measurements (background magnetic field of ~0.25 nT is aim). The Cluster launch is scheduled for December 1995 with Ariane 5 from Kourou.

49)    ISTP Global GEOSPACE Science - Energy Transfer in Geospace, ESA/NASA/ISAS brochure, 1992

Cluster operations are performed by ESA with support provided by NASA's DSN (Deep Space Network).

Orbit:
Near-polar orbits of the spacecrafts with an apogee of 20 $R_E$ into the solar wind, and a perigee of 3 $R_E$. Six months later the apogee swings through the geotail. The inter-spacecraft distances are smaller ($<R_E$) on the dayside orbital phase, e.g. when the apogee is turned into the solar wind, and up to 3 $R_E$ in the nightside orbit phase, e.g. when the apogee swings through the geotail.

**Sensors:**[50]
A Wave Experiment Consortium (WEC) was formed to get maximum scientific return from the available spacecraft instruments. WEC comprises five coordinated experiments designed for measuring electric and magnetic fluctuations, and small-scale structures within critical layers in the Earth's magnetosphere. These WEC experiments are: STAFF, EFW, WHISPER, DWP, and WBD.

- **FGM** = Fluxgate Magnetometer. (PI: A. Balogh, Imperial College, London, UK)
  Study of small scale structures and processes in the Earth's environment. Objective: Provision of inter-calibrated measurements of the magnetic field vector **B** at the four Cluster S/C. Identical instrumentation on all S/C (two tri-axial fluxgate sensors and a data processing unit). Measurement of $\underline{B}$ wave from DC to 10 Hz, resolution $\geq 6$ pT. Four magnetometer ranges:

  |   |   |
  |---|---|
  | $\pm 256$ nT | resolution = $\pm 0.015$ nT |
  | $\pm 1024$ nT | resolution = $\pm 0.061$ nT |
  | $\pm 8192$ nT | resolution = $\pm 0.5$ nT |
  | $\pm 65536$ nT | resolution = $\pm 4$ nT |

- **STAFF** = Spatio-Temporal Analysis of Field Fluctuations (N. Cornilleau-Wehrlin, CRPE/CNET, France)
  Objectives: Study of wave-particle interaction in the region where the solar wind meets the Earth's magnetosphere. Measurement of $\underline{B}$ wave from up to 10 Hz, compressed data up to 4 kHz, cross-correlator for $<\underline{E}, \underline{B}>$.
  A STAFF instrument is provided on all four S/C, it consists of a three axial search coil magnetometer and of a spectrum analyzer to perform onboard each satellite the auto and cross correlations between electric and magnetic components.

- **EFW** = Electric Fields and Waves (G. Gustavson, Swedish Institute of Space Physics, Uppsala, Sweden)

  Objectives: measurement of the electric field and density fluctuations, determination of the electron density and temperature, study of nonlinear processes that result in acceleration of the plasma, study of large-scale phenomena from data of all four S/C. The instrument consists of double probes, mounted on two pairs of wire booms, each 100 m tip to tip. Measurement of $\underline{E}$ wave from 10 Hz, compressed data up to 100 kHz, sensitivity $< 50$ nV/m $(Hz)^{1/2}$. EFW measurement modes:

  - Instantaneous spin plane components of the electric field vector, over a dynamic range of 0.1 - 700 mV/m, and with variable time resolution down to 0.1 ms.
  - Low energy plasma density, over a dynamic range at least 1 to 100 $cm^{-3}$
  - Electric field and density fluctuations in double layers of small amplitude, over dynamic ranges of 0.1 to 50 mV/m for the fields, and 1 - 50% for the relative density fluctuations, time resolution of 0.1 ms on some occasions
  - Waves, ranging from electrostatic ion cyclotron emissions having amplitudes as large as 60 mV/m at frequencies as low as 50 mHz, to lower hybrid emissions at several hundred Hz and with amplitudes as small as a few $\mu$V/m.

---

50)  The Cluster Mission - Scientific and Technical Aspects of the Instruments, esa SP-1103, ISSN 0379-6566, Oct. 1988

- Time delays between signals from up to four different antenna elements on the same S/C, with a time resolution of 25 µs on some occasions.
- The S/C potential

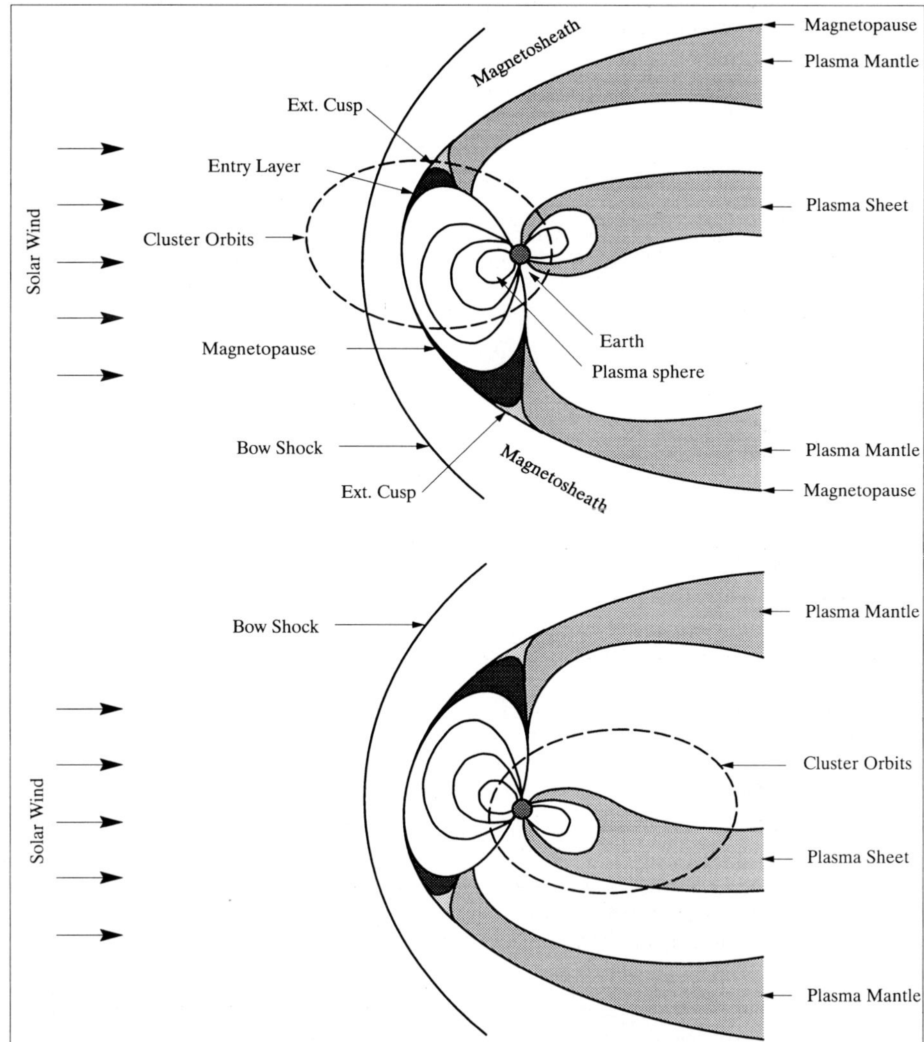


**Figure 17:**     **Cluster Orbits in Relation to the Magnetosphere at 6 Month Intervals**

- **WHISPER** = Waves of High Frequency and Sounder for Probing of Density by Relaxation (P. M. E. Décréau, Laboratoire de Physique et Chimie de l'Environnement, Orléans, France)
  Objectives: accurate measurement of the total plasma density within the range from 0.2 - 80 cm$^{-3}$ (prime objective), continuous survey of one electric component of plasma waves in the frequency range from 4 - 80 kHz with an accuracy of about 160 Hz (secondary objective).

WHISPER employs the method of relaxation sounding (using parts of the EFW wire booms). The analysis of density variations is performed via active sounding of plasma resonances. Expected measurement results:

-   identification of regions in space and mass transport
-   spatial extension and drift speed
-   density fluctuations
-   wave mode identification
-   coldest component of the electron density

Passive measurements of natural plasma waves up to 400 kHz.

- **WBD** = Wide Band Data (D. A. Gurnett, Univ. of Iowa, USA)
  Objectives: Study of E-field wave form up to ~ 100 kHz in the Earth's magnetosphere. A wideband receiver system measures electric and magnetic fields over a frequency range from 10 Hz to 600 kHz. The Cluster wideband receiver is similar to the instruments flown on ISEE-1 and DE-1. WBD makes use of the EFW sensors (two electric dipole antennas, two search coil magnetometers); conversion frequencies: 0, 125 kHz, 250 kHz, 500 kHz; bandpass filter ranges: 1 kHz-100 kHz, 50 Hz-25 kHz, 10 Hz-10 kHz; frequency resolution: limited by FFT (75 Hz typical); time resolution: 10-20 ms (per FFT spectrum);
  The wideband technique involves transmitting band-limited waveform data to a ground station using a high-rate data link (250 kBit/s R/T or 100 kBit/s in burst mode).

- **DWP** = Digital Wave Processor (L. J. C. Woolliscroft, Univ. of Sheffield, UK)
  Objectives: Correlation of wave/particle phenomena and wave/particle interactions. DWP is the onboard control system (multi-processor unit) of all WEC instruments performing data compaction and compression, event selection, particle/wave correlation, and control of WHISPER.

- **EDI** = Electron Drift Instrument (Pi: G. Paschmann, MPE, Garching, Germany)
  Objectives: accurate and highly sensitive measurements of the electric field. EDI measures the drift of a weak beam of test electrons which for certain emission directions return to the S/C after one gyration (the drift is related to the electric field and to the gradient in the magnetic field).
  EDI consists of two emitter/detector assemblies, each with a $2\pi$ FOV. E = (0.1 - 10 mV/m, <100 Hz), $\nabla \underline{B}$, $|\underline{B}|$ = 5-1000 nT, emission and tracking of 2 electron beams.

- **CIS** = Cluster Ion Spectrometry (Pi: H. Rème, Centre d'Etudes Spatial des Rayonnement, Toulouse, France)
  Objectives: Study of the dynamics of magnetized plasma structures in the vicinity of the Earth's magnetosphere (physics of the bow shock, the magnetopause boundary, the polar cusp, the geomagnetic tail, and the plasma sheet).

  CIS consists of two instruments, CODIF and HIA to measure both the cold and hot ions from the solar wind, the magnetosheath, and the magnetosphere (including the ionosphere) with sufficient angular, energy, and mass resolution.
  **CODIF** = Composition and Distribution Functions analyzer (energy range = 0.02 - 40 keV/q) to measure the 3-D distribution of major ion species ($H^+$, $He^{2+}$, $He^+$, and $O^+$). The instrument is a symmetrical hemispherical analyzer with RPA (Retarding Potential Analyzer) and TOF (Time-of-Flight) electronics, FOV = $2\pi$ x 8°; split geometric factor.
  **HIA** = Hot Ion Analyzer for high time resolution of the solar wind (energy range = 3 eV/q - 40 keV/q). The instrument is a symmetric quadri-spherical analyzer; FOV = $2\pi$ x 8° with high resolution. ($\geq 2.8°$).

- **PEACE** = Plasma Electron and Current Analyzer (Pi: A. D. Johnstone, Mullard Space Science Laboratory, Holmbury, UK)

Objectives: Study of the 3-D velocity distribution of electrons.
PEACE consists of two electron sensor devices. LEFA = Low Energy Electron Analyzer (energy range = 0 - 100 eV). The instrument is a spherical electrostatic analyzer with a FOV of $\pi$ x 3.8°.
HEFA = High Energy Electron Analyzer (energy range = 0.1 - 30 keV). The instrument is a troidal electrostatic analyzer with a FOV = $2\pi$ x 4.6°.

- **RAPID** = Research with Adaptive Particle Imaging Detectors (PI: B. Wilken, MPI für Aeronomie, Lindau/Harz, Germany)
  Objectives: Study of suprathermal plasma distributions in the energy range from 20 - 400 keV and 2 keV/n - 1500 keV for electrons and ions, respectively.
  RAPID provides complete coverage of the unit sphere in phase space. It is made up of two instruments, the IIMS and IES. IIMS = Imaging Ion Mass Spectrometer for ion distribution. IIMS is a position-sensitive instrument with solid state detectors and TOF (Time-of-Flight) electronics. IES = Imaging Electron Spectrometer for the measurement of energetic electrons. IES has a position-sensitive solid state detector.

- **ASPOC** = Active S/C Potential Control (PI: W. Riedler, Institut für Weltraumforschung, Graz, Austria)
  Objective: control of the S/C charge potential to insure effective, complete measurement of the ambient plasma distribution functions.
  The basic approach involves the emission of charges from the S/C sufficient to balance the excess of charge accumulating on the satellite by the environment (photo emission of electrons drives the S/C potential positive relative to the plasma potential).
  The method employed is field ionization with a Liquid Metal Ion Source (LMIS) of the type "solid needle". LMIS uses Indium as charged material. The instrument consists of an electronics box and two cylindrical emitter modules with 5 needles each.

## A.16   CORONAS-I

Coronas [Complex of Orbital (Near-Earth) Observations of the Activity of the Sun][51),52) is a Russian Intercosmos project with the cooperation of several international partners [Ukraine (Dnepropetrovsk) for satellite construction, Poland and CSFR in sensor development, Germany for science data reception and data interpretation (DARA, DLR-DFD, and Astrophysikalisches Institut Potsdam (AIP)].

Main objectives: study of the internal structure of the sun and the different processes of solar flares. Coronas-I (with IZMIRAN as project coordinator) is the first of three planned spacecraft, the second S/C is named Coronas-F (with FIAN as project coordinator) to be launched in late 1994. Satellite mass = 2200 kg, diameter = 2 m, height = 6 m. The launch of Coronas-I is planned for early 1994 from Plesetsk. There are a total of 10 scientific instruments onboard. The lifetime of a mission is expected to be 1.5 years. The CORONAS project is viewed as an important successor mission to the most successful Solar Maximum Mission (SMM of NASA).

The data of the solar flare observations is considered for the study of:
- the energy transfer from the solar interior to the surface and the subsequent release in solar flare events,
- the explosive flare processes on the Sun,
- the characteristics of solar cosmic rays, exit conditions, their propagations in the IMF and subsequent influences on the magnetosphere of the Earth,

51)   V. N. Oraevsky, Yu. D. Zhugzhda, 'Project CORONAS-I - Orbital Observations of the Solar Activity and Oscillations, 1991, Coronas Information Series, paper by AIP, DLR-DFD and IZMIRAN, K Pflug (editor)
52)   I. Sobelman, I Zhitnik et al., " XUV and Optical Observations of the Sun by Means of the TEREK-C Telescope/ Coronograph and the RES-C Spectroheliometer aboard the CORONAS-I satellite",1992, Coronas Information Series, paper by AIP, DLR-DFD and IZMIRAN, K Pflug (editor)

- development of a theory of flares and techniques of forecasting of their geophysical effects.

Orbit: Circular polar orbit, altitude of 500 km, inclination = 83°; the orbit will be quasi-synchronous to assure 20 day observation periods in which the satellite will be outside the Earth's shadow.

**Sensors:**

- **TEREK-C** = Solar X-Ray Telescope/Coronograph. PIs: I. A. Zhitnik, I. Sobelman. Instrument is of TEREK heritage flown on PHOBOS-1 in 1988.
  Objectives: studies of large and fine structures in the solar atmosphere, determination of the characteristics of the hot solar plasma in active regions, studies of coronal holes, etc. TEREK records images of the solar corona in the following spectral ranges:
  - 5 - 25 Å, X-ray channel;   Lines of ions: O VII-VIII, Mg XI, Fe XVII, Si XIII (Wolter-1 type grazing incidence objective with filters mounted on a 4-position wheel for subchannel selection)
  - 170 - 180 Å,              Lines of ions: Fe IX-XI
  - 290 - 320 Å,              Lines of ions: He II and Si XII
  - near 52 Å,                Lines of: Si XII
  - near 130 Å,               Lines of: Fe XXIII-XXIV
  - 4000 - 6000 Å, channel K   optical coronograph

  The detectors in all channels are CCD frame transfer image sensors with a dynamic range of $\sim 10^5$. Spatial resolution of $\sim 7$".

- **RES-C** = Solar X-Ray Spectral Polarimeter. Objective: Study of hot solar plasma structures in the range of $1\text{-}3 \times 10^7$ K. RES-C provides monochromatic images of the sun x-ray radiation and linear polarization simultaneously in three spectral ranges:
  - 180 - 205 Å (XUV channel) consisting of two subchannels with perpendicular dispersion planes. Each channel contains a grazing incidence plane grating (2400 lines) with multilayer coating, spherical mirrors working at normal incidence and intensified cooled CCD frame transfer image sensors. The spectral resolution is 0.03 Å, the angular resolution 6" x 90".
  - 8.41 - 8.43Å (Mg XII channel). The Mg XII channel comprises the bent quartz (d=4.25 Å; 5 cm) and crystal spectrometer with CCD image sensor. The spectral resolution is about $7 \times 10^{-4}$Å, the angular resolution is 6" x 6".
  - 1.85 - 1.87Å (Fe XXV channel). This channel contains the bent spherical quartz (d=1.18Å, R=30 cm) crystal with CCD image sensor. The spectral resolution is $2 \times 10^{-4}$ Å, the angular resolution is 2'.

- **DIOGENESS** = Diagnostic of Energy Sources and Sinks in Flares. Spectrometer and Photometer within the soft X-ray region. Objectives: provision of energy source and sink measurements that enable the study of the balance of the solar flare energy contained in its main thermal reservoir.
  The DIOGENESS instrument consists of three independent units: a Bragg high-resolution spectrometer (BS), a X-ray spectrophotometer (BF), and a microcomputer for the control of BS and BF.
  BS spectral range: 2.835 - 3.356Å. The most important lines are CaXVIII, CaXIX and CaXX. The BS time resolution is 0.1 - 10 s.
  BF spectral range: 2 - 160 keV. The energy channels are: 2-4, 4-8, 10-20, 20-40, 4-80, and 80-160 keV. The time resolution is 1 second in the range of 2-8 keV and 0.1 second in the range of 10-160 keV.

- **HELICON** = Solar X-ray and γ-ray Scintillation Spectrometer. Helicon has the following characteristics:
  - number of detectors: 2 (solar and antisolar directions)
  - energy range: 10 keV - 8 MeV

- energy windows for the investigation of time histories: 10-50 keV, 50-250 keV, 250-800 keV
- sensitive area of each detector: $314\,\mathrm{cm}^2$ which corresponds to a sensitivity for flares with fluencies $> 10^{-7}\,\mathrm{erg\ cm}^{-2}\,\mathrm{s}^{-1}$
- statistical criterion for identification of flare: $G\sigma$ above the background
- number of PHA (Pulse Height Analyzer) channels: 64 for each detector
- number of spectra registered during one flare: up to 64 (depending on flare power)
- accumulation time for single spectra: 2 ms to 8 s (depending on the magnitude of the flare)
- number of time analysis channels: 4096
- time resolution: from 2 ms to 0.25 s
- background spectrum control: each 2 min
- duration of time history recording: up to 4 min

- **IRIS** = Solar Burst Spectrometer. Objective: provision of measurements for the study of integral intensities and X-ray spectra with the following characteristics:
  - energy channels: 2-4, 4-8, 8-12, 12-20, 20-30, 30-40, 40-60, 60-90, 90-120, 120-150, 150-200, and 200-300 keV
  - studies of X-ray precursors of solar flares in the range of 2-20 keV with a sensitivity of $10\,\mathrm{erg\ cm}^{-2}\,\mathrm{s}^{-1}$
  - studies of the dynamics of hard X-ray spectra during the flare impulse phases in the range of 30-120 keV with a time resolution of about 0.01 s

- **SUFR-Sp-C** = Solar UV Radiometer. Objectives: measurement of the solar radiation intensity in the UV range of $\lambda < 130$ nm. SUFR records the solar EUV radiation with the use of thermoluminescent material (24 specimens arranged on a rotating wheel). The thermoluminescent material $CaSO_4$ (Mn) is practically insensitive to radiation in the range of $\lambda > 130$ nm.
Dynamic range of the SUFR-Sp-C measurements: $0.1\text{-}30\,\mathrm{erg\ cm}^{-2}\,\mathrm{s}^{-1}$
The instrument contains 24 screens: one without filters measuring the total EUV radiation flux for $\lambda < 130$ nm, others with different filters (MgF, Al foil, Mylarfilm). The filters isolate the wave bands of soft X-rays (0.3-12 nm) and the Lyman alpha hydrogen line (121.6 nm). The absolute error does not exceed 15%. The exposure time for one screen and measuring time of its thermoluminescent signal is $18.5 \pm 0.5$ s. The total measurement cycle last 8 minutes.

- **VUSS** = Vacuum UV Solar Spectrometer. Objective: long-term measurements of the solar EUV radiation. The instrument performs absolute measurements of solar line intensities in the wave lengths region of 20-58 nm. Its operation is based on the concept of collision photoionization spectroscopy. The maximal spectral resolution is $\lambda/\Delta\lambda \approx 400$. The threshold sensitivity is $5 \times 10^6$ quants $\times\,\mathrm{cm}^{-2}\,\mathrm{s}^{-1}$.

- **DIFOS** = Solar Flux Optical Photometer (PI: Yu. D. Zhugzhda). Objective: observation of global solar oscillations in the spectral range of 0.4-1.0 μm. A three-channel instrument at 0.520 μm (.1 μm band), 0.710 μm (0.12 μm band), and 0.4-1.0 μm range. Intensity measurements with a relative accuracy of about $10^{-5}$. The time resolution is 16 seconds.

- **SORS** = Solar Radiospectrometer. Objectives: investigation of solar radiobursts, diagnosis of surrounding ionospheric plasma near the satellite.
SORS works with two sweeping receivers in the frequency ranges of 50 kHz - 30 MHz, and 25 MHz - 300 MHz. A wide-band channel ($\Delta f = 4$ MHz) is used for the study of detailed structures of radiobursts in time scales of ms.
The SORS measurements of solar radio-emission in the low frequency range $< 25$ MHz will complement the ground-based radioastronomical measurements with high time resolution. Direct measurements of UV emission and of local plasma density on the same satellite will improve ionospheric modelling and forecasting.

- **SKL** = Solar Cosmic Ray Spectrometer Complex (PI: S. N. Kuznetsov). Objectives: measurement of charged and neutral components of energetic particle fluxes (energy spectra and of the elemental, isotopic and charge composition). SKL consists of three devices: SKE-3, MKL, and SONG.

  - SKE-3 - measures the fluxes of nuclei with charges ranging from $z=1$ to 10, and with energies from 1.5-19 MeV / nucl for $^4$He and 3.7-4.6 MeV / nucl for $^{20}$Ne.
  - MKL - is a cosmic ray monitor for measuring the spectra of electrons (0.5 - 12.0 MeV), protons (1 - 200 MeV), and alpha-particle fluxes (120 - 240 MeV), which includes 5 detectors.
  - SONG - measures the solar neutrons, gamma rays (continuum and lines) and detects charged particles, mainly electrons and protons. A Cs (Te) detector (200 mm diameter, h = 100 mm) surrounded by a $4\pi$ anticoincidence shielding of a plastic scintillator is used. The SONG instrument permits the detection of:
    - neutrons with E $\geq$ 30 MeV (5 channels)
    - gamma rays in the range of 0.1-100 MeV (9 chan.) and 0.3-16 MeV (236 chan.)
    - high-energy protons in the ranges 200-500 MeV and 11-108 MeV (6 chan.).
    SONG operates in two modes ("flare" and "monitoring") with a time resolution of 0.04 s, 2.5 s, or 120 s.

**Data Transmission Parameters:**

- Transmission power  +9 dBW
- Antenna  2 V-dipole
- Antenna gain  +0.5 dB
- Polarization  right circular
- EIRP  +5.5 dBW
- Carrier frequency  136.35 MHz / 137.45 MHz (VHF)
- Bandwidth  120 Hz
- Modulation  BPSK, bi-phase
- Data rate  32 kbit/s / 64 kbit/s
- Word length  8 bit
- Frame length  128 byte

Coronas-I has an onboard data recording capability. The ground segment consists of Russian S/C control (TT&C function) and science data reception by the DLR-DFD ground station Neustrelitz (only science data reception station in the project). Neustrelitz is responsible for all science data reception, for intermediate archiving of all data, for science data distribution to all parties involved. DLR-DFD Neustrelitz processes/evaluates also data from the sensors SUFR and VUSS. The AIP (Astrophysikalisches Institut Potsdam) receives data of the following sensors for evaluation: DIFOS, HELICON, SORS, and SKL.

| Nr. | Instrument/Subsystem | Acronym | Mass (kg) | Power (W) | Principle Institut |
|---|---|---|---|---|---|
| 1 | Solar XUV Telescope/ Coronograph | TEREK-C | 45 | 40 | FIAN |
| 2 | Solar X-ray Spectral Polarimeter | RES-C | 35 | 30 | FIAN |
| 3 | Diagnostic of Energy Sources and Sinks in Flares | DIOGENESS | 15.4 | 24 | AI CSAV, CBK |
| 4 | Solar X-ray and $\gamma$-ray Scintillation Spectrometer | HELICON | 35 | 10 | LFTI |
| 5 | Solar Burst Spectrometer | IRIS | 30 | | LFTI |
| 6 | Solar UV Radiometer | SUFR-Sp-C | 3.6 | 12 | IPG |
| 7 | Vacuum UV Solar Spectrometer | VUSS | 18.5 | 20 | IPG |
| 8 | Solar Flux Optical Photometer | DIFOS | 6 | 18 | IZMIRAN, GAO, FMI |
| 9 | Solar Radiospectrometer | SORS | 18.5 | 21.5 | IZMIRAN, CBK |
| 10 | Solar Cosmic Ray Spectrometer Complex | SKL | 63 | 25 | NIIYaF, IEP,SAV |
| 11 | Processing unit for amplitude and time analysis of input signals | AVS | 12 | 22 | MIFI |
| 12 | System of compilation of scientific information | SSNI | 60 | 80 | OKB MEI |

**Table 10:** **Overview of CORONAS-I Scientific Payload**

| IZMIRAN | Institute of Terrestrial Magnetism, Ionosphere and Propagation of Radio Waves, Troitsk near Moscow, IZMIRAN is coordinating the project Coronas-I |
|---|---|
| LFTI | Physical Technical Institute (St. Petersburg) |
| GAO | Main Astronomical Observatory of the Ukrainian Academy of Sciences, Kiev |
| OKB MEI | Special Research Bureau of Moscow, Power Engineering Institute |
| MIFI | Moscow Engineering Physical Institute |
| NIIYaF | Institute of Nuclear Research of Moscow State University |
| IPG | Institute of Applied Geophysics, Moscow |
| FIAN | Lebedev Physical Institute of the Russian Academy of Sciences, Moscow (Coronas-F coordinator) |
| FMI | Physical Mechanical Institute, Lvov (Ukraine) |
| | Scientific Production Facility for Space Devices (Dnepropetrovsk, Ukraine) - Coronas satellite development and integration |
| AI CSAV | Astronomical Institute of the Academy of Sciences of the CSFR, Prague |
| AI SAV | Astronomical Institute of the Slovak Academy of Sciences |
| IEP Kosice | Institute of Experimental Physics of the Slovak Academy of Sciences |
| CBK | Space Research Center of the Polish Academy of Sciences |
| | |
| DLR-DFD | Germany, Neustrelitz Station - reception of all science data |
| AIP | Astrophysikalisches Institut, Potsdam, Germany - processing/interpretation of some sensor data |

**Table 11:     Principle Institutes involved in the cooperative Project Coronas**

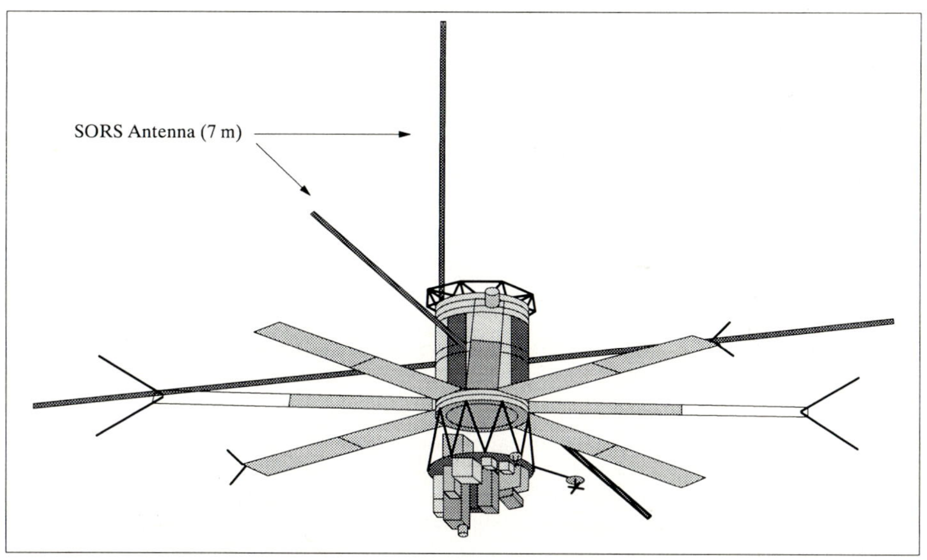


**Figure 18:     The CORONAS-I Spacecraft Model**

## A.17     CRRES (Combined Release and Radiation Effects Satellite)

The CRRES program is a joint NASA (MSFC)/DoD (USAF) mission in the field of Solar-Terrestrial energy transport with the objective to study ionospheric and magnetospheric transport phenomena and to study the effects of the natural radiation environment on microelectronic components (DoD). The satellite mass is 1753 kg (payload = 678 kg including the chemical canisters), nominal spin rate = 2 rpm. The spacecraft was launched on an At-

las/Centaur vehicle from Cape Canaveral on July 25, 1990. The nominal mission duration is 3 years, the extended mission is planned for 6 years (in support of the GGS mission). The joint NASA/DoD mission is scheduled to last one year after launch, this is followed by a two year DoD science mission. After the third year of operation the CRRES mission will become part of the GGS (Global Geospace Science) program.[53),54)]
The satellite failed on October 9, 1991.

Orbit: Elliptical orbit of 350 km x 35584 km, referred to as the Geosynchronous Transfer Orbit (GTO), inclination = 18.15°, orbit period = 10.56 hrs.

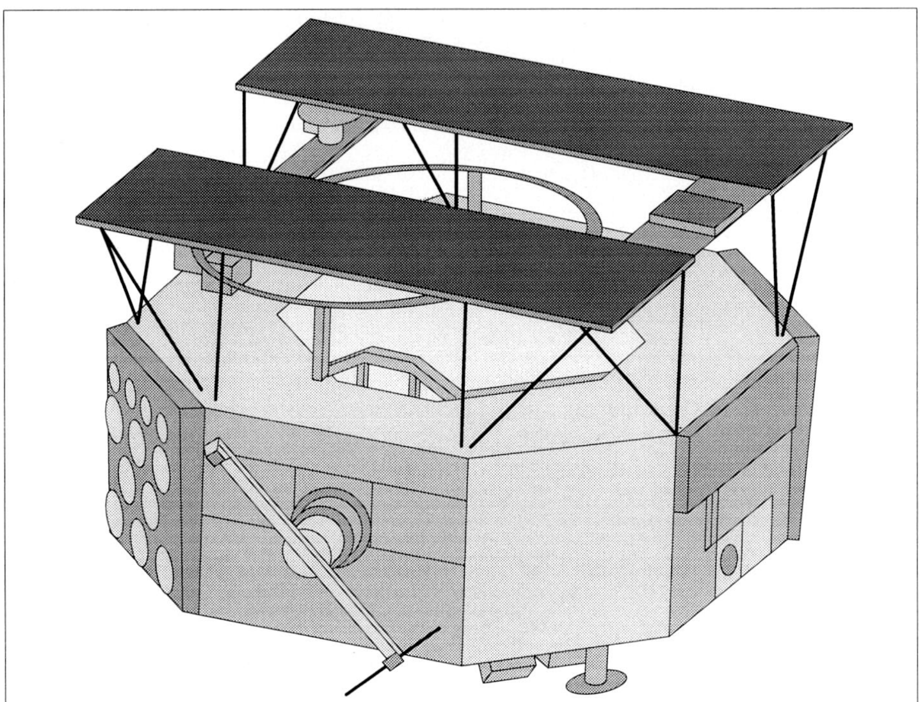

**Figure 19:    The CRRES S/C Model**

CRRES carries a complement of chemical release canisters, to be released at certain times over ground observation sites, and diagnostic facilities. These releases form large clouds of metal vapor that interact with the ionospheric and magnetospheric plasma and the Earth's magnetic field. These interactions will be studied with optical, radar, plasma wave and particle instruments from the ground, from aircraft, and the CRRES spacecraft.

The experiments performed near perigee are intended to further the understanding of the interaction of plasmas with magnetic fields, the coupling of the upper atmosphere with the ionosphere, the structure and chemistry of the ionosphere, and the structure of low-altitude electric fields. Those experiments which are performed near apogee in the Earth's magnetosphere are intended to study the formation of diamagnetic cavities, coupling between the magnetosphere and ionosphere, and the effects of artificial plasma injections upon the stability of the trapped particles in the radiation belts.

53)   R. A. Cooper, D. H. Burks, "Space Physics Missions Handbook", NASA, Office of Space Science and Applications, Feb. 1991
54)   Special Section on CRRES, Journal of Spacecraft and Rockets, Vol. 29, No. 4 July-Aug. 1992, pp. 555-617

The onboard scientific experiments consist of 46 electronic boxes, 10 booms, and 24 chemical canisters (Ba, Li, Ca, and Sr). There are 16 PI's responsible for the 16 separate chemical releases (24 canisters) at various times during the CRRES mission. The three chemical release campaigns were:

1. low-altitude releases (near perigee) over the South Pacific in September of 1990.
2. high-altitude releases (from 6000 to about 33500 km) over North America in January and February of 1991.
3. low-altitude releases over the Caribbean in July and August of 1991.

| Investigation | Experiment Code | Mass (kg) | Data Rate (bit/s) | Principal Investigator |
|---|---|---|---|---|
| Chemical Releases | NASA | 890 | | 16 PIs from 13 institutions |
| Low Altitude Satellite Studies of Ionosphere Irregularities (LASSII) | NRL-701 | 28.5 | 13616 | P. Rodrriguez, NLR |
| Space Radiation Experiment | AFGL-701 | 168.6 | 11027 | E. G. Mullen, AFGL |
| Energetic Particles and Ion Composition (EPIC) | ONR-307 | 34.4 | 2674 | R. Vondrak, Lockheed |
| Solar Flares II | ONR-604 | 15 | 750 | J. Simpson, U. of Chicago |
| High Efficiency Solar Panel (HESP) | AFAPL-801 | 5.2 | 63 | T. Trumble, Wright Patterson AFB |

**Table 12:    The CRRES S/C Science Payload**

**Sensors:**

- LASSII = Low-Altitude Satellite Studies of Ionosphere Irregularities (PI: P. Rodriguez, NRL). Objective: monitoring of naturally occurring and artificially produced ionospheric perturbations and the effects of ionospheric perturbations on communication paths. Measurements are conducted near perigee of selected orbits. LASSII is composed of three plasma experiments:

  - $P^3$ = Pulsed Plasma Probe. A Langmuir probe experiment that operates as two pulsed plasma probes, measuring ionospheric electron densities and temperatures.
  - QIMS = Quadrupole Ion Mass Spectrometer. Measurement of the densities of positive ions in both the natural ionosphere and in the chemical release experiments. Energy range from 500 keV to 100 MeV.
  - ELFWA = Extremely Low Frequency Wave Analyzer. Measurement of the spectrum of the electric and magnetic field fluctuations (electric field spectra from 10 to 250 Hz, and single-axis magnetic field spectra from 10 Hz to 125 Hz.

- EPIC = Energetic Particles and Ion Composition Experiment. EPIC consists of four instruments:

  - IMS-HI = Ion Mass Spectrometer - High/Medium Energy. Measurement of energy spectra and pitch angle distributions of magnetospheric ions and neutral particles.
  - IMS-LO = Ion Mass Spectrometer - Low Energy. Two instruments, measure the composition of plasmas that are the sources of radiation belt particles, and to provide data on the origin and acceleration processes of these plasmas. E/q = 0.11-35 keV/e, and M/q from 1-32 amu/e.
  - EPAS = Electron and Proton Wide-Angle Spectrometer. It covers the energy range from 21-285 keV, and for protons from 37 keV - 3.2 MeV. EPAS consists of two units to measure electrons simultaneously in ten directions and ions in four directions over a total angle of $\approx 100°$.
  - SEP = Spectrometer for Electron and Protons. SEP measures the energy and pitch angle distribution of energetic electrons and protons throughout the S/C orbit.

## A.18    Dynamics Explorers (DE-1 and DE-2)

NASA/GSFC program with international cooperation on the payload.[55] Combined DE-1 and DE-2 launch on Aug. 3, 1981 with Delta vehicle from Vandenberg. Objectives: Study of the flow of solar energy and matter from space through the Earth's magnetic field into the upper atmosphere (coupling of energy, electric currents, electric fields, and plasmas between the magnetosphere, ionosphere, and the atmosphere).
The DE S/C have a spin rate of 10 rpm, mass = 424 kg (at launch), payload mass = 105 kg

Orbit: Coplanar polar orbits;
- DE-1 (high altitude mission) 464 km x 23370 km highly elliptical orbit ; inclination = 90°
- DE-2 (low altitude mission) of 309 x 1012 km; inclination = 90°; period = 101 min; DE-2 re-entered the atmosphere on Feb. 19, 1983.

The orbits of the two spacecraft rotate in local time, precess and decay throughout the mission. This permits investigations in a multitude of constellations.

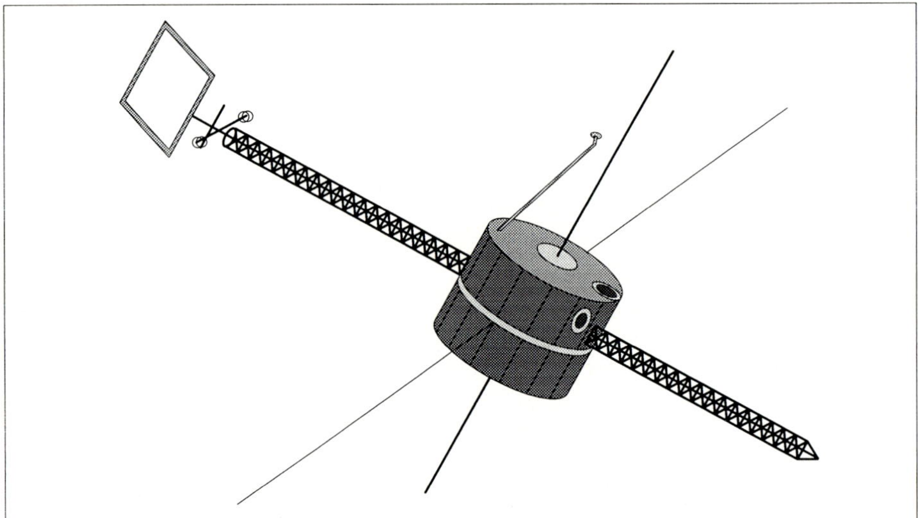

**Figure 20:**    **The Dynamics Explorer S/C Model**

### A.18.1    DE-1 Instruments (High Altitude Mission)

- **MAG-1** = Magnetometer (PI: Seguria, MSFC).[56]
  Measurement of the vector magnetic field. Objectives include: observation of field-aligned currents, magnetospheric equatorial currents, and ULF waves.
  The instrument is a triaxial fluxgate magnetometer. Resolution = ±1.5 nT in basic mode. In addition there are two modes of higher resolution: ±0.25 nT and ±20 pT. The sampling rate is 16 vector samples per second in all modes.

- **PWI** = Plasma Wave Instrument (PI: S. D. Shawhan, NASA).[57]
  Measurement of electromagnetic waves and electrostatic waves (PWI can distinguish


55)    R. A. Hoffman, G. D. Hogan, R. C. Maehl, "Dynamics Explorer Spacecraft and Ground Operating Systems", Space Science Instrumentation, 5, 1981, pp. 349-367
56)    W. H. Farthing, L. J. Cahill, et al., "Magnetic Field Observations on DE-A and -B", Space Science Instrumentation, 5, 1981, pp. 551-560

between electromagnetic and electrostatic phenomena and determines wave polarization and wave propagation direction). The quasi-static electric fields are measured parallel to the spacecraft spin axis in a range of 2 mVm$^{-1}$ to 2 mVm$^{-1}$, and perpendicular to the spin axis 0.5 mVm$^{-1}$ to 2 mVm$^{-1}$ at 16 samples per second. The AC electric field sensors include a 200 m tip-to-tip long wire antenna. Frequency range of AC fields: 1 Hz - 2 MHz. Integrated into this instrument is a Linear Wave Receiver (LWR, from Stanford) measuring wave amplitude in the frequency range 1.5 - 16 kHz.

- **HAPI** = High Altitude Plasma Instrument (PI: J. L. Burch, SWRI).[58]
  Objectives: measurement of the velocity-space distributions of electron and positive ions. Information on the hot-plasma effects of the coupling of energy, mass, and momentum within the Earth's atmosphere, ionosphere, and magnetosphere.
  HAPI consists of five electrostatic analyzers mounted in a fan-shaped angular array at angles of 45°, 78°, 90°, 102°, and 135° with respect to the S/C spin axis, each making simultaneous measurements of electrons and positive ions at up to 64 energy steps from 5 eV to 32 keV.

- **RIMS** = Retarding Ion Mass Spectrometer (PI: C. R. Chappel, MSFC).[59]
  Measurement of the thermal ion density, temperature bulk velocity, and ion composition. RIMS consists of three nearly identical sensor heads which contain a retarding potential analyzer, followed by a magnetic ion mass spectrometer with two channel electron multiplier detectors.

- **EICS** = Energetic Ion Mass Spectrometer (heritage of ISEE and GEOS-1,-2 spectrometers, PI: E. G. Shelley, Lockheed)[60]
  Objective: measurement of the ion composition [energy and pitch angle distributions of the principal mass constituents ($O^+$, $H^+$) of the upward flowing ions from the auroral acceleration region].
  EICS is a high sensitivity ($\approx 1$ cm$^2$ -sr-eV) high resolution (M/$\Delta$M $\geq 10$ FWHM at focus) instrument. It covers the entire mass range from < 1 amu/e to >150 amu/e in 64 mass channels at each of 32 energy per charge steps covering the range from 0 to $\sim 17$ keV/e. Data are used to investigate the coupling between the magnetosphere and the ionosphere via the auroral acceleration region.

- **SAI** = Spin-scan Auroral Imager (PI: L. A. Frank, Univ. of Iowa).[61]
  Measurement of visual and UV global auroral images. The instrument consists of three spanning photometers to view the dim emissions from Earth in the presence of strong stray light sources. The rotation of the S/C and the instrument scanning mirror provide a 2-D array of pixels comprising an image frame.

### A.18.2 DE-2 Instruments (Low Altitude Mission)

- **MAG-2** = Magnetometer.
  Measurement of the vector magnetic field (see MAG-1).

- **VEFI** = Vector Electric Field Instrument (PI: N. C. Maynard, GSFC)[62]
  Measurement of the vector electric field.
  Instrument: Six 11 m cylindrical antennas are deployed along 3 orthogonal axis to form the sensor array. Measurements of the DC electric field are made 16 times per second.


57) S. D. Shawhan, R. A. Helliwell, et al., "The Plasma Wave and Quasi-Static Electric Field Instrument (PWI) for Dynamics Explorer-A", Space Science Instrumentation, 5, 1981, pp. 535-550
58) J. L. Burch, R. A. Hoffman, et al., "High-Altitude Plasma Instrument for Dynamics Explorer-A", Space Science Instrumentation, 5, 1981, pp. 455-463
59) C. R. Chappell, J. H. Hoffman, et al., "The Retarding Ion Mass Spectrometer on Dynamics Explorer-A", Space Science Instrumentation, 5, 1981, pp. 477-491
60) E. G. Shelley, et al., "The Energetic Ion Composition Spectrometer (EICS) for the Dynamics Explorer-A", Space Science Instrumentation, 5, 1981, pp. 443-454
61) L. A. Frank et al., "Global Auroral Instrumentation for the Dynamics Explorer Mission", Space Science Instrumentation 5, 1981, pp. 369-393

A 20 channel comb filter spectrometer monitors the AC electric fields, concentrating on frequencies below 1 kHz.

- **NACS** = Neutral Atmosphere Composition Spectrometer (of Pioneer heritage, PI: G. R. Carignan, Univ. of Michigan)[63]
  Measurement of the abundances of neutral species (atomic oxygen, helium, argon, molecular nitrogen) over an altitude range of at least 100 km to 300 km upward in the Earth's atmosphere. The NACS instrument is a quadrupole mass spectrometer with electron bombardment ionization (an electron multiplier in the counting mode for detection is used).

- **WATS** = Wind and Temperature Spectrometer (PI: N. W. Spencer, GSFC).[64]
  Measurement of vertical, zonal and meridional components of neutral wind. In addition, measurements of the concentration and velocity of the ambient thermal ions is possible. Instrument accuracies: concentration of neutral particles $\sim 20\%$; their temperature $\sim 5$ K;

- **FPI** = Fabry-Perot Interferometer (PI: P. B. Hays, Univ. of Michigan).[65]
  Objective: Study of cross-sectional views of the dynamic and thermodynamic state of the thermosphere. Measurement of drift and temperature of neutral and ionic atomic oxygen in the thermosphere. FPI has a FOV of $0.53^\circ$ (semi-cone angle) defined by a doublet objective lens and a field stop.

- **IDM** = Ion Drift Meter (PI: R. A. Heelis, Univ. of Texas).[66]
  Measurement of the components of the ion drift normal to the spacecraft velocity. Data are used in studies of the convection pattern and energy deposition in the ionosphere.

- **RPA** = Retarding Potential Analyzer (PI: W. B. Hansen, Univ. of Texas).[67]
  Measurement of the bulk ion velocity in the direction of the spacecraft motion, the constituent ion concentrations and the ion temperature along the satellite path. In addition, the spectral characteristics of irregularities in the total ion concentration are determined.

- **LANG** = Langmuir Probe (PI: L. H. Brace, GSFC).[68]
  Measurement of electron temperature and density and ion density in the Earth's ionosphere.(heritage of Pioneer Venus mission).

- **LAPI** = Low Altitude Plasma Instrument (PI: R. Hoffman, GSFC) .[69]
  Objectives: measurement of high-resolution velocity-space distributions of positive ions and electrons from 5 eV to 30 keV, and from $0^\circ$ to $180^\circ$ in pitch angle.
  The instrument contains an array of 15 parabolic electrostatic analyzers and two Geiger-Mueller counters mounted on a one-degree of freedom platform.

Data:
Onboard tape recorder (2 for each S/C). Science data transmission rate of 16.384 kbit/s.


62)  N. C. Maynard et al., "Instrumentation for Vector Electric Field Measurements from DE-B", Space Science Instrumentation, 5, 1981, pp. 523-534

63)  G. R. Carignan, et al., "The Neutral Mass Spectrometer on Dynamics Explorer B", Space Science Instrumentation, 5, 1981, pp. 429-441.
64)  N. W. Spencer, et al., "The Dynamics Explorer Wind and Temperature Spectrometer", Space Science Instrumentation, 5, 1981, pp. 417-428
65)  P. B. Hays, et al., "The Fabry-Perot Interferometer on Dynamics Explorer", Space Science Instrumentation, 5, 1981, pp. 395-416
66)  R. A. Heelis, W. B. Hanson, et al., "The Ion Drift Meter for Dynamics Explorer-B", Space Science Instrumentation, 5, 1981, pp. 511-521
67)  W. B. Hanson et al., "The Retarding Potential Analyzer for Dynamics Explorer-B", Space Science Instrumentation, 5, 1981, pp. 503-510
68)  J. P. Krehbiel, L. H. Brace, W. H. Pinkus, R. B. Kaplan, et. al., "The Dynamics Explorer Langmuir Probe Instrument", Space Science Instrumentation, 5, 1981, pp. 493-502
69)  J. D. Winningham, R. A. Hoffman, et al., "The Low Altitude Plasma Instrument (LAPI)", Space Science Instrumentation, 5, 1981, pp. 465-475

## A.19 DMSP (Defense Meteorological Satellite Program)

Background[70],[71],[72]: DMSP is the meteorological program of the US Department of Defense (DoD) which originated in the mid-1960s with the objective to collect and disseminate worldwide cloud cover data on a daily basis.

From mid-1965 to early 1970, the series of DMSP satellites was known as the Block 4 series. S/C of this series were a spin-stabilized octagon with a mass of about 50 kg. The payload consisted of two vidicon cameras for the collecting of high-resolution TV pictures (resolution: 2.4 km at image center to about 16 km at the edge of a picture). Some later S/C also had a low-resolution VIS (.4 - 4 μm) and infrared (8.0 - 12 μm) sensor. From these, better sensors emerged with first attempts to combine reflected and emitted energy in the cloud analysis problem.

Beginning in early 1970, a new generation of satellites was inaugurated, the Block 5 series. It was a three-axis stabilized spacecraft that initially employed a scanning radiometer as its imaging sensor (four channels). The VIS channels had a spectral band from 0.4 - 1.1 μm and the lower resolution channel (approx. 3.6 km at nadir) also had a low-light level amplification system. These two design factors considerably enhanced the utility of the system.

DMSP employs the concept of total system integration from the spacecraft operation through data reduction and interpretation. The space segment is controlled by the Air Force Space Command (AFSPCMD) through command sites centered in Colorado Springs, Co. By the end of 1992, 22 Air Force terminals, all equipped to receive Block 5 image data, were deployed worldwide.

| Series | Satellite | Launch Date / End Mission | Sensor Complement |
|---|---|---|---|
| Block 5D-1 | F-1 | 11.9.1976 / 17.9.1979 | OLS, SSH, SSJ/3, SSB, Contamination Monitor |
|  | F-2 | 4.6.1977 / 19.3.1978 | OLS, SSH, SSJ/3, SSB, SSB/0, IFM, SSI/E,SSI/P |
|  | F-3 | 30.4.1978 / Dec. 1979 | OLS, SSH, SSJ/3, SSB, GFE-3R |
|  | F-4 | 6.6.1979 / 29.8.1980 | OLS, SSH, SSJ/3, SSI/E, SSM/T, SSC, SSD |
|  | F-5 | 14.7.1980 (failed) | OLS, SSH-2, SSJ/3, SSI/E, SSB/O, SSR |
| Block 5D-2 | F-6 | 20.12.1982 / 24.8.1987 | OLS, SSH-2, SSI/E, SSJ/4, SSB/A |
|  | F-7 | 18.12.1983/ 17.10.1987 | OLS, SSM/T, SSI/E, SSJ/4, SSB, SSJ*, SSM |
|  | F-8 | 18.6.1987 / 13.8.1991 | OLS, SSM/I, SSM/T, SSI/ES, SSJ/4, SSB/X-M |
|  | F-9 | 3.2.1988 | OLS, SSM/T, SSI/ES, SSJ/4, SSB/X |
|  | F-10 | 1.12.1990 | OLS, SSM/I, SSM/T, SSI/ES, SSJ/4, SSB/X-2 |
|  | F-11 | 28.11.1991 | OLS,SSM/I,SSM/T-2,SSJ/4,SSI/ES-2,SSB/X-2 |
|  | S-12 | projected | OLS, SSM/I, SSM/T,  SSJ/4, SSI/ES-2, SSB/X-2, SSM |
|  | S-13 | projected | OLS, SSM/I, SSM/T, SSM/T-2, SSJ/4, SSI/ES-2, SSB/X-2, SSM, SSZ, |
|  | S-14 | projected | OLS, SSM/I , SSM/T, SSM/T-2, SSJ/4, SSI/ES-2, SSB/X-2, SSM, SSZ |
| Block 5D-3 | S-15 | projected | OLS, SSM/I , SSM/T, SSM/T-2, SSJ/4, SSI/ES-2, SSB/X-2, SSM, SSF |
|  | S-16 | projected | OLS, SSMIS , SSI/ES-2, SSJ/4, SSM, SSF, SSY |
|  | S-17 | projected | OLS, SSMIS , SSI/ES-2, SSJ/4, SSM, SSULI, SSUSI |

**Table 13:    Recent DMSP Series Satellites with their Sensor Complements**

DMSP satellites are launched from Vandenberg AFB. Real-time commands and stored command files are relayed via commercial communication satellites to Command Readout Stations (CRSs) which then relay these command to the DMSP satellites. Data may also be


70)   W.D. Meyer, DMSP: Review of its Impact, in Monitoring Earth's Ocean, Land, and Atmosphere from Space, Volume 97, Progress in Astronautics and Aeronautics, AIAA, 1985, pp. 131- 147
71)   S. Ferry, 'The Defense Meteorological Satellite System Sensors: An historical Overview', May 1989
72)   R. B. Gomez, M. C. Colton, D. Boucher, F. P. Kelly, "The Defense Meteorological Satellite Program (DMSP)", ISSSR, Maui, Hawaii, 16-20 Nov. 1992

relayed via tracking stations (located in Hawaii, New Hampshire and Thule (Greenland)). The meteorological data are processed by centralized users at the Air Force Global Weather Center, Fleet Numerical Oceanography Center, and deployed tactical assets.

The Block 5D-1 satellites were flown from 1976 to 1980. The first Block 5D-2 satellite was launched in 1982. Block 5D-2 satellites are the designated satellites until the mid-1990s when the Block 5D-3 satellite  series is anticipated. The first Block 5D-3 satellite is S-15 (last S-20). The block 5D-3 satellites are projected to carry similar complement of sensors as the 5D-2 configurations. The required pointing accuracy of 0.01° is achieved by 3 orthogonal gyroscopes measuring short-term variations in the S/C attitude.

**Orbital Parameters of DMSP:**[73]
Sun-synchronous orbits, altitude = 811 - 853 km, inclination = 98.9°, period = 101.6 minutes, there are normally two satellites in operation at any one time (one in a morning and one in an afternoon equatorial crossing time).

**Block 5D Sensors (operational)**

- **OLS** = Operational Linescan System (primary sensor on each satellite). Objective: day and night cloud cover imagery. The OLS instrument consists of an eight-inch telescope which oscillates back and forth six times per second.
  The OLS measures in VIS and TIR bands. Swath width = 1600 nmi (3000 km) from a nominal 450 nmi (833 km) orbit altitude. OLS provides global coverage in both visible (L data) and IR (T data) modes. Fine resolution data with a nominal linear resolution of 0.3 nmi (0.56 km) are collected as needed, day and night, by the IR detector, and as needed, during daytime only, by a segmented silicon diode detector (LF data). A high resolution photometer tube is used for nighttime visible imagery.
  - Band 1: VIS wavelength = 0.4 - 1.1 μm,
  - Band 2: TIR wavelength = 10.5 - 12.6 μm (8 - 13 μm, old prior to 1979), resolution = 0.56 km for fine resolution data (stored data is smoothed to 2.7 km resolution), Continuous data collection, polar stereographic image products have ground resolution of 5.4 km. Swath = 3000 km
    The IR system counts are automatically calibrated to very between 190 and 310 K of effective blackbody or brightness temperatures.

- **SSM/I** = Special Sensor Microwave Imager.[74]
  The SSM/I is a passive microwave radiometer. It detects thermal energy emitted by the Earth's atmosphere in the microwave portion of the spectrum (similar to NIMBUS-7 SMMR). SMM/I data is also used to determine ocean surface wind speed, ice coverage and age, areas and intensity of precipitation, cloud water content, and land surface moisture. Swath width = 753 nmi (1394 km).
  - Channel 1: frequency = 19,35 GHz, wavelength = 1.55 cm, antenna beam width = 1.98°, geometric footprint dimension 70 x 45 km, Earth incidence angle = 53.1°. The swath is organized into 64 pixels for the lower frequency channels, and 128 pixels for the 85.5 GHz channels.
  - Channel 2: frequency = 22.235 GHz, wavelength = 1.35 cm
  - Channel 3: frequency = 37.00 GHz, wavelength = 0.81 cm
  - Channel 4: frequency = 85.50 GHz, wavelength = 0.35 cm
    The SSM/I sensor executes a 45° conical scan of the Earth's surface, atmosphere and space in the direction of, or opposite to the Earth and atmosphere. The instrument records both horizontally and vertically polarized electromagnetic emission near 19, 22, 37, and 85 GHz.

73)   R. Massom, 'Satellite Remote Sensing of Polar Regions', Applications, Limitations and Data Availability, Belhaven Press, London

74)   J. W. Sherman , "The Near-Term Suite of Satellite Sensors to Support Developing Countries' Climate and Global Change Photograms", Proceedings of the Twenty-Fourth International Symposium on Remote Sensing of Environment, ERIM Ann Arbor Mi., Volume I, 27-31 May 1991, pp. 27-28

| Wavelength (mm) | Frequency (GHz) | Polarization | Resolution (km) | Environmental Response |
|---|---|---|---|---|
| 15.5 | 19.35 | V/H | 50 | Ocean surface wind, land surface moisture |
| 13.5 | 22.235 | V | 50 | Integrated water vapor content, Ocean surface wind |
| 8.1 | 37.0 | V/H | 25 | Rain, cloud water content, ice cover |

**Table 14:**     **Some SSM/I Sensor Characteristics**

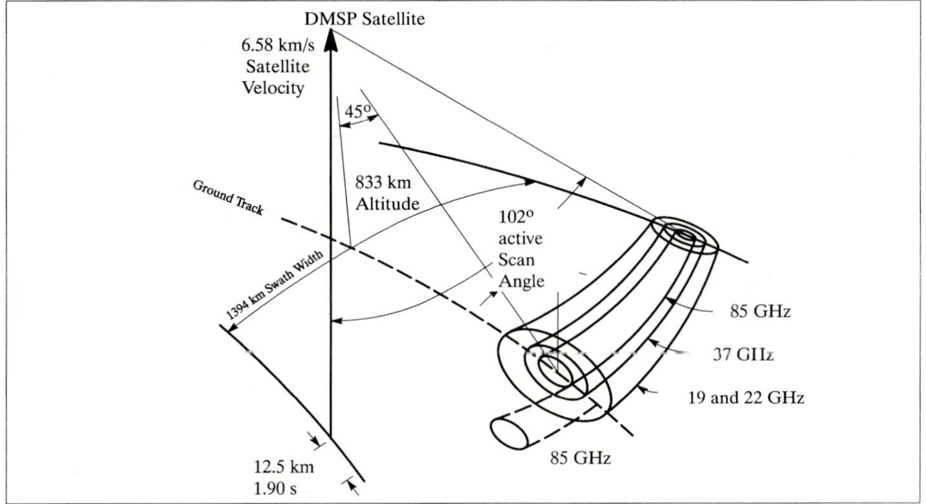


**Figure 21:**     **The Scan Geometry of the DMSP SSM/I Sensor**[75)]

- **SSM/T-1** = Special Sensor Microwave Temperature Sounder. (in operation since 1985)
  The SSM/T is a seven channel passive microwave sounder. It measures the Earth surface and atmospheric emission in the 50 to 60 GHz oxygen band. The SMM/T is a cross-track nadir scanning radiometer having a FOV of 14.4°. At nominal altitude (833 km) the subtrack spatial resolution is an app. circle of 174 km diameter at nadir and an ellipse of 213 x 304 km toward the limb. There are seven total cross-track scan positions separated by 12° with a maximum cross-track scan angle of 36°. Swath width = 1500 km (data coverage gap between successive orbits).
  Scan across-track through nadir. Frequencies: 50.5, 53.2, 54.35, 54.9, 58.4, 58.825 and 59.4 GHz. The daily data volume is 1.5 MByte per satellite.

- **SSM/T-2** = Special Sensor Microwave Water Vapor Profiler-2
  The SSM/T-2 is a modification of SSM/T for water vapor sounding. It detects electromagnetic radiation at 91.5, 150, and 183 GHz and has footprints ranging from 46 to 120 km (depending on frequency). First flight of instrument on F-11 in 1991.

- **SSJ/4** = Auroral Electron and Ion Spectrometer.
  The SSJ/4 is a next generation sensor of the SSJ/3. Objective: Measurement of transfer energy, mass, and momentum through the magnetosphere - ionosphere in the Earth's magnetic field. - The SSJ/4 sensor consists of four electrostatic analyzers that record the

---

75)   The rotating antenna sweeps the surface in two alternating modes - one in which all four frequencies are recorded, and another in which only 85 GHz data are recorded. The use of a single antenna results in different ground resolutions for each frequency.

flux of precipitating ions or electrons at 20 fixed energy channels between 50 eV and 30 keV. The primary source of the particles precipitating into the upper atmosphere is the northern and southern auroral zone. The daily data volume is approximately 1 Mbyte per satellite.

- **SSI/ES** = Special Sensor Ionospheric Plasma Drift/Scintillation Meter.
  The SSI/ES is an improved version of the SSI/E. In addition to the Langmuir probe and planar collector which make up the SSI/E, the SSI/ES has a plasma drift meter and a scintillation meter. An upgraded version of the SSI/ES (SSI/ES-2) is planned. The data volume of SSI/ES is 12 MByte/satellite-day.

- **SSM** = Triaxial Fluxgate Magnetometer.
  The SSM measures geomagnetic fluctuations associated with geophysical phenomena (i.e., ionospheric currents flowing at high latitudes). In combination with the SSI/ES (or SSI/ES-2) and the SSJ/4, the SSM will provide heating and electron density profiles in the high-latitude ionosphere.

- **SSB/X-2** = Gamma Ray Particle Detector.
  The SSB/X-2 is an array-based system which detects the location, intensity, and spectrum of X-rays emitted from the Earth's atmosphere. The array consists of four identical and independent directional detectors.

- **SSZ** = (classified sensor)
  SSZ is a static Earth-viewing sensor, monitoring electromagnetic radiation.

- **SSMIS**= SSM/I + SSM/T-1 + SSM/T-2 (Block 5D-3 sensor). In the future (i.e. in the Block 5D-3 satellite era) the Block 5D-2 passive microwave sensor suite will be combined into a single new sensor package - the SSMIS (see Ref..72)).

**Old Sensors of DMSP (not operational any more, for Reference only)**

- **SSL** = Lightning Detector (old series).
  The SSL was a "one-of-a-kind" experiment. Night operation to detect lightning flashes in the 0.4 - 1.1 µm range. FOV = 1200 x 1200 nmi.

- **SSB** = Gamma Tracker. (old series)
  Used to track fallout and nuclear debris entrained in the atmosphere from 10 to 15 km. The sensor detected the fission gammas emitted by the debris.

- **SSE** = Temperature Sounder. (old series)
  Eight channel scanning filter radiometer. Six of the channels measure in the 15 µm carbon dioxide band, one in the 12 µm window, and the last in the rotational water vapor band near 20 µm. Radiance measurements of the Earth's atmosphere are processed to obtain vertical temperature profiles.

- **SSJ** = Auroral Electron and Ion Spectrometer
  The SSJ counted ambient electrons with energies ranging from 50 eV - 20 keV. It determined the number of electrons having energies within certain sub-ranges of the 50 eV to 20 keV spectrum by utilizing a time-sequenced variable electrostatic field to deflect the particles toward the channeltron detector.

- **SSJ/3** = Auroral Electron and Ion Spectrometers.
  The SSJ/3 is a next generation sensor of the SSJ. The SSJ/3 was flown on all Block 5D-1 spacecraft with the exception of F-1.
  Objective: Measurement of transfer energy, mass, and momentum through the magnetosphere - ionosphere in the Earth's magnetic field.

- **SSI/E** = Topside Ionospheric Plasma Monitor.
  The SSI/E measures the ambient electron density and temperatures, the ambient ion

density, and the average ion temperature and molecular weight at the DMSP orbital altitude. The instrument consists of an electron sensor (Langmuir probe) and an ion sensor mounted on a 2.5 m boom. The ion sensor is a planar aperture, planar collector sensor oriented to face into the spacecraft velocity vector at all times.

- **SSJ*** = Space Radiation Dosimeter. The SSJ* measures the accumulated radiation dose produced by electrons in the 1 - 10 MeV energy range, protons of greater than 20 MeV, and the effects of the occasional nuclear interactions produced by energetic protons. Accumulated dose was measured of a period of one year (minimum).

- **SSD** = Atmospheric Density Sensor.
The SSD provides a measure of major atmospheric constituents (nitrogen, oxygen, and ozone) in the Earth's thermosphere by making the Earth limb observations of the ultraviolet radiation from the thermosphere. The sensor measured the radiation emitted from excitation of molecular nitrogen by impinging solar radiation.

- **SSC** = Snow Cloud Discriminator.
The SSC is  sensor push broom scan radiometer. Wavelength 1.51 - 1.63 µm (IR). A binary system, it was used to determine the presence of snow versus clouds. It is a proposed concept sensor intended to help determine if machine processing could make the snow/cloud determination. The along-track scan is provided by the forward motion of the S/C, while the 40.2° cross-track scan is provided by a linear array of 48 detector elements at the image plane of a wide-angle lens.

- **SSB** = Gamma X-Ray Detector.

- **SSB/A** = X-Ray Spectrometer.
The SSB/A detects X-rays and gamma rays from bomb debris or those X-rays produced by the Bremsstrahlung process when electrons precipitate from the Earth's radiation belts. By sensing these X-rays, the SSB/A provides location of the aurora as it orbits the Earth.

- **SSB/O** = Omnidirectional Gamma Detector.
The SSB/O was an experiment to determine if more accurate atmospheric measurements could be obtained by measuring the co-orbiting particles and the upward flux and subtracting it from the sub-satellite scene. The experiment was extremely successful. The SSB/O was sensitive to X-rays in the energy range of approximately 1.5 keV.

- **SSB/S** = Scanning X-Ray Detector.
The SSB/S detects the location, intensity and spectrum of X-rays emitted from the Earth's atmosphere. The detector consists of three sensors.

- **SSB/X, SSB/X-M** = Gamma Ray Detectors.
The SSB/X is an array-based system which detects the location, intensity, and spectrum of X-rays emitted from the Earth's atmosphere. The array consists of four identical and independent directional detectors. The SSB/X-M and SSB/X-2 follow from the SSB/X and are also gamma-ray and particle detectors. The SSB/X-M and SSB/X-2 consist of two identical and independent gamma-ray detectors and three particle detectors (capable of detecting gamma-ray bursts).

- **SSH** = Infrared Spectrometer.
The IR spectrometer is a multispectral sounder for humidity, temperature, and ozone measurements. The SSH provides soundings of temperature and humidity and a single measurement of ozone for vertical and slant paths lying under and to the side of the sub-satellite track.
The SSH made a set of radiance measurements in narrow spectral channels lying in the absorption bands of carbon dioxide, water vapor, and ozone. The radiance measurements were mathematically inverted to yield vertical temperature profiles, water vapor, and total ozone content.

- **SSH-2** = Infrared Temperature and Moisture Sounder.
  The SSH-2 provides soundings of temperature and humidity for vertical and slant paths lying under and to the side of the sub-satellite track. SSH-2 is physically identical to SSH, but with different (and tighter) filter bands. 16 frequency bands in the range of 3.7 - 30 μm. Resolution = 60 km. Swath width = 2240 km.

## A.19.1   DMSP Data Availability - Visible and Infrared Imagery

NOAA/NESDIS-NSIDC[76] and -NGDC[77] (Boulder Co.) have established a collection of digital satellite imagery acquired from the DMSP series of the US Air Force under NOAA-NESDIS contract. Hence, DMSP imagery data are available for the general user community (among them the SSM/I, SSM/T and SSM/T-2 sensor data). These data are prepared from a global, digital intensity file used operationally by the Air Force in forecasting and are subsequently archived at NGDC [DMSP imagery are archived after operational use (usually 45 to 60 days)]. A digital archive of DMSP data is operational at NGDC since April 1992 (NGDC receives two 5 GByte tapes in compressed format every day). Archival services are continually upgraded.

The imagery collection consists of three positive transparency products produced by USAF.

1. Limited coverage (at 0.6 km resolution) is available for areas surrounding direct-readout sites operated at several locations around the world, principally the Western United States, Europe, and Southeast Asia.

2. Global coverage is available on single-orbit strips (at 2.7 km resolution)

3. Mosaics compiled from several orbits, with latitude and longitude grids added, are available globally (at a resolution of 5.4 km).

Mosaics are the only pre-gridded product available. Single-orbit strips can be custom gridded for an additional fee. Each of these products has been produced since 1973, except the mosaics, which are available from December 1975 on.

| Parameter | Spatial Resolution (km) | Range of Values | Quantization Levels | Absolute Accuracy |
|---|---|---|---|---|
| Ocean Surface Wind Speed | 25 | 3 - 25 | 1 | $\pm$ 2 m/s |
| Ice: area covered<br>Ice: age<br>Ice: edge location | 25<br>50<br>25 | 0 - 100<br>1. year, multiyear<br>N/A | 5<br>1 yr, >2 yr<br>N/A | $\pm$ 12%<br>None<br>$\pm$ 12.5 km |
| Precipitation over land ares | 25 | 0 - 25 | 0,5,10,15,20, $\succeq$25 | $\pm$ 5 mm/hr |
| Cloud water | 25 | 0 - 1 | 0.05 | $\pm$ 0.1 kg/m$^2$ |
| Integrated water vapor | 25 | 0 - 80 | 0.10 | $\pm$ 2.0 kg/m$^2$ |
| Precipitation over water | 25 | 0 - 25 | 0,5,10,15,20,$\succeq$25 | $\pm$ 5 mm/hr |
| Soil moisture | 50 | 0 - 60% | 1 | None |
| Land surface temperature | 25 | 180 - 340 K | 1 | None |
| Snow water content | 25 | 0 - 50 cm | 1 | $\pm$ 3 cm |
| Surface type | 25 | 12 types | N/A | N/A |
| Cloud amount | 25 | 0 - 100% | 1 | $\pm$ 20% |

**Table 15:**      **Environmental Products of the SSM/I Sensor**[78]


76)   'Defense Meteorological Satellite Program, Visible and Infrared Imagery Collection', NOAA-NSIDC, Feb. 1984
77)   Data Management Plan for the Archive of DMSP Digital Data at NGDC, April 28, 1992, Draft; Courtesy of W. Kroehl, NGDC
78)   J. Holloinger, "DMSP Special Sensor Microwave/Imager Calibration/Validation", Final Report, Vol. I and II, NRL, 1989

| NOAA/AVHRR Sensor | System Parameter/ (Applications) | DMSP/OLS Sensor |
|---|---|---|
| 1.1 km LAC<br>4.0 km GAC (degraded at edges)<br>good<br>good | spatial resolution<br><br>(sea ice leads)<br>(meteorology) | 0.55 km 'fine'<br>2.7 km 'smooth' (const. across swath)<br>better<br>better |
| 5 narrow channels<br>Ch1: .55-0.68 µm<br>Ch2: .725-1.0 µm<br>Ch3: 3.55-3.93 µm<br>Ch4: 10.30-11.30 µm<br>Ch5: 11.50-12.50 µm<br>good<br>good | spectral resolution<br><br><br><br><br><br>(sea surface temp.)<br>(vegetation index) | 2 broad-band channels<br>VIS: 0.4-1.1 µm<br>IR: 10.5-12.5 µm<br><br><br><br>marginal<br>N/A |
| IR yes (10 bit)<br>VIS no (pre-launch, drifts) | absolute calibration | IR yes (8 bit)<br>VIS no (continued gain adjustment) |
| no<br><br>N/A<br>developmental<br><br>N/A | Visible-Band dynamic range/nighttime operation<br>(auroral characteristics)<br>(biomass burning,)<br>NESDIS, NASA<br>(moonlit clouds/snow) | yes<br><br>yes<br>added potential with unique visible band<br><br>good |
| no<br><br>N/A<br>N/A | coincident passive micro-wave<br>(snow/ice, rain rate)<br>(sfc. wind/soil moisture) | yes<br><br>good<br>good |
| no<br><br>no | coincident space environ-ment measurements<br>(auroral image feature + electron flux) | yes<br><br>yes |

**Table 16:** **Comparison of two Sensors - NOAA/AVHRR and DMSP/OLS**[79]

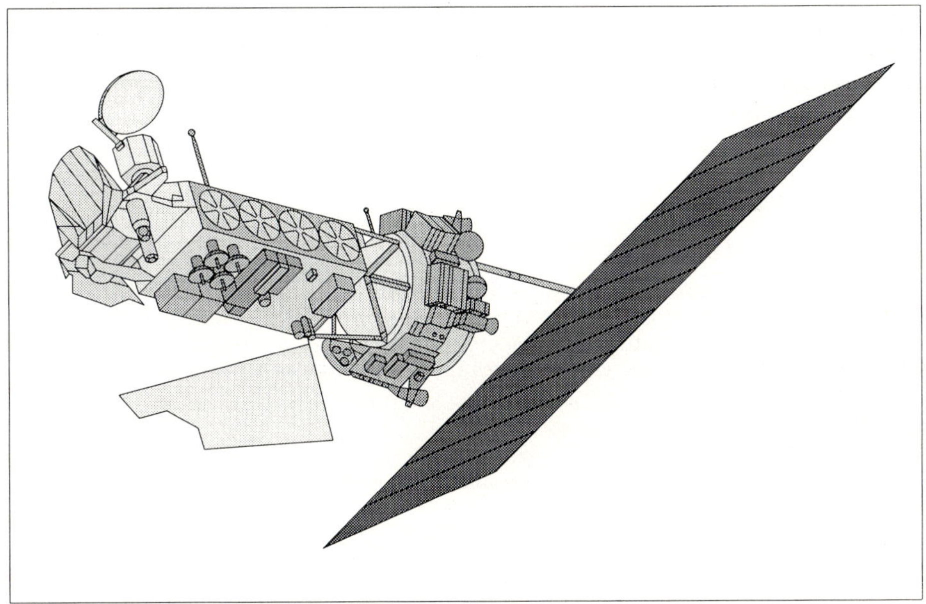

**Figure 22:** **DMSP S/C Model**

79)  Courtesy G. Scharfen, NOAA/NESDIS/NSIDC

## A.20    EGS (Experimental Geodetic Satellite, Ajisai)

Japanese (NASDA) geodetic mission. Launch: Aug. 12, 1986 from Tanegashima Space Center with the H-1 launch vehicle. The primary objective was performance confirmation of the H-1 launch vehicle. Objectives of EGS: provision of long-range geodetic applications aimed at rectifying Japan's domestic geodetic triangular network, determining the exact position of many Japanese islands and establishing Japan's geodetic point of origin.[80]

EGS is a passive spherical satellite of 2.15 m diameter and a mass of 685.2 kg, carrying 318 mirrors and 120 laser reflector assemblies (1436 corner cube reflectors) for precise satellite laser ranging (SLR) measurements from ground-based laser ranging stations. The reflectors and mirrors on the satellite sphere reflect laser beams back to the source, regardless of the angle of incidence. (Solid Earth studies, crustal movements, plate tectonics,). The reflection of solar light is used to determine the direction to the satellite from an observation site.

The body of the satellite is a hollow sphere made of glas-fiber-reinforced plastics. The surface is covered with corner cube reflectors (CCR's) and solar light reflectors. Twelve CCR's form a set of laser reflectors (LR's), and 120 LR's are distributed almost uniformly over the surface. The remaining part of the surface is covered with 318 solar reflectors. The base of the mirrors is made of an alloy of aluminum and the surface is coated with a silicon oxide for protection.

Orbit: circular orbit with an altitude of 1500 km, inclination = 50°, period = 116 minutes.

Observation method of EGS: The basic concept for geodetic use of the EGS is to simultaneously determine directions through photography and measuring distances by laser ranging. The principle is to determine the relative geodetic location of two observation sites by simultaneously measuring the distance and direction to the EGS at a known station (named the Base Station) and an unknown station (Primary Station) of which the precise geodetic location is unknown. The distance from a site to the EGS is measured by SLR. To determine the location of a number of unknown stations around the base station, at least one Transportable SLR (TSLR) is used.

The laser ranging surveys are conducted by the 'Geophysical Survey Institute' of the Ministry of Construction, and the 'Hydrography Department of the Maritime Safety Agency', Ministry of Transport.


## A.21    EOS - Original Program Version

EOS[81],[82],[83]( = Earth Observation System) is the planned and coordinated Remote Sensing Program (Mission to Planet Earth) of the USA (NASA, NOAA, USGS, NSF, DOE, USDA, EPA, DOD, etc. in the Global Change Program). The overall objective is to provide a comprehensive set of measurements on many different aspects of global change. In particular:

* Documenting global change (observations)
* Enhancing the understanding of key processes (process research)
* Predicting global and regional environmental change (integrated modeling and prediction)

The considered program time frame is 25 years (from 1990). Two families of polar platforms are considered, which are complementary and closely coordinated, namely the EOS-A se-

---


80)    M. Sasaki, H. Hashimoto, " Launch and Observation of the Experimental Geodetic Satellite of Japan', IEEE Transactions on Geoscience and Remote Sensing, Volume 25, Nr. 5, Sept. 1987

81)    "Earth Observing System", Reference Handbook 1990, and 1991, NASA-GSFC

82)    'Optical Remote Sensing of the Atmosphere', 1990 Technical Digest Series of the Optical Society of America, Volume 4, pp. 23-58

83)    G. Asrar, D. J. Dokken, "EOS Reference Handbook", March 1993, NASA

ries and the EOS-B series with TITAN-IV launch vehicles (large observatory platforms). The planned EOS program consists of 4 to 6 satellites and a very large infrastructure to be built for the dissemination of the wealth of data. The first EOS launch is planned for 1998. There is an extensive and coordinated preparatory program underway in support of all EOS activities, this applies for the ground segment (data centers, communication networks, facilities, etc.), as well as for the space segment (selection of candidate instruments for particular payloads, etc.).

- Space Segment: large observatories, all similar. Six Titan IV launch vehicles
- Sensors: large number of instruments to cover all aspects of global change
- Ground Segment: Integrated EOS Data and Information System (EOSDIS) providing a comprehensive infrastructure in the flight-, data- and user segments (mission control, data reception, communication, management, data processing, data storage, etc.), and long-term database access services to a large user community for data interpretation and utilization.

Note: The Original Program version of EOS has been replaced by the **'Restructured Program Version'** (next chapter), a major shift of NASA policy in its Earth Observation program. In the Summer and Fall of 1991 there were overall program reviews, in particular by CEES (Committee on Earth and Environmental Sciences), in order to harmonize the major objectives of 'Mission to Planet Earth' with the intended EOS concepts and in particular with very tight budget projections for long-term support capability by all affected agencies. The original EOS program version is kept in this text for reasons of continuity and context to the reader. All EOS sensors are described and updated under A.21.

**EOS Sensors:** (Candidates for EOS payloads)

- **ACRIM** = Active Cavity Radiometer Irradiance Monitor (PI:R. C. Willson, JPL). The ACRIM sensor points toward the sun; the assembly is mounted on a two-axis tracker to observe the solar disk during each orbit.
  The objective of ACRIM is to monitor the variability of total solar irradiance with state-of-the-art accuracy and precision, thereby extending the high-precision database compiled by NASA since 1980 by other ACRIM experiments as part of the Earth radiation budget in the National Climate Program. Note: The first ACRIM sensor was flown on the Solar Maximum Mission (Launch: Feb. 14, 1980)
  The ACRIM instrument contains 3 independent active cavity radiometer (ACR) solar-monitoring sensors and a sun-position sensor. One ACR monitors the solar irradiance, the other two are used to calibrate the optical degradation of the first instrument. FOV = $\pm 2.5°$; mass = 39 kg; duty cycle = 100% (daylight only); power = 35 W; data rate = 1 kbit/s.

- **AIRS** = Atmospheric Infrared Sounder (NASA-JPL instrument). Objective: High-resolution measurement of global temperature profiles in the atmosphere. Measurement of the Earth irradiation in the range of 3.74 - 15.4 µm, simultaneously in 2300 high-spectral resolution channels. This is supplemented with 6 channels in the visible/IR range between 0.4 and 1.7 µm. AIRS has an IFOV of $1.1°$ and FOV = $\pm 49°$ scanning capability perpendicular to the spacecraft ground track (swath width = 1650 km, 13.5 km horizontal resolution in nadir, 1 km vertical). The data rate of AIRS is 1.42 Mbit/s, mass = 140 kg, power = 224 W.

- **Altimeter**. This is a nadir-looking dual-frequency radar altimeter (TOPEX heritage) that maps the topography of sea surface and polar ice sheets. The shape and strength of the radar return pulse also provide measurements of ocean wave height and wind speed, respectively. The primary objective is to provide information on the ocean surface current velocity (for ocean circulation models).
  The altimeter is a dual frequency radar altimeter measuring at 13.6 GHz (Ku-Band) and at 5.3 GHz (C-Band). Mass = 265 kg, power = 250 W, data rate = 85 kbit/s.

- **AMSU-A/MHS** = Advanced Microwave Sounding Unit (NOAA Instrument on EOS)/Microwave Humidity Sounder (MHS is provided by EUMETSAT). The AMSU-A (see also MTS, new name) provides temperature soundings whereas the MHS provides humidity soundings. AMSU-A is a 15-channel instrument; FOV = $\pm 49.5°$; IFOV = $3.3°$ (beamwidth); temperature resolution: 0.25 - 1.3 K; spatial resolution = 40 km horizontal at nadir; mass = 100 kg; data rate = 3.2 kbit/s; power = 125 W; duty cycle = 100%.
  MHS is a passive radiometer for humidity profiling, consisting of 5 channels: 1 at 89 GHz, 1 at 166 GHz, and 3 at 183.3 GHz. Coverage is about $50°$ on both sides of the suborbital track with an IFOV of $1.1°$. Data rate = 4.2 kbit/s; mass = 66kg; duty cycle = 100%; power = 85 W; swath width = 1650 km; spatial resolution = 13.5 km horizontal at nadir.

- **ASTER** = Advanced Spaceborne Thermal Emission and Reflection Radiometer (NASDA sensor). Previous name: ITIR = Intermediate Thermal Infrared Radiation. Objective: Surface (land and water) and cloud imaging with spatially high resolutions and with multi-spectral channels from the visible to the thermal infrared spectrum. Spectral bands:
  - Visible and Near Infrared (VNIR) 3 bands (0.5 - 0.9 μm, 15 m resolution. Stereoscopic viewing capability along track.
  - Short wavelength infrared (SWIR) 6 bands (1.6 - 2.5 μm), with 30 m resolution
  - 5 Thermal Infrared (TIR) channels from 8 - 12 μm, with 90 m resolution

  The swath width is 60 km. Pointing capability is provided for the cross-track direction ($\pm 106$ km for SWIR and TIR, and $\pm 314$ km for VNIR). Expected nominal life is 5 years. Data rate: 8.3 Mbit/s (89.2 Mbit/s peak); mass = 400 kg; instrument IFOV: =43 μrad (at nadir for SWIR), = 128 μrad (at nadir for TIR), = 21 μrad (at nadir for VNIR).
  ASTER will provide surface radiative (brightness) temperature and the multispectral TIR data can be used to derive surface kinetic temperature and spectral emissivity. ASTER will also provide cloud measurements such as cloud amount, type, spatial distribution, morphology, and radiative properties.

- **CERES** = Clouds and Earth's Radiant Energy System (PI: B. Barkstrom, NASA Langley). CERES is of ERBE heritage; instrument mass = 45 kg (1 instrument), power = 42 W, data rate = 10 kbit/s, duty cycle = 100%). Long-term measurement of the Earth's radiation budget. CERES will also provide global measurements of atmospheric radiation from the top of the atmosphere to the surface. The CERES instruments are a pair of broadband scanning radiometers (one cross-track mode, one rotating plane) based on ERBE. CERES measures longwave and shortwave infrared radiation using thermistor bolometers to determine the Earth's radiation budget. The first instrument (cross-track scanning) will essentially continue the ERBE mission. There are three channels in each radiometer. The second instrument (biaxially scanning) will provide angular flux information to improve the accuracy of current models.
  - Total radiance in the range of 0.3 - 50 μm.
  - Shortwave: 0.3 to 5 μm
  - Longwave: 8 to 12 μm

  Limb-to-limb scanning with nadir IFOV of 14 mrad, FOV = $\pm 78°$ cross-track, $360°$ azimuth. Spatial resolution = 21 km at nadir.

- **EOS-Color** = EOS Ocean Color Instrument (NASA/GSFC).
  EOS-Color is a second generation instrument, based on CZCS and SeaWiFS. EOS-Color has the same basic specifications as SeaWiFS (see chapter A.92).

- **EOS-SAR** (EOS Synthetic Aperture Radar). Objective: Monitoring of global deforestation and its impact on global warming; soil, snow, and canopy moisture and flood inundation, and their relationship to the global hydrologic cycle; sea ice properties and their impact on polar heat flux.

The EOS-SAR (JPL) is a three-frequency (L-, C-, and X-Band) multipolarization instrument providing HH and VV plus cross-polarization and phase measurements. The X-Band will be provided by Germany and Italy (heritage of SIR-C missions, see also A.96 on page 261). The instrument uses electronic beam steering in the cross-track direction to acquire images at selectable incidence angles from 15 to 50°. The EOS-SAR has a varying spatial resolution and swath width capability in three distinct modes:

- 20 to 30 m resolution with a swath width of 30 to 50 km (local high-resolution mode)
- 50 to 100 m resolution with a 100 to 200 km swath width (regional mapping mode)
- 250 to 500 m resolution with a swath width of up to 500 km (global mapping mode). The EOS-SAR is scheduled to fly on a dedicated platform. Data rates: 180 Mbit/s (max), 15 Mbit/s average

- **EOSP** = Earth Observing Scanning Polarimeter (PI: L. D. Travis, GSFC).
A photopolarimeter providing simultaneous measurements of radiance and linear polarization in 12 spectral bands from 0.41 to 2.25 µm. A cross track scanning mirror sweeps the 10 km FOV (at nadir) from limb to limb, generating global maps. EOSP will determine cloud properties (such as optical thickness and phase), global aerosol distribution, atmospheric corrections, and land and vegetation characteristics.
Spectral bidirectional reflectance distribution function accurate to 5%, polarization accurate to 0.2%; swath = ± 65° (limb-to-limb scan); spatial resolution = 10 km at nadir; mass = 19 kg; duty cycle = 100%; power = 14 W; data rate = 44 kbit/s (88 kBit/s peak); FOV = ± 65°; instrument IFOV = 14.2 mrad.

- **GGI** = GPS Geoscience Instrument (JPL). GGI is a high-performance GPS receiver-processor. It will include 18 dual-frequency satellite channels and three distributed GPS antennae. The antennae will be oriented to provide full-sky coverage for precise orbit determination and Earth limb coverage for radio occultation measurements. GGI will serve four principal science objective:

- cm-level global geodesy
- high-precision atmospheric temperature profiling
- ionospheric gravity wave detection and topographic mapping
- precise positioning in support of other science instruments

Occultation tracking will provide several hundred daily atmospheric temperature profiles to better than 1 K accuracy from 5 to 50 km altitude, with better than 1 km resolution. Occultation measurements will also contribute to the determination of global atmospheric energy balance and possible long-term trends. The instrument is derived from the GPS flight receiver developed for Topex/Poseidon.

Data rate = 50 kbit/s; power = 105 W; duty cycle = 100%, mass = 60 kg; Measurement approach: measures pseudo-range, carrier phase, and amplitude data from up to 18 GPS satellites using three antennae looking for ward, zenith, and aft; data from satellites are later combined with ground GPS receiver data to yield cm-level positioning.

- **GLAS** = Geoscience Laser Altimeter System (former name GLRS; PI: B. E. Schutz, NASA/GSFC).
GLAS measures ice sheet geodynamics, cloud properties, and geological processes and features. In addition, operation of GLAS over land and water will provide along-track topography.
The GLAS laser is a frequency-doubled, cavity-pumped, solid-state Nd:YAG laser with energy levels of 120 mJ (1.064 µm) and 60 mJ (0.532 µm). The pulse repetition rate is 40 pulses/s, and the beam divergence is approximately 0.2 mrad. The infrared pulse is used for surface altimetry, the green pulse is used for atmosphere measurements. The altimeter uses a 90 cm diameter telescope.

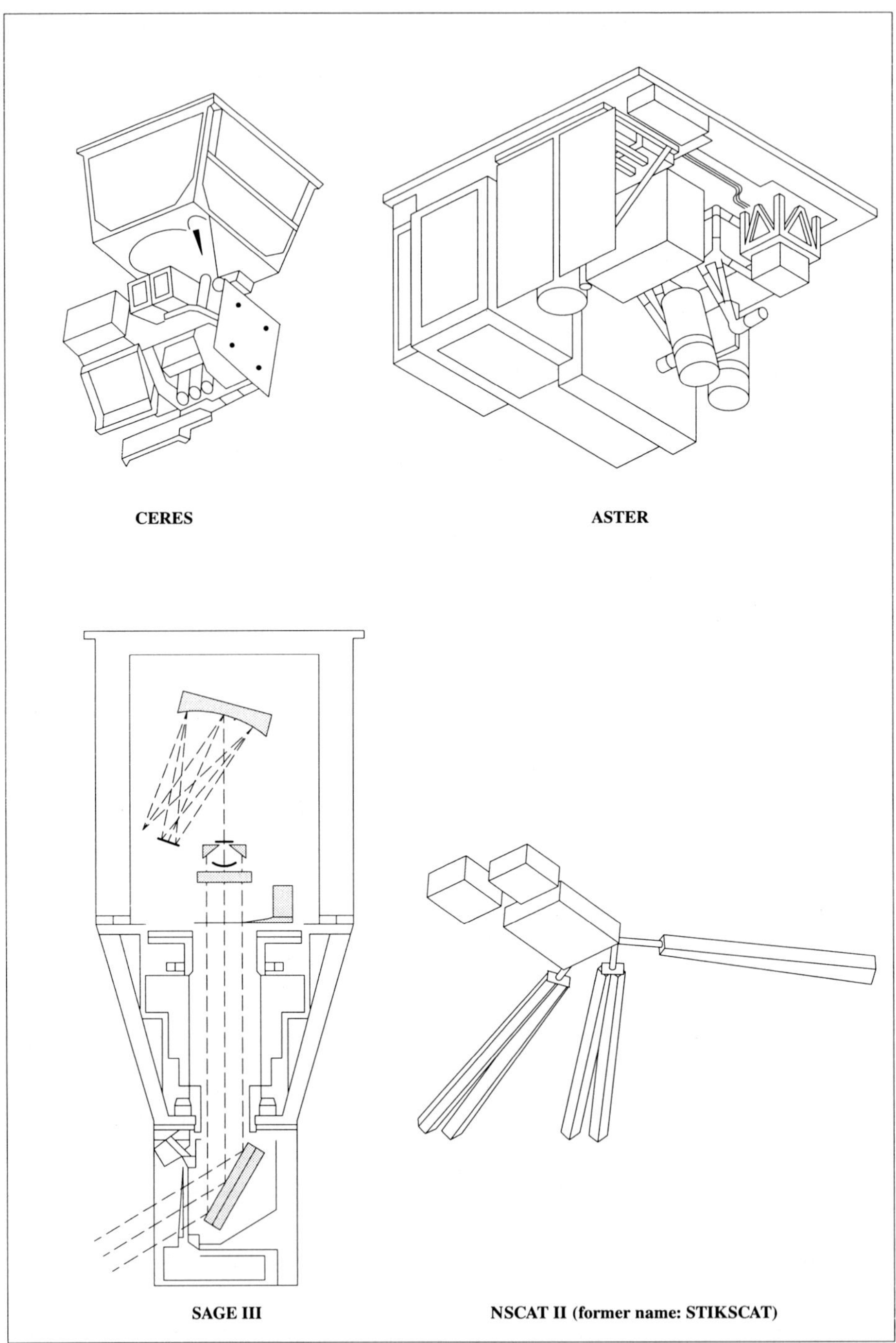


**Figure 23:** **Schematic Models of some EOS Sensor**

The GLAS altimeter is required to measure range (for ice-sheet applications) with an intrinsic precision of better than 10 cm with a 70 m surface spot size (footprint). Along-track cloud and aerosol height distributions will be determined with a vertical resolution of 75-200 m and a horizontal resolution from 150 m for dense clouds to 50 km for aerosol structure and planetary boundary layer height.

The data rate of the instrument is < 200 kbit/s; mass = 125 kg; duty cycle 50% (average); power = 175 W; FOV: nadir only; Instrument IFOV = 70 m laser footprint.

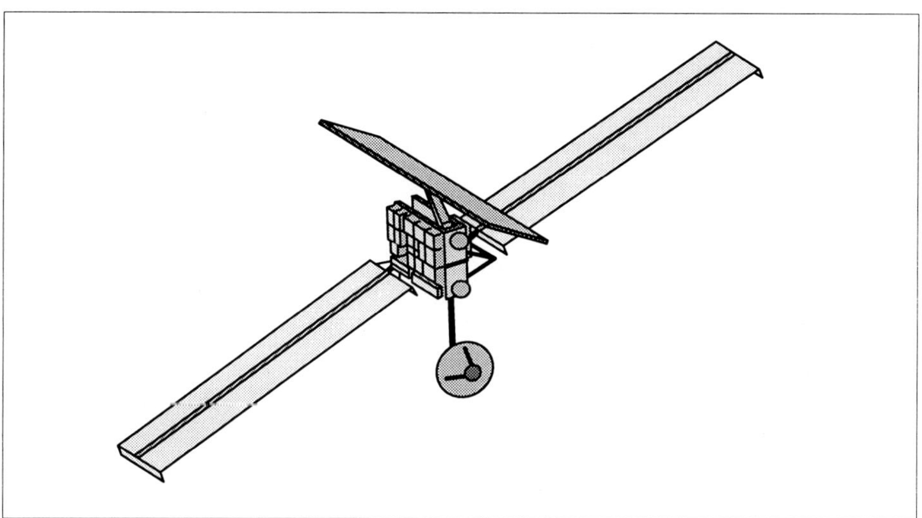

**Figure 24:**     **EOS-SAR S/C Model (dedicated Platform)**

- **HiRDLS**   = High-Resolution Dynamics Limb Sounder (NASA and UK sensor, PIs: J. Barnett, Oxford University; J. Gille, NCAR).
  Spectral range: 6 to 18 µm. HiRDLS observes global distributions of temperature and concentrations  of $O_3$, $H_2O$, $CH_4$, $N_2O$, $HNO_3$, $NO_2$, $N_2O_5$, $CFC_{11}$, $CFC_{12}$, $ClONO_2$ from the upper troposphere, stratosphere, and mesosphere plus water vapor, aerosol, and cloud tops; 2000 - 3000 km swath width (typically six profiles across swath). Spatial resolution: profile scanning 400 x 400 km horizontally (equivalent to 4° long x 4° lat) x 1 km vertically; averaging volume for each data sample 1 km vertical x 10 km across x 400 km along line-of-sight.
  HiRDLS is an infrared limb-scanning radiometer designed to sound the upper troposphere, stratosphere, and mesosphere. HiRDLS performs limb scans in the vertical at multiple azimuth angles, measuring infrared emissions in 21 channels (temperature distribution) ranging from 6.12 - 17.76 µm. Winds and potential vorticity are determined from spatial variations of the height of geopotential surfaces.
  Data rate = 40 kbit/s; mass = 150 kg; duty cycle = 100%; power = 180 W; FOV(scan range): elevation ±2.5° about -25.3° below horizontal, azimuth: -20° (sun side) to +50° (anti-sun side). Detector IFOV: 1 km vertical x 10 km (2.5°) horizontal.

- **HIRIS** = High-Resolution Imaging Spectrometer. HIRIS provides both high spectral and spatial resolution images of the Earth, and it can sample any point on the surface at a minimum of every 2 days. HIRIS measures in the 0.4 to 2.45 µm wavelength region in 192 spectral bands (first spaceborne hyperspectral imager!). The nadir pixel size is 30 m, the swath width = 24 km. The peak data rate of HIRIS is 100 Mbit/s, the average data rate is 3 Mbit/s. Objective: Improvement of models for the biosphere.

- **LAWS** = Laser Atmospheric Wind Sounder. LAWS is a 'Doppler lidar system' for direct tropospheric wind measurements (flux, distribution, transport phenomena). Instrument view angle = 45° off nadir. Data rate = 1.5 Mbit/s.

- **LIS** = Lightning Imaging Sensor
  8.5 km spatial resolution; 1 ms temporal resolution for retrieving lightning flashes; views at 777.4 nm on 128 x 128 ccd array; swath width: 1100km x 1100km.

- **MIMR** = Multi-frequency Imaging Microwave Radiometer (ESA sensor, see also MIMR description in chapter A.78.2, METOP-1). Applications: atmosphere, ocean, land and cryosphere parameters.
  Measures precipitation rate, cloud water, water vapor, temperature profiles, sea surface roughness, sea surface temperature, ice, snow, and soil moisture.
  Measurement approach: passive microwave radiometer; 6 frequencies (6.8 to 90 GHz), each band with H and V polarization; swath of 1400 km at incidence angle of 50°; FOV varies with frequency: 60 x 40 km at 6.8 GHz to 4.8 x 3.1 km at 90 GHz. Spatial resolution:
  - 4.86 km (90 GHz)
  - 11.62 km (36.5 GHz)
  - 22.3 km (23.8 GHz)
  - 38.6 km (10.65 GHz)
  - 60.3 km (6.8 GHz)

  Data rate = 62 kbit/s; mass = 273 kg; duty cycle = 100%; power = 270 W;

- **MISR** = Multi-Angle Imaging Spectro-Radiometer (PI: D. J. Diner, NASA/JPL)
  Nine CCD cameras fixed fore-, aft-, and nadir-looking, view angles ± 70.5°; 4 spectral bands centered at 0.443, 0.555, 0.67, and 0.865 μm; each of the 36 instrument data channels (i.e. four spectral bands for each of the 9 cameras) is individually commandable to provide ground (nadir) sampling of 240 m, 480 m, 960 m, or 1920 m; swath = 356 km; multi-angle coverage of the entire Earth in 9 days at the equator, and two days at the poles.
  Application: MISR provides global maps of planetary and surface albedo, and aerosols and vegetation properties. Monitoring of global and regional trends in radiatively important optical properties (i.e. opacity, single scattering albedo, and scattering phase function) of natural and anthropogenic aerosols.
  Instrument duty cycle = 50%; mass = 106 kg; power = 67 W; data rate = 3.8 Mbit/s (average), 6.5 Mbit/s (peak); FOV = ± 59° down-track by ± 15° cross-track.

- **MLS** = Microwave Limb Sounder (UARS heritage, PI: J. W. Waters, NASA/JPL).
  Measures thermal emission from the atmospheric limb in submillimeter and millimeter wavelength spectral bands. Spectral bands centered at: 215 GHz, 440 GHz, 640 GHz, and 2.5 THz. Spectral resolution = 1MHz.
  The MLS investigation will study and monitor atmospheric processes that govern stratospheric and mesospheric ozone. Emphasis is on chlorine and nitrogen destruction of ozone.
  Measurements are performed continuously, at all times of day and night, altitude range from the upper troposphere to the lower thermosphere. The vertical scan is chosen to emphasize the lower stratosphere and upper troposphere. Complete latitude coverage is obtained each orbit. Pressure (from $O_2$ lines) and height (from a gyroscope measuring small changes in the FOV direction) are measured to provide accurate vertical information for the composition measurements.
  Measurement approach: passive limb sounder; thermal emission spectra collected by offset Cassegrain scanning antenna system; Limb scan = 0 - 120 km; spatial resolution = 3 x 300 km horizontal x 1.2 km vertical
  Data rate: 5 kbit/s; FOV: Borsight 62-74° relative to nadir; IFOV = ± 2.5° (half-cone, along-track); mass = 500 kg, power = 540 W; duty cycle = 100%.

- **MODIS** = Moderate-Resolution Imaging Spectrometer (PI: V. Salomonson, GSFC). MODIS is an imaging spectrometer, consisting of a cross-track scan mirror and collecting optics, and a set of linear detector arrays with spectral interference filters located in four focal planes. The optical arrangement provides imagery in 36 discrete bands between 0.4 and 15 µm. The spectral bands provide a spatial resolution of 250 m, 500 m, or 1 km at nadir. (MODIS heritage: AVHRR, HIRS, Landsat TM, Nimbus-7 CZCS). Objective: Measurement of biological and physical processes.

  MODIS polarization sensitivity is less than 2% for the visible out to 2.2 µm; SNR = 830:1 (443 nm), = 745:1 (520 nm), = 503:1 (865 nm); absolute irradiance accuracy of 5% for < 3 µm and 1% for > 3 µm; absolute temperature accuracy of 0.2 K for oceans and 1 K for land; daylight reflection and day/night emission spectral imaging;  swath width of 2300 km at 110° FOV; mass = 250 kg; duty cycle = 100%; power = 225 W; data rate = 8.3 Mbit/s (average), 11 Mbit/s (day), 2.5 Mbit/s (night). Instrument IFOV = 250 m (2 bands), =500 m (5 bands), = 1000 m (29 bands).

  MODIS provides global coverage every 1 to 2 days. It will provide specific global survey data, which include the following:
  - Surface temperature with 1 km resolution, day and night, with absolute accuracy of 0.2 K for oceans and 1 K for land surfaces.
  - Ocean color, defined as ocean-leaving spectral radiance within 5% from 415-653 nm, based on adequate atmospheric correction from NIR sensor channels
  - Chlorophyll fluorescence within 50% at surface water concentrations of 0.5 mg/m$^{-3}$.
  - Vegetation/land surface cover, conditions, and productivity
  - Cloud cover with 250 m resolution by day and 1 km resolution at night
  - Cloud properties, characterized by cloud droplet phase, optical thickness, droplet size, cloud-top pressure, and emissivity
  - Aerosol properties defined as optical thickness, particle size, and mass transport
  - Fire occurrence, size, and temperature
  - Global distribution of total precipitable water

- **MOPITT** = Measurement of Pollution in the Troposphere (PI: J. Durmond, Canadian sensor, CSA). Mopitt measures emitted and reflected infrared radiance in the atmospheric column. Analysis of these data permit retrieval of CO profiles and total column $CH_4$. Mopitt will measure tropospheric CO and $CH_4$ concentrations to study how these gases interact with the surface, ocean, and biomass systems.

  Measurements: Correlation spectroscopy utilizing both pressure-modulated and length-modulated gas cells, with detectors at 2.3, 2.4, and 4.7 µm. Vertical profile of CO and total column of $CH_4$ to be measured; CO concentration accuracy is 10%; $CH_4$ column abundance accuracy is 1%. Swath = 616 km, spatial resolution = 22 x 22 km; Data rate = 6 kbit/s; duty cycle = 100%

  MOPITT is designed as a scanning instrument. FOV = 1.8° (22 km at nadir). The instrument scan line consists of 29 pixels, each at 1.8° increments. The maximum scan angle is 26.1° off-axis (swath width of 640 km).

- **NSCAT II** = NASA Scatterometer II (PI: M. Freilich, NASA/JPL). Scatterometer with six slotted waveguide Ku-Band "Stick" fan-beam antennas, previous designation was Stikscat. Objective: acquire accurate, high-resolution, continuous, all-weather measurements of near-surface vector winds over the ice-free global oceans. Application: studies of tropospheric dynamics and air-sea momentum fluxes. NSCAT II is selected for flight on ADEOS II. As of Fall 1993 NSCAT II has been renamed to **SeaWinds**.[84] NASCAT II is an active microwave radar scatterometer using a classical fan-beam design (pulse transmissions at 14 GHz and measurement of backscattered signals). The sensor acquires data from two 600 km wide swaths separated by a 325 km wide gap at

---

84)    M. King, Editor's Corner of The Earth Observer, Vol. 5, Nr. 4, July/August 1993, p. 2

nadir. All six antennae transmit and receive vertically polarized radiation, one antenna on each side also transmits/receives horizontal polarization. Instrument mass = 270 kg, power = 290 W, data rate = 5.1 kbit/s.

NASCAT II data products consist of global multi-azimuth normalized radar cross section measurements; 25 km$^2$ resolution ocean vector winds ($\sim 12\%$ speed and 20° direction accuracies for wind speeds of 3-50 m/s) in each of the swaths; and spatially/temporarily averaged wind field maps with 1° spatial resolution and 2-day temporal resolution.

- **SAGE III** = Stratospheric Aerosol and Gas Experiment III (PI: M. P. McCormick, NASA/LaRC).
  SAGE III is an Earth limb-scanning grating spectrometer (heritage: SAM II, SAGE I, SAGE II). SAGE III will measure global profiles of aerosols, $O_3$, $NO_2$, $NO_3$, OClO, $H_2O$, temperature, and pressure between cloud tops and the upper mesosphere with 1- to 2-km vertical resolutions.

  Measurement approach: Self-calibrating solar and lunar occultation, with nine spectral channels, from 290 to 1550 nm, to study aerosols, ozone, OClO, $NO_2$, $NO_3$, water vapor, temperature, and pressure. Swath: n/a (looks at sun through the Earth's limb); spatial resolution = 1 - 2 km vertical; duty cycle: during solar and lunar occultation; data rate = 100 kbit/s for 8 minutes (three times per orbit); power = 15 W (60 W peak); FOV = ± 180° azimuth, 19 to 29° elevation; mass = 40 kg.

- **SOLSTICE II** = Solar Stellar Irradiance Comparison Experiment (PI: G. Rottman, NCAR). SOLSTICE II is composed of an ultrahigh-resolution spectrometer, low resolution spectrometers, and an extreme UV photometer.
  Objective: daily measurements of the full-disk solar UV irradiance between 5 and 440 nm. Application: studies in the changes of photochemistry, dynamics and energy balance in the middle atmosphere, model the penetration of solar radiation down into the Earth's atmosphere.

  The SOLSTICE II instrument consists of a five-channel spectrometer together with the required gimbal system to point the instrument at the sun and selected stars. Photometric accuracy is better than 5% absolute (1% relative). The spectral resolution is 0.2 nm and 0.0015 nm. Instrument mass = 99.5 kg; power = 34 W; data rate = 5 kbit/s; FOV = 1.5°. The SOLSTICE II data product will be a daily average of the solar UV irradiance from 5 to 440 nm.

- **SAFIRE** = Spectroscopy of the Atmosphere Far Infrared Emission.
  The objective of the SAFIRE experiment is to improve the understanding of the middle atmosphere ozone distribution by conducting global-scale measurements of the chemical, radiative, and dynamical processes that influence ozone changes.
  SAFIRE is a passive limb emission instrument that combines the advantages of far-infrared Fourier transform spectroscopy and space-proven mid-infrared broadband radiometry. The sensor provides simultaneous observations of key Oy, HOy, NOy, ClOy, and BrOy gases, coupled with dynamical tracer measurements, including vertical profiles of temperature, $O_3$, OH, $HO_2$, $H_2O_2$, $H_2O$, HDO, $CH_4$, $NO_2$, $HNO_3$, $N_2O$, $N_2O_5$, HCl, HOCl, HBr, and HF.
  Measurement approach: swath: limb viewing from 0 - 106 km to within 4° of the poles; spatial resolution = 3 km vertical (infrared), 1.5 km vertical (mid-infrared)
  Data rate = 8.7 Mbit/s; mass = 407 kg; power = 465 W; FOV = 1 x 1° square swept over a depression angle of 17 to 29° (10° from the orbital plane)

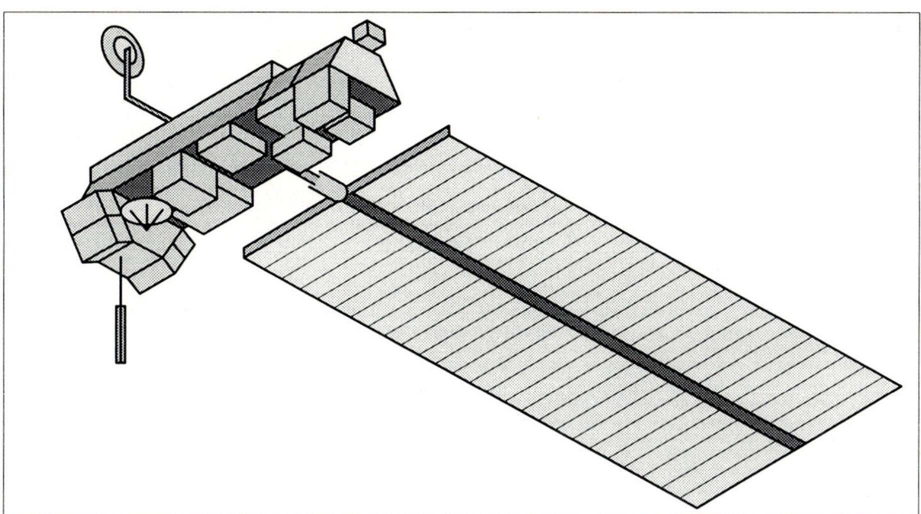

**Figure 25:    Model of the EOS-A Platform (old concept)**

- **TES** = Tropospheric Emission Spectrometer (PI: R. Beer, NASA/JPL).
  TES is a high spectral resolution infrared imaging Fourier transform spectrometer with spectral coverage from 2.3 - 15.4 µm and spectral resolution of 0.025 cm$^{-1}$. Maximum sampling time of 8 seconds with a SNR of up to 600:1. TES has the capability to make both limb and nadir observations. Limb mode: height resolution = 2.3 km, height coverage = 0 - 30 km. In the nadir modes, TES has a spatial resolution of 50 x 5 km (global) or 5 x 18 km (local). TES is a pointable instrument, it can access any target within 45° of the local vertical, or produce regional transects up to 1700 km without any gaps in coverage. TES addresses four of the most pressing issues in global change:

Observations from TES will further the understanding of long-term variations in the quantity, distribution, and mixing of minor gases in the troposphere, including sources, sinks, troposphere-stratosphere exchange, and the resulting efforts on climate and the biosphere. TES will provide global maps of tropospheric ozone and its photochemical precursors. Other objectives:

- Simultaneous measurements of $NO_y$, CO, $O_3$, and $H_2O$, determination of the global distribution of OH.
- Measurements of $SO_2$ and $NO_y$ as precursors to the strong acids $H_2SO_4$ and $HNO_3$
- Measurements of gradients of many tropospheric species
- Determination of long-term trends in radiatively active minor constituents in the lower atmosphere.

TES data generate 3-D profiles on a global scale of virtually all infrared-active species from the Earth's surface to the lower stratosphere.

Data rate = 3.24 Mbit/s (average operation), 19.6 Mbit/s (peak operation); FOV = + 45° to - 71° along-track, ±71° cross-track; IFOV = 24 x 7.5 mrad (narrow angle) 240 x 75 mrad (wide angle); swath = n/a; mass = 340 kg; power = 430 W

## A.22    EOS - Restructured Program Version (1992)

The restructured EOS program represents an integrated monitoring program of measurements of key climate change variables, coupled with a comprehensive data and information system (EOSDIS). 'EOS restructured' permits, however, only limited observations in the field of atmospheric chemistry and solid Earth geophysics.[85] The continuity of global measurements remains a high priority. In addition, 'EOS restructured' depends more strongly on collaboration with foreign partners. In a sense, tight money everywhere may be a vigorous catalyst and a harbinger for more intensified cooperation and data sharing on a truly global scale to make more use of all the resources available to everyone!

The restructured EOS program is based on the following assumptions:[86),87),88),89)]

- Three classes of small and intermediate spacecraft are defined for EOS payloads (a result of the restructuring is the shift from "large observatories" to intermediate and smaller spacecraft).

- Six launch vehicles are needed through FY 2002 in three classes:
  a.   Pegasus (max. S/C mass of 600 lbs = 272 kg)
  b.   Delta II (max. S/C mass of 7040 lbs = 3200 kg)
  c.   Atlas II AS (max. S/C mass of 13800 lbs = 6270 kg)

- The first launch in the EOS program is scheduled for June 1998

- The restructured EOS program considers only a total of about 20 instruments (considerably smaller than before)

- The program is based on a 5-year life per mission. A total of 3 mission cycles are considered over a 15-year period.

- The ground segment with the EOSDIS core system remains conceptually unaffected by the restructuring. The overall data input will be smaller, however, due to expected smaller source data amounts (a matter of scale).

### A.22.1   Mission to Planet Earth

NASA's Mission to Planet Earth is a Presidential initiative and represents the agencies contribution to the U.S. Global Change Research Program and to related international efforts to better understand the planet Earth and how humans may be affecting it.

Mission to Planet Earth builds on earlier missions to study the Earth's global ozone changes, atmospheric dynamics and ocean circulation. The strategy for Mission to Planet Earth consists of several interrelated elements, each of which builds on or complements the others:

1.   Near-term missions. These include:
     - UARS, Topex/Poseidon
     - Shuttle Spacelabs (such as the ATLAS series)
     - Landsat series
     - Explorer-class Earth Probes series (ozone, ocean winds, tropical rainfall, gravity, topography, etc. as well as support for aircraft and in situ research
     - Cooperative missions with international partners and/or other agencies

2.   Research Base for Scientific Analysis. This includes:
     - EOSDIS (Earth Observing System Data and Information System). The prototype EOSDIS will be on-line in 1994 for near-term science operations.


85)) "Our Changing Planet: The FY 1993 U.S. Global Change Research Program", A Report by the Committee on Earth and Environmental Sciences, p. 66
86)) Papers were provided by ESAD of NASA HQ's, Washington D.C.; Title: "Restructured Earth Observing System (EOS) Program Review", (with updates from the Review), Lennard Fisk, Dec. 11, 1991
87)) Vision 21, The NASA Strategic Plan, January 1992
88)) 'The Restructuring of the Earth Observing System', NASA paper
89)) 'Report of the Earth Observing System (EOS) Engineering Review Committee', Sept. 1991

3. A core Earth Observing System (EOS). EOS is designed to gather a 15-year data set on the Earth's coupled systems. See Fig. 26 and A.22.2 to A.22.7 for definition. Note, the sensors are defined under the 'EOS Original Version'.

4. Follow-on EOS Missions.
   - Follow-on Earth Probes Series
   - EOS-SAR
   - Geostationary platforms

As in the original EOS program, most spacecraft will fly in sun-synchronous polar orbits - but with different equatorial crossing times - the AM and PM series. The AM- and PM S/C (the principal EOS S/C) will be repeated twice on five-year centers for at least fifteen years' data coverage. In addition, the program includes two series of smaller spacecraft, one dedicated to precision altimetry, and to other providing for occultation measurements from non-synchronous orbits. The restructured EOS also includes an extension of ocean color data purchase as well as contributions to the Earth radiation budget and lightning instruments for flight on TRMM and its follow-ons.

## A.22.2  EOS AM-1 Mission

The AM [Ante Meridian (10:30 AM equator crossing time)] platform is scheduled as the first EOS core program payload because:

- the proposed AM instrument payload fits the early funding profile and is less costly overall, thereby meeting the required development schedule for launch in June 1998

- the proposed PM instrument payload has increased schedule risk caused by addition of AIRS and MIMR along with MODIS.

EOS AM-1 objectives: Observation of clouds, aerosols, and radiative balance. Characterization of the terrestrial surface.

Orbit: Sun-synchronous polar orbit, altitude 705 km, 10:30 AM equatorial crossing (descending node); launch on Atlas II AS vehicle

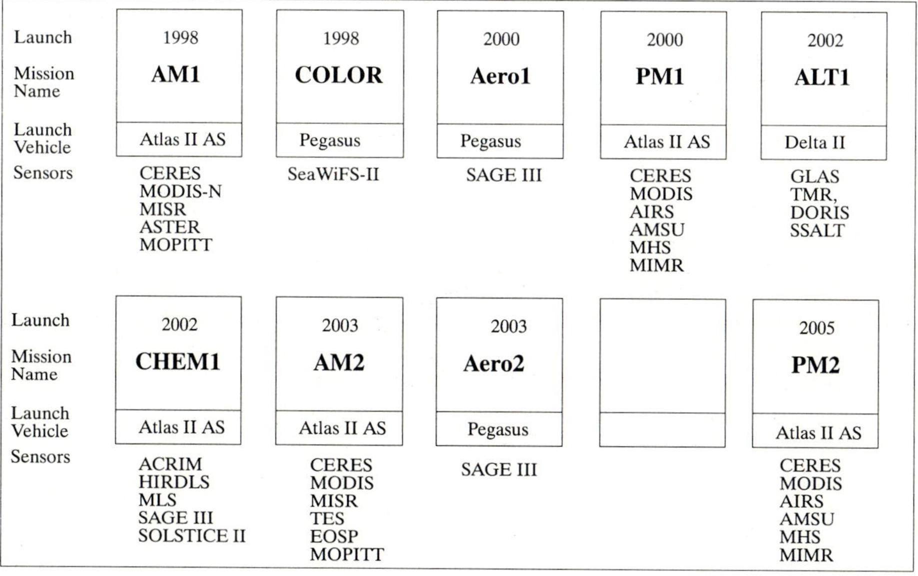

| | | | | | |
|---|---|---|---|---|---|
| Launch | 1998 | 1998 | 2000 | 2000 | 2002 |
| Mission Name | **AM1** | **COLOR** | **Aero1** | **PM1** | **ALT1** |
| Launch Vehicle | Atlas II AS | Pegasus | Pegasus | Atlas II AS | Delta II |
| Sensors | CERES MODIS-N MISR ASTER MOPITT | SeaWiFS-II | SAGE III | CERES MODIS AIRS AMSU MHS MIMR | GLAS TMR, DORIS SSALT |
| Launch | 2002 | 2003 | 2003 | | 2005 |
| Mission Name | **CHEM1** | **AM2** | **Aero2** | | **PM2** |
| Launch Vehicle | Atlas II AS | Atlas II AS | Pegasus | | Atlas II AS |
| Sensors | ACRIM HIRDLS MLS SAGE III SOLSTICE II | CERES MODIS MISR TES EOSP MOPITT | SAGE III | | CERES MODIS AIRS AMSU MHS MIMR |

**Figure 26:**   **Definition of Rescoped EOS Program Launch Profile**

Sensors: **CERES, MODIS (-N), MISR, ASTER,   MOPITT**

Radiative Fluxes: CERES, MODIS, and MISR will measure radiative and physical properties of clouds and aerosols.

Land-Atmosphere Interaction:
- MODIS and MISR will observe terrestrial surface, land cover, and productivity
- ASTER will provide fine spatial/spectral observations to link global/regional/local scales
- ASTER and MISR will observe volcanic effects

Tropospheric Chemistry: MOPITT will initiate global observations of CO and $CH_4$

*The EOS-AM1 spacecraft will also include the following communication features:*
- *Direct Access System (DAS), which is composed of:*
    - *the Direct Playback (DP) subsystem*
    - *the Direct Broadcast (DB) subsystem, and*
    - *possibly the Direct Downlink (DDL) subsystem.*

*While it is planned that all EOS data will be recorded and played back via TDRSS, DAS will provide a backup option for direct transmittal on onboard data to ground receiving stations via an X-Band transmitter subsystem (DP subsystem). DAS will also support transmission to ground stations of qualified EOS users around the world who require direct data reception.*


## A.22.3   EOS Ocean Color Mission

The EOS ocean color mission is simply referred to as 'EOS Color' and is the continuation of the SeaStar program with the sensor SeaWiFS II. The policy of 'Data Purchase', as introduced by the SeaStar-1 mission for commercial customers, will be continued for 'EOS Color' until the second MODIS-N is launched on EOS PM-1 spacecraft. An overlap of the end of SeaWiFS I is planned for cross calibration and reliable continuity of observation.

Orbit: Sun-synchronous polar orbit, altitude about 700 km; launch on Pegasus vehicle in second half of 1998.

Sensor: **SeaWiFS II**
SeaWiFS II observations of phytoplankton and chlorophyll will be used to estimate the oceanic biomass and primary production.


## A.22.4   EOS-Aerosol Mission

The objective of the 'EOS-Aero1' mission is the observation of atmospheric aerosols. The aerosol series involves 5 S/C launched on 3-year centers to provide a 15 year coverage.

Orbit: altitude = 705 km, 57° inclination, launch on a Pegasus vehicle in early 2000.

Sensor: **SAGE III**
- SAGE III primarily will observe aerosols in the troposphere and stratosphere, respectively, for radiative effects and studies of cloud-aerosol interactions. SAGE III also observes ozone and water vapor in the upper troposphere

Comments:
- The Aero1 observations are to complement SAGE III data on the CHEM1 spacecraft in 2002
- The Aero-1 observations are to complement MISR aerosol data on the AM1 spacecraft


## A.22.5   EOS-PM1 Mission

Plans call for EOS PM-1 to be launched in 2000. The overall objectives of the PM-1 mission are as follows:

- Clouds, precipitation, and radiative balance
- Terrestrial snow and sea ice
- Sea surface temperature and ocean productivity

Orbit: Sun-synchronous polar orbit, altitude = 705 km, 13:30 (PM) equatorial crossing (ascending node); launch on Atlas II AS class vehicle in Dec. 2000

Sensors: **CERES, MODIS, AIRS, AMSU,  MHS,  MIMR**

- Observation of radiative fluxes
  CERES (also on AM-1 and TRMM-1) and MODIS (also on AM-1) should reduce errors in radiative fluxes due to inadequate diurnal sampling
- Observation of the physical climate system
  AIRS, AMSU, and MHS will provide improved atmospheric temperature and humidity profiles
  MIMR will estimate sea surface temperature, atmospheric water, precipitation, and snow and ice content
  These data and scatterometer winds from ADEOS-2 or Chem-1 will address air-sea fluxes

The plan calls for all EOS PM satellites to be identical, except that CERES will consist of two scanners on EOS PM-1 and a single scanner on the follow-on flights. The PM series will include the DB and DP communication capabilities of DAS, with the DB system transmitting all instrument data, except CERES.

## A.22.6   EOS-Altimeter Mission

The objectives of the altimeter mission are the study of ocean circulation, and the ice sheet mass balance. A launch is planned on a Delta II vehicle in early 2002. The EOS-ALT payload was dramatically reconfigured during the rescoping exercise, with the ALT and GGI instruments eliminated to reduce total program cost. The instrument complement that resulted consists of the following:

Sensors: **GLAS, TMR, DORIS, and SSALT**

In addition to precise orbit tracking (DORIS and TMR, see A.110) and altimeter calibration and orbit determination (SSALT), the EOS ALT series will provide measurements of sea ice and glacier surface topography, cloud heights, and aerosol vertical structure (GLAS).

## A.22.7   EOS-Chemistry Mission

The rescoping of the EOS program removed STIKSCAT and TES from this platform. A polar orbit is considered. A launch is planned on an Atlas II AS vehicle in 2002.

Sensors: **ACRIM, HiRDLS, MLS, SAGE III, SOLSTICE II**

The EOS CHEM instruments will provide measurements of solar energy flux (ACRIM); solar UV radiation (SOLSTICE II); atmospheric aerosols, ozone, and water vapor (SAGE III); atmospheric trace gases (HiRDLS); and ozone based on chlorine monoxide, bromine oxide, and water vapor (MLS).

## A.22.8   Deferred or Deselected Sensors as a result of Restructuring

- HiRDLS (= High Resolution Dynamics Limb Sounder) and STIKSCAT (Stick Scatterometer) have been delayed until 2002.

Affected science field: ability to initiate timely observations of stratospheric/tropospheric chemical constituents and ocean surface stress.

- HIRIS (High-Resolution Imaging Spectrometer) was deselected.

- MODIS-T (Moderate -Resolution Imaging Spectrometer - Tilt) was deselected. Affected science field: ability to spectrally characterize ocean color.

- GLRS-A (Geoscience Laser Ranging System - Altimeter) is deferred to 2002. Affected science field: ability to initiate observations of ice-sheet mass balance. Note: GLRS-A has been renamed into GLAS = Geoscience Laser Altimeter System, to reflect the retention of the GLRS altimeter component and the deselection of the GLRS ranging component.

- LAWS (Laser Atmospheric Wind Sounder). There will be no funding for the development and flight of LAWS. Affected science field: ability to make direct observations of atmospheric winds.

- SWIRLS (Stratospheric Wind Infrared Limb Sounder) and either MLS (Microwave Limb Sounder) or SAFIRE (Spectroscopy of the Atmosphere Far Infrared Emission) were deselected.
  Affected science field: ability to fully characterize the stratosphere during a period of rapid anthropogenic change.

- IPEI (Ionospheric Plasma and Electrodynamics Instrument), XIE (X-Ray Imaging Radar), and GOS (Geomagnetic Observing System) were all deselected.
  Affected science field: ability to monitor and characterize the near-space environment, including particle fluxes and magnetic field measurements.

- GLRS (Geoscience Laser Ranging System) is to be descoped, eliminating laser ranging. Affected science field: ability to observe solid Earth deformation.

| Launch | Space-craft | Lifetime | Instrument Complement |
|---|---|---|---|
| 1998 | AM1 | 5 | MODIS, MISR, CERES (2), MOPITT, ASTER |
| 2003 | AM2 | 5 | MODIS, MISR, CERES, EOSP, TES, (MOPITT) |
| 2008 | AM3 | 5 | MODIS, MISR, CERES, EOSP, TES |
| 1998 | COLOR | 3 | SeaWiFS-Type |
| 2000 | AERO1 | 3 | SAGE III |
| 2003 | AERO2 | 3 | SAGE III |
| 2006 | AERO3 | 3 | SAGE III |
| 2009 | AERO4 | 3 | SAGE III |
| 2012 | AERO5 | 3 | SAGE III |
| 2000 | PM1 | 5 | MODIS, AMSU, MIMR, AIRS, MHS, CERES(2) |
| 2005 | PM2 | 5 | MODIS, AMSU, MIMR, AIRS, MHS, CERES |
| 2010 | PM3 | 5 | MODIS, AMSU, MIMR, AIRS, MHS, CERES |
| 2002 | ALT1 | 5 | GLAS, TMR, SSALT, DORIS |
| 2007 | ALT2 | 5 | GLAS, TMR, SSALT, DORIS |
| 2012 | ALT3 | 5 | GLAS, TMR, SSALT, DORIS |
| 2002 | CHEM1 | 5 | HiRDLS, SOLSTICE II, ACRIM, MLS, SAGE, TBD(Jap. sensor) |
| 2007 | CHEM2 | 5 | HiRDLS, SOLSTICE II, ACRIM, MLS, SAGE, TBD |
| 2012 | CHEM3 | 5 | HiRDLS, SOLSTICE II, ACRIM, MLS, SAGE, TBD |

**Table 17:**     **Overview of Rescoped EOS Program**

## A.23    Equator - S

Equator-S ('small') is a Solar-Terrestrial energy transport mission of MPE-Garching for plasma and field measurements in the magnetospheric boundary regions, the auroral source region, and the ring-current at near-equatorial latitudes. Equator - S is considered a complement to the 'Global Geospace Science (GGS) Program'. A launch is planned in April 1996 with (Ariane V launch from Kourou). The S/C and payload is provided by Germany (PI for mission: G. Haerendel, MPE).[90]

Within the GGS/ISTP context simultaneous measurements in the solar wind (WIND) and in the magnetospheric boundary regions (EQUATOR-S) will allow a more quantitative understanding between input parameters and transfer efficiencies. The scenario of simultaneous measurements considers the following S/C: EQUATOR-S at equatorial latitudes, POLAR at polar latitudes, and GEOTAIL in the deep tail of the magnetosphere.

The Equator-S spacecraft is spin-stabilized, it has a spin rate of 60 rpm to allow for high temporal resolutions for the measurement of the particle distributions functions. S/C total payload mass $\approx$ 170 kg.

Orbit: A near-equatorial orbit with a 10-12 $R_E$ geocentric distance apogee and low perigee ($\approx$ 500 km).

Sensors:

- **Magnetometer** (Lead Investigators: H. Lühr, TU Braunschweig, J. Rustenbach, DLR Berlin, F. Primdahl, Danish Space Research Institute, Lynby). A ring core fluxgate instrument which has been flown successfully on sounding rocket payloads. Instrument characteristics:
  - Ranges: $\pm$ 2000 nT / $\pm$ 30 000 nT
  - Digitization: 16 bit
  - Sample rate: 50 vectors/s
  - Mass of sensor: 500 g
  - Mass of electronics: 2500 g
  - Power consumption: 3 W

- **EDI** = Electron Drift Instrument (Lead Investigators: G. Paschmann, MPE, R. Tobert, Univ. of New Hampshire, C. E. McIlwein and R. W. Fillius, Univ. of Ca. at San Diego). This instrument is identical in concept and design to the EDI sensor on CLUSTER. EDI measures the drift of a weak beam of test electrons which is related to the ambient electric field and the gradient in the magnetic field. EDI is capable of measuring the electric field, local gradients in the magnetic field, and the magnetic field itself. EDI consists of two emitter/detector assemblies, each with a $2\pi$ FOV. E = (0.1 - 10 mV/ m, <100 Hz), $\nabla \underline{B}$, $|\underline{B}|$ = 5-1000 nT, emission and tracking of 2 electron beams.
  - Mass: 9.5 kg
  - Power consumption: 9 W
  - Data rate: 3.5 kbit/s

- **CODIF** = Composition and Distribution Functions analyzer, consisting of the following instruments (to save weight):
  - **3D-Electron Analyzer** (of WIND 3D Plasma instrument heritage; Lead Investigators: G. K. Parks, M McCarthy, Univ. of Washington, Seattle) Measurement of electrons from 10 eV to 30 keV; $\Delta E/E$ = 0.2; geometric factor = $10^{-3}$ cm$^2$ sr; FOV = 180° x 12°; angular resolution = 5.6° x 5.6°. Instrument mass = 9.3 kg, power = 6.7 w, data rate = 5 kbit/s.

---

90)    "A Small Equatorial Satellite to Complement the Global Geospace Science Program - Equator - S", MPE Equator-S proposal, 30 Sept. 1991

- **Ion Composition Instrument** (Lead Investigator: E. Möbius, Univ. of New Hampshire). This instrument is essentially a copy of CODIF on the CLUSTER mission (heritage of SULEICA and the 3-D Plasma Instrument flown on AMPTE). CODIF is a high sensitivity mass resolving spectrometer with an instantaneous $360° \times 8°$ field of view to measure the 3-D distribution function of the major ion species ($H^+$, $He^{2+}$, $He^+$, $O^+$, and a combination of $O_2^+$ and $NO^+$). CODIF covers the energy range from about 0 to 40 keV/q. The instrument is a symmetrical hemispherical analyzer with RPA (Retarding Potential Analyzer) and TOF (Time-of-Flight) electronics.
  Instrument mass = 7.5 kg; power = 6 W, data rate = 5 kbit/s.

- **Energetic Particle Detectors** (Lead Investigators: P. Wenzel and T. Sanderson, ESTEC; R. Lin, UC Berkeley; H. Rème, CESR; G. K. Parks, U. of Washington, Seattle). The instrument consists of a pair of double-ended semiconductor detector telescopes provising 3-D measurements of electrons and ions with high sensitivity and high energy resolution. Measurement ranges: electrons from 400 keV to > 1 MeV; instrument mass = 3.6 kg, power = 2.7 W, data rate = 1 kbit/s.

- **Potential Control Device** (Lead Investigators: W. Riedler and K. Torkar, Space Research Institute, Graz; A Perdersen, R. Grard and R. Schmidt, ESTEC). The instrument reduces the S/C potential (caused by the ambient plasma charging of the outer surfaces) to acceptable levels. The method employs beams of 6 keV indium ions to neutralize the S/C charging (see also 'ASPOC' of the CLUSTER mission). Instrument mass = 1.9 kg, power = 3 W, data rate = 0.1 kbit/s.

Data:
An onboard solid state mass memory of 64 MByte is provided allowing data storage for about 7 hours. Nominal data rate = 20 kbit/s.


## A.24    ERBS  (Earth Radiation Budget Satellite)

ERBS = Earth Radiation Budget Satellite. NASA Earth Radiation Budget Experiment (ERBE) Research Program (at GSFC). Launch of the free-flyer ERBS satellite from Space Shuttle Challenger: October 5 1984 (STS-13). The ERBS is a three-axis momentum-biased S/C. Communication via TDRSS. See also Table 50.

Objective: Measurement of reflected and emitted energy at various spatial levels. Observations provide useful data for studies of geographical-seasonal variations of the Earth's radiation budget. Data will be correlated with NOAA polar-orbiting TIROS S/C.[91]

Orbit: Non-sun-synchronous circular orbit. Altitude = 610 km, Inclination = 57°, Period = 96.8 min. As of 1/1994 the ERBS mission is still operational. However, S/C is operational on only 1 battery. SAGE-II is operational, ERBE is only operational in periods of enough sunlight.

Sensors:

- **ERBE** (Earth Radiation Budget Experiment). The ERBE dual instrument provides measurements on several spatial and temporal scales. Spectral range: 0.2 - 50 μm.
  ERBE Nonscanner (NS) sensor: 5 channels with cavity radiometer detectors.. Four of the channels are primarily nadir-pointing, with the fifth channel used for solar viewing. Data rate: 160 Bit/s.
  ERBE Scanner: 3 channels (6,7,and 8) with radiometric thermistor bolometers. Chan-

91)   J.A.Dezio, C.A. Jensen, "Earth Radiation Budget Satellite", in Monitoring Earth's Ocean, Land, and Atmosphere, Vol. 97 by AIAA, 1985, pp. 261-292

nel 6 isolates shortwave radiation at 0.2 - 5 µm; channel 7 covers the longwave radiation at 5-50 µm; and channel 8 provides total radiation measurements at 0.2 - 50 µm. Data rate = 960 Bit/s.

- **SAGE-II** (Stratospheric Aerosol and Gas Experiment II). Earth limb-scanning grating spectrometer. Monitoring of concentrations and distributions of stratospheric aerosols, nitrogen dioxide, and other constituents. SAGE-II is a 7 channel radiometer. Spectrum Range: 0.385 - 1.020 µm. Data rate = 6.3 kbit/s.

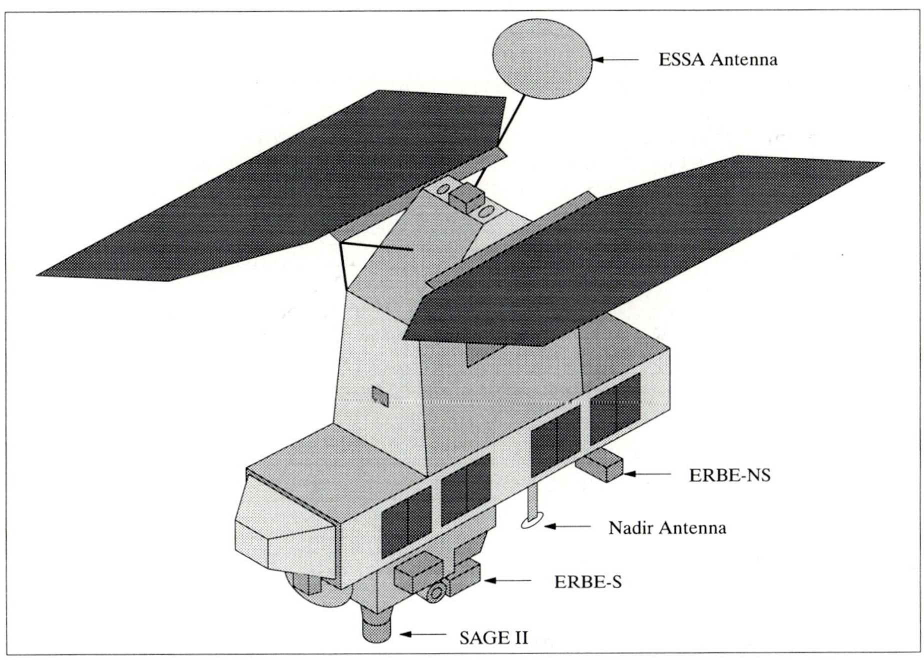


**Figure 27:    The ERBS S/C Model**

## A.25    ERS-1 (European Remote-Sensing Satellite)

ERS [92],[93],[94],[95] = European Remote Sensing Satellite (ESA Program). ERS-1 launch: July 17. 1991 from Kourou (Ariane IV vehicle, Mission Control at ESOC). The ERS program was initiated in 1981, it includes also the build-up of a corresponding ground segment for the exploitation of all data.

There is a world-wide cooperation in the area of data services (reception and exchange). DLR-DFD provides, among other European nation's agencies, a contribution in the area of science data reception (O'Higgins station in Antarctica), and in the area of science data processing with the 'Processing and Archiving Facility' (D-PAF). The other PAFs in the ground segment are provided by France (F-PAF), Italy (I-PAF) and the United Kingdom (UK-PAF).


92)   ESA Bulletin No. 65 Feb. 1991
93)   W. Markwitz, "Das ERS-1 Bodensegment, Empfang, Verarbeitung und Archivierung von SAR Daten", Die Geowissenschaften, 9. Jahrgang, Heft 4-5, April-Mai 1991, pp. 111-115
94)   D. Gottschalk, "ERS-1 Mission and System Overview", Die Geowissenschaften, 9. Jahrgang, Heft 4-5, April-Mai 1991, pp. 100-101
95)   M.F. Buchroithner, J. Raggan, D. Strobl "Geokodierung und geometrische Qualitätskontrolle", Die Geowissenschaften, 9. Jahrgang, Heft 4-5, April-Mai 1991, pp. 116-112

Application: A broad range of disciplines is covered. Observation of oceans, polar ice, land ecology, geology, forestry, wave phenomena, bathymetric (water depth), atmospheric physics, meteorology, etc. Scientific research: PIPOR (Program for International Polar Ocean Research); PISP (Polar Ice Sheet Proposal). Demonstration of concept and technology for space and ground segments (performance and operational capability).

Orbit: Sun-synchronous polar ($98.52°$ inclination), 785 km altitude; Nominal mission duration of 2 years, 3rd year reduced mission; there are the following orbit coverage cycles:
1. Reference orbit - 3 day repeat cycle (high repetition change monitoring with dedicated calibration sites; this orbit has been used during the commissioning phase)
2. Ice-Orbit - similar 3 day repeat cycle with a slightly different longitudinal phase. The main limitations of a 3 day cycle are the restricted global coverage for the imaging SAR and the wide separation of the RA-1 tracks.
3. Mapping-Orbit - 35 days repeat cycle (guaranteeing full earth coverage). This enables SAR imaging of every part of the Earth's surface, with at least twice the frequency of the coverage at middle and high latitudes.

**Sensors:**

*   **AMI** (Active Microwave Instrument)[96] Synthetic Aperture Radar (SAR) . Two separate radars are incorporated within the AMI, a SAR for 'Image and Wave mode' operation, and a scatterometer (SCAT) for 'Wind mode' operation. This instrument can operate in either one of the following modes:

    a.    AMI in **Imaging mode**. Measurement in C-Band (Frequency = 5.3 GHz (equivalent to 5.66 cm wavelength), Bandwidth = 15.55 MHz) in VV polarization; Look Angle = $23°$; radiometric resolution = 5 Bit on raw data (SAR mode), this corresponds to about 30 m spatial resolution; swath width = 100 km. Data rate = 105 Mbit/s.
    Imaging mode operating time per orbit = 12 minutes ((12% duty cycle)including 4 minutes in eclipse).

    b.    AMI in **Wave Mode**. Measurement of the changes in radar reflectivity of the sea surface due to surface waves. Provision of images (5 km x 5 km), also referred to as "imagettes", at regular intervals of 200 km along track. These imagettes are transformed into spectra providing information about the lengths and directions of the ocean wave systems. Characteristics: Freq. = 5.3 GHz, Polarization = Linear Vertical (LV); Incidence (look) angle = $23°$. Wave direction: 0 - $180°$. Resolution = 30 m. Data rate = 370 kbit/s. Duty cycle of 70 %

    c.    AMI in **Wind Scatterometer Mode (AMI-SCAT)**[97]. Use of 3 separate sideways looking antennas (fore, mid and aft beams, see Figure 30) for the measurement of sea surface wind speed and direction. Characteristics: Wind direction range = $0\text{-}360°$; accuracy = $\pm20°$; Wind speed range = 4-24 m/s; Spatial resolution = 50 km; Grid spacing = 25 km; Swath width = 500 km (same side as SAR imaging); Frequency = 5.3 GHz; Polarization = LV. Data rate = 500 kbit/s. Operation over all oceans; (note: the AMI-SCAT cannot be operated in parallel with the AMI SAR imaging mode; however, parallel operation of the wind and waves modes is possible).
    The three antenna beams continuously illuminate a swath of 500 km each measur-

---

96)    E. P. W. Attema, "The Active Microwave Instrument On-Board the ERS-1 Satellite", Proc. IEEE, Vol. 79, No.6, June 1991, pp. 791- 799
97)    ERS-1 User Handbook, esa SP-1148, May 1992, pp. 6-7

ing the radar backscatter from the sea surface for overlapping 50 km resolution cells using a 25 km grid spacing. The result is three independent backscatter measurements relating to cell center nodes on a 25 km grid (three different viewing directions, separated by a very small time delay). This permits surface wind vector determination using 'triplets' within the mathematical model.

AMI is an instrument providing data for a wide range of research disciplines such as: climatology, oceanography, glaciology, land processes, operational meteorology, etc.

- **Radar Altimeter (RA-1),** operating in K-Band. RA is a nadir-pointing pulse radar taking precise measurements of the echos from the ocean and ice surfaces. Frequency = 13.8 GHz; Pulse length = 20 µs; Pulse repeat frequency = 1020 Hz; Chirp bandwidth = 330 MHz (sea), 82.5 MHz (ice). RA-1 operates in 2 modes: Ocean mode and Ice mode. RA-1 provides measurements leading to the determination of:[98),99),100)]
    - altitude (ocean surface elevation for the study of ocean currents, the tides and the global geoid)
    - significant wave height
    - ocean surface wind speed
    - various ice parameters (surface topography, ice types, sea/ice boundaries)
- **ATSR** (Along-Track Scanning Radiometer and Microwave Sounder). ATSR (build by CRPE, France and RAL, UK) consists of two instruments: the MWS (Microwave Sounder) and the IRR (Infrared Radiometer)

MWR Characteristics:
ATSR-MWS is a nadir-looking passive microwave radiometer measuring microwave-equivalent temperatures in the 23.8 and 36.5 GHz spectrum with an IFOV of 20 km (=footprint), each channel has a bandwidth of 400 MHz.. Measurements of the atmospheric water-vapor and liquid content (prime objective of MWR). This is used to improve the accuracy of the sea surface temperature measurements and also to provide accurate tropospheric range correction for the RA-1.

IRR Characteristics:
4 Spectral channels: 1.6, 3.7, 11 and 12 µm. Spatial resolution = 1 km x 1 km (IFOV at nadir). Radiometric resolution < 0.1 K. Predicted accuracy = 0.5 K over a 50 km x 50 km area with 80 % cloud cover. Swath width = 500 km. The scanning technique enables the Earth's surface to be viewed at two different angles (0° and 52°) in two curved swaths 500 km wide and separated, along track, by about 700 km. Measurements of:
    - cloud-top temperature and cloud cover
    - sea-surface temperature (prime objective of IRR)
- **LRR** (Laser Retro-Reflector). A passive optical device for accurate satellite tracking from ground (laser ranging stations of SLR network) to support instrument data evaluation. LRR Characteristics: Wavelength = 350 - 800 nm (optimized for 532 nm). Efficiency: greater than 0.15 end-of-life. Reflection coefficient: greater than 0.8 end-of-life. FOV: elevation half-cone angle 60°, azimuth 360°. Diameter: < 20 cm.

- **PRARE[101)]** (Precise Range and Range Rate Equipment) see also chapter A.81. The precise satellite range determination will lead to higher-accuracy altitude measurements that will extend the mission to ocean circulation studies and geodetic applications such as sea-surface topography and crustal dynamics.


98)  G. Schreier, K. Maeda, B. Guindon, "Three Spaceborne SAR Sensors: ERS-1, JERS-1, and RADARSAT- Competition or Synergism?", Geo Informationssysteme, Heft 2/1991, Wichmann Verlag, Karlsruhe, pp. 20 - 27
99)  R. Winter, D. Kosmann "Anwendungen von SAR-Daten des ERS-1 zur Landnutzung", Die Geowissenschaften, 9. Jahrgang, Heft 4-5, April-Mai 1991, pp. 128-132
100) W. Kühbauch, "Anwendung der Radarfernerkundung in der Landwirtschaft", Die Geowissenschaften, 9. Jahrgang, Heft 4-5, April-Mai 1991, pp. 122-127

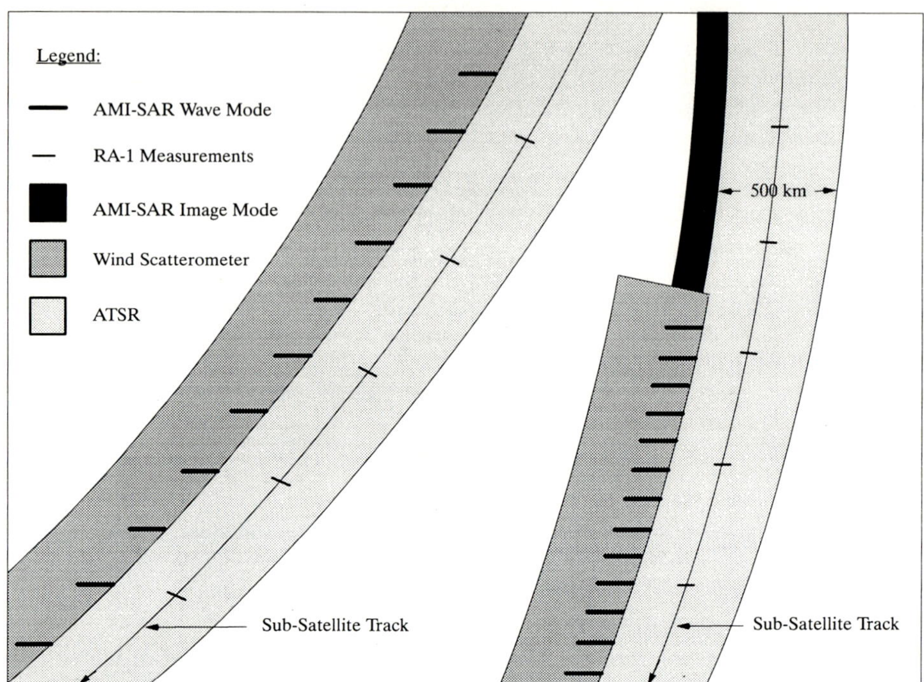


**Figure 28:    Schematic Swath Coverages for ERS-1 Sensors**

**Data:**

Image size SAR: 100 km x 100 km (4 Looks Full Digital Image Processing)

Transmission: Frequency = X-Band Link (SAR); data rate = 105 Mbit/s; there is no on-board recorder for SAR image data, hence, **the data generated in SAR mode can only be transmitted in real-time during passes over receiving stations** [ESA stations at Kiruna, Fucino, Maspalomas (Canary Islands), and Gatineau (Canada), and national facilities, like the Canadian station 'Prince Albert', the DLR-DFD stations "O'Higgins" (in Antarctica) and a portable station which can be setup anywhere (DTXS = DFD Transportable X-Band Station), the Japanese stations "Hatoyama", "Kumatomo" and "Syowa", the Indian (ISRO) station Hyderabad, the Alaska SAR Facility (ASF), Alice Springs (Australia), Tromsö (Norway), Cuiaba (Brazil), Cotopaxi (Ecuador), Beijing (China), Saudi Arabia, Thailand, etc.].

All ground stations are equipped with "Fast Delivery" SAR processors, capable of generating quicklook images after reception of the pass. These "Fast Delivery Products" (FDP) are directly mailed to the national PAF's (Processing and Archiving Facility).

- D-PAF     = DLR-DFD in Oberpfaffenhofen
- F-PAF     = CERSAT, Brest France
- I-PAF     = ASI, Matera, Italy
- UK-PAF  = RAE, Farnborough, UK

**Earthnet ERS Central Facility (EECF)**

---

101) **Note: The PRARE onboard instrument of the ERS-1 payload could not achieve an operational status after launch. The instrument had worked nominally for 5 days after launch (five contacts with the command station showed nominal telemetry). A thorough failure analysis came to the conclusion, that the most likely cause of the PRARE failure is a RAM damage due to radiation (destructive RAM latch-up).**

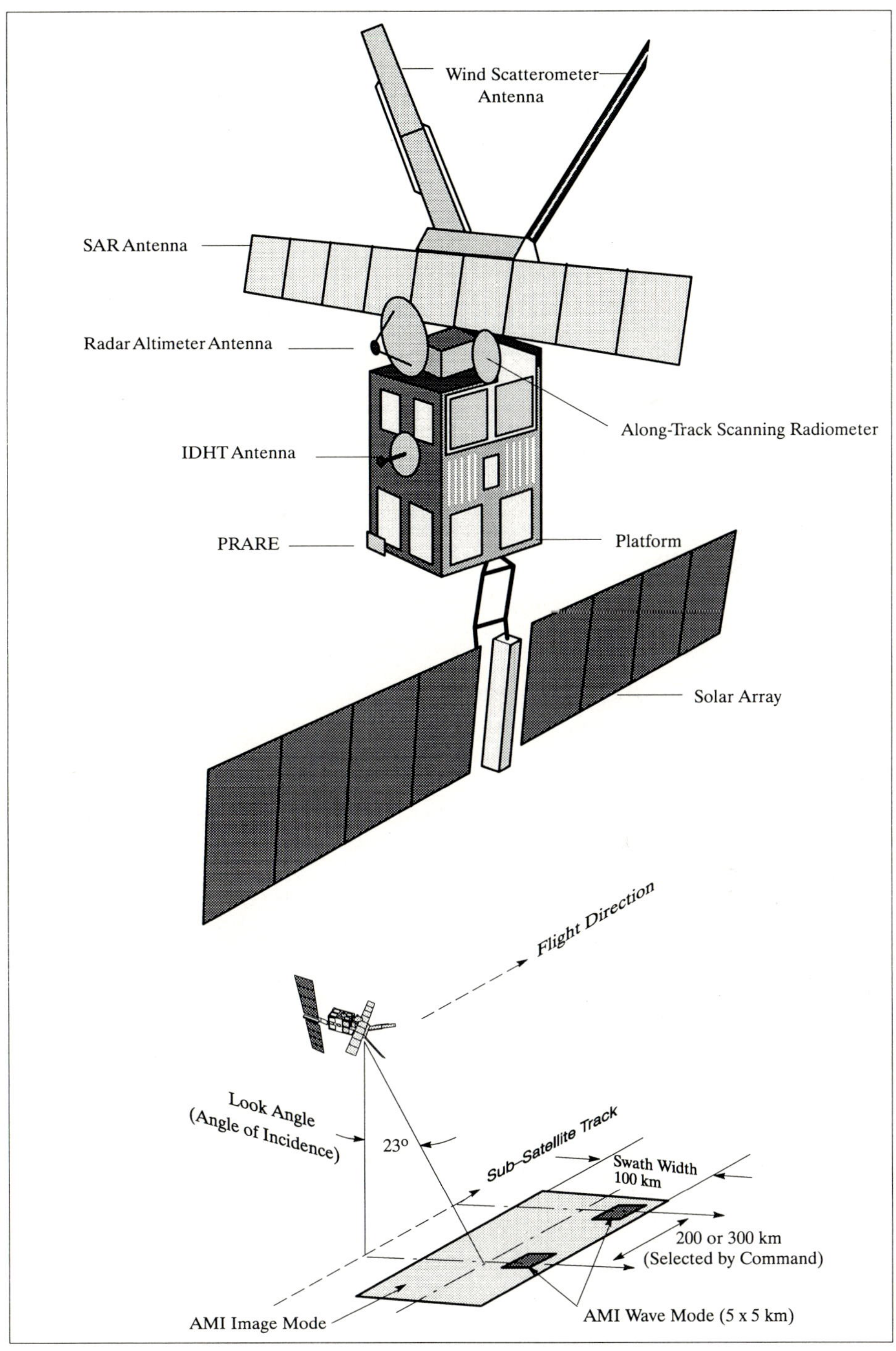


**Figure 29:    The ERS-1 Satellite Model and SAR (AMI) Observation Geometry**

### Commercial ERS-1 Data Distributors

In 1992 ESA selected three distributors for ERS-1 Data. These are[102]:

-   Radarsat International of Ottawa, Canada (responsible for commercial sales in Canada and USA)
-   EURIMAGE of Rome, Italy (market in Europe, North Africa, and the Middle East)
-   SPOT Image of Toulouse, France (rest of the world)

| SAR Image Products | SAR Wave Mode Products | Altimeter Products | Wind Scatterometer Products |
|---|---|---|---|
| Annotated Raw Data<br>Fast-Delivery Image<br>Fast Delivery Image Copy<br>Single-Look Complex Image<br>Precision Image<br>Ellipsoid Geocoded Image<br>Terrain Geocoded Image | Annotated Raw Data<br>Fast Delivery Product<br>Fast Delivery Product Copy<br>Intermediate Product Copy<br>Complex Imagette<br>Detected Imagette and its Spectrum<br>Imagette Precise Spectrum | Annotated Raw Data<br>Fast-Delivery Product<br>Fast-Delivery Product Copy<br>Offline Intermediate Product<br>Sea Surface Height<br>Sea Surface Topography<br>Ocean Geoid | Fast-Delivery Product<br>Fast-Delivery Product Copy<br>Extracted Wind Copy<br>De-aliased Offline Wind Fields |
| Orbit Products | Earth Gravity Products | ATSR Products |  |
| Preliminary Orbit<br>Precise Orbit | ERS-1 Gravity Model<br>1. Generation<br>ERS-1 Gravity Model<br>2. Generation |  |  |
|  |  |  |  |

**Table 18:    ERS-1 Data Products**

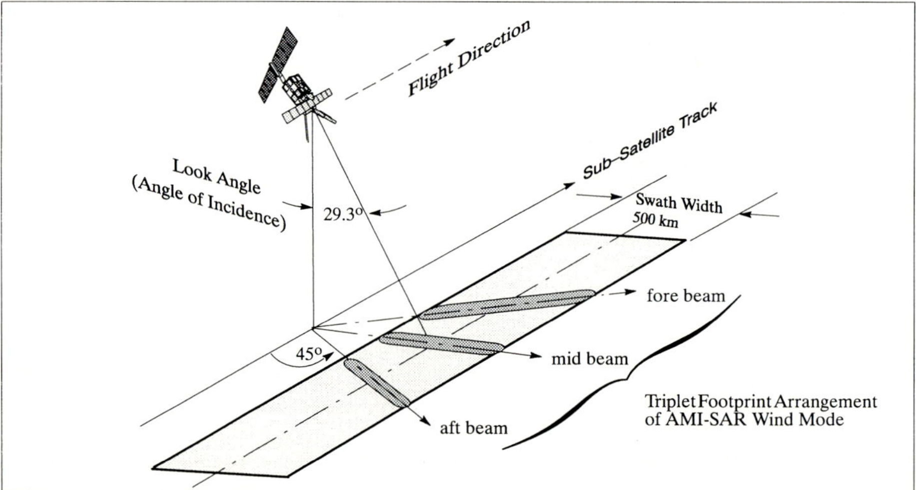


**Figure 30:    ERS-1 Wind Scatterometer Observation Geometries**

## A.26    ERS-2

ERS-2 is the follow-up mission of ERS-1, currently scheduled (and approved) for a launch in Jan. 1995 (3 year mission). ERS-2 has the same mission objectives as ERS-1, plus an atmospheric chemistry mission objective (with GOME) and an advanced ATSR instrument (AATSR).

---

102) 'ESA Signs Long-awaited Imagery Sales Deal', Space News, Feb. 10-16 1992, p. 4

Orbit: Sun-synchronous polar orbit. Altitude = 824 km; (local crossing time of the equator = 9:30 AM). Orbital period is about 100 minutes. Repeat cycle = 3 days.

**Sensors:**

Same as ERS-1 (AMI, AATSR, RA, PRARE, LRR)+ GOME (atmospheric chemistry)

**GOME** = Global Ozone Monitoring Experiment[103],[104],[105].

GOME is a passive nadir-looking double spectrometer which observes solar radiation transmitted through or scattered from the atmosphere or from the earth's surface.

The GOME double spectrometer (1. stage: prism, 2. stage: grating) is operating over the spectral range of 240 nm to 790 nm (i.e. ultra-violet/visible) at moderately high spectral resolution (0.2 nm to 0.4 nm). In addition to the improved backscattering technique, GOME will exploit the full capabilities of the enhanced ATSR. The GOME measurement concept is based on 'Differential Optical Absorption Spectroscopy' (DOAS). DOAS is a proven method in balloon flights.

GOME objectives: observation of ozone and other important trace gas species in the troposphere as well as in the stratosphere.

Its main mode of operation is nadir looking, but it will also be able to look at the sun and the moon for calibration purposes. For onboard calibration, GOME does not depend on one technique, but seeks to exploit several. Both absolute radiometric and wavelength calibrations will be possible. This means, that GOME will be able to retrieve ozone distributions by exploiting the traditional backscatter approach, as well as with the more novel differential optical absorption spectroscopy.

GOME measures across-track in 3 steps (normal mode)

Pixel size on the ground = 40 x 1.7 km.

Spatial resolution: between 40 x 40 km and 40 x 320 km.

GOME spectral bands:
- Channel 1 (240 -295 nm), 512 channels
- Channel 2 (290 - 405 nm), 1024 channels
- Channel 3 (400 - 605 nm), 1024 channels
- Channel 4 (590 - 790 nm), 1024 channels


Data (GOME):

A continuous data stream of 40 kbit/s is generated (requirement for global coverage).


**AATSR** = Advanced Along Track Scanning Radiometer. Imaging radiometer operating in 4 infrared channels (1.6, 3.7, 11.0, and 12.0 μm) and with 4 visible/reflected channels (0.65, 0.85, 1.27 and 1.6 μm). AATSR retains the full capability of ATSR (of ERS-1)

The different viewing direction are achieved by employing a plane inclined mirror in the instrument which rotates and produces a conical scan of the instrument IFOV, the scan being centered about a line of sight approximately 23°, i.e. along-track of nadir.

Pixel size for the ocean IR channels is 1km x 1 km, and 0.5 km x 0.5 km for the land visible channels.

Objectives of AATSR: precise measurement of sea surface temperatures; extension of measurements to land applications.

AATSR is intended to operate continuously. Data from channels 3 (3.7 μm, night measurements) and channel 4 (1.6 μm, day measurements) are not used together. The overall data rate = 1 Mbit/s.


103) C.J. Readings, 'The Interim GOME Science Report', Feb. 1990,
104) 'The Global Ozone Monitoring Experiment (GOME) and ERS-2', earth Observation quarterly, ESA periodical Nr. 32 Dec. 1990
105) 'ERS-2 and Beyond', p. 95 in ESA Bulletin No. 65 February 1991

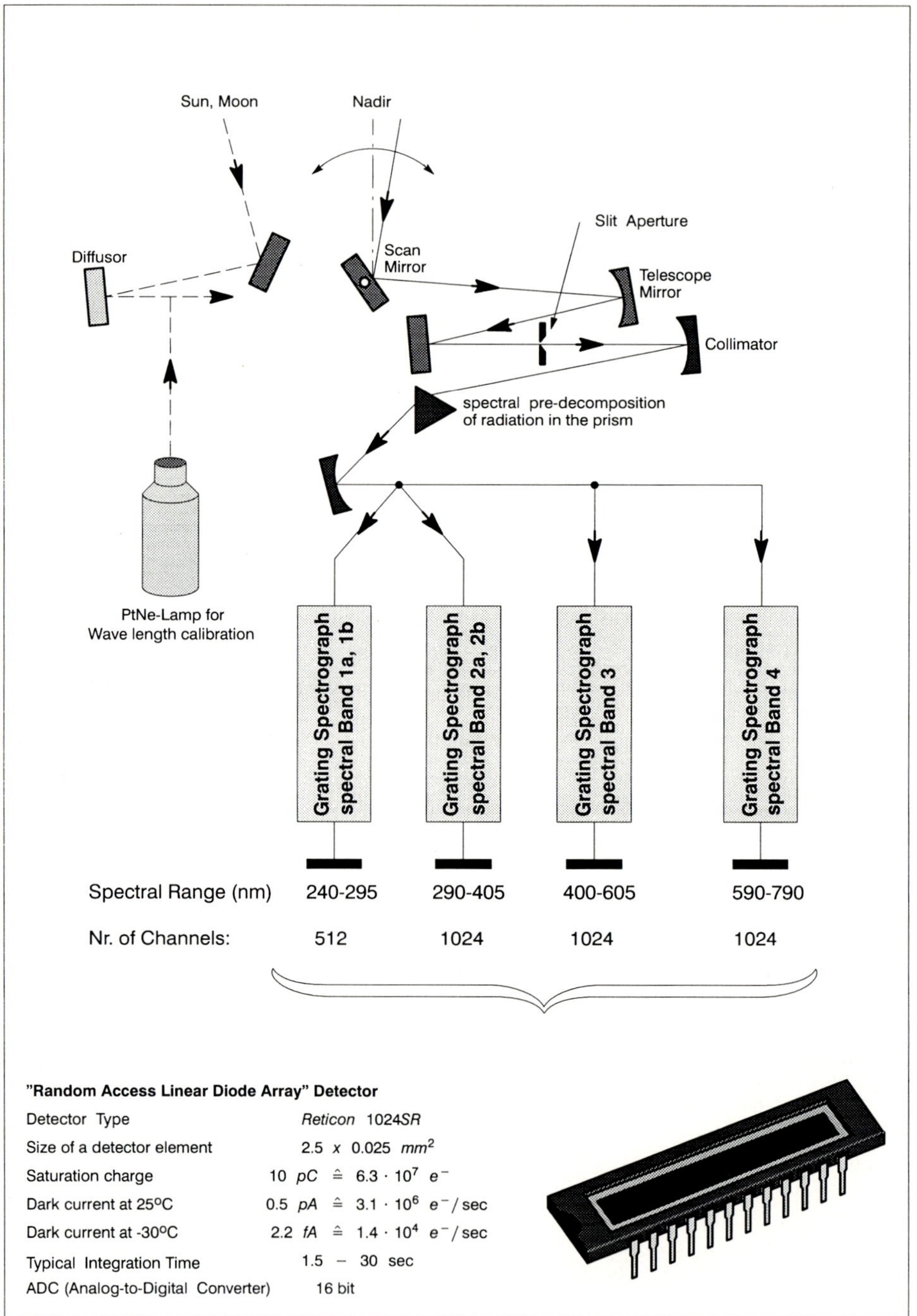



**Figure 31:     Optical Principle of the GOME Double-Spectrograph**[106)]

---

106)  Courtesy of W. Schneider, DLR

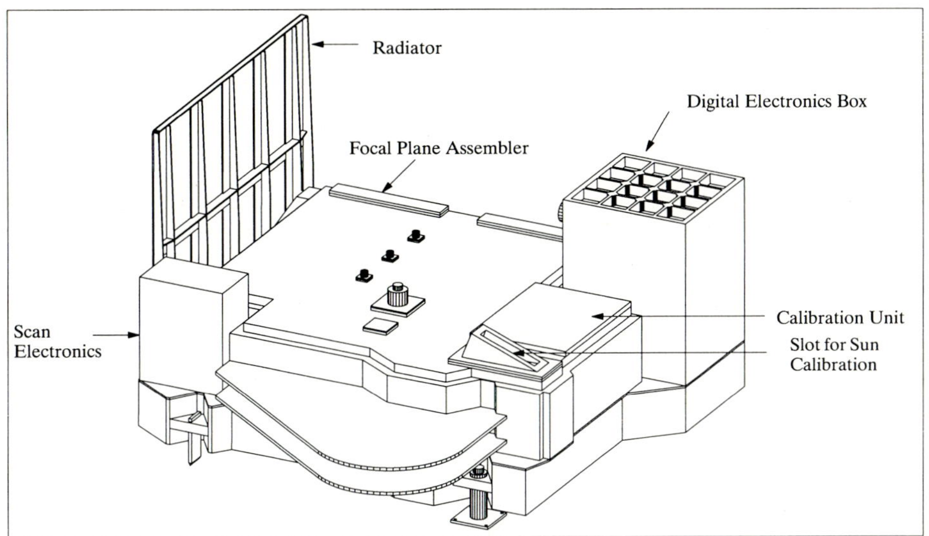


**Figure 32:     The GOME Instrument Model**

## A.27    ETALON

ETALON[107](="standard") = Soviet Geodynamical Satellites. ETALON-1 was launched on January 10, 1989 from Baikonur together with 2 GLONASS satellites. (ETALON-1 is also known as Cosmos-1989). ETALON-2 (identical S/C) was launched on May 31, 1989.

Orbits (ETALON-1 and -2): Near-circular orbit, altitude = 19130 km, eccentricity = 0.00068, inclination = 64.8°, nodal period = 10874.7 days (29.8 years), perigee period = 94402.9 days (258.6 years), one revolution period = 675 min (11.25 h)

Features of ETALON S/C.
Shape = sphere with a diameter of 1.294 m, mass = 1415 kg, laser reflectors = 2140, distance from the geometrical center to the plane of probable reflection = 558 mm

Objectives: ETALON is a passive satellite system dedicated entirely to laser ranging [see also LAGEOS (A.54 )]. Solid Earth studies: geodynamic processes, development of high-accuracy global references, long-period disturbances, geopotential modelling, etc.

GFZ of Potsdam is the data collection and distribution center of ETALON laser measurements in Europe.


## A.28    EURECA (European Retrievable Carrier)

An ESA spacecraft/platform[108] designed and developed to be recovered from orbit after completion of a particular mission, returned to Earth, refitted with a new sensor complement, and to be relaunched for another extended/new mission. A typical mission duration is of the order of 6 months to one year. The concept allows a complete payload/sample recovery.

107)  S.K. Tatevian, A.N. Zakharov, "The Geodynamical Satellite ETALON", CSTG Bulletin No. 11, Title: New Satellite Missions for Solid Earth Studies, 1989, pp. 3-9
108)  ESA Press Release, ESA/ESTEC, 1991

### A.28.1   EURECA-1 Mission

Launch of EURECA-1 by Shuttle (Atlantis, STS-46) from KSC on July 31, 1992. Retrieval of EURECA-1 by Shuttle (Endeavour) during the STS-57 mission from June 21 - July 1, 1993. Mission operations of EURECA-1 was provided by ESOC in Darmstadt; some payload operations was provided by MUSC (Microgravity User Support Center) of DLR in Cologne, Germany. The spacecraft was developed and integrated at DASA/ERNO, Bremen, Germany.[109),110)]

Orbit: Low Earth Orbit (LEO) circular; altitude = 502 km, inclination = 28°, retrieval altitude of 476 km.

**Sensor Complement**

Only those sensors are described and indexed that fit into the scope of the book, namely: "Observation of the Earth and its Environment", the other sensors (in the life- and material sciences) are simply listed with a short commentary.

ERA = Exobiology and Radiation Assembly. A multi-user life science facility for experiments on the biological effects of space radiation. PI: H. Bücker, DLR, Germany.

AMF = Automatic Mirror Furnace. Optical radiation furnace for the growth of single, uniform crystals from the liquid or vapor phases, using the travelling heater or Bridgman methods. PI: K. W. Benz, Universität Freiburg, Germany.

SGF = Solution Growth Facility. A multi-user facility for the growth of monocrystals from solution. PI: J. C. Legros, Université Libre de Bruxelles, Belgium.

PCF = Protein Crystallisation Facility. A multi-user solution growth facility for protein crystallisation in space. PI: W. Littke, Universität Freiburg, Germany.

MFA = Multi-Furnace Assembly. A multi-user facility dedicated to material science experiments. PI: A. Passerone, National Research Council, Genova, Italy

HPT = High Precision Thermostat. An instrument for long-term experiments requiring microgravity conditions and high precision temperature measurement and control (typical experiments are: 'caloric', 'critical point', and 'phase transitions'). PI: G. Findenegg, Ruhr Universität, Bochum, Germany.

SFA = Surface Forces Adhesion. Study of the dependence of surface forces and interface energies on physical and chemical-physical parameters such as surface topography, surface cleanliness, temperature, and the deformation properties of the contacting bodies. PI: G. Poletti, Università di Milano, Italy.

RITA = Radio-Frequency Ionization Thruster Assembly. Study of the use of electric propulsion in space. PI: H. Bassner, DASA-MBB, Munich, Germany.

IOC = Inter-Orbit Communication Instrument. A technological experiment to provide a pre-operational in-flight test and demonstration of all required instrument functions and services. PI: R. Tribes, CNES, France.

ASGA = Advanced Solar Gallium Arsenide Array. A technological experiment for the performance testing of future solar arrays. PI: C. Flores, CISE SPA, Segrate, Italy.

WATCH = Wide Angle Telescope. Objective: detection of celestial gamma-ray burst sources and x-ray sources with photon energies in the range of 5 - 200 keV, measurement of source direction. PI: N. Lund, Danish Space Research Institute, Lyngby, Denmark.

---


109)   P. Ferri, H. Hübner, S. Kellock, W. Wimmer, "The Joint ESA-NASA Operations for Eureca's Deployment and Retrieval", ESA bulletin, Number 76, November 1993, pp. 81-90

110)   F. Dreger, J. Fertig, D. Gawthrope, S. Martin, et. al., "Eureca: The Flight Dynamics of the Retrieval", ESA bulletin, Number 76, November 1993, pp. 92-99

**TICCE** = Timeband Capture Cell Experiment. Study of microparticle population distributions in near-Earth space, typically Earth debris, meteoroids, and cosmic dust. TICCE captures micron-dimensional particles with velocities in excess of 3 km/s and stores the debris for retrieval and post-mission analysis. PI: J. A. M. McDonnell, University of Kent, UK.

Particles detected by the instrument pass through a front foil and into a debris collection substrate positioned 100 nm behind the foil. Each perforation in the foil has a corresponding debris site on the substrate. The foil is being moved during the mission time in 50 discrete steps. The phase shift between the debris site and the perforation enables the determination of the impact time. The instrument mass = 8 kg, dimensions: 690 x 690 x 80 mm.

**ORA** = Occultation Radiometer Instrument. Objective: measurement of aerosols and trace gas densities in the Earth's mesosphere and stratosphere (vertical profiles between 20 -100 km of: ozone, nitrogen, dioxide, water vapor, carbon dioxide, and background and volcanic aerosols) . PI: E. Arijs, BIRA (Belgisch Instituut voor Ruimte Aeronomie), Brussels, Belgium.

ORA measures the intensity of solar radiation during the sunrise and sunset phases of each orbit. It uses the SUN pointing capabilities of EURECA and makes measurements in ten narrow wavelength bands (UV/VIS,NIR) in the spectral region 250 - 1100 nm. ORA consists of a UV/VIS unit, an NIR unit and a control/electronics unit. The UV/VIS unit has 8 similar modules which measure the relative solar irradiance in narrow bands. Each module contains a quartz window, an interference window to select the appropriate wavelength, optics to limit the detector viewing angle, and a photodiode.

**SOVA** = Solar Constant and Variability Instrument. Objective: measurement of the solar constant, its variability and its spectral distribution. PIs: D. Crommelynck, IRMB, Bruxelles, Belgium; and C. Fröhlich, PMOD/WRC Davos, Switzerland.

The instrument is composed of three boxes, two of which (SOVA1 and SOVA2) house the various devices dedicated to functions such as Sun-pointing, Sun photometers, and two different types of absolute radiometers. The absolute radiometers are very accurate instruments for the measurement of total irradiance. They are essentially cavities, in which the solar radiation is trapped and absorbed. Heat flux transducers convert the radiative power into electrical power and return a calibrated measurement of the Sun's radiation. The high precision radiometer (RELOS) is used to measure solar oscillations in total irradiance. The Sun-pointing resolution of SOVA2 is 2 arc seconds. The Sun photometers use interference filters to select the wavelength band and silicon diode devices to measure the radiation. The complete set of photometers covers the spectrum from 303 nm to 865 nm.

**SOSP** = Solar Spectrum Instrument. Objective: study of solar physics and the solar-terrestrial relationship in aeronomy and climatology. Measurement of the absolute solar irradiance and its variations in the spectral range 170 - 3200 nm (accuracy of 1% in VIS/IR and 5% in UV). PI: G. Thullier, CNRS, Verriers le Buisson, France.

SOSP has three spectrometers, one for each spectral domain (UV/VIS/IR). The spectrometers select the spectral range using a double monochromator; irradiance measurement with a photomultiplier in UV and VIS, in IR with a cooled lead sulfide cell. Spectral resolution: 1 nm in UV/VIS, 20 nm in IR. In-flight calibration is provided.


## A.29   EXOS (Exospheric Observations)

Japanese (ISAS) Solar-terrestrial interaction program with a series of four satellites for the study of the magnetosphere and auroral phenomena.

### A.29.1   EXOS-A (Kyokko)

ISAS mission with the objective to study the magnetosphere (auroral activity), plasma environment and VLF (Very Low Frequency) emissions. S/C mass = 126 kg. Launch: Feb. 4, 1978 by M-3H-2 vehicle from Kagoshima. End of EXOS-A mission in November 1979 due to the degradation of solar cells by radiation belt particle bombardment.

Orbit: apogee = 3980 km, perigee = 640 km, inclination = 65°.

Sensors:

- **UV-TV** Camera. This wide-angle TV camera takes an instantaneous auroral picture that covers almost the entire polar region, in every 128 seconds during its operation over the northern polar region. About 10 to 20 pictures were obtained during one pass. More than 20,000 pictures of aurorae in UV were obtained in all.

- UV Glow Spectrophotometer. Spectrophotometry of ultraviolet glows from the thermosphere, magnetosphere, and interplanetary space in the spectral range of 300 - 1300 Å.

- Electron Spectrometer. Measurement of the upward and downward electron fluxes along the magnetic field line of force in the energy range from several ev up to around 10 keV. Measured data show precipitation patterns with longitudinal as well as latitudinal variations.

- Ion Mass Spectrometer. Objective: study of the light ion trough, atmospheric composition of ionized species of H, He, and O atoms in the thermosphere.

- Ionospheric Plasma Probes. Measurements of density and temperature of ambient electrons by plasma probes, and plasma waves especially with reference to the emission associated with auroral phenomena.

ISAS also conducted with Kyokko international cooperation with a Scandinavian group during the rocket and balloon campaigns at Kiruna, Sweden.

### A.29.2   EXOS-B (Jikiken)

ISAS mission with the objective to study the magnetosphere plasma resonance and echo phenomena as part of IMS (International Magnetosphere Study) program.[111]
Launch of EXOS-B on Sept. 16, 1978. The mission lasted over 3 years.

Orbit: quasi-equatorial orbit, apogee = 30,051 km, perigee = 227 km, inclination = -31°, period = 523.6 minutes. within 3 years the apogee has gradually changed from the initial value of 30.051 - 28,000 km, while the perigee is oscillating in the range from 227 to 290 km.

**Sensors:**

- **SPW** = Stimulated Plasma Wave Experiment (H. Oya, Tohoku University) in the plasmasphere and in the magnetosphere across the plasmapause.
  Objective: study of the plasma wave resonance mechanisms (upper hybrid resonance, plasma resonance, harmonics of the electron cyclotron resonance, electrostatic resonance, sequence of the diffuse resonance). Plasma waves are exited by RF pulses transmitted from the S/C to study nonlinear wave-plasma interactions.

  SPW instrumentation consists of four systems: the signal generation system (high power RF transmitter for the simulation of the plasma waves in the frequency range from 10 kHz to 3 MHz), the power amplifier, antenna system, and the receiver. The signal gen-

---

111)  Journal of Geomagnetism and Geoelectricity including Space Physics, Volume 33, No. 1, 1981, featuring
      EXOS-B, pp. 1-160

erator can produce two types of frequency variations that are swept frequency mode: Tx sweep, and the fixed frequency mode: Tx-Fix-n (Tx-Fix-1 = 21 kHz, Tx-Fix-2 = 90 kHz, Tx-Fix-3 = 390 kHz, and Tx-Fix-4 = 1.7 MHz).

- **NPW-A** = Natural Plasma Wave Astronomy Mode (H. Oya, Tohoku University). Objective: observation of the planetary (terrestrial kilometric radiation) and solar radio waves in the frequency range from 10 kHz - 3 MHz.

  Instrumentation: the satellite is equipped with two sets of long dipole antennas which are arranged orthogonally to each other in the perpendicular plane of the S/C spin axis (length = 102 m tip-to-tip and 72 m tip-to-tip). The 102 m antenna is mainly used for plasma and radio wave reception, while the 72 m antenna is used for plasma relaxation sounder and the impedance probe.

- **VLF** = Very Low Frequency Wave Detectors (I. Kimura, Kyoto University). Objective: observation of wave particle interactions in the magnetosphere (electromagnetic and electrostatic waves in a frequency range between 150 Hz and 9.5 kHz, the measured range of electron is from 5 eV to 11 keV). The VLF receiver is normally connected to the 102 m dipole antenna.

- **DPL** = Doppler Shift Measurement. Objective: detection of ionization ducts in the magnetosphere. DPL observes the Doppler shift (intensity, phase shift, and antenna capacitance at the same frequency) of an NWC (22.3 kHz) signal transmitted from North West Cape in Australia.

- **IEF** = Impedance and Electric Field Measurement. Objectives: measurement of an antenna impedance , DC electric field and AC electric field (low frequency) using a long cylindrical antenna system. Study of electron density and wave phenomena in the frequency range from 1 Hz to 450 Hz.

  IEF has a cross-dipole antenna system with originally designed length of 120 m; these antenna elements were not fully deployed due to an unstable satellite attitude. The actual length of each antenna element is as follows: A1 = 33.4 m, A2= 36.2 m, B1 = 51.7 m, B2 = 51.3 m. IEF consists of four instruments:
  - IEF-1 is a frequency swept impedance probe using the A1 antenna, a frequency range from 10 kHz to 3 MHz with a swept period of 2 seconds or 8 seconds. An equivalent capacitance range measured is from 0.5 pF to 5000 pF.
  - IEF-S and D are a single Langmuir probe and a double probe using A1 and A2 antennas respectively.
  - IEF-C is an electric field detector using A1 and A2 or B1 and B2 antennas with 1 Hz to 450 Hz electric field measurements.

- **MGF** = Triaxial Fluxgate Magnetometer. Objective: measurement of the geomagnetic vector field. Study of relation between field-aligned currents and dynamics of the magnetosphere, and the generation and propagation mechanisms of hydromagnetic waves near the plasmapause. Resolution from 2 nT to 380 nT in four ranges. Sampling rate of either 1 second or 4 seconds. The dynamic range of MGF covers 247 nT to 46,840 nT.

- **ESP** = Charged Particle Detectors. Objectives:
  - investigate the dynamics of electrons and ions in the inner magnetosphere
  - study of wave-particle interaction in the inner magnetosphere together with the wave measurements
  - observe a response of magnetospheric plasma when an active experiment, that is SPW (high power radio wave emission) and/or CBE is performed.

  ESP is composed of an electron sensor, an ion sensor and electronic circuits. Measurements of electron and ion fluxes in the energy ranges from a few eV up to 10keV and of ion fluxes from 10 eV up to 30 keV, respectively in the magnetosphere of L-shells from $L \cong 2.5$ to $L \cong 8$. Energy resolution $\Delta E/E = 3\%$ for electrons and 5% for ions.

- **CBE** = Controlled Beam Experiment. Objectives: control of the satellite potential by an electron beam emission. Study of the wave excitation (linear and non-linear wave phenomena due to the beam-plasma interaction).

### A.29.3   EXOS-C (Ohzora)

ISAS mission with the objective of electron temperature measurement in high latitude regions of the plasma bubble and to study the anomalous electron heating in the upper atmosphere.
Launch: Feb. 14, 1984 by a M-3S-4 vehicle from Kagoshima. S/C mass = 208 kg. The S/C is stabilized stabilized with the aid of a momentum wheel, and the spin axis of the wheel is automatically controlled so that one side of the S/C faces the solar disk. Power = 160 W.

Orbit: apogee = 865 km, perigee = 354 km, inclination = 74.6°, period = 97 min.

**Sensors:**

- Limb Scanning Radiometer (passive device) for the 1.27 µm infrared atmospheric band airglow to deduce the ozone density in the altitude range of 70 - 90 km.

- UV Spectrometer for nadir observation of the backscattered UV radiation (2500 - 3500 Å) to obtain the ozone profile in the altitude range of 25 - 60 km.

- Solar Image Radiometer. Measurements in several visible and near-infrared bands to detect the limb absorption by the stratospheric aerosols and ozone.

- IR Solar Spectrometer. Monitoring the limb absorption by the stratospheric water vapor, methane, carbon dioxide, and ozone.

- Topside Ionospheric Plasma Sounder (including a receiver). Measurement of the plasma waves due to the plasma turbulence associated with the precipitating particles.

- Receiver for global monitoring of electromagnetic waves radiated from the power line networks and high-power electric power apparatus on the ground.

- Electron Temperature Probe (for electron density and temperature measurements), consisting of 4 sensor electrodes (S1, S3, S5, S7) mounted at the ends of the solar cell paddles. The electrode plane is parallel to the satellite spin axis. Dayside cusp observation. Characteristics: frequency of the oscillator = 30 kHz; amplitude of the signal = 500 mV, 250 mV, 0 mV; input impedance of DC amplifiers = 110 MΩ [112]

- Energy spectrum analyzer for precipitating particles.

### A.29.4   EXOS-D (Akebono)

ISAS mission for 'exospheric observations' with the objective to study the generation mechanism of auroral charged particle precipitation in the polar cusp in comparison with the simultaneously observed electric field. The S/C is spin-stabilized at 7.5 rpm, the spin axis points toward the sun, attitude control by 3-axis magnetic torquing. Payload mass = 295.4 kg, science instruments = 97 kg (including antennas and booms). Surface conductivity: 98.9% of the S/C surface, including the solar cells, is made conductive so as to minimize the electrostatic disturbances around the satellite. [113],[114]
Launch: 21. Feb. 1989 with M-3SII-4 vehicle from Kagoshima, Japan.

Orbit: elliptical polar orbit with an initial apogee into the southern hemisphere, apogee = 10,482 km, perigee = 272 km, inclination = 75.1°, period = 212 minutes.

---


112) K. I. Oyama et al, "Electron Temperature Probe on Board Japan's 9th Scientific Satellite Ohzora", J. Geomagnetism and Geoelectricity, Volume 37, 1985, pp. 413-430

113) EXOS-D (Akebono) - Japan's 12th Scientific Satellite - A Study of auroral particle acceleration processes, ISAS brochure

**Sensors:**

- **EFD** = Electric Field Detector (PI: H. Hayakawa, ISAS)
  Objective: study of the quasi-static vector electric field along the spacecraft orbit, in particular: a) the electric field structure in the auroral ionosphere and its relationship to ion dynamics and auroral ion acceleration, b) global dynamics of the polar ionosphere, c) plasma wave instability, d) electric field in the "plasma bubble".

  EFD measures the electric field by two techniques: namely the double probe and the ion beam technique. Double probe measurement from DC to 10 Hz; the ion beam measures the electric field twice per spin period.
  Ion beam technique: the instrument consists of 4 lithium ion guns, 4 lithium ion detectors, data control unit, an ion beam direction control unit, and a peripheral interface module.
  Double probe technique: measurement of the electric field in the plane perpendicular to the satellite spin axis using two pairs of wire antennas (30 m). The sampling rate is 32, 8, and 2 samples per second.

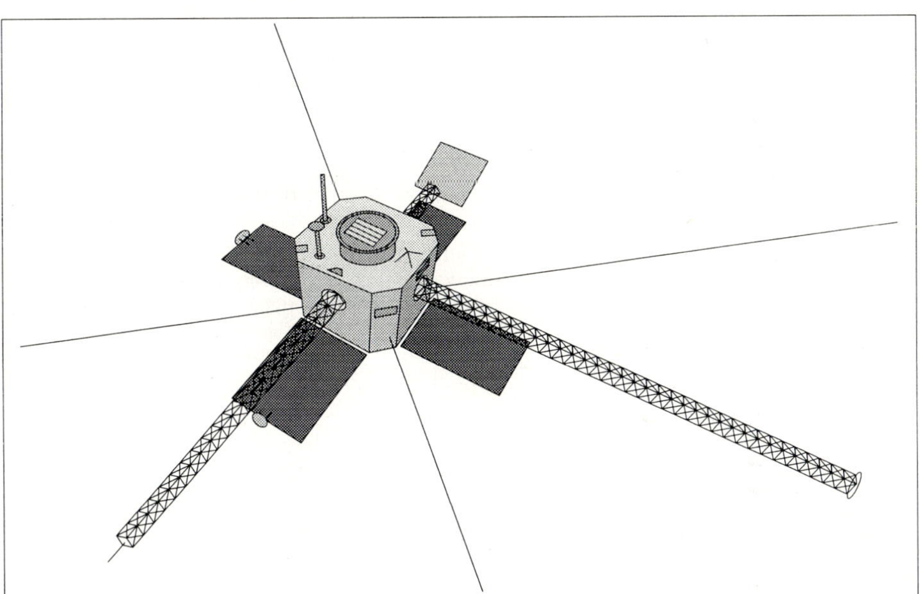

**Figure 33:**    The Exos-D (Akebono) S/C Model

- **MGF** = Magnetic Field Detectors (Pi: H. Fukunishi, Tohoku University) Objectives:
  - study the relationship between fine structures of field-aligned currents and auroral forms by comparing magnetometer data with ATV data and ground-based auroral TV data.
  - determine the current carriers using simultaneous magnetic field and particle data
  - study the driving mechanism of field-aligned currents from simultaneous measurements of magnetic fields and electric fields
  - identification of plasma wave modes in the frequency range < 100 Hz

  MGF is a triaxial fluxgate magnetometer (mounted at the tip of a 5 m mast) and a triaxial search coil magnetometer (mounted at the tip of a 3 m mast). The fluxgate mag-

---

114) Selected papers on EXOS-D (Akebono) Observations in Geophysical Research Letters, Volume 18, Nr. 2, Feb. 1991, pp. 293-352

netometer measures within four ranges: $\pm 65{,}536$ nT, $\pm 16{,}384$ nT, $\pm 4{,}096$ nT, and $\pm 1024$ nT with resolutions of: 2 nT, 0.5 nT, 0.125 nT, and 0.031 nT respectively. Sensor drive frequency = 15 kHz; frequency response = DC to 50 Hz.

Search coil magnetometer: triaxial search coil with $10^5$ turns and permalloy cores, frequency response = 1-1000 Hz; sensitivity = 1 pT at 10 Hz

- **VLF** = Very Low Frequency Wave Detectors (Pi: I. Kimura, Kyoto University) Objectives: study of the behavior of plasma waves associated with auroral particles, wave particles interaction mechanisms, and propagation characteristics of whistler mode and electrostatic mode waves in the magnetospheric plasma.

The instrument consists of the following subsystems: uses a pair of dipole wire antennas, 60 m tip-to-tip.

  - WBA (wide band receiver for observation of VLF spectra), 50 Hz - 10.5 kHz (5 kHz) E or B component
  - MCA (multi-channel analyzers), 16 channels for E, 16 channels for B
  - PFX (step frequency receiver for the measurement of wave normal direction and pointing flux), band width = 50 Hz
  - ELF (ELF band receiver, 4 channels), 100 Hz, 50 Hz
  - VIP (Vector impedance measurement of the wire antennas)
  - DUP (Data processing unit)

Joint experiments between VLF and the HIPAS (High Power Active Stimulation) facility located in Fairbanks, Alaska, were made in 1989. The ground VLF observations were successful in the sense that the VLF signals associated with the amplitude modulated HF HIPAS transmissions were clearly detected.

- **PWS** = Plasma Wave Detectors and Sounder (Pi: H. Oya, Tohoku University) Objectives: measurement of high frequency plasma waves (observe naturally occurring plasma and electromagnetic waves, conduct stimulated plasma wave experiments).

The instrument consists of:

  - a dynamic spectrum analyzer (frequency range from 10 kHz to 5 MHz, frequency resolution = 1 kHz, dynamic gain is about 100 dB)
  - a polarization measurement system (frequency range from 20 kHz to 5 MHz)
  - a pointing vector measurement system (measurement of wave amplitudes and phases of two electric field components and three magnetic field components, the analyzed wave frequency is stepped in frequency from 60 kHz to 1.3 MHz)
  - a high power transmitter/sounder (coverage of a wide frequency range at high power, frequency range from 20 kHz to 12 MHz, also operational modes to detect the plasma density with high spectral resolution)
  - an impedance measuring system (swept frequency impedance of the antenna in the plasma is measured in the same frequency range as the sounder)

the PWS sensors employ two sets of 60 m tip-to-tip antennas and a 70 cm diameter three dimensional loop antenna.

- **LEP** = Low Energy Particle Detectors (Pi: T. Mukai, ISAS) Objectives: comprehensive observations of energy and pitch angle distributions of electrons, ions, and mass per unit charge of the ions in the auroral magnetosphere. In particular:

  - study of the characteristics of charged particles associated with various types of auroras
  - study of particle acceleration due to electrostatic potential drops along magnetic field lines
  - identification of charge carriers of the field-aligned currents
  - study of wave-particle interactions
  - study of particle modulations (<16 Hz) and determination of source locations

- direct detection of flux modulations of electrons and ions in the VLF and HF ranges
- study of the plasma transport between the polar ionosphere and the distant magnetosphere

LEP-S1 instrument (3-D energy/charge analyzer). Energy range = 10 eV - 16 keV for electron measurement and 13 eV/Q -20 keV/Q for ion measurement; scanning rate = 32 steps/second; energy resolution $\Delta E/E$ = 12%; FOV = $8°$ x $10°$ centered at $180°$, $150°$, $120°$, $90°$, $60°$ with respect to the solar direction.

LEP-S2 (3-D energy/charge analyzer). Identical to the LEP-S1. The view directions of LEP-S1 and LEP-S2 are symmetrical with respect to the spin axis of the satellite.

LEP-M (3-D energetic ion-mass spectrometer). Energy range = 1 - 25 keV/Q. Type of measurement = $135°$ spherical electrostatic analyzer, $40°$ magnetic analyzer and MCP; simultaneous measurements of mass and pitch angle distributions, differential energy spectra (32 steps) for ion species $H^+$, $He^{++}$, $He^+$, $O^{++}$, and O. Scanning rate = 16 steps/second for energy scanning; FOV = $3°$ x $34°$ with the longer dimension parallel to the spin axis of the satellite.

- **SMS** = Suprathermal Ion Mass Spectrometer (Pi: B. Wahlen, NCR, Canada) Objectives: measurement of the distribution functions of the major as well as the minor ion constituents of the magnetospheric ion population.

SMS is a radiofrequency type ion mass spectrometer to measure the mass, energy and angular distributions of positively charged ions species. The instrument uses a time-of-flight technique to measure the ion velocities and electrostatic deflection to define the ion energy per unit charge, thereby determining the ion mass per unit charge. Instrument characteristics:

- energy/unit charge (E/Q) range: -0 < E/Q < 4 kV
- mass/unit charge (m/Q) range: - 0.8 < m/Q < 60 AMU/Q
- energy resolution ($\Delta E$): - 0.05 < $\Delta E$ < 0.2 kV (programmable)
- mass resolution ($\Delta m/m$): - 0.05 < $\Delta m/m$ < 0.2 (programmable)
- angular resolution in plane to S/C spin axis ($\Delta\theta$): $-3°$ < $\Delta\theta$ < $90°$ (FWHM, programmable and dependent on energy and geometric factor)
- angular resolution in plane to the scan ($\Delta\theta$): $-3°$ < $\Delta\theta$ < $90°$ (FWHM, programmable and only available for low energy E/Q < 0.1 kV ions)

- **TED** = Thermal Electron Detectors (Pi: K. Oyama, ISAS). Objectives: measurement of the behavior of the background plasma electrons associated with auroral phenomena. In particular:

- Detection of the field-aligned current carriers responsible for the downward current in the auroral region.
- Observation of the behavior of thermal electrons in the trough region and their relationship to double layers or the V-shaped electric fields.
- Observation of the velocity distribution function anisotropies of thermal electrons in the polar region.
- Study of the heating mechanism of thermal electrons due to the wave-particle interaction.
- Observation of the behavior of thermal electrons at low altitudes in the South Atlantic geomagnetic anomaly.
- Observation of the equatorial "plasma bubble" and study of the non-thermal properties associated with this phenomenon.

TED-1 instrument characteristics (electron temperature $T_e$ and floating point potential $V_f$): energy range: $T_e$ = 0 - 1.0 eV, -5V < $V_f$ < 5V; type of measurement: two planar probes with a shape of two semi-circular disks, each mounted on the tip of a solar paddle. The surfaces of the two probes are arranged to be perpendicular to each other. Superposition of AC signal of 1 kHz.

TED-2 instrument characteristics (velocity distribution and electron density): energy range: 0 - 2.5 eV, 0 - 5 eV; density range: $10^2$ - $10^6$ cm$^{-3}$; type of measurement: two planar probes of circular shape with their surfaces mutually perpendicular.

- **ATV** = Visible and UV Auroral Television (Pi: T. Oguti, University of Tokyo)
  Objectives: imaging of the global aurora through two spectral windows in the VIS and UV ranges with high sensitive devices which are essential to the observation of faint polar cusp auroras.

  ATV produces auroral images in the VIS (557.7 nm) and UV (115-160 nm) spectral ranges using a despun mirror system. ATV uses a CCD imaging system of 488 (vertical) x 376 (horizontal) pixels with a FOV of 30° (vertical) x 40° (horizontal) for VIS, and 36° (vertical) x 36° (horizontal) for UV, respectively. The integration time of the CCDs may be set to 100 ms, 200 ms, 400 ms, or 600 ms.

Data:
Onboard storage capability of 67 Mbit (bubble recorder). Data transmission in S-Band PCM at 65, 16, and 4 kbit/s; and UHF analog data transmission at 10 kHz and at 5 kHz. The EXOS-D (Akebono) science data will be placed in the public domain some years after acquisition.


## A.30    FAST (Fast Auroral Snapshot Explorer)

NASA/GSFC mission within the 'Small Explorer Program' with the objective to measure and study the rapidly varying electric and magnetic fields and the flow of electrons and ions in the aurora regions of the Earth, study of the physical causes of complex auroral displays, investigations of how electrical and magnetic fields accelerate electrons, protons, and other ions in the auroral regions. The S/C instrument data will be correlated with observations by other S/C as well as by geomagnetic ground stations from Earth. The S/C spins in a cartwheel mode with a period of about 5 seconds, has a mass of 162 kg, length = 0.86 m, diameter = 1.17 m, design life = 1 year, payload of 4 instruments. The projected launch date is August 1994 (placing FAST into the ISTP program) with an enhanced Pegasus vehicle from Vandenberg Ca.. Mission duration = 1 year. PI for mission: C. W. Carlson of UCB (University of California at Berkeley). [115],[116]

Orbit: Non-Sun-synchronous orbit, perigee = 350 km, apogee = 4200 km (apogee maximum at northern latitude), inclination = 83°, period = 133 minutes

Sensors:

- **ESA** = Quadrispherical Electrostatic Electron Analyzer (UCB). The objective is to measure electrons and ions with a high time resolution. ESA consists of an Ion Spectrometer and an Electron Spectrometer to make detailed distribution function measurements; in addition there is an Electron Stepped ESA to make very high time resolution electron measurements with lower energy and angle resolution. The Electron Stepped ESA has three different energy sampling modes.

- **TEAMS** = Toroidal Energy Angle Mass Spectrograph (Uni. of New Hampshire and Lockheed Palo Alto Research Lab). The objective is to measure the full 3-D distribution of the major ion species with 1/2 spin period. TEAMS typically looks for H$^+$,

115)  D. Baker, G. Chin, R. Pfaff, "NASA's Small Explorer Program", Physics Today, Dec. 1991, pp. 44-51
116)  C. W. Carlson, "The Fast Auroral Snapshot Explorer", EOS, Vol. 73, Nr. 23, 1992, pp. 249, 253, 254

$He^{++}$, $He^+$, $O^+$, and a combination of $O^{++}$ and $NO^+$ over an energy range of a few eV to 5 keV. TEAMS combines the selection of ion by energy/charge by electrostatic deflection and subsequent time-of-flight analysis.

- **EFLPI** = Electric Field / Langmuir Probe Instrument (UCB). The objective is to measure the vector electric field from DC to 2 MHz, the thermal plasma density and temperature, density fluctuations, propagation velocity of structures, and wavelength measurements using 3 orthogonal boom pairs, two of which have spherical probes at two locations. EFLPI includes a wave/particle correlator.

- **MFI** = Magnetic Fields Instrument (UCLA). The objective is to measure vector DC and AC magnetic fields using a 3-axis fluxgate magnetometer and a 3-axis search coil magnetometer.

Data: Onboard storage capability of 1 Gbit solid-state memory. The instruments acquire data at a max. rate of 8 Mbit/s (snapshot data collection, at max. rate about 100 s of data can be acquired per auroral pass). Onboard data preprocessing capability. Continuous synoptic data are collected at a slower rate to put the "snapshots" into proper context. Real-time data will be acquired at stations in the auroral zone, memory dumps can be acquired at other NASA stations. Prime focus is the Northern Apogee Campaign (January - March 1995) when simultaneous ground measurements will be conducted.

Note: in the 'sample array' column of Table 19, the designation '4Mx48Ex64Ω' refers to: 4M (mass) x 48E (energy) x 64 Ω (solid state array);

| Measurement (ESA) | Coverage | Energy Range (keV) | ΔE/E | Sampling Resolution | FOV | Angular Resolution | Sample Array |
|---|---|---|---|---|---|---|---|
| Ion Mass Spectrometer | 3-D | 0-12 | 0.13 | 2.5 s (3D) 78 or 156 ms (2D) | 360° x 11° | 10x22.5° 3D 5.6 or 11.2° | 4Mx48Ex64Ω (3D) 4Mx48Ex16α (2D) |
| Ion Spectrometer | 2-D | 0-24 | 0.20 | 78 ms | 360° x 12° | 11.2x12° | 48Ex32α |
| Electron Spectrometer | 2-D | 0-30 | 0.15 | 78 ms | 360° x 10° | 11.2x10° | 48Ex32α |
| Electron Stepped ESA | 2-D | 0-30 | 0.15 | 1.6 ms | 360° x 10° | 22.5x10° | 6Ex16α |

**Table 19:    FAST Particle Detectors**

| Measurement | Components | Frequency Range | Sampling Resolution | Range | Resolution | Data Rate (max.) kbit/s |
|---|---|---|---|---|---|---|
| DC E-Field | 3-axis | 0-300 Hz | 30 μs | ± 1.6 V/m | 0.05 mV/m 16 bit | 5000 |
| Wave E-Field | 3-axis | 0.3-20 kHz | 30 μs | ± 200 mV/m | 0.05 mV/m 16 bit | 5000 |
| Swept Freq. spectrum analyzer | 3-axis | 0-2 MHz | 32 μs | 70 dB | 8 bit | 246 |
| E-Field Rectifier Filters | 3-axis | 0.1-2 MHz | 30 μs | 0.1-1 mV/m | 8 bit | 960 |
| AC E-Field burst memory | 2 - 4 | 0 - 1000 kHz | 0.5 μs | selectable impulse | 10 bit | 2000 |
| Density | 3-axis | 0.2 - 1 MHz | 0.5 ms | $1\text{-}10^5$ cm$^{-3}$ | 8 bit | 8 |
| Temperature | - | 0 - 2 kHz | 1 s | 0.1 eV-1 keV | 8 bit | 0.08 |
| MFI Fluxgate Magnetometer | 3-axis | 0-50 Hz | 2 ms | $10^{-5}$ - 0.6 G | 16 bit | 48 |
| MFI Search Coil Magnetometer | 3-axis | 10 Hz - 2.5 kHz | 0.1 ms | - | 16 bit | 240 |

**Table 20:    FAST Field Instruments[117]**

---

117)  Information provided by C. Cattell of UCB

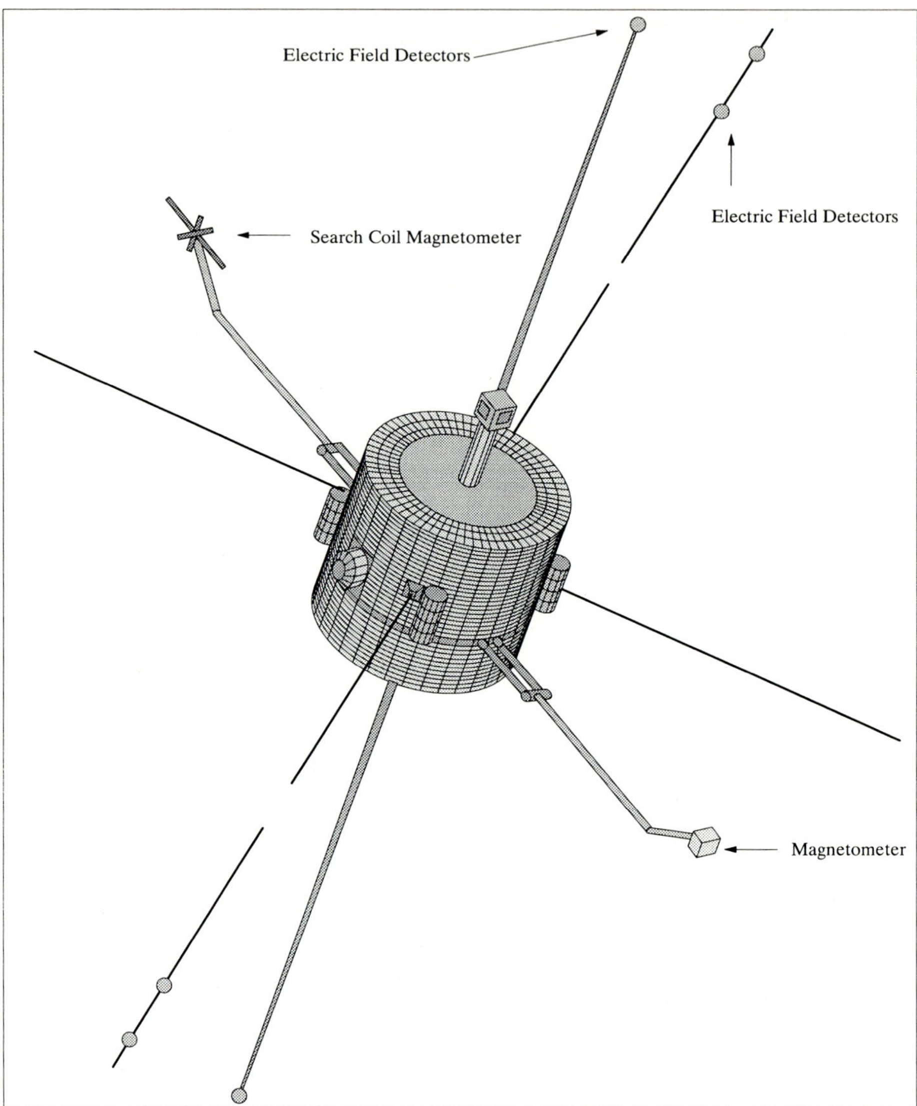


**Figure 34:      The FAST S/C Model**

## A.31     Fengyun-1

Fengyun-1 or FY-1 (Fengyun = wind and cloud) is a meteorological satellite series of PRC (Peoples Republic of China) organized by SMA (State Meteorological Administration and SOA (State Oceanic Administration). Launch vehicle: Long March 4. Launch of FY-1A: Sept. 7. 1988; Launch of FY-1B:  Sept. 3, 1990.
The satellite is a hexahedron of 1.4 x 1.4 x 1.2 m; mass = 750 kg. There are three solar cell arrays mounted on the two sides of the satellite respectively. The attitude of the satellite is three-axis stabilized.  Note: FY-1A suffered serious malfunctions in space.

Orbit: Sun-synchronous polar orbit with an altitude of 900 km; orbital period = 102.86 minutes; inclination = 99.1°, eccentricity < 0.005; equatorial crossing at 7:45 AM.

**Sensors:**

- **VHRSR**[118] = Very High Resolution Scanning Radiometer (VIS and IR). Two identical instruments onboard. A 5-channel instrument: IFOV of 1.2 mrad and 1.1 km spatial resolution; FOV=55.4°; swath = 3235 km; the following spectral bands are defined:

| Channel | Wavelength ($\mu$m) | Objectives |
|---|---|---|
| 1 | 0.58 - 0.68 | daytime cloud and surface image |
| 2 | 0.725 - 1.10 | daytime cloud image and water, ice, snow, and vegetation observation |
| 3 | 0.48 - 0.53 | ocean color image |
| 4 | 0.53 - 0.58 | ocean color image |
| 5 | 10.5 - 12.5 | diurnal cloud image surface observation and sea surface temperature |

The satellite provides data in three modes: HRPT (High Resolution Picture Transmission), APT (Automatic Picture Transmission), and DPT (Delayed Picture Transmission). The data formats of HRPT and APT are very similar to those of the NOAA series satellites (satellite attitude is included). The Satellite Meteorological Center of PRC broadcasts the FY-1 orbital prediction via GTS so that users throughout the world may receive HRPT and APT data in real time.[119]

| Carrier Frequency | Data Rate Subcarrier Frequency | Modulation Mode | Power |
|---|---|---|---|
| HRPT: 16.95.5, 1704.5 MHz | 0.6654 Mbit/s | PCM/PSK | 5 W |
| APT: 137.035, 137.795 MHz | 2.4 kHz | AM/FM | 8 W |

**Table 21:    HRPT/APT Transmission Channel Characteristics of FY-1**

## A.32    Fengyun-2

Fengyun-2 or FY-2 is a Geostationary meteorological satellite series of the Peoples Republic of China. Spin-stabilized S/C at 100 rpm. Planned Launch in 1994 with a Long March 3 vehicle[120].

Location: 105° Eastern Longitude.(spares: 122° E, 79°E)

Objectives:

- Acquiring daylight visible cloud map, day-and-night infrared cloud map and water vapor distribution map
- Data collection from meteorological, oceanic and hydrological observational platforms
- To broadcast a stretched digital cloud map, a low resolution cloud map and weather map information
- To obtain the cloud top and sea surface temperature as well as the wind field distribution by data processing.

Sensor: **Scanning Radiometer**


118) Q. B. Zheng, X. R. Xue, 'Optical Design of the Remote Sensing Instrument for FY-1 Meteorological Satellite', Chinese Journal of Infrared & Millimeter Waves, Volume 9, Number 2, 1989
119) 'The Data Format and the calibration Parameters of FY-1 Meteorological Satellite', Satellite Meteorology Center, SMA
120) 'China Launches first Fengyun II', Flight International, 11-17 Dec. 1991, p. 20

| Parameter | Visible | Infrared | Water vapor |
|---|---|---|---|
| Spectral band ($\mu$m) | 0.55 - 1.05 | 10.5 - 12.5 | 6.3 - 7.6 |
| No. of channels | 4 (+4 spare) | 1 (+1 spare) | 1 (+1 spare) |
| Nadir Resolution | 1.43 km<br>IFOV=40 $\mu$rad | 5.73 km<br>IFOV=160 $\mu$rad | 5.73 km<br>IFOV = 160 $\mu$rad |

**Table 22:**  **Spectral Characteristics of the Scanning Radiometer**

Scanning modes: a) normal scanning; b) optional scanning; c) single line scanning. A picture frame (normal scanning) = 30 minutes, a scan (North-South) = 2500 steps

Digital and cloud map transponder:
- Frequency: 1.7/2.0 GHz
- EIRP ($dB_w$) original cloud map and stretched cloud map: 57.5
- EIRP ($dB_w$) weather map broadcast: 46

Data Collection Transponder:
- Frequency:    401/468 MHz
- EIRP:     47.3 $dB_w$

No. of Channels:  domestic: 100   abroad: 33

Channel band width: 331 kHz    122 kHz

## A.33  FORTE (Fast On-Orbit Recording of Transient Events)

FORTE is a planned mission of Los Alamos National Laboratory (LANL) and Sandia National Laboratories (SNL) with an advanced radio frequency (RF) impulse detection and characterization experiment. The prime objective is the measurement of electromagnetic impulses (EMP), primarily due to lightning, within a noise environment dominated by continuous wave carriers, such as TV and FM stations. The goal is to develop an understanding of the correlation between the optical flash and the VHF emissions from lightning.[121]

FORTE will also conduct ionospheric physics experiments. The effects of large scale structures within the ionosphere will be studied [such as traveling ionospheric disturbances and horizontal gradients in the total electron content (TEC) on the propagation of broad bandwidth signals].

Orbit: altitude = 800 km, inclination = 65°

The S/C has a total mass of 180 kg and an average power of 50 Watt (body-mounted solar cells). The S/C is nadir-pointing with a biased momentum wheel stabilization (three-axis stabilized). The satellite will be operated from SNL in Albuquerque, New Mexico. A launch on Pegasus is planned at the end of 1995. The FORTE payload consists of the following instruments:

- **RF System.** The RF system provides the following features:
    - Three broad bandwidth RF receivers (20 and 100 MHz bandwidth) covering the frequency range from 30 - 300 MHz
    - High-speed, low power digitizers (300 Msamples/s)
    - Broad bandwidth, dual polarization-selective VHF antenna
    - Extensive on-board signal processing (adaptive discrimination, signal categorization by attributes)
- **Optical System.** The optical system consists of a coarse imager with a ground resolution of 10 x 10 km for lightning flash location (500 frames/s) and a fast photo-detector (50 k samples/s) for the recording of individual light curves.

---

121) Paper provided by LANL (D. C. Cobb)

- Segmented array to image lightning
- Simultaneous fast radiometer for enhanced time resolution
- The optical system augments the RF system to develop high performance lightning discriminator/locator capability.

- **Event Classifier**. The event classifier provides on-orbit characterization of impulsive RF events which satisfy the trigger criteria (extensive digital signal processing is involved).

## A.34    FREJA

**Swedish**/German small-satellite mission with international cooperation in the instrumentation (follow-up to Viking, see A.113). The objectives are: magnetospheric research, to make high-resolution measurements in the upper ionosphere and lower magnetosphere (auroral phenomena). The satellite was built by SSC (project management: Swedish Board of Space Activities) and weighs 256 kg (spinning disk with a diameter of 2.2 m and with a spin axis solar-orientated, spin rate = 10 rpm). Nominal lifetime = 2 years. Project scientists: R. Lundin of Swedish Institute of Space Physics, and G. Haerendel, MPE Garching.
Piggyback launch of Freja on Oct. 6. 1992 with a Long March (CZ-2C) vehicle from Jiuquan Spacecraft Launch Center, China. Operational mission as of 1/1994.

Orbit: apogee = 1790 km, perigee = 650 km, inclination = 65°

**Sensors**:[122],[123] The science payload includes six radial wire booms (1 < 15 m) and two stiff radial booms (1m and 2m).

- **F1**: Electric Field Experiment (PI: Göran Marklund, Royal Institute of Technology, Stockholm). The objective is to further explore (and estimate the relative roles of) various suggested mechanisms for particle acceleration either by quasi-dc parallel electric fields (e.g. double layers, anomalous resistivity and magnetic mirror effects) or by wave-resonance.
  Measurement of the electric field as potential difference between opposite probes. Two components can be measured since the probes are all in the spin axis of the satellite (the $3^{rd}$ components is deduced). The measurement parameters are as follows: maximum field = 1V/m; minimum field = 0.03 mV/m; accuracy $\approx$ 0.5 mV/m; sampling rate (normal mode) = 768 s$^{-1}$; sampling rate (burst mode) = 6144 s$^{-1}$. F1 measures in addition the satellite potential.
  The electric field is measured with the double probe technique. Six spherical probes are used, extended on three wire boom pairs in the spin plane.

- **F2**: Magnetic Field Experiment (PI: L. Zanetti, APL, Johns Hopkins University, Md, USA). Objectives: static dc structure measurements of large and small scale size currents with full orbit data and 50 m resolution; vector wave measurements up to Nyquist frequencies of 64 Hz and spin axis measurements up to 256 Hz in the normal 256 kbit/s telemetry mode.
  Instrument characteristics:

  - Triaxial ring core fluxgate sensor, 2m boom mounted
  - low noise (10 µV) analog, 16 bit A/D, S/C mounted
  - internal 1.3 Mbit RAM, event trigger, FFT S/W, oversampling
  - real-time data output: 14.3 kbit/s for 256 kbit/s TM rate, and 28.6 kbit/s for 512 kbit/s TM rate
  - DC - B: 128 vector samples/s (vs/s), range ± 65,000 nT, ± 1 nT
  - AC - B: 128 vs/s, bandpassed 1.5-128 Hz, range ± 500 nT


122)  M. André (editor) and the Freja Science Team, 'The Freja Scientific Payload', Swedish Institute of Space Physics, Kiruna, May 1991
123)  'The Freja Scientific Satellite', brochure of Swedish Space Corporation

- Spin axis: 1.5 - 256 Hz FFT, 2 Hz resolution
- instrument mass: 3.5 kg (excluding boom and mount)
- instrument power: 3.7 W (including dc/dc)

- **F3C**: Particles Experiment - Cold Plasma Analyzer (PI: B. A. Wahlen, NRC, Canada). Objectives: quantitative measurements of the cold ionospheric plasma distribution; small scale auroral plasma structure investigation; identification of the plasma wave mode; investigation of low-altitude energization mechanisms; direct observation of the clod plasma drift velocity; coordinated ground-based observations.
Instrument characteristics: boom-mounted sensor; technique: electro-static analyzer;

  - Species detected: negatively and positively charged particles (electrons and ions)
  - Energy range: $0 \leq E/Q \leq 300$ eV
  - Dynamic range: $10^2 \leq n \leq 10^5$ cm$^{-3}$
  - Energy resolution: $0.05 \leq \Delta E/E \leq 0.20$ (programmable)
  - Angular resolution: azimuth, $10^{\circ} \leq \Delta\phi \leq 90^{\circ}$ (programmable)
  - Angular resolution: elevation, $5^{\circ} \leq \Delta\alpha \leq 90^{\circ}$ (programmable)
  - Temporal resolution: $\sim 10$ ms (typical), maximum $\sim 0.1$ ms (on one selected parameter)
  - Instrument mass: 5.5 kg (including boom)
  - Instrument power: 8.4 W (normal)
  - Telemetry rates: 16.384 kbit/s (nominal; 32.768 kbit/s (maximum)

- **F3H**: Particles Experiment - Hot Plasma Analyzer (PI: L. Eliasson, Swedish Institute of Space Physics, Kiruna). Objectives: study of ion distributions of ionospheric origin (mainly $H^+$ and $O^+$) in the energy range from a few eV up to several keV traverse to the geomagnetic field; electron distributions observed on auroral field lines; study of auroral acceleration mechanisms;
The instrument consists of three units: MATE (an electron spectrometer), TICS (a 3-D ion composition spectrometer, and DPU (data processing unit).
**MATE** characteristics:
measurement of the angular and energy distributions of electrons with high temporal and spatial resolution; energy range: 0.1 keV - 120 keV; angular segments: 32; energy levels: 16; FOV/sensor head: $2^{\circ}$ x $10^{\circ}$; energy resolution: 15-30% FWHM; minimum sampling time: 10 ms/energy-angle matrix; maximum data rate: 51.2 kByte/s (no data compression); normal data rate: $\approx 20$ kbit/s; mass = 2.7 kg; power = 3.8 W.
**TICS** characteristics:
measurement of 'hot' magnetospheric and 'cold' ionospheric ion distributions, study of the heating/acceleration of ions perpendicular to the magnetic filed lines. TICS consists of a $90^{\circ}$ spherical 'top hat' electrostatic analyzer with $360^{\circ}$ FOV followed by a cylindrical sector magnet momentum analyzer. Energy range: 0.5-15000 eV/q; mass range: 1-40 AMu/q; angular segments: 32; energy levels: 32; energy resolution: 10% FWHM, FOV/sector head: $5^{\circ}$ x $10^{\circ}$; sampling time: 10ms/mass-angle matrix (32 x 32) FOV: $360^{\circ}$; geometric factor:$1_{-10}^4$ (cm$^2$ s sr keV/keV) per $11^{\circ}$ opening; time resolution: 0.5 spin period; maximum data rate: 102.4 kByte/s (no data compression); normal data rate: $\approx 20$ kbit/s; mass: $\approx 3.3$ kg; power: 4.5 W.

- **F4**: Waves Experiment (PI: B. Holback, Swedish Institute of Space Physics, Uppsala). Objectives: study of wave modes and energy, measurement of electric, magnetic, density and electron temperature wave fields and turbulence, background plasma density and electron temperature.
The F4 instrument utilizes a number of sensors mounted on booms in the satellite spin plane. Three pairs of spherical probes (P1-P4 and P8-P9) are mounted on three wire boom pairs (switchable for electric field and plasma density measurements). A Cylindrical Langmuir Probe (CYLP) is mounted at the outer end of a stiff boom. The High Frequency (HF) experiment uses the P1 and P2 signals or the dedicated HF probes

(HFa, HFb). The Search Coil Magnetometer (SCM) assembly consists of 3 search coils mounted at the end of one stiff boom. Measured quantities of F4 instrument:

- electric wave fields ($\Delta E$)
- magnetic wave fields ($\Delta B$)
- plasma density ($\Delta n/n$) and temperature ($\Delta T_e/T_e$) variations
- Langmuir (DC) current to provide the cold plasma characteristics ($N_e$ and $T_e$)

| Branch | Sensor Signal | Sampling rate k-samples/s | Word (bit) | Telemetry (Word/s) | Duty Cycle (%) |
|---|---|---|---|---|---|
| HF 1 channel | HF Probes $e_1 - e_2$ | 8000.000 | 8 | 1024 | 0.01 |
| MF 1 channel | $e_{i,k}$ $b_{x,y,z}$ | 32.768 | 16 | 512 | 1.6 |
| LF 4 channels | $e_{i,k}$; $b_{x,y,z}$; $\Delta n/n_{1,2,3,4}$ | 4.096 | 16 | 768 | 18.8 |
| n up to 4 probes | $n_{1,2,3,4}$ | 0.128 | 16 | 128 | 100 |

**Table 23:**     **Summary of Key Parameters for the F4 Wave Analyzer**

- **F5**: Auroral UV Imager (PI: J. S. Murphree, University of Calgary, Canada). Objectives:

  - Determination of the growth rate of distortions in discrete arcs
  - Determination of the local time extent of optical substorm onsets
  - Characteristics of discrete arcs from Maxwellian and non-Maxwellian source distributions (in conjunction with the particle observations)
  - Determine whether episodic expansion or continuous propagation is the dominant form of substorm spiral motion
  - Characterize substorm onset precursor activity.

  F5 is of Viking mission heritage (see V5 in chapter A.113). F5 is an inverted Cassegrain Burch type camera [mass = 10.8 kg; power = 8.4 W (high); TM rate = 88.064 kbit/s or 44.032 kbit/s; image storage capability: 952 kByte]. F5 Measurement characteristics:

  - Camera optics: inverse Cassegrain, speed F/1; focal length = 22.4 mm; FOV = $25^\circ$ x $20^\circ$; optical axes are $90^\circ$ from spin axis.
  - Spectral Passband: camera 0: 134 nm - 180 nm ($BaF_2$ + CsI); camera 1: 125 nm - 160 nm($CaF_2$ + KBr);
  - Detector: intensified CCD; image size (max) 385 x 4598 pixels; pixel size = 22 µm x 22 µm; full well potential: $3 \times 10^5$ e⁻
  - Resolution: angular= $0.0783^\circ$ x $0.0783^\circ$; spatial (from apogee with 2 x 2 pixels) ~ 5 km
  - Exposure: 0.37 s (for 6 second spin period)

- **F6**: Electron Beam Experiment (PI: G. Paschmann, MPE, Garching). Objectives: measurement of electric fields associated with the auroral acceleration region (method based on sensing the drift of weak electron beams). F6 is of GEOS heritage, see also EDI instrument on Cluster missions (A.15). F6 has 3 electron guns (3 components of E-field, up to 700 samples/s) consisting of the following components:

  - an electron gun with a magnetic deflection system
  - a position-sensing detector
  - the analog detection electronics
  - the gun and deflector supplies
  - interfaces with the electronics box

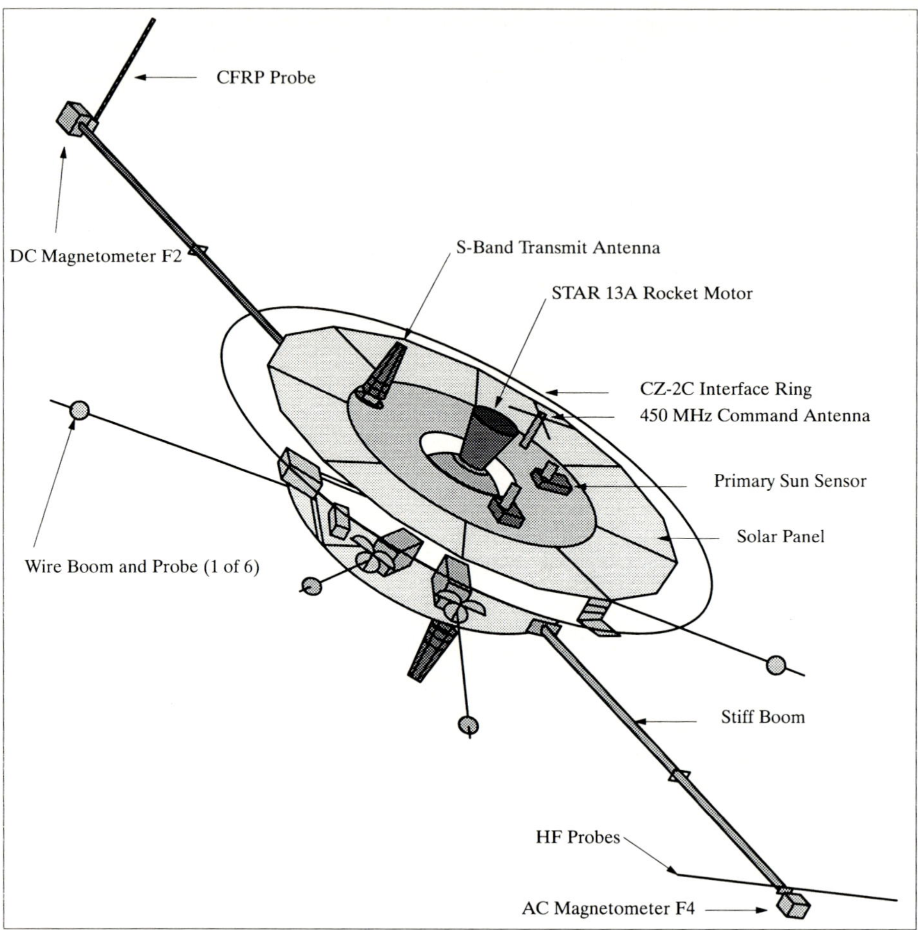


**Figure 35:**      **The Freja S/C Model**

- **F7**: Particle Correlator Experiment (PI: M. H. Boehm, MPE, Garching). Objectives: measurement of electron distributions, electron-electric field correlations, particle-particle correlations.

  The F7 TESP/correlator instrument consists of a swept-energy electrostatic analyzer to measure the full electron distribution function at 64 ms time resolution, and wave/particle and particle/particle correlations (correlation with F4) measuring variations in the distribution function at frequencies up to several MHz. TESP measurement parameters:

  - Angular channels: 32, evenly spaced
  - Energy range: $\sim 10$ eV - $\sim 30$ keV
  - Typical number of energy steps: 32 or 64
  - Typical sampling time per step: 1 ms
  - Sweep retrace time: < 1 ms
  - Maximum entrance aperture deflection angle: $\sim 30°$, partially blocked by the satellite
  - Total geometric factor (numerically calculated): 0.025 cm$^2$ ster keV/keV

- Approximate energy width (retarding grid off): 15% FWHM
- Approximate energy width (retarding grid on): $\sim 7\%$ FWHM at $\sim 0.006$ cm$^2$ ster keV/keV.

Freja Data:
Onboard data storage capability. S-Band downlink with data stream phase-modulated onto the carrier, data rates (direct transmission) = 262 kbit/s or 524 kbit/s, >2 Mbit/s (burst memory mode). Esrange ground station at Kiruna for S/C control and science data reception, Prince Albert station in Saskatchewan also for data reception.

Temporal resolution of Freja data:

- Fields: $\leq 0.1$ ms
- Particles: $\approx 10$ ms
- Images: $\approx 6$ s

Spatial resolution of Freja data:

- Fields: $\leq 1$ m
- Particles: $\approx 100$ m
- Images: $\approx 2$ km

## A.35  GEMINI

NASA Program with a total of 10 missions between 23. March 1965 and 15. Nov. 1966.[124]

Orbit: Altitudes between 160 km and 320 km, inclination = 28 to 29°, period between 88 and 95 minutes. Note: Gemini-11 at altitude of 640 to 1360 km.

Sensors:

| | | | | |
|---|---|---|---|---|
| • Camera | Hasselblad 500 C | Focal length = 80 mm | Mission | 3 - 8 |
| Camera | Hasselblad 500 C | Focal length = 250 mm | | 7 |
| Camera | Hasselblad SWA | Focal length = 38 mm | | 9 - 12 |
| Maurer Space Camera | | Focal length = 80 mm | | 9 - 12 |
| Zeiss Contarex | | Focal length = 35 mm | | 4 |
| Zeiss Contarex | | Focal length = 250 mm | | 5 |

## A.36  GEO-IK

GEO-IK[125] = Space Geodetic Complex. A Soviet S/C for Solid Earth Research, launched May 30, 1988. Builder and operator of GEO-IK S/C = NPO PM, Krasnojarsk. The mission is planned for an operational life of 5 years. Two more GEO-IK satellites were launched in 19989 and 1990 respectively.

Orbit: Near-circular orbit, altitude = 1500 km, inclination = 74°, period = 116 minutes

The satellite is equipped with a Doppler transmitter, a flashlight system and laser corner reflectors. The S/C is oriented with the long axis toward the Earth's mass center, stabilization is realized with binormal towards the orbital axis.
The frequency range of the Doppler transmitter is 150 and 400 MHz

Objectives: geodetic measurements (geodetic connection of islands, development and control of local geodetic networks, improvement of the general ellipsoid parameters, connection of the initial geodetic dates, etc.). The scientific program is carried out by the Astronomical Council of the USSR Academy of Sciences.

---

124) Systeme und Sensoren' p. 45, Taschenbuch zur Fernerkundung, Wichmann, 1990
125) S.K. Tatevian, "The Space Geodetic Complex GEO-IK", CSTG Bulletin No. 11, Title: New Satellite Missions for Solid Earth Studies, 1989, pp. 9-11

The geodetic complex GEO-IK consists of the satellite itself, the ground-based tracking sites, and the S/C control center.


## A.37    GEOS (GEOstationary Satellite)

The GEOS missions[126] (1 and 2) were intended to be reference S/C for IMS (International Magnetosphere Study, 1976-79).


### A.37.1   GEOS-1

GEOS-1 mission (ESA). Launch April 20 1977 by Delta vehicle from Cape Canaveral (a booster failure meant that geostationary orbit could not be achieved). Orbit: 2050 km perigee, 38000 km apogee. The S/C performed measurements in this orbit for 14 months. S/C mass = 574 kg (S/C integration by British Aircraft Corporation). Seven experiments aboard.
Application: Detailed study of the magnetosphere's radial distribution of plasma, particles and waves (in conjunction with ground-based observations in Scandinavia, Iceland, Antarctica, and Alaska). Magnetospheric transport phenomena.

Sensors:[127] (simultaneous waves and particle experiments along with electric and magnetic fields and total plasma density)

- **S 300** = Search Coil Magnetometer (CRPE, France; ESTEC; Danish Space Research Institute). Objectives: study of magnetospheric wave phenomena in both electric and magnetic domains; measurement of AC-magnetic fields up to 30 kHz; DC/AC electric fields and plasma resonances up to 80 kHz; mutual and self-impedance

- **S 302** = 2 Electrostatic Analyzers (Mullard Space Science Lab, UK). Objectives: measurement of thermal plasma (electron, protons) up to 500 eV.

- **S 303** = Combined Electrostatic and Magnetic Analyzer (Univ. of Bern; MPI, Garching). Objectives: measurement of the composition (1-140 amu) and energy spectra of ions up to 16 keV.

- **S 310** = 10 Electrostatic Analyzers (Kiruna Geophysical Observatory, Sweden). Objectives: measurement of the pitch-angle distribution of electrons and protons in the 0.2-20 keV energy range.

- **S 321** = Magnetic Deflection System followed by Solid-state Detectors (MPI, Lindau). Objectives: measurement of the pitch-angle distribution for electrons (20-300 keV) and protons (20 keV - 3 MeV).

- **S 329** = Tracing of Electron Beam over one or more Gyrations (MPI Garching). Objectives: measurement of DC electric field grad $|\mathbf{B}|$.

- **S 331** = Fluxgate Magnetometer (CNR, Frascati, Italy). Objectives: measurement of the three components of the DC and ULF magnetic field.

Data: Transmission rate of 110 kbit/s containing high-speed data (wide-band analog and correlator data from the wave experiment), and low-speed data, consisting of low-frequency data from the wave experiment and the data from all other experiments.

---

126)  JANE's Spaceflight Directory, 1988-89, pp. 332-333
127)  GEOS - Projects under Development, ESA Report to COSPAR, Jan. 1977, pp. 112-123

## A.37.2   GEOS-2

The GEOS-2 satellite (ESA) carried the same payload (sensor complement) as GEOS-1 and is regarded a replacement for GEOS-1 (GOES-2 was originally considered as backup). GEOS-2 was launched July 14 1978 from Cape Canaveral. Orbit: 25,640 km perigee, 35,592 km apogee at 0.77° inclination, positioned at 37° East.
Objective: Measurement of fluctuations in the magnetic field. Fluctuations in the magnetic field are picked up by small antennae (about 25 cm). GEOS-2 had six antennae to measure the three components of the field in two frequency bands.

GEOS-2 provided two years of data, was then placed in hibernation for 8 months, revived for 8 months in 1981 to support the EISCAT program of upper atmosphere motion measurements. GEOS-2 remained in use until the end of 1983. Periodic monitoring support (1984) of the chemical releases of the AMPTE mission.[128]

## A.38   GEOS-3 (Geodynamics Experimental Ocean Satellite)

GEOS-3 is a NASA mission[129]. Launch: April 9 1975 with a Thor-Delta launch vehicle from Vandenberg. Operation of satellite for over 3 1/2 years.

Orbit: altitude = 843 km, inclination = 115°;

Application: Determination of oceanographic and geophysical parameters, satellite altimetry. Estimation of significant wave height. Geometric, gravimetric and other geodetic investigations. First estimations of surface wind speeds with altimeter data. Provision of the first comprehensive data set in most areas of the world's oceans.

**Sensors:**

* **Radar Altimeter**. Multimode radar system with two distinct data-gathering modes (global and intensive). First radar altimeter to provide surface height measurements (50 cm range precision in the global mode and 20 cm in the intensive mode. One measurement per second. Sensor frequency = 13.9 GHz,

* **Laser System,** consisting of the spaceborne laser retroreflector subsystem and the ground-based laser-ranging systems (provision of precision satellite-ranging data). The laser reflector consists of 264 quartz cube corner reflectors mounted on a 45° conic frustum.

* **Doppler System**, consisting of 2 spaceborne transmitters and ground Doppler receiving stations. Dual frequencies (162 and 324 MHz) are coherently related. Data rate = 15.6 kbit/s or 1.56 kbit/s.

## A.39   GEOSAT (Geodetic/Geophysical Satellite)

US Navy Altimeter Mission[130],[131]. Mission control, data handling and processing, archiving and distribution at APL, Navy and NOAA. GEOSAT consisted of two missions: The first was the classified Geodetic Mission (GM); launch: March 1985; mission duration = 18 months (until Sept. 1986). The second mission is known as the 'Exact Repeat Mission' (ERM), which was unclassified, it started Oct. 1, 1986 and ended in January 1990.

Objectives: Provision of a dense global grid of altimeter data for Navy use in the areas of geodesy (Earth's gravitational models), the study of fronts and eddies, winds, waves and ice topography, physical oceanography in the 'Exact Repeat Mission'.

128)  "GEOS", interavia Space Directory 1992-93, pp. 155-156
129)  H.R. Stanley, 'The GEOS 3 Project', Journal of Geophysical Research, July 30, 1979, pp. 3779-3783
130)  "The Navy GEOSAT Mission: An Overview", Johns Hopkins APL Technical Digest, Volume 8, No. 2, 1987
131)  'The Navy GEOSAT Mission Radar Altimeter Satellite Program', in Monitoring Earth's Ocean, Land, and Atmosphere from Space, Volume 97, 1985 AIAA, pp. 440-463

Orbit:
1. GM mission: Sun-synchronous polar orbit, inclination=108°. Altitude = 800 km. Orbit of non-repeating ground tracks in order to obtain a densely sampled map of the marine geoid.
2. ERM mission: Starting Oct. 1, 1986 the orbit was changed to an exact repeat cycle of 17 days for the observation of geodetic parameters of the oceans.

**Sensor:**

**Radar Altimeter.** Frequency = 13.5 GHz; range precision = 5 cm. Range measurement between satellite and subsatellite point (at nadir) of the orbit with high measurement precision. Orbit determination with 2-frequency Doppler Tracking System (TRANET).

**Data:**

Onboard recording up to 12 hours. S-Band Downlink to APL Tracking Facility.
Data products: Raw data were processed into Geophysical Data Records (GDRs) by APL and NOAA. The GDRs include height data derived from average echoes at a rate of 10 Hz, and mean height values at 1 Hz (every 6.7 km on the surface).
Although the primary objective of the Geosat mission was to operate over oceans, echoes were also collected over ice and land surfaces.[132)

NOAA/NODC in Washington DC is providing Geosat data of the ERM period on CD-ROM (Geosat altimeter crossover differences) as of Dec. 1992.


## A.40    GFO (Geosat Follow-On Program)

The objective of the GFO program is to provide operational altimetry data for the US Navy. The launch of the first GFO satellite is planned for the end of 1995. There will be subsequent launches to provide a longterm altimetric data source. Expected lifetime = 5 years for each satellite.

Areas of application: The altimetry data from the GFO program will be used to obtain ocean topography measurements which can be used to derive the location of fronts, eddies, and current data. This information is pivotal for the development of global structure models.

Orbit: Sun-synchronous polar orbit, inclination=108°. Altitude = 800 km. Orbit of non-repeating ground tracks in order to obtain a densely sampled map of the marine geoid.

**Sensors:**

* Radar Altimeter. Frequency = ; range precision = . Range measurement between satellite and subsatellite point (at nadir) of the orbit with high measurement precision.

* Radiometer. Measurement of the water vapor content along the altimeter pulse path. A two-channel instrument that provides water vapor time corrections for the altimeter.

* GPS = Global Precision System. A GPS receiver for direct orbit measurement.

**Data:**
All data links to/from GFO are encrypted. The science data will be made available to NOAA who in turn are able to provide unclassified data to the general user community.

---

132)  D. R. Mantripp, J. K. Ridley, C. G. Rapley, "Antarctic map from the Geosat Radar Altimeter Geodetic Mission", ESA earth observation quarterly, No. 37-38, May-June 1992, pp. 6-10

## A.41   GEOTAIL

Japanese (ISAS)/NASA collaborative mission within ISTP (S/C built by ISAS, launch by NASA/GSFC). Geotail inaugurates the Collaborative Solar-Terrestrial Research Program (COSTR). COSTR defines the NASA contributions to the Geotail, SOHO, and Cluster missions. Geotail objectives: study of the structure and dynamics of the geomagnetic tail. In particular:

- Determine the overall plasma, electric and magnetic field characteristics of the distant and geomagnetic tail.
- Determine the role of the distant and near-Earth tail in substorm phenomena and in the overall magnetospheric energy balance and relate these phenomena to external triggering mechanisms.
- Study the processes that initiate magnetic field reconnection in the near-Earth tail and observe the microscopic nature of the energy conversion mechanism in the reconnection region.
- Study plasma entry, energization, and transport processes in interaction regions such as the inner edge of the plasma sheet, the magnetopause and the bow shock, and investigate associated boundary layer regions.[133]

Geotail launch on July 24, 1992 from Cape Canaveral with a Delta 2 vehicle.[134] S/C mass = 970 kg (330 kg propellant, 105 kg science payload). The S/C is a 20 rpm spin-stabilized cylinder of 2.2 m   diameter and 1.6 m in height.

Orbit: Geotail uses the gravity of the moon to assist its orbit on the night side of the Earth, where the magnetotail is stretched out as a result of the impact of the solar wind encountering the Earth. Geotail's orbit extends from 220 $R_E$ (1,401,620 km) at its farthest point to 8 $R_E$ (50,960 km) at its nearest point.

- Distant tail orbit: 1.75 years in distant tail configuration (double lunar swingby to an 8 x 220 $R_E$ orbit)
- Near tail orbit: 1.45 years in near tail configuration (reduced to an 8 x 30 $R_E$ orbit, 7.5° inclination).

Science background: The solar wind, emanating from the Sun, injects plasma into the magnetosphere and transfers energy to it. Several times a day the magnetosphere undergoes a disturbance called a substorm. As the substorm grows, most of the solar energy is dissipated within the magnetosphere, ionosphere and upper atmosphere. This disturbance ultimately causes auroral displays, the acceleration of charged particles to high energies, the emission of intense plasma waves, and the generation of strong ionospheric currents that produce significant changes in the upper atmosphere. These waves and currents often result in severe problems on Earth with regard to communications, power supplies, and spacecraft electronics.

**Sensors:**

- **EFD** = Electric Field Detector (PI: K. Tsuruda, ISAS). Objectives: study of the coupling of the E-Filed in the near-Earth magnetosphere and in the ionosphere (in particular during substorms).
  EFD uses electric-field antennas sampling at 64 samples per second, and an electron beam technique at 2 samples per spin.

- **MGF** = Magnetic Fields Measurement (PI: S. Kokubun, Uni. of Tokyo, R. Lepping, GSFC, instrument sponsored by ISAS). Objectives: study of the transport dynamics of mass, momentum, and energy between the magnetospheric and ionospheric plasma. Study of the merging in the magnetotail.
  Instrument: MGF contains also the Geotail Inboard Magnetometer provided by the US.

133)  'The Geotail Mission', in NASAFacts, GSFC, June, 1992
134)  'Delta Launches Geotail', Space News, July 27-Aug. 9 1992, p. 12

- **HEP** = High Energy Particles Experiment (T. Doke, Waseda University, Tokyo, instrument sponsored by ISAS). Objectives: measurement of high energy particles up to 25 MeV for electrons, 35 MeV for protons, and 210 MeV/charge for ions. Measurements may indicate the plasma boundary surfaces and reflect whether magnetic field lines are open or closed.

- **LEP** = Low Energy Particles Experiment (PI: T. Mukai, ISAS). Objectives: study of the dynamics of the magnetotail plasmas, plasma circulation and its variability in response to fluctuations in the solar wind and in the interplanetary magnetic field. Measurement of electrons from 6 eV to 36 keV, and ions from 7 eV to 42 keV/ charge.

- **PWI** = Plasma Waves Investigation (PI: H. Matsumoto, Kyoto University, instrument sponsored by ISAS). Objectives: study of the wave phenomena related to plasma dynamics in the different regions on various scales (phenomena include magnetic-field-line merging, moving plasmoids, and particle acceleration). Measurement of plasma waves in the frequency range of 5 Hz - 800 kHz. PWI contains also the Multi-Channel Analyzer provided by the US.

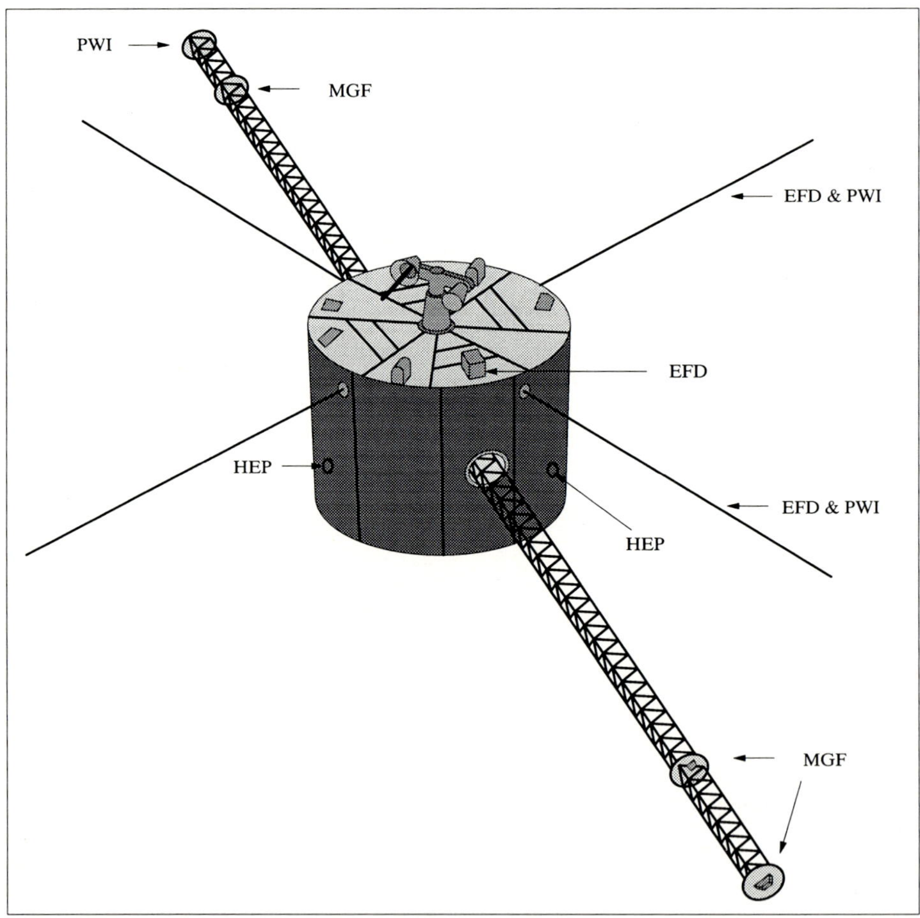


**Figure 36:**    **The Geotail S/C Model**

- **EPIC** = Energetic Particle and Ion Composition Experiment (PI: D. J. Williams, APL, John Hopkins University, instrument sponsored by NASA). Objectives: measurement of the charge, mass, and energy of ions. Study of the relative importance of ion sources and mechanisms for acceleration, transport and loss of particles, the formation and dynamics of magnetospheric boundary layers.
  Instrument: ion composition spectrometer and a telescope. Measurement of ions from 50 keV to 3 MeV and from 10 keV to 230 keV.

- **CPI** = Comprehensive Plasma Investigation (PI: L. A. Frank, Uni. of Iowa, instrument sponsored by NASA). Objectives: measurement of the 3-D plasma in the Earth's magnetotail. The plasma data will be correlated with the magnetic field, plasma waves, electric particles, and auroral imaging data to determine the magnetotail plasma dynamics.
  Instrument: measurement range of 1 eV - 50 keV for the Hot Plasma and Ion Composition Analyzer, and 150 eV - 7 keV energy /unit charge for the Solar Wind Analyzer. Plasma parameters, including heat flux and field-aligned current density, are measured.

Data: Two onboard recorders at 450 Mbit each. Real-time/playback transmission rates at 16.384 kbit/s, 65.536 kbit/s, or 131.072 kbit/s. Ground data reception at Usuda and Kagoshima stations (Japan) and NASA DSN.

## A.42   GLONASS

GLONASS [135] = Global Orbiting and Navigation Satellite System. GLONASS is an operational satellite navigation system of Russia in an experimental phase. The system consists of several satellites capable of providing high-accuracy position information on a global scale for users on sea, land, or in space.

The first GLONASS satellite series was launched Oct. 12, 1982. Further launches were in 1983 (2), 1984 (2), 1985 (2), 1986 (1), 1987 (2) and 1988 (1). Each successful launch has always placed a total of 3 satellites into orbit. As of September 1993 there were 14 operational GLONASS satellites. The GLONASS constellation will eventually be completed with a total of 27 satellites with three spares.

| Int. Satellite ID | Cosmos Nr. | Glonass Nr. | Nr. of Channels | Orbital Plane |
| --- | --- | --- | --- | --- |
| 1989-1A | 1987 | 40 | 9 | 1 |
| 1990-45A | 2079 | 44 | 21 | 3 |
| 1990-45B | 2080 | 45 | 3 | 3 |
| 1990-45C | 2081 | 46 | 15 | 3 |
| 1990-110A | 2109 | 47 | 4 | 1 |
| 1990-110B | 2110 | 48 | 13 | 1 |
| 1990-110C | 2111 | 49 | 19 | 1 |
| 1991-25B | 2140 | 51 | 11 | 3 |
| 1992-5A | 2177 | 53 | 22 | 1 |
| 1992-5B | 2178 | 54 | 2 | 1 |
| 1992-5C | 2179 | 55 | 17 | 1 |
| 1992-47A | 2204 | 56 | 1 | 3 |
| 1992-47B | 2205 | 57 | 24 | 3 |
| 1992-47C | 2206 | 58 | 8 | 3 |

**Table 24:**    **Overview of Active Glonass Satellites (Status: July 1992)** [136]

Measurement principle: The user makes range and Doppler measurements to a number of navigation satellites within visibility. Position and velocity are part of the message content from the navigation satellite. Orbit: Nearly-circular orbit, altitude about 19100 km.

---

135) "Understanding Signals from GLONASS Navigation Satellites", International Journal of Satellite Communications', Vol. 7 11-12, 1989, pp. 11-22
136) A launch of 3 new Glonass satellites took place on Feb. 17, 1993, GPS World, March 1993, p. 24

GLONASS and NAVSTAR-GPS have a lot of communalities or similarities with respect to: orbits, frequencies, and message formats. GLONASS employs an Earth-centered, Earth-fixed Cartesian coordinate frame of reference different from that used by GPS: GLONASS uses SGS 85; GPS uses WGS84 (World Geodetic System).[137]

| Parameter | | GLONASS | NAVSTAR/GPS |
|---|---|---|---|
| Orbital Parameters | | | |
| Period (minutes) | | 675.73 | 717.94 ($\sim$ 12 hours) |
| Inclination (degree) | | 64.8 | 55.0 |
| Semi-major axis (km) | | 25.510 | 26.560 |
| Orbital plane separation (degree) | | 120 | 60 |
| Ground track repeat (orbits) | | 17 | 2 |
| Transmission Signal Parameters | | | |
| Signal separation technique | | FDMA | CDMA |
| L-Band Carrier Frequencies (MHz) | | L1= 1602.5625-1615.5 | L1 = 1575.42 |
| | | L2= 1246.4375-1256.5 | L2 = 1227.60 |
| PRS clock rate (MHz) | C/A code | 0.511 | 1.023 |
| (PRS=Pseudorandom Sequence) | P code | 5.11 | 10.23 |
| PRS length (chips) | C/A code | 511 | 1.023 |
| | P code | $5.11 \times 10^6$ | $6.187104 \times 10^{12}$ |
| Navigation Message | | | |
| Superframe duration (minutes) | | 2.5 | 12.5 |
| Superframe capacity (bits) | | 7,500 | 37,500 |
| Word duration (seconds) | | 2.0 | 0.6 |
| Word capacity (bits) | | 100 | 30 |
| Number of words per frame | | 15 | 50 |
| Satellite ephemeris specification | | Geocentric Cartesian coordinates & their derivatives | Kepler elements and perturbation factors |
| Time reference | | UTC (SU) | UTC (USNO) |
| Position reference | | SGS 85 | WGS 84 |

**Table 25:     Selected NAVSTAR/GLONASS Parameters[138]**

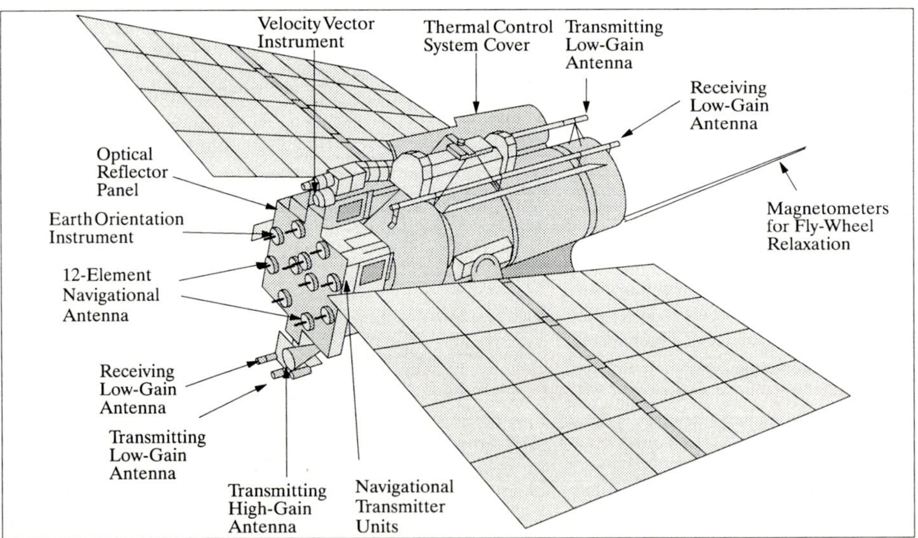


**Figure 37:     The GLONASS S/C Model[139]**


137)  P. N. Misra, E. T. Bayliss, R. R. LaFrey, M. M. Pratt, R. A. Hogaboom, R. Muchnik, "GLONASS Performance in 1992: A Review", GPS World, May 1993, pp. 28-38

138)  Y. Gouzhva, I. Koudryavtsev, V. Korniyenko, I. Pushkina, "GLONASS Receivers: An Outline", GPS World, January 1994, pp. 30-36

139)  Courtesy of A. Selivanov, ISDE and B. Zhukov, IKI, Moscow

## A.43    GPS (NAVSTAR-GPS)

NAVSTAR[140], [141],[142] = NAVigation System with Time And Ranging. **GPS = Global Positioning System**. Operator: DOD (US Department of Defense), in particular the US Airforce. GPS is a space-based radio positioning/navigation system that provides three-dimensional position, velocity, and time information to suitably equipped users anywhere on or near the surface of the Earth. The total system consists of 21 operational satellites in near-circular orbit in six orbital planes, so that observers anywhere on the surface of the earth have at least four satellites in view. In addition, there will be 3 backup satellites in orbit. Initially the orbital inclination was defined to be 63°, but this figure has been changed to 55° to allow shuttle launches. The orbital altitude is 20,200 km. The operational phase of GPS is scheduled to start in 1994.

GPS/NAVSTAR is based on direct one-way range measurements with synchronized time references. The satellites share a common time system known as 'GPS time' and transmit (broadcast) a precise time reference as a spread spectrum signal at two frequencies in L-Band: L1 = 1575.42 MHz, L2 = 1227.6 MHz. Two spread spectrum codes are used: a civil coarse acquisition (C/A) code and a military precise (P) code. L1 contains both a P and a C/A code, while L2 contains only the P code.

The accuracy of both codes is different. The receiver of the civil code cannot decode the military P code when the security status 'Selective Availability' in GPS satellites is turned on. (Note: With selective availability turned on, military users determine their location within 17.8 m, while civilian users determine their position within an accuracy of 100 m; hence, selective availability degrades the navigation information to all civil users). In the meantime some civil users get centimeter-type accuracy through phase tracking of the GPS carrier (even with the degraded GPS signals), this is also referred to as "realtime kinematic carrier-phase tracking".

| Predictable Accuracy | Repeatable Accuracy | Relative Accuracy |
|---|---|---|
| PPS = Precise Positioning Service (reserved for military applications) | | |
| Horizontal = 17.8 m<br>Vertical = 27.8 m<br>Time = 90 ns | Horizontal = 17.8 m<br>Vertical = 27.7 m | Horizontal = 7.6 m<br>Vertical = 11.7 m |
| SPS = Standard Positioning Service (available to general user community) | | |
| Horizontal = 100 m<br>Vertical = 156 m<br>Time = 175 ns | Horizontal = 100 m<br>Vertical = 156 m | Horizontal = 28.4 m<br>Vertical = 44.5 m |

**Table 26:    GPS Accuracy Characteristics**

P.S. - During the Persian Gulf war (Jan./ Feb. 1991) military receivers were not available in sufficient numbers, so Selective Availability was turned off for friend and foe (all observers could receive the precision position on the civil frequency). After the Gulf war GPS was brought back to the old segregation policy. There is no limit to irony!

Measurement principle: The user makes a range measurement and a range delay measurement, resulting in a three-dimensional position (with four satellites this can be longitude, latitude, and elevation).. The visibility of up to 3 satellites allows a two-dimensional position determination (for instance longitude and latitude which is sufficient for most earth surface movement).

140)  "The NAVSTAR GPS System", AGARD Lecture Series No. 161, ISBN 92-835-04771, Sept. 1988
141)  "Understanding Signals from GLONASS Navigation Satellites", International Journal of Satellite Communications, Vol. 7 11-12, 1989, pp.11-22
142)  'Navstar', Jane's Spaceflight Directory 1988-89, 4. Edition, pp. 404-405

NAVSTAR-GPS Service: The oldest NAVSTAR satellite is in orbit since 1978. Three-dimensional coverage is available globally and daily for a duration of 16 to 18 hours. Two-dimensional coverage is available for 21 to 22 hours daily. GPS employs the following reference system: DOD World Geodetic System (WGS).

The control segment of GPS will include a number of monitoring stations and ground antennas throughout the world. The monitor stations will use a GPS receiver to passively track all satellites in view and thus accumulate ranging data from the satellite signals. The information from the monitor stations will be processed at the master control station (MCS) to determine satellite orbits and to update the navigation message to each satellite. This updated information will be transmitted to the satellites via ground antennas, which will also be used for transmitting and receiving satellite control information.

| PRN (Pseudo Random Noise) | SVN (Satellite Vehicle Navstar) | Launch Date | Date of Service Availability | Orbital Plane | Comment Block | Deactivated |
|---|---|---|---|---|---|---|
| 6 | 3 | 6. Oct. 78 | | A3 | I | 25 Apr. 92 |
| 11 | 8 | 14. Jul. 83 | | C3 | I | 04 May 93 |
| 13 | 9 | 13. Jun. 84 | | C1 | I | |
| 12 | 10 | 8. Sep. 84 | | A1 | I | |
| 3 | 11 | 9. Oct. 85 | | C4 | I | |
| 2 | 13 | 10. Jun. 89 | 11. Jul. 89 | B3 | II | |
| 14 | 14 | 14. Feb. 89 | 15. Apr. 89 | E1 | II | |
| 15 | 15 | 1. Oct. 90 | 15. Oct. 90 | D2 | II | |
| 16 | 16 | 18. Aug. 89 | 14. Oct. 89 | E2 | II | |
| 17 | 17 | 11. Dec. 89 | 6. Jan. 90 | D3 | II | |
| 18 | 18 | 24. Jan. 90 | 16. Feb. 90 | F3 | II | |
| 19 | 19 | 21. Oct. 89 | 26. Nov. 89 | A4 | II | |
| 20 | 20 | 24. Mar. 90 | 18. Apr. 90 | B2 | II | |
| 21 | 21 | 2. Aug. 90 | 22. Aug. 90 | E2 | II | |
| 23 | 23 | 26. Nov. 90 | 10. Dec. 90 | E4 | IIA | |
| 24 | 24 | 4. Jul. 91 | 30. Aug. 91 | D1 | IIA | |
| 25 | 25 | 23. Feb. 92 | 24. Mar. 92 | A2 | IIA | |
| 28 | 28 | 10.. Apr. 92 | 25. Apr. 92 | C2 | IIA | |
| 26 | 26 | 07. Jul. 92 | 23. Jul. 92 | F2 | IIA | |
| 27 | 27 | 09. Sep. 92 | 30. Sep. 92 | A3 | IIA | |
| 32 | 32 | 22. Nov. 92 | 11. Dec. 92 | F1 | IIA | |
| 29 | 29 | 18. Dec. 92 | 5. Jan. 93 | F4 | IIA | |
| 22 | 22 | 3. Feb. 93 | 4. Apr. 93 | B1 | IIA | |
| 31 | 31 | 4. Apr. 93 | 13. Apr. 93 | C3 | IIA | |
| 07 | 37 | 13 May 93 | 12 Jun. 93 | C4 | IIA | |
| 09 | 39 | 26 Jun. 93 | 20 Jul. 93 | A1 | IIA | |
| 05 | 35 | 30 Aug. 93 | 28 Sep. 93 | B4 | IIA | |
| 04 | 34 | 26 Oct. 93 | 29. Nov. 93 | D4 | IIA | |
| 06 | 36 | NET 03 Mar. 94 | | C1 | IIA | |
| | | NET = No Earlier Than | | | | |

**Table 27:**     **GPS Launch Dates and Constellations of Active Satellites**

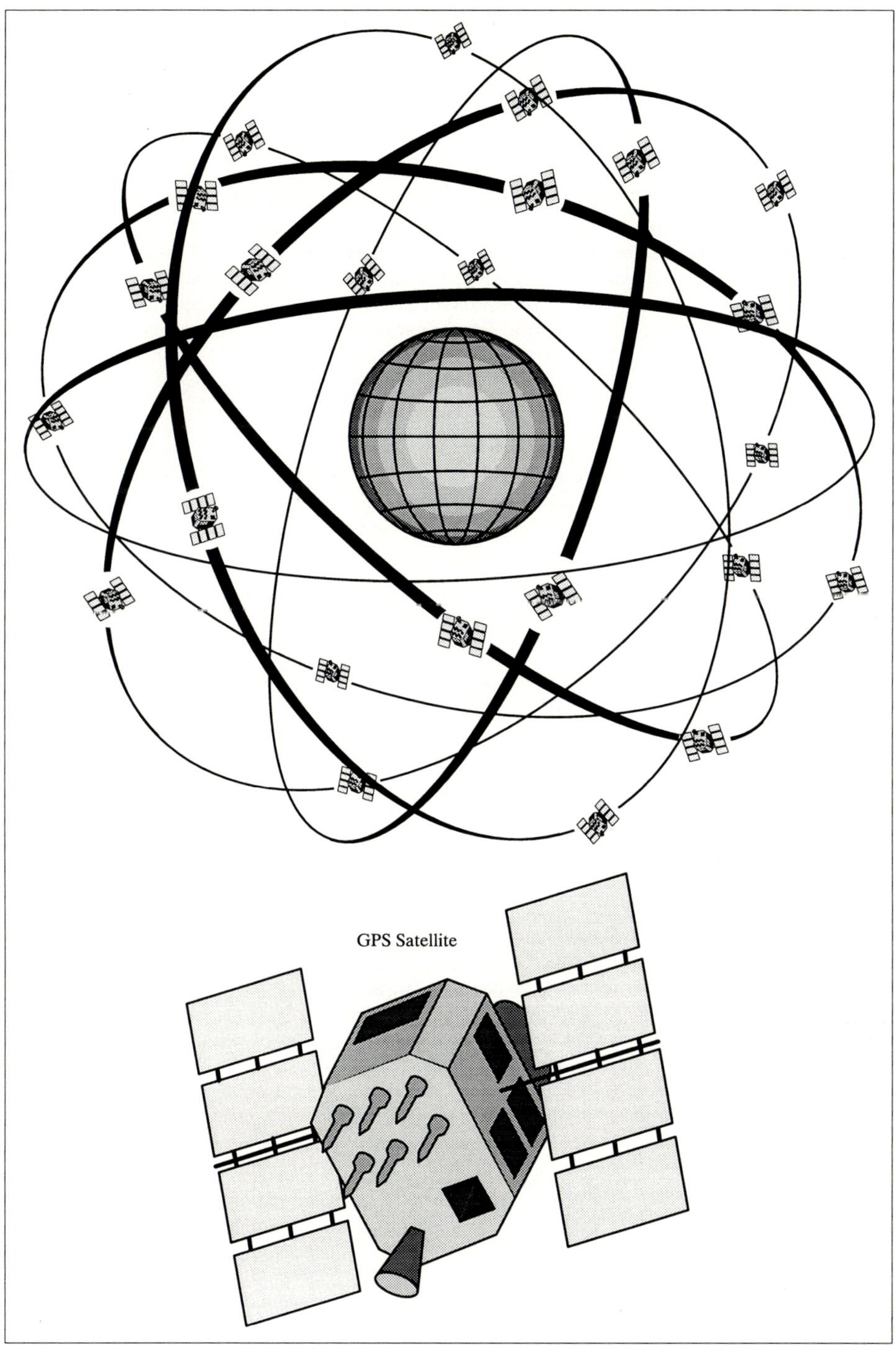


**Figure 38:      Global Positioning System (GPS)**

### A.43.1  Availability of GPS/GLONASS Systems

The GPS system was offered to general civil use, worldwide - with no user fees attached for the next 10 years - by the FAA administrator James B. Busey at the Air Navigation Conference (ANC) in Montreal in Sept. 1991. This makes the GPS services available to anyone who can live with the 100 m accuracies of the SPS mode. The 10 year period starts at the end of 1993 when GPS is fully deployed (21 operational satellites plus three spares)[143]. GPS can provide a nearly 100% service availability (three-dimensional) with a 21 satellite constellation.

At the same conference, the Soviet Union (now Russia) offered the GLONASS service to the general worldwide civil user community for a period of 15 years after full deployment at no charge. The commitment to the SPS service accuracies of 100 m latitude/longitude and 150 m altitude were matched by Russia. As of Sept. 1991 there are 12 GLONASS satellites in orbit, full orbital capability of 24 satellites will be achieved by 1995.

The general civil availability of GPS/GLONASS services solves certainly some basic regulatory issues for the immediate future. Long-range service provision of GPS/GLONASS for the civil sector beyond the years 2003/2010 is, however, questionable and requires an unprecedented level of international cooperation. Nevertheless, the current commitments of service give rise to consider GPS/GLONASS for many important applications, including the development of new navigation receiver systems for civil aviation.

The international community realizes also the need for an integrated long-range program of a Global Navigation Satellite System (GNSS). GNSS is proposed by the ICAO (International Civil Aviation Organization). GPS and GLONASS will certainly be key elements in this constellation.[144]

| Item | Block IIA | Block IIR (replenishment) |
| --- | --- | --- |
| Weight (kg) | 1670 | 2037 |
| Design Life (years) | 7.5 | 10 |
| Cost/Spacecraft (million) | $ 48 | $ 28 |
| Contractor | Rockwell International | GE Astro Space |

**Table 28:**     **Comparison of future Block IIA and Block IIR GPS Satellites**[145]

### A.43.2  GPS Applications

GPS and GLONASS are both basic service providers to large military and civil user communities by providing instantaneous position, velocity, and time. These parameters are fundamental for navigation in general, they are also implicit to virtually all applications in the field of Earth observation and surveying. Current interest in GPS/GLONASS applications lies in the potential of the system's capabilities, the number of uses will be limited only by the imagination. Only civil applications are considered in this context.

**Research Applications**

GPS lends itself to many different scales of application[146], from the global level to the most discrete and site-specific. In between are regional and national projects. At the global level GPS applications are planned to help define the dimensions of the planet itself: creating geoidal models, establishing the terrestrial reference frame and earth rotation parameters, mapping the dynamic topography of the oceans and monitoring plate tectonics. The list of ongoing and possible research applications of GPS services is long. Some examples are:

143)  "GPS - the Next Generation", GPS World, Nov. Dec. 1991, pp. 12-16
144)  H. Montgomery, "Organizing the Technology", GPS World, April 1992, pp. 18-20
145)  'GPS - the Next Generation', GPS World, Nov./Dec. 1991, p. 12
146)  Glen Gibbons, "What in the World!?!" GPS WORLD, April 1991, p. 21-24

- Measurement of crustal motion in Central and South America (CASA)
- Tide gauge positioning along the Baltic Sea (unified height system to study the relative land uplift and sea level variations). Poland, Finland, Sweden, etc.
- Development of regional differential GPS (DGPS) systems to aid maritime vessel traffic services along coastlines and in harbors (US Coast Guard and serval Scandinavian agencies)[147]
- CNRS of France started a project using GPS positioning to help georeference early photos of glaciers in the French Alps. Growth and recession of ice sheets.
- NASDA is planning to integrate GPS into the navigation system of its H-II orbiting plane.
- Russian scientists use GPS for oceanic work. Models for earthquake predictions, etc.
- A new earthquake forecasting research program in the USA is for instance based on GPS technology (1991). The proposed concept uses an array of GPS monitoring stations positioned along a fault, all of these stations are tied into a computer center, which calculates earth movements in the order of 1 cm.[148]
- NASA/JPL with its extensive experience in interplanetary navigation and corresponding planet atmosphere interpretation is investigating its occultation techniques to be applied directly for GPS receiver measurements. This would allow for the determination of the Earth's major atmospheric constituents and other parameters. See also GGI sensor definition under EOS sensors, and GPS-MET of Microlab-1.
- There are numerous projects and campaigns in Solid Earth Research. GIG '91 (GPS Experiment for the International Earth Rotation Service and Geodynamics).
- The Space Shuttle has been flown GPS twice in 1993. NASA/JSC is considering to fly two GPS attitude determination experiments in 1995 (platforms deployed from Shuttle).
- In the future GPS will be used for attitude determination on commercial satellites.
- GPS is being used for precision time transfer and geodetic positioning
- There are also numerous questions that involve standards and procedures.
- etc.

## Commercial Applications

In the commercial world the potential of GPS hasn't been tapped yet, it is just getting to know the technology and gearing up for new navigation and surveying products. The following applications could be supported:[149]

- Civilian air traffic control. GPS could serve as a guidance system for instrument landings. This would be very advantageous to all especially to 3. World Nations who have practically no guidance systems installed at all. The completion of the GPS and GLO-NASS constellations will be of utmost importance to the civil aviation community. The year 1994 will probably see the first *certified* GPS receiver for non-precision landing.
- Field demonstrations have been conducted of runway and approach guidance of aircraft with carrier-phase DGPS, and precision landing using a DGPS system combined with an inertial navigation system.
- Navigation on the oceans. Marine navigators are asking for a DGPS (see A.43.5 for DGPS) network around coast lines.
- Navigation systems for automobiles. In Sept. 1991 public transportation agencies in Denver and Dallas signed contracts to install GPS-based automatic vehicle location (AVL) systems (used for route guidance, fleet management, costumer information services, etc.). As of 1993, new systems, called IVHS (Intelligent Vehicle/Highway Systems) are on the drawing boards.
- Navigation systems for outdoors (tracking, boating, expeditions, etc.)

147) B. Tryggö, R. Bäckström, "Threading the Needle: Differential GPS on the Baltic Sea", in GPS World Sept. 1991, pp. 22-26
148) "GPS is Newest Aid in Earthquake Forecasting", Space News, March 18-24 1991, pp. 22
149) "Smart Policy: Make Best GPS Data Available to All", Space News, April 1-7 1991, pp. 15

- Navigation systems for the blind people!
- Commercial surveying. We are just beginning to explore the huge potential of the GPS surveying technique in terms of accuracy, efficiency, and productivity. Example: an integrated GPS tachymeter system is being developed with a realtime datalink and direct output to a Geographic Information System (GIS).
- As of 1993 surveying systems in USA with dual (or more) video cameras and GPS receivers map roads and the surrounding infrastructure (all video imagery is tagged with GPS coordinates), the data is fed into GIS databases for later retrieval.
- There is seemingly not a single application in surveying and navigation that cannot be improved using GPS/GLONASS.
- etc.

As of 1993 the market offers a wide variety of GPS receivers/products to serve an increasingly diverse range of GPS applications. the list of manufacturers would be too long to be considered here. In general the GPS applications technology is now both operationally capable and also becoming more affordable to the consumer market.

There are also investigations and assessments under way for a joint service provision of GPS and GLONASS systems. The combination of both systems increases considerably the number of simultaneous satellite visibility to the user[150].

### A.43.3   IGS (International GPS Service for Geodynamics)

As of 1993 an IAG (International Association of Geodesy) workshop is in the process of defining and installing a GPS satellite tracking service along with databases for remote access. The program is called "IGS", its primary goal is to "provide the science community with high-quality GPS orbits on a rapid basis, to provide Earth rotation parameters of high resolution as a byproduct, to expand geographically the current International Terrestrial Reference Frame (ITRF) maintained by the International Earth Rotation Service (IERS), and to monitor the global deformations of the Earth's crust."[151]

A number of campaigns (since 1992) by an IAG subgroup (G. Beutler, I. Mueller, etc.) have been conducted to test the feasibility of the program. Initial tracking service products (pilot program) are provided by the following data centers: JPL (central bureau), NASA's Crustal Dynamics Data Information System (CDDIS), Institut Geographique in France, and the Scripps Institution of Oceanography.

### A.43.4   CIGNET

CIGNET = Cooperative International GPS Network of IAG (International Association of Geodesy). CIGNET is continuously tracking the GPS satellites providing a fiduciary data set that may be used for orbit computation and research. The data from all stations are sent weekly to the National Geodetic Survey (Rockville, Md.) where they are archived. The following SLR (Satellite Laser Ranging) ground stations are participating in the initial CIGNET program, a final operational network is considered with about 150 stations:[152]

- Kokee-Orbit, Hawaii, USA
- Mojave, California, USA
- Onsala, Sweden
- Richmond, Florida, USA
- Tromsö, Norway
- Tsukuba/Kashima, Japan

---


150) N.E. Ivanow, V. Salistchew, "GLONASS and GPS: Prospects for a Partnership", GPS WORLD, April 1991, p. 36-40

151) G. Beutler, E. Brockmann, "Proceedings of the International GPS Service for Geodynamics (IGS) -Workshop", 25-26 March, 1993, Astronomical Institute, University of Bern

152) CIGNET Report, CSTG Bulletin No. 11, Title: New Satellite Missions for Solid Earth Studies, June 1989, pp. 235-256

- Westford, Massachusetts, USA
- Wettzell, Germany
- Yellowknife, Canada

While the current CIGNET provides improved ephemeris data, the more expansive geodetic system would also provide positioning data. As for costs, operating a couple of commercially available GPS/GLONASS receivers at a site that is part of a global CIGNET is orders of magnitude less expensive than building a new laser or VLBI facility.

Note: The objectives of IGS and CIGNET seem to be fairly identical. Both systems are IAG-sponsored initiatives. In 1993 it looks like IGS will make the race to a full-service capability.

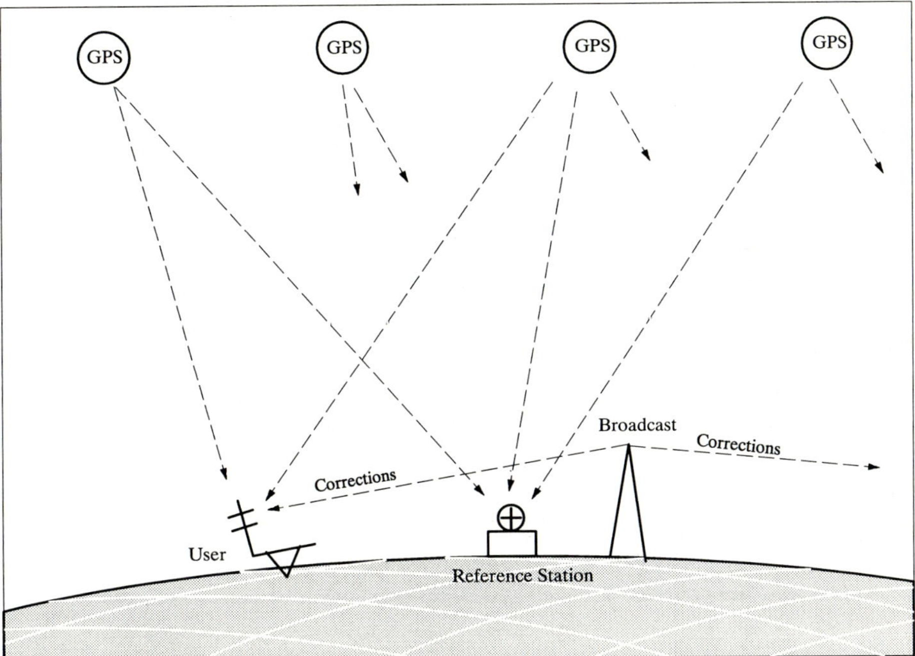


**Figure 39:**    **Differential Operation of GPS with a Reference Station**

## A.43.5  DGPS

A DGPS (Differential GPS)[153] system employs a local reference station, which has a high-quality GPS receiver and an antenna at a known, surveyed location. The reference station estimates the slowly varying components of the GPS satellite range measurement errors, and transmits them as corrections to users within communication range of the station. With this concept higher position accuracies are achieved for users in the vicinity of a DGPS reference station (the accuracy reduces with increasing radial distance).

A surveying application: By using GPS equipment on an aircraft - typically dual-frequency, 24 channel receivers with the antenna as close as possible to the camera lens itself - and a DGPS base station on the ground, the position of the plane itself can be recorded to an accuracy level of about 2 meters (this obviates the need to send in survey crews in advance to set visible ground markers).

---

153) P.K. Enge, R.M. Kalafus, M.F. Ruane, "Differential Operation of the Global Positioning System", IEEE Communications Magazine, July 1988, Vol. 26, No.7, pp. 48-59

In the USA there are plans to modify more than 200 radiobeacons to enable broadcasts of DGPS signals along coastal regions. A full DGPS network is planned to be operational by 1996. The network will broadcast corrections to ranging data received from GPS satellites to improve vessel navigation up to a range of a few hundred km (from the reference station).

Many radiobeacon networks of other countries are also being upgraded for the broadcast of differential GPS corrections to marine users. The corrections will improve the accuracy of a mariner's GPS fix to <10 m which allows harbor and harbor entrance navigation. The first commercial DGPS radiobeacon receivers made it to the market at the end of 1991.

A DGPS positioning service was developed by Magnavox and CUE Network in the USA. This positioning service provides real-time, 1-5 m accuracy DGPS corrections, broadcast over FM subcarriers from selected area radio stations.[154] The broadcast standard is referred to as RBDS (Radio Broadcast Data System), RBDS is compatible with the pager-based DGPS service, called ACC-Q-POINT (formerly known as:Pinpoint).
Another commercial service provider company 'Differential Corrections Inc. of Cupertino, Ca. has plans to rapidly establish a worldwide network of FM RBDS stations. In 1993 there are 8 US cities with this service (Los Angeles, Chicago, Washington D. C., San Francisco, Detroit, San Diego, Salt Lake City, and Albuquerque), a total of 25 US cities are planned for the end of 1993, the service is also offered in Europe.[155]

Wide-area DGPS corrections coverage will be broadcast from geostationary satellites ( Inmarsat 3, launch of Inmarsat 3 in 1994) at or near the GPS L1 frequency. Inmarsat 3 carries a transponder to broadcast the L1 lookalike GPS signal with a C/A code assigned for geostationary satellite operation. The signal will be generated through ground station control and will provide a civil adjunct service to the DOD-operated service. Conventional GPS receivers can track this signal with minor software modifications to select the assigned C/A code and to demodulate the data broadcast by the geostationary satellite. Ultimately, four Inmarsat geostationary satellites will be available. Other satellites are also planned to carry similar payloads.[156]

### A.43.6   Future of GPS/GLONASS Services

New satellites of GPS/GLONASS that are under construction will not only ensure a seamless continuance of GPS/GLONASS service but will also introduce several new features when put into operation. One such feature will be the GPS integrity channel (GIC). It will be used to broadcast GPS and GLONASS health information to aircraft users. The GIC monitors the satellite signals and broadcasts "use/don't use" signals as well as coarse differential corrections to users. The GIC will use geostationary satellites to warn aircraft worldwide in realtime (<10 seconds).

### A.43.7   GPS Information Services for Civil User Community

In the USA a GPS Information Center (GPSIC) and a Civil GPS Service Interface Committee (CGSIC) are provided. The GPSIC communicates update GPS information to the national information centers [such as GIBS (GPS Informations- und Beobachtungssystem of DGFI, Frankfurt) in Germany] for easier access to the user community. The objective is to further civil applications of the GPS system as a whole. Direct access is available to the BBS (Bulletin Board Service)

- USA computer (modem) access to BBS: 703-866-3890
- Germany GIBS-BBS: Internet access (Telnet Nr. 141.74.240.25)

154)  B. McGarigle, " 'Top 40' Hydrography: Surveying with FM-based DGPS", GPS World April 1993 pp. 37-40
155)  'California-Based Firms Offer Highly Accurate GPS Services', Space News, Nov. 29-December 5, 1993, p. 7
156)  GPSWORLD, Dec. 1993, p. 39

## A.44    GMS (Geostationary Meteorological Satellite)

GMS = Geostationary Meteorological Satellite of JMA (Japan Meteorological Agency) and NASDA. GMS is also known by the name of Himawari. Operational meteorological program of Japan with the following  satellites:

- GMS-1 F1    Launch: July 14 1977  Delta 2914 vehicle of NASA Position: de-orbit
- GMS-2 F2    Launch: Aug. 11 1981 N-II vehicle of NASDA    Position: de-orbit
- GMS-3        Launch: Aug. 3 1984   N-II vehicle of NASDA    Position: 120° East
- GMS-4        Launch: Sept. 6. 1989 H-I  vehicle of NASDA    Position: 140° East
- GMS-5        Launch: projected for 1994 with H-II vehicle

The mission objectives of the GMS program are:

1.  Weather watch, to obtain and transmit pictures of the Earth and its cloud cover using VISSR

2.  Collection of weather data (ships, buoys, aircraft, weather stations)

3.  Distribution of weather data (geostationary repeater for the collection of weather data)

4.  Monitoring of solar particles

GMS-4 is a spin-stabilized satellite weighing 725 kg. The S/C consists of a despun Earth-oriented antenna assembly and a spinning section rotating at 100 rpm. Despun S-Band and UHF antennas provide high gain for on-orbit communications with ground stations[157].

**Sensors:**

- **VISSR** (Visible and Infrared Spin-Scan Radiometer). Spectral bands: 0.50 - 0.75 µm (visible), and 10.5 - 12.5 µm (infrared). Resolution: 1.25 km (visible) and 5 km (infrared). The VISSR is used to obtain visible and infrared spectrum mappings of the Earth and its cloud cover.

   VDM (VISSR Digital Multiplexer). VDM is a high-speed pulse code modulation (PCM) encoder. It converts VISSR signals into digital form and multiplexes them with image and message synchronization data.

- SEM (Space Environment Monitor, NOAA sensor). Spectral bands: 1 - 500 MeV (protons); 8 - 390 MeV (alpha particles); >2 MeV (electrons). SEM measures three kinds of solar particles: protons, alpha, and electrons.

- DCS (Data Collection System) from ground platforms (see METEOSAT and NOAA-GOES)


## A.45    GOMS (Geost. Operational Meteorological Satellite)

GOMS[158] = Geostationary Operational Meteorological Satellite. Russian program series in the construction phase at VNIIEM, Moscow. The first GOMS satellite (GOMS-1) is scheduled to be launched in March/April 1994 (by a Proton carrier vehicle). International cooperation with other nations meteorological programs is invited.

Satellite: mass = 2300 kg (including payload of 550 kg), GOMS is designed as a three-axis stabilized spacecraft with orientation to the Earth and along the velocity vector. Nominal life = 2-3 years. Orientation accuracy = 5-10 arc/min.

---

157)  M. Homma, M. Minowa, M. Kobayashi, M. Harada, "Geostationary Meteorological Satellite System in Japan" in
     'Monitoring Earth's Ocean, Land, and Atmosphere from Space', Volume 97 AIAA, 1985, pp. 570 - 583
158)  'Space System with Geostationary Meteorological Satellite (GOMS)', Paper of NPO Planeta, Moscow, Nov.
     1990

Orbit: Geostationary, position = 76° East

Objectives:

- to acquire television images in real time of the Earth's surface and cloud cover within a radius of 60° centered at the subsatellite point in the visible and IR regions of the spectrum
- to measure temperature profiles of the Earth surface (land and ocean) as well as top of cloud cover temperatures
- to measure the radiation state and magnetic field of the space environment at the geostationary position
- to transmit via digital radio channels television images, temperature and radiation as well as magnetometric information to the main and regional data receiving and centers
- to acquire the information from Russia and international data collection platforms (DCPs), located in the GOMS visibility, and to transmit the obtained information to all receiving and processing centers (RPCs)
- to call for the DCPs to transmit the information to the satellite
- to retransmit the processed meteorological data in the form of facsimile or alphanumerical information from the receiving processing centers to the receiving stations of the independent data users via GOMS
- to provide the exchange of high-speed digital data (retransmission via GOMS) between all regional centers of Russia (Federal Service for Hydrometeorology and Environmental Monitoring).

**Sensors:**

- **STR** = Scanning **TV R**adiometer providing imagery in VIS and IR bands. Objectives: observation of clouds and underlying surface in VIS and IR spectra and temperature data of underlying surface, determination of top of clouds.
  Spectral bands: 0.4 - 0.7 µm (VIS), 10.5 - 12.5 µm (IR-1); Number of scan lines per frame = 8000 (VIS) and 2500 (IR); IFOV = 31.5 µ rad (VIS), =160 µ rad. (IR); FOV = 13500 x 13500 km; spatial resolution = 1.5 km (VIS) and 6.5-8.0 km (IR). The imaging session frequency is not less than 30 min, the length of one take (frame time) is 15 minutes. Direct data transmission rate = 2.56 Mbps.

- **RMS** = Radiation Measurement System (Instruments for helio-geophysical measurements, radiation and magnetic parameters). FOV = 13500 x 13500 km. Objective: registration of particles (protons, electrons, α-particles), measurement of the X-ray radiation from the sun, measurement of the magnetic field vector components.
- density of electron fluxes with energies in four bands from 0.04 - 1.7 MeV
- density of proton fluxes with energies in four bands from 0.5 - 90 MeV
- density of alpha particles with energies from 5 - 12 MeV
- intensity of the galactic cosmic radiation with energies > 600 MeV
- solar X-ray radiation intensity with energies from 3 - 8 keV
- intensity of solar UV radiation in four wave bands up to 1300 Å
- magnetic induction vector component quantities along three axis with ± 180 nT interval

## A.45.1   Radio Complex for Data Collection, Transmission and Relay

- GOMS Communication Characteristics
- transmission of TV images and heliogeophysical information from the satellite to the receiving and processing centers - RPCs (Radio channels I and II).
- carrier frequencies: 1685 MHz (S-Band) and 7465 MHz (X-Band)
- data transmission rate = 2.56 Mbps
- Radio channel III: transmission of data from DCPs; frequencies: 401 - 403 MHz; transmission of data is possible through 33 international and 100 Russian channels at rates of 100 b/s.

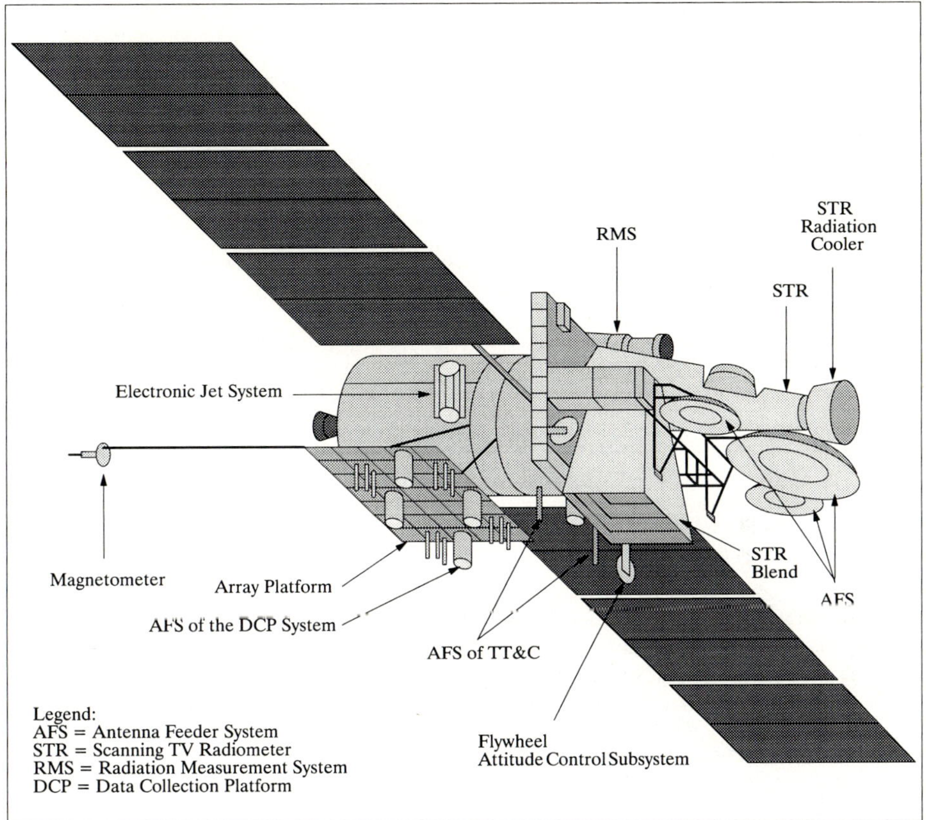


**Figure 40:    The GOMS S/C Model**

- Radio channels IV and V: retransmission of information obtained from DPCs to RPCs; frequencies: 1697±1 MHz (S-Band) and 7482 ±1 MHz (X-Band)
- Radio channels VI and VII: transmission of facsimile information in standard WEFAX format and alphanumerical data from RPCs to GOMS. Frequencies: 2115±1.5 MHz and 8195±1.5 MHz. Data transmission rate = 1200 b/s
- Radio channel VIII: retransmission of facsimile and alphanumerical data from GOMS to independent receiving stations. Frequency: 1691±1.5 MHz
- Radio channel IX: transmission of high-speed digital information from RPC to GOMS. Frequency: 8190±5 MHz; data rate up to 0.96 Mbit/s
- Radio channel X: transmission of high-speed digital information from GOMS to the RPCs. Frequency: 7.465±2.5 MHz; data rate up to 0.96 Mbit/s
- Radio channel XI: calling for DCP from GOMS. Frequency: 469±1 MHz
- Radio channel XII: transmission of DCP request from RPC to GOMS. Frequency: 2119± 1 MHz

- GOMS Data Receiving and Transmitting Modes

- Channels I and II operate 24 - 48 times per day, each session lasts 15 minutes
- Channels III, IV and V operate under the commands to call for the DCP information
- Channels VI, VII and VIII operate continuously
- Channels IX and X function in the interval when channels I and II are not active
- Channels XI and XII operate on request

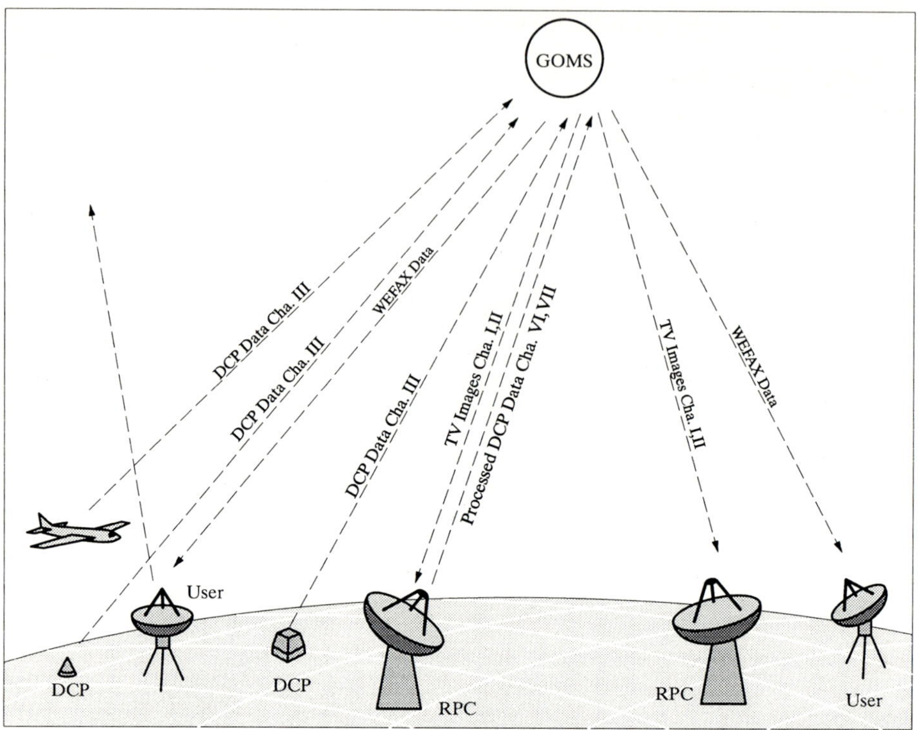


**Figure 41:    Data Collection and Distribution Scenario for GOMS**

## A.46    HCMM (Heat Capacity Mapping Mission)

NASA Mission, also known as **AEM-1** (Application Explorers Mission-1). Launch: 26. April 1978, end of operation: Aug. 1980; Scout launch vehicle from Wallops, payload weight = 134 kg.[159),160)]

Orbit: Sun-synchronous near-polar orbit, altitude = 620 km, inclination = 97.6°, period = 97 minutes, repetition cycle = 16 days, local equator crossing at 2 pm (ascending node). On Feb. 21-23 1980 the HCMM orbit was lowered from 620 km to 540 km altitude.

Application: Measurement of heat budget, surface temperature, soil moisture, urban heat islands, etc. HCMM was an experimental remote sensing mission to test thermal inertia to help discriminate between different surface materials, and to identify different states, such as degree of soil wetness.

Sensor: **HCMR** (Heat Capacity Mapping Radiometer)
Spectral range      0.55 - 1.11 μ (for solar reflection, pixel size = 500 x 500 m)
Spectral range      10.5 - 12.5μ (for thermal emission, pixel size = 600 x 600 m)
Image size          2016 lines (1680 pixels/line), 8 bit resolution
IFOV                0.83 milliradians

HCMM data were stored and processed at GSFC. Data products: geometrically corrected and calibrated images of radiance or equivalent black-body temperature.


159)  P. Slater, 'Remote Sensing' Optics and Optical Systems, Addison-Wesley, 1980, pp. 462 - 465
160)  HCMM System in 'Manual of Remote Sensing', Second Edition, American Society of Photogrammetry, 1983,
      pp. 663-670

# A.47    HIROS (High Resolution Observing Satellite)

A planned Japanese (NASDA) mission for a launch in 2000 with a mission design life of 3 years. S/C mass of ~3500 kg. HIROS objectives/applications: regional land observation.

- Target of mapping scale is 1 : 25,000. Image data can be used for the generation of topographic maps.
- Monitoring of desertification, deforestation, pollution of coastal zones. Contributions to IGBP and WCRP.
- Contributions to disaster monitoring, such as the eruption of volcanoes, earthquakes, tidal waves, floods, etc.

Orbit: Sun-synchronous sub-recurrent polar orbit, altitude ~700 km,

Sensors:

- **AVNIR-2** = Advanced Very High Resolution Radiometer (successor to instrument on ADEOS, NASDA sensor). A spatial resolution in the order of 2.5 m (of panchromatic spectral band) is required. A sensor pointing function is required to increase the observation capability/frequency.

- **VSAR** = Variable off-nadir angle Synthetic Aperture Radar (NASDA sensor). The design goal is a spatial resolution of 10 m, and a variable off-nadir pointing function.

- **DCS** = Data Collection System.


# A.48    IMP-8 (International Monitoring Platform)

This NASA/GSFC S/C is also known by the names of Explorer 50 and IMP-J. IMP-8 is in a series of ten IMP missions with the primary objective to perform detailed and near-continuous observations of the Sun -Earth environment (monitoring of the solar wind, measuring the plasma/field environment of the magnetosheath and the magnetotail). Launch on Oct. 26 1973 from the Eastern Test Range by a Delta vehicle.

Orbit: Geocentric elliptical, apogee = 288 672 km , perigee = 141050 km, inclination = 11.9°, period = 12.6 days, spin vector direction= normal to ecliptic, spin rate = 22 rpm.

Sensors:[161]

1.    **Field Investigations**

- GNF = Magnetic Fields Experiment (PI: N. F. Ness, GSFC). Boom-mounted triaxial fluxgate magnetometer (mass=3.2 kg) in the range of ± 36 nT per sensor. The resolution per sensor is ± 0.3 nT. Vector measurements are performed every 40 ms.

- GAF = DC Electric Fields Investigation (PI: T. L. Aggson, GSFC). Instrument (mass=11.5 kg) utilizes two 60 m wire antennas to measure dc electric fields in the solar wind and magnetosheath.

- IOF = AC Electric and Magnetic Fields Experiment (PI: D. A. Gurnett, University of Iowa). The plasma wave instrumentation (mass=12 kg) consists of a 121.8 m tip-to-tip dipole antenna for electric field measurements (in the S/C spin plane) and a triaxial search coil for magnetic field measurements. Magnetic measurements are made in 7 channels in the frequency range 40 Hz to 1.78 kHz (cycle time of 10.24 s for one set of measurements). The electric field is measured by a 15 channel spectrum analyzer operating between 40 Hz and 178 kHz, and a wide band receiver tunable to 2000, 500, 125, and 31.1 kHz.

161)  J. H. King, "Availability of IMP-7 and IMP-8 Data for the IMS Period", The IMS Source Book, GSFC, pp. 10-20,

**2.  Plasma Investigations**

- LAP = Los Alamos Plasma Experiment (PI: S. J. Bane, Los Alamos National Laboratory). Solar Plasma Electrostatic Analyzer Experiment. The instrument (mass=6.3 kg) consists of a hemispherical plate electrostatic analyzer for the measurement of ion and electron distributions. Resolution time < 2 minutes.

- IOE =  Low Energy Particles Investigation (PI: L. A. Frank, University of Iowa). The instrument (mass=2.6 kg) consists of a Low-Energy Proton and Electron Differential Energy Analyzer (LEPEDA) and a Geiger tube. LEPEDA measures fluxes of ions and electrons, separately and simultaneously, in each of 16 energy per charge channels (between 50 eV and 45 keV) in each of 16 azimuthal directions around the S/C spin vector. The Geiger tube measures electrons in the range > 45 keV. A full measurement cycle = 82 seconds.

- MAP = Solar Plasma Faraday Cup Experiment (PI: H. S. Bridge, MIT).  The instrument (mass=6.5 kg) consists of a split collector Faraday cup measuring ions in the energy range of 50 eV/q to 7 keV/q in 24 steps, and electrons in the range from 22 eV/q to 2 keV/q in 21 steps. An angular distribution is taken at each energy step. The measurement sequence is controlled by the measured flux. Typical measurements are 30 seconds in the solar wind and magnetosheath.

**3.  Energetic Particles Investigations**

- GWP =  Energetic Particles Experiment (PI: D. J. Williams, NOAA). The instrument (mass=3.3 kg) is a solid-state telescope which measures fluxes of ions in four energy channels ranging from 0.05 - 0.20 MeV to 2.1 - 4.5 MeV, and electrons in the ranges 30-90 and 100-200 keV.

- MAE = Ion and Electron Investigation (PI:G. Gloeckler, University of Maryland). The experiment consists of two detector systems (mass=7 kg). An electrostatic deflection spectrometer (EDS) measures the energy per charge of incident ions in several ranges between 37 and 1200 keV/q.

- APP = Charged Particles Experiment (PI: S. M. Krimigis, JHU/APL). The instrument is a solid-state telescope (mass=3.9 kg) measuring fluxes of protons in 11 energy channels between 0.29 and 140 MeV, and alpha particles in 6 channels between 0.64 and 52 MeV/n. Time resolution for the measurement cycle is 10.24 s.

- GME = Solar and Cosmic Ray Particles Investigation (PI: F. B. Mc Donald, GSFC). The experiment (mass=11 kg) consists of three telescopes intended to measure the energy spectra and composition of solar and galactic electrons, protons, and heavier nuclei. The lowest energy, solid-state telescope measures electrons (>0.15, 0.35, 0.75 MeV) and protons (between 0.05 and 25 MeV) and particles $2 \leq Z \leq 28$ (1.6-12 MeV/n). The second telescope measures protons (0.8-4 MeV) and 4-20 MeV/n nuclei with $1 \leq Z \leq 26$. The third telescope uses CsI scintillator elements to measure electrons (2-12 MeV) and Z=1 to 30 nuclei (20-500 MeV/n).

- CAI = Electron Isotopes Investigation (PI: E. C. Stone, JPL). The instrument (mass=8 kg) consists of an 11 element solid-state telescope for composition and spectra measurements of galactic and solar cosmic rays in the energy range of 1-40 MeV/n.

- CHE = Cosmic Ray & Solar flare Isotopes Investigation (PI: J. A. Simpson, U. of Chicago). The instrument (mass=7.4 kg) consists of a pair of solid-state telescopes. The main telescope measures nuclei in the energy range of 10 to 100's of MeV/n, and electrons in the range of $\sim 2$ to $\sim 25$ MeV. The second telescope measures protons and alpha particles in the 0.5-1.8 MeV/n range.

**Data:** Telemetry primary transmitter (PCM) at frequency of 137.980 MHz, secondary transmitter at frequency 136.800 MHz. Science Data is archived at World Data Center A of GSFC.

# A.49   INSAT

INSAT[162] = Indian National Satellite system is a joint venture of the following Indian Departments: DOS (Department of Space), DOT (Department of Telecommunications), IMD (Indian Meteorological Department), AIR (All India Radio). DOS is responsible for the operation of the INSAT space segment. IMD operates a Meteorological Data Utilization Center (MDUC, at Delhi) for the dissemination and distribution of the INSAT meteorological images and ancillary data.

INSAT is a multipurpose operational satellite system series (geostationary) employed for meteorological observation over India and the Indian Ocean, as well as for domestic telecommunications (nationwide direct TV broadcasting, TV program distribution, meteorological data distribution, etc.). The INSAT-1 series satellites were built by Ford Aerospace Corporation (USA) to Indian specifications.

Sensor: (Meteorological Package)
**VHRR** = Very High-Resolution Radiometer (ISRO sensor). Spectral ranges: 0.55 - 0.75 µm (VIS) and 10.5 - 12.5 µm (IR). Resolutions: 2.75 km for VIS and 11 km for IR channels. Scanning time/image = 30 minutes. Scanned Field: normal mode (20° E-W and 14° N-S); full frame mode (20° E-W and 20° N-S); sector scan mode (20° E-W and 4.5° N-S).

Data Collection System (**DCS**): This system gathers and relays environmental data (meteorological, hydrological and oceanographic) from unattended land- and ocean-based automatic data collection platforms (DCPs). The global data receive frequency (platforms to satellite) is 402.75 MHz (UHF-Band). As of 1991 over a 100 DCPs have been installed.

**INSAT-1 Satellite Series:**

- INSAT-1B
  Launch: August 30 1983; position: 74° East;  The satellite is used as a standby for IN-SAT-1D.

- INSAT-1C
  Launch: July 22 1988 (Ariane vehicle). The satellite lost earth lock on Nov. 22 1989 and is inoperable now.

- INSAT-1D
  Launch: June 12 1990 (US Delta vehicle); position: 74° East.

**INSAT-2 Satellite Series:**

- This satellite family is considered the second generation INSAT series for the nineties. INSAT-2A is an operational satellite (launch on July 9, 1992 with Ariane from Kourou). INSAT-2B was launched on July 22, 1993 (Ariane vehicle). The INSAT-2 operational satellites (2C, 2D and 2E) will have enhanced capabilities. The INSAT-2 meteorological sensor VHRR has a 2 km resolution in VIS and 8 km resolution in the IR spectral ranges.

  INSAT-2 has a data collection system (DCS) with a data relay transponder (DRT) for environmental data. It will also be furnished with **SAS&R** (Satellite Aided Search and Rescue) system. SAS&R provides an emergency alert capability for the Indian subcontinent and beyond as part of the international satellite aided search and rescue program (see also chapter A.74.3). India has signed an agreement with the international COS-PAS-S&RSAT council for the use and operation of LUTs (Local User Terminals) and an MCC.

**INSAT VHRR Data Distribution:**
The INSAT-1 VHRR data is distributed in near realtime to/from 22 SDUCs (Secondary Data Utilization Centers) throughout the country.

---

162) 'Space Applications', DOS Annual Report 1990-91. pp. 13-23

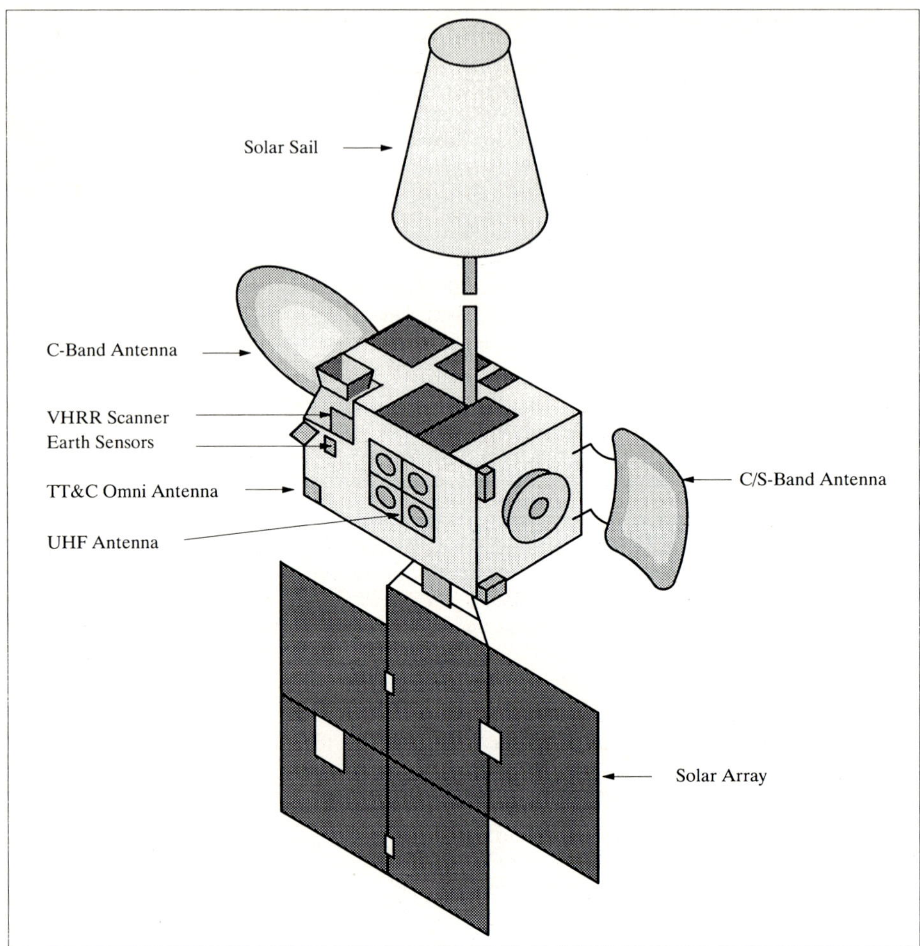


**Figure 42:    The INSAT S/C Model**

## A.50    INTERBALL

Russian multi-spacecraft mission (IKI, Project Scientist: L. Zelenyi) with international cooperation on payload instruments within ISTP. Objectives: Study of the solar-terrestrial transport phenomena (solar wind energy and interaction with the Earth's magnetosphere, its storage there and subsequent dissipation into the tail and auroral regions of the magnetosphere, ionosphere, and atmosphere during magnetospheric substorms).

The program[163],[164] consists of 4 satellites (Prognoz series), a "Tail Probe" configuration and an "Auroral Probe"configuration, each of the satellite configurations is provided with a subsatellite; namely: "Tail C2-X" and "Auroral C2-A". Nominal mission duration: 1 year.

163) "INTERBALL - Study of Magnetospheric Plasma and Solar-Terrestrial Relations", Academy of Sciences of the USSR Space Research Institute, 1987
164) J. Büchner, L. M. Seljenyi, "Interbol erforscht die Magnetosphäre", Astronomie und Raumfahrt, GDR, 25. Jahrgang, 1987, Heft 3, pp. 77-80

Projected launch date for "Tail Probe" is mid-year 1994, and for the "Auroral Probe" end of 1994.[165)]

INTERBALL S/C data: Spin stabilized spacecraft with nominal spin rate of 0.5 rpm. Mass of Auroral Probe payload = 212 kg; mass of Tail Probe payload = 150 kg. Each satellite has 3 telemetry systems ; RTK (16 kbit/s max.), SSNI (32 kbit/s max.), and STO (60 kbit/s max.).
RTK =    Radio Telemetrii Komplex
SSNI =   System Sbora Nutschnoi Informasii
STO =    System Telemetrii

Earth Orbits:

- Auroral Probe (coplanar with subsatellite C2-A): Apogee = 20000 km; perigee = 750 km; inclination = 65°; distance between Auroral Probe and C2-A = 10 - 1000 km
- Tail Probe (coplanar with subsatellite C2-X): Apogee = 200000 km; perigee = 500 km; inclination = 65°; distance between Tail Probe and C2-X = 10 - 10000 km.

The study objective of the subsatellites is aimed to distinguish the space and time dependence of the measured plasma parameters for moving satellite systems in moving plasmas (transport phenomena). Measurement of the correlation vector of the magnetic field, plasma 3-D flux and ion 3-D flux from two points separated between 1 and 1000 km.

The simultaneous measurements of the same plasma parameters on the mother satellite and the subsatellites permits the determination of the space and time variations. The plasma measurements are done with a very high time resolution (0.2-1 s) providing the study of the fine structure of the magnetospheric boundaries. The overlapping plasma sensors allow further the study of 3-D distribution functions of ions (separately for several species) and electrons in a wide range of energy with good resolution and experiment reliability.

### A.50.1  "Auroral Probe" Sensors

Project manager: R. A. Kovrazhkin, IKI, Moscow

**1.   Auroral Plasma Experiments:**

- **SKA-3** = Electron/Proton Distribution Experiment (PIs: A. K. Kuzmin, F. K. Shuiskaya, IKI); objective: measurement of electron and proton distribution. Energy range = 0.03 - 15 keV; electron and ion anisotropy (M=1, 4, 16), E=30-500 keV/Q; time resolution = 1.5 and 0.2 s/sample; data rate = 3 kbit/s; mass = 32 kg

- **ION** = Ion Spectra Experiment (PIs: J. A. Sauvaud, CESR, France, R. A. Kovrazhkin, IKI, Moscow); objective: measurement of ion spectra and anisotropy (M=1, 4, 16), E=0.005-20 keV/Q; time resolution = 3.2 s/sample; data rate = 3 kbit/s; mass = 17 kg

- **PROMICS-3** = 3-D Spectrometer - Ion Composition Experiment (I. Sandahl, IRF, Kiruna, Sweden; N. Pissarenko, IKI, Moscow); energy range < 100 keV/Q; time resolution = 1.2 and 2.6 s/sample; data rate = 2 kbit/s; mass = 12.9 kg

**2.   Magnetic, Electric Fields and Wave Experiments**

- **IMAP-3** = Investigation of Magnetic Fields  (PIs: I. Arshinkov, SDS Lab. Bulgaria, L.Zhuzgov, IZMIRAN, Russia); measurements of three components, range = + 60000 nT; resolution = 1 nT; DC to 10 Hz; time resolution = 0.12 s/sample; data rate = 1 kbit/s; mass = 4.6 kg

- **IESP-2** = Electric Field and Intensity Experiment (PIs: V. Chmyrev, IZMIRAN; G. Stanev, CLKI Bulgaria); measurement range = 0 - 50 Hz; time resolution = 0.03 s/sample; data rate = 7.2 kbit/s; mass = 6.7 kg

---

165) "Interball Project - Magnetospheric System of 4 Spacecraft", The Solar-Terrestrial Science Project of the Inter-Agency Consultative Group for Space Science, esa SP-1107, November 1990, pp. 61-73

- **NVK-ONCH** = VLF Electromagnetic Waves Experiment (PIs: Yu. Mikhailov and A. Goljavin, IZMIRAN); measurement range = 20 Hz - 20 kHz; data rate = 2 x 12 kHz; mass = 17 kg

- **MEMO** = Analyzer of Magnetic Waves (PIs: F. Lefeuvre, LPCE, France; M. Mogilevsky, IKI); measurement range < 2 MHz; time resolution = 0.25 and 120 s/sample; data rate = 20 kbit/s; mass = 14.5 kg

- **POLRAD** = Polish Radiometer - Auroral Kilometric Radio-radiation Experiment (PIs: I. Hanash, Poland, I. Lishin, IRE, M. Mogilevsky, IKI, Moscow); measurement range = 20 kHz - 2 MHz; data rate = 2.3 kbit/s; mass = 22.5 kg

3.  **Thermal Plasma Experiments**

- **HYPERBOLOID** = Ion Mass-Analyzer (PIs: J. J. Berthelier, CRPE, France; T. Muliazchik, IKI); E = 0.0 - 100 eV; v = 0.1 - 20 km/s; species: $H^+$, $He^+$, $O^+$, $O^{++}$, $N^+$, $N_2^+$, $NO^+$, $O_2^+$; time resolution = 1 s/sample; data rate = 8 kbit/s; mass = 15 kg

- **KM-7** = Cold Plasma Experiment (PIs: J. Smilauer, Prague, CSFR; V. V. Afonin, IKI); measurement of of plasma electrons; temperature range < 10 eV; time resolution = 0.1 s/sample; data rate = 0.3 kbit/s; mass = 2.7 kg

- **ALPHA-3** = Trapped Ion Experiment (PI: V. Bezrukikh, IKI); $N > 1\ cm^{-3}$; energy < 25 eV; time resolution = 1 s/sample; data rate = 0.16 kbit/s; mass = 3.5 kg

- **RON** = Ion Emitter Experiment (PIs: W. Riedler, Space Research Institute, Graz, Austria; R. Schmidt, ESA-ESTEC; Yu. Galperin, IKI); $N_2^+$, $In^+$; current = 1 - 10 µA; data rate = 0.2 kbit/s; mass = 7.5 kg

4.  **Energetic Particles Experiment**

- **DOK-2A** = Electron and Proton Experiment (PIs: K. Kudela, Institute of Exp. Physics, Kosice, CSFR; V. M. Lutsenko, IKI); $E_e$ = 10-400 keV; $E_p$ = 15-1000 keV; time resolution = 1 s/sample; data rate = 0.8 kbit/s; mass = 5.5 kg

5.  **Auroral Oval Image Experiments**

- **UFSIPS** = Radiation Emission Experiment (PIs:A. K. Kuzmin, IKI; K. Palazov, IKI-BAN Bulgaria; ); measurement in the lines: 1304Å, 1356Å, and 1493Å; time resolution = 1 line per rotation; data rate = 0,2 kbit/s; mass = 25 kg;

- **UVAI** = UV Auroral Imager (PIs: L. L. Cogger, Uni. of Calgary, Canada; Yu. I. Galperin, IKI); range = 1400 - 1600Å; time resolution = 1 picture per rotation; data rate = 3 kbit/s; mass = 20 kg

6.  **Subsatellite C2-A**

- C2-A = Measurement of electric and magnetic fields, VLF waves, plasma and energetic particles (PIs: P. Triska, Ya. Voita, Geophysics Institute, Prague); data rate = 40 kbit/s; mass = 50 kg


## A.50.2   "Tail Probe" Sensors

Project manager: M. Nozdrachev, IKI, Moscow

1.  **Plasma Experiments**

- **SKA-1** = 3-D Ion Distribution Measurement (O. Vaisberg, A. Leibov, IKI); range = 0.1 - 5 keV/Q; time resolution = 3.7 s/sample; data rate = 2 kbit/s; mass = 26.1 kg

- **VDP** = 3-D Ion Faraday Cups (PIs: G. Zastenker; Z. Nemechek, Charles University Prague, CR); $E_i > 0$ eV; time resolution = 0.1 s/sample; data rate = 1 kbit/s; mass = 4.9 kg

- **ELECTRON** = 3-D Electron Distribution Function (PIs: J. A. Sauvaud, CESR, France; O. Vaisberg, N. Borodkova, IKI); range = 0.01 - 30 keV; time resolution = 120 or 3.7 s/sample; data rate = 1 kbit/s; mass = 6.5 kg

- **CORALL** = Wide-Range 3-D Ion Spectrometer (PIs: R. Himenez, Cuba; Yu. Yermolaev, IKI); range = 0.1 - 30 keV/Q; time resolution = 120 s/sample; data rate = 1 kbit/s; mass = 5 kg

- **AMEI-2** = Energy-Mass Analyzer (PIs: R. Koleva, STIL, Bulgaria; V. Smirnov, IKI); (M = 1-16); E = 0.1 - 10 keV/Q; time resolution = 120 s/sample; data rate = 0.2 kbit/s; mass = 9 kg

- **MONITOR-3** = Solar Wind Analyzer (PIs: A. Feodrov, IKI; Ya. Shafrankoya, Charles University, Prague, CR); E = 0.4 - 15 keV/Q; time resolution = 1 s/sample' data rate = 8 kbit/s; mass = 7.8 kg

- **PROMICS-3** = 3-D Ion Composition Spectrometer (PIs: I. Sandahl, IRF, Sweden; N. Pissarenko, IKI); (M = 1, 32); E = < 100 keV/Q; time resolution = 1.2 or 2.6 s/sample; data rate = 2 kbit/s; mass = 12.9 kg

2. **Thermal Plasma Experiment**

- **ALPHA-3** = Ion Trap Experiment (PI: V. Bezrukikh, IKI); $N > 1$ cm$^{-3}$; E < 25 eV/Q; time resolution = 16 s/sample; data rate = 1 kbit/s; mass = 3.5 kg

3. **Energetic Particles and X-Ray Experiments**

- **SKA-2** = Spectrometric Device Complex (PIs: E. Morozova, IKI, S. Fisher, CR); low and energetic charged particle composition and anisotropy; $E_e$ = 40 - 200 keV; $E_i$ = 50 keV - 150 MeV; time resolution = 120 s/sample; data rate = 0.25 kbit/s; mass = 22.4 kg

- **DOK-2X** = Electron and Proton Experiment (PIs: K. Kudela, Institute of Exp. Physics, Kosice, Slovakia; V. Lutsenko, IKI); measurement of fluxes and anisotropy; $E_e$ = 10 - 400 keV; $E_i$ = 15 - 1000 keV; time resolution = 120 or 1 s/sample; data rate = 2 kbit/s; mass = 5.5 kg

- **RF-15** = Solar X-Ray Experiment (PIs: O. Likin, IKI; F. Farnik, Astr. Institute, CR; J. Silvester, Poland); range = 2 - 200 keV; time resolution = 0.1 s/sample; data rate = 0.01 kbit/s; mass = 10 kg

4. **Fields and Wave Experiments**

**Complex ASPI** = Analysis of Plasma Spectra Instabilities (Project manager: S. Romanov, Science manager: S. Klimov, IKI). ASPI consists of the following instruments: OPERA, MIF-M, IFPE, ADS, PRAM, also the subsatellite C2-X wave experiment is included.

- **OPERA** = Onde di Plasma Et Radiazioni Aurorali (PIs: E. Amata, CNR/IFSI, Italy; S. Savin, IKI); measurement in frequency range of 0 - 150 kHz; time resolution = 2 or 64 sample/s; data rate = 3 kbit/s; mass = 3.7 kg

- **MIF-M** = Multi-component Investigations of Fluctuations of the Magnetic Field (PIs: S. Romanov, M. Nozdrachev, IKI); frequency range = 0 - 40 kHz; time resolution = 1 or 64 sample/s; data rate = 2.5 kbit/s; mass = 8.5 kg

- **IFPE** = Investigations of Fluctuations of Protons and Electrons (PIs: J. Büchner, MPE, Berlin; H. Lehmann, DLR Berlin; S. Romanov, IKI); frequency range = 0.1 - 1000 Hz; time resolution = 1 or 64 sample/s; data rate = 2.5 kbit/s; mass = 4.9 kg

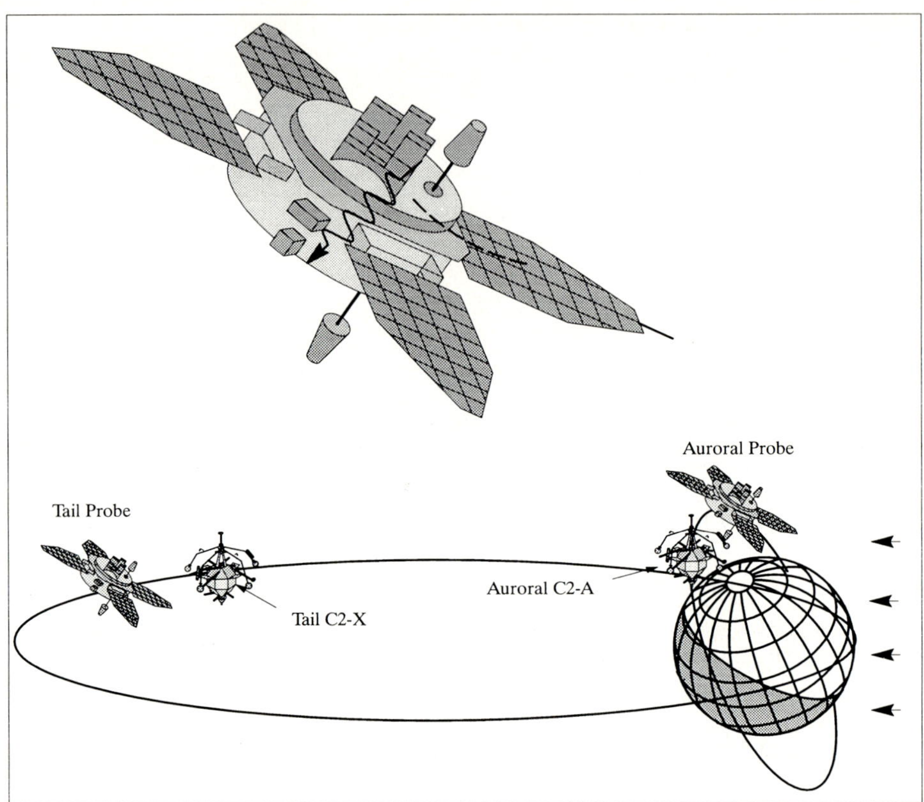


**Figure 43:    The INTERBALL S/C Models and Trajectories**

- **ADS** = Analyzer of Dynamic Spectra (PIs: J. Juchnievicz, Space Res. Institute, Poland; S. Klimov, IKI); frequency range = 0.5 Hz - 40 kHz, time resolution = 8 or 1 s; data rate = 1 kbit/s; mass = 5 kg

- **PRAM** = Adaptive Processing of Wave Information (PIs: S. Romanov IKI); data rate = 4 kbit/s; mass = 2.4 kg, digital processing of MIF-M, IFPE and RF-15 information.

- **IMAP-2** = Magnetometer (PIs: I. Arshinkov, SDS Lab., Bulgaria; L. Zhuzgov, IZMI-RAN); measurement range = $\pm$ 200 nT; resolution = $\pm$0.05 nT; time resolution = 0.12 s/sample; data rate = 0.5 kbit/s; mass = 4.9 kg

- **AKR-2** = Analyzer of Kilometric Radiation (PIs: L. Fisher, Komensky, University of Bratislava, Slovakia; V. Grigorjeva, Astron. Inst. of Moscow University); frequency range = 100 kHz - 1.5 MHz; time resolution = 0.06 s/sample; data rate = 1.8 kbit/s; mass = 2.1 kg

## 5.    Subsatellite C2-X

- **C2-X** = Measurement of electric and magnetic fields, waves, plasma and energetic particles (PIs: P. Triska, Ya. Voita, Geophysics Inst. Prague, CSFR); data rate = 20 kbit/s; mass = 50 kg; wave experiment (magnetic field: 0.1 Hz - 20 kHz; electric field: 0.1 Hz-250 kHz; plasma current: 0.1 Hz - 256 kHz). Note, C2-X is is incorporated into the ASPI Complex.

## A.51    IRS-1A (Indian Remote Sensing Satellite)

IRS-1A[166],[167] = Indian Remote Sensing Satellite. Launch: March 17, 1988 by ISRO (Soviet launch vehicle Vostok from the Baikonur Cosmodrome. The IRSO S/C control center is in Bangalore. TT&C function is provided by ISTRAC (ISRO Tracking Network), supported by DLR (GSOC, Weilheim), NOAA (Fairbanks), ESA (Malindi) and the USSR (Bearslake)  ground stations. IRS-1A is operational as of 1/1994.

IRS-1A is a 3-axis stabilized satellite in sun-synchronous orbit, nominal altitude = 904 km, inclination = 99.49°; Swath width: LISS 1 = 148 km, LISS 2 = 2 x 74 km; period = 103.2 minutes; the repeat cycle = 22 days. equator crossing at 10:26

Application: Land use, agriculture, forestry, hydrology, soil classification, coastal wetland mapping, natural resources (in particular pinpointing likely groundwater locations), disaster monitoring, cartography, etc.

**Sensors:**

**LISS** = Linear Imaging Self-Scanning Sensor. LISS-I and LISS-II are 2 multispectral cameras (CCD Detector Arrays). Resolution on the ground = 73 m with the first system, the second system has a resolution of 36.5 m. Both imaging cameras scan the same region.

**Data:** Image size: LISS 1 = 148 km; LISS 2 = 145 km
Data Rates:          LISS 1: S-Band,  5.2 Mbit/s, PCM/BPSK modulation
                     LISS 2:  X-Band, 2 x 10.4 Mbit/s, PCM/QPSK modulation

| Wavelength Ranges (μm) | Pixelsize (m) LISS 1 | LISS 2 | Radiometric Resolution (bit) |
|---|---|---|---|
| 0.46 - 0.52 (blue) | 36 | 72 | 7 |
| 0.52 - 0.59 (green) | 36 | 72 | 7 |
| 0.62 - 0.68 (red) | 36 | 72 | 7 |
| 0.77 - 0.86 (NIR) | 36 | 72 | 7 |

**Table 29:**    **Specification of the LISS Instruments**

After 3 years of service (March 1991) the sensor LISS-1 acquired > 57500 scenes, while LISS-2 acquired > 230,000 scenes. IRS-1A is still fully operational.

IRS-1A Data products are being sold nationally and internationally by the ISRO NRSA (National Remote Sensing Agency) Data Centre, Hyderabad. These products compete directly with Landsat TM and MSS data as well as with Spot image data.

## A.51.1    IRS-1B

**IRS-1B** is a follow-up satellite of IRS-1A.  The launch of the S/C was on August 29, 1991 (same tracking support configuration as for IRS-1).  IRS-1B is fully operational as of Sept. 16, 1991. The IRS-1B instruments are practically identical with  those of IRS-1A (Satellite mass = 975 kg).[168],[169]

Orbit: Polar sun-synchronous orbit; altitude = 904 km, inclination = 99.49°, period = 103.2 minutes. Repeat cycle: 22 days.

ISRO's policy allows ground stations from other countries to have direct access to the Indian satellite imagery. Any existing ground station equipped to receive data from SPOT or from Landsat will be able to receive the IRS-1B satellite's data with very minor changes (ISRO can supply the upgrade).


166) "Indian Remote Sensing Satellite and Associated Data Products", A.K.S. Gopalan, Proceedings of the Twenty-Third International Symposium of Remote Sensing of Environment, Vol. I, p. 71, ERIM, Ann Arbor Mich., 1990
167) IRS NewsLetter, ISRO, Vol. 2 No. 1, March 1991
168) 'India Expands Access to Imagery', Space News Aug. 26 - Sept. 8, 1991, p. 22
169) 'India Calls IRS-1B Launch a Success', Space News, September 9-15, 1991, p. 12

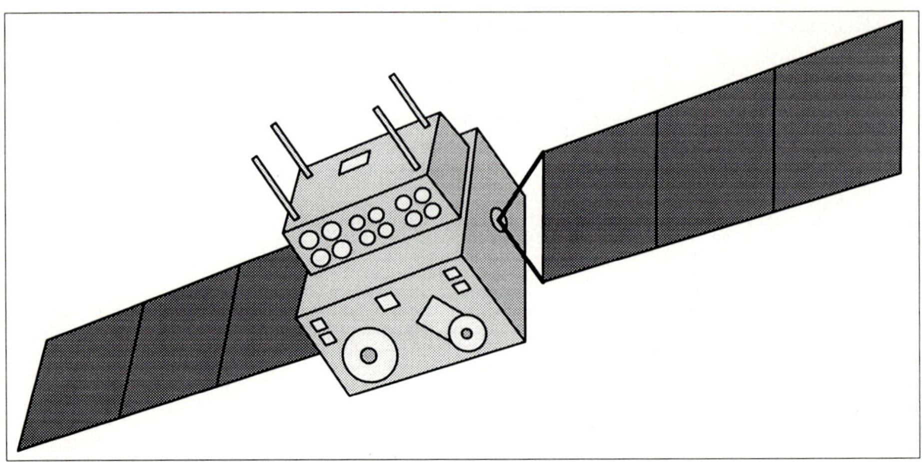

**Figure 44:**     **The IRS-1B S/C Model**

## A.51.2   IRS-1C/1D

IRS-1C[170] is (1994) being built by ISRO (Indian Space Research Organization) as a second generation remote sensing satellite series with enhanced capabilities in terms of spatial resolution and spectral bands. Onboard recorder.

Orbit: Polar sun-synchronous orbit; altitude = 817 km (±5 km); equatorial local crossing time = 10:30 A.M, descending node; inclination = 98.69°; period = 101.35 min;

Launch of IRS-1C is scheduled for mid 1995 with a Russian launcher (Vostok)[171]. The IRS-1D launch is planned for 1997. Both platforms are 3-axis stabilized.

Sensors:

The three sensors are cameras operating in the pushbroom scanning mode using solid state charge-coupled device (CCD) detectors.

- **PAN** = Panchromatic Camera.

  | | |
  |---|---|
  | Spectral bands: | 0.5 - 0.90 µm |
  | Spatial resolution: | $\leq$ 10 m |
  | Swath width: | 70.5 km |
  | Radiometric resolution: | 64 grey levels |
  | Repetition cycle: | 24 days (revisit is possible in 5 days) |
  | Off-nadir viewing capability | ± 26° (with this a revisit cycle of 5 days can be achieved) |
  | Image data rate: | 84.9 Mbit/s |

- **LISS-III** = Linear Imaging Self-Scanning Sensor. Continuous service multispectral imagery. Application: Land and water resources management.

  | | |
  |---|---|
  | Spectral bands: | 0.52 - 0.59 µm |
  | | 0.62 - 0.68 µm |
  | | 0.77 - 0.86 µm |
  | | 1.55 - 1.75 µm |
  | Spatial resolution: | 23.5 m |
  | Swath width: | 142 km |

---

170) IRS-1C Executive Summary,IRS-1C/1D Project, May 1990, ISRO
171) 'India's IRS-1C Satellite to offer sharper Images', Space News, May 25-31, 1992 p. 11

|                          |                    |
|--------------------------|--------------------|
| Radiometric resolution:  | 128 grey levels    |
| Repetition cycle:        | 24 days            |
| Data rate:               | 42.45 Mbit/s       |

- **WiFS** = Wide Field Sensor (camera). Application: Vegetation index mapping.

|                          |                    |
|--------------------------|--------------------|
| Spectral bands:          | 0.62 - 0.68 μm     |
|                          | 0.77 - 0.86 μm     |
| Spatial resolution:      | 188 m              |
| Swath width:             | 770 km             |
| Radiometric resolution:  | 128 grey levels    |
| Repetition cycle:        | 5 days             |
| Data rate:               | 2 Mbit/s           |

Transmission Data Rate: 125 Mbit/s (X-Band), TT&C in S-Band (Mission control centre at Bangalore). A great variety of data products are already defined for the IRS-1C mission.

## A.51.3   IRS-1E (P1)

Technological mission with the primary objective to test an ISRO-developed launch vehicle by the name of PSLV (Polar Space Launch Vehicle). The launch occurred on Sept. 20 1993, but the satellite failed to achieve orbit (**PSVL failure of 2. stage**).Orbit: Sun-synchronous polar orbit with equatorial crossing at 10:30 AM descending node, altitude = 904 km, period = 103 min, repeat cycle = 22 days.[172)]

Sensors:

- **LISS-I** = Linear Imaging Self-Scanning System. Same definition as under IRS-1A.

- **MEOSS** = Monocular Electro-Optical Stereo Scanner (DLR sensor) with push-broom CCD technology. MEOSS is a stereo camera system capable of recording 3 images simultaneously with a single lens by means of linear scanning. The scanner is operating in the spectral range of 0.57-0.7 μm. Resolution = 50 m along track, 158 m across track, 45 m in vertical direction; swath width = 510 km, 8 bit quantization. Application: stereo view capability to study topography, geology, terrain analysis and modelling, snow/ice mapping, meteorology (cloud height and movement), etc..

## A.51.4   IRS-P2

Cooperative ISRO/DLR mission with the objective to acquire remote sensing data for oceanographic-, land-, and atmospheric applications. The projected launch date is 1995 from SHAR (East coast launch facility of India) with a PSLV launcher. Three-axis stabilized S/C of 890 kg.[173)]

Orbit: Sun-synchronous circular orbit with an equatorial crossing at 10:30 AM descending node, altitude = 817 km, inclination = 98.7°, repeat cycle = 24 days, period = 101 min.

Sensors:

- **LISS-II** = Linear Imaging Self-Scanning System. Same definition as under IRS-1A (A.51 on page 139).

- **MOS** = Multispectral Optoelectronic Scanner (see also MOS-OBSOR under PRIRO-DA, A.82 on page 231). An imaging spectrometer for the visible and near-infrared spectrum (VNIR). MOS is provided by DLR Berlin.
  Objective: Image generation of the Earth surface (surface - atmosphere interaction, ocean color, phytoplankton, regional and global distributions of man-made aerosols

---

172) IRS-1E MEOSS Utilization Plan, ISRO, July 1991
173) Document on Configuration of IRS-P2 and MOS and their Interfaces, ISAC, Bangalore, Nov. 1992

and its links to gaseous admixtures, spectral and spatial cloudiness characteristics, etc.) in the VNIR region of 0.4 - 1.01 µm.

The sensor apparatus consists of three complementary instruments: (see also Figure 80). The MOS operation requires at least one calibration per month (with respect to the Sun).

- **MOS-A**: Spectral range = 755 - 768 nm (4 channels); nadir scan, FOV = 13.6° (swath direction), 0.34° (flight direction), swath width = 200 km, spatial resolution = 5.8 km x 1.3 km, radiometric resolution: dynamic range = 12 bit quantization; data rate = 37.5 kbit/s.
  Measurement of atmospheric turbidity. The data from MOS-A are used for correction of the atmospheric influence (scattering) on the multispectral data of MOS-B.

- **MOS-B**: Spectral range = .4 - 1.01 µm (13 channels in VNIR); nadir scan; FOV = 13.4° (swath direction), 0.1° (flight direction); swath width = 195 km; spatial resolution = 1.5 km x 1.5 km, radiometric resolution: dynamic range = 12 bit quantization; data rate = 121.5 kbit/s.

- **MOS-C**: Spectral range = 1.6 and 2.3 µm (2 channels); nadir scan; FOV = 13.3 ° (across-track) and 0.1° along-track; swath width = 195 km; spatial resolution = 1.5 km x 1.5 km; radiometric resolution: dynamic range = 12 bit quantization; data rate = 18.7 kbit/s

Data: No onboard data storage capability. Downlink broadcast of R/T science data in X-Band (8.316 GHz) to a dedicated ground station network and to a general user community. TT&C operations in S-Band. The following ground receiving stations are considered: Hyderabad and Mauritius (ISRO), Weilheim (DLR), Mas Palomas (ESA), Natal (INPE), Djakarta (LAPAN).

## A.52    ISEE (International Sun-Earth Explorer)

A NASA/ESA cooperative program involving 3 satellites. The ISEE-1 and ISEE-3 S/C were the principal contributions of NASA, while ISEE-2 was built and managed by ESA.

### A.52.1    ISEE-1 and -2 Mission

Objectives: Observation of the near-Earth magnetosphere and its boundaries, better understanding of many phenomena, such as the Earth's bow shock, the magnetosheath and magnetopause, interactions between the tail and aurorae, and particle populations and flows in the tail.

Joint launch of ISEE-1 and -2 S/C on Oct. 22. 1977 ( by NASA). Highly eccentric orbit around the Earth with an apogee of 23 $R_e$. Both S/C in the same orbit plane with a controllable separation distance, inclination = 28.9°. Both S/C reentered during Sept. 1987. The S/C were built by NASA (ISEE-1) and ESA (ISEE-2).

Sensors:[174](only a very brief overview is given, the reader is referred to the references). Note: the ISEE sensor acronyms are 3-letter words: two letters refer to the PI followed by one letter designating the S/C: ISEE-1 = M (Mother), ISEE-2 = D (Daughter), ISEE-3 = H (Heliocentric).

- **ANM/AND = Electrons & Protons Instrument** (ISEE-1 and -2, PI: K. A. Andersen, UCB). Objectives: Study of the varies energetic particle phenomena found in the Earth's magnetosphere, magnetopause, magnetosheath, bow shock, and upstream me-

---

174) **Special issue on 'Instrumentation for the International Sun-Earth Explorer Spacecraft' in IEEE Transactions on Geoscience Electronics, Volume 16, Nr.3, July 1978**

dium. Measurement over a wide range of energies, from ~ 1.5 to 300 keV for both electrons and protons.

- **BAM/PAD = Fast Plasma Experiment** (on ISEE-1 and -2, PIs: S. J. Bame, Los Alamos Scientific Lab, G. Paschmann, MPI Garching). Three electrostatic analyzers (with 90° spherical section) provide electron and proton measurements. Each instrument uses a divided secondary emitter system to intercept the analyzed particles. Measurement ranges:
  - Protons:   5 eV - 40 keV        ISEE-1
  - Electrons: 5 eV - 20 keV        ISEE-1
  - Ions:       50 eV - 40 keV       ISEE-2
  - Electrons: 5 eV - 20 keV        ISEE-2

  ISEE-1 carries also a solar wind experiment (SWE) to measure solar wind ions with high resolution.

- **FRM/FRD = Low Energy Protons & Electrons** (ISEE-1 and -2, PI: PI: L. A. Frank, Univ. of Iowa) Objective: study of directional intensities of positive ions and electrons over a large solid angle. Energy range: $1 \, eV \leq E/Q \leq 50 \, keV$ in 63 bands with 17% resolution

- **GUM/GUD = Plasma Wave Investigation** (ISEE-1 and -2, PI: D. A. Gurnett, Univ. of Iowa). Objective: Study on wave/particle interaction in the Earth's magnetosphere and in the solar wind. The instrument on ISEE-1 uses three electric dipole antennas with lengths of 215 m, 73.5 m, and 0.6 m for the electric field measurements, and a triaxial search coil antenna for magnetic field measurements. The ISEE-2 instrument uses two electric dipoles with lengths of 30 m and 0.6 m, and a single-axis search coil antenna for magnetic field measurements.
  - Magnetic field levels:          10 - 100 kHz (3 axis, 16 channels)
  - Electric field levels:          10 Hz - 10 kHz (3 axis, 12 channels).
  - The sweep frequency spectrum analysis of the electric field signals: 10 kHz - 200 kHz (128 steps)

- **HAM = Plasma Density Experiment** (ISEE-1, PI: C. C. Harvey, ESTEC)
  Objective: Study of the total electron density by means of radio techniques.
  The propagation experiment measures the phase velocity of a radio wave of frequency 683 kHz and 272.5 MHz (transmitted from the Mother S/C ISEE-1 and received on the daughter S/C, ISEE-2).

- **RUM/RUD = Fluxgate Magnetometer Experiment** (ISEE-1 and -2, PI: C. T. Russell, UCLA). Objective: Study of the dynamic plasma (magnetospheric phenomena) and field environment of the Earth.
  Magnetometer is mounted on a 3 m boom. The instrument has two commandable ranges of $\pm 256 \, \gamma$ and $\pm 8192 \, \gamma$ with an accuracy of 0.025%.

- **EGD = Solar Wind Ion Experiment** (ISEE-2, PIs: E. Egidi, G. Moreno, CNR Frascati). Objective: Study of the transient phenomena in the solar wind to obtain a spatial gradients of the interplanetary plasma.
  The instrument is based on two identical hemispherical electrostatic energy selectors for the measurement of positive ions in two different energy windows.
  - Ions:       50eV/q - 25 keV/q
  - Electrons: 35 eV - 7 keV

- **HEM = VLF Wave Propagation Experiment** (ISEE-1, PI: R. A. Helliwell, Stanford)
  Objective: Study of VLF-wave-particle interactions in the magnetosphere (note: VLF = Very Low Frequency in the 10 - 30 kHz range). A second goal is the determination of the effects upon energetic particles in the magnetosphere of electrical power transmission line radiation.
  The instrument setup consists of three separate elements:

- a broadband VLF receiver on ISEE-1
- a broadband VLF transmitter located at Siple station in the Antarctic
- ground stations in the Antarctic and Canada

- **HPM = DC Electric Field Experiment** (ISEE-1, PI: J. P. Heppner, GSFC)
  Objective: Study of the transfer mechanisms (mass, momentum, and energy at the magnetopause), in particular the spatial extent and variability of the zone of strong electric fields, or fast convection in adjacent magnetospheric regions.
  Instrument: 8 channel spectrum analyzer. Measurement ranges: 0.1 Hz - 3200 Hz in 9 steps.

- **HOM = Low Energy Cosmic Ray Experiment** (ISEE-1, PI: D. Hovestadt, MPI Garching). Objective: Measurement of elemental abundances, charge state composition, energy spectra, and angular distributions of energetic ions in the energy range of 2 keV/charge to 80 MeV/nucleon, and of electrons between 75 - 1300 keV.
  The instrument consists of three sensor systems:
  - ULECA is an electrostatic deflection analyzer, its energy range from $\sim 3$ to 560 keV/charge
  - ULEWAT is a double dE/dX versus E thin-window flow-through proportional counter/solid-state detector telescope covering the energy range from 0.2 to 80 MeV/nucleon (Fe).
  - the ULEZEQ sensor consists of a combination of an electrostatic deflection analyzer and a thin-window proportional counter. The energy range is 0.4 MeV/nucleon to 6 MeV/nucleon. Objective: collection of composition data in the trapped radiation zone.

- **MOM = Quasi-Static Electric Field Experiment** (ISEE-1, PI: F. S. Mozer, UCB)
  Objectives:
  - study of the quasi-static electric field over a dynamic range of 0.1 - 200mV/m
  - study of wave electric fields at frequencies $<1000$ Hz with a sensitivity $<1\mu$V/m $(Hz)^{1/2}$ at all frequencies
  - study of plasma density and temperature

  Measurements are made of the potential difference between a pair of 8 cm diameter vitreous carbon spheres which are mounted on the ends of wire booms and are separated by 73.5 m in the spin plane of the satellite.

- **OGM = Fast Electron Spectrometer Experiment** (ISEE-1, PI: K. W. Ogilvie, GSFC)
  Objective: Study of three-dimensional plasma distribution in the solar wind, magnetosheath, outer magnetosphere, and near tail regions.
  Instrument with three energy ranges: 7.5-512 eV, 11-2062 eV, and 109-7285 eV.

- **SHM = Ion Composition Experiment** (ISEE-1, PI: R. D. Sharp, Lockheed, Palo Alto)
  Objective: Study of the composition of the hot magnetospheric plasma. Ion composition of the ring current, the plasma sheet, the plasmasphere, the magnetosheath, and the solar wind in order to establish the origin of the plasmas in the various regimes of the magnetosphere and to identify mass and charge dependent acceleration, transport, and loss processes.
  The instrument consists of two ion mass spectrometers which can be operated independently. The spectrometers point 5° above and 5° below the ISEE-1 spin plane. Measurement ranges: 1 AMU to > 150 AMU in 64 channels at each of 32 energy channels covering the energy per charge range from 0 to $\sim 17$ keV/e.

- **WIM/KED = Medium Energy Particles Experiment** (ISEE-1, -2, PI: D. J. Williams, NOAA, Boulder Co.). Objective: Study and identify the physical mechanisms of medium energy particles associated with acceleration, source and loss processes, and boundary and interface phenomena throughout the orbits of ISEE-1 and -2.
  The experiment consists of the WIM instrument (Wide Angle Particle Spectrometer

and a Heavy Ion Telescope) on ISEE-1 and the KED instrument (five sensor systems mounted at various angular positions with respect to the S/C spin axis) on ISEE-2.

- Protons: 20 keV - 2 MeV in 8 channels, in 16 channels          ISEE-1
- Electrons: 20 keV - 1.2 MeV in 8 channels, in 16 channels      ISEE-1
- Protons: 20 keV - 2 MeV in 12 channels                         ISEE-2
- Electrons: 20 keV - 300 keV (to 1.2 MeV for 90° unit)          ISEE-2

## A.52.2   ISEE-3 Mission

Objective: Measurement of the ingredients of the interplanetary medium: the properties of the solar-wind plasma, magnetic and electric fields, solar and heliospheric charged particles and galactic cosmic rays. Exploration of the distant geomagnetic tail.[175]
ISEE-3 S/C: Launch Aug. 12 1978 (NASA) by Delta vehicle from Cape Canaveral.

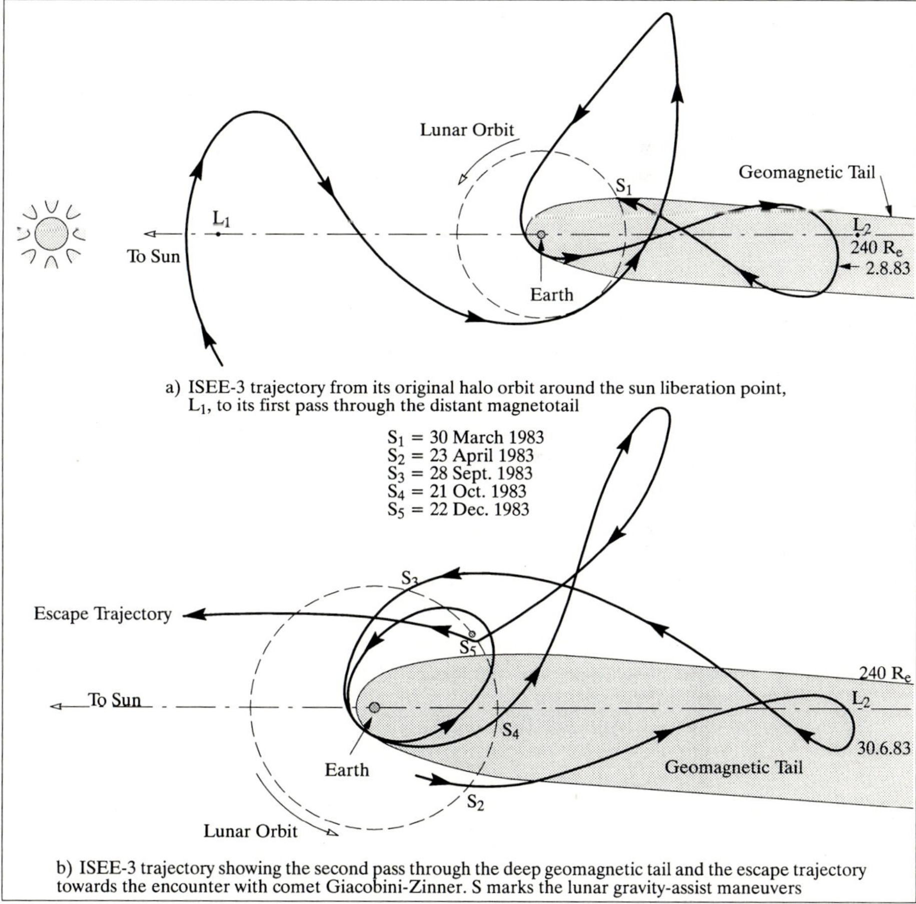


**Figure 45:    ISEE-3 Spacecraft Trajectory Overview**

175) K. P. Wenzel, "Earth's Distant Geomagnetic Tail Explored by ISEE-3 Spacecraft", ESA Bulletin 37, 1984 pp. 46-50

Orbit: ISEE-3 was first placed 1.5 million km ($\sim 240$ Earth radii, $R_e$) sunward from the Earth in a halo orbit around the so-called Lagrangian Point $L_1$ (also referred to as liberation point). At $L_1$ the S/C co-rotated around the Sun with the Earth during the course of each year.

ISEE-3 was redirected in mid-1982 from its $L_1$ position (which it had occupied for more than 4 years) to study the distant Earth's magnetic tail throughout most of 1983. After that ISEE-3 was redirected into a heliocentric orbit to fly past the Giacobini Zinner comet on Sept. 11, 1985, and upstream of Halley's comet on March 28, 1986. The S/C was renamed ICE (International Cometary Explorer) in 1984.

**Sensors:**

The ISEE-3 payload consisted of 13 instruments provided by both US and European groups.

- **ANH = X-Rays and Electrons Instrument** (PI: K. A. Anderson, UCB)
  - Measurement of solar flare X-ray bursts and transient cosmic gamma-ray bursts. A proportional counter and scintillation detector cover the energy range from 5 - 228 keV.
  - Measurement of electrons from $\sim 2$ keV to $\sim 1$ MeV with high energy and angular resolution. (Study of interplanetary and solar electrons in the energy range between the solar wind and galactic cosmic rays).

- **BAH = Solar Wind Plasma Experiment** (PI: S. J. Bame, Los Alamos Scientific Lab) Two electrostatic analyzers ( with 135° spherical section) provide electron and ion measurements. Each instrument uses a divided secondary emitter system to intercept the analyzed particles.

- **HKH = High Energy Cosmic Ray Experiment** (PI: H. H. Heckman, UCB) Multidetector cosmic ray experiment to identify the charge and mass of incident cosmic ray nuclei from H through $F_e$ species (over energy ranges from 20 to 500 MeV/nucleon).

- **HOH = Low Energy Cosmic Ray Experiment** (PI: D. Hovestadt, MPI, Garching) Objective: Study of nuclear and ionic composition of solar, interplanetary, and magnetospheric accelerated and trapped particles.
  Measurement of elemental abundances, charge state composition, energy spectra, and angular distributions of energetic ions in the energy range of 2 keV/charge to 80 MeV/nucleon, and of electrons between 75 - 1300 keV.

- **DFH = Low Energy Proton Experiment** (PI: R. J. Hynds, Imperial College, London).[176] Objective: Study of low energy protons from a solar flare to relate particle fluxes measured near the Earth to fluxes in the upper corona (investigation of the gross scale of coronal control) .
  DFH experiment to measure low energy protons in the energy range from 35-1600 keV.

- **MEH = Cosmic Ray Electrons and Nuclei** (PI: P. Meyer, Univ. of Chicago). Objective: Study of the long and short-term variability of cosmic ray electrons and nuclei.
  Measurement of the energy spectrum of cosmic electrons in the range of 5-400 MeV. In addition, determination of the energy spectra and relative abundances of nuclei from protons in the iron group (energies from 30 MeV/n to 15 GeV/n).

- **OGH = Plasma Composition Experiment** (PI: K. W. Ogilvie, GSFC) Objective: Study of the dynamics and energetics of the solar wind acceleration region. Ion mass spectrometer for the measurement of ionic composition of the solar wind.

- **SCH = Plasma Wave Instrument** (PI: F. L. Scarf, TRW, Los Angeles) Objective: Study of interplanetary wave-particle interactions in the spectral range from 1 Hz to 100 kHz.

176) A. Balogh, R. J. Hynds, J. J. van Rooijen, G. A. Stevens, T. R. Sanderson, K. P. Wenzel, "Energetic Particles in the Heliosphere - Results from the ISEE-3 Spacecraft", ESA Bulletin 27, 1981, pp. 4-12

Measurements of magnetic field and electric field components on long booms (90 m tip to tip). Magnetic field levels: 8 channels, 60 dB range, 20 Hz - 1 kHz. Electric field levels: 16 channels, 80 dB range, 20 Hz - 100 kHz.

- **SBH = Radio Mapping Experiment** (PI: J. L. Steinberg, Meudon Obs., Paris)
  Objectives: a) monitoring of the solar wind flow and perturbations of the magnetic field in conjunction with simultaneous measurements on ISEE-1 and -2 (bow shock, magnetopause, neutral sheet), and b) propagation studies of particle fluxes and shock waves in the solar wind (large scale structure of the magnetic field).
  Measurement of the interplanetary scintillation of natural radio sources with the use of two dipole antennas, one in the spin plane (90 m tip to tip) and one along the spin axis (15 m tip to tip). Each of these antennas drives two radiometers (10 kHz bandwidth and 3 kHz bandwidth).

- **SMH = Helium Vector Magnetometer** (PI: E. J. Smith, JPL)
  Objective: Continuous observation of the interplanetary magnetic field near 1 AU (structure, direction, polarity north-south component, magnitude, dynamic phenomena).
  Boom-mounted magnetometer sensor (3 m) with the following characteristics:
  - 8 dynamic ranges of: $\pm 4$, $\pm 14$, $\pm 42$, $\pm 144$, $\pm 640$, $\pm 4000$, $\pm 22000$, $\pm 140000\,\gamma$
  - frequency response: 0 - 3 Hz within three bands (0.1 - 1, 1 - 3, and 3 - 10 Hz) for measurements of fluctuations parallel to the S/C spin axis.

- **STH = Heavy Isotope Spectrometer Telescope, HIST** (PI: E. C. Stone, CIT)
  Objective: measurement of the isotopic composition and energy of solar, galactic, and interplanetary cosmic ray nuclei for the elements Li through Ni in the energy range from $\sim 5$ to 250 MeV/nucleon.
  Charge, Isotope, and energy range: Z 3 - 28 (Li to Ni); A 6 - 64 (($^6$Li to $^{64}$Ni)
  Mass resolution: Li 0.065 - 0.83 proton masses; Fe 0.18 - -0.22 proton masses.

- **TYH = Medium Energy Cosmic Ray Experiment** (PI: T. Y. von Rosenvinge, GSFC)
  Objective: measurement of the charge composition of nuclear energetic particles over the energy ranges from $\sim 1$ - 500 MeV/ nucleon, and charges from Z=1 to Z=28.
  The experiment consists of two telescopes. The combined charge, mass, and energy intervals covered by these two telescopes are as follows:
  - Nuclei charge of energy spectra: Z = 1-30, energy range 1-500 MeV/nucleon
  - Isotopes:      Z=1, $\Delta$M=1, from 4-70 MeV/n
         Z=2, $\Delta$M=1 from 1-70 MeV/n
         Z=3-7, $\Delta$M=1 from 30-140 MeV/n
  - Electrons:      $\sim$2-10 MeV
  - Anisotropies:      Z=1-26 (1-150 MeV/n for Z=1,2); Electrons: 2-10 MeV

## A.53     JERS-1 (Japanese Earth Resources Satellite)

JERS-1 = "Japanese Earth Resources Satellite". JERS-1 is a joint NASDA/MITI Project (NASDA developed the satellite, MITI sponsored the instruments). The NASDA launch (on H-I carrier) of JERS-1 occurred on Feb. 11, 1992 (from Tanegashima). The initial SAR antenna unfolding problems were overcome in the first half of April 1992[177]. Nominal mission duration = 2 years. Satellite mass = 1400 kg.

Application: Survey of geological phenomena, land usage (agriculture, forestry), observation of coastal regions, geologic maps, environment, disaster monitoring, etc.

Objectives: Generation of global data sets with SAR and OPs sensors in order to survey resources; to establish an integrated Earth observation system, to verify instrument/system performances.

---

177) "Japanese Elated as JERS-1 Rescue Works", Space News, April 13-19, 1992, p. 1 and p. 20

Orbit: Sun-synchronous polar orbit; Inclination = 97.7°; repeat cycle = 44 days (westward); altitude = 568 km; period = 96 minutes, local mean time = 10:30 - 11:00 AM

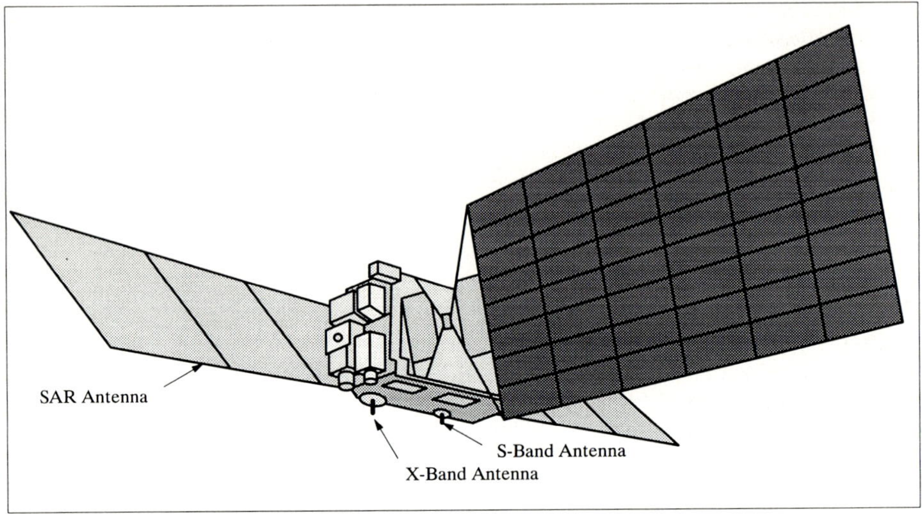


**Figure 46:    JERS-1 Spacecraft Model**

**Sensors:**

- **SAR** (Synthetic Aperture Radar). Measurement in L-Band (1.275 GHz, 15 MHz bandwidth) in HH polarization; Look Angle = 35.21° (off-Nadir). Spatial resolution of 18 m (3 Looks); Swath width = 75 km. Resolution = 18 (range) x 18 m (azimuth, 3 looks);

- **OPS = Optical Sensor**. Measurement of reflected sun light in 8 bands (color imagery), 3 are in the VIS and NIR bands, 4 SWIR bands, (and 1 stereoscopic band). Swath width = 75 km, Resolution = 18.3 m (scanning direction) x 24.2 m (flight direction). Band 4 is for off-nadir viewing (15 + 33/60 degree); band 3 and 4 make a stereo pair.
  - Band 1 = 0.52 - 0.60 μm
  - Band 2 = 0.63 - 0.69 μm
  - Band 3 = 0.76 - 0.86 μm
  - Band 4 = 0.76 - 0.86 μm (forward viewing)
  - Band 5 = 1.60 - 1.71 μm
  - Band 6 = 2.01 - 2.12 μm
  - Band 7 = 2.13 - 2.25 μm
  - Band 8 = 2.27 - 2.40 μm

**Data:**

SAR data and OPS data are recorded onboard and transmitted to the stations Hatoyama and Fairbanks during a pass. Two downlink frequencies: 8.15 GHz and 8.35 GHz (X-Band). Data rate: = 60 Mbit/s per carrier

Ground stations:
NASDA, EOC (Hotoyama, Japan)
Tokai University (Kumamoto, Japan)
National Institute of Polar Research (Syowa, Antarctica)
Alaska SAR Facility (ASF, Fairbanks, Alaska)
Canada, CCRS (Gatineau, Prince Albert)

ESA (Kiruna, Fucino, Maspalomas)
National Research Council of Thailand (Bangkok, Thailand)
ACRES (Australia)
etc.

JERS-1 SAR Data Products:[178]
Level 0    Unprocessed signal data
Level 1    Partially processed signal data products, range compression only
Level 1.1  Basic image product
Level 2.0  Bulk image product
Level 2.1  Standard geocoded image product
Level 3.0  Precise correct image product
Level 4    Geocoded with terrain correction

JERS-1 OPS Data Products:
Level 0    Unprocessed data
Level 1    Radiometrically processed image
Level 2    geometrically corrected image
Level 3    Precisely corrected image using GCPs
Level 4    Precisely corrected image using registered marks
Level 5    Stereo image


## A.54   LAGEOS-I (Laser Geodynamics Satellite)

LAGEOS-I[179] = Laser Geodynamics Satellite (NASA). Launch: May 4 1976 with a Delta
launch vehicle from Vandenberg.
Orbit: Near-circular orbit, altitude = 5950 km, inclination = 110°.

Objective: First NASA satellite dedicated wholly to laser ranging. LAGEOS was designed
to act as a permanent reference point so that the Earth's progress could be tracked relative
to the satellite (in contrast to the traditional system of tracking satellites relative to the
Earth). The USGS uses LAGEOS to measure continental drift (plate tectonics, crustal deformations).

The first 4 years until 1980 were devoted to determining Lageos precise orbit and to building
up a global network of 14 Earth stations. By accurately measuring the time for a laser pulse
to travel to the satellite and return, the position of the laser system could be determined to
about 10 cm. Under NASA's Crustal Dynamics Project (started in 1979), 56 investigators
from 12 countries were making repeated measurements between their locations and Lageos.

The Lageos satellite is an aluminum sphere with a brass core. Its 426 prisms, called cube-
corner reflectors, give it an appearance of a golf ball (60 cm diameter and 411 kg weight).
The three-dimensional prisms reflect laser beams back to the source, regardless of the angle
from which they come.

The LAGEOS-I satellite is known to extremely high accuracy, the location of a laser ranging
station on the surface of the Earth can be determined to a precision of less than 1 cm (by
measuring the time for a laser pulse to travel from the laser ranging station to the satellite
and return).

---

178) K. Maeda, M. Nakai, O. Ryuguji, 'JERS-1/ERS-1 Verification Program and Future Verification Program', Ad-
       vanced Space Research, Vol. 12, No. 7, pp. 327-331, 1992
179) Jane's Spaceflight Directory 1988-89, Fourth Edition, pp. 83-84

### A.54.1   LAGEOS-II

**LAGEOS-II**[180],[181],[182] is a collaborative NASA-ASI mission (ASI built LAGEOS-II based on the same design as the NASA-produced LAGEOS-I), a follow-up of LAGEOS-I. Shuttle (Columbia) launch from Cape Canaveral: Oct. 22 1992 (NASA).

Italy developed and provided the IRIS (Italian Research Interim Stage), a solid-fueled booster, which carried the satellite from the Shuttle's parking orbit into the required Lageos II orbit.

Orbit: LAGEOS-II and LAGEOS-I are deployed in prograde (LAGEOS-II: 52° inclination) and retrograde (LAGEOS-I: 110° inclination) orbital planes. Near-circular orbit, altitude = 5950 km.

Objectives: LAGEOS-II is an integral part of the Crustal Dynamics Project (CDP). Study of the Earth's crust in the Mediterranean region. Research in solid Earth geophysics [study of global and local tectonic processes, polar motion and Earth rotation, determination of Universal Time (UT-1), the recovery of Earth and ocean tidal parameters, and geopotential modelling].

LAGEOS-II is an identical S/C to LAGEOS-I. LAGEOS-I and -II are passive satellites dedicated exclusively to laser ranging.

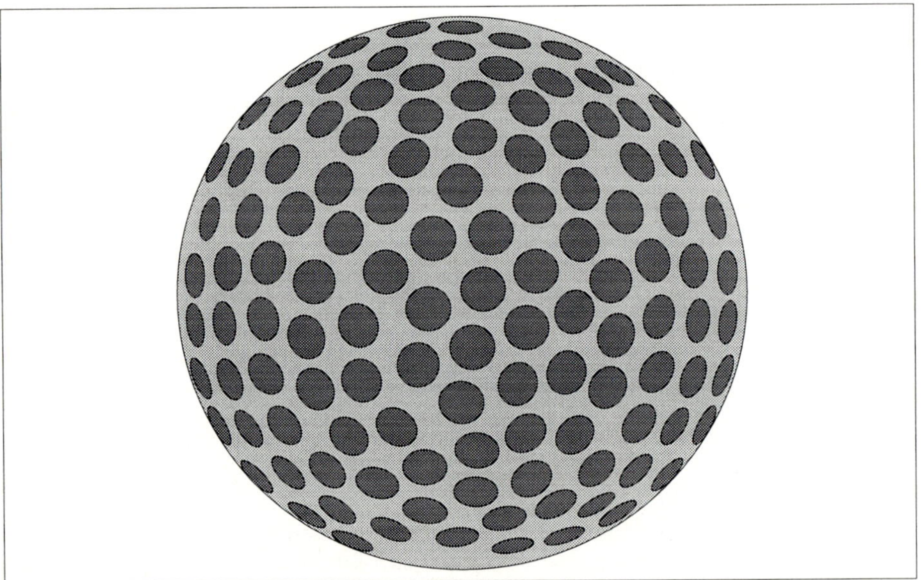

**Figure 47:**    **Model of the LAGEOS Sphere with its Reflectors**

Both LAGEOS satellites will be tracked by a global network of fixed and transportable lasers from some 65 sites. The current precision for laser systems varies from 15 cm to less than 1 cm for single-shot range measurements.

Data to be made available to investigators consist of both preprocessed and analyzed data (i.e. station positions, baselines, and Earth rotation parameters as a function of time). Data


180)  R. Kolenkiewicz, S. Zerbini, "LAGEOS-II: A collaborative NASA-ASI Mission", CSTG Bulletin No.11, Title: New Satellite Missions for Solid Earth Studies,, June 1989, pp. 13-18
181)  'Columbia Successfully Lofts Italian Lageos Satellite', Space News, Oct. 26-Nov. 1, 1992, p. 13
182)  NASA/ASI Lageos II brochure

will be archived in the Crustal Dynamics Data Information System (CDDIS) at GSFC. The facility is accessible for interactive log-ins through SPAN, INTERnet, etc.


## A.55    LANDSAT

Background: NASA launched ERTS-1 (Earth Resources Technology Satellite) on July 23 1972. This satellite was subsequently renamed Landsat-1. The Landsat system remained an experimental NASA program until 1983. The system was then declared operational and its management was turned over to NOAA. The Land Remote-Sensing Commercialization Act of 1984 authorized a phased commercialization of remote sensing data.

The NOAA[183),184),185)] Landsat satellite family [built by GE Astro Space of East Windsor, N.J. (S/C), and Hughes Santa Barbara Research Center (sensors)] is operational since 1972 in the field of Earth observation ( land observation in the visible spectral range). Starting in 1984 the Landsat operation and data handling was handed over to a commercial contractor company: **Eosat** (Earth Observation Satellite Company, Lanham, Md; Eosat is a joint venture of GE Co. and Hughes Aircraft Co.). The science data (of LS-4 and LS-5) are processed and archived by EOSAT and commercially distributed. The LS-4 and LS-5 S/C and sensors are regarded second generation technology in the LS program. While LS-6 (with the enhanced Thematic Mapper) will be built and operated by Eosat (including data archiving and distribution)[186)] it looks that the future Landsat program moves back to NASA control.

Application: Land Use, Agriculture, Forestry, Geology, Water Resources, Mapping, etc. Landsat data are particularly suited for long-term estimation and monitoring of standing vegetation biomass, biological productivity, and the movement of fragile ecosystem boundaries.

- LS-1 Launch: July 23, 1972, altitude = 907 km,; End of operation: 1978
- LS-2 Launch: Jan. 22, 1975, altitude = 908 km, ; End of operation: 1983
- LS-3 Launch: March 5, 1978, altitude = 915 km, ; End of operation: 1983
- LS-4 Launch: July 16, 1982, altitude = 705 km, ; is still in operation as of 1993 (put on standby July 1993)
- **LS-5 Launch:  March 1, 1984; altitude = 705 km, is (1994) in operation;**
  (Provides MSS and TM data to foreign ground stations via S-Band and X-Band)

The commercial purchase of the science data is very expensive for the research community (a Landsat tape costs about $ 5000). NOAA and Eosat made an agreement in Nov. 1990, that all Landsat data older than 2 years return into the category "Public Domain", hence are affordable for research. The "Public Domain" data is provided and distributed by the EROS Data Center (EDC), Sioux Falls S.D..
Effective February 1993 - all MSS data is in the public domain at EDC

Orbit for LS-1 to LS-3: Sun-synchronous polar orbit (AM Orbit), Altitude = 907- 915 km, Inclination = 99.2º, Period = 103 minutes, Repeat coverage = 18 days

Orbit for LS-4 and LS-5: Sun-synchronous polar orbit (AM Orbit), Altitude = 705 km, Inclination = 98.2º, Period = 99 minutes, Repeat coverage = 16 days

**Sensors:**

- **MSS** (Multi-Spectral Scanner) on LS 1-5; Radiometric resolution = 8 Bit; spatial resolution = 79 m; swath width = 185 km; Spectral ranges for Landsat-1 to Landsat-3


183)  "Fernerkundung, Daten und Anwendungen", W. Markwitz/R. Winter, Wichmann Verlag, 1989, S. 32 - 36
184)  "Taschenbuch zur Fernerkundung", F. Strathmann, Wichmann Verlag, 1990
185)  Monitoring Earth's Ocean, Land, and Atmosphere from Space, Volume 97, AIAA, 1985, Chapter 3
186)  'Landsat Moves Under Control of NASA, Hesitant DoD', Space News, Dec. 2-8, 1991, p. 5

Channel 4: 500 - 600 nm
Channel 5: 600 - 700 nm
Channel 6: 700 - 800 nm
Channel 7: 800 - 1100 nm
Channel 8: 10.4 - 12.6 μ (chan. 8  only on LS-3 until July 11, 1978)

Spectral ranges: Landsat-4 to Landsat-5
Channel 1: 500 - 600 nm
Channel 2: 600 - 700 nm
Channel 3: 700 - 800 nm
Channel 4: 800 - 1100 nm

Data:
Image size: 2583 Lines x 5500 Pixel (EDIPS Format)
Image size: 2286 Lines x 3600 Pixel (Telespazio Format)

Transmission: frequency = 2229.5 MHz (LS 1-3), 2265.5 MHz (LS 4-5)
Data rate: 15.06 Mbit/s

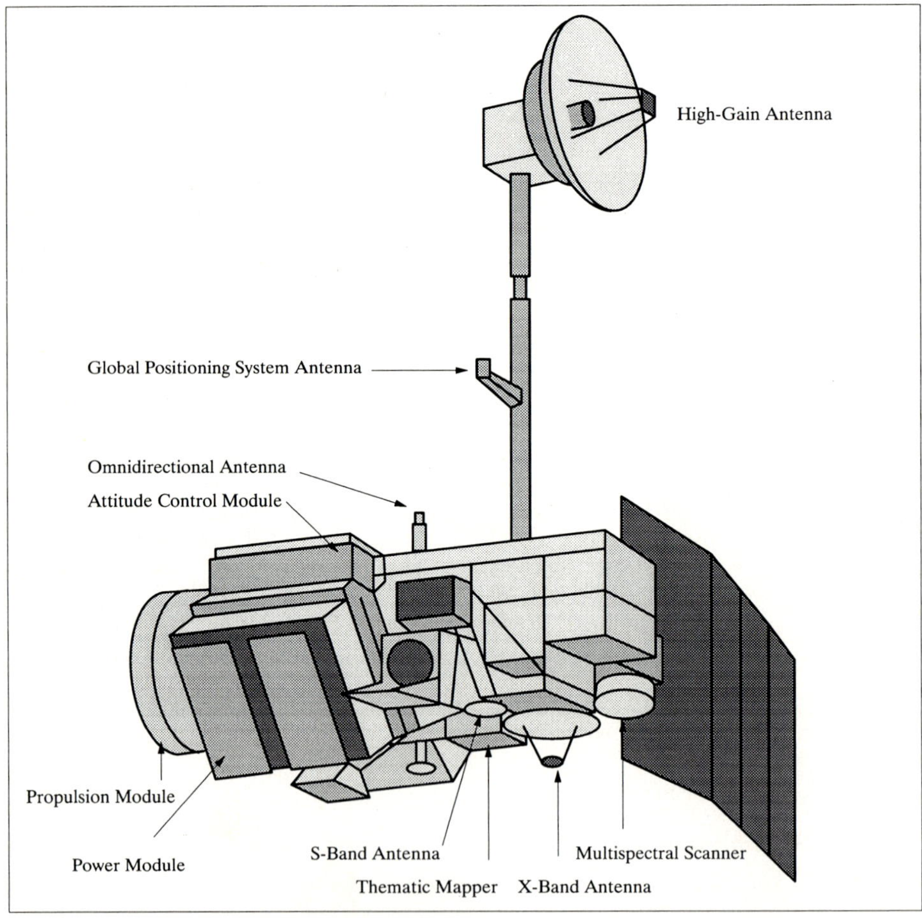


**Figure 48:     Model of the Landsat-4 and 5 Spacecraft**

- **TM** (Thematic Mapper); TM is a scanning optical sensor operating in the visible and infrared ranges.
  Operational since 1984 (first operation on LS-4). Geometric resolution = 30 m (120 m for channel 6), swath width = 185 km.

  Spectral Ranges:
  - Channel 1: 450 - 520 nm (VIS, blue)
  - Channel 2: 520 - 600 nm (VIS, green)
  - Channel 3: 600 - 690 nm (VIS, red)
  - Channel 4: 760 - 900 nm (NIR)
  - Channel 5: 1550 - 1750 nm (SWIR)
  - Channel 7: 2080 - 2350 nm (SWIR)
  - Channel 6: 10.4 - 12.5 $\mu$ (TIR)

  Data:
  Image size: 185 x 172 km; 5760 Lines x 6928 Pixels
  Transmission: frequency = 8215.5 MHz; Data rate = 84.9 Mbit/s (246 MByte per scene).

- **RBV** (Return Beam Vidicon Camera). Geometric resolution = 40 m. Spectral range: 505 - 750 nm. Ground image size = 98 x 98 km.

Data Transmission from NOAA S/C to ground stations via S-Band, X-Band, and Ku-Band.


## A.55.1  Landsat-6

The LS-6 payload was launched on October 5, 1993 with a Titan 2 booster from Vandenberg Air Force Base. S/C builder and integrator: Martin Marietta Astro Space (formerly General Electric Astro Space). **The satellite failed to achieve its orbit**, communication with the satellite was never established. A formal review is conducted by NOAA to investigate the failure.[187]

Orbit: Altitude = 705 km, polar sun-synchronous orbit, inclination = 98.2°, Period = 99 min, repetition cycle (repeat coverage) = 16 days, equatorial crossing time: 9:45 AM. Nominal life = 5 years.

**Sensor:**
LS-6 is designed to carry a single sensor, the Enhanced Thematic Mapper (ETM) which includes several new features that will significantly improve the data quality.

- **ETM (Enhanced Thematic Mapper)**

  LS-6 Capabilities:  ETM (seven TM Bands, plus an additional panchromatic band)
  Two onboard recorders (playback to ground in X-Band), each recorder is capable of recording/reproducing at a rate of 85 Mbit/s, and each can store $\approx$ 15 minutes worth of image data, or 29 scenes.
  Three pointable antennas
  Simultaneous acquisition of TM and Pan Data
  ETM generates 3 different data streams:
  - Mode 1    Seven Spectral bands
  - Mode 2    Panchromatic band (P) with channels 4,5, and 6
  - Mode 3    Panchromatic band (P) with channels 4,6, and 7

---

187) 'Satellite Loss Raises Questions for Eosat's Future', Space News, October 11-17, 1993, p. 3

| Band Nr. | Wavelength (μm) | Detectors | IFOV (μrad) | Ground Res. (m) |
|---|---|---|---|---|
| PAN | 0.50 - 0.90 | SiPD (32) | 18.5 x 21.3 | 13 x 15 |
| 1 | 0.45 - 0.52 | SiPD (16) | 42.5 | 30 |
| 2 | 0.52 - 0.60 | SiPD (16) | 42.5 | 30 |
| 3 | 0.63 - 0.69 | SiPD (16) | 42.5 | 30 |
| 4 | 0.76 - 0.90 | SiPD (16) | 42.5 | 30 |
| 5 | 1.55 - 1.75 | InSb (16) | 42.5 | 30 |
| 7 | 2.08 - 2.35 | InSb (16) | 42.5 | 30 |
| 6 | 10.4 - 12.5 | HgCdTe (4) | 170 | 120 |

**Table 30:**     **Summary of Landsat-6 ETM Bandwidth Specification**

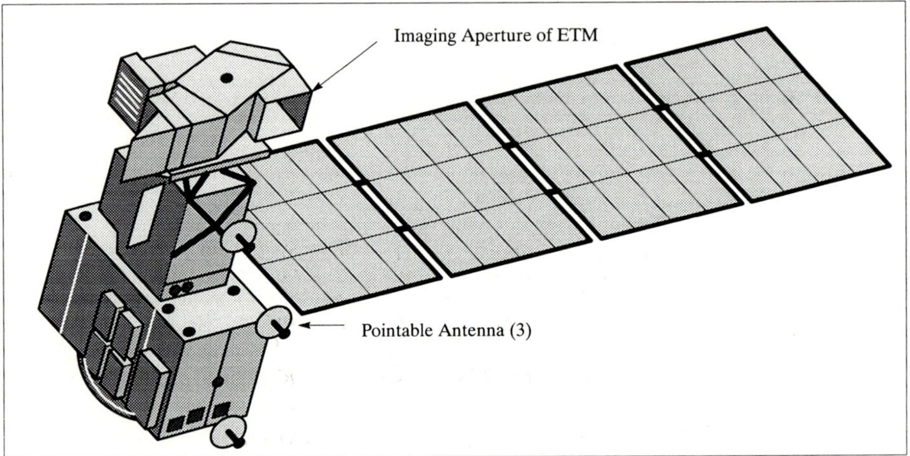


**Figure 49:**     **The Landsat-6 S/C Model**

Spectral ranges:

- Channel P: 500 - 900 nm (panchromatic), Appl.: 15 m resolution, cartography
- Channel 1: 450 - 520 nm (VIS, blue), Appl.: Water penetration, bathymetry (water depth), Chlorophyll absorption, distinguishes deciduous/coniferous
- Channel 2: 520 - 600 nm (VIS, green), Appl.: Matches green reflectance, peak of healthy vegetation, assessment of plant vigor
- Channel 3: 600 - 690 nm (VIS, red), Appl.: Chlorophyll absorption, plant type discrimination
- Channel 4: 760 - 900 nm (NIR), Appl.: Plant cell structure, plant vigor, complete absorption by water, shoreline mapping
- Channel 5: 1550 - 1750 nm (SWIR), Appl.: Moisture content, soil mapping, thin cloud penetration
- Channel 7: 2080 - 2350 nm (SWIR), Appl.: Hydroxyl ion absorption, geology
- Channel 6: 10.4 - 12.5 μ (TIR), Appl.: Brightness temperature, soil moisture, plant heat stress.

Ground pixel resolution: 15 m (panchromatic), 30 m (ch. 1-5, ch.7) 120 m (ch.6)
Ground image size: 185 x 185 km

Data:
Landsat-6 has three X-Band frequencies for downlink (8082.5, 8212.5, and 8342.5 MHz) to allow combinations of up to two real time data links (TM and/or a panchromatic mode) and two tape recorder dumps. The ETM outputs two 85 Mbit/s serial composite streams, each

consisting of the digital image area, internal ETM calibration data, image-related timing data, S/C time code, ephemeris, and attitude data. The ancillary timing, calibration, and attitude data are used by the Landsat-6 Image Data Processing System S/W to perform geometric image corrections. The X-Band downlink communications subsystem permits the simultaneous downlinks to three stations using pointable antennas.[188]

## A.55.2  Landsat-7

A Landsat-7 payload is being considered and in the process of being ordered as a follow-up satellite for a planned launch period in 1997 (procurement responsibility for LS-7 by DOD and NASA)[189]. The LS−7 program considers the design and development of a new sensor, referred to as HRMSI (High Resolution Multispectral Imager).

## A.55.3  Landsat Archival Data (Status)[190]

Landsat data archived by the US (about 820,000 scenes as of 1992) exists in several formats which, in most cases, can only be processed by unique hardware/software systems. Some of these data become irretrievable due to magnetic tape degradation and obsolescence of processing systems. To date, only a small fraction of this data have been converted to usable and maintainable media. Only 15% of of the approx. 310,000 scenes of MSS data acquired between 1972 and 1978 have been converted to usable CCT (Computer Compatible Tape) format. .... Similarly, since 1979, approximately 350,000 additional MSS scenes and 190,000 TM scenes have been acquired and stored on HDTs (High Density Tape).[191]

- ~ 310,000 Landsat 1-2-3 MSS scenes from 1972-78 are on WBVTs (Wide Band Video Tape) at the EROS Data Center (EDC). As of October 1993 150,000 scenes on 7,247 HDT's have been transcribed onto 48 GB DCRSi's (Digital Cassette Recording System-incremental) cassettes, with a scheduled completion of all scenes by June 1994.[192]

- ~ 350, 000 Landsat 3-4-5 MSS scenes from 1979 to 1991 are on HDTs at EDC. The EDC EDIPS system, which is over 14 years old, is available to process these data.

- ~ 214, 000 Landsat 4-5 TM scenes from 1982 to 1993 on HDTs are at the GSFC tape staging and storage facility. TM data conversion started in November 1993 scheduled to be completed by June 1996.

Upon completion of data conversion activities, approximately 820,000 Landsat scenes, or 70 Terabyte of data, that currently exist on ≈ 60,000 tapes will be placed on ≈ 2000 cassettes.

## A.56  LDEF (Long Duration Exposure Facility)

A NASA-LaRC[193] freeflying spacecraft which accommodates technology-, science-, and applications experiments for long-term exposure to the space environment. with the objective to collect small meteoroids and space debris in Low Earth Orbit (LEO) for post-mission impact analysis of surfaces and materials that have been exposed to the space environment. Directional resolution of the flux of meteoroids and space debris particles.

---

188)  EOSAT Landsat Technical Notes, September 1992
189)  'Competitors to challenge Landsat Incumbent Team', Space News, Feb. 24 - March 1, 1992, p. 13
190)  'USGS EROS Data Center, Global Change Data Management and Information System Related Activities', CEOS Summary Report of WGD 10 Meeting, Annapolis Md. April 16-18, 1991
191)  'Satellite and Earth Science Data Management Activities at the U.S. Geological Survey's EROS Data Center', Report to CEOS, WGD-13, Canberra, Australia, Oct. 27-29, 1992
192)  'Satellite and Earth Science Data Management Activities at the U. S. Geological Survey's EROS Data Center', Report to CEOS, WGD-15, Frascati, Italy, Oct. 18-20, 1993
193)  A. S. Levine (editor), "LDEF - 69 Months in Space, First Post-Retrieval Symposium", NASA Conference Publication 3134 (Part 1 and Part 2), Proceedings of a symposium sponsored by NASA at Kissimmee, Florida, June 2-8, 1991

| Exp. Nr. | Experiment Title (Sponsoring Institution) | Tray Numbers (Exp. Location) |
|---|---|---|
| A0015 | Free-Flyer Biostack Experiment, DLR (Inst. für Flugmedizin) | C2, G2 |
| A0019 | Influence of Extended Exposure in Space on Mechanical Properties of High-Toughness Graphite-Epoxy Composite Material (U. of Michigan) | D12 |
| A0023 | Multiple Foil Micro-Abrasion Package,MAP (U. of Kent, UK) | C3, C9, D12, E6, H11 |
| A0034 | Atomic Oxygen Stimulated Outgassing (Southern University/NASA-MSFC | C3, C9 |
| A0038 | Interstellar Gas Experiment (NASA-JSC/ University of Bern) | E12, F6, H6, H9 |
| A0044 | Holographic Data Storage Crystals for LDEF (Georgia Institute of Technology) | E5 |
| A0054 | Space Plasma High Voltage Drainage (TRW Space and Technology Group) | B4, D10 |
| A0056 | Exposure to Space Radiation of High-Performance Infrared Multilayer Filters and Materials Technology Experiments (University of Reading/British Aerospace) | B8, G12 |
| A0076 | Cascade Variable Conductance Heat Pipe (McDonnell Douglas Astronautics Co.) | F9 |
| A0114 | Interaction of Atomic Oxygen with Solid Surfaces at Orbital Altitudes (U. of Alabama in Huntsville/NASA-MSFC) | C3, C9 |
| A0133 | Effect of Space Environment on Space Based Radar Phased Array Antenna (Grumman Aerospace Corporation) | H7 |
| A0134 | Space Exposure of Composite Materials for Large Space Structures (NASA-LaRC) | B9 |
| A0135 | Effect of Space Exposure on Pyroelectric Infrared Detectors (NASA-LaRC) | E5 |
| A0138-1 | Study of Meteoroid Impact Craters on Various Materials (CERT/ONERA) | B3 |
| A0138-2 | Attempt at Dust Debris Collection with Stacked Detectors (CERT/ONERA) | These experiments are also referred to as: 'Frecopa' |
| A0138-3 | Thin Metal Film and Multilayers Experiment (CNRS/LPSP) | |
| A0138-4 | Vacuum Deposited Optical Coatings Experiment (Matra S. A, Optical Division) | |
| A0138-5 | Ruled and Holographic Gratings Experiment (Inst. SA/JOB IN-YVON Division) | |
| A0138-6 | Thermal Control Coatings Experiment (CERT/ONERA, CNES/CST) | |
| A0138-7 | Optical Fibers and Components Experiment (CERT/ONERA) | |
| A0138-8 | Effect of Space Exposure of some Epoxy Matrix Composites on Their Thermal Expansion and Mechanical Properties (Matra S. A., Space Division) | |
| A0138-9 | The Effect of the Space Environment on Composite Experiments (Aerospatiale) | |
| A0138-10 | Microwelding of Various Metallic Materials Under Ultravacuum (Aerospatiale) | |
| A0139A | Growth of Crystals from Solutions in Low Gravity (Rockwell Int. Science Center/ Technical University of Denmark) | G6 |
| A0147 | Passive Exposure of Earth Radiation Budget Experiment Components (The Eppley Laboratory, Inc.) | B8, G12 |
| A0171 | Solar Array Materials Passive LDEF Experiment (NASA-MSFC, NASA-LeRC, NASA-GSFC,NASA-JPL) | A8 |
| A0172 | Effects of Solar Radiation on Glasses (NASA-MSFC, Vanderbilt University) | D2, G12 |
| A0175 | Evaluation of Long-Duration Exposure to the Natural Space Environment on Graphite-Epoxy Mechanical Properties (Rockwell Int. Corp., Tulsa Facility) | A1, A7 |
| A0178 | A High Resolution Study of Ultra-Heavy Cosmic Ray Nuclei, 'UHCRE' (Dublin Institute for Advanced Studies, Ireland, ESA/ESTEC) | A2, A4, A10, B5, B7, C5, C6, C8, C11, D1, D5, D7, D11, E2, E10, F4 |
| A0180 | The Effect of Space Environment Exposure on the Properties of Polymer Matrix Composite Materials (University of Toronto) | D12 |
| A0187-1 | Chemistry of Micrometeoroids (NASA-JSC, University of Washington, Rockwell Int. Science Center | A3, A11 |
| A0187-2 | Chemical and Isotopic Measurements of Micrometeoroids by Secondary Ion Mass Spectrometry 'SIMS' (McDonnell Center for the Space Sciences, MPI für Nuclear Physics Heidelberg, Munich Technical University, Dornier Co.) | C2, E3, E8 |
| A0189 | Study of Factors Determining the Radiation Sensitivity of Quartz Crystal Oscillators (Martin Marietta Laboratories) | D2 |
| A0201 | Interplanetary Dust Experiment, IDE (Inst. for Space Science and Technology, NASA-LaRC, North Carolina University) | B12, C3, C9, D6, G10, H11 |
| M0001 | Heavy Ions in Space (Naval Research Laboratory, Washington) | H3, H12 |
| M0002-1 | Trapped Proton Energy Spectrum Determination (AF Geophysics Laboratory) | D3, D9, G12 |
| M0002-2 | Measurement of Heavy Cosmic-Ray Nuclei on LDEF (U. of Kiel, Germany) | E6 |
| M0003 | Space Environment Effects on Spacecraft Materials (The Aerospace Corporation) | D3, D4, D8, D9 |
| M0004 | Space Environment Effects on Fiber Optics Systems (AF Weapons Laboratory) | F8 |
| M0006 | Space Environment Effects (AF Technical Applications Center) | C2 |
| P0003 | LDEF Thermal Measurement System (NASA-LaRC) | Center ring |
| P0004-1 | Seeds in Space Experiment (George W. Park Seed Company, Inc.) | F2 |
| P0004-2 | Space-Exposed Experiment Developed for Students, SEEDS, (NASA-HQ) | F2 |
| P0005 | Space Aging of Solid Rocket Materials (Morton Thiokol, Inc.) | Center ring |
| P0006 | Linear Energy Transfer Spectrum Measurement (U. of San Francisco, NASA-MSFC) | F2 |

| Exp. Nr. | Experiment Title (Sponsoring Institution) | Tray Numbers (Exp. Location) |
|---|---|---|
| S0001 | Space Debris Impact Experiment (NASA/LaRC) | A5, A6, A12, B1, B2, B6, B8, B11, C4, C7, D2, D6, E1, E4, E7, E11, F1, F3, F5, F7, F10. F11, G4, G8, H5 |
| S0010 | Exposure of Spacecraft Coatings (NASA-LaRC) | B9 |
| S0014 | Advanced Photovoltaic Experiment (NASA-LeRC) | E9 |
| S0050 | Investigation of the Effects of Long Duration Exposure of Active Optical System Components (Eng. Exp. Station, Georgia Institute of Technology) | E5 |
| S0050-1 | Investigation of the Effects of Long Duration Exposure on Active Optical Materials and UV Detectors (NASA-LaRC) | E5 |
| S0069 | Thermal Control Surfaces Experiment (NASA-MSFC) | A9 |
| S0109 | Fiber Optic Data Transmission Experiment (JPL) | C12 |
| S1001 | Low Temperature Heat Pipe (NASA-GSFC, NASA-ARC) | F12, H1 |
| S1002 | Investigation of Critical Surface Degradation Effects on Coatings and Solar Cells Developed in Germany (MBB) | E3 |
| S1003 | Ion Beam Textured and Coated Surfaces Experiment (NASA-LeRC) | E6 |
| S1005 | Transverse Flat Plate Heat Pipe Experiment (NASA-MSFC, Grumman Aerospace Corporation) | B10 |
| S1006 | Balloon Materials Degradation (Texas A&M University) | E6 |

**Table 31:**     **Summary of LDEF Experiment Complement**

LDEF was deployed on April 7 1984 by Shuttle, STS-11 (41 C). Retrieval was planned after 10 months. Due to the Challenger accident (STS-25), it was eventually recovered 69 months after launch on January 12, 1990 (STS-32). As a consequence of the delay, much more data had been gathered than planned. Post mission deintegration in SAEF-II (Spacecraft Assembly and Encapsulation Facility) at KSC.

The LDEF spacecraft is an open-grid, 12-sided, cylindrical structure. Length of cylinder = 9.1 m, diameter = 4.3 m. Gravity gradient stabilized spacecraft, with the longitudinal axis pointing toward the center of the Earth. Surface elements are fixed relative to LDEF's velocity vector. Magnetic actuators control the rotation around the longitudinal axis. The LDEF structure is configured with 72 equal-size rectangular openings on the sides and 14 openings on the ends (six on the Earth-facing end, and eight on the space-facing end) for mounting experiment trays. Total mass = 9700 kg.

Orbit: almost circular orbit, altitude = 477 km, inclination = 28.5°. At the time of retrieval the orbital altitude had decreased to 335 km.

**Experiments**: All LDEF experiments are self-contained in trays that are clamped to the facility structure. The LDEF has 72 peripheral and 14 end experiment trays. The 12 sides of the LDEF structure are numbered rows 1 through 12 in a clockwise direction when facing the end with the support beam (the Earth-facing end in orbit). The six longitudinal locations are identified alphabetically as Bay A through Bay F, starting at the end with the support beam. A tray location is designated by the Bay and Row: A-1, B-5, F-8, etc. The Earth-facing end is designated by a G identifier, the locations have even-number clock-position identifications (G-12, G-2, G-4, G-6, G-8, and G-10). The space-facing end is designated by an H identifier, the locations follow also a clock-position convention (H-12, H-1, H-3, H-5, H-6, H-7, H-9, and H-11).

The 57 LDEF experiments are extensively described in the references, only a few experiments follow with short descriptions.[194]

**Frecopa** = French Cooperative Payload (PI: J. C. Mandeville, CERT/ONERA and others). Objective: determination of the number of impacts, and the size and chemical composition of the impacting cosmic dust and debris. A collection area of about 2000 cm$^2$ is exposed to the space environment (multilayer thin foil detectors). In addition a large variety of materials placed on the same tray (8500 cm$^2$) is exposed.

---

194) W. Flury, "Europe's Contribution to the Long Duration Exposure Facility (LDEF) Meteoroid and Debris Impact Analysis", ESA bulletin, Number 76, November 1993, pp. 112-118

**SIMS** = Chemical and Isotopic Measurements of Micrometeoroids by Secondary Ion Mass Spectrometry (PI's: E. Zinner Washington University, St. Louis, E. K. Jessberger, MPI Heidelberg, and others). Objective: chemical and isotropic measurements of micrometeoroids by secondary ion mass spectrometry. The experiment consists of 237 capture cells, each measuring 8.6 x 9.4 cm, located on three different rows. The target material (germanium wafers) is of very high purity, essential for the determination of the composition of the deposit. The SIMS technique is sufficiently sensitive to allow analysis of deposits that are 0.5 to 5 monolayers thick.

**MAP** = Multiple-Foil Micro-Abrasion Package (J. A. M. McDonnell, U. of Kent, UK). Objective: measurements of impactor velocity, density, angle of incidence, and chemical composition. A capture cell experiment. Each detector consists of two foils and a pure, polished stop plate to catch fragments of impacting particles. Deployment of MAP detectors on leading and trailing surfaces, and the surfaces normal to them, of LDEF.

**UHCRE** = A High Resolution Study of Ultra-Heavy Cosmic Ray Nuclei (PI: D. O'Sullivan, Dublin Institute for Advanced Studies, Ireland, and others). The experiment consists of 18 m$^2$ of thermal blankets used to collect a large number of impact records of various sizes. The thermal blankets of UHCRE provide an impact record of meteoroids and debris.


# A.57    LFC (LARGE FORMAT CAMERA)

Shuttle Missions (STS-2 and STS-13) 12. Nov. 1981 and Oct. 5-13, 1984 (part of OSTA-3 payload)[195]

| Orbit: | 001 - 022 | 352 km altitude |
|---|---|---|
| (STS-13) | 023 - 036 | 272 km altitude |
| | 037 - 128 | 225 km altitude |

Inclination: 57°.

Application: High-resolution mapping camera. Cartographic mapping to achieve imagery at a scale of 1:50,000. LFC (408 kg) was used, among other tasks, at China's request, to try to obtain data on geological faults in Asia.

**Sensor:**
LFC = Large Format Camera (NASA Experiment, )

| | |
|---|---|
| Spectral range | 400 - 900 nm |
| Focal length | 305 mm |
| Image Format | 23 x 46 cm |
| Image scale | 1 : 738 000 |
| FOV | 73.7° across track; 41.1° along track |
| Swath width | 170 x 340 km (at 225 km altitude) |
| Kodak B/W (Neg.) | Panatomic-X Aerocon 3412 |
| Kodak B/W (Neg.) | High definition aerial 3412 |
| Kodak Color (Pos.) | High definition Aerochrome SO-131 |
| Kodak Color (Pos) | Aerial Color SO-242 |
| Data rate | 225 frames/film |

---

195) B.B. Schardt, B.H. Mollberg, 'The Orbiter Camera Payload System's Large-Format Camera and Attitude Reference System' in Monitoring the Earth's Ocean, Land, and Atmosphere from Space, Volume 97 , AIAA, 1985, pp.684 - 709

## A.58    LITE (Lidar In-Space Technology Experiment)

LITE[196)] = Lidar In-Space Technology Experiment (NASA/LaRC). The LITE-1 payload is scheduled to fly on a Shuttle mission (STS-64) in June 1994. A follow-up Shuttle flight with LITE-2 is planned for 1995, a third mission, LITE-3 is planned for 1996.

Orbit: Altitude = 296 km; inclination = 28.5°, the orbit allows lidar measurements over both land and water.

Objective: Detection of stratospheric and tropospheric aerosols; measurement of the planetary boundary layer, cloud top heights, atmospheric temperature and density in the range from 10 to 40 km. Regional and diurnal studies.

Sensor: The instrument consists of a nominal 1 m diameter telescope receiver, a three-color neodymium: YAG laser transmitter, a boresight unit, and the system electronics. LITE makes extensive use of the Shuttle resources for electrical power, thermal control, and command and data handling. The laser transmits energy at three harmonically related wavelengths into the atmosphere. The receiver collects the energy backscattered from the atmosphere and brings it to focus on three detectors.

- Laser transmitter module:

| | 1064 nm | 532 nm | 355 nm |
|---|---|---|---|
| Output wavelengths: | 1064 nm | 532 nm | 355 nm |
| Output energy (millijoules): | 486 | 460 | 196 |
| Beam divergence (milliradians): | 0.9 | 0.6 | 0.6 |
| Beam quality (X diff. limited): | 5.0 | 4.5 | 5.5 |
| Pulse Rate Freq. (all wavelengths): | 10 pulses per second | | |
| Pulse width (nonoseconds): | 27 | 27 | 31 |

Telescope:

| | |
|---|---|
| Primary mirror diameter: | 0.956 m |
| Secondary mirror diameter: | 0.311 m |
| Focal length: | 4.825 m |
| Focal ratio: | f/5.1 |
| Obscuration ratio: | 0.11 |

- Aft Optics (Nighttime measurements):

| | | | |
|---|---|---|---|
| Optical throughput (%) | 58 | 49 | 49 |
| Color filter BW (nm) | 675 | 265 | 60 |
| Detector quantum efficiency (%) | 33 | 14 | 21 |

- Aft Optics (Daytime measurements):

| | | | |
|---|---|---|---|
| Interference filter transmission (%) | 50 | 40 | 12 |
| Overall optical throughput (%) | 29 | 20 | 6 |
| Interference filter BW (nm) | 1 | 0.3 | 1 |
| Detector quantum efficiency (%) | 33 | 14 | 21 |

- Field of View (all wavelengths): Selectable: 1.7 mrad; 3.4 mrad; 5.6 mrad; and blocked

- Signal processing Electronics (all channels)

| | |
|---|---|
| Baseband amplifier bandwidth: | 2 MHz |
| Filter characteristics: | 7 pole Bessel Lowpass |
| Digitizer resolution: | 12 bits |
| Digitizing rate: | 10 MHz during sample period |
| Data sample period: | 660 microseconds |

196)  'Lidar In-Space Technology Experiment (LITE): NASA's first In-Space Lidar System for Atmospheric Research', Optical Engineering, Jan. 1991, Vol. 30 No. 1 pp. 88-95

## A.59   MAGSAT

MAGSAT is a Johns Hopkins APL/NASA/USGS satellite mission for the survey of the Earth's Magnetic Field.[197),198),199)] Launch: Oct. 30, 1979. Objectives: collection of data for improved modeling of the time-varying magnetic field generated within the core of the Earth, and to map variations in the strength and vector characteristics of crustal magnetization.

Orbit: Sun-synchronous orbit; inclination = 96.76°; perigee = 352 km, apogee = 561 km. The satellite remained in orbit for seven and a half months until June 11, 1980.

**Sensors:** the sensors were mounted on an instrument platform at the end of the 6 m magnetometer boom to eliminate the effect of spacecraft fields. The basic MAGSAT mission required knowledge in the magnetic field orientation to a total system accuracy of < 20 arcseconds.

- **Scalar Magnetometer** (built in Canada), Cesium vapor type, with an accuracy of about 1.5 nT. The sensor developed internal oscillations in its lamp circuitry shortly after launch which prevented full data recovery and which, at times, slightly degraded its accuracy. Its data were, however, sufficient to provide in-flight calibration for the vector magnetometer, except during the last few weeks of operation.

- **Vector Magnetometer** (built at GSFC), fluxgate type, with an accuracy of < 3 nT for each component <3 nT (after in-flight calibration). The instrument consisted of three fluxgate ring-core sensors mounted on a MACOR (machinable ceramic) structure.

The onboard attitude determination system required two star cameras, two Attitude Transfer System (ATS) optical heads (for pitch/yaw and roll). Two ATS (pitch/yaw and roll) mirrors were mounted on the back of the vector magnetometer at the end of the boom, providing a reflected beam of light for accurate magnetometer axis determination. A precision sun sensor was mounted in addition on the vector magnetometer for redundant measurements.

Data: Magsat carried two onboard tape recorders to allow 10 hours of recorded data to be dumped to a NASA ground station during a pass. The data are available in several formats from the National Space Science Data Center (NSSDC) at GSFC and at NOAA/NGDC (subset of NASA data).

## A.60   MAPS (Measurement of Air Pollution from Satellites)

MAPS is a NASA/LaRC Shuttle payload with the objective to measure (nearly) global tropospheric carbon monoxide (CO) distributions in the middle troposphere in order to evaluate the CO sources and chemistry, and to determine the seasonal variations of this key atmospheric trace gas.

The MAPS instrument is a gas filter correlation radiometer consisting of an electro-optical sensor, an electronics module, a digital tape data recorder, and an aerial camera. MAPS is mounted onto MPESS (Multi-Purpose Experiment Support Structure) in the forward end of the Shuttle cargo bay. The instrument makes nadir-viewing measurements from Shuttle orbits of approximately 220 km altitude.[200)]

The electro-optical sensor incorporates two gas cells, one containing 350 hPa of CO, and another containing 149 hPa of $N_2O$; their corresponding detectors, a direct radiation detector, an external balance and gain check system, and an internal balance system.


197)  F. F. Mobley, L. D. Eckard, G. H. Fountain, G. W. Ousley, "Magsat - A New Satellite to Survey the Earth's Magnetic Field', IEEE Trans. on Magnetics, Vol. Mag. 16, No. 5, September 1980, pp. 758-760

198)  R. Langel, G. Ousley, J. Berbert, 'The MAGSAT Mission', Geophysical Research Letters, Vol. 9, No.4, April 1982, pp. 243-245

199)  R. Langel, "The Magnetic Earth as Seen from Magsat, Initial Results', Geophysical Research Letters, Vol. 9, No.4, April 1982, pp. 239-242

200)  Information provided by V. Connors and D. O. Neil of NASA/LaRC

| MAPS sensor | Gas filter correlation radiometer |
|---|---|
| Spectral band | 2080 to 2220 cm$^{-1}$ (4.5 - 4.8 μm) |
| Resolution (sensitivity) | 1.7 10$^{-7}$ W cm$^{-2}$ sr$^{-1}$ noise-equivalent radiance in the radiometer channel<br>5.7 10$^{-9}$ W cm$^{-2}$ sr$^{-1}$ noise-equivalent radiance in the difference channel |
| Swath width | footprint is 20 km at ground surface (corresponds to spatial resolution) |
| Data rate | 43 bit/s (on-board recorded and downlinked during flight) |

**Table 32:     Specification of some MAPS Parameters**

MAPS measures the absorption lines of CO and $N_2O$ of the incoming radiation. The spectrum is split into three beams, one beam passes through a cell containing CO and falls onto a detector - the CO gas cell acts as a filter. A second beam falls directly onto a detector. The observed signal differences are used for CO determination in the altitude range of 0-18 km. The third beam of the incident radiation passes through a cell containing $N_2O$ and is measured by another detector. Again, the $N_2O$ gas cell acts as a filter for the effects of $N_2O$ present in the atmosphere. Since the global distribution of $N_2O$ is well known, the measured values of $N_2O$ may be used as an indicator for the presence of clouds in the field of view and to correct the simultaneous CO measurement for systematic errors.

The aerial camera mounted next to the MAPS electro-optical head provides information on cloud cover and terrain over which the sensor data are gathered.

MAPS was first flown on STS-2 (12.-14. November 1981) as part of the OSTS-1 payload, then on STS-13 (5. -13. October 1984) as part of the OSTA-3 payload. These flights revealed longitudinal variations in CO distributions and identified biomass burning in the Southern Hemisphere as a major global CO source (equal in magnitude to the amount of CO generated in the Northern Hemisphere by fossil fuel combustion). MAPS is scheduled to fly again on Shuttle in March/April 1994 for a 10 day Earth observing mission.


## A.61     METEOR-1 Series

"Meteor" is the Soviet/Russian generic name for its Low-Earth-Orbiting (LEO) meteorological satellite family which started operation in 1969 (built by VNIIEM, Moscow). Sponsoring agency: ROSGIDROMET. Prior to this series there was an experimental Cosmos series with the first orbiting meteorological satellite, "Cosmos-44", which was launched on August 25, 1964, and was followed by nine analogous Cosmos satellites until 1969, when the series was officially named 'Meteor-1'. The designations of the series Meteor-1, -2, and -3 define different payload configurations and different orbits, the Meteor-3 series is presently (1993) the most advanced payload generation and the operational series. A total of 25 Meteor-1 satellites were launched from 1969 to 1977 (3 to 4 launches per year, see Table 34)[201],[202]. Note: Meteor-1-18 is also referred to as Meteor-Priroda-1, Meteor-1-25 is also named Meteor-Priroda-2, and Meteor-1-28 is the satellite Meteor-2-2.

Background:[203] ROSGIDROMET, the Russian Federal Service for Hydrometeorology and Environmental Monitoring, is the national central agency providing relevant environmental and climate information to the public, various industrial organizations and decision-making bodies. Hydrometeorological services together with the provision of information on environmental pollution are based on the performance of the National Observing System for the State of the Environment and Climate (GSNP). GSNP operates data acquisition facilities (ground network of synoptic, hydrological and upper-air stations, weather radars, aircraft, rocket sounding stations, satellites and ocean ships), telecommunication and processing facilities, analysis and forecasts based on advanced atmospheric and hydrological models.

---

201)  'The Cambridge Encyclopedia of Space', Cambridge University Press 1990, p. 235
202)  The original text was reviewed by Y. V. Trifonow of VNIIEM, Moscow
203)  COSPAR-90-Paper by A. Karpov, USSR State Committee for Hydrometeorology, Moscow. Title of paper: "Hydrometeorological, Oceanographic and Earth-Resources Satellite Systems operated by the USSR".

Application: Meteorological sciences, operational meteorology

Orbit: Non-Sun-synchronous polar orbit, average altitude of 650 km for the first 9 satellites and 900 km for the rest of the series, inclination = 81-82°.

Sensors:

- **TV** = TV optical instrument (MR-600A). Framing (vidicon technique). Spectral range: 0.4 - 0.8 μm; swath width = 1000 km; resolution = 1.25-3 km;

- **IR** = TV IR instrument Lastocha. Spectral range: 8 - 12 μm; swath width = 1000 km; resolution = 15 km;

- **AC** = Radiation Budget Sensor. AC is a narrow-sector device for the measurement of the Sun's radiation fluxes, the thermal radiation of the Earth's surface, cloud cover and the atmosphere. Spectral range: 0.3 - 30 μm; swath width = 2500 km; resolution = 40 km x 50 km.

The Meteor series, somewhat similar to the NOAA-TIROS series, provides a daily weather review for more than two-thirds of the globe, observation of clouds, ice cover and atmospheric radiation; they also permit the study of weather fronts and jet stream currents, monitoring of clear-air turbulence (for warning of the intercontinental airliners, etc.).


## A.62    METEOR-2 Series

Soviet/Russian Earth observation satellite series in the field of meteorology. Sponsoring agency: ROSGIDROMET. The Meteor-2 series (developed and built by VNIIEM) came into use in 1975. Until 1990, 21 satellites of this type have been launched. The system comprises 2-3 satellites continuously operating in near polar orbit with an average altitude of 900 km.

Orbit: Non-Sun-synchronous polar orbit, altitude between 850 and 950 km, inclination = 81-82°.

Sensors:
The on-board instrument package includes three television-type (frame technique) VIS and IR scanners, a 5-channel scanning radiometer and a device (RMK) for measuring radiation flux densities in the near-Earth space (see sensor definition of Meteor-3 series).


## A.63    METEOR-Priroda Series

USSR experimental mission series considered as predecessors of the Resurs-O1 series (see A.85). Sponsoring agency: ROSGIDROMET. First Earth resources satellite series for continuous imaging from space. First launch: 9. July 1974. In 1983 Meteor-Priroda data was used to determine snow depths on all Soviet agricultural regions and to assist in planning the spring work season and forecasting the harvest. The Meteor-1 series universal platform was also used for the Meteor-Priroda series up to Nr. 4. Starting with Meteor-Priroda-5, a Meteor-2 platform was employed.

Orbit: Sun-synchronous polar orbit, altitude = 630 km, inclination = 98°, period = 97-98 minutes, repetition cycle = 16 days

Sensors:

- **Fragment 2** = Experimental multi-channel opto-mechanical system of high resolution [analogous in function to MSS (Multispectral Scanner System) on Landsat].

Swath width = 85 km
Resolution = 80 m for spectral range of 0.4-1.1 µm; = 240 m for spectral range of 1.2-1.8 µm; = 480 m for the spectral range of 2.1-2.4 µm

- **MSU-S** = Multispectral Scanner - moderate resolution (2-channels). Spectral range: of 0.58 - 0.7 µ, and 0.7 - 1.0 µ, resolution = 140-240 m, swath = 1380 km.

- **MSU-SK** = Multispectral Scanner of moderate resolution with Conical Scanning. Spectral ranges: 0.5-0.6, 0.6-0.7, 0.7-0.8, 0.8-0.9 µm. Spatial resolution = 175-243 m (in VIS) and 500 m (in the IR). Swath = 600 km

- **MSU-E** = High-Resolution Multispectral Scanner with Electronic Scanning. Spectral ranges (3) of 0.5 - 0.7 µm, 0.7 - 0.8 µm, 0.8 - 1.0 µm. Spatial resolution = 28 x 28 m; swath = 28 km.

- **MSU-M** = Multispectral Scanner of low resolution; swath width = 2000 km; resolution = 1 km x 1.7 km; spectral ranges (4): 0.5 - 0.6, 0.6 - 0.7, 0.7 - 0.8, 0.8 - 1.0 µm.

- **SHF** = Passive MW Radiometer measuring the emission radiation of the atmosphere/ ocean system at the following microwave lengths/frequencies: 0.8 cm (37.5 GHz), 1.35 cm (22.22 GHz), 1.55 cm (19.4 GHz), and 4 cm (7.5 GHz). The swath width = 800 - 900 km depending on orbit height. Resolution = 1.2 - 3.0°.

- **SI-GDR** = Spectrometer/Interferometer (German Democratic Republic). An infrared Fourier spectrometer (original designation: PM = Profile Module) with the objective of temperature profile measurements in the atmosphere. Spectral range: 6.25 - 25 µm; spectral resolution = 5 cm$^{-1}$ ; nadir pointing; resolution = 2.2 x 2.4°.

- **R10M** = Radiation Measurement Complex, analogous to RMK-2

| Name of Satellite | Launch Date | Period (min) | Perigee (km) | Apogee (km) | Inclina-tion (°) | Remarks / Sensor Complement |
|---|---|---|---|---|---|---|
| Meteor-P.-1 | 9.7.1974 | 102.6 | 877 | 905 | 81.2 | MSU-M, MSU-S, TV, IR,, SHF |
| Meteor-P.-2 | 15.5.1976 | 102.4 | 866 | 908 | 81.2 | MSU-M, MSU-S, IR, SI-GDR, R10-M, SHF |
| Meteor-P.-3 | 29.6.1977 | 97.5 | 602 | 685 | 97.9 | MSU-M, MSU-S, SI-GDR, R10-M, SHF |
| Meteor-P.-4 | 25.1.1979 | 97.5 | 628 | 656 | 98.0 | MSU-M, MSU-S, SI-GDR, R10-M, |
| Meteor-P.-5 | 18.6.1980 | 97.3 | 589 | 678 | 98.0 | MSU-M, MSU-S, Fragment-2, R10-M, MSU-SK, MSU-E |
| Meteor-P.-6 | 10.7.1981 | 97.6 | 611 | 688 | 97.9 | MSU-M, MSU-S, R10-M, SHF |

**Table 33:    Overview of the Meteor-Priroda Missions**


## A.64    METEOR-3 Series

Russian Earth observation satellite series. Sponsoring agency: ROSGIDROMET. The first Meteor-3 launch was on 24 Oct. 1985 from Plesetsk using the F2 launch vehicle. The Meteor-3 series is the current generation of Russian meteorological satellites. Nominal lifetime = 2 years. An advanced Meteor series (Meteor-3M) is being planned for launch in 1996 (note: the launch of built and stored satellites occurs on a basis of overall operational requirements).

Both types of satellites (Meteor-2 and Meteor-3) provide main and regional centers of Russia and CIS with global data on the distribution of clouds, snow and ice in VIS and IR bands, radiation flux data at least twice daily, atmospheric temperature - humidity sounding data, data on the temperature of the underlying surface, cloud-top heights and sea surface temperature.

| Name of Satellite | Launch Date | Period (min) | Perigee (km) | Apogee (km) | Inclination (°) | Remarks / Sensor Complement |
|---|---|---|---|---|---|---|
| Meteor-1-1 | 23.3.1969 | 97.9 | 644 | 713 | 81.2 | TV, IR, AC |
| Meteor-1-2 | 6.10.1969 | 97.7 | 630 | 690 | 81.2 | TV, IR, AC |
| Meteor-1-3 | 17.3.1970 | 96.4 | 655 | 643 | 81.2 | TV, IR, AC |
| Meteor-1-4 | 28.4.1970 | 98.1 | 637 | 736 | 81.2 | TV, IR, AC |
| Meteor-1-5 | 23.6.1970 | 102 | 863 | 906 | 81.2 | TV, IR, AC |
| Meteor-1-6 | 15.10.1970 | 97.5 | 633 | 674 | 81.2 | TV, IR, AC |
| Meteor-1-7 | 20.1.1971 | 97.6 | 630 | 679 | 81.2 | TV, IR, AC |
| Meteor-1-8 | 17.4.1971 | 97.2 | 620 | 646 | 81.2 | TV, IR, AC |
| Meteor-1-9 | 17.7.1971 | 97.3 | 618 | 650 | 81.2 | TV, IR, AC |
| Meteor-1-10 | 29.12.1971 | 102.7 | 880 | 905 | 81.2 | TV, IR, AC |
| Meteor-1-11 | 30.3.1972 | 102.6 | 878 | 903 | 81.2 | TV, IR, AC |
| Meteor-1-12 | 30.6.1972 | 103 | 897 | 929 | 81.2 | TV, IR, AC |
| Meteor-1-13 | 27.10.1972 | 102.6 | 893 | 904 | 81.2 | TV, IR, AC |
| Meteor-1-14 | 20.3.1973 | 102.6 | 882 | 903 | 81.2 | TV, IR, AC |
| Meteor-1-15 | 29.5.1973 | 102.5 | 867 | 909 | 81.2 | TV, IR, AC |
| Meteor-1-16 | 5.3.1974 | 102.2 | 853 | 906 | 81.2 | TV, IR, AC |
| Meteor-1-17 | 24.4.1974 | 102.6 | 877 | 907 | 81.2 | TV, IR, AC |
| **Meteor-P-1** | **9.7.1974** | **102.6** | **877** | **905** | **81.2** | **MSU-M, MSU-S, TV, IR, RMK, SHF** |
| Meteor-1-19 | 28.10.1974 | 102.5 | 855 | 917 | 81.2 | TV, IR, AC |
| Meteor-1-20 | 17.12.1974 | 102.4 | 861 | 910 | 81.2 | TV, IR, AC |
| Meteor-1-21 | 1.4.1975 | 102.6 | 877 | 906 | 81.2 | TV, IR, AC |
| Meteor-2-1 | 11.7.1975 | 102.5 | 872 | 906 | 81.3 | Exp. launch of Meteor-2/TV, IR, SM, RMK-2 |
| Meteor-1-22 | 18.9.1975 | 102.3 | 867 | 918 | 81.2 | TV, IR, AC |
| Meteor-1-23 | 25.12.1975 | 102.4 | 857 | 913 | 81.3 | TV, IR, AC |
| Meteor-1-24 | 7.4.1976 | 102.3 | 863 | 906 | 81.2 | TV, IR, AC |
| **Meteor-P-2** | **15.5.1976** | **102.4** | **865** | **907** | **81.2** | **MSU-M, MSU-S, IR, SI-GDR, RMK, SHF** TV, |
| Meteor-1-26 | 16.10.1976 | 102.5 | 871 | 904 | 81.3 | IR, AC |
| Meteor-2-2 | 7.1.1977 | 103 | 893 | 932 | 81.3 | TV, IR, SM, RMK-2 |
| Meteor-1-28 | 5.4.1977 | 102.5 | 869 | 909 | 81.2 | TV, IR, AC |
| Meteor-2-3 | 14.12.1977 | 102.5 | 872 | 906 | 81.2 | TV, IR, SM, RMK-2 |
| Meteor-2-4 | 1.3.1979 | 102.3 | 857 | 908 | 81.2 | TV, IR, SM, RMK-2 |
| Meteor-2-5 | 31.10.1979 | 102.6 | 877 | 904 | 81.2 | TV, IR, SM, RMK-2 |
| Meteor-2-6 | 9.9.1980 | 102.4 | 868 | 906 | 81.2 | TV, IR, SM, RMK-2 |
| Meteor-2-7 | 15.5.1981 | 102.5 | 868 | 904 | 81.3 | TV, IR, SM, RMK-2 |
| Meteor-2-8 | 25.3.1982 | 104.2 | 954 | 976 | 82.5 | TV, IR, SM, RMK-2 |
| Meteor-2-9 | 15.12.1982 | 102 | 836 | 904 | 81.3 | TV, IR, SM, RMK-2 |
| Meteor-2-10 | 28.10.1983 | 101 | 780 | 901 | 81.2 | TV, IR, SM, RMK-2 |
| Meteor-2-11 | 5.7.1984 | 104 | 954 | 974 | 82.5 | TV, IR, SM, RMK-2 |
| Meteor-2-12 | 7.2.1985 | 104 | 950 | 975 | 82.5 | TV, IR, SM, RMK-2 |
| Meteor-3-1 | 24.10.1985 | 110.3 | 1235 | 1263 | 82.5 | MR-2000M, MR-900B, IR, SM, RMK-2 |
| Meteor-2-13 | 26.12.1985 | 104 | 952 | 975 | 82.5 | TV, IR, SM, RMK-2 |
| Meteor-2-14 | 27.5.1986 | 104.1 | 953 | 974 | 82.5 | TV, IR, SM, RMK-2 |
| Meteor-2-15 | 5.1.1987 | 104 | 950 | 973 | 82.5 | TV, IR, SM, RMK-2 |
| Meteor-2-16 | 18.8.1987 | 104.1 | 954 | 974 | 82.5 | TV, IR, SM, RMK-2 |
| Meteor-2-17 | 30.12.1987 | 104 | 952 | 975 | 82.5 | TV, IR, SM, RMK-2 |
| Meteor-2-18 | 30.1.1988 | 104.1 | 947 | 973 | 82.5 | TV, IR, SM, RMK-2 |
| Meteor-3-3 | 26.7.1988 | 109.4 | 1198 | 1221 | 82.5 | MR-2000M,MR-900B,Klimat,SM,RMK-2, |
| Meteor-3-4 | 25.10.1989 | 109.5 | 1191 | 1228 | 82.6 | MR-2000M,MR-900B,Klimat,SM,RMK-2,IR |
| Meteor-2-19 | 28.2.1990 | 104.1 | 951 | 974 | 82.5 | TV, IR, SM, RMK-2 |
| Meteor-2-20 | 28.6.1990 | 104 | 950 | 974 | 82.5 | TV, IR, SM, RMK-2 |
| Meteor-2-21 | 28.9.1990 | 104.1 | 953 | 975 | 82.5 | TV, IR, SM, RMK-2 |
| Meteor-3-5 | 24.4.1991 | 109.5 | 1195 | 1230 | 82.5 | MR-2000M,MR-900B,Klimat,SM,RMK-2, |
| Meteor-3-6 | 15.8.1991 | 109.5 | 1195 | 1230 | 82.5 | MR-2000M,MR-900B,Klimat,SM,RMK-2, IR, TOMS |
| Meteor-2-24 | 31.8.1993 | 104.1 | 979 | 941 | 82.5 | TV, IR, SM, RMK-2 (spare satellite) |
| Meteor-3-7 | 25.1. 1994 | 109.3 | 1186 | 1201 | 82.5 | MR-2000M,MR-900B,Klimat,SM,RMK-2, SCARAB, PRARE |
| | | | | | | |
| | | | | | | |
| Cosmos-1066 | 23.12.1978 | 102.2 | 848 | 908 | 81.2 | Exp. payload on Meteor-1 satellite technology to carry out special geophysical experiments |
| Intercosmos-Bulgaria 1300 | 7.8.1981 | 101.9 | 825 | 906 | 81.2 | Bulgarian sensors on Meteor-2 satellite technology (ionospheric plasma and high-energetic fluxes of charged particles, electric and magnetic fields, upper atmospheric glow in UV and VIS) |

**Table 34:**      **Russian Environmental/Meteorological Satellites (Chronological Order)**[204]

---

204) Courtesy of B. S. Zhukov (IKI), Y. V. Trifonov (VNIIEM), and S. N. Voyakin (NPO Planeta), Moscow

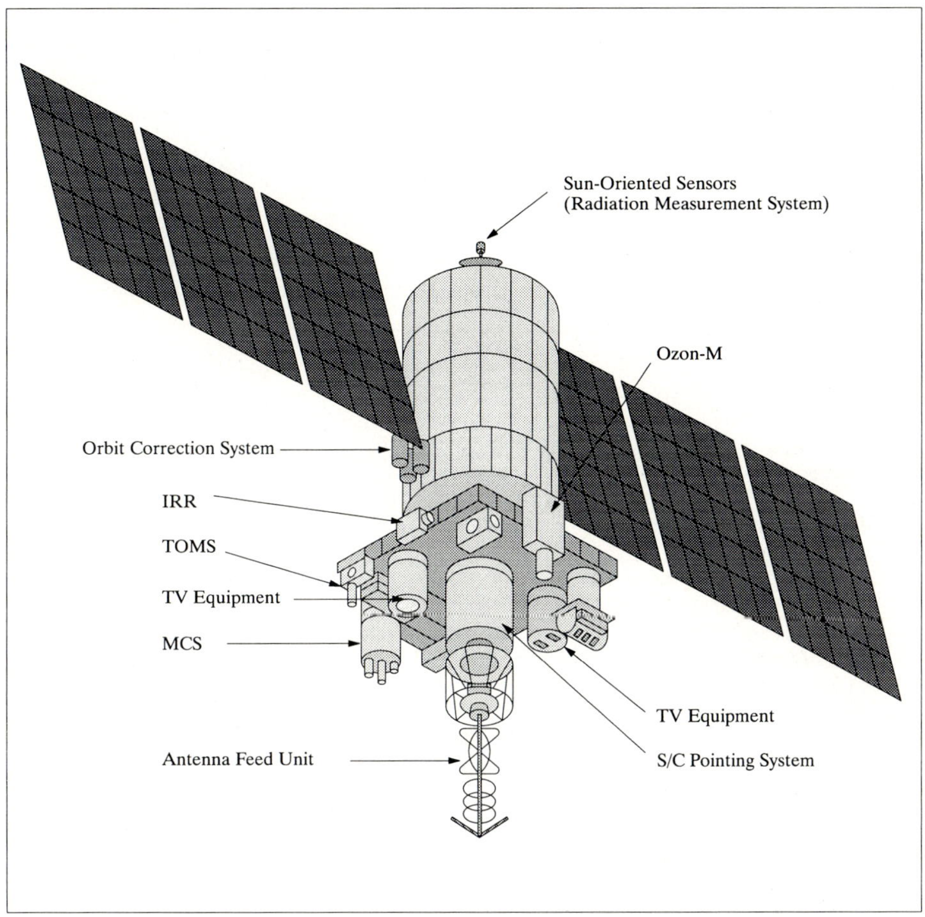


**Figure 50:    The Meteor-3 Series S/C Model**

| Parameter | Meteor-3 Series | Meteor-3M Series |
|---|---|---|
| Payload Mass (kg) | 700 | 900 |
| Orbital Altitude (km) | 1200 - 1250 | 900 - 950 |
| Orbit Inclination ($^o$) | 82.5 | 82.5 |
| Satellite pointing accuracy for all 3 axis (arc/min) | 20 -30 | 10 |
| Stabilization accuracy ($^o$/s) | 0.05 | 0.001 |
| Power, average/day, (W) | 600 | 1000 |
| Payload voltage range (V) | 24 - 34 | $27 \pm 1$ |
| Nominal Life (year) | 2 | 3-5 |
| Geometric correction accuracy (km) | 15 | 5 |
| Transmitter frequency (MHz)<br>real-time mode<br>store-and-forward mode | 137-138 (analog)<br>466.5 (analog) | 137-138 (analog)<br>1690-1710 (digital) |
| Orbit correction engine | available | available |

**Table 35:    Meteor-3 Series Satellite/Observation Characteristics**

National Russian/CIS plans envisage the development of an integrated hydrometeorological satellite system in the 90s, comprising geostationary and low-orbit satellites. Onboard instruments of the geostationary satellites will provide continuous observations of the Earth's disk within 60° with respect to the stationary point over the equator.

Orbit: Non-sun-synchronous near-polar orbit, drifting slowly with local time; altitude = 1230 km; inclination = 82.5°; Period = 109 minutes; the higher altitude of Meteor-3 enables to extend the instrument swath width, providing therefore complete coverage of the Earth surface.

As of 1/1994 the following Meteor satellites are operational: Meteor-2-21, Meteor-3-5, Meteor-3-6, Meteor-2-24 (last spare of series), and Meteor-3-7. Meteor-3-8 has been ordered.

**Sensors:**

**1. Television Systems**

- TV Camera System **MR-2000M** and **MR-900B**; Objective: Observation of the daytime Earth cloud cover in the visible spectrum (0.5 - 0.7 µm) at a local solar angle not less than 5°. Spatial resolution of 0.7-1.4 km for MR-2000M and 1-2 km for MR-900B. The swath width = 3100 km (MR-2000) and 2600 km (MR-900). The camera MR-2000M provides storage and direct transmission operation. The camera MR-900B has no storage operation mode (only direct transmission). Stable range of information reception within a radius of 3000 km.
  Output products: individual images, photomosaics of images from 2-3 passes over receiving station within 300 km in radius. Global photomosaics of images of various regions of the globe (twice daily), cloud-free photomosaics of arctic and antarctic oceans once in 5 days.

| Parameter | Real-time Mode | Store-and-forward Mode |
|---|---|---|
| Swath width (km) | 2600 | 3100 |
| Number of Pixels/scan line | 900 | 2000 |
| Number of reproduced brightness levels | 50 | 50 |

**Table 36: TV System Parameters**

**2. Optical Scanning Systems**

- **Klimat**= Infrared Radiometer[205].Electromechanical device with a scan angle of ± 48°, a swath width of 3100 km, a spatial resolution of 3 km; surface temperature range = 223-313 K; temperature difference at 300 K, background = 0.2 K. Measurement spectrum: 10.5 to 12.5 µm.
  Output products: Global photomosaics of Northern and Southern Hemispheres, tropical zone, individual images; digital SST and top of cloud height charts, tropical cyclone coordinates, cloud amount data on regular grid over the globe.

- **SM** = Multichannel Spectrometer, also known as **"Device 174 K"** (optical instrument for atmospheric observation). Electromechanical device, scan angle =± 24°. Scanning 10-channel IR radiometer for atmospheric thermal sounding. Spectral range: 9.65 - 18.7 µm (9.65, 10.60, 11.10, 13.33, 13.70, 14.25, 14.43, 14.75, 15.015, 18.70). Resolution = 42 km; swath width = 1000 km; Output products: SATEM messages with atmospheric thermal sounding data (total ozone content)

205) Y. V. Trifonov, "Meteor-3 space system for hydrometeorological observation", VNIIEM, Moscow, 1991

### 3. Radiation Measurement System

- **RMK-2** = Radiation Measurement Complex. Objectives: Registration of flux densities of proton in the 5-90 MeV and electrons in the 0.15-3.0 MeV energy regions.
    - Electron and proton flux density. Ranges: 2.5 -1 x $10^5$ particles/cm$^2$s, and 0.111 - 4.4 x $10^3$ particles/cm$^2$s
    - Radiation dose exposition in the range: 1 x $10^{-7}$ - 5 x $10^{-4}$ particles/s
    - Number of channels: 12

- **MIVZA-M** = Low Resolution Microwave Scanner. The instrument is being considered for Meteor-3-8 and for the follow-up Meteor-3M series. Objective: measurement of the atmospheric and background radiation at three wavelengths for the calculation of the following parameters:
    - for the atmosphere: an integral humidity and its water content, effective temperature of clouds, and the rain rate
    - for the land: height and boundary of snow
    - for the ocean: wind speed value, ice coverage, and estimation of age and boundary.

The instrument consists of a thermostatic microwave unit (MWU), a calibrator, a control unit, a data processing unit (DPU), and a data storage unit (DSU). Observation is performed by a conical scanning motion of the antenna system.

The MWU is made up of the following components: a two-mirror antenna, receivers at 1.5 cm, 0.86 cm, and 0.32 cm wavelengths, a Faraday-plane polarization switch, and a data control and collection unit (DCCU) to generate control signals needed for the microwave components, to transform an analog signal from the receivers to a digital one, and to generate a consistent data flow. The Faraday switch of the receivers enables the detection of sequential radiation at vertical and horizontal polarizations. The MWU is mounted on the scanning platform that provides the polarization stability whenever the antenna sight direction varies. The calibrator provides a line-by-line calibration from the space background radiation as well as from a hot body reference. The DPU performs the digital processing, including radiometric correction, straight line transformation, and side-lobe antenna pattern accounting. Data are registered in the DSU whose memory accounts for the recording of one complete pass. The data will be transmitted to the ground stations in the form of radio brightness temperature charts.

Meteor Data: There is a prime network of ground stations in Moscow, Novosibirsk, Chabarowsk, and Tashkent (operator/coordinator: NPO Planeta). There are also 80 APT (Automatic Picture Transmission, compatible with NOAA APT) stations in Russia/CIS receiving Meteor data (on 137-138 MHz). The APT stations provide data reception for Meteor and for NOAA series satellites.

A new feature of the Meteor program is the incorporation of sensors and instruments from other space agencies through international cooperation. The following foreign sensors/payloads are flown or planned on Russian missions:

- **TOMS** (NASA sensor)
  A Meteor-3 series S/C (Meteor-3-5) was launched successfully on August 15, 1991 with NASA sensor participation [206],[207]. A refurbished **TOMS** sensor (of the original engineering model for Nimbus 7) is part of this payload. The instrument transmits its data to ground stations in the US and Russia on a daily basis (US archive at GSFC, Russian archive in Dolgoprudny outside of Moscow). The orbital life time is projected at 2 years. As of Fall 1993 the Meteor S/C and TOMS are performing well. Life expectancy through 1994/5. Objective: Mapping of the vertical ozone profiles. TOMS has a swath with of 3100 km. Spectral bands (6):0.3125, 0.3175, 0.3313, 0.3398, 0.3600, and 0.3800 µm. The ground resolution is 47 x 47 km at nadir and 62 x 62 km overall.

206) 'Soviets to Launch U.S. Ozone Mapper', Space News Aug. 5-18 1991, p. 14
207) 'TOMS Arrives Successfully in Space', Space News Aug. 19-25, 1991, p. 2

| Parameter / Channel Number | 1, 2 | 3, 4 | 5 |
|---|---|---|---|
| Wavelength | 1.5 cm | 0.86 cm | 0.32 cm |
| Frequency | 20.0 GHz | 35.0 GHz | 94.0 GHz |
| Polarization | V, H | V, H | H |
| RF Bandwidth | 2 GHz | | |
| Dynamic range | 0 - 350 K | | |
| NE$\Delta$T | 0.1 K | | |
| Absolute accuracy | 1.5 K | 1.5 K | 2.5 K |
| Antenna beamwidth | 3.08° | 1.83° | 0.69° |
| IFOV (altitude = 850 km) | 80 km | 55 km | 20 km |
| Look angle | 40 - 45° | | |
| Swath width | 1500 km | | |
| Scanning type | conical scanning | | |
| Conical scan angle | ± 60° | | |
| Scan period | 12-14 s | | |
| Data rate | 0.28 kbit/s | | |
| Mass of instrument | 45 kg | | |
| Power | 75 W | | |

**Table 37:**    **Specification of the MIVZA-M Sensor Parameters**

- **SCARAB** (French/Russian/German sensor).
  SCARAB is a joint development of a scanning radiometer. It's objective is the collection of data on short and long wave radiation to estimate the Earth's radiation budget at the top of the atmosphere. The spatial resolution of SCARAB is 60 km, the swath width = 3000 km. Applications: climatology and atmospheric research. Two launch dates of the Scarab sensor are scheduled on Meteor-3-7: Jan. 1994 and 1995. Spectral bands:
  - 0.2 - 50 µm
  - 0.2 - 4 µm
  - 0.5 - 0.7 µm
  - 10.5 - 12.5 µm
- **PRARE** (German Microwave Tracking System, see chapter A.81). PRARE flies on Meteor-3-7 as a passenger instrument.

## A.65    METEOR-3M Series

Advanced satellite series starting with launches in 1996/7 (some S/C parameters are given in Table 35).

Orbit: Sun-synchronous orbit, altitude =900 km, inclination = 82.5°

Sensors: the sensor complement of the 3M series is yet to be defined as of end of 1993.

## A.66   METEOSAT

METEOSAT[208],[209],[210],[211] = Operational meteorological ESA program (initiated in 1972, since Nov. 1983 official MOP) covering Europe, Africa and the Atlantic Ocean. The Meteosat series provides also Earth observation (science) data. The program consists of 3 satellites. All METEOSAT satellites are in geosynchronous orbit (geostationary).

Meteosat itself is part of a global network of geostationary meteorological satellites distributed around the equator. All activities regarding satellite and operations are coordinated by the CGMS (Coordination of Geostationary Meteorological Satellites) committee represented by WMO and the countries providing the satellites (Europe (ESA), USA (NOAA), USSR, Japan (JMA), India (ISRO)).

Payload: High-Resolution Radiometer (image generation in the infrared region (IR), in the water vapor absorption bands (WV), and in the visible range (VIS). The resolution at nadir is 5 km for the IR- and WV channels and 2.5 km for the VIS-channel).

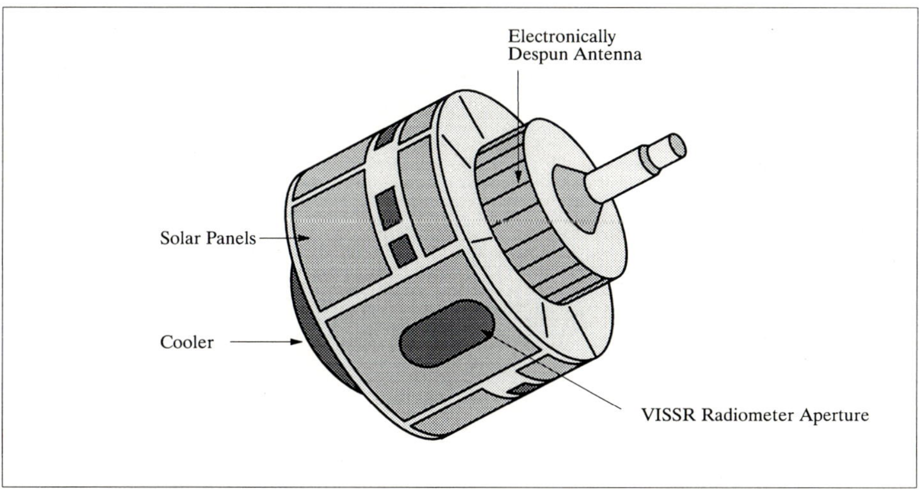


**Figure 51:**    **The Meteosat S/C Model**

**Application:** Weather forecast, Meteorology, Climatology (sea surface temp. monitoring, cloud cover statistics, wind and rainfall assessment).

An image is provided in 1/2 hour periods in the visible range and in the IR- and WV regions.

Satellites:
-   METEOSAT-1 (Prototype of the preoperational program).
    Launch: Nov. 23, 1977; service operation until Nov. 1979
-   METEOSAT-2 (Prototype of the preoperational program)
    Launch: June 19, 1981; service operation from 12 Aug. 81 until 11 Aug. 1988 as prime satellite, re-orbiting in Dec. of 1991.
-   METEOSAT-3 (Prototype of the preoperational program, initially designated as P2)
    Launch: June 1988 (2 years); expected mission duration until 1994. Note: this satellite is presently (1992) assigned to the West Atlantic as a temporary replacement for GOES.

208)  ESA Information Note to the Press No. 4, 11 Feb. 1991, "MOP-2 Ready for Launch"
209)  "Current and Planned European Operational Meteorological Satellite Systems", John Morgan, Proceedings of the Twenty-Third International Symposium on Remote Sensing of Environment, Bangkok, Thailand, April 18-25, 1990, ERIM, Ann Arbor Mich, Vol. I, pp. 107-116.
210)  'The Meteosat Operational Programme - From Experiments to Exploitation', earth observation quarterly, no. 25, march 1989
211)  Introduction to the METEOSAT Operational System, esa BR-32 ISSN 250-1589, Sept. 1987

MOP (Meteosat Operational Program); the MOP phase extends to 1998. MOP is considered the 1. Meteosat generation. Starting in 1987 the weather satellite program 'Meteosat' is being managed and operated by **Eumetsat** (an independent agency since 1986 in Darmstadt). ESA builds and operates the MOP series as a EUMETSAT contractor.

EUMETSAT (MOP) Missions:

- METEOSAT-4       MOP-1 (1. op. Satellite) Launch: March 6, 1989 (Nominal life=2 years for METEOSAT-4 to 6)
- METEOSAT-5       MOP-2, Launch: March 2, 1991
- METEOSAT-6       MOP-3, Launch: Nov. 20, 1993 (last satellite of 1. series)
- METEOSAT-7       MOP-4, projected launch: Dec. 1995 (Meteosat-7 is a new procurement initiated in 1991; there is an option for a second launch in mid 1997 (METEOSAT-8); nominal life = 5 years of each)

Sensor:
**VISSR** (Visible and Infrared Spin scan Radiometer, Platform spin: 100 rpm), also referred to as **MSR** (Multispectral Radiometer)
Spectral Ranges:       0.5 - 0.9 μm (VIS), visible band,
                   5.7 - 7.1 μm (WV), emission of water vapor,
                   10.5 - 12.5 μm (IR),

| | |
|---|---|
| FOV | 18° |
| IFOV | 0.065 mrad (VIS), 0.14 mrad (IR, WV) |
| Ground pixel size | 2.5 x 2.5 km for VIS and WV (5 x 5 km for IR) |
| Image lines | 5000 (VIS), 2500 (IR,WV) |
| Number of pixels/image | 25 Mega pixels (VIS), 6.25 Mega pixels (IR, WV) |
| Scanning time/image | 30 minutes (duty cycle = 100%) |

Applications: Earth and atmospheric monitoring, operational meteorology, climatology. Basic climatological data sets and precipitation index are derived daily. Measurements: cloud coverage, cloud motion winds, cloud top heights, upper tropospheric humidity, precipitation

**DCS** = Data Collection System. This system gathers and relays environmental data transmission that originate at remote automatic data collection platforms (DCPs for ground truthing data). DCPs may be installed at fixed sites on land or on mobile supports (buoys, ships, balloons or aircraft). These platforms can be set to transmit data in four modes: at fixed times (self-timed); in response to DCS interrogation; in an emergency mode during which the platform signals for a DCS interrogation within 1 min; and in an adaptive random reporting mode. The total number of DCPs in the Meteosat coverage area is = 800 (registered by ESOC), 400 of which are active on average. Status of 1991.

**All geostationary meteorological satellites series (METEOSAT, NOAA-GOES, GMS (NASDA), etc. use DCS (or variations thereof; note: the ARGOS-DCS concept on Polar-orbiting NOAA-POES S/C with the corresponding PTTs (ground) is different in uplink frequency, message length and format).**

The principle Meteosat missions can be combined by the user to provide an integrated data access system. This concept is known as 'Meteosat Operational Systems for data Acquisition and InterChange (MOSAIC). Users can readily combine satellite imagery, local reports from the data collection system, as well as conventional meteorological data and forecasts. A total of 66 channels are provided in the 402 MHz range. Meteosat acts in this setup simply as a relay for most data distribution services.

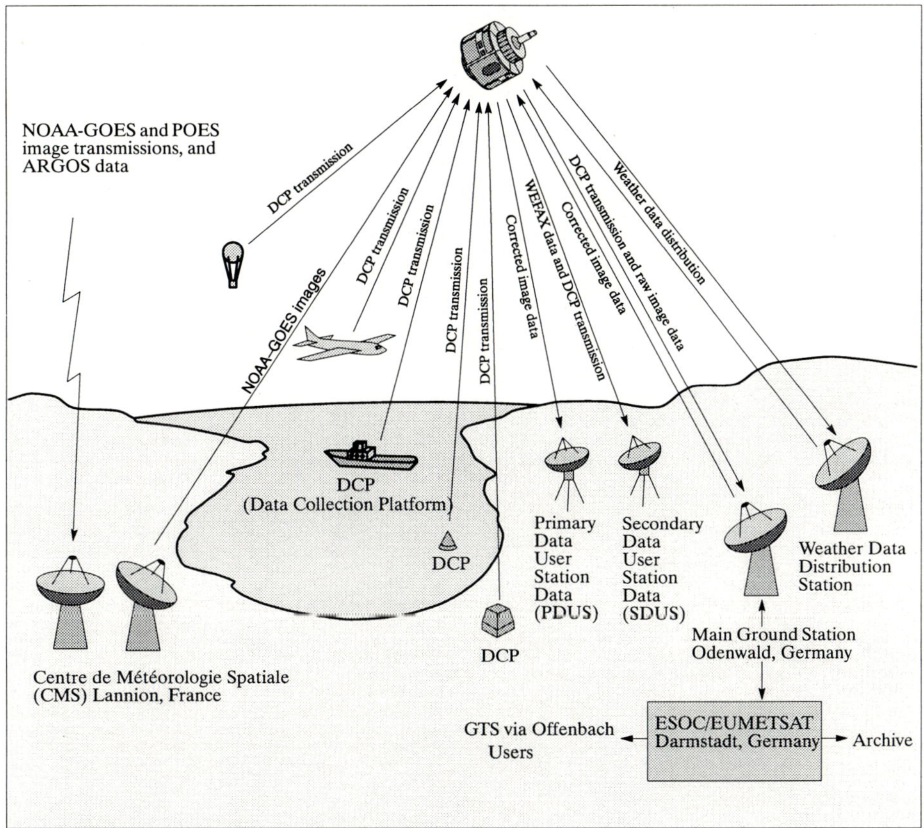


**Figure 52:    Basic Configuration of the Meteosat Operations System**

Data Transmission: Date rate = 333 kbit/s; Frequencies: 1691.0 MHz (channel 2) and 1694.5 MHz (channel 1), 1686.833 MHz (raw image data to the main ground station), 1695.77 ± 0.165 MHz (for Meteorological Data Dissemination MDD). The majority of channel 1 and channel 2 transmissions originate from ESOC (to user community).
S-Band is used for the provision of the following services:

-    to transmit the raw image data to the ground
-    to relay processed image data from the central ground station to other user stations
-    for ranging signals
-    for S/C commanding
-    to transmit telemetry
-    WEFAX (Weather Facsimile) transmissions are in a format compatible with the APT (Automatic Picture Transmission) service of the NOAA-GOES and POES series[212),213)].

The UHF-Band is used to receive environmental data from the DCPs. DCP data may be relayed to Meteosat on up to 33 international and 33 regional channels; regional DCP operate in the 402.1 - 402.2 MHz band; the international channels operate in the 402.0 - 402.1 MHz band.

---

212)  'Meteosat WEFAX Transmissions', ESOC paper, March 1990
213)  'Meteosat High Resolution Image Dissemination', ESOC paper, Oct. 1989

## A.66.1  MSG (METEOSAT Second Generation)

The current 1. generation METEOSAT series is planned to be replaced by a 2. generation series (MSG) with a probable launch of the first prototype satellite in 2000. MSG program start in 1993. The basic requirements on MSG call for significant improvements in the quality of observations and services over the 1. generation series. MSG satellites are planned for operations over a period of at least 10 years. ESA/EUMETSAT take a two-phase approach to program implementation:[214]

1. Pre-operational program or demonstration phase (with two satellites MSG-1 and MSG-2)

2. Operational program phase (MSG-O)

MSG satellites are spin-stabilized (cylindrical in shape, similar to the 1. generation S/C).

Orbit: geosynchronous at 0.0° longitude. Two satellites are simultaneously in orbit (one operating and one in cold redundancy) for reasons of availability.

Application: Operational meteorology and climate monitoring.

| Channel | | 99% energy bandwidth ($\mu$m) | Sampling Distance (NS /EW) (km) | Dynamic Range $*W/(m^2\ sr\ \mu m)$ | NE$\Delta$R or NE$\Delta$T | Absolute Accuracy |
|---|---|---|---|---|---|---|
| Name | Center of $\lambda$ ($\mu$m) | | | | | |
| VIS 0.6 | 0.635 | 0.15 | 3 x 3 | 0 - 620 * | 1.7 W/(m$^2$ sr $\mu$m) | no requirement |
| VIS 0.8 | 0.83 | 0.24 | 3 x 3 | 0 - 380 * | 1.0 W/(m$^2$ sr $\mu$m) | |
| IR 1.6 | 1.615 | 0.35 | 3 x 3 | 0 - 110 * | 0.3 W/(m$^2$ sr $\mu$m) | |
| IR 3.8 | 3.8 | 0.80 | 3 x 3 | 0 - 350 K | 0.25 K at 300 K | < 0.5 K at 'hot' for reference temperatures 'cold' and 'hot' |
| IR 8.7 | 8.7 | 0.80 | 3 x 3 | 0 - 335 K | 0.25 K at 300 K | |
| IR 10.8 | 10.8 | 2.00 | 3 x 3 | 0 - 335 K | 0.25 K at 300 K | |
| IR 12.0 | 12.0 | 2.00 | 3 x 3 | 0 - 335 K | 0.30 K at 300 K | |
| IR 6.2 ,WV | 6.25 | 1.80 | 3 x 3 | 0 - 270 K | 0.25 K at 250 K over 9 x 9 km | < 1.0 K at 'hot' for ref. temperatures at 'hot' and 'cold' |
| IR 7.3,WV | 7.35 | 1.00 | 3 x 3 | 0 - 300 K | | |
| IR 13.4 | 13.4 | 0.72 | 3 x 3 | 0 - 300 K | 0.70 K at 270 K over 18 x 18 km | < 0.5 K at 'hot' for ref. temp. 'cold' and 'hot' averaged over 18 x 18 km |
| IR 9.7 | 9.7 | 0.48 | 3 x 3 | 0 - 335 K | 0.25 K at 255 K over 18 x 18 km | |
| HRV | Silicon | 0.40 | 1 x 1 | 0 - 500 * | .25 W/(m$^2$ sr $\mu$m) | no requirement |

**Table 38:**     **Requirements and Channel Definitions for the SEVIRI Instrument**

Sensors:

- **SEVIRI** = Spinning Enhanced Visible and Infra-red Imager. Principal onboard instrument for imaging and sounding (12 channel instrument as defined in Table 38). The channels VIS 0.6, VIS 0.8, IR 1.6 and HRV are referred to as "warm", while the channels IR 3.8 to IR 13.4 are referred to as "cold". The cylindrical instrument has a diameter of about 1 m and a height of 2.1 m along the spin axis of the satellite. Mass = 133 kg, power = 115 W. The major sub-elements of SEVIRI are:
  - Scan assembly with scan mirror and calibration sources
  - Ritchey-Chretien telescope
  - Focal plane with detectors and passive cooler

  Imaging parameters:
  - Earth frame East-West          17.40° (303.7 mrad.)
  - Earth frame North-South        17.28° (301.7 mrad.)
  - Scan range North-South         20.0° (349.1 mrad., 1389 steps)

---

214) 'Meteosat Second Generation Programme Proposal', ESA/PB-EO, Nov. 1992

|   |   |   |
|---|---|---|
| - | Scan line step | 51.88 arcsec (251.5 mrad., 9 km at SSP) |
| - | Scan mechanism step | 25.94 arcsec (125.8 mrad., 9 km at SSP) |
| - | Spin rate | 100 rpm |
| - | Line cycle | 0.6 s |
| - | Imaging time per line | 30 ms (5%) |
| - | Earth imaging time | 12 minutes |
| - | Calibration + stabilization | 3 minutes |
| - | Repeat cycle | 15 minutes |

- S&R payload within the S&RSAT/COSPAS system. Provision of transparent relay function for search and rescue operations.

- DCP onboard collection system.

- WEFAX

| Channel Groups | Scanning Parameters | Data rate before stretching | Data rate after stretching |
|---|---|---|---|
| 3 VNIR Channels (2 VIS +1 IR) | 3 detectors per channel 3750 pixels per line 10 bits per pixel | 11.25 Mbit/s | 0.5625 Mbit/s |
| 8 IR Channels | 3 detectors per channel 3750 pixels per line 10 bits per pixel | 30.00 Mbit/s | 1.50 Mbit/s |
| 1 HRV (High Resolution Visible) Channel | 9 detectors per channel 5625 pixels per line 9 bits per pixel | 30.375 Mbit/s | 0.7594 Mbit/s |
| Total | | 71.625 Mbit/s | 2.8219 Mbit/s |

**Table 39:    Projected Data Rates of the SEVIRI Instrument**


## A.67    Microlab-1

Microlab is a commercially built and operated small satellite platform concept of Orbital Sciences Corporation (OSC), Dulles Va., offering fully integrated services of payload integration, Pegasus launch services from an L1011 aircraft, S/C operation, communication services, data processing, etc. to the science community at large.[215]

Microlab-1 is the first small satellite mission in this series which is scheduled for launch in Spring 1994. The mission design life is two years with a goal of four years. The primary payload of the Microlab-1 spacecraft is the OTD (Optical Transient Detector), a lightning instrument package provided by NASA/MSFC. The secondary payload on Microlab-1 is GPS-MET (GPS-Meteorology), an instrument provided by UCAR (University Consortium for Atmospheric Research, in addition collaboration with JPL, Stanford University, and U. of Arizona), which uses the GPS signals to measure Earth atmospheric properties. The science data processing of OTD and GPS-MET is done at MSFC and UCAR, respectively. Microlab-1 mass = 69 kg, power = 42 W average, data rate = 2 Mbit/s (one OSC tracking station support). The S/C shape in orbit resembles a butterfly (a cylinder of 1 m diameter and 37 cm in height, whose covers are opened like wings of a bird, the gravity gradient boom is pointing toward the Earth's center).

Orbit: circular orbit, altitude = 785 km, inclination = 70°. The attitude control system is an active magnetic gravity-gradient system for nadir pointing within $\pm 5°$ (1 $\sigma$). The nominal pointing accuracy of Microlab pointing knowledge is $\pm 2°$ (1$\sigma$), dependent on the modeling accuracy of the Earth's magnetic field. The short-term pointing stability is 0.10°/s. Yaw steering is used to maximize sun exposure of solar arrays.

---

215)  Information provided by G. Moody and D. Finn of OSC, and by W. J. Koshak of NASA/MSFC

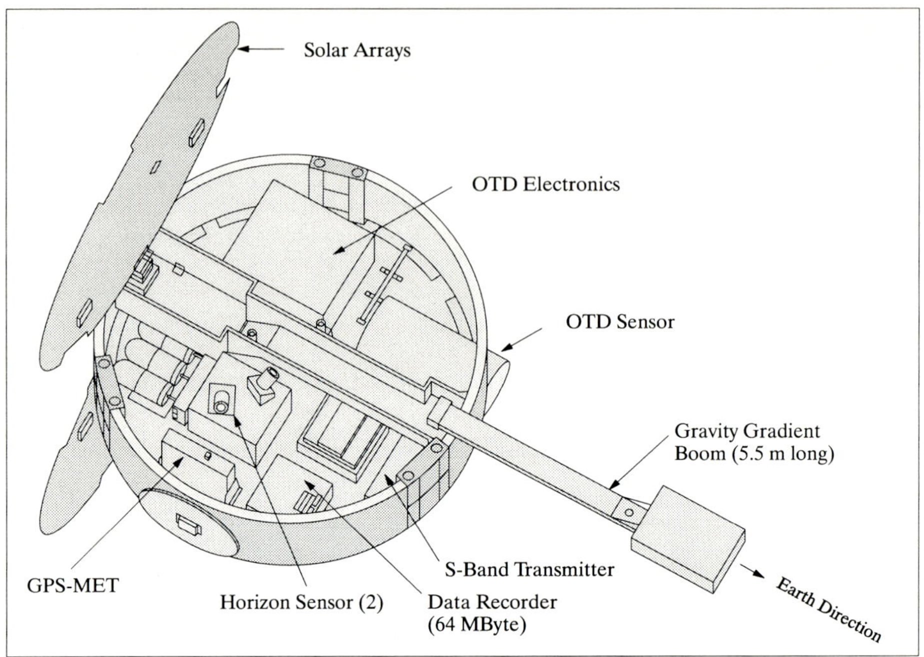


**Figure 53:** **The Microlab-1 S/C Model**

Sensors:

**OTD** = Optical Transient Detector (MSFC instrument). Objectives: observation of the global distribution of lightning leading to the formation of a climatic database of the spatial and temporal distribution of thunderstorms and lightning. The goal is to better understand the thunderstorm activity on a global scale. OTD is a staring imager for the detection of lightning over a large region of the Earth's surface with storm scale resolution, it marks the time of occurrence and location of the lightning and measures its radiant energy. Both intra-cloud and cloud-to-ground discharges will be detected during daytime and nighttime conditions. OTD images a scene much like a television camera; however, daytime detection of highly transient lightning sources against a bright cloud-top background makes actual data handling and processing much more involved than that required by a simple imager. The OTD instrument design is based on LIS of TRMM.

OTD is composed of six major subsystems: an imaging system, a focal plane assembly (including a CCD array detector, pre-amplifiers, and multiplexers), a Real-Time Event Processor (RTEP) and background remover, an event processor, and formatter, power supply, and interface electronics. The imaging system is a simple telescope consisting of a beam expander, an interference filter, and re-imaging optics. The filter is narrow band (8.4Å) and is centered about a prominent neutral oxygen emission triplet in the lightning spectrum to optimize SNR in the presence of a bright solar-lit cloud top.

**GPS-MET** = GPS Meteorology (proof-of-concept-system). Objective: global remote sensing of the atmosphere using the radio occultation measurement method. GPS-MET is a modified commercial high-precision GPS receiver with the capability of simultaneously receiving GPS signals on L1 and L2 frequencies. The receiver can track both C/A and P codes and also implements a codeless carrier recovery technique. All observables are sampled 50 times per second. A single low-gain antenna (a standard 'patch' on a ground plane) points roughly in the anti-velocity direction.

High resolution atmospheric soundings can be retrieved, when the radio path between the LEO GPS receiver and one GPS satellite traverses the Earth's atmosphere. When the path of the GPS signal begins to transect the mesopause at about 85 km altitude, it is sufficiently retarded that a detectable delay in the order of 1 mm in the dual frequency carrier phase observations is obtained by the LEO GPS receiver. As the signal path descends through successively denser layers of the atmosphere, the delay increases to approximately 1 km at the Earth's surface. Thus the atmosphere creates a unique signal with over 6 orders of magnitude in dynamic range. A single LEO GPS receiver can observe more than 500 such occultations per day. The key observables are atmospheric temperature and moisture distributions.

GPS-MET measurement resolutions:

- Precision of phase measurements: order of 1 mm
- Velocity accuracy: $10^{-4}$ m/s
- Position accuracy: 1 m (vertical: $<$ 1 km; horizontal: $<$ 200 km)

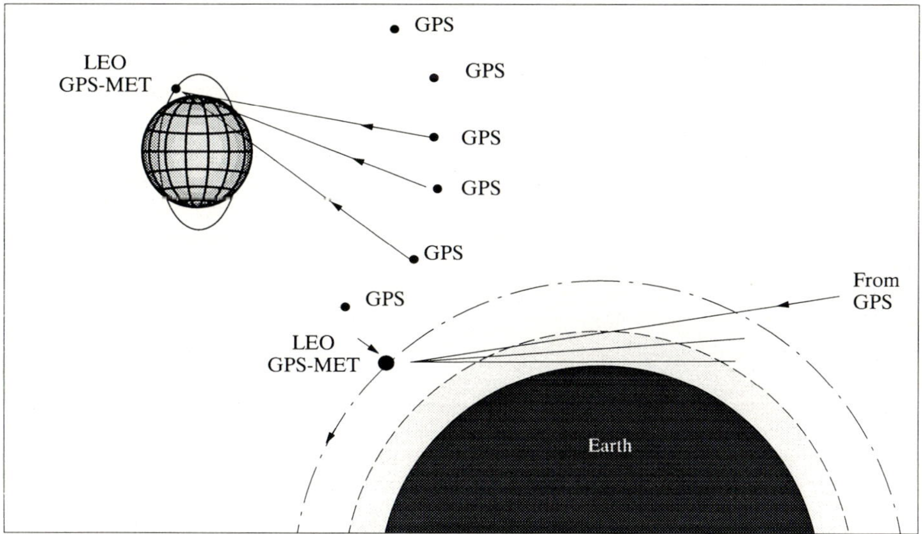


**Figure 54:      Observational Geometries of the GPS-MET Occultation Measurements**

# A.68    MIR-1 Orbital Station

MIR (=Peace) is the Soviet Union's (operator: NPO Energia) modular orbital station (launch of the MIR core module: Feb. 20 1986) of considerable dimensions and payload capacities. MIR is comprised of the following separate modules and a Soyuz (Sojus) TM spacecraft:

1. the **MIR Core module**, which contains the cosmonaut crew quarters (mass = 20,000 kg, 4.2 m diameter), existing

2. the **Kvant 1 module** for astrophysics  (mass = 11,000 kg), existing

3. the **Kristall module**, which is used to manufacture new materials in micro-gravity (mass = 20,000 kg), existing

4. the **Kvant 2 module**, which houses the airlock used by cosmonauts to exit MIR for space walks (mass = 20,000 kg), existing

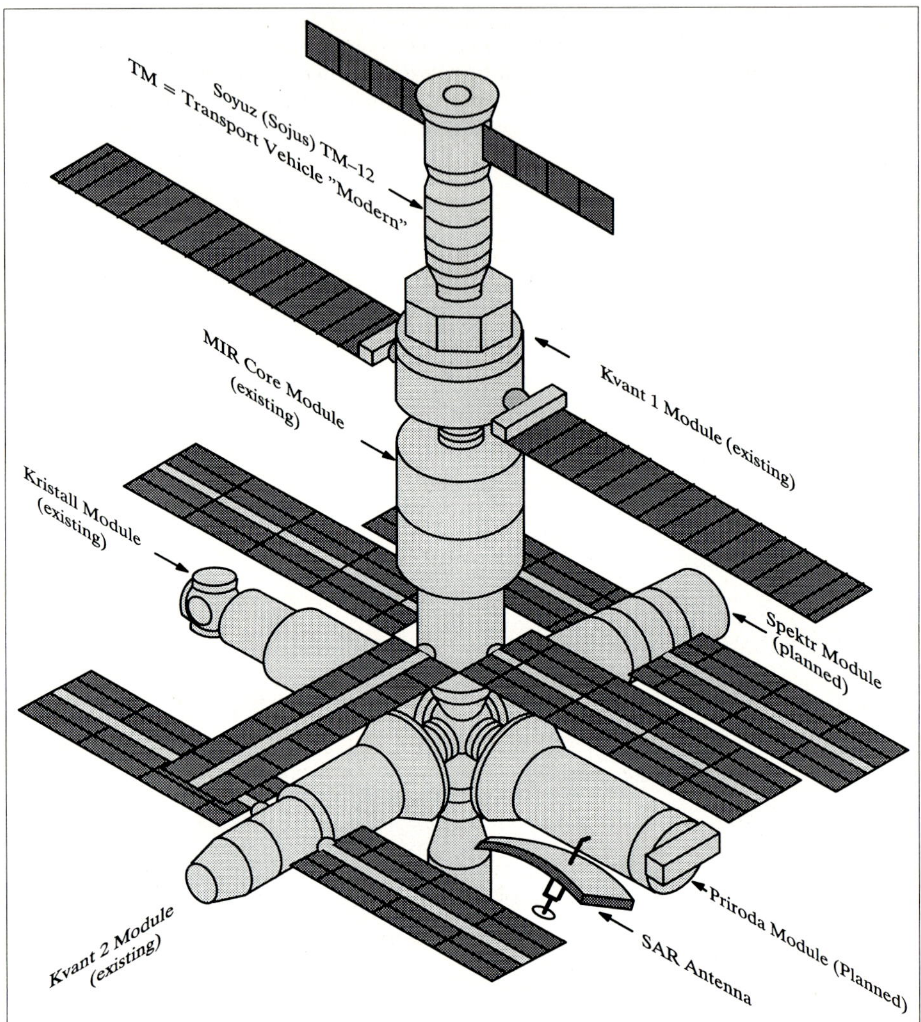


**Figure 55:     Overall Configuration of the MIR Orbital Station**

5.   the Soyuz TM-12 spacecraft (service vehicle)

6.   the *Priroda module*, which is used for Earth observation (this module will be added to MIR in early 1995 (see chapter A.82 for Priroda)

7.   the *Spektr (Spectrum) module*, planned extension (launch in Mid 1994)

The station has been occupied continuously by cosmonaut crews since Sept. 1989.[216] A successor to MIR-1 is planned with MIR-2 in 1997 (planned orbit inclination = 65°).

**One of the MIR objectives is Earth observation (resources, environment).** The presence of the onboard crew is a great advantage for the maintenance and repair of all onboard equip-

216) 'Soviets to set Record Pace for MIR Repairs', Space News June 10-16, 1991, p. 12

ment and for the collection of meaningful correlative information. The crew makes actually observations possible that require very intricate and complicated operational sequences or methods, not easily susceptible for algorithmic definition. For the time frame of 1990 to 1993 there is a considerable amount of Earth observation (resources) scheduled and also the introduction of new and more advanced sensor equipment.

Orbit: Altitude = 350 - 410 km; Inclination = 51.6°;

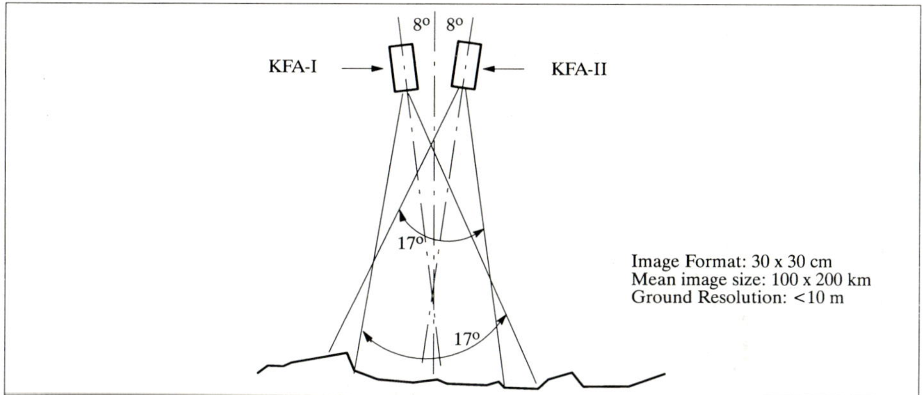


**Figure 56:     Optical Geometries of the KFA-1000 Camera (Priroda-5) on MIR**

**Sensors:** (Earth Observation equipment (sensors, etc.) is located in the MIR core module)

- Module Kvant 2 - Camera KAP-350 (350 mm focal length). Provision of high-quality images (Format 180 x 180 mm,  scale 1 : 1000 000, resolution = 30 m)

- Module Kristall (was introduced in 1990)
  Contains two cameras of the type KFA-1000, which are mounted fan-like for extra-wide swath width (see A.84 for a KFA-1000 description). The two cameras KFA-1000 are also known under the designation: 'Priroda-5'

- Module "I" is the complex "PRIRODA", scheduled to be introduced at MIR in 1993 (see A.82 for PRIRODA definition).

- Hasselblad camera 500-EL.
  Hand-held (portable) cameras for the crew to document their visual observations.

- Spectrometer - System "Gemma" (hand-held)
  Prototype testing in MIR as of 1989. Measurement of 456 narrow channels in the spectral range from 0.4 - 1.1 μ.

- MKS-M = Multichannel Spectrometer of DLR, consisting of the following systems:
  - MKS-M-AS = Atmospheric spectrometer (atm. corrections in the VIS range)
  - MKS-M-BS = Bio spectrometer (determination of ocean phytoplankton)

**Data:**[217]. MIR data are being offered commercially to the user community by 'Energia USA' of Henderson Va., a US subsidiary of NPO ENERGIA, Moscow

---

217) "Earth Imagery from MIR offered to Commercial Buyers", Space News, April 25-May3, 1992, p. 27

**Photographic Complex**

| Name of Instrument | Spectral Range (nm) | Observation Surface | Spatial Resolution | MIR Module | Image Format/Nr. of Images per film |
|---|---|---|---|---|---|
| Priroda-5 | 400-800 | 32° x 16° 100 x 200 km | 5-10 m | Kristall | =30 x 30 cm$^2$/1500 |
| KAP-350 | 400-800 | 40° | 30-40 m | Kvant-2 | = 18 x 18 cm$^2$/600 |
| MKF-6MA | 480-840 | 32° x 24° | 30 m | Kvant-2 | = 56 x 81 mm$^2$/2400 |
|  |  |  |  |  |  |

**TV - Complex**

| Name of Instrument | Spectral Range (nm) | Observation Surface | Spatial Resolution | MIR Module | Remarks |
|---|---|---|---|---|---|
| KL-103 W | 400-800 | 16° x 20° | 250 m | Kvant-2 | Color Camera |
| Atlas | 400-800 | 20° x 41° | 200-10 m | Kvant-2 | |

**Optical Complex - Spectrometer**

| Name of Instrument | Spectral Range (nm) | Field of View (FOV) | Spatial/Spectral Resolution | MIR Module | Remarks |
|---|---|---|---|---|---|
| Telespectrometer Phase | 445-2200 | 10' | 2-/15, 100, 200, 250 m | Kvant-2 | Impulse Frequency = 10 GHz |
| IR Spectrometer Istok-1 | 3.6-16 μm 64 chan. | 12' x 48' | 0.8x6 km$^2$/ 125,250 nm | Priroda | $\theta v$ =0-90°, Nadir and Occultation |
| Telespectrometer ITS-7D | 4000-8000 nm | 0.2 x 0.2° 0.2 x 0.8° 60 x 240 km$^2$ | -/150 and 300 nm | Kvant-2 | Spectrum - 1 s |
| Phönix | 2.61-2.63 μm | 1°24' x 25' | -/0.4 cm$^{-1}$ | Spektr | Spectrum - 1 s |
| Volchov-1 | 5-22 μm | 10° | -/10 cm$^{-1}$ | Spektr | Spectrum - 2 s |
| Volchov-2 | 5-22 μm | 20' | -/16 cm$^{-1}$ | Spektr | Spectrum - 2 s |
| Skif | 400-1200 | 0.87 x 0.17° | -/14 a. 3nm | Core | |
| MKS-M-AS | 757-770 nm | 1.15 x 0.09° | 6.3 x 0.5 km$^2$/1.5nm | Core | |
| MKS-M2-BS | 415-830 | 0.46 x 0.46° | 2.4x2.4 km$^2$ /10 nm | Core | |
| MKS-M2-AS | 757-770 | 1.15 x 0.09° | 6.3 x 0.5 km$^2$/1.5nm | Kvant-2 | Spectrum - 20 ms on pointable platform |
| MKS-M2-BS | 415-1030 nm | 0.46 x 0.46° | 2.4x2.4 km$^2$ /10 nm | Kvant-2 | Spectrum - 20 ms on pointable platform |
| Spektr 256-Z | 450-830 nm | 8° | 65x120 km$^2$ /1.5 nm | | |
| Ozon-M | 0.25-0.29μm 0.37-0.39μm 0.60-0.64 μm 0.99-1.03 μm | 2' x 25' | 15 km / 2-7Å 1 km height | Priroda | Occultation |
| IR Radiometer Jausa-100-3 Neva-3 Neva-5 | 1.8-3.0μ 1.8-3.0μ 3.0-5.0μ | 50' x 50' 10' x 10' 10' x 20' | -/- -/- -/- | Spektr Spektr Spektr | |
| Matrix Spectrometer Zwet | 1.5-3.3 μ | 20' x 20' | -/0.02 μm | Spektr | Spectrum - 0.5 s |

**Optical Complex - Scanner**

| Name of Instrument | Spectral Range (nm) | Channels | Swath Width/ FOV | Spatial/Spectral Resolution | MIR Module | Remarks |
|---|---|---|---|---|---|---|
| MOS-Obsor-A | 757-768 | 4 | 80 km | 1.4nm /2.7km | Priroda | Imaging Spectrometer |
| MOS-Obsor-B | 457-1030 | 13 | 80 km | 10nm /0.6km | Priroda | Imaging Spectrometer |
| MSU-SK | .5-12.5 μm | 5 | 350 km | 100nm ./120x300m | Priroda | conical Scanner (39°) |
| MSU-E | 0.5-0.9 μm | 3 | 2 x 27 km | 100nm/25m | Priroda | CCD-Pushbroom camera |

**Optical Complex - Lidar**

| Name of Instrument | Spectral Range (nm) | Channels | Swath Width/ FOV | Spatial/Spectral Resolution | MIR Module | Remarks |
|---|---|---|---|---|---|---|
| Balkan-1 | 532 nm | | 90" | 3m vertical | Spektr | Impulse = 0.15 J Freq.=1 x in 5.5 s |
| Alissa (France) | 532 nm | | 3' | 150m vertical | Priroda | Impulse = 40 mJ Frequency = 8 Hz |

| Name of Instrument | Wave Length (cm) | Swath (km) | Pixel Size (km) | View Angle | MIR Module | Resolution |
|---|---|---|---|---|---|---|
| Microwave (MW) Complex - Radiometer | | | | | | |
| KR-05 ($\lambda$ scan) | 0.46-0.55 ($\Delta f = 1GHz$) | | 1.5 | | Spektr | |
| Complex Ikarus IKAR-N,R-30 (Nadir) R-80 R-135 R-225 R-600 | 0.3-6 0.3 0.8 1.35 2.25 6.0 | 60-750 60 | 5-75 60 | Nadir | Priroda | 0.15 K |
| IKAR-P RP-225 | 2.25 (3 polar) | 750 | 75 | 40° | Priroda | 0.15 K |
| RP-600 (Panorama) | 6.0 (3 polar) | 750 | 75 | 40° | Priroda | 0.15 K |
| IKAR-Delta (scanning) RD-30 RD-80 RD-135 RD-400 | 0.3 0.8 1.35 4.0 | 400 400 400 400 | 5 8 15 50 | 40° 40° 40° 40° | Priroda | 1.5 K 0.5 K 0.4 K 0.15 K |
| Active Microwave (MW) Instruments | | | | | | |
| SAR Travers | 9.3 23 | 50 50 | 0.15 | 35° | Priroda | |
| Altimeter Greben | 2.25 | 2.5 | 2.5 x 9.75 | Nadir | Priroda | min. ±10 cm |

**Table 40:**     **Earth Observation Instruments on the Soviet Space Station MIR**[218),219),220)]

# A.69    MOMS-01

**MOMS**[221),222)] = Modular Optoelectronic Multispectral Scanner (German sensor, built by MBB, payload-/experiment of DLR, University of Munich, etc..). MOMS-01 was a Shuttle payload (mounted on SPAS = Shuttle Pallet Satellite) on two missions (STS 7: June 18 1983 - 6 day mission; and STS 10 (41-B): Feb. 3, 1984 - 8 day mission).

MOMS-01 is a two-channel system which uses the spectral ranges of 575 - 625 nm for geologic observations, and 825 - 975 nm for vegetation detection. A double optical system was used per band for illuminating the CCD array consisting of 4 groups with a length of 6900 pixels. Pixel size on ground = 20 x 20 m. The total experiment time was 26.5 minutes on STS-7, and 30 minutes on STS-10.

Application: Observation of the Earth with configurable sensors. The stereoscopic monitoring capability provides a very good interpretability of the data for topographic and geographic features.

Orbit: Shuttle Orbit, Inclination = 28.5°; STS 7 = 292 km altitude; STS 11 = 289 - 300 km altitude; Swath width = 140 km

Sensor: **RETICON** CCPD 1728, 4 CD's per Line; Radiometric resolution = 7 Bit; Measurement in the red and infrared regions. The RETICON sensor is regarded the first imaging charge-coupled device (CCD) instrument that was operated in outer space.

Data: Data rate = 40 Mbit/s; the data were recorded onboard. The MOMS-01 missions yielded 450 individual scenes for geoscientific thematic evaluation.

218)  Overview Paper provided by G. Zimmermann of DLR (IKF) Berlin, Aug. 1991
219)  Note: The sensors of existing modules are operational (Priroda and Spektr modules are planned)
220)  MIR Earth Images are sold  by 'Energia Deutschland GmbH', a joint venture of NPO Energia, Moscow and Kayser-Threde of Munich, Germany - see Space News, Aug. 17-23, 1992, p. 13
221)  'MOMS-02 - Ein multispektrales Stereo-Bildaufnahmesystem für die zweite deutsche Spacelab-Mission D2', F. Ackermann, J. Bodechtel, F. Lanzl, D. Meissner, P. Seige, H. Winkenbach; Geo-Informations-Systeme, Zeitschrift für interdisziplinären Austausch innerhalb der Geowissenschaften, Wichmann Verlag, Jahrgang 2, Heft3/1989, S. 5 - 11.
222)  J. Bodechtel, D. Meißner, P. Seige, H. Winkenbach, J. Zilger, "The MOMS Experiment on STS-7 and STS-11 - First Results and Further Development of the Modular Optoelectronic Multispectral Scanner", Proceedings of the Eighteenth International Symposium on Remote Sensing of Environment, Volume I, 1984, pp. 77-85

## A.70    MOMS-02

**MOMS-02** on the Spacelab D-2 Mission (10 day flight). Launch: April 26, 1993. MOMS-02 is an advanced version of MOMS (or MOMS-01). The D-2 orbit has an inclination of 28.5° and an altitude of 296 km (this means: observation of equatorial regions).

The data will be used in the following applications: Land cover (vegetated areas: land use, biomass estimation; unvegetated areas: lithology, mineral prospecting, tectonic investigations); geomorphology, pedology, ecology, basic research in the spectral signatures of rocks, soil and vegetation, etc.[223]

Objectives:

* Stereoscopic visual observation (high degree of interpretability)
* Provision of high-quality topographic regional maps (scale 1:50000) and digital terrain models ( <5 m pixel size on ground)
* Test of a digital photogrammetric observation technique and processing system (Prototype)
* Correlation of high-resolution panchromatic data with multispectral data.

Experiment:        Triple-Stereoscopy
                       Along-Track-Stereoscopy
                       High-resolution Imagery
                       Multispectral observation (refined modelling of MS classification)
                       Combination of Stereo- and Multispectral Imagery

The MOMS-02 payload has a total of 5 optical systems: three are used for stereoscopic imagery, two are employed for multispectral imagery.

**MOMS-02 Sensor:** Spectral Ranges: (MS)

| | |
|---|---|
| Channel 1: | 440 - 505 nm |
| Channel 2: | 530 - 575 nm |
| Channel 3: | 645 - 680 nm |
| Channel 4: | 770 - 810 nm |

Panchromatic Bands (high resolution (HR) + stereo (ST))

| | |
|---|---|
| Channel 5: | 520 - 760 nm (nadir looking) |
| Channel 6: | 520 - 760 nm (+21.4°, forward tilt) |
| Channel 7: | 520 - 760 nm (-21.4°, backward tilt) |

| | |
|---|---|
| Ground pixel size | 13.5 m x 13.5 m (ch. 1-4, 6,7); 4.5 m x 4.5 m (ch. 5) |
| Swath width | 78 km (ch. 1-4, 6,7); 37 km (ch. 5) |
| Focal length (f) | 220 mm (ch. 1-4); 237.2 mm (ch. 6,7), 660 mm (ch. 5) |
| IFOV/TFOV | 45.45 μrad / 10.0° (ch. 1-4) |
| | 42.16 μrad / 15.0° (ch. 6,7) |
| | 15.15 μrad / 8.0° (ch. 7) |

Quantization: 8 bit uncompressed for MS and Stereo, 6 bit compressed for Stereo

Seven variable observation modes of different band combinations are defined. The modes can be summarized as follows:

* Mode 1    cha. 5,6,7          HR + ST, full stereo
* Mode 2    cha. 1-4            MS, full multispectral
* Mode 3    cha. 3,4,6,7        2MS / 2 ST
* Mode 4    cha. 1,3,4,6        3 MS / 1 ST
* Mode 5    cha. 1,3,4,7        3 MS / 1 ST

223) J. Bodechtel, S. Lutz, "Neue Wege der Erderkundung", aus Einsichten, Forschung an der LMU, pp. 38-43, 1992

- Mode 6   cha. 2,3,4,5          3 MS / HR
- Mode 7   cha. 1,3,4,5          3 MS / HR

Data: Onboard recording onto HDT, max. recording time is 5.5 hours; this amounts to about $2.5 \times 10^{12}$ bit of data. The max. observation data rate = 100 Mbit/s.

There are also investigations and advanced plans under way to fly MOMS-02 along with PRIRODA-1 on MIR in 1995[224].

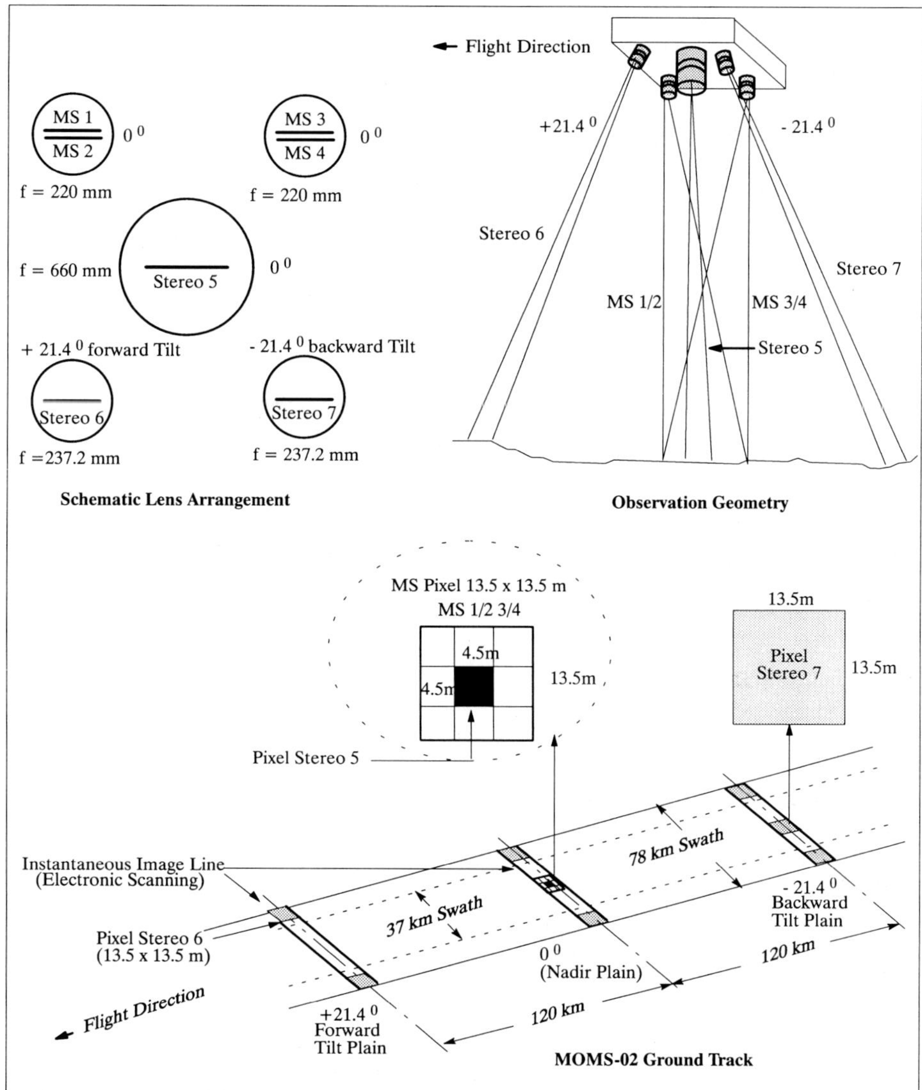


**Figure 57:**    **MOMS-02 Optical Layout and Geometries for the D-2 Mission**

224) Deutsch-russische Kooperation in der Erdbeobachtung, Nutzungskonzept für die präoperationelle Mission des Modularen Optoelektronischen Stereo-Scanners MOMS-02 auf dem Modul PRIRODA der sowjetischen Raumstation MIR (MOMS-02P), DARA, Nov. 1991

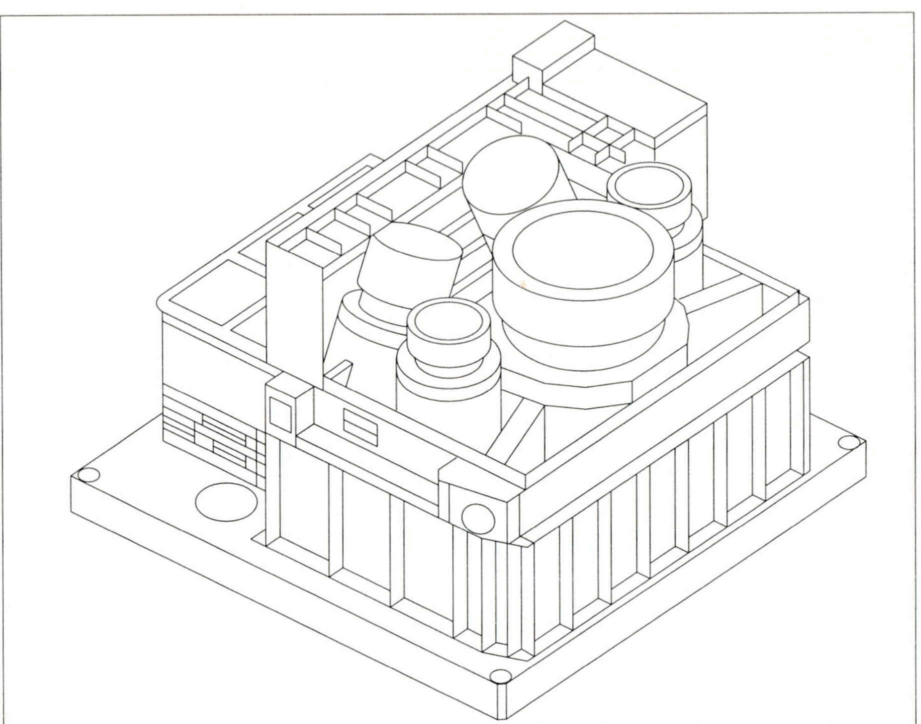

**Figure 58:**    **The MOMS-02 Sensor Model**

## A.71    MOS (Marine Observation Satellite)

MOS = Marine Observation Satellite. MOS-1 is Japan's (NASDA) first experimental Earth observation mission, also referred to as MOMO-1. MOS-1 is operational since Feb. 1987 (launch Feb. 19, 1987). ESA/Earthnet is a data distributor of MOS-1. The follow-up mission is **MOS-1B**, launch: Feb. 7, 1990.

Application: Observation of the ocean surfaces (color), vegetation, land ecology, measurement of water vapor in the stratosphere, measurement of surface temperatures, etc.

Orbit: Sun-synchronous with a repeat cycle of 17 days. Inclination = 99.1°; altitude = 908 km;. period = 103 min; Local sun time = 10:15 for MOS-1 and 10:33 for MOS-1b

**Sensors: (MOS-1B)**

- **MESSR** (Multispectral Electronic Self-Scanning Radiometer) 2 units; 50 m resolution on ground, measurement in 4 channels, visible range and near-infrared range. Bands: 0.51-0.59 µm, 0.61-0.69 µm, 0.73-0.80 µm, 0.80-1.10 µm. Swath width = 100 km (200 km when both camera systems are operating)

- **VTIR** (Visible and Thermal Infrared Radiometer, mechanically scanning mirror instrument); resolution = 900 m on ground in the visible range and a 2700 m resolution within the three infrared ranges. Bands: 0.5-0.7 µm, 6.0-7.0 µm, 10.5-11.5 µm, 11.5-12.5 µm. Swath width = 1500 km

- **MSR** (Microwave Scanning Radiometer, mechanical); MSR is a Radio Sensor, measuring (scanning) the Earth surface in flight direction. Spectral bands: 23.8 GHz, 31.4 GHz. Resolution: 23 km (at 31 GHz) and 32 km (at 23 GHz). Swath width = 320 km

**Data:**

- MESSR = 9 Mbit/s
- VTIR = 0.8 Mbit/s
- MSR = 2 kbit/s

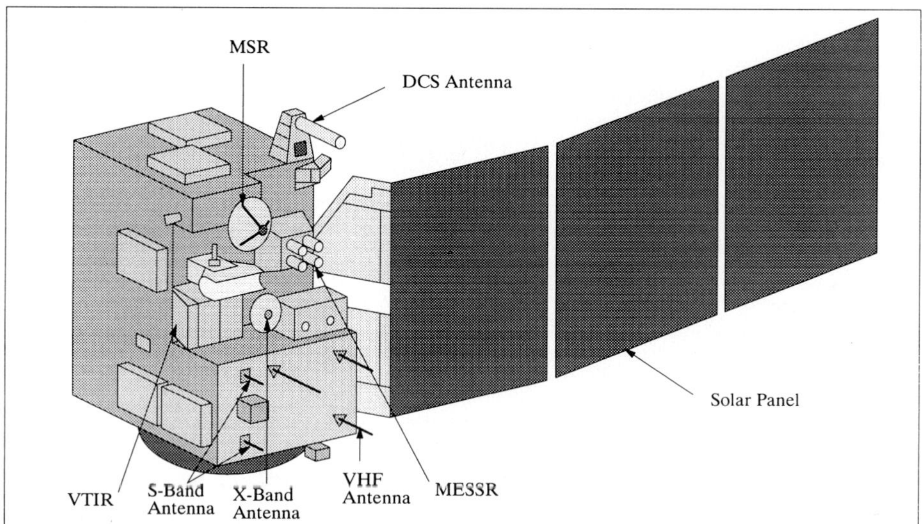


**Figure 59:** **The MOS-1B S/C Model**

## A.72 NIMBUS

NASA/GSFC Earth Observation program[225],[226]. NIMBUS-7 was launched in 1978. (longest Earth observation mission so far; the Nimbus series S/C with their orbits and sensor complements are listed in Table 50). The S/C is three-axis stabilized and has a mass of 965 kg. The attitude control subsystem maintains S/C alignment with the local orbital reference axis to within 0.7° of the pitch axis and 1° of the roll and yaw axis.
Status 1/1994: NASA stopped operation of SAM II and ERB on January 4, 1994. TOMS failed in May 1993.

Application: Oceanography, Pollution of the Atmosphere, Meteorology. Measurement of trace gases and particles in the atmosphere. Distribution phenomena of air pollution. Observation of the ocean surface color and of the temperature. Ice observation, in particular in the coastal regions.

Orbit: Sun-synchronous polar orbit, altitude = 955 km, period = 104.16 minutes.

**Sensors:** (2 of the nine instruments are operational as of Fall 1993: the non-scanning portion of ERB and SAM II).

- **CZCS** = Coastal Zone Color Scanner. Measurement of temperature, chlorophyll concentration, sediment distribution, Gelbstoff-concentrations as a salinity indicator, and the temperature of coastal waters and open ocean. Channel wavelengths: o.433-0.453 µm, 0.51-0.53 µm, 0.54-0.56 µm, 0.66-0.68 µm, 0.70-0.80 µm and 10.5-12.5 µm. Spatial resolution: 825 m x 825 m (each band) at nadir, swath width = 1600 km. Data rate = 800 kbit/s. Data Products: global maps of chlorophyll concentration, sediment distribution,

225) "The NIMBUS-7 User's Guide", NASA/GSFC, Prepared by The Landsat/Nimbus Project, Aug. 1978
226) "NIMBUS-7, Observing the Atmosphere and Oceans", NASA pamphlet Dec. 1983

gelbstoff concentrations as a salinity indicator, and temperature of coastal waters and the open ocean. CZCS was operational for nearly 8 years before it fell silent in mid-1986.

- **ERB** = Earth Radiation Budget. Global measurement of Earth fluxes, Solar fluxes, and zonal insolation. 22 optical channels - 10 solar viewing channels covering 0.2 - 50 μm ; 4 Earth viewing channels with fixed wide-angle FOV's to measure radiation from the entire Earth disk (0.2 - 50 μm, 0.2 - 50 μm, 0.2-3.8 μm, and 0.695-2.8 μm); and 8 Earth-viewing channels to scan from nadir to horizon in several vertical planes with narrow-angle FOVs. Spatial resolution ~ 150 km (scanner) in each band.
  Swath width: unencumbered FOV for all solar channels is 10°, with a maximum FOV of 26° for all except channel 9 which is 28°; unencumbered FOV for the fixed-angle Earth-viewing channels is 121°, with a maximum of 133.3° (channel 12 also has a setting for 89.4° and 112.4°, respectively); the 8 Earth-viewing scanning channels have a FOV of 0.25 x 5.12°.
  Data rate < 4 kbit/s. Data products: determination of the radiation budget of the Earth on both synoptic and planetary scales by simultaneous measurement of incoming solar radiation and outgoing Earth-reflected (shortwave) and emitted(longwave) radiation.

- **LIMS** = Limb Infrared Monitor of the Stratosphere. Measurement of vertical gas concentrations and temperature profiles in the stratosphere. Observables: $O_3$, $H_2O$, $NO_2$, $HNO_3$ and temperature. Channel wavelengths: 6.25, 6.75, 9.65, 11.35, 15.25 μm and 1 broad channel from 13.3 - 17.2 μm.

- **SAM II** = Stratospheric Aerosol Measurement II. The instrument is a sun photometer that views a small portion of the sun through the Earth's atmosphere during S/C sunrise and sunset. Measures aerosol extinction and extinction ratio profiles, and stratospheric optical depth as a function of altitude, latitude and longitude. Channel wavelength: 0.98-1.02 μm. Spatial resolution is 1 km (vertical); swath width is 5 - 40 km vertical. Measurement calibration technique: the Langley Technique is used to calculate zero air mass solar intensity. By measuring the intensity of the sun over a several-hour period on a clear, optically stable day and using the calendar day and latitude/longitude of the measurement location, one can plot intensity versus air mass.
  Data rate: < 500 bit/s. Data products: global maps of the concentration and optical properties of stratospheric aerosols as a function of altitude, latitude, and longitude, which prove valuable for studies on radiative transfer and climatic effects; aerosol transport sources and sinks in the stratosphere; seasonal variations and sudden warming phenomena; and volcanic injection phenomena. When no clouds are present in the instruments IFOV then tropospheric aerosols are also mapped.

- **SAMS** = Stratospheric and Mesospheric Sounder. Measurement of vertical gas concentrations ($H_2O$, $CH_4$, CO and NO) and temperature profiles in the stratosphere and mesosphere. Observation of 'resonant scattering of solar radiation'. Channel wavelengths: 9 channels defined by gas cell modulation 4.1 to 15 μm and 25 - 100 μm. SAMS was operational until April 9, 1985.

- **SBUV/TOMS** = Solar Backscatter Ultraviolet/Total Ozone Mapping. Measurement of vertical $O_3$ profiles, total atmospheric $O_3$, solar irradiance, and terrestrial radiances. SBUV is a double Ebert-Fastie spectrometer and filter photometer - TOMS is a single Ebert-Fastie scanning spectrometer. SBUV serially monitors 12 selected narrow wavelength bands in the spectral region from 0.250 - 0.340 μm or continuously scans from 0.160 - 0.400 μm while the photometer measures the light in a fixed band centered at 0.343 μm. Spatial resolution of SBUV = 11.3° in each band. Swath width = 200 km, separated by the 26° longitude interval between successive orbits.
  TOMS - measures the backscattered radiation sampled at six wavelengths from 0.312 - 0.340 μm. Spatial resolution is 3° (scanned through the satellite subpoint and perpendicular to the orbital plane). Swath width = 2700 km.

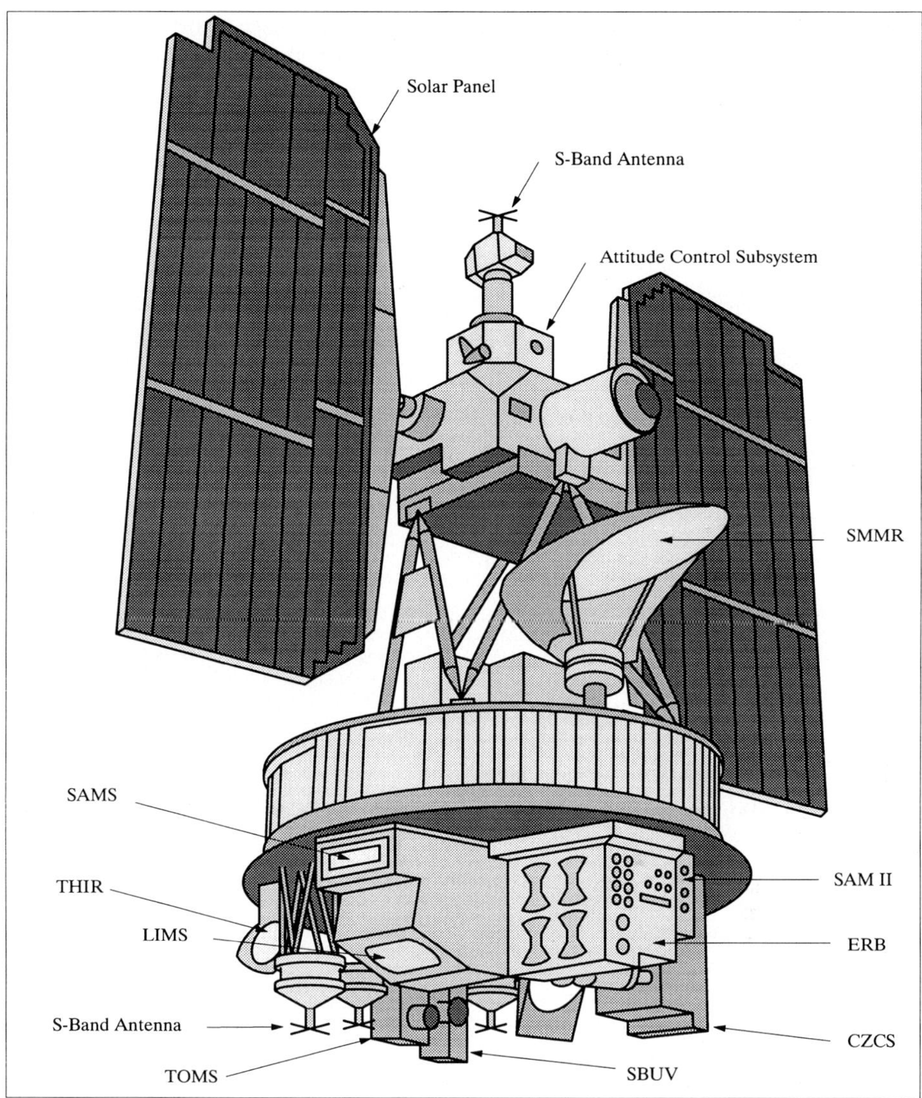


**Figure 60:** **The Nimbus-7 Spacecraft Model**

Measurement/calibration technique: SBUV - both channels simultaneously view identical fields of solar radiation scattered by the terrestrial atmosphere in the nadir of the solar flux scattered from the instrument diffuser plate that is deployed on command. TOMS - measurements are made sequentially in $3^\circ$ steps in the $\pm 51^\circ$ cross scan from nadir, creating a contiguous map of total ozone (scans of consecutive orbits overlap).

Data rate: < 1 kbit/s. Data products: The vertical distribution of ozone, global maps of total ozone and 200-mb height fields; incident solar ultraviolet irradiance and ultraviolet radiation backscattered from the Earth. SBUV helps to determine the total amount of atmospheric ozone in a vertical column above the subsatellite point; vertical profile of ozone above the ozone maximum; measurements of the ultraviolet solar spectral irradiance and its temporal variability over the 160 - 400 nm range (with a spectral resolution of 1 nm). TOMS exploits the polar orbit to yield global, contiguous maps of total ozone.

**TOMS** is scheduled or flying on the following missions (4 improved instruments are being built with new diffusers and considering a larger wave spectrum):
- Meteor 3-6 (TOMS Engineering Model onto a Soviet satellite, launch: 1991)
- Earth Probes, Launch: 1993
- ADEOS, Launch: 1995
- Earth Probes, Launch: 1997

- **SMMR** = Scanning Multichannel Microwave Radiometer. Observation of sea-ice parameters, ocean surface conditions, atmospheric conditions, land parameters, glacial features. Channel wavelengths: 6.6, 10.7, 18.0, 21.0, 37.0 GHz (frequency). SMMR ops ended on July 6, 1988.

- **THIR** = Temperature Humidity Infrared Radiometer. Observation of surface temperature and cloud top temperature. Channel wavelengths: 6.75 µm and 11.5 µm. THIR operation ended on Nov. 30, 1987.

Data: S-Band downlink. Onboard recording of data. All data are preprocessed in NSSDC (National Space Science Data Center) archived and distributed.

S/C Status (Fall 1992): overall the S/C is in good health, though its orbit is drifting to earlier equator crossing times ( ~ 10:40 AM local time) making data analysis more difficult (an equator crossing time of noon would be optimal). There is no longer sufficient fuel to prevent this drift. Only the TOMS sensor remain fully operational. The SUBV, SAM II, and ERB sensors are functioning in a reduced capacity. All other sensors are non-operational.


## A.73    NOAA-GOES

GOES = Geostationary Operational Environmental Satellite. A NASA/NOAA weather satellite series (NASA manages the program, NOAA operates the satellites and provides the services to the user community). The original weather satellite program became operational in 1974 and was known under the term: SMS = Synchronous Meteorological Satellites. All GOES satellites are planned for an operational life of nominally 5 years (2 operational satellite configuration, GOES-East and GOES-West).

*According to NOAA convention all preoperational satellites are designated with a letter - with the launch and the start of the operational phase, the letter name is converted into a number, for instance: GOES-A (preoperational) is changed to GOES-1 (operational).*

| | | | |
|---|---|---|---|
| SMS-1 | Launch: May 17, 1974 | SMS-A | (Name prior to launch) |
| SMS-2 | Launch: Feb. 6, 1975 | SMS-B | (Name prior to launch) |
| GOES-1 | Launch: Oct. 16, 1975 | GOES-A | (Name prior to launch) |
| GOES-2 | Launch: Jun. 16, 1977 | | |
| GOES-3 | Launch: Jun. 16, 1978 | | |
| GOES-4 | Launch: Sept. 9, 1980 | GOES-4 VAS failed in Nov. 82 | |
| GOES-5 | Launch: May 22, 1981 | | |
| GOES-6 | Launch: Apr. 28, 1983 | | |
| **GOES-7** | **Launch: 1987 (Feb. 26), currently operational (1994)** | | |

The GOES program of the future has the following tentative schedule:[227]

GOES-I    Launch: 4/1994 (GOES-I is also referred to as GOES-Next)
GOES-J    Launch: 4/1995

---

[227] "Construction of Lightning Mapper Snagged in GOES Delays", Space News, March 4-10, 1991, pp. 18

GOES-K  Launch: 1999
GOES-L  Launch: 2000
GOES-M Launch: 2004

Note: The GOES-I-M series is undergoing a change from spinning S/C to a three-axis stabilized S/C to increase the observational dwell times and to achieve other improvements[228].

Application: Weather forecast, Meteorology. Derivation of temperature and moisture profiles. Monitoring severe storms and tropical cyclones (hurricanes), etc. .

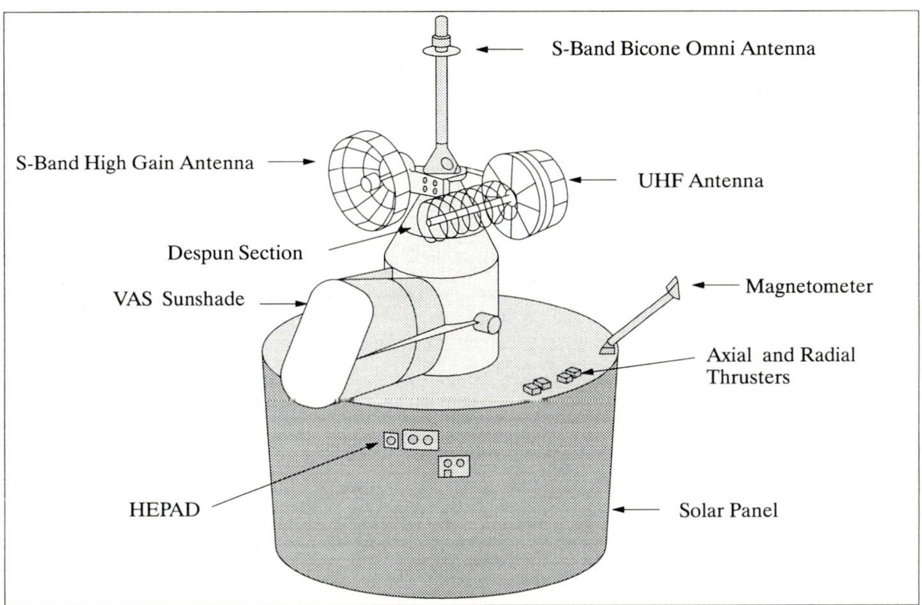


**Figure 61:    The GOES First Generation S/C Model**

**Sensors:** (see also Table 50 for GOES sensor configurations)

**VISSR** = Visible Infrared Spin Scan Radiometer. Measurement (scanning) in 8 channels in the visible range and 2 channels in the infrared spectrum.

**VAS**[229],[230] = VISSR and Atmospheric Sounder (successor to VISSR). VAS is a radiometer with 8 visible channel detectors and 6 thermal detectors that detect IR radiation in 12 spectral bands (two of the detectors are used primarily for imaging and four for atmospheric sounding). The spatial resolution is 0.9 km in the VIS range and 7-14 km in the IR range. Full Earth-disk coverage is accomplished by spinning in the west to east direction at 100 rpm. The VAS sensor has three operational modes:

1.  The VISSR operating mode (same as that of the original VISSR)

2.  The multispectral imaging (MSI) operating mode. Measurement of relatively frequent (e.g. half hourly) full Earth-disk imagery of the atmospheric vapor, temperature, and cloud distribution, as well as variations in the surface skin temperature of the Earth.

228)  E. P. Mercanti, "Need for Expanded Environmental Measurement Capabilities in Geosynchronous Earth Orbit", Proceedings of the Twenty-Fourth International Symposium on Remote Sensing of Environment', ERIM, Volume I, pp. 45-55
229)  R. Koffler, L. Spayd, "30 Years of Operational Environmental Satellites: A Retrospective and Future View of the United States Program", presented at the Twenty-Third International Symposium on Remote Sensing of the Environment, Bangkok, Thailand, April 18-25, 1990, pp. 95-97
230)  J.R. Greaves, W.E. Schenk, 'The Development of the Geosynchronous Weather Satellite System', in Monitoring Earth's Ocean, Land, and Atmosphere from Space, Volume 97, 1985, pp. 150-181

3.  In the dwell sounding mode (DS) up to 12 spectral filters can be positioned in sequence into an optical train while the scanner is dwelling on a single N-S scan line.
    Note: the DS mode cannot be used simultaneously with either the VISSR or MSI mode.

**DCS** = Data Collection System. This onboard system gathers and relays environmental data transmission that originate at remote automatic Data Collection Platforms (DCPs for ground truthing data). These platforms can be set to transmit data in four modes: at fixed times; in response to DCS interrogation; in an emergency mode during which the platform signals for a DCS interrogation within 1 min; and in an adaptive random reporting mode. The DCS collected data are retransmitted from satellite to small ground-based regional data utilization centers. The system also allows for the retransmission of narrow-band **WEFAX** (Weather Facsimile) data to existing small, ground-based APT (Automatic Picture Transmission) receiving stations from a larger facility.

## A.73.1   NOAA-GOES Data Collection System (DCS)

The Data Collection System (DCS) on NOAA-GOES satellites provides an operational data collection service for a large number of land- and sea-based data collection platforms (DCPs). Geostationary satellites (like GOES) offer the advantage of continuous coverage, however, their coverage is not global but limited to their large area of view, in addition they cannot provide a DCP location service by Doppler techniques (due to their geostationary orbit). The overall system is composed of the following subsystems:[231),232),233)

1.  The space segment: a DCS platform on 2 operational GOES satellites (GOES-East, and GOES-West). Each GOES S/C is equipped with two DCS transponders, one active and one as backup.
2.  The deployed Data Collection Platforms (DCPs) in the ground segment
3.  The NESDIS DCS ground receive system CDA (Command and Data Acquisition Station) at Wallops Island, Virginia.
4.  The DAPS (Data Collection System Automatics Processing System) at CDA.

**The Space Segment: DCS**

The DCS onboard GOES uses the GOES S/C for the relay of data from remotely located in-situ sites at or near the Earth's surface to properly equipped receiving stations in radio view of the GOES satellites. Each GOES provides an RF link between the DCP and CDA. They up-convert DCP data from UHF (401.9 MHz) to S-Band (1694.5 MHz) for transmission to CDA, and down-convert the CDA -DCP interrogate signal from S-Band (2034.9 MHz) to UHF (468.8 MHz).

The DCS uses 400 kHz of satellite transponder bandwidth . This bandwidth is subdivided (by frequency division multiplexing) into 200 domestic channels with a 1.5 kHz channel separation (channels 1-200), and 33 international channels (channels 202 - 266, even numbered only) with a 3 kHz channel separation. The 33 international channels are common with the METEOSAT and GMS spacecraft.

**Data Collection Platforms in the Ground Segment**

All DCPs serviced by GOES must be type-certified by NOAA-NESDIS. The types available to the user community have the following functional capabilities:

*   Self-timed DCPs (this is the most common type of platform in use)
*   Self-timed and Random Reporting DCPs

231)  "The Geostationary Operational Satellite Data Collection System", NOAA Technical Memorandum NESDIS 2, June 1983
232)  "Users Guide for Random Reporting - An Introduction to GOES Random Reporting Services", NOAA, April 1985
233)  User Interface Manual, Version 1.1, for the 'Data Collection System Automatic Processing System (DAPS)', Integral Systems Inc., Sept. 1990

- Random Reporting DCPs (relatively simple and inexpensive platforms)
- Interrogated DCPs

**1) Self-timed DCPs.** - These are simple platforms containing a transmitter and a pre-programmed self-timer to report at a specific time and intervals to the DCS. The sensor measuring cycle is independent of the DCP reporting cycle. Data of the sensor may be recorded at the DCP prior to a message transmission to DCS. The reporting frequency from a DCP to the DCS may vary from a few minutes to a few hours.

**2) Self-timed and Random Reporting DCPs.** - Same functionality of 1) and in addition has the capability of transmitting over a secondary channel when environmental conditions require more frequent reporting than offered under self-timed operation.

**3) Random Reporting DCPs.** - The platform contains a transmitter that broadcasts at random time (threshold reporting, set by the user). - During normal periods a random reporting DCP is expected to report up to 3 times per day that it is properly functioning.

**4) Interrogated DCPs.** - Scheduled transmissions initiated from the CDA at NESDIS (the schedule is stored at CDA). This type of DCP contains a receiver and transmitter. The receiver detects its own DCP ID upon message reception from CDA. Upon detection of its ID the DCP will transmit all data accumulated since the last reporting sequence. Interrogated DCPs may be interrogated as often as every 5 minutes or as infrequently as once per day. In addition some interrogated DCPs have a second or alert reply channel that may be used for threshold reporting. - Interrogated DCPs may be reconfigured by CDA command when requested by the user.

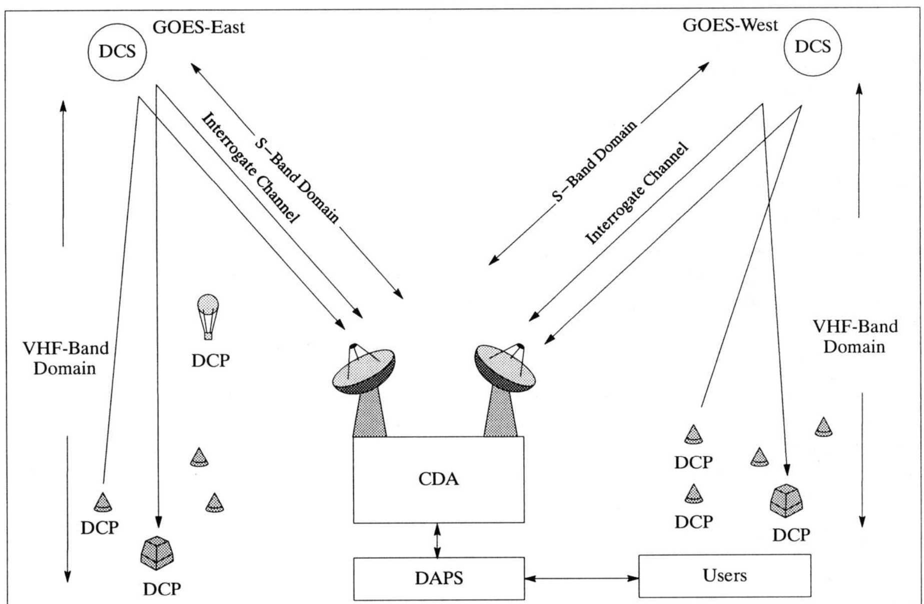


**Figure 62:** **Schematic Overview of the GOES Data Collection System**

## NOAA-NESDIS Ground Segment

The NOAA ground system consists of the RF front-end equipment and the DCS computer equipment or DAPS. The received messages from the DCPs are routed to the DAPS which provides all required processing and dissemination functions including data storage and distribution to the user community.

**DCS Access Method**

A data transmission from a DCP in the ground segment to the DCS on-board GOES is generally initiated (i.e. self-triggered) by the ground segment DCPs (except for the few interrogated DCPs). All (or most) of the 200 channels of the onboard DCS are in listening mode trying to lock onto any message that is being offered to them by the DCPs on a **random access basis**. The assigned transmission times of the DCPs are actually time intervals (slots) of 1 minute duration.

As of the end of 1993, NOAA service records list 7728 activated DCPs being serviced by GOES S/C, out of a total of 11927 DCPs that are assigned to the GOES user community at large.[234] - There are 6 DCPs of the type 'interrogated' that are being serviced by NOAA/ NESDIS, they are being used for seismic and tsunamic warning. In addition there are another 15 DCPs which use the interrogate mode as backup to the self-timed mode and to command these DPCs to perform specific functions under specific conditions.

## A.73.2   NOAA-GOES SEM Instruments

The **SEM** = Space Environment Monitor instrument package on each of the GOES satellites provides data which are processed by SEL (Space Environment Laboratory, Boulder Co.) of NOAA in near real-time for space environment forecasts. The SEM instruments survey the sun, they monitor the solar-terrestrial electromagnetic environment that affects a number of Earth activities [operational reliability of ionospheric radio (satellite communications, navigation systems), over-the-horizon radar, electric power transmission, crews of high-altitude aircraft and of space stations].

Each SEM package is composed of for separate sensor systems: two energetic particle sensors, a magnetometer, and a solar soft X-Ray sensor. Starting with GOES-L the SEM package will be extended by a fifth instrument, the SXI. Note: the GOES series SEM package differs considerably from the POES series SEM package.

- **EPS** = Energetic Particle Sensor. Objective: monitoring of solar protons and alpha particles produced during flares. Measurement in 7 channels for high proton fluxes, 6 channels for alpha particle fluxes, and in 3 channels for electron flux.

- **HEPAD** = High Energy Proton and Alpha Particle Detector. Objective: monitoring of the very high energy protons and alpha particles in large solar flares. HEPAD uses the phenomenon of Cerenkov radiation which occurs when an incident particle travels at speeds greater than the local speed of light in a high refractive index material, in this case fused silica.

- **Magnetometer** of SEM package. Objective: measurement of the magnitude and direction of the ambient magnetic field. Three boom-mounted  orthogonal sensors for the three components. Sensitivity of 0.2 nT, measurement range of $\pm 1000$ nT.

- **XRS** = Solar X-Ray Sensor. Objective: measurement of the solar X-ray spectrum in two broad energy bands: 1-8 and 0.5-4 Å. The X-ray sensor is continuously looking into the sun direction. The collimator and ion chamber assembly is mounted on a single axis positioner which in turn is mounted on the S/C solar yoke.

- **SXI** = Solar X-Ray Imager (starting with GOES-L). Objective: data for the study of the structure of energetic processes in the upper atmosphere of the sun (positions of flares), improvement in the prediction of solar flares. SXI uses a grazing incidence mirror to image the sun in the spectral range from 6 - 60 Å. Image resolution: 512 x 512 pixels. Each pixel has a FOV of 5 arc seconds. Data rate: 100 kbit/s. SXI produces normally an image every 10 minutes with an option of producing one image per minute. A filter wheel permits the pass band of the system to be modified to provide additional spectral information. SXI is under construction at MSFC as of 1993.

---

234)  Information provided by M. J. Nestlebush of NOAA/NESDIS

| Channel Designation | Particle Type | Nominal Energy Range (MeV) |
|---|---|---|
| P1 | Proton | ≤ 0.8 - 4 |
| P2 | | 4 - 9 |
| P3 | | 9 - 15 |
| P4 | | 15 - 40 |
| P5 | | 40 - 80 |
| P6 | | 80 - 165 |
| P7 | | 165 - 500 |
| A1 | Alpha | 4 - 10 |
| A2 | | 10 - 21 |
| A3 | | 21 - 60 |
| A4 | | 60 - 150 |
| A5 | | 150 - 250 |
| A6 | | 300 - 500 |
| E1 | Electron | ≥ 0.6 |
| E2 | | ≥ 2.0 |
| E3 | | ≥ 4.0 |

**Table 41:**   **Summary of Energetic Particle Sensor Outputs**

| Channel Designation | Particle Type | Nominal Energy Range (MeV) |
|---|---|---|
| P8 | Proton | 350 - 420 |
| P9 | | 420 - 510 |
| P10 | | 510 - 700 |
| P11 | | > 700 |
| A7 | Alpha | 2560 - 3400 |
| A8 | | >3400 |

**Table 42:**   **Summary of HEPAD Channel Outputs**

## A.73.3   NOAA-GOES Second Generation

GOES-I-M Mission Sensors: Starting with GOES-I (→ GOES-8) each satellite will be furnished with 2 so-called second generation instruments (with simultaneous imaging and sounding operational capabilities). In addition it is planned that all GOES (starting from GOES-I) will be provided with 3-axis spin stabilization.[235],[236]

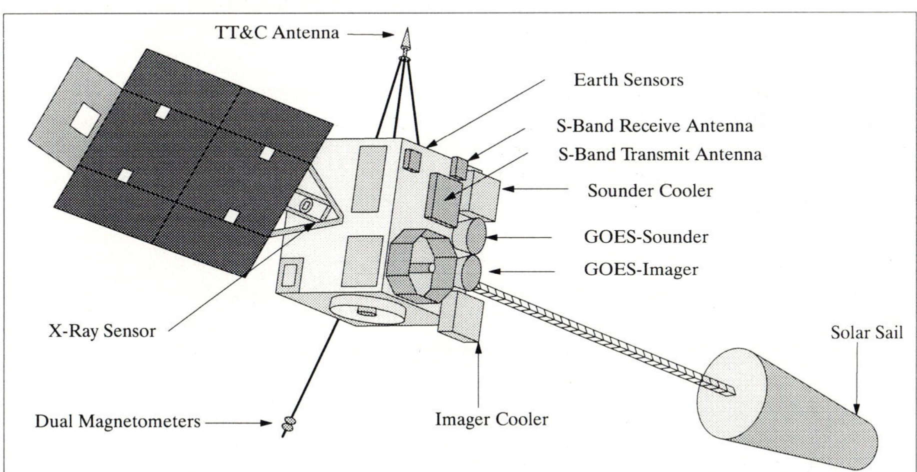


**Figure 63:**   **The GOES Second Generation S/C Model**


235)  J. Savides, "Geostationary Operational Environmental Satellite GOES I-M", System Description, Space Systems/Loral, Palo Alto, Ca., Dec. 1992
236)  "The GOES I-M Series Satellites - A brief description and Status Report", NOAA draft paper, March 1993

A single-wing solar array on the S/C rotates about the satellite pitch axis to track the sun during orbital motion generating a minimum of 1057 W at summer solstice, at the end of five years. A conical shaped solar sail on top of the 16 m boom of the S/C is used to balance the torque caused by the solar radiation pressure.

Orbit: Geosynchronous at an altitude of 35770 km above the equator. GOES-East will be positioned at 75° West longitude, GOES-W(est) at 135° West longitude.

**Sensors:**

**GOES-Imager** = Multispectral Imaging Radiometer for operational meteorology and climatology (in the VIS and IR ranges). The GOES-Imager measures the cloud cover and the cloud motion vectors. Resolution: = 1km in the VIS range and 4 km in the IR region. A normal full Earth disk scan (18° x 18°) is done in 25 minutes (a 3000 km x 3000 km area can be scanned in 3.1 minutes). The key features of the imager are summarized in Tables 43 and 47). The pointing accuracy of the GOES-Imager is expected to be within a 4 km radius at nadir.

**GOES-Sounder** = Infrared Sounder for operational meteorology and climatology (19 channel discrete filter radiometer). Objective: atmospheric soundings (data products are: vertical temperature and moisture profiles, layer mean temperature and moisture, total precipitable water, and lifted index). Resolution: 8 km. The GOES-Sounder is of HIRS/2 heritage providing similar instrument operation and performance. A filter wheel rotating ten times per second provides for data sampling at ten steps per second. The detector filter arrangement makes use of four spectral bands (see Table 44). Four detectors are simultaneously irradiated in each band, providing an output from four Instantaneous Geometric Field of Views (IGFOVs). In this manner the system can sample four 8 km IGFOV each 0.1 seconds.

The data products of the Sounder are used to augment data from the Imager to provide information on atmospheric temperature and moisture profiles, surface and cloud top temperatures, and the distribution of atmospheric ozone.

Both Imager and Sounder employ a servo-driven, two-axis gimballed mirror system in conjunction with a 31 cm Cassegrain telescope (separate sensors with independent operation). Each has a flexible scan control, enabling coverage of small areas as well as global scenes (Earth's full disk), and close-up, continuous observations of severe storms.

An onboard 'Image Navigation and Registration' (INR) system provides good pointing accuracy i.e. geographical location capability of the Imager and Sounder pixels in near real-time. In this context image navigation refers to the determination of the location of a pixel within an image in terms of Earth longitude and latitude, registration refers to pixel stability of maintaining pointing of each pixel to a specific Earth location within an image and between repeated images.

In addition the GOES I-M missions will fly the **SEM** subsystem (see description under A.73.2) which consists of the following instruments (and will provide **DCS** and **WEFAX** services).

- Magnetometer, similar to the versions on the previous GOES satellites, consisting of the three separate orthogonal probes.

- Solar XRS (X-Ray Sensor), similar to the previously flown designs; however, it is continuously pointed at the sun, by virtue of being mounted on the solar array yoke, and on top of a positioner that tracks the sun in the north-south direction.

- The EPS and HEPAD are functionally identical to those flown on the previous GOES satellites.

- The SXI will be part of the SEM subsystem starting with GOES-L.
  **SXI** = Solar X-Ray Imager. Objective: Solar storm warning. SXI measures the solar X-ray emissions.

| Spectral Channels | 1 (VIS) | 2 (IR) | 3 (IR) | 4 (IR) | 5 (IR) |
|---|---|---|---|---|---|
| Prime Measurement Purpose | Cloud Cover | Nighttime Clouds | Water Vapor | Surface Temp. | SST & Water Vapor |
| Wavelength (µm) | 0.55 - 0.75 | 3.80 - 4.00 | 6.50 - 7.00 | 10.20 - 11.20 | 11.5 - 12.50 |
| S/N or NE$\Delta$T | 150:1 | 1.4 K at 300 K | 1.0 K at 230 K | .35 K at 300 K | .35 K at 300 K |
| Detector Type | Silicon | InSb | HgCdTe | HgCdTe | HgCdTe |
| Spatial Resolution (µrad) | 28 ($\approx$1 km x 1 km at Nadir) (+0-10%) | 112 ($\approx$4 km x 4 km at Nadir) (+0-10%) | 224 ($\approx$8 km x 8 km at Nadir) (+0-25%) | 112 ($\approx$4 km x 4 km at Nadir) (+0-25%) | 112 ($\approx$4 km x 4 km at Nadir) (+0-25%) |

**Table 43:    GOES-Imager Performance Requirements**

| Detector Channel (Band) | | Central Wavelength (µm) | NE$\Delta$N (mW/m$^2$/sr/cm$^{-1}$) | Prime Measurement Purpose |
|---|---|---|---|---|
| HgCdTe (Longwave) | 1 | 14.71 | 0.66 | Temperature Sounding |
| | 2 | 14.73 | 0.58 | Temperature Sounding |
| | 3 | 14.06 | 0.54 | Temperature Sounding |
| | 4 | 13.96 | 0.45 | Temperature Sounding |
| | 5 | 13.37 | 0.44 | Temperature Sounding |
| | 6 | 12.66 | 0.25 | Temperature Sounding |
| | 7 | 12.02 | 0.16 | Surface Temperature |
| HgCdTe (Midwave) | 8 | 11.03 | 0.16 | Surface Temperature |
| | 9 | 9.71 | 0.33 | Total Ozone |
| | 10 | 7.43 | 0.16 | Water Vapor Sounding |
| | 11 | 7.02 | 0.12 | Water Vapor Sounding |
| | 12 | 6.51 | 0.15 | Water Vapor Sounding |
| InSb (Shortwave) | 13 | 4.57 | 0.013 | Temperature Sounding |
| | 14 | 4.52 | 0.013 | Temperature Sounding |
| | 15 | 4.45 | 0.013 | Temperature Sounding |
| | 16 | 4.13 | 0.008 | Temperature Sounding |
| | 17 | 3.98 | 0.0082 | Surface Temperature |
| | 18 | 3.74 | 0.0036 | Surface Temperature |
| Silicon (Visible) | 19 | 0.70 | 0.10% Albedo | Cloud |
| Silicon (Star Sense) | | 0.65 | 6:1 SNR | 4th magnitude stars |

**Table 44:    GOES-Sounder Performance Requirements**

| Operating Mode | Imaging & Sounding | SEM Package | Communications | Attitude Control |
|---|---|---|---|---|
| Sunlight (Normal) | on | on | on | Earth pointing with INR |
| Eclipse | off | on, except XRS off | on, except WEFAX and MDL off | Earth pointing without INR |
| Launch operations (Transfer orbit) | off | off | off | Earth pointing during maneuvers, otherwise Sun pointing |
| On-Orbit Storage | off | off | off | Earth pointing without INR |

**Table 45:    GOES Second Generation Satellite On-Orbit Operating Modes**

| Data Link Type | Source | Uplink (MHz) | Downlink (MHz) | Destination |
|---|---|---|---|---|
| Command | CDA/DSN | 2034 | | S/C |
| Telemetry + SEM | S/C | | 1694 | CDA, SEL-Boulder Co., DSN |
| Ranging | DSN, STDN | 2034 | 2209 | S/C, DSN |
| WEFAX | CDA | 2033 | 1691 | Users/APT |
| DCPI | CDA | 2034 | 468 | DCP |
| DCPR | DCP | 401 | 1694 | CDA/Users |
| S&R | ELT/EPIRB | 406 | 1544 | S&R Ground Station |
| MDL (diagnostic data) | S/C | | 1681.5 | SOCC/SEL |
| PDR (Processed Data Relay at 2.111 Mbit/s) | CDA | 2027.7 | 1685.7 | Users (via S/C) |
| Image Raw Data (2.62 Mbit/s) | S/C | | 1676 | CDA |

**Table 46:    Overview of GOES Communication Links**

**GOES Data Transmission:**  Direct Broadcast and Relay in S-Band.

UHF-Band at 401, 406 and 468 MHz. The UHF-Band is used to receive environmental data from the DCPs. It is also used for S&R service provision.

The GOES satellites are operated and controlled via CDA and SOCC of NOAA, the sensor data are processed and distributed by the Data Processing Center of NOAA (see Fig. 67). The GOES ground segment uses also 7 satellite field service stations, located at: Anchorage, Honolulu, Kansas City, Miami, New Orleans, San Francisco, and Washington D. C..

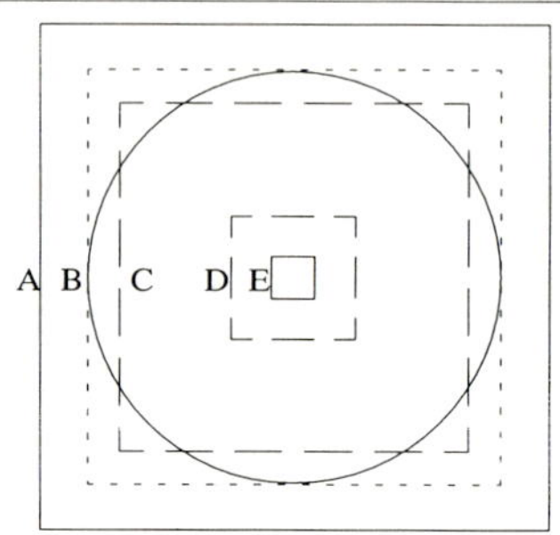


GOES-Imager Scan Filed

A.  Maximum Frame Field 19.2° E-W by 19° N-S
 Maximum Star Field 23° E-W by 21 N-S
B.  Full Earth Limb Scan in 25 minutes
C.  Earth Scan 60° N-S by 60° E-W in 22 min
D.  Area Scan 3000 km x 3000 km in 3.1 min
E.  Small Area Scan 1000 km x 1000 km in 40 s

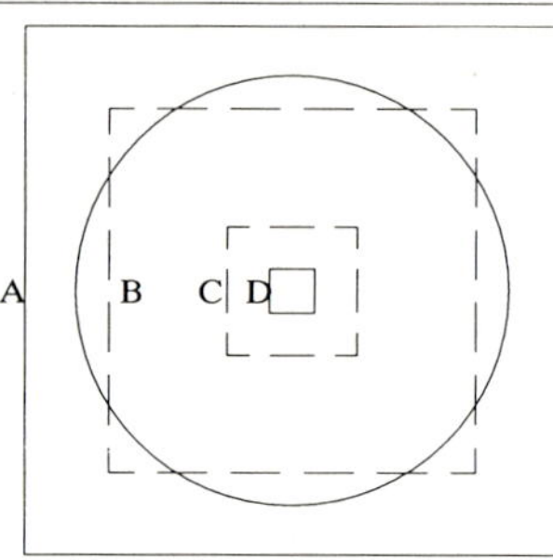


GOES-Sounder Scan Field

A.  Maximum Frame Field 23° E-W by 21° N-S
B.  Full Earth Sounding 60° N-S x 60° E-W
 464 minutes
C.  Area Sounding 3000 km x 3000 km in 42 minutes
D.  Small Area Sounding 1000 km x 1000 km in 5 minutes

**Figure 64:**    **GOES Second Generation Scan Operations**

| Feature | GOES-Imager | GOES-Sounder |
|---|---|---|
| Optical Aperture | 31.1 cm | 31.1 cm |
| Total Step & Sample Time | N/A | 0.1 s (0.2 s and 0.4 s optional) |
| Methods of Scan | 2-axis, linear E-W, line step N-S | 2-axis, step & dwell E/W 280 µrad steps N/S 1120 µrad steps (optional 2240 µrad steps, 0.2 s dwell) |
| Scan Rate | 20°/s optical | 40 soundings/s |
| Slew Rate | 10°/s mechanical | 10°/s mechanical |
| Spatial Resolution (µrad) | VIS = 28, Ch. 2,4, & 5 = 112, Ch. 3 = 224 | all channels = 242 µrad (diameter) |
| Sampling | 1.75/IGFOV VIS, Ch. 2, Ch. 4 & Ch. 5 3.5/IGFOV channel 3 | 4 IGFOVs sampled simultaneously |
| Sampling Rate | 183.3 µs/pixel (IR) 45.8 µs/pixel (VIS) | 0.1 s |
| Channel Co-registration | ± 28 µrad | Within 22 µrad of channel 8 |
| Star sensing | Uses visible array SNR 6 for 4th mag stars (400 samples) | Separate visible array SNR 6 for 4th mag stars (each sample) |
| Data Output | 10 bit quantization | 13 bit quantization |
| Data Rate | 2.6208 Mbit/s | 40 kbit/s |
| Data Format | NRZ-S, PN code | NRZ-S, PN code |
| Patch Temperature | Regulated at 94 K, 101 K or 104 K | Regulated at 94 K, 101 K or 104 K |
| Time between Space Looks | 2.2 s large frame 9.2 or 36.6 s smaller frame | 2 min |
| Time between Black Body Calibrations (Nominal) | 10-30 min (can override or inhibit) | 20 min (can override or inhibit) |
| Priority Frame Select | 1 level normal 2 levels priority 1 level star sense | 1 level normal 2 levels priority 1 level star sense |

**Table 47:**    **GOES Second Generation Instrument Parameters**

## A.74   NOAA-POES

The NOAA-POES series (Polar-Orbiting Operational Environmental Satellites) and the NOAA-GOES series are regarded as the backbone of the US meteorological program. Both types of satellites carry data collection systems for the readout of in situ environmental data. In addition to providing standard meteorological readings in remote locations, the data collection systems allow the monitoring of sea conditions through measurements on ocean buoys, readout of data from aircraft and balloons, and the relay of hydrological data on river levels, snow thickness, etc.
Objective: Provision of continuous meteorological soundings (systematic observations), measurement of energy budgets (with energetic particle sensors, etc.) .

The current POES series satellites are named simply NOAA-9 through NOAA-12 in the order of launches. The program has evolved over several generations of satellites (TIROS, ESSA, TIROS-M to the TIROS-N series), starting in 1960 with TIROS-1 to the most recent operational satellite of NOAA-12 (see Table 50). Along with the general program evolution, there has also been an enormous sophistication in sensor technology, hence in observation capability. A large number of daily atmospheric profiles or "soundings" find their way into global forecasting models.

Orbit: Sun-synchronous polar, Altitude = 800 - 850 km, Inclination = 98-99°; Period = 102 minutes, repeat cycle  = 12 hours, 11 days.
A continuous S-Band broadcast is provided for an international user community. **Any user with a receiving station has access to the data.**
The following POES/TIROS-N satellites can be received at the present time (1992):

- NOAA-9  Launch: Dec. 12, 1984 (see Table 50 for sensor configuration)
  First operation of SBUV/2 and ERBE instruments
  Status 1992 (S/C in standby operation, the power system is considered as very marginal): Digital tape recorder failed 2 months after launch. ERBE stopped operation in Jan. 1987. AVHRR has at times exhibited anomalous behavior in its synchronization in MIRP. MSU channels 2 and 3 have failed, and the power system is degraded. SBUV/2 is operating satisfactorily. The satellite is collecting, processing, and distributing SBUV/2 and ERBE-Nonscanner data. It is also providing S&R data.
  Status 1/1994: S/C standby operation. Operational instruments: AVHRR, SSU, DCS, HIRS, S&R, SBUV/2.

- NOAA-10 Launch: Sept. 17, 1986           (7:32 AM Descending Orbit)
  Status 1992 (S/C in standby operation utilizing STIP record mode): The ERBE-nonscanner mode is performing well, while the ERBE-Scanner mode failed, which has exhibited a scan sticking anomaly that is apparently generic to the instrument. The S&R processor (406 MHz) has failed.
  Status 1/1994: S/C in standby operation. Operational instruments: AVHRR, MSU, DCS, HIRS, S&R.

- NOAA-11 Launch: Sept. 24, 1988           (1:49 PM Ascending Orbit)
  Status 1993 (operational, primary and backup grating encoder malfunction during sun sweeps): SBUV/2 instrument is operational. Two gyros have failed (Y and Z gyros). Attitude control is being maintained through the use of new reduced gyro flight software. Orbit is degrading, overpass time is later each day.
  Status 1/1994: S/C operational. Operational instruments: AVHRR, SSU, MSU, HIRS, S&R, SBUV.

- NOAA-12 Launch: May 14 1991 (AM Orbit)
  Status 1/1994: All instruments are operational.

- NOAA-13 Launch: August 9, 1993 (PM Orbit). NOAA lost contact on August 21, 1993 with the spacecraft. Catastrophic failure, S/C lost.

**Sensors:**

**AVHRR** Scanner (Advanced Very High Resolution Radiometer). Radiometric resolution = 10 Bit, spatial resolution in the order of 1.1 km
Measurement of cloud coverage, sea surface temperature, vegetation, aerosols. The benefit of AVHRR data lies in its high temporal frequency of coverage, global coverage at least once per day (the disadvantage of AVHRR data is its coarse spatial resolution: 1.1 km at subsatellite point). Applications: Operational meteorology, oceanography, climatology, vegetation monitoring, land and sea ice observation.
Scanning type: cross-track, IFOF 1.4 milliradians (average), scanning rate = 360 per minute, Sampling rate = 2048 per scan, sample step = 0.95 milliradians (1.36 samples per IFOV), scan angle (max) = 55.4° off nadir, swath width = 3000 km (approx.).

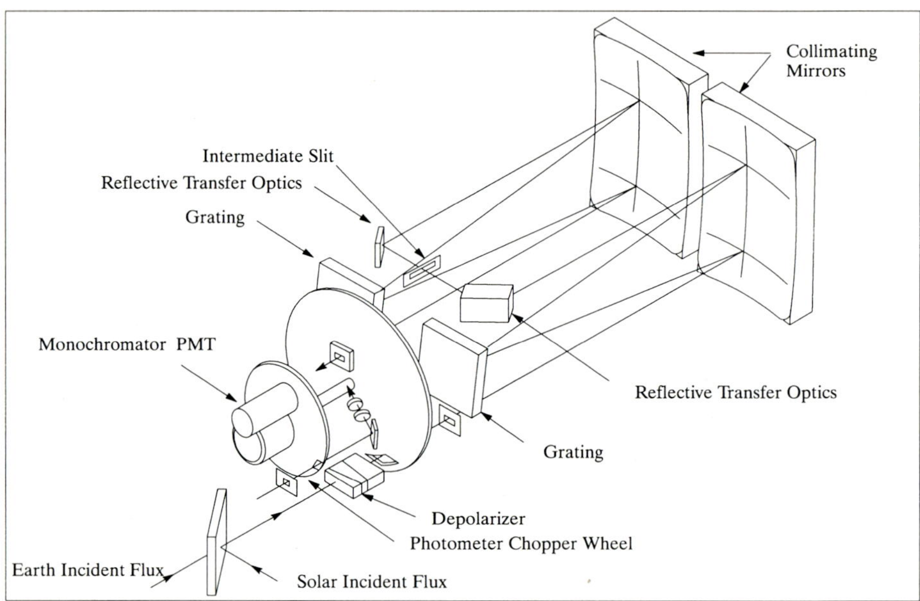


**Figure 65:**    **Schematic Optical Diagram of the SBUV Instrument**

Spectral Ranges:      580 - 680 nm
                      725 - 1100 nm
                      3550 - 3930 nm
                      10.3 - 11.3 µm (10.5 - 11.5 on NOAA-10)
                      11.4 - 12.4 µm (not on NOAA-10)

**HIRS/2I** (High Resolution Infrared Radiation Sounder)
Measurement of atmospheric temperature and humidity. This instrument detects and measures energy emitted by the atmosphere to construct a vertical temperature profile from the Earth's surface to an altitude of about 40 km. Measurements are made in 20 spectral regions in the infrared band (one frequency lies at the high frequency end of the visible range)
Spectral range =      6.72 - 14.95 µm (channels 1-12)
                      3.76 - 4.57 µm (channel 13 - 19)
                      0.69 µm (channel 20)
Nadir scan, swath = 2240 km. Spatial resolution = 20.4 km.

**MSU** = Microwave Sounding Unit (JPL sensor)
MSU detects and measures the energy from the troposphere to construct a vertical temper-

ature profile to an altitude of about 20 km. Measurements are made by radiometric detection of microwave energy divided into 4 frequency channels. Each measurement is made by comparing the incoming signal from the troposphere with the ambient temperature reference load. Because its data are not seriously affected by clouds, the MSU is used along with HIRS/2I to remove measurement ambiguity when clouds are present. Data output: 12-bit binary at 0.32 kb/s.

Spectral ranges:　　　　50.3, 53.74, 54.96, and 57.95 GHz (4 channels)
Nadir scan
FOV　　　　　　　　　95°
Swath width　　　　　　2348 km
Spatial resolution　　　105 km at nadir

**SSU** = Stratospheric Sounding Unit, (UK sensor, not on NOAA-12)
Temperature measurements in the upper atmosphere are derived from radiance measurements made in 3 channels using a pressure-modulated gas ($CO_2$) to accomplish selective bandpass filtration of the sampled radiances. Measurements of temperature profiles, top of atmosphere radiation (from 25 km to 50 km in altitude)
Spectral ranges ($cm^{-1}$): 669.99, 669.63, 669.36
Resolution (km at nadir): 147.3
IFOV: 10°
Data output: 12-bit binary sampled at 0.48 kb/s.

**ERBE** = Earth Radiation Budget Experiment (on NOAA-9 and 10 only). Measurement of Earth radiation gains and losses. Duty cycle: 100%. Spectral bands: 0.5-0.7 µm, 0.2-4.0 µm, 0.2-50 µm, 10.5-12.5 µm. Ground resolution of 50 km in nadir. Swath width = 3000 km.

**SBUV/2** = Solar Backscatter Ultraviolet (on PM satellites only)
The SBUV/2 sensor is a spectrally scanning UV radiometer (nadir-viewing sensor). Measuring of solar irradiance and scene radiance (backscattered solar energy) over a spectral range of 0.16 to 0.40 µm. Duty cycle = 100%. Resolution = 170 km.
Application: atmospheric chemistry (measurement of trace gases). Sensor objectives:

-　　to make measurements from which total ozone concentration in the atmosphere can be determined to an absolute accuracy of 1 percent.
-　　to make measurements from which the vertical distribution of atmospheric ozone can be determined to an absolute accuracy of 5%.
-　　to measure the solar spectral irradiance from 0.16 to 0.40 µm.

**SEM** = Space Environment Monitor (POES series instruments)
SEM is a multichannel, charged-particle spectrometer that measures the population of the Earth's radiation belts and the particle phenomena resulting from solar activity (both of which contribute to the solar/terrestrial energy interchange). SEM consists of two or three separate sensor units and a common DPU (data processing unit). The sensor units are TED, MEPED, and HEPAD (occasionally). The lower-energy sensors (TED, plus the proton and electron telescopes of MEPED) have pairs of sensors with different orientations because the direction of the particle fluxes is important in characterizing the energy interchanges taking place. Objectives:

-　　to determine the energy deposited by solar particles in the upper atmosphere
-　　to provide a solar storm warning system

**TED** = Total Energy Detector. Performance: Proton: 0.3 to 20 keV in 11 bands. Electron: 0.3 to 20 keV in 11 bands.

**MEPED** = Medium Energy Proton and Electron Detector. Performance: Proton: 30 to 2500 keV in 5 bands. Electron: >30 to > 300 keV in 3 bands. Ions: > 6 MeV. Omniproton: > 16 MeV, > 36 MeV, > 80 MeV.

**HEPAD** = High Energy Proton and Alpha Particle Detector (see A.73.2).

**EHIC** = Energetic Heavy Ion Composition (U. of Chicago and NRC HIA of Canada) EHIC measures the chemical and isotopic composition of energetic particles between hydrogen and nickel over the energy range of 0.5 to 200 MeV/nucleon. The experiment will measure energetic solar flare particles in the polar regions where the Earth's magnetic field connects to the interplanetary field carried in the solar wind. It will also measure trapped energetic particles in the magnetosphere.
The primary objective of EHIC is to obtain elemental and isotopic composition data, which can be used to test models for solar flare ion acceleration, ion transport in interplanetary space, ion entry into the magnetosphere, and nucleosynthetic processes leading to the elemental and isotopic mix found at the Sun.

**MAXIE** = Magnetospheric Atmospheric X-Ray Imaging Experiment (Lockheed and University of Bergen, Norway). MAXIE maps the intensities and energy spectra of x rays produced by electrons that precipitate the atmosphere. With mechanical scanning, MAXIE will obtain new high-resolution x-ray imaging data on auroral and substorm processes with a temporal resolution and repetition rate so far not available.
Combined with other data, a more complete analysis of the radiation environment is obtained. Derived products include vertical profiles of ionization and electrical conductivity, as well as the imaging of the aurora in sunlight. These products are used to determine the atmospheric scale height and other synoptic information.

**RAIDS** = Remote Atmospheric and Ionospheric Detection System (NOAA & USAF). RAIDS is an USAF space test program instrument. It uses 8 UV detectors (extreme UV to near IR). Applications: simultaneous observations of the neutral and ion composition of the day and night side of the Earth and auroral region. Objectives: measurement of ionospheric electron density through limb scanning, detection of natural airglow emissions from molecular and atomic constituents (measurement range: 100-700 km altitude). Duty cycle: 100%.

**Data:** Image size: 2577 Lines, 2048 Pixels. Transmission: frequency = 1070 MHz; Data rate = 665 kbit/s (S-Band real time data); = 2.6616 MBit/s (S-Band playback). SEM data are processed and archived by NOAA/NGDC/SEL, Boulder Co.

### A.74.1   Planned POES series of NOAA

Policy issues: NOAA, ESA and EUMETSAT have an agreement in principle to collaborate in the collection of global environmental and meteorological observations using polar orbiting satellites.[237]
[The current NOAA polar program is based on the services of two operational satellites flying in complementary sun-synchronous orbits, one in a "morning or AM" orbit, and the second in an "afternoon or PM" orbit. It so turns out that the data from the PM mission is primary in the USA, with the AM mission providing supplementary and back-up coverage. In Europe the converse is true, with the AM mission providing the most timely coverage.]

One element of the joint program is the agreement, that Europe (ESA, EUMETSAT) will assume responsibility for providing the AM segment, while NOAA will continue to provide the future PM segment of the mission as well as the majority of the payload sensors for both AM and PM segments of the program. A key element of the European meteorological polar program will be the POEM missions (which are undergoing a redefinition as of 1992).

It is NOAA policy[238],[239] to continue its proven downlink broadcast transmission of science data in S-Band, at no cost, to all equipped users. The current downlink rate of science data is 667 kbit/s, with a NOAA-specific data format and data handling.

Beginning with NOAA-O, an improved transmission technique will be substituted using the CCSDS-Standard with its greater range of support capabilities. NOAA wants to manage the

237) A. F. Durham, "Future Polar Satellite Program Plan for Global Environmental Observations", IAF 92-0083, 43rd Congress of the International Astronautical Federation, Aug. 28-Sept. 5, 1992 Washington D. C.

expected larger data volumes by effective onboard data compression techniques. Future S-Band data rates are planned with up to 3.5 Mbit/s (HRPT).

Note: The instrumentation of NOAA-13, that failed in orbit shortly after launch, was identical with the instrumentation on the earlier TIROS Series (NOAA-9, 10, 11, 12, 13), but with the addition of EHIC and MAXIE. These latter instruments will not be available for NOAA-J.

**NOAA-J   (PM Orbit) Projected Launch: Fall 1994**
Sensors: AVHRR, HIRS, MSU, SSU, SEM, ARGOS, S&R,

NOAA-J is also referred to as NOAA-Next (it is next in line for launch). NOAA-J is in the ATN (Advanced TIROS-N) program, which provides for additional instruments of opportunity, such as the S&R. The ATN program is a cooperative effort between NASA, NOAA, the UK, and France for providing day and night global environmental and associated data for regular daily operations.

NOAA-J orbit: Circular Sun-synchronous near-polar orbit; altitude = 870 km; inclination =98.86°; orbital period =102.12 min; Equator crossing at about 1:40 PM northbound and 1:40 AM southbound local solar time.

**Candidate Sensors:**
AVHRR, HIRS, AMSU-A (MTS), AMSU-B (MHS), SEM, ERBE, SBUV,MCP, ARGOS, S&R[240]

The instrumentation on NOAA K, L, M and N is almost identical as before with the following upgrades:

AVHRR : has an additional channel (total of 6)
The sensors MSU and SSU are being replaced by AMSU (Advanced Microwave Sounding Unit).

Orbit: Sun-synchronous polar, altitude of 700 - 850 km,
- NOAA-K (AM Orbit)          Projected Launch: Jan. 1996
- NOAA-L (PM Orbit)          Projected Launch: May 1997
- NOAA-M (AM-Orbit)          Projected Launch: Jan. 1999
- NOAA-N (AM/PM-Orbit)       Projected Launch: May 2000
- NOAA-N' (PM-Orbit)         Projected Launch: Jan. 2002

- NOAA-O Study phase         Launch: 2004  "All are PM Satellites"
- NOAA-P Study phase         Launch: 2007
- NOAA-Q Study phase         Launch: 2010

**Instrument candidates for the NOAA POES O,P,Q series:**[241]

**VIRSR** = Visible Infra-Red Scanning Radiometer is considered (old name AVHRR/4). Applications: operational meteorology, oceanography, climatology, vegetation monitoring. Objectives: measurements of land and sea surface temperatures, cloud cover and precipitation, snow and ice cover, vegetation index, soil moisture, and albedo. Duty cycle: 100%. Resolution = 1.1 km at nadir. Swath width = 3000 km.
VIRSR features:

238) Bruce H. Needham, "Instrumentation and Services for the NOAA Polar-Orbiting Operational Environmental Satellites (POES) in the 21 st Century", NOAA/NESDIS, Office of System Development, Washington D.C., '90
239) "Pre-Phase-A Study of NOAA O,P,Q Spacecraft and Ground Segment LRPT and HRPT Data Handling and Transmission Subsystems" Draft Final Report, Oct. 16 1990, Atlantic Research Corp. prepared for NASA-GSFC
240) **Note: Although the original acronym for 'Search and Rescue' is SAR in the context of NOAA missions, it was changed in this document consistently to S&R in order to distinguish it from the other widely-used meaning of SAR, namely 'Synthetic Aperture Radar', a sensor type. A consequence of S&R is also S&RSAT (instead of SARSAT)**
241) CEOS Summary Report, WGD-10 Meeting, Annapolis Md., April 16-18, 1991

- 7 full-time VIR channels versus 6 previously
- 12 bit quantization versus 10 bit previously
- NE$\Delta$T increased to 0.10 K versus 0.12 K
- Addition of in-orbit visible channel calibration
- Scan detection reversed to same as sounders
- Data rate is estimated to about 1.3 Mbit/s

**MTS** = Microwave Temperature Sounder is considered (old name of AMSU-A)
MTS features:
MTS provides atmospheric temperature measurements from the surface up to 40 km in 15 channels (23.8 GHz, 31.4 GHz, 12 channels between 50.3 to 57.3 GHz and 89 GHz). Coverage is about 50° on both sides of the suborbital track with an IFOV of 3.3°. Temperature resolution: 0.25 - 1.3 K.

- 6 new channels (60 GHz) are added (in addition to the 15) to increase the sounding capability to 73 km versus 45 km before

All channels have a spatial resolution of about 40 km at nadir. Swath width approx. 2200 km. Data rate of 2.2 kbit/s.

**MHS** = Microwave Humidity Sounder is considered (EUMETSAT sensor). MHS is basically identical to the AMSU-B flown on the NOAA K,L, M, N series. MHS has 5 channels (at 89 GHz, 157 GHz, and 183 GHz) spanning the height range from the surface to about 42 km. MHS can also be used to monitor precipitation.
All channels will have a spatial resolution of about 15.4 km at nadir. Cross-track scanning capability of ± 48.5° from nadir. Swath width approx. 2200 km. Data rate = 4 kbit/s. MHS ensures a twice daily full global coverage.

**SBUV/3** = Solar Backscatter Ultraviolet Radiometer (further development from Nimbus). This instrument is pointed toward nadir, there are 12 channels between 255 nm and 340 nm, measurement of ozone profiles, in addition measurement of backscatter. Sensor on PM missions only.

**TOMS** = Total Ozone Mapping Spectrometer. Measures the 'total column ozone' in nadir direction in 6 channels. Sensor on PM missions only.

**IRTS** = Infrared Temperature Sounder. (old name: HIRS/4). Applications: operational meteorology and climatology. Objectives: measurements of atmospheric temperature profiles, humidity soundings, water vapor, total ozone content, surface temperature, cloud parameters. Duty cycle: 100%. Resolution = 19.5 km. Swath = 2240 km. IRTS features:

- IFOV increased to 19.5 km versus 21 km at nadir
- Scan time increased to 8 seconds includes calibration

**SEM** = Space Environment Monitor (old name: SEM (Aug))

**LEFI** = Local Electric Field Instrument (new addition to SEM package). Objectives: measurement of the ambient vector electric field.

**ARGOS** and **S&RSAT**

**Growth Instruments under Consideration** (if available)

AIRS = Advanced Infrared Sounder (NASA prototype operational instrument replaces ITRS, PM missions only).

HiRDLS = High Resolution Dynamic Limb Sounder. (NASA prototype operational instrument replaces SBUV, PM missions only).

CERES = Clouds and Earth's Radiant Energy System. (NASA prototype instrument, PM missions only).

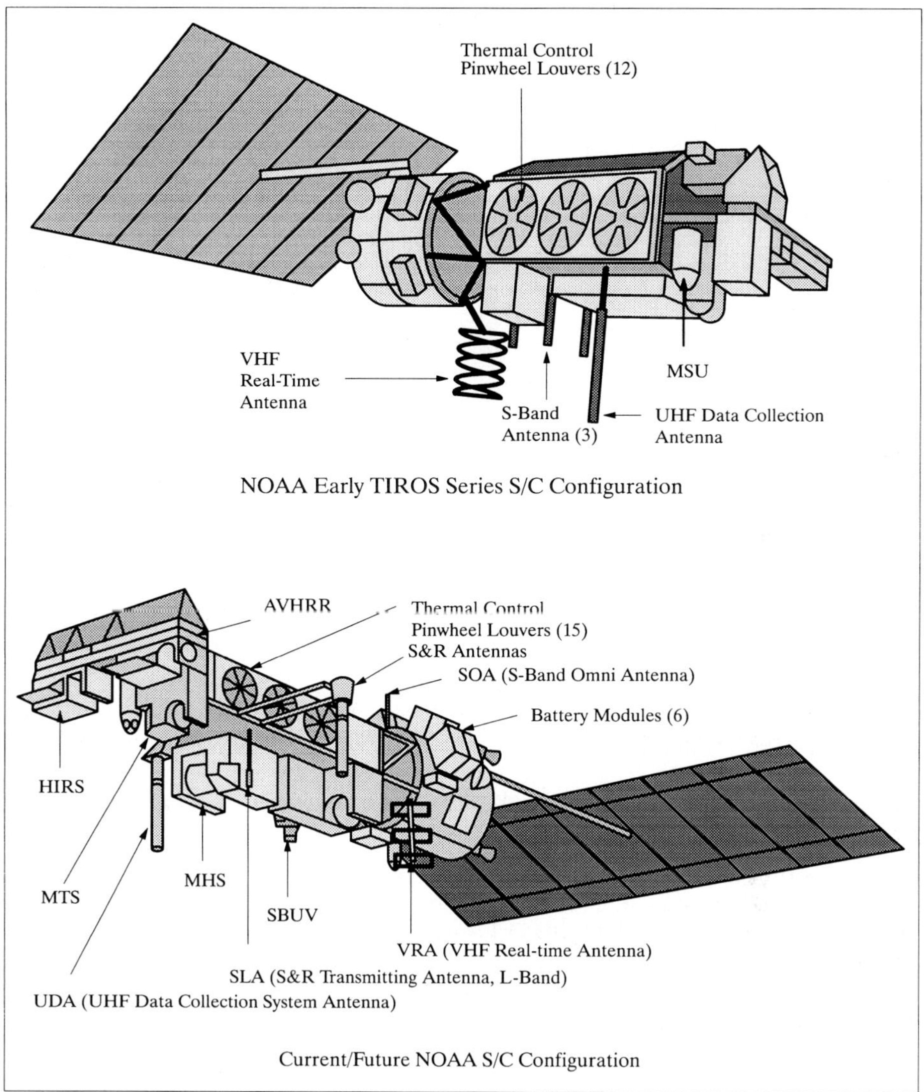


**Figure 66:    Typical NOAA POES Series S/C Models**

| Type of Service | NOAA-9 through NOAA-N | NOAA-O,P,Q (proposed) |
|---|---|---|
| Global Recorded Data | Reduced Resolution AVHRR, Full Resolution Soundings, All other Data at Full Resolution | Full Resolution Imager Full Resolution Soundings All other Data at Full Resolution |
| Local Area Coverage | Approximately 10 min/orbit | Global Availability |
| HRPT | Full Res. AVHRR & TIP Data, 1698 or 1707 MHz, Data Rate = 665.4 kbit/s, Split Phase PSK | Full Res. Imagery & TIP Data, 1698 to 1710 MHz, Data Rate $\leq$ 3.5 Mbit/s , QuadraPhase (I=1.5 Mbit/s, Q=2 Mbit/s) |
| APT (LRPT in OPQ era) | Low Res. 2 Channel AVHRR Analog, 137.50 & 137.62 MHz | 2-3 Channels Imager Digital @ 72 kbit/s, Full Res. Soundings, 137-138 MHz |
| Beacon | HIRS, SSU, MSU, SEM, DCS, SBUV, & Eph | No equivalent, Data included in LRPT |

**Table 48:    Real-Time and Global Data Services Current/Future NOAA POES Missions**

| Sensor Type | Current Series (84-95)NOAA-9,10,11,12, I, J | Next Series ('96-'00) NOAA-K,L,M,N | Future Series ('04 - 2010) NOAA-O,P,Q | Prototypes (1998 + NASA EOS) |
|---|---|---|---|---|
| VIR Imagers | AVHRR/2 | AVHRR/3 | VIRSR | |
| IR Sounders | HIRS/2; SSU (UK) | HIRS/3 | IRTS | AIRS |
| μ-Wave Sounders | MSU | AMSU-A1,A2 AMSU-B (UK) | MTS; MHS (Europe) | MIMR |
| Ozone | SUBV/2 (PM satellite only) | SUBV/2 (PM satellite only) | SBUV; TOMS | HiRDLS |
| Earth Radiation Budget | ERBE (NOAA-9 and 10 only) | | | CERES |
| Space Environment | TED; MEPED | TED; MEPED | TED; MEPED; LEFI; | |
| Data Collection | ARGOS/2 (CNES) | ARGOS/3 | ARGOS/3+ | |
| Search and Rescue (S&R) CNES,CRC | S&RSAT | S&RSAT | S&RSAT | |
| Winds | | | | STIKSCAT |
| Sea State | | | | ALT |

**Table 49:    Evolution of NOAA's Polar Orbiting Satellite Sensor Suite**[242]

## A.74.2   ARGOS on NOAA-POES Satellites

**ARGOS** Data Collection System (DCS) on NOAA S/C (provided by CNES/France, installed on all TIROS-N family satellites since 1978, first satellite equipped with ARGOS was TIROS-N).

The ARGOS/DCS supports NOAA in its overall environmental mission objectives, collecting (ground and space) truthing data. The concept uses many ground segment platforms (fixed and moving), i.e. buoys, free-floating balloons, and remote weather stations, and equips them with a **Platform Transmitter Terminal (PTT)** package. These PTTs collect and process relevant environmental data and transmit these to the NOAA-POES satellites. The onboard ARGOS DCS receives the incoming signal and measures both the frequency and relative time of occurrence of each transmission. The S/C retransmits these data via the CDA (Command and Data Acquisition) stations (one at Wallops Island Va., the other at Gilmore Creek, Alaska; there is in addition a downlink station (CMS) at Lannion, France, (see Figure 52) to a central processing facility. The DCS information is decommutated and sent to the ARGOS processing center where it is processed, distributed to the user community, and stored on magnetic tape for archival purposes. (see also Figure 67).

Each ARGOS PTT transmits encoded messages at regular intervals on a 401.650 MHz uplink. Messages transmitted by the various platforms within satellite visibility are received and selected for processing on a random access basis. The satellite DCS computes the Doppler shift on the receive frequency and generates the ARGOS telemetry message which includes PTT identification, sensor data, measured frequency, and time and date of measurement. A small portion of the S/C downlink is reserved for ARGOS data. Each time a satellite is within visibility of one of the 3 receiving stations, it downlinks the recorded data.

Some ARGOS System Characteristics: (see also chapter A.10)

- Minimum platform/satellite elevation angle of visibility . . . . . . 5°
- Percentage of platforms with four Doppler Measurements per day . . . . . >85%
- Measured location accuracy . . 350 m
- Message capacity for sensor data . . . . . . 32 to 256 bits
- Messages are of duration <1 s and are transmitted at regular intervals by any PTT
- Uplink operational frequency   401.650 MHz
- Typical power of PTT uplink  is 200 mW at intervals of 90-15 seconds for location (drifting) PTTs and 200-300 seconds for data collection-only (fixed) platforms.

---

242)  CEOS Summary Report, WGD-10 Meeting, Annapolis Md., April 16-18, 1991

In 1988 an 'Argos World Service' was introduced providing five times daily location reports on vehicles and freight carrying a standard transmitter. The NOAA Argos packages receive all messages within a 5000 km diameter visibility circle at any instant; four PTTs can be processed simultaneously by NOAA-1 to 12 (and up to J) S/C. The capacity will be increased to eight for NOAA-K S/C and its successors. The data are formatted and stored, then dumped each time the satellites moves within visibility of one of the three ground stations (Wallops, Gilmore Creek, or CMS). VHF and S-Band transmitters also perform real-time relay (broadcast) for any user station within visibility.

### A.74.3   S&RSAT on NOAA-POES Satellite Series

The **S&RSAT**[243] (CNES/CRC) - **COSPAS** (Russian) program is an international cooperative satellite-based radio location system for search and rescue operations (established in 1979). It includes the NOAA program the Russian/CIS merchant navy, the Canadian defense and communications ministries, and CNES/France (as of 1992 around 30 countries joined the radio location search and rescue program). S&RSAT relays emergency radio signals from aviators, mariners, and land travellers in distress to ground stations. S&RSAT equipment onboard NOAA polar orbiting satellites is provided by Canada and France. (Note: the first S&R equipment was provided for NOAA-8 launched in March 1983).

**S&R** = Search and Rescue (S&RR = S&R Repeater, CRC/Canada; S&RM = S&R Memory, CNES/France).
The S&R instruments consist of a 3-band (121.5, 243, and 406.05 MHz) repeater S&RR and a 406.025 MHz processor S&RM. The S&RR downlink is at 1544.5 MHz and, besides the three repeated bands, also includes the 2400 b/s bit stream S&RM output. The 121.5 and 406 MHz bands are also serviced by two USSR COSPAS satellites which, together with the NOAA satellites (currently on NOAA-9, -10, and -11), provide improved timeliness of response.
The 121.5 -referred to as ELT 121.5 (Emergency Locator Transmitter) - and 243 MHz bands service emergency beacons are required on many aircraft, with a smaller number carried on maritime vessels. The 406 MHz band presently services the US fishing fleet, which is required to carry emergency beacons, and large international ships, which soon will be required to carry them. The beacons are also carried by some aircraft and smaller vessels and are being used by terrestrial carriers.

The 406 MHz emergency beacon signals (also referred to as EPIRB-406 - Emergency Position Indicating Radio Beacon) are immediately processed and stored on board the satellite and are transmitted to the ground from a continuous memory dump, providing complete worldwide coverage. Around the world, ground station LUTs (Local User Terminal) acquire the processed data and unique beacon identification and send these located and identified alerts to MCCs (Mission Control Centers), which forward the alerts to appropriate Rescue Coordination Centers for action. The 406 MHz beacons are designed to work well with the satellite, the system nominally provides better than 4 km accuracy, 90% ambiguity resolution on first pass, and better than 90% location probability on one pass. Note: the US S&R operational ground system facilities consist of S&RSAT, SOCC at Suitland Md. as the MCC, and three LUTs. In addition to the US facilities, many other cooperating nations operate their own LUTs and MCCs.

By late 1988, there were 15 LUTs in seven countries: Ottawa, Goose Bay, Edmonton and Fort Churchill in Canada; Toulouse, France; Tromsö, Norway; Lasham, UK; Kodiak, San Francisco, and St. Louis, USA; Moscow, Arkangelsk, Novosibirsk and Vladivostok, USSR; Bangalore, India; and Sao Paulo, Brazil.

The 121.5/243 MHz emergency beacons, whose use predates the satellite system, have not been specified to work with the satellite; consequently the results are variable, depending

---

243)  **Note: In compliance with the S&R designation the original acronym for SARSAT was changed to S&RSAT in this document.**

on the quality of the beacon. Nominally, location accuracy is about 20 km. All the processing is accomplished within the LUT, and because the satellite does not store these data, only beacons with mutual view of the satellite and LUT will be detected. No identification is included with the 125.5/243 MHz transmissions. Consequently, many non-beacon sources are also detected as beacons, increasing the difficulty of using these alerts. Even with these problems, the large number of beacons in the field have provided an impressive performance history. More than 1700 people have been saved by the S&R/COSPAS forces using satellite-derived locations, and for more than 300 of these people, the satellite provided the only means of alert[244),245)].

The S&RSAT system is also considered for geostationary meteorological satellites (like future Meteosat) as well as for the POEM satellite series. GOES-7, launched in May 1986, carried an experimental 406 MHz relay into geostationary orbit.

The USSR began deploying the space segment with the launch of Cosmos 1383 in June 1982 from Plesetsk into a 989 x 1028 km, 83° inclination orbit (Tsikada navigational satellite series). Designated **COSPAS 1** (= Space System for Search of Vessels in Distress'), the 121.5 MHz band remained operational until December 1987, with 406 MHz utilized primarily for interference monitoring. Cosmos 1447 (launched March 24, 1983) and 1574 (June 21, 1984) adopted the roles of COSPAS 2 and 3, with a third vehicle available as replacement.

As of 1993 the COSPAS-S&RSAT system has become the main system for emergency signalling at sea. The ICAO has also approved the COSPAS-S&RSAT system. In its 1993 program, airplanes will begin to be equipped with a new generation of radio beacons operating on both 121.5 and 406 MHz.

There are also plans under consideration to augment the COSPAS-S&RSAT space segment with geostationary satellites. Their use means that the EPIRB-406 signal will be received almost instantly. However, in order to automatically determine the coordinates of the emergency signal, it will be necessary to wait for the system's low-orbit satellite (position determination can only be provided from a system that moves relative to the Earth). The system thus works in two stages. In the first stage only the emergency signal is received via the geostationary satellite (GOES, INSAT and GMS satellites are planned to be equipped for this service). This signal is transmitted to the search and rescue service to prepare for the operation. In the second stage, the site of the signal origin is determined by the low-orbit satellite.

LUTs in 1993: Ottawa, Goose Bay, Edmonton and Fort Churchill in Canada; Toulouse, France; Tromsö, Norway; Lasham, UK; Kodiak, San Francisco, St. Louis, Hawaii, and Puerto Rico, USA; Arkangelsk, Novosibirsk, Nakhodka, Tilichiki, and Vladivostok, Russia; Alice Springs, Australia; Bangalore, and Lucknow, India; Hong Kong; Sao Paulo, Brazil; Santiago, Chile; Lahore, Pakistan; Yokohama, Japan; Wellington, New Zealand; Bari, Italy; Singapore; Guam; Maspalomas, Canary Islands; King George Island, Ant.[246)]

244)  Advanced TIROS-N (ATN) NOAA-I, NASA /NOAA bulletin 1991
245)  'Proceedings of the Twenty-Third International Symposium of Remote Sensing Environment', Vol. I, Bangkok,
      Thailand, April 18-25, 1990,, Erim, P.O. 8618 Ann Arbor Mich. p. 94
246)  Y. G. Zurabov, "The COSPAS-S&RSAT System: Results and Prospects", Space Bulletin, Vol. 1, Nr. 1 1993, pp.
      11-13

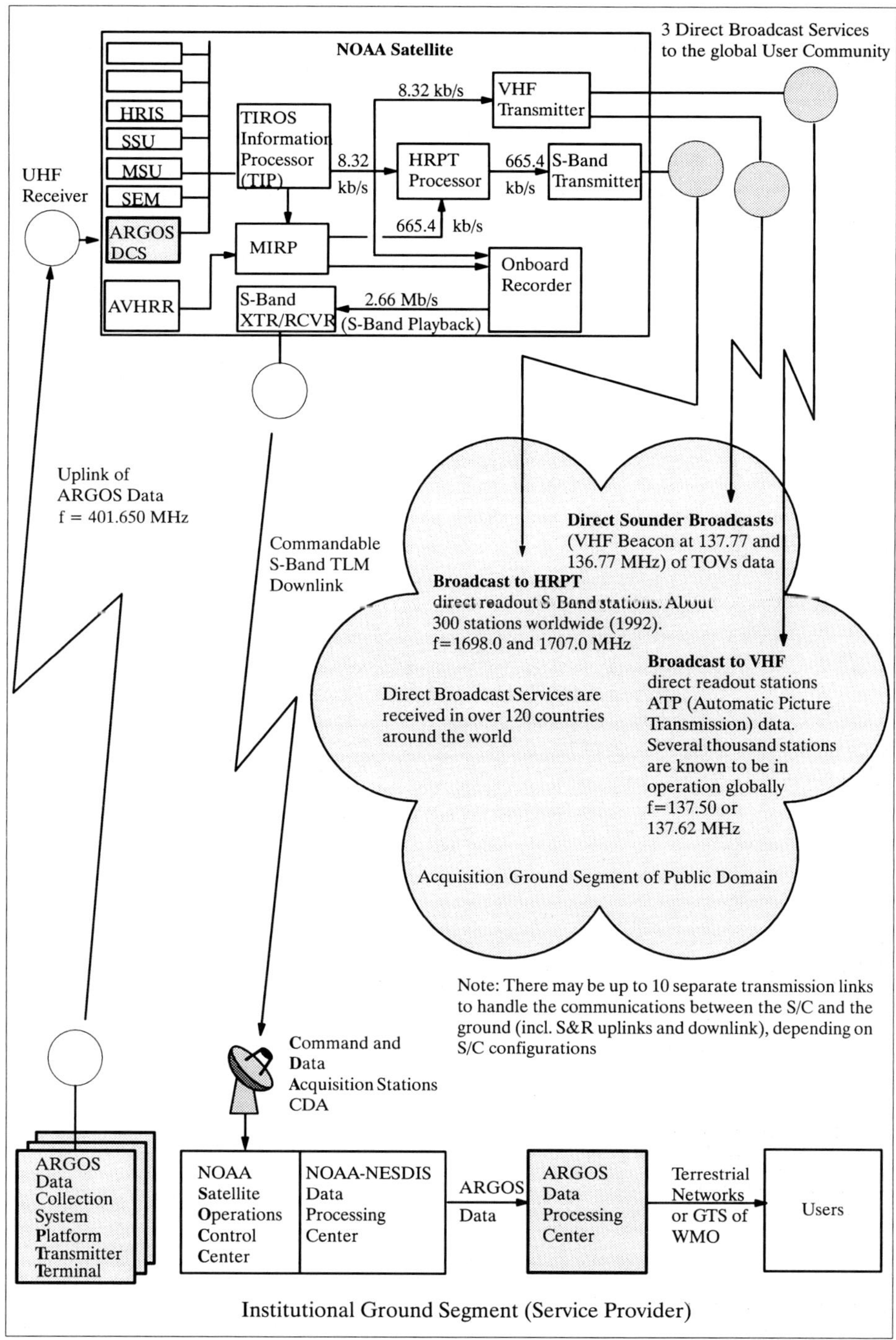


**Figure 67:**    **The ARGOS System Concept within the NOAA POES/TIROS Family**

| Satellite Name | Launch Date | Period (min) | Perigee (km) | Apogee (km) | Inclination (°) | Remarks/Sensor Complement |
|---|---|---|---|---|---|---|
| TIROS-1 | 01 Apr. 60 | 99.2 | 796 | 867 | 48.3 | 1 TV-WA and 1 TV-NA |
| TIROS-2 | 23 Nov. 60 | 98.3 | 717 | 837 | 48.5 | 1 TV-WA, 1 TV-NA, passive & active IR scan |
| TIROS-3 | 12 Jul 61 | 100.4 | 854 | 937 | 47.8 | 2 TV-WA, HB, IR, IRP |
| TIROS-4 | 08 Feb 62 | 100.4 | 817 | 972 | 48.3 | 1 TV-WA, IR, IRP, HB |
| TIROS-5 | 19 Jun 62 | 100.5 | 680 | 1119 | 58.1 | 1 TV-WA, 1 TV-MA |
| TIROS-6 | 18 Sep 62 | 98.7 | 783 | 822 | 58.2 | 1 TV-WA, 1 TV-MA |
| TIROS-7 | 19 Jun 63 | 97.4 | 713 | 743 | 58.2 | 2 TV-WA, IR, ion probe, HB |
| TIROS-8 | 21 Dec 63 | 99.3 | 796 | 878 | 58.5 | 1st APT TV direct readout & 1 TV-WA |
| NIMBUS-1 | 28 Aug 64 | 98.3 | 487 | 1106 | 98.6 | 3 AVCS, 1 APT, HRIR, 3-axis stabilization |
| TIROS-9 | 22 Jan 65 | 119.2 | 806 | 2967 | 96.4 | First "wheel", 2 TV-WA, global coverage |
| TIROS-10 | 02 Jul 65 | 100.6 | 848 | 957 | 98.6 | Sun synchronous, 2 TV-WA |
| ESSA-1 | 03 Feb 66 | 100.2 | 800 | 965 | 97.9 | 1st operational system, 2 TV-WA, FPR |
| ESSA-2 | 28 Feb 66 | 113.3 | 1561 | 1639 | 101.0 | 2 APT, global operational APT |
| NIMBUS-2 | 15 May 66 | 108.1 | 1248 | 1354 | 100.3 | 3 AVCS, HRIR, MRIR |
| ESSA-3 | 02 Oct. 66 | 114.5 | 1593 | 1709 | 101.0 | 2 AVCS, FPR |
| ATS-1 | 06 Dec. 66 | 24 hr | 41,257 | 42,447 | 0.2 | Spin scan camera |
| ESSA-4 | 26 Jan 67 | 113.4 | 1522 | 1656 | 102.0 | 2 APT |
| ESSA-5 | 20 Apr. 67 | 113.5 | 1556 | 1635 | 101.9 | 2 AVCS, FPR |
| ATS-3 | 05 Nov 67 | 24 hr | 41,166 | 41,222 | 0.4 | Color spin scan camera |
| ESSA-6 | 10 Nov 67 | 114.8 | 1622 | 1713 | 102.1 | 2 APT TV |
| ESSA-7 | 16 Aug 68 | 114.9 | 1646 | 1691 | 101.7 | 2 AVCS, FPR, S-Band |
| ESSA-8 | 15 Dec 68 | 114.7 | 1622 | 1682 | 101.8 | 2 APT TV |
| ESSA-9 | 26 Feb 69 | 115.3 | 1637 | 1730 | 101.9 | 2 AVCS, FPR, S-Band |
| NIMBUS-3 | 14 Apr. 69 | 107.3 | 1232 | 1302 | 101.1 | SIRS A, IRIS, MRIR, IDCS, MUSE, IRLS |
| ITOS-1 | 23 Jan 70 | 115.1 | 1648 | 1700 | 102.0 | 2 APT, 2 AVCS, 2 SR, FPR, 3-axis stabilization |
| NIMBUS-4 | 15 Apr. 70 | 107.1 | 1200 | 1280 | 99.9 | SIRS B, IRIS, SCR, THIR, BUV, FWS, IDCS, IRLS, MUSE |
| NOAA-1 | 11 Dec 70 | 114.8 | 1422 | 1472 | 102.0 | 2 APT, 2 AVCS, 2 SR, FPR |
| NOAA-2 | 15 Oct 72 | 114.9 | 1451 | 1458 | 98.6 | 2 VHRR, 2 VTPR, 2 SR, SPM |
| NIMBUS-5 | 11 Dec 72 | 107.1 | 1093 | 1105 | 99.9 | SCMR, ITPR, NEMS, ESMR, THIR |
| NOAA-3 | 06 Nov 73 | 116.1 | 1502 | 1512 | 101.9 | 2 VHRR, 2 VTPR, 2 SR, SPM |
| SMS-1 | 17 May 74 | 1436.4 | 35,605 | 35,975 | 0.6 | VISSR, DCS, WEFAX, SEM (EPS,MAG,XRS) |
| NOAA-4 | 15 Nov 74 | 101.6 | 1447 | 1461 | 114.9 | 2 VHRR, 2 VTPR, 2 SR, SPM |
| SMS-2 | 06 Feb 75 | 1436.5 | 35,482 | 36,103 | 0.4 | VISSR, DCS, WEFAX, SEM |
| NIMBUS-6 | 12 Jun 75 | 107.4 | 1101 | 1115 | 99.9 | ERB, ESMR, HIRS, LRIR, T&DR, SCAMS, TWERLE, PMR |
| GOES-1 | 16 Oct 75 | 1436.2 | 35,728 | 35,847 | 0.8 | VISSR, DCS, WEFAX, SEM (EPS,MAG,XRS) |
| NOAA-5 | 29 Jul 76 | 116.2 | 1504 | 1518 | 102.1 | 2 VHRR, 2 VTPR, 2 SR, SPM |
| GOES-2 | 16 Jun 77 | 1436.1 | 35,600 | 36,200 | 0.5 | VISSR, DCS, WEFAX, SEM (EPS,MAG,XRS) |
| GOES-3 | 15 Jun 78 | 1436.1 | 35,600 | 36,200 | 0.5 | VISSR, DCS, WEFAX, SEM (EPS,MAG,XRS) |
| TIROS-N | 13 Oct 78 | 98.92 | 849 | 864 | 102.3 | AVHRR,HIRS-2,SSU,MSU,HEPAD, MEPED, TED |
| NIMBUS-7 | 24 Oct 78 | 99.28 | 943 | 955 | 104.09 | LIMS, SAMS, SAM-II, SBUV/TOMS, ERB, SMMR, THIR, CZCS |
| NOAA-6 | 27 Jun 79 | 101.26 | 807.5 | 823 | 98.74 | AVHRR,HIRS-2,SSU,MSU,HEPAD,MEPED,TED |
| GOES-4 | 09 Sep 80 | 1436.1 | 35,600 | 35,600 | 0.5 | VAS,DCS,SEM(EPS,MAG,XRS,HEPAD),WEFAX |
| GOES-5 | 22 May 81 | 1436.1 | 35,600 | 35,600 | 0.5 | VAS,DCS,SEM(EPS,MAG,XRS,HEPAD),WEFAX |
| NOAA-7 | 23 Jun 81 | 101.92 | 852 | 869 | 98.9 | AVHRR,HIRS-2,SSU,MSU,HEPAD,MEPED,TED |
| NOAA-8 | 28 Mar 83 | 101.2 | 801 | 826 | 98.2 | AVHRR,HIRS-2,SSU,MSU,MEPED, TED,S&R |
| GOES-6 | 28 Apr. 83 | 1436.1 | 35,803 | 35,771 | 0.1 | VAS, DCS, WEFAX, SEM (EPS,MAG,XRS) |
| ERBS | 05 Oct 84 | 96.8 | 393 | 608 | 57 | ERBE, SAGE-II |
| NOAA-9 | 12 Dec 84 | 102.0 | 842 | 862 | 98.9 | AVHRR,(HIRS-2,SSU,MSU)=TOVS, SEM, S&R, SBUV, ERBE, |
| NOAA-10 | 17 Sep 86 | 101.27 | 803 | 824 | 98.66 | AVHRR,(HIRS-2,SSU,MSU)=TOVS, MEPED, TED, S&R, SBUV, ERBE, |
| GOES-7 | 26 Feb 87 | 1436.1 | 35,759 | 35,826 | 0.05 | VAS, DCS, WEFAX, SEM (EPS,MAG,XRS) |
| NOAA-11 | 24 Sep 88 | 102.14 | 845 | 863 | 98.91 | AVHRR,(HIRS-2,SSU,MSU)=TOVS, MEPED, TED, S&R, SBUV, ERBE, |
| NOAA-12 | 14 May 91 | 101.34 | 806 | 825 | 98.7 | AVHRR, HIRS, MSU,SEM (MEPED,TED) |
| NOAA-13 | 9 Aug. 93 | a S/C failure (power loss) occurred on Aug. 21, 1993 | | | | AVHRR,TOVS,SBUV,SEM,ARGOS,S&R,EHIC, MAXIE |
| | | | | | | |

**Table 50:**　　**Summary of U.S. Civilian Env./Met. Satellites (Chronological Order)[247]**

---

247) Proceedings of the Twenty-Third International Symposium of Remote Sensing of Environment, Vol. I, Bangkok, Thailand, April 18-25, 1990, p. 89, Erim, Ann Arbor, Mich.

## A.75    ODIN

Odin is a planned Swedish small-satellite mission for astronomical and atmospheric research (previously named MOSES = Molecules in Outer Space and Earth Stratosphere) by the Swedish National Space Board (SNSB). Launch is planned for Mid-1996 (on a Pegasus vehicle). The anticipated mission duration is 3 years (2 years design life). SNSB plans to carry out the project in cooperation with other countries.[248]

Applications: astronomy and aeronomy (atmospheric research: stratospheric ozone chemistry, mesospheric ozone science, summer mesospheric science, coupling of atmospheric regions).

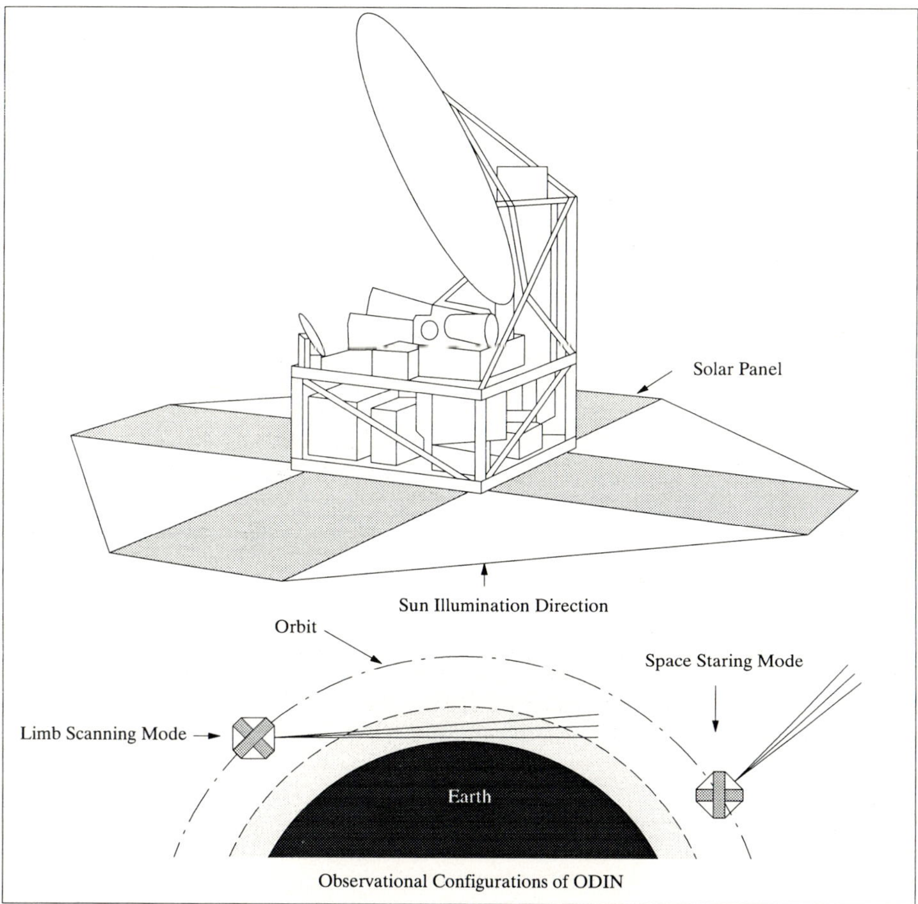


**Figure 68:    The ODIN S/C Model and Observational Configurations**

Objectives: Measurements in the wavelengths of 0.5 - 0.6 mm and 2.5 mm. These contain emission lines from important molecules such as water vapor, molecular oxygen, ozone and carbon monoxide for the study of atmospheric processes as well as in the study of astronomical objects. Complementary information on the Earth's atmosphere come from spectral lines at UV and VIS wavelengths. Major scientific issues relate to star formation processes, interstellar chemistry and atmospheric ozone balance.

---

248)  ”ODIN - A Small Satellite for Astronomy and Atmospheric Research”, SSC/SNSB brochure

Orbit: Sun-synchronous polar orbit; altitude = 700 km; inclination = 97.7°; with ascending node at 18:00 hrs ; period = 98.7 minutes.

Spacecraft: type = 3-axis stabilized with reaction wheels, star trackers and gyros; mass = 205 kg (140 kg bus plus 65 kg payload); power = 280 W; pointing = $\pm 15$ arcsec in staring mode ($\pm 1.2$ arcmin scanning); Datalink: > 500 kbit/s. Onboard storage = 100 Mbyte in solid state memory.

For aeronomy the spacecraft follows the Earth limb -scanning the atmosphere up and down from 15 km to 120 km at a rate of up to 40 scans per orbit. When observing astronomical sources Odin is continuously pointing towards the object for up to 60 minutes. Note: Both sensors point into the same direction.

**Sensors:**

**SMR** = Sub-Millimeter-wave Radiometer. This is a passive microwave limb sounder with one receiver at a wavelength $\lambda$ = 2.5 mm and 4 bands within the submillimeter range (0.5 - 0.6 mm). Antenna reflector type: offset Cassegrain telescope (1.1 m diameter, surface: 10 $\mu$m rms, material: carbon fiber reinforced plastic).

| | | |
|---|---|---|
| - | Type | Single sideband heterodyne receivers, |
| - | Frequencies: | 119 GHz, 488 GHz, 535 GHz, 553 GHz, and 575 GHz |
| - | Coverage: | 15 - 20 GHz in each submillimeter band |
| - | Bandwidth: | 100 MHz to 1 GHz |
| - | Resolution: | 0.1 MHz to 1 MHz |
| - | Sensitivity: | 1 K in 1 MHz with S/N = 5 after 15 minutes |
| - | Mixers: | Cooled Schottky mixers |
| - | Local Oscillators: | LO based on Gunn diodes and frequency multipliers |
| - | LNA | Cooled HEMT low noise amplifiers |
| - | Spectrometer | 1000 channel hybrid autocorrelator |

The 5 single sideband receivers in SMR are continuously switched between a reference source of known signal strength and the signal from the telescope. The telescope is periodically targeted towards well-known celestial objects. These procedures ensure both stability and good calibration.

**OS** = Optical Spectrograph. This is, similar to SCIAMACHY, a UV/VIS limb sounder. OS is a purely atmospheric measurement device of the type: grating spectrometer. OS measurements are in the spectral region from 200 to 800 nm and at 1270 nm, the resolution is about 2-3 nm below 300 nm, 0.5-1 nm above 300 nm, and 10 nm in the IR. Aperture = 10 $cm^2$. FOV = 0.02° x 2°. OS is co-aligned with SMR. The OS instrument design features gratings coupled to arrays with a combination of Si-based CCD detectors for VIS, and InGaAs detectors for IR.

Spectral lines:
The SMR covers transitions of aeronomical interest from the following molecules: $ClO$, $CO$, $NO_2$, $N_2O$, $H_2O_2$, $HO_2$, $H_2O$, $H_2^{18}O$, $NO$, $HNO_3$, $O_3$, and $O_2$; and atomic and molecular transitions of astrophysical interest from:

$Cl$, $H_2^{18}O$, $H_2O$, $H_2S$, $NH_3$, $H_2CO$, $O_2$, $CS$, $^{13}CO$, $H_2CS$, $SO$, $SO_2$
The Optical Spectrometer (OS) is aimed at studying the following species in the Earth's atmosphere: Aerosols, $ClO$, $O_3$, $O_2$, $O_4$, $NO$, and $NO_2$.


## A.76  OKEAN-O

A Ukrainian/Russian satellite series (OKEAN=Ocean) for the operational monitoring of ocean surfaces (sea surface temperatures, wind speed, sea color, status of ice coverage, cloud coverage and precipitation) day and night operation. The Okean-O series with their polar orbits provides valuable complementary data of the arctic and antarctic regions, which are not visible from geostationary meteorological satellites.

Orbit: Sun-asynchronous polar orbit, altitude = 660 km, inclination 82.5°, period = 98 minutes.

The OKEAN program started in 1979 with the pre-operational (i.e. experimental) Cosmos program[249]. Mission objective: a major interest was in the operation of active and passive microwave sensors. The active MW sensor provides an all-weather observation capability.

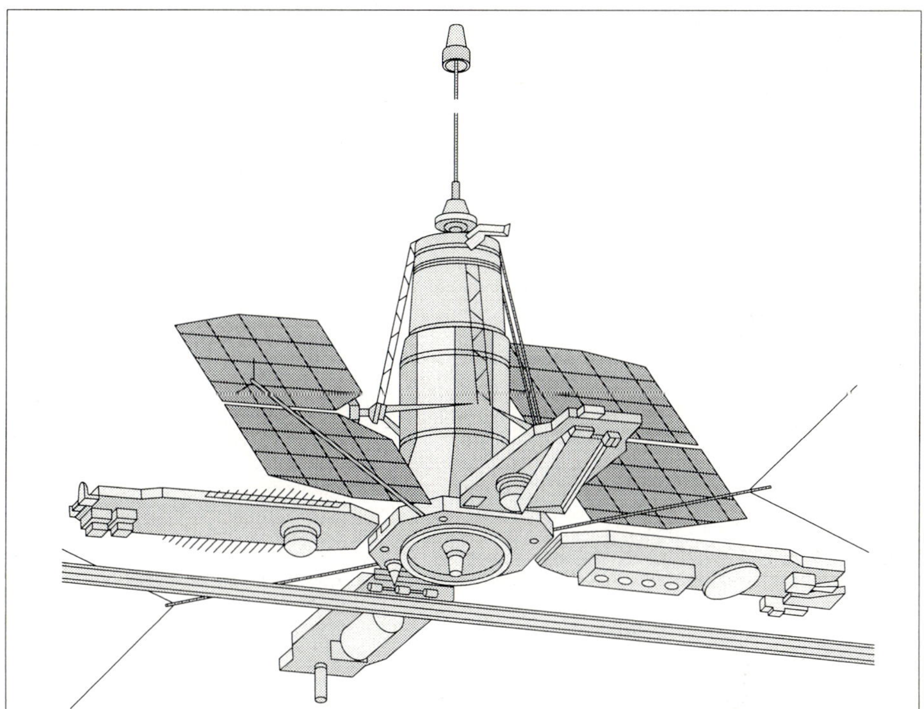

**Figure 69:**    **The OKEAN S/C Model**

## A.76.1   Experimental Cosmos Program

The predecessor series of the Okean series was the experimental Cosmos series.

- Cosmos 1076;   Launch: 1979; Orbit: altitude = 660 km; inclination = 82.6°
- Cosmos 1151;   Launch: 1981; Orbit: altitude = 660 km; inclination = 82.6°

- Cosmos 1500;   Launch: 28 Sept. 1983; Orbit: 649 x 679 km; Inclination = 82.6°
- Cosmos 1602;   Launch: 28 Sept. 1984; Orbit: 629 x 664 km; Inclination = 82.5°
- Cosmos 1766;   Launch: 28 July 1986; Orbit: 640 x 696 km; Inclination = 82.5°
- Cosmos 1869;   Launch: 16 July 1987; Orbit: 635 x 688 km; Inclination = 82.5°

The Cosmos satellites 1076 and 1151 provided the following sensors:

- **Device** ν =  Passive MW Radiometer (NPO Vega) measuring the emission radiation of the atmosphere/ocean system at the following (micro) wave lengths / frequencies: 0.8

---

249)  Verbal information provided by B. Kutuza of IRE (Russian Academy of Sciences), Moscow

cm (37.5 GHz), 1.35 cm (22.22 GHz), 3.2 cm (9.37 GHz), and 8.5 cm (3.53 GHz). Nadir pointing.

- **Device** $\pi$ = a polarimeter measuring radiation at wavelength 3.2 cm (with 2 orthogonal polarizations at a look angle of 53° relative to nadir, and with one polarization at nadir).

- **Device 174 K** = IR spectrometer.

- Device NN with 4 visible channels for monitoring of ocean color.

The simultaneous spectral measurements in several wavelengths provide a means of evaluating the hydro-physical parameters of the sea surface, the meteorological parameters of the atmosphere as well as the . The parameters are: sea surface temperature, wind speed at the sea surface boundary, liquid vapor content in clouds, integrated water vapor, etc.

The Cosmos satellite series (1500 - 1869, see above) was equipped with a subset of instruments of the first series, i.e. one or two passive MW radiometers per satellite measuring the emission radiation of the atmosphere/ocean system. Cosmos 1500 was the first Soviet satellite equipped with a side-looking, all-weather radar, namely **RLSBO**.


## A.76.2   OKEAN-O1 Operational Series

Operational series:

-   Okean-O1-1;   Launch: 5 July 1988; Orbit: 635 x 666 km; Inclination = 82.5°
-   Okean-O1-2;   Launch: 28 Feb. 1990; Orbit: 639 x 666 km; Inclination = 82.5°
-   Okean-O1-3;   Launch: 4 June 1991; Orbit: 664 x 684 km; Inclination = 82.5°
    (conditional operation since Jan. 1992, RLSBO not working!)
-   Okean-O1-4;   Launch: Projected for January 1994

Sensors (Okean-O1 Series):

- **RLSBO** = Side Looking Real Aperture Radar (Kharkov IRE, Ukraine, PI: Kalmykov), prime sensor of the Okean series (11.1 m antenna length). Wavelength/frequency: 3.2 cm/9.7 GHz, X-Band; resolution = 2.1 km - 2.8 km in flight direction, = 1.2 km - 0.7 km in cross track direction; swath = 450 km.

  -   impulse duration = 3 µs
  -   impulse peak power = 100 kW
  -   pulse repeat frequency = 100 Hz
  -   polarization = V (vertical)

- **RM-08** = Passive Microwave Scanning Radiometer (Kharkov IRE, Ukraine). Wavelength/Frequency: 0.8 cm / 36.6 GHz; resolution = 15 x 20 km, swath = 550 km. Objective: monitoring of atmospheric vapor, sea ice, ocean surface temperature with an accuracy of 1-2 K.

- **MSU-M** = Multispectral Scanner of low resolution (ISDE, Moscow). Resolution = 1.0 x 1.7 km, swath = 1900 km. Objective: cloud monitoring and sea surface temperature. Spectral ranges: 0.5 - 0.6 µm; 0.6 - 0.7 µm; 0.7 - 0.8 µm; 0.8 - 1.1 µm

- **MSU-SK** = Multispectral Scanner of moderate resolution (with conical scanning, (ISDE, Moscow). Spectral ranges: 0.5-0.6, 0.6-0.7, 0.7-0.8, 0.8-0.1 µm. Spatial resolution = 175-243 m (in VIS) and 500 m (in the NIR). Swath = 600 km

- **MSU-S** = Multispectral Scanner of moderate resolution (ISDE, Moscow). Spectral ranges: 0.6 - 0.7 µm, 0.7 - 0.9 µm; resolution = 370 m; swath = 1100 km

- **Kondor** = DCPs data collection system

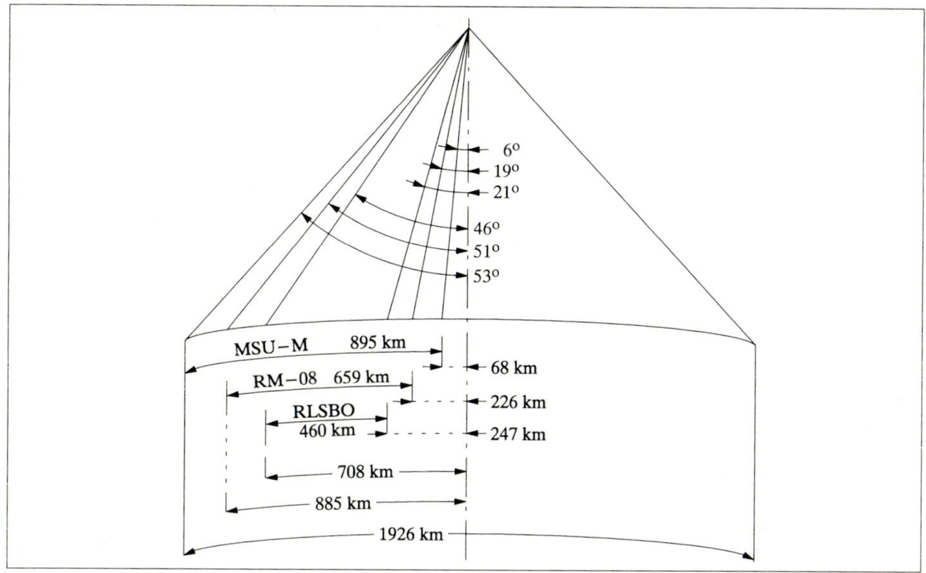


**Table 51:    Observation Geometries for OKEAN Instruments (shown to one side only)**

One of the features of Okean type satellites is the possibility of direct data reception by various users having simplified hardware and software for data processing. The downlink frequency of 137.4 MHz (carrier frequency with 2.4 kHz subcarrier frequency) is used for direct broadcasts of RLSBO and MSU-M imagery with reduced resolution to any of the CIS APT stations as well as to foreign users. The second science data downlink: carrier frequency of 466.5 MHz.

Data collected are transmitted to the receiving stations at Moscow, Novosibirsk, and Khabarovsk (Pacific region). Maps can be readied in 10 days and ice forecasts sent directly to ships at sea.

### A.76.3   OKEAN-O Series

An upgraded series with regard to the sensor complement over the OKEAN-O1- series. The first launch of this series (OKEAN-O-1) is projected for 1995.

Sensors:

- **MSU-V** = High-Resolution Multispectral Scanner (Kazan). CCD detector. The instrument has a swath width of 180 - 200 km. Objectives:
  Spectral ranges:
  - Band 1    0.45 - 0.52 µm (VIS)            resolution = 50 m
  - Band 2    0.52 - 0.62 µm (VIS) resolution = 50 m
  - Band 3    0.62 - 0.74 µm (VIS) resolution = 50 m
  - Band 4    0.76 - 0.90 µm (VIS) resolution = 50 m
  - Band 5    0.90 - 1.10 µm (NIR) resolution = 50 m
  - Band 6    1.55 - 1.75 µm (SWIR)          resolution = 100 m
  - Band 7    2.10 - 2.35 µm (SWIR)          resolution = 275 m
  - Band 8    10.3 - 12.6 µm (TIR) resolution = 275 m

- **MSU-SK** = Multispectral Scanner of Moderate Resolution with conical scanning (ISDE, Moscow). Swath width = 600 km; Spectral bands: 0.5 - 0.6 µm, 0.6 - 0.7 µm, 0.7 -

0.8 µm, 0.8 - 1.1 µm, 10.3 - 11.8 µm. Spatial resolution = 160 m (in VIS) and 600 m (in TIR).

- **RTVK-M** = Radio and TV-Complex as part of MSU-M (VNIIOPhI, Moscow). Objective:
  MSU-M swath width of 1900 km, spatial resolution of 1.6 - 2.0 km; 4 spectral bands: 0.5 - 0.6 µm, 0.6 - 0.7 µm, 0.7 - 0.8 µm, 0.8 -1.1 µm

- **RLSBO** = Side Looking Real Aperture Radar. There are two instruments, one looking to either side of the ground track. The near angle of incidence is ± 20°. Note: RTVK-M and RLSBO instruments are integrated into one complex.

- **Trasser** = Polarization Spectroradiometer.
  - Swath width
  - Spatial resolution:
  - Spectral range:      0.427 - 0.809 µm; 62 spectral lines of width = 6 nm????
  - Spectral resolution   2 Å
  - Polarization:       H and V

- **Delta-2** = Scanning Microwave Radiometer (MEI, Moscow). Objective:
  The instrument is nadir looking and provides 4 frequency bands at the following wavelengths (frequencies): 0.8 cm (37.7 GHz), 1.35 cm (22.2 GHz), 2.2 cm (13.6 GHz), and 4.5 cm (6.6 GHz). Swath width = 800 km; spatial resolution = 20 - 100 km.

- R-600 = Passive Microwave Radiometer (sounder ???, built by MEI, Moscow). Objective: measurement of emissive microwave radiation at the atmosphere/sea surface interface.
  Wavelength (frequency) = 6 cm (5 GHz); spatial resolution = 130 x 130 km.

- R-225 = Passive Microwave Sounder (built by MEI). Objective:
  Wavelength (frequency) = 2.25 cm (13.3 GHz); spatial resolution = 130 x 130 km.

- Kondor-2M = DCPs data collection system


## A.77   Ørsted

A Danish research satellite [named in honor of the Danish scientist Hans Christian Ørsted (1777-1851) who discovered electromagnetism in 1820] developed by a consortium organizations, including the University of Copenhagen, the Technical University of Denmark (TUD), the Danish Meteorological Institute, the Danish Space Research Institute, and Computer Resources International. The mission objective is to perform highly accurate and sensitive measurements of the geomagnetic field and global monitoring of the high energy charged particles in the Earth's environment. The data is used for improvement of geomagnetic models and the study of auroral phenomena, correlation with Earth-based measurements to study the relationship between the external field and the energy coupling of the magnetosphere-ionosphere system.[250],[251]

Ørsted is in the class of a low-cost microsatellite with a projected launch as a secondary payload on a US Air Force Delta II rocket in September 1995. S/C mass = 60 kg, power = 50 W (22 W average load), S/C stabilization by a gravity gradient boom (8 m boom containing the star imager and magnetometers) and active magnetic torquing (3-axis magnetorquer coils). Box-like shape of the satellite (34 x 45 x 68 cm) that is covered by solar panels. Design life of at least 1 year with a possible extension to 3 years. Ørsted is supposed to fill the gap in magnetic measurements between Magsat (launch in 1979) and GAMES (NASA mission, projected for 1998).

---

250) Information provided by F. Primdahl of TUD, Lyngby, Denmark
251) P. Donaldson, "Mapping Magnetism", Space, April 1993

Data: Science data sampling rates: 1 to 100 samples/s. Onboard data storage and downlinking every 12 hours or less. Uplink and downlink in S-Band (2.114 GHz and 2.296 GHz respectively). Maximum downlink rate is 256 kbit/s. Three downlink stations in Denmark.

Orbit: Sun-synchronous circular polar orbit, altitude = 830 km, inclination = 98.7°.

Sensors:

**Scalar Magnetometer** (LETI and CNES, the sensor is provided free of charge by France). Overhauser proton-precession magnetometer [coils for proton resonance excitation and detection, and a resonator for Electronic Spin Resonance (ESR) pumping of a nitro-oxide solution ] for measuring the magnetic field scalar values with an absolute measurement error of less than 1 nT. Measurement range: 16,000 - 64,000 nT; sampling rate = 1 Hz. Calibration is verified in-flight by comparison with the vector magnetometer measurements. Instrument mass = 1 kg. The instrument is boom-mounted at 8 m distance.

**Vector Magnetometer** (TUD and Danish Research Institute). The instrument is a compact spherical coil (CSC) triaxial fluxgate magnetometer to measure the the magnetic field vectors with an angular resolution of 20 arcsec and an absolute measurement error of less than 3-5 nT. Dynamic range: ± 65,536 nT; resolution: 0.5 nT, 100 vector samples/s burst mode, 10 vector samples/s normal mode. Instrument mass = 300 g. The sensor is boom-mounted at a distance of 6 m from the satellite.

**CPE** = Charged Particle Experiment (Danish Meteorological Institute). The instrument consists of six solid-state particle detectors for high energy measurements of electrons (30 keV - 1 MeV), protons (200 keV - 30 MeV), and α-particles (1-100 MeV). Detectors look in different directions with a FOV of 15-45°

**Star Imager** (TUD). The instrument is a CCD star imager for determining the pointing vector for the CSC fluxgate magnetometer with a resolution of 20 arcsec or less. Boom-mounted close to the position of the vector magnetometer. The camera of the star imager is a 604 x 576 pixel CCD device.

**GPS receiver** (NASA). 3-D position, velocity and time tag information for all science data.

## A.78    POEM (Polar Orbit Earth-Observation Missions)

POEM-1[252],[253],[254],[255],[256],[257]="Polar Orbit Earth-Observation Missions". This is the first Earth observation mission within the  ESA Observation Program utilizing the Polar Platform (PPF) being developed in the Columbus Programme.

Background: Proposed were originally two parallel and staggered mission series. The first series was planned with a Morning Orbit (M), the second series with a (After) Noon Orbit (N)]. The objectives  of the M-series were dedicated for the study of Meteorology/Atmosphere/Ocean/Ice/Environment, while the N-Series was considered for the study of Land Resources/Atmosphere/Environment.
This scenario was followed in 1991 by the POEM-1 mission (a single satellite mission) which fulfilled both environmental and meteorological mission objectives.

The updated POEM programme of 1992/93 considers a split scenario of two dedicated mission series called ENVISAT (Environmental Satellite) and METOP (Operational Me-


252)  "The ESA Earth Observation Programme and its Role in Global Remote Sensing", P. Goldsmith, Proceedings of the Twenty-Third International Symposium of Remote Sensing of Environment", Vol. I, ERIM, Ann Arbor, Mich., pp. 125-137.

253)  Programme Proposal for the first Polar Orbit Earth-Observation Mission using the Polar Platform, Part 1 , ESA Paper, 31-08-89

254)  Objectives and Strategy for the Earth-Observation Programme of the European Space Agency,ESA,, Oct. 88

255)  Polar Platform Concept Evaluation, ESA Paper, 25 Sept. 1989

256)  Programme Proposal for the first ESA Polar Platform, ESA/PB-EO (89) 32, 1. Sept. 1989

257)  Programme Proposal for the Development and Exploitation of the First Polar Orbit Earth-Observation Mission (POEM-1) using the Polar Platform, ESA/POEM 1, Issue 1, 28.10.91 Part 1, Issue 1, 30.10.91 Part 2

teorology Programme). The key element of the split scenario is the transfer of the meteorological instrument package onto a separate platform with a morning orbit devoted primarily to operational meteorology and climate monitoring. This allows the ENVISAT mission to focus on environmental issues that are more research-oriented with a package of essentially pre-operational instruments.

### A.78.1   ENVISAT-1

Overall objectives: Studying and monitoring of the Earth's environment on various scales, from local through regional to global. Monitoring and management of the Earth's resources, both renewable and non-renewable. Continuation and improvement of the services provided to the worldwide operational meteorological community. Contribution to the understanding of the structure and dynamics of the Earth's crust and interior.

Major disciplines covered: meteorology, climatology, environment, atmosphere, vegetation, hydrology, land use, ocean and ice processes.

Projected Launch: 1998 on Ariane 5 launch vehicle, nominal life time of 4 years. The ENVISAT-1 payload is integrated into PPF.[258]

Orbit: Near-circular sun-synchronous orbit; altitude = 800 km; ascending node 10 AM local time.

Data: two onboard tape recorders, each with a capacity of 30 Gbit, a record rate capability of 5 Mbit/s and a replay rate of 50 Mbit/s.

Payload data transmission: 2 Ka-Band channels at 50/100 Mbit/s via DRS; 2 X-Band channels at 50/100 Mbit/s (direct transmission to ground). One high-rate channel of 100 Mbit/s is considered for ASAR; 10 low/medium rate channels at up to 32 Mbit/s.

Attitude pointing: better than $0.1°$ (3 sigma) is required; attitude measurement: better than $0.03°$ (3 sigma) is required.

**Sensors:**

1.   Instruments provided by ESA

-   MERIS (Medium Resolution Imaging Spectrometer)
-   MIPAS (Michelson  Interferometer for Passive Atmospheric Sounding)
-   RA-2 (Radar Altimeter-2, a Laser Retro-Reflector (LRR) is included)
-   MWR (Microwave Radiometer)
-   ASAR (Advanced SAR)
-   GOMOS (Global Ozone Monitoring by Occultation of Stars), chemistry and dynamics of upper atmosphere

2.   Announcement of Opportunity (AO) Instruments

-   SCIAMACHY (Scanning Imaging Absorption Spectrometer for Atmospheric Cartography, provided by Germany and The Netherlands)
-   AATSR (Advanced Along Track Scanning Radiometer, provided by the UK), see also ATSR (ERS-1,2)
-   DORIS (see A.110.1) provided by France
-   SCARAB (Scanner for the Radiation Budget, provided by France)

**MERIS** = Medium Resolution Imaging Spectrometer[259]. MERIS is a passive optical push-broom instrument (CCD technology) intended to measure reflected radiation from the Earth's surface and from clouds with high spectral resolution (high radiometric resolution in the VIS range). MERIS will be able of changing gain and offset settings to adept performances to targets and different albedos (sea, clouds, land, ).

258)  B. Pfeiffer, B. Gardini, J. Cendral, "Envisat-1 and the Polar Platform: The Concept and the History", esa Bulletin, Nr. 76, November 1993, pp. 8-13
259)  "MERIS Medium Resolution Imaging Spectrometer", ESA brochure

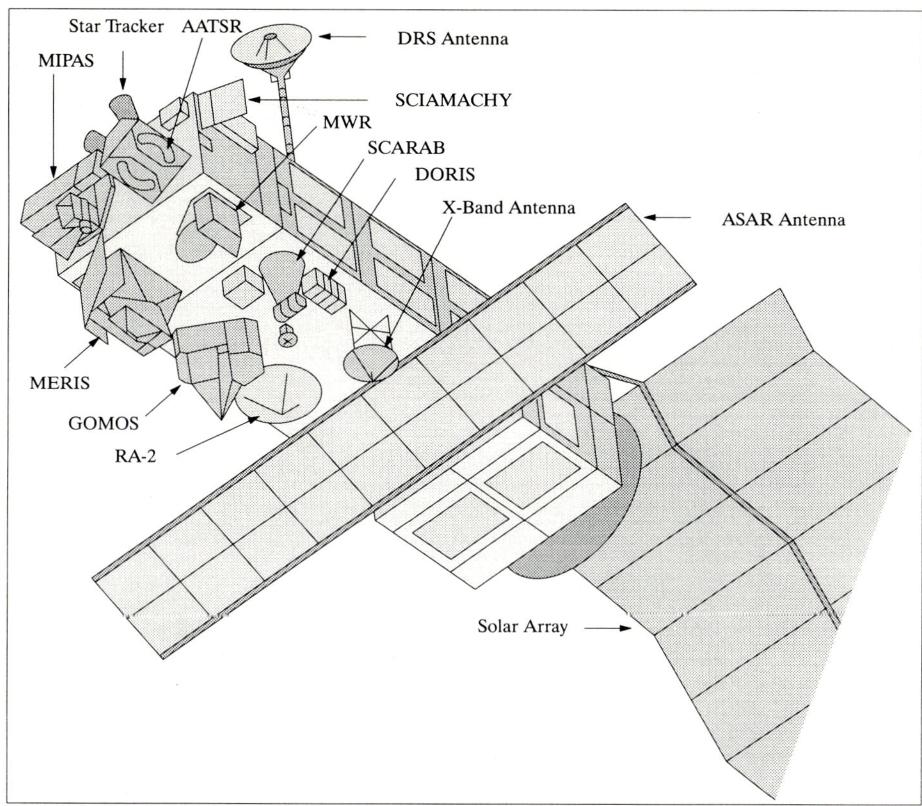


**Figure 70:     Model of the Envisat-1 S/C**

- • MERIS Objectives/Applications:[260]

- Ocean: monitoring of marine biophysical and biochemical parameters (chlorophyll, Gelbstoff concentrations, suspended particles). Mapping of sea pollution, coastal erosions, sea ice.
- Atmosphere: monitoring cloud distribution, cloud altitude, water vapor column content, aerosols
- Land: monitoring of vegetation/biomass, inland water, agriculture/forestry. Mapping of snow and ice, vegetation stress/calamities (fires)

- • Instrument: a programmable, high-spectral resolution imaging spectrometer.

- Total swath is covered by 6 identical modules (cameras) each having a 14° FOV, with 0.4° overlap between adjacent cameras. The cameras view the Earth through 6 depolarizing windows and common depointing prism.
- Pushbroom mode. An image is formed at the entrance slit of a spectrometer, which disperses the image spectrum on an area array CCD matrix (576 lines x 770 columns).
- Integration time/frame ≈ 40 ms (250m)
- TFOV= ±41°
- Mass = 155 kg
- Power = 165 W

260) M. Morel , J. L. Bézy, F. Montagner, A, Morel, J. Fischer, "Envisat's Medium-Resolution Imaging Spectrometer", esa bulletin, Nr. 76, November 1993, pp. 40-46

- Spectral characteristics:
- Spectral range:                   0.40 - 1.05 µm
- Spectral sampling interval         1.25 nm
- Spectral bandwidth:                2.5 - 25 nm (programmable)
- Registration between bands:        <0.1 pixel
- Band transmission capability:      15 bands programmable in position and width
- Band center knowledge              <1 nm

- Radiometric characteristics:
- Polarization sensitivity           <0.3% over the full spectral range
- Radiometric accuracy:              <2% relative to the sun
- Spectral band to spectral band accuracy: <0.05%
- Pixel to pixel accuracy:<NEDL = Noise Equivalent Detected Radiance at top of atmosphere with input signal Locean (upwelling radiance from ocean surface)
- Dynamic range                      up to albedo 1.0

- Spatial resolutions:
- Open ocean observation:            1 km at nadir
- Land and coastal zones:            250 m at nadir

- Swath width: 1450 - 1500 km, providing global coverage every 3 days.

- Operational Modes:
- Full resolution (250 m nadir); data rate = 24.6 Mbit/s (no onboard recording is considered), duty cycle of about 10%
- Low resolution (1000 m nadir); data rate = 1.7 Mbit/s, full coverage duty cycle

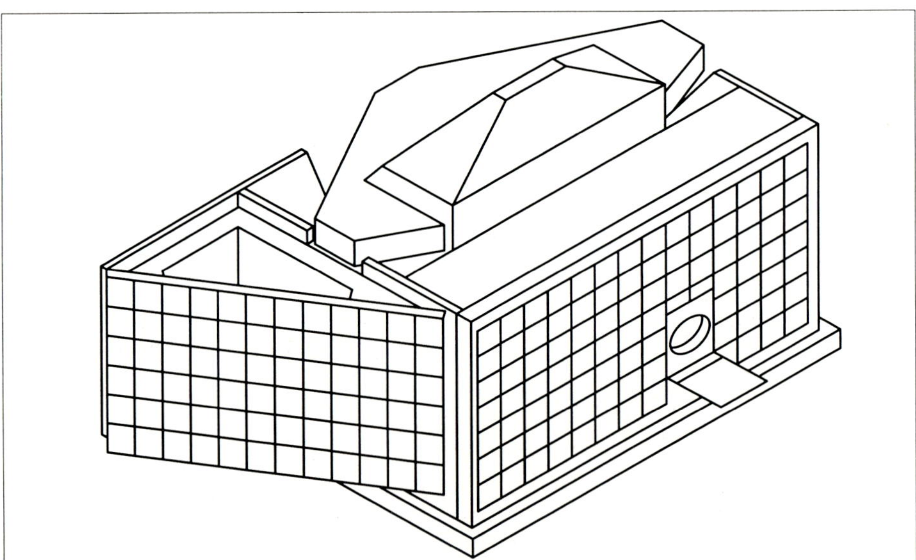

**Figure 71:**    **Schematic Model of the MERIS Instrument**

**RA-2** (extended capability of ERS-1 and 2). RA-2 is an adaptive pulse limited radar altimeter (nadir-pointing). Frequencies: 13.575 GHz (K-Band) and a second frequency at 3.2 GHz (S-Band) for the measurement and correction of ionospheric delays. Adaptable width/resolution windows (measurements over ice surfaces), autonomous resolution control. A LRR is considered part of RA-2 (see ERS-1).

The main objectives of RA-2 are the high-precision measurements of distance from the satellite to the Earth surface (topography) and the measurement of the power and the shape of the radar echoes from ocean, ice and land. The ground processing of the radar echo power and shape data enables the determination of wind speed and significant wave height in the observed target area. Sea ice monitoring through frequent temporal coverages.[261]

RA-2 Instrument parameters:
- Operating frequency: 13.575 GHz (Ku-Band), and 3.2 GHz (S-Band)
- Bandwidth: 320, 80, 20 MHz CW (Ku-Band), 160 MHz (S-Band)
- Transmit peak power: 60 W Ku-Band, 60 W S-Band
- PRF: 1800 Hz (Ku-Band), 450 Hz (S-Band) interleaved operation
- Pulse width: 20 µs
- FFT resolution: 128 complex points (samples) at 16 bit
- Echo averaged on board: 100 (Ku-Band), 25 (S-Band)
- Antenna beamwidth: 1.84° (Ku-Band), 6.2° (S-Band)
- Antenna diameter: 1.2 m
- Total mass: 106 kg
- Total power consumption: 168 W
- Data rate: 64/100 kbit/s (nominal/max)

- **ASAR** (Advanced SAR)[262],[263]
  ASAR measures the radar back-scatter of the Earth's surface at C-Band in either V or H polarization. The instrument will be designed to operate in three principal operating modes (image, wide swath and wave mode) and in two other modes.
- 'Image mode' and 'Alternating Polarization mode': ASAR operates as an imaging radar collecting data from relatively narrow swaths (up to 120 km) with high spatial resolution (30 m). The image mode provides continuous coverage over a single swath that can be pointed anywhere with 15-45° incidence angles.
- Alternating polarization mode: this technique provides vertically and horizontally polarized imaging of the same scene by interleaving looks with each polarization along track within the synthetic aperture.
- Wide swath mode: continuous coverage over a swath width of 406 km is imaged, which is divided into 5 subswaths ranging from 60 to 100 km in width; spatial resolution = 100 m.
- Wave mode: imaging of small areas of 5 x 5 km, in frequent intervals over along-track distances of 100 or 200 km on ocean surfaces. Measurement of the change in radar backscatter from the sea surface due to ocean surface waves.
- Global monitoring mode: provides continuous along-track sampling across a 400 km swath. Allows ASAR to be operated in a reduced spatial resolution mode of 1000 m (with a corresponding reduced data rate for on-board recording). Global monitoring of features such as ice or snow coverage, deforestation, desertification or humidity.

The design of ASAR incorporates a flexible swath position capability, this means that in contrast to the single swath of ERS, the image mode of ASAR offers the choice of up to seven swaths at various distances from the subsatellite track at incidence angles from 15° to 45° ,mid-swath (also referred to as **ScanSAR** technique).

Application: The main objective of the ASAR is to provide information on: ocean waves, sea ice extent and motion, snow and ice extent, surface topography, land surface properties, Earth's bio-mass (especially deforestation in equatorial zones), surface soil moisture and wetland extent.

261) A. Resti, "Envisat's Radar Altimeter: RA-2", esa bulletin, Nr. 76, November 1993, pp. 58-60
262) "ASAR Advanced Synthetic Aperture Radar", ESA brochure
263) S. Karnevi, E. Dean, D. J. Q. Carter, S. S. Hartley, "Envisat's Advanced Synthetic Aperture Radar: ASAR", esa bulletin, Nr. 76, November 1993, pp. 30-35

| Support modes | |
|---|---|
| Off | Instrument electrically disconnected from the platform |
| Standby | Instrument communications active and capable of receiving commands and transmitting telemetry on demand. Internal monitoring is active. |
| Heater | Ovens of all units requiring temperature stabilization are active and stable |
| Pre-operation | All units are active, but no transmissions are occurring |
| **Operation modes:** | |
| Wide swath | This mode operates with a wide swath and reduced spatial resolution |
| Image | High resolution, selectable swath position |
| Wave | Sampled imaging mode, low data rate |
| Global monitoring | Wide swath, low spatial resolution, low data rate |
| Alternating polarization | Interleaved VV and HH polarized imaging at high resolution, selectable swath position |
| Calibration mode | External calibration: External characterization to ground receivers |
| **Test/Health-check modes** | |
| Module stepping | Individual health check on each T/R (transmitter/receiver) module |
| Test | On-ground testing facility |

**Table 52:    Summary of ASAR Operating Modes**

| Operating Mode / Parameter | Image Mode | Alternating Polarization | Wide Swath Mode | Global Monitoring | Wave Mode |
|---|---|---|---|---|---|
| Polarization | VV or HH | VV + HH | VV or HH | VV or HH | VV or HH |
| Spatial Resolution | ~30 m | ~30 m | ~100 m | ~1000 m | ~30 m |
| Radiometric Resolution | 1.8-3.0 dB | 2.9-4.7 dB | 2.6-3.0 dB | 1.6 dB | 1.6-2.5 dB |
| Ambiguity Ratio point target / distributed target | ≥ 27 dB / 7-21 dB | ≥ 20 dB / 8-23 dB | ≥ 23 dB / 7-25 dB | ≥ 25 dB / 9-26 dB | ≥ 27 dB / 7-22 dB |
| Swath Width | 56-120 km 7 sub-swaths | 56-120 km 7 sub-swaths | 406 km 5 sub-swaths | 400 km 5 sub-swaths | 5 x 5 km 2 sub-swaths |
| Incidence angle range | 15-45° | 15-45° | | | 15-45° |
| Localization accuracy | <0.35 km | <0.35 km | <0.35 km | <0.35 km | <0.35 km |
| DC power consumption | 1200 W | 1200 W | | | |
| Data rate (mean) | | | | | |
| Radiometric Accuracy | ≤ 1.4 dB | ≤ 1.4 dB | ≤ 1.4 dB | ≤ 1.3 dB | ≤ 1.7 dB |

**Table 53:    ASAR Predicted In-Orbit Performance**

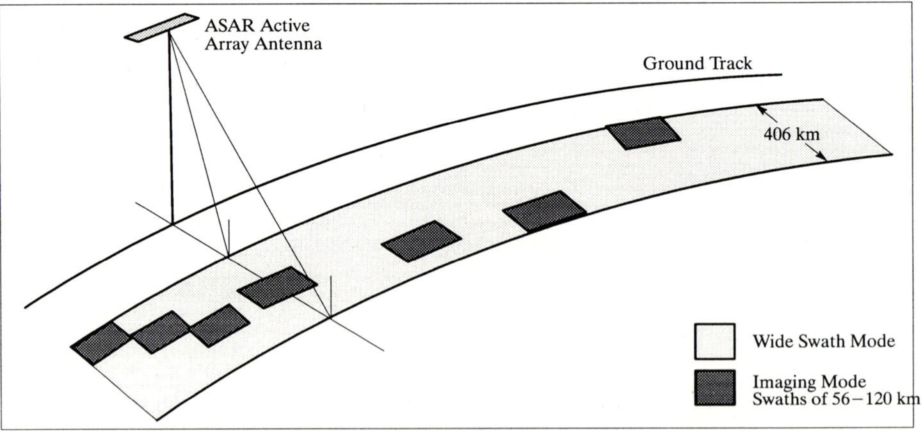


**Figure 72:    ASAR Observation Geometries**

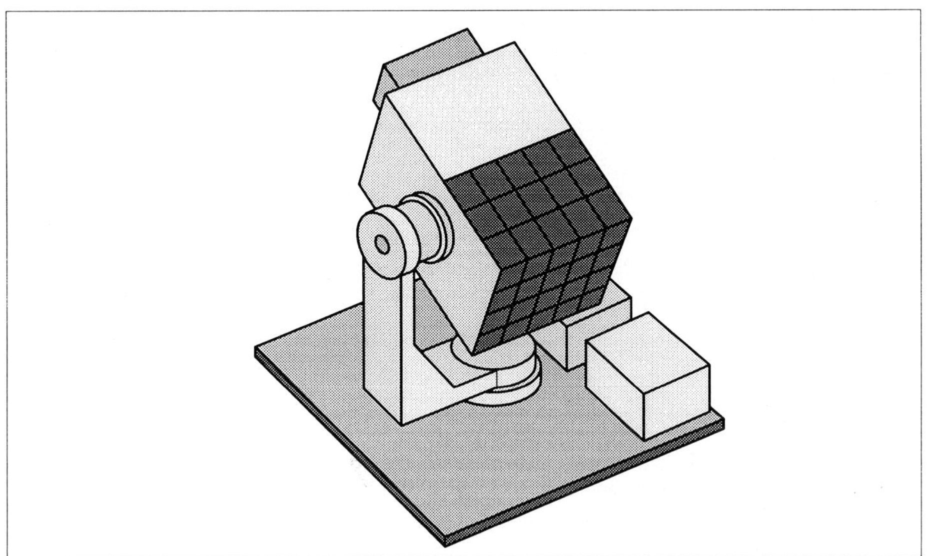

**Figure 73:      Schematic Model of the GOMOS Instrument**

**GOMOS**[264),265)] = Global Ozone Monitoring by Occultation of Stars. GOMOS is a UV/ Visible/Near-infrared limb viewing spectrometer, operation in star occultation mode. GOMOS relies on the self-calibrating occultation method (which minimizes the impact of instrument degradation). Using stars rather than Sun or Moon as light sources, it achieves a permanent, homogeneous global coverage through the measurement of 25 to 30 stars per orbit. The optical configuration is based on two grating spectrometers (range: 250 - 950 nm) permits to measure stratospheric profiles of $O_3$, $H_2O$, $NO_2$, $NO_3$, aerosols and temperatures with a 1.7 km vertical resolution.

Objectives: High accuracy and global long-term monitoring of stratospheric species ($O_3$, $NO_2$, $H_2O$, OClO, BRO, ClO) aerosols and vertical temperature distributions (Principle: star occultation). The measurement target lies between 20 and 100 km.

GOMOS Instrument:[266)]
The instrument optical design is based on a single telescope concept: a Cassegrain telescope simultaneously feeding a UV/VIS medium resolution spectrometer, the NIR high resolution spectrometers, two photometers (monitoring the input signal scintillation) and two (redundant) star trackers. All sensors and their associated front-end electronics are mounted on a thermally controlled CFPR optical bench.
The two-stage Steering Front Mechanism (SFM) has an angular steering range of 100° in azimuth and 8° in elevation, and a pointing accuracy of some 10 µrad at a bandwidth of 5 Hz. The SFM relies on a combination of linear motors and flexible joints for its operation. The accurate tracking function is performed via a digital closed control loop using star tracker information read at 100 Hz.

The occultation-mode operation of GOMOS involves pointing the instrument toward preselected stars, acquiring and then tracking them with high accuracy. The instrument has 3 additional calibration modes to meet the stringent pointing requirements, these are:

264)  A. Popescu, P. Ingmann, ""Envisat's Global Ozone Monitoring by Occultations of Stars Instrument: GOMOS",
       esa bulletin, Nr. 76, November 1993, pp. 36-39
265)  GOMOS handout from 'Atmospheres Panel Meeting' in Washington DC, Feb. 26-27, 1991
266)  "GOMOS  - Global Ozone Monitoring by Occultation of Stars", Esa brochure

- A linearity monitoring mode, allowing to calibrate for possible non-linearity of the sensor chains within the large dynamic range.
- A spatial-spread monitoring mode, allowing to optimize the position of the CCD read-out regions with respect to that of the stellar spectrum.
- A uniformity monitoring mode, allowing the measurement of the sensitivity difference between the various CCD pixels of a spectral sampling interval.

| Parameter | UV/VIS Spectrometer | IR1 Spectrometer | IR2 Spectrometer | Photometer |
|---|---|---|---|---|
| Spectral range | 250-675 nm | 756-773 nm | 926-952 nm | 650-700 & 470-520 |
| Spectral Resolution | 0.9 nm | 0.12 nm | 0.16 nm | N/A |
| SNR | >12 | >6 | >3 | >15 |
| Linearity | 1% | 1% | 1% | N/A |
| Stability | 1% | 1% | 1% | N/A |

**Table 54:**　**GOMOS Performance Characteristics**

Some GOMOS parameters:

- Telescope: 150 - 300 mm aperture Cassegrain
- cooled 2D CCD Array
- 2-Axis stabilized Platform
- Star Tracker: $\pm 0.3^\circ$ FOV
- Total mass: < 150 kg
- Power: < 180 W operating, 30 W standby
- Data rate: 220 kbit/s

**SCARAB** (Scanner for the Radiation Budget)[267]
Global measurement of radiative budget. SCARAB is a 4-channel mechanical cross-track scanner. SCARAB begun as initially as a French-Soviet project, it has now also German participation (DARA/GKSS). The four channels on SCARAB include broad-band, total and SW (short wave) channels for radiation budget determination, and narrow-band IR window, and visible channels for improved cloud detection (0.5 - 0.7 µm; 0.2 - 4.0 µm; 0.2 - 50 µm; and 10.5 - 12.5 µm). All 4 channels will have relatively coarse spatial resolution (60km). This is regarded as adequate for providing information on cloud radiation forcing. FOV = 100°; IFOV = 48 x 48 mrad (60 km at sub-satellite point). Data rate = 1 kbit/s; power = 35 Watt; mass = 30 kg

The SCARAB sensor is in the family of ERBE (NOAA sensor, ERBE scanners ceased operation in 1990) and CERES (EOS sensor). Status: 2 flight models are under construction to be flown on the Meteor-3 satellite series.

**MWR** = Microwave Radiometer. A nadir-viewing, 2-channel passive sensor operating at 23.8 and 36.5 GHz. Measurement of radiation reflected by the Earth. Derivation of surface temperature and atmospheric water content. FOV = 20 km diameter. The MWR measurements serve as input for RA-2 (correction of tropospheric effects by water vapor). MWR is of the same design as on ERS-1 and ERS-2.

Instruments planned for ENVISAT-2

- ATLID (Atmospheric Lidar). This sensor is intended to monitor cloud and aerosol layers in the atmosphere.

- ALADIN = Atmospheric Laser and Doppler Instrument (Doppler and Wind Lidar)

---

267) Robert Kandel, "Radiation and the Energy Balance", Paper presented at the ESA 'Earth Observation User Consultation Meeting', ESTEC, May 1991, The Consultative Document Collection of Preprints for the Meeting.

## A.78.2   METOP-1

European needs for meteorological observations in polar ("morning" and "afternoon") orbits have been generously provided by NOAA S/C and payloads (including some instruments developed in Europe) over the last two decades and will be provided until the year
2000. However, there is a longterm NOAA plan to confine its future coverage to the "afternoon" orbit after the year 2000. Europe is interested and also required to pick up this service
gap, to share the burden of service provision with the USA. Hence, the prime objectives of
the METOP mission series are as follows:

- to ensure continuity and availability for operational purposes of polar meteorological
  observations from the "morning" orbit to the global user community.

- to provide enhanced monitoring capabilities (complimentary to ENVISAT) to fulfil the
  requirements to the study of the Earth climate system as expressed by a number of international cooperative programs such as: GCOS (Global Climate Observing System),
  IGBP (International Geosphere and Biosphere Program), and WCRP (World Climate
  Research Program). The aim is the provision of continuous long-term data sets.

Orbit: Near-circular sun-synchronous morning orbit (9 AM), altitude = 850 km,

The METOP-1 S/C will be developed by ESA/EUMETSAT joint venture and operated by
EUMETSAT. The operational meteorology payload is provided by EUMETSAT, while the
climate monitoring payload is furnished by ESA. A launch is planned for the year 2000.

### 1.   Operational Meteorological Sensor Payload (Eumetsat/NOAA)

| Payload Sensor | Operational Met. Mission Objectives | Climate Mission Objectives (in addition to the Met. parameters) |
|---|---|---|
| VIRSR = Visible Infra-Red Scanning Radiometer | global imagery, global op-erational sounding | ocean measurements (SST), clouds and Earth radiation budget, land measure-ments |
| MTS = Microwave Tempera-ture Sounder (old AMSU-A) | global operational sound-ing | sea ice |
| MHS = Microwave Humidity Sounder (old AMSU-B) | global operational sound-ing | clouds and Earth radiation budget, sea ice |
| IASI = Improved Atmospheric Sounder Interferometer | global operational sound-ing | ocean measurements (SST), clouds and Earth radiation budget, some atmos. trace constituents |
| ARGOS = Remote Data Col-lection System (DCS-3 on-board, PTTs and DCPs in the ground segment) | data collection and location | collection of climate data |
| MCP = Meteorological Com-munications Package (acquir-ing and broadcasting data from the sensors) | global data access, local data access | global data access, local data access |
| S&R = Search and Rescue System | Cooperative satellite-based radio location system for search and res-cue operations. Relay of emergency radio signals from aviators, mari-ners and land travellers in distress to ground stations. | |
| SEM = Space Environment Monitor | Monitoring of the S/C environment (solar-terrestrial) | |

**Table 55:**   **METOP-1 Core Operational Instruments**

## 2.  Climate Instrument Payload (ATSR, MIMR, SCARAB, GOME, ASCAT)

| Climate Mission | Objectives |
| --- | --- |
| Operational meteorological observations | Continuity of 30 years of atmospheric observations and enhanced capacity to measure atmospheric constituents |
| Ocean measurements (incl. surface winds) | To monitor the exchange of energy between the ocean and atmosphere |
| Clouds and Earth radiation budget | Radiation is the primary energy source of the climate system and principle heat input to the oceans |
| Sea ice | The extent of sea ice is an important variable in connection with both ocean and heat budget and radiation balance |
| Atmospheric minor constituents | The concern over the depletion of stratospheric ozone suggests that the maintenance of a continuous data set of global total column ozone and vertical profiles is vital |
| Ocean color | Ocean color (chlorophyll content) is important for monitoring primary ocean productivity |
| Land measurements | Monitoring of spectral albedo and soil moisture variations |

**Table 56:     METOP-1 Climate Monitoring Requirements**


**MIMR** = Multifrequency Imaging Microwave Radiometer[268]. MIMR is an Earth-viewing, multifrequency passive microwave radiometer operating at 6 different frequencies (6.8, 10.65, 18.7, 23.8, 36.5, and 89 GHz) with a dual (V, H) polarization for each frequency. MIMR uses a conical-scan concept with a mechanically rotating antenna of 1.6 m diameter. Effective swath width = 1400 km (daily coverage of more than 82% of the Earth's surface). Spatial resolution < 5 km for the highest frequency channel.

Applications: Monitoring of atmospheric parameters, oceanography, ice/snow and land processes, soil moisture, precipitation, sea surface temperature, surface wind speed, sea ice concentration and properties, water vapor and liquid water content.

MIMR Instrument Parameters:

- Mass = 200 kg
- Power consumption = 190 W
- Duty cycle = 100%
- Data rate = 64 kbit/s

| Channels | 1 | 2 | 3 | 4 | 5 | 6 |
| --- | --- | --- | --- | --- | --- | --- |
| Frequency (GHz) | 6.8 | 10.65 | 18.7 | 23.8 | 36.5 | 89 |
| Pixel size - along track (km) | 60 | 38 | 22 | 20 | 11.6 | 4.9 |
| Pixel size - across track (km) | 60 | 38 | 22 | 20 | 11.6 | 4.9 |
| Sampling interval - along track (km) | 15.5 | 15.5 | 15.5 | 15.5 | 7.75 | 3.87 |
| Sampling interval - across track (km) | 15.5 | 15.5 | 15.5 | 15.5 | 7.75 | 3.87 |
| Radiometer sensitivity (K) | 0.2 | 0.4 | 0.5 | 0.5 | 0.5 | 0.7 |
| Radiometric stability (K) | 0.2 | 0.4 | 0.5 | 0.5 | 0.5 | 0.7 |
| Radiometric accuracy (K) | 1 | 1 | 1.5 | 1.5 | 1.5 | 1.5 |

**Table 57:     MIMR System Performance for a 705 km Orbit**


**ASCAT** (Advanced Wind Scatterometer)[269],[270] Objective: Determination of the wind vector fields at sea surface by measurement of the radar backscattering cross section $\sigma^o$.

The ASCAT is an active radar measuring system exploiting the Doppler principle. ASCAT operates in C-Band at 5.255 GHz with a long pulse with linear frequency modulation. Mea-


268)  "MIMR Multifrequency Imaging Microwave Radiometer on POEM", ESA brochure F-31
269)  'ASCAT Advanced Scatterometer' ESA brochure
270)  H. Ebner, H. R. Schulte, H. Hölzl, D. Miller, P. Hans, "ASCAT - Advanced Wind Scatterometer", IGARSS '92
       Volume I, pp. 435-439

surement of the radar reflectivity of the sea surface over two 500 km wide swaths, one on either side of the satellite ground track, this feature results in a much faster global coverage capability. Heritage of ERS-1 AMI-SCAT sensor. The major differences between ASCAT and AMI-SCAT are:

- enhanced coverage due to double swath operation
- use of solid state technology (improved radiometric performance)
- reduced downlink data rate due to onboard data processing

In each swath a regular 25 km grid of points (nodes) is defined where the $\sigma^0$ values are determined. The wind data at each node are extracted from a wind model. The measurements are accomplished by consecutive antenna operations, covering all three directions of viewing: fore, mid and aft beam with 45°, 90°, and 135° respectively relative to ground track. The antenna subsystem consists of six radiating waveguide antennas - two fore, two mid, and two aft antennas (V shaped configuration).

System Requirements:

- spatial resolution = 50 km (along and across track)
- localization accuracy = 5 km (along and across track)
- radiometric resolution $k_{pe}$ = 3-8% (varies with wind speed and incidence angle)
- radiometric accuracy = 0.46 dB (Interbeam)
- polarization = VV
- wind measurement accuracies $\leq$ ±2 m/s (wind speed)
- wind measurement accuracies $\leq$ ±20° (wind direction)
- downlink data rate < 60 kbit/s

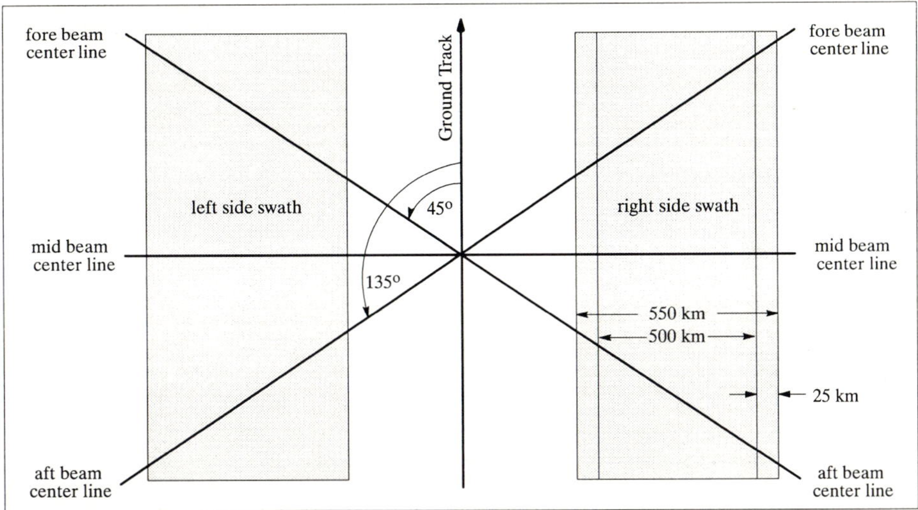


**Figure 74:**    **Top View of ASCAT Observation Geometries**

**IASI** (Improved Atmospheric Sounder Interferometer, CNES/ASI)[271]
IASI is a Fourier Transform nadir-viewing imaging interferometer operating in the thermal infrared spectrum from 3.4 (2940 cm$^{-1}$) to 15.5 μm (645 cm$^{-1}$), at moderate spectral resolution (0.25 cm$^{-1}$ unapodized). Objectives: sounding the tropospheric moisture and temperature, measuring the column-integrated $O_3$, CO, $CH_4$, and $NO_2$, and measurement of some trace gases which drive the budget of tropospheric chemistry and contribute to the greenhouse effect.

---

271) "Improved Atmospheric Sounding Infrared", ASI/CNES brochure, April 1991

**Data**: DRS (European Data Relay Satellite System) is the baseline for payload data communications along with classical direct downlinking of payload data to existing ground stations.

Service Provision: The METOP program, as successor to the NOAA POES morning series, is required to provide continuous broadcast of its meteorological data to the world-wide user community, so that any ground station in any part of the world can receive local data when the satellite passes over that receiving station. This implies continued long-term provision of HRPT (frequency: 1701 MHz, data rate: 3.5 Mbit/s) and LRPT (frequency: 137-138 MHz, data rate: 72 kbit/s) downlink services.


## A.79    POLAR

NASA/GSFC Solar-Terrestrial mission within GGS and ISTP with the objective to study the ionospheric role in substorm phenomena and in the overall magnetospheric energy balance (measurements of plasma entry and transport into the northern dayside cusp regions at high altitudes, and over the southern polar cap at low altitudes, global imaging of the northern auroral zone).[272],[273]

Projected launch in Fall 1994 with a Delta II vehicle from Vandenberg Air Force Base. Spacecraft mass = 1297 kg (269 kg propellant, 264 kg science payload). Nominal lifetime = 3 years. S/C is spin-stabilized at 10 rpm (a smaller platform of the S/C is despun and can be pointed to maintain the viewing field of certain instruments), 2.4 m $\oslash$, 1.8 m height, cylinder with conductive surfaces. Solar arrays provide 440 W, including 186 W for payload.

Orbit: Polar orbit with an apogee of 9 $R_E$ and a perigee of 2 $R_E$; period = 18 hours,

**Sensors:**

**1.    Observation of local electromagnetic fields in the low frequency range**

**MFE** = Magnetic Field Experiment (PI: C. Russell, UCLA). Objectives: Study of the coupling of the solar wind and the magnetosphere through currents driven in the polar cusp (energy and momentum exchange with the magnetosphere at the cusp-magnetosheath interface).
Instrument: two fluxgate magnetometers; sampling of the magnetic field with in the frequency range from 0 - 50 Hz. Magnetic field strengths between $10^{-6}$ and 0.6 Gauss are detectable.

**EFI** = Electric Fields Instrument (PI: F. Mozer, U. of Ca., Berkeley). Objectives: Study of the electric-field structure of the high-latitude magnetosphere, the cusp, and the plasma mantle.

EFI is a dual-probe instrument. Sampling of the electric field between 0 and 20 kHz. Electric field strengths between 0.1 and > 1000 mV per meter at a rate of 40 samples per second in the normal mode, more than 1000 samples per second in the burst mode. The burst will be coordinated with HYDRA and TIDE instruments.

**PWI** = Plasma Wave Instrument (PI: D. Gurnett, U. of Iowa). Objectives: Study of the wave/particle processes mediated by the electromagnetic turbulence (momentum transfer in the geospace system, particularly in the boundaries).
The PWI instrument samples the electric-field noise above the highest EFI frequencies well into the radio band. Magnetic-loop and search coils are used to sample the magnetic fluctuations above the highest frequencies detectable by MFE (identification of the characteristic modes of plasma behavior).

---

272)  'The Solar-Terrestrial Science Project of the Inter-Agency Consultative Group for Space Science', esa SP-1107, November 1990, pp. 11-15
273)  'ISTP Global GEOSPACE Science - Energy Transport in Geospace', ESA/NASA/ISAS brochure, 1992 of GSFC

**HYDRA** = Hot Plasma Analyzer Experiment (PI: J. Scudder, GSFC). Objectives: Study of low-energy electrons in the global magnetic topology. Study of electron and ion signatures that accompany geomagnetic substorms, auroral arcs, field-aligned currents, and particle precipitation.

The HYDRA ensemble of electrostatic analyzers is one part of the HYDRA instrument, which measures electrons and ions in energy per unit charge between 1 eV and 30 keV, and observes them in 12 directions simultaneously. In addition, another device, called a parallel plate analyzer, with a position-sensitive array, images the electrons within and near the magnetic field with a 1.5° angular resolution, energy ranges between 10 eV and 10 keV. Sampling rates: 2/second.

**TIDE/PSI** = Thermal Ion Dynamics Experiment / Plasma Source Investigation (PI: T. Moore, MSFC). Objectives: Study of low-energy ions (transport mechanisms) to evaluate the ionosphere as a source of plasma for the magnetosphere.
TIDE samples ions extracted from the ionosphere (whose mass is determined by a time-of-flight scheme). Energy range: 0.1 eV per charge to 100 eV per charge.

## 2.   Observation of the particle populations associated with the electromagnetic fields

**TIMAS** = Toroidal Imaging Mass-Angle Spectrograph (PI: E. Shelley, Lockheed Palo Alto Research Lab.). Objectives: Study of the properties, location, and morphology of the polar cusp, which is the principal source region for entry of solar wind plasma and the hot ionospheric plasma into the magnetosphere.
The TIMAS instrument samples ions of resolved mass that are either energized ions of ion ospheric origin or stored particles of solar wind origin. Energy range: 50 eV - 30 keV. Sampling rate at 10 times per minute (one per satellite spin). TIMAS data is used in combination with data from TIDE and from SWICS on the WIND satellite.

**CAMMICE** = Charge and Mass Magnetospheric Ion Composition Experiment (PI: T. Fritz, Los Alamos National Lab.). Objectives: Study the mechanisms that control the energization, storage, and precipitation of particles in the high-latitude magnetosphere.
CAMMICE permits to determine the composition of major ion constituents in the near-Earth plasma sheet and in the ring current. Energy ranges between 6 keV per charge and 60 MeV per ion. For full angular distribution, CAMMICE measures at a rate of 10 samples per minute, or once per spin of the S/C. The angular resolution approaches 0.2°.

**CEPPAD/SEPS** = Comprehensive Energetic-Particle Pitch Angle Distribution / Source Loss Cone Energetic Particle Spectrometer (PI: B. Blake, Aerospace Corp. Ca.). Objectives: Investigation of the quantitative information on the sources, energization, transport, and losses of energetic particles in the magnetosphere. Measurement of the rate of particle precipitation into the Earth's upper atmosphere.
The CEPPAD investigation uses a variety of techniques to provide detailed energy spectra and angular distributions of energetic particles. SEPS is on the despun platform exploits the pointing relative to the ambient magnetic field to greatly enhance its angular resolution and measurements of energetic particle precipitation. The sensors, which cover the energy range from 15 keV to 10 MeV, separate protons and electrons. The complete 3-D particle distribution is measured every 6 seconds.

## 3.   Imaging instruments for spatial observations

**UVI** = Ultraviolet Imager (PI: M. Torr, MSFC). Objectives: Study of the spatial and temporal descriptions of the aurora and images of the total particle flux, characteristic energy, thermospheric neutral composition, and ionospheric conductances.
UVI images the dayside and nightside auroras in the vacuum UV range using five specially designed filters. The detector is an intensified CCD used in conjunction with a fast reflective optical system to image an 8° FOV at a nominal rate of two frames per minute.

**VIS** = Visible Imaging System (PI: L. Frank, U. of Iowa). Objectives: Quantitative assessment of the dissipation of magnetospheric energy into the auroral ionosphere. Development of an energy flow model within the magnetosphere using VIS data in three ways: to illustrate the topology of the magnetosphere, to delineate the response of the magnetosphere and the magnetotail to substorms and solar-wind conditions, and to identify the locations and mechanisms for suprathermal charged-particle acceleration.

The VIS instrument uses an image intensifier readout through 12 visible narrow-band filters producing 5 separate auroral images per minute.

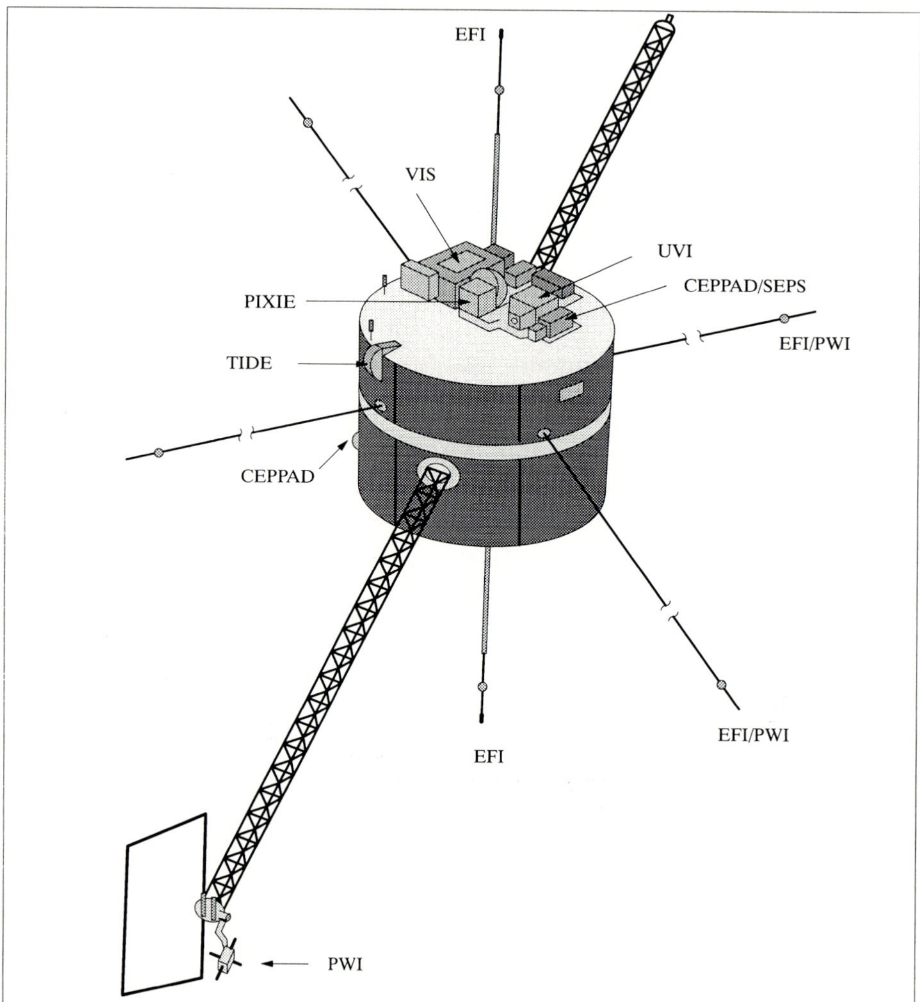


**Figure 75:** **The POLAR S/C Model**

**PIXIE** = Polar Ionospheric X-ray Imaging Experiment (PI: W. Imhof, Lockheed Palo Alto Research Lab.). Objectives: Study of the morphology and spectra of energetic electron precipitation and the effect upon the atmosphere. Derivation of the total electron energy deposition rate, the energy distribution of the precipitating electrons, and the altitude profile of ionization and electrical conductivity.

PIXIE uses a pin-hole camera concept to measure the spatial distribution and temporal variation of x-ray emissions from the Earth's atmosphere. The time resolution of images is 60 seconds or better (the optimal radial distance to obtain the best images is 6 $R_E$).

Data: Onboard recording capability of science data (1.3 Gbit digital tape recorder). Science data transmission of 56 kbit/s (RT) and 512 kbit/s (PB) to DSN (four 60 minute nominal contact periods per day). The Polar S/C provides onboard interconnection of instrumentation for data communication. Data sharing among the instruments can be triggered by pattern recognition schemes in onboard computers.

| Investigation | Experiment Acronym | Mass (kg) | Data Rate (Bit/s) |
|---|---|---|---|
| Magnetic Field Experiment | MFE | 5.0 | 500 |
| Electric Field Instrument | EFI | 31.9 | 2500 |
| Plasma Wave Instrument | PWI | 18.4 | 2520 |
| Hot Plasma Analyzer | HYDRA | 14.4 | 4400 |
| Thermal Ion Dynamics Experiment | TIDE | 33.3 | 2520 |
| Toroidal Ion Mass Spectrograph | TIMAS | 16.5 | 3600 |
| Charge and Mass Magnetospheric Ion Composition Experiment | CAMMICE | 12.9 | 1280 |
| Comprehensive Energetic Particle Pitch Angle Distribution | CEPPAD | 14.4 | 4380 |
| Ultraviolet Imager | UVI | 18.0 | 12000 |
| Visible Imaging System | VIS | 24.0 | 11000 |
| Polar Ionospheric X-ray Imaging Experiment | PIXIE | 24.5 | 3500 |

**Table 58:     Instrument Summary of the Polar S/C Payload**


# A.80    PoSAT-A (Portuguese Satellite)

PoSAT is the first satellite of Portugal built by Surrey Satellite Technology Ltd. (SSTL) at the University of Surrey (UK) within a Technology Transfer Programme between UK and Portugal. The primary objective of the programme is to stimulate a Portuguese space industry, the purpose of the PoSAT-A mission is to generate a nucleus of engineers with first-hand space technology experience for Portugal's possible future satellite programmes.[274],[275]

PoSAT-A was launched on Ariane, September 26, 1993 as an auxiliary payload on the Spot-3 launch from Kourou. Satellite construction at SSTL by a joint team of Portuguese and SSTL engineers. PoSAT-A uses the modular UoSAT microsatellite platform. The S/C is stabilized by an Earth-pointing gravity boom and by a 3-axis magnetic torquing system. S/C mass = 50 kg, power = 20 W, downlink data rate = 9.6 or 38.4 kbit/s, VHF and UHF bands, ground receiving stations are in Portugal (Sintra) and the UK. S/C operation via Sintra station. S/C design life = 5 years. Onboard data recording capability of 32 MByte into solid-state memory.

Orbit: Sun-synchronous, altitude = 820 km, inclination = 98°, period = 100 minutes.

Sensors:

**EIS** = Earth Imaging System. Two CCD imaging cameras with two optical assemblies. The cameras are aimed in the same direction but offer a different FOV. The wide angle image has a ground resolution of 2 km and a FOV of 1000 km (NIR 800-900 nm filter; 4.8 mm focal length). The narrow angle camera has a ground resolution of 200 m and a FOV (swath width) of about 100 km (Red 600-700 nm filter, 50 mm focal length). The cameras feature a monochromatic design with optical filters chosen to contrast ground properties such as soil moisture content. The image is read form each CCD by a Transputer Data Processing Ex-

274)  Information provided by J. Radbone of SSTL, University of Surrey, UK
275)  'First PoSAT images', Space, Vol. 9, Nr. 9, December 1993, p. 6

periment (TDPE) and after processing, the image is being provided to the On-Board Computer (OBC) and recorded. Camera sensor (both cameras): EEV CCD04-06 image sensor + chip set; 578 x 576 array (interleaved fields); anti-blooming & electronic integration control. Operational Earth imaging is performed on a scheduled snapshot basis.

**Star Sensor** (attitude control). The sensor is based on the EIS camera but with suitable optics for imaging the faint light from stars. The FOV is 15° (in the Y-direction of the S/C). The Star Sensor image is analyzed by the TDPE and measurement data returned to the OBC.

**GPS Receiver** (based on the Trimble TRANS-II receiver). GPS data provides on-board position, velocity and time reference. The data is used by OBC to generate an orbital element set and to provide scheduling and synchronization to other on-board computers, and to allow ground stations equipped with a GPS receiver to experiment with applications for real-time DGPS.

**CRE** = Cosmic Ray Experiment. CRE monitor the space radiation environment and its effects on the S/C semiconductor electronics. CRE contains a PIN diode of 900 mm$^2$ and a multi-channel analyzer capable of detecting energetic particles (5-600 MeV) with a wide range of Linear Energy Transfer. The large number of channels and the fast response time of the sensor allow the CRE to build up the spectrum of observed energies of particles.

**TDE** = Total Dose Experiment. TDE is part of CRE monitoring the total accumulated ionizing dose in a solid-state RADFET dosimeter. The larger memory devices in the computers are regularly 'washed' to detect and log Single Event Upset (SEU) information.

**DSPE** = Digital Signal Processing Experiment. The DSPE consists of 2 Texas Instruments processors (C25 and C30 from TM320 series). The DSPE can be used as a programmable communications modem to modulate downlink data from, or uplink data for the OBC, thus enabling experiments with new modulation techniques to be optimized for LEO satellite mobile communications.


## A.81    PRARE Tracking System

PRARE (Precise Range and Range Rate Equipment)[276],[277] is a microwave ranging system operating at centimeter accuracy levels for the measurement of satellite to ground range and range rate. The concept is based on an autonomous spaceborne two-way, dual frequency microwave tracking system with its own telemetry, telecommand, data storage, timing and data transmission capability.

PRARE is flown on missions with requirements for high-precision orbit determination. Such missions are: ERS series, GPS series, Polar Platform, etc.

PRARE consists of three major components:

1. Space Segment
   This is a small self-contained hardware unit with a mass of 17 kg, a power consumption of 39 VA (operational), 8 VA in standby mode, and host satellite independent communication links.

2. Ground Station Network
   The network consists of small automated and transportable ground stations who's positions are known. The currently (1991) installed network consists of 25 stations positioned around the globe.

276) "The Precise Range and Range Rate Equipment PRARE: Status Report on System Development, Preparations for ERS-1 and Future Plans", Submitted by F. Flechtner, K. Kaniuth, Ch. Reigber, H. Wilmes of DGFI, Second International Symposium on Precise Positioning with the Global Positioning System (GPS '90), Sept. '90, Ottawa
277) P. Hartl, C. Reigber "Das PRARE-System der ERS-1 Mission", Die Geowissenschaften, 9. Jahrgang, Heft 4-5, April-Mai 1991, pp. 156-162.

3.  Control Segment (in the case of ERS-1 it consists of:)
-   a command station
-   a master station (with a time control unit and a preprocessing computer in connection with a user center, e.g. DFD). The master station may also command the space segment.
-   a station for calibration, and an additional dumping station

**PRARE Measurement Principle**

Two continuous signals are emitted from the space segment to the ground, one of which is in S-Band (2.2 GHz), the other in the X-Band (8.5 GHz) frequency range. Both signals are modulated with a PN-code (pseudo random noise, 1 MChips/s for the S-Band and 10 MChips/s for the X-Band) used for the distance measurement and containing data signals (broadcast information) for the ground station operation (prediction of orbit visibility, etc.). The time delay in the reception of the two simultaneously emitted signals is measured at the ground stations with an accuracy of better than 1 ns, the result is retransmitted and stored in the onboard memory of PRARE for later ionospheric correction of the data.

In the ground station the received X-Band signal is transposed to 7.2 GHz, coherently modulated with the regenerated PN code (or with one of three orthogonal copies for code multiplexing) including the two-frequency delay data, as well as housekeeping and meteorological ground data, and is retransmitted to the PRARE space segment.

**PRARE Control Segment**

The PRARE control segment for ERS-1 consists of a command station in Stuttgart, a master station at DLR/DFD in Oberpfaffenhofen, a calibration station in Wettzell, and an additional dumping station in Tromsoe, Norway.

*   Command Station. A command station is required for the maiden flight of PRARE on ERS-1 in order to control eventualities and contingency situations. The functions are:
    -   Monitoring and commanding of the PRARE space segment
    -   Simulation capabilities
    -   Has the ability to establish spacecraft contact in contingency situations during visibility periods, even without receiving a downlink signal.

*   Master Station. Functions as a central receiving and preprocessing station for all PRARE tracking data from the global network.
    -   Planning and scheduling of all tracking activities
    -   Update of orbit prediction elements (Uplink)
    -   Referencing of the PRARE space clock time to the UTC time standard.
    -   Processing of the PRARE tracking data and comparison with predicted data.

*   Calibration Station. Provides the capability to calibrate PRARE inflight with a precise ranging standard (Laser station).

*   Dumping Station. The large number of tracking stations generates so much environmental and other data so that onboard recording could be a problem (with 80 kByte storage capacity). Hence, an additional high-latitude dumping station (for about ten data dumps per day) was selected parallel to the master station in Oberpfaffenhofen in order to improve the downlink capability.

**PRARE Ground Stations**

A PRARE ground station is mobile and consists of the following components:
    -   an antenna unit with an offset antenna of 60 cm diameter, fronted electronics and tracking system
    -   an electronics unit including RF-modules, station processor and power supply
    -   a monitor and test computer (PC) as user interface.

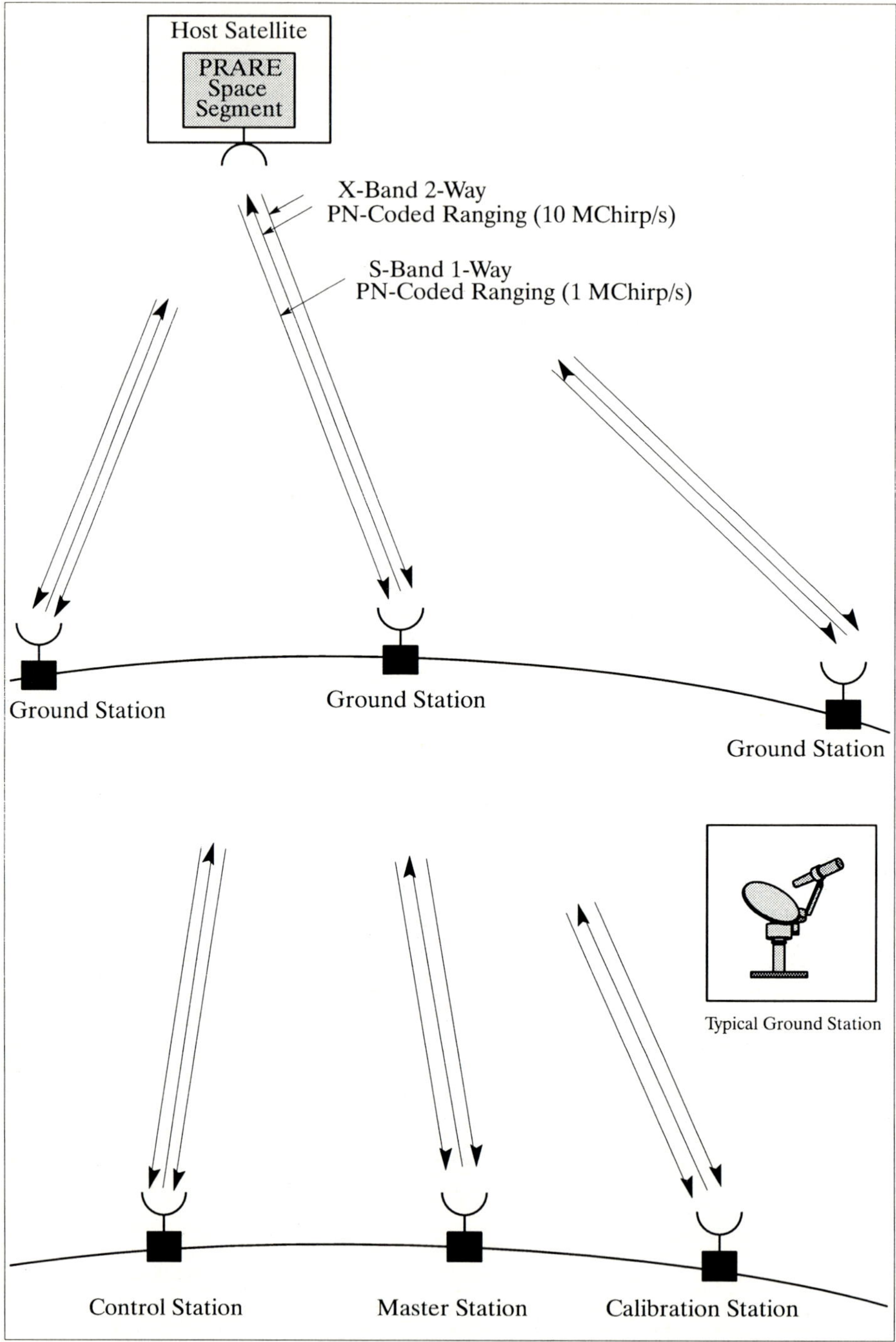



**Figure 76:      PRARE Measurement Principle with Space- and Ground Segment**

**Functional Extensions for PRARE ♦PRAREE**

Further extensions of PRARE functions are considered after a successful flight on ERS-1. The prime feature of future flights could be the real-time onboard availability of the actual measured orbit. This requires more onboard processing. It would allow a source referencing of the science data with the actual orbit.

PRAREE[278] is an extended version of ERS-1,2 PRARE. PRAREE is an autonomous two-way, 3 microwave frequency (Ku, S and UHF bands) tracking system with its own telemetry, telecommand, data storage, timing and data transmission capability, which is complemented by two uplink laser channels in different colors, modulated by ranging codes. The optical channels are used for the analysis and correction of atmospheric propagation effects.

# A.82   PRIRODA-1

PRIRODA-1[279],[280],[281] (=Nature) is a  multisensor research module series of the Soviet Union's orbital station MIR for the observation of the environment (the former Institut für Kosmosforschung = IKF, Berlin, was a partner and instrument provider in this mission, now DLR). Planned international PRIRODA-1 Mission: Launch in early 1995. Nominal mission life = 2-3 years.

Application: Geoecology (radiophysical methods for the study of the environment, infrared-spectroradiometric methods for the study of spatial IR-radiation of the atmosphere, spectroradiometric methods for the study of reflected solar irradiation (VNIR) in particular in the region 'Atmosphere - Earth surface'), Oceans, Hydrology, Meteorology, Geology, etc..

An extensive science program goes along with the Priroda mission. The basic mission objectives are of a research nature in the following fields:[282]

- Land surface exploration

  - Snow cover
  - Investigations of soils
  - Vegetation
  - Earth surface mapping at different wavelength ranges for solid Earth research

- Ocean exploration

  - Global Earth survey by passive sensing techniques
  - Sea surface temperature determination
  - Wind speeds and direction
  - Surface roughness due to the wind
  - Ocean color characteristics
  - Ocean bioproductivity investigations
  - Ice cover determination

- Atmosphere investigations

  - Large-scale atmospheric processes
  - Atmosphere - ocean interactions
  - Lower stratosphere and troposphere physical parameters


278) Programme Proposal for the Development and Exploitation of the First Polar Orbit Earth-Observation Mission (POEM-1) using the Polar Platform, ESA/POEM 1, Issue 1, 28.10.91 Part 1, Issue 1, 30.10.91 Part 2, An. 5, p. 2

279) "PRIRODA", Ein Forschungsmodul der sowjetischen Orbitalstation MIR zur Fernerkundung der Erde, Wissenschaftliche Nutzlast Technische Beschreibung, Institut für Kosmosforschung (IKF), Berlin, 1990

280) "PRIRODA-Experimente", Programm zur Beschaffung, Verarbeitung, Bewertung und Anwendung von Daten des Multisensorsystems PRIRODA der sowjetischen Orbitalstation MIR, 1992 -94, DARA, Berlin, Mai 1991

281) "Complex for Remote Sensing of the Earth", Science Program, DLR paper 1991

282) Orbital Station MIR, Complex of Remote Sensing of the Earth "PRIRODA", Scientific Program, Brochure of IRE, Moscow, 1991

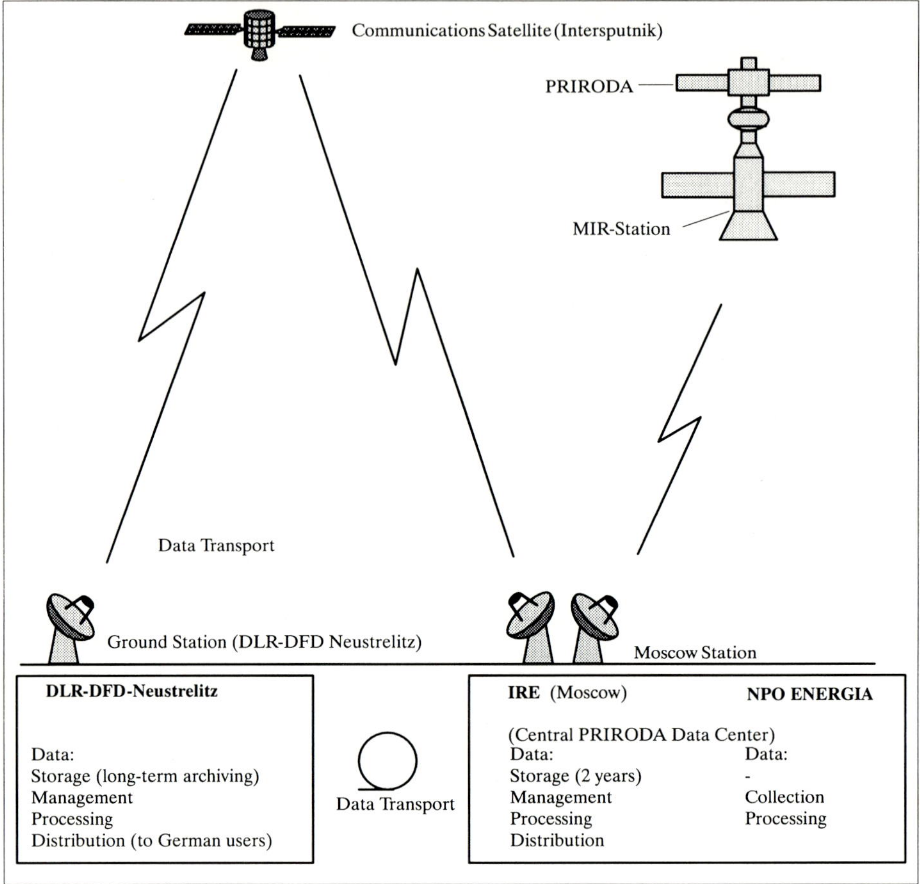


**Figure 77:      PRIRODA-1 Data Distribution in the Ground Segment**

- Ozone experiment
- Ecological investigations

The great variety of observation techniques by the various instruments provides synergies on many levels in the evaluation and interpretation of the data (for instance: the combination of SAR and radar altimeter data with MW radiometer data in ocean research).

Orbit (same as that of MIR): Altitude = 300 to 400 km, 51.6° inclination

**Sensors:**

1. **PRIRODA passive microwave instruments, consisting of the following instrument blocks: IKAR-N, IKAR-D, IKAR-P, and a scanning radiometer[283].**

- **IKAR-N:** - 5 microwave radiometers (R-30, R-80, P-135, R-225P, and RP-600). IKAR-N (Control and data collection module). Wavelengths of 0.3 cm, 0.8 cm, 1.35 cm, 2.25 cm, and 6.0 cm in nadir direction. R-225P can also measure different polarizations. Objective: Measurement of microwave radiation (emission) at the atmosphere/sea surface interface.

283)  G. Zimmermann,"Mission PRIRODA", German Proposals to Scientific Program, DARA Bulletin, Dec. 1991

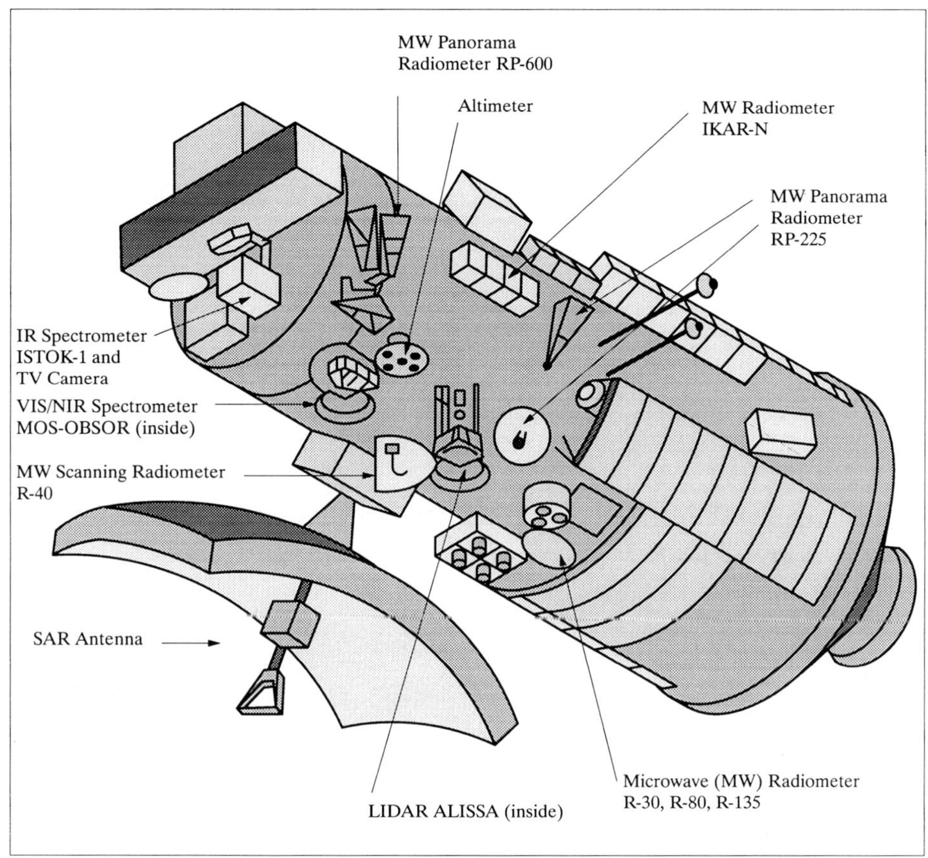


**Figure 78:**     **Model of the Priroda Module on the MIR Station**

- **IKAR-D** (DELTA): - a 3 channel scanning radiometer at wavelength 0.3, 0.8, and 1.35 cm and a look angle of 40° from nadir

- **IKAR-P**: - (Panorama) 3-channel scanning microwave radiometer (RP-225) at wavelength of 2.25 cm, and a 5-channel panorama radiometer RP-600 at 6.0 cm. Both with a look angle of 40° against nadir.
  Objective: Measurement of ocean surface and temperature parameters.

- **R-400** - a one-channel polarization scanner microwave radiometer. Wavelength of 4.o cm (7.5 GHz). Look angle also 40° against nadir.
  Objective: Measurement of the thermal radiation of the Earth surface (radio wave range).

## 2. PRIRODA active MW-Equipment, consisting of a SAR and a radar altimeter

- **Travers** (RSA) = Sideview radar system with synthetic aperture (SAR, measurement in two frequencies). Russian sensor, built by MEI (Moscow Power Engineering Institute).
  Objective: generation of radar images of the Earth's surface (land and ocean).

  - Wavelengths of scanning beam: 9.2 cm (3.28 GHz, S-Band), and 23 cm (1.28 GHz, L-Band).

| Parameter | S-Band SAR | L-Band SAR |
|---|---|---|
| Frequency | 3.28 GHz | 1.28 GHz |
| Wavelength | 9.2 cm | 23 cm |
| Bandwidth | 5 MHz | |
| Peak power | 300 W | 1000 W |
| Impulse power | 250 W | 600 W |
| Compressed pulse width | 0.4 ms (effective pulse), 25 ms (total duration) | |
| Pulse repetition frequency | 3 kHz (depends on the swath range) | |
| Antenna gain | 37 dB | 33 dB |
| Antenna size | 2.8 m x 6 m | |
| Polarization | VV or HH | |
| Quantization | complex 4 bit | |
| Swath width (H=400 km) | 50 km | |
| Resolution in azimuth:<br>Resolution in azimuth:<br>Range resolution: | 150 m (for satellite processing)<br>20 m (for ground processing)<br>100 m | |
| Look angle | 35° | |
| Data rate | 16 Mbit/s (ground processing) | 16 Mbit/s (ground processing) |
| Data rate | 2.56 Mbit/s (onboard processing) | 2.56 Mbit/s (onboard processing) |
| Orbit repetition cycle | 6 days (coverage) | |
| Operation mode | L- and S-Band operate in parallel | |

**Table 59:**     **Specification of SAR Travers Parameters**

- **Greben = Precision radar altimeter** (PRV). Russian sensor (built by MEI).
  Objective: precision measurements of oceanic altitudes, determination of wave heights, etc.

  - Altitude region:     300 - 400 km
  - Spatial resolution at ground     2.5 km across track; 9.75 km along track
  - Field of view (FOV), 350 km     2.5 km
  - Antenna nadir pointing     0°
  - Data rate (output)     8 kbit/s

  Greben provides a range resolution of 10 cm, observation frequency = 13.76 GHz; the uncompressed pulse width employs double modulation (frequency modulation + Sherman code), the width one pulse sweep is 2000 µs, while the uncompressed pulse width is 1.7 µs. Beam limited footprint = 13 km (on 3 dB level).

3. **PRIRODA optical equipment, consisting of ISTOK-1, MOS-A, MOS-B, MSU-SK, MSU-E, TV camera, Ozone-M, and a lidar ALISSA.**

- **ISTOK-1** = IR Spectroradiometric System (sounder)
  The apparatus of ISTOK-1 consists of the following modules:

  - the opto-mechanical block (OMB),
  - the electronic block of the IR-spectrometer system (EBS),
  - the data processing block (digital processing of video data),
  - the computer block (built in Rumania)

  Objective: Measurement of atmospheric irradiation for different observation angles.

  - Vertical temperature profiles of the atmosphere (up to 30 km with an accuracy of 4 - 5 K) and its $CO_2$ band radiation.
  - Altitude profile of pressure and temperature (with 0.2 km resolution).
  - Humidity profiles from the measurement of the water vapor band (6.3 µm)
  - Measurement of ozone content in the band 9.6 µm
  - Measurement of the ocean surface temperature
  - Atmospheric chemistry
  - Study of convective processes in the atmosphere

The IR-spectrometer consists of two identical diffraction polychromatographs for the spectral ranges of 4 - 8 µm, and 8 - 16 µm.

- Number of spectral channels:    64
- Spectral resolution (band width): 0.125 µm (4-8 µm region), 0.025 µm (8 - 16 µm region)
- Sensitivity: = 1 x 10 (W/cm/sr/m) for range 1; = 5 x 10 (W/cm/sr/m) for range 2
- Radiometric resolution: 14 bit
- Spatial resolution    0.75 - 3 km (in flight direction)
- Look angles: -15° to 132° in orbit plane; 0° to 90° in azimuth plane
- Maximum rotation rate of instrument: 2.4 degrees/s
- Data rate: 10 kbit/s


- **MSU-SK** = Optical Multispectral Scanner - moderate resolution. (heritage: from Meteor and Cosmos 1639, the sensor is being built by the Institute of Space Device Engineering, ISDE). MSU-SK measures in 4 channels of the visible spectrum: (0.5 to 0.6 µm, 0.6 to 0.7 µm, 0.7 to 0.8 µm, and 0.8 to 1.1 µm) with a resolution of about 120 m; MSU-SK measures also in the infrared spectrum (10.4 to 12.6 µm, TIR) with a resolution of about 400 m. The swath width is 320 km (H=400 km).

  - Look angle: 39°
  - SNR for channel 1 = 70
  - SNR for channels 2-5 = 100
  - Noise level for channel 5 = 0.5 K
  - Range of radio temperatures for channel 5 = 210 - 320 K (TIR)
  - Data rate for channels 1-4 = 2.56 Mbit/s
  - Data rate for channel 5 = 1.28 Mbit/s

- **MSU-E** = Optical Multispectral Scanner (electronic scanning, CCD technology, built by ISDE). Measures in three visible channels (0.5 to 0.6 µm, 0.6 to 0.7 µm, and 0.8 to 0.9 µm); the resolution is 20 m; the average swath width is 2 x 27 km (2 devices measure in parallel, one to each side of the ground track).

  - Look angles: ± 32° (the instrument may be pointed in the cross-range direction thereby extending the swath range)
  - Radiometric resolution: 8 bit
  - Data rate for each channel: 3.84 Mbit/s

Both scanners (MSU-SK and MSU-E) are operationally coupled (for instance: two channels may be switched on MSU-SK plus 3 channels on MSU-E).


- **MOS-OBSOR** = Multispectral Optoelectronic Scanner. An imaging spectrometer for the visible and near-infrared spectrum (VNIR). MOS is provided by DLR, Berlin. Objective: Image generation of the Earth surface (surface - atmosphere interaction, ocean color, phytoplankton, regional and global distributions of man-made aerosols and its links to gaseous admixtures, spectral and spatial cloudiness characteristics, etc.) in the VNIR region of 0.4 - 1.01 µm.
  The sensor apparatus consists of two complementary instruments: (see also Figure 80).


- **TV Camera**
  Objective: surveying the region of measurement with 256 x 256 pixels of about 900 x 900 m edge. Spectral ranges: 0.4 - 0.75 µm, look angle = 15°, spatial resolution = 300 m.


- **Ozon-M** = Sounder apparatus for the measurement of the ozone concentration profile and other trace gas elements in the atmosphere. (Russian sensor, Organization: INTEGRAL)

| Parameter | MOS-A | MOS-B |
|---|---|---|
| Spectral range | 755 - 768 nm | 400 - 1010 nm |
| Nr. of channels | 4 | 13 |
| Spectral channels | 756.7, 760.6, 763.5, 766.4 nm | 408, 443, 485, 520, 570, 615, 650, 685, 750, 815, 870, 945, 1010 nm |
| Spectral resolution | 1.4 nm | 10 nm |
| Swath width (H = 400 km) | 80 km | 82 km |
| Spatial resolution | 2.8 km | 0.7 km |
| Quantization | 12 bit | 12 bit |
| Measurement range | 0.1 - 25 W/cm$^2$/ nm/sr | 0.2 - 25 W/cm$^2$/ nm/sr |
| Data rate | 4 kbit/s | 210 kbit/s |

**Table 60:	Specification of some MOS-A and MOS-B Parameters**

The apparatus is a 4-channel diffraction-scanner spectrometer. An autonomous graded shading mechanism permits the direct measurement of solar radiation. Objective: Measurement of the solar radiation; spectral transparency of the atmosphere; determination of ozone profiles.

- Total number of spectral channels (bands): 188
- Spectral range:	257 - 330 nm	Band width:	0.3 nm
-	360 - 462 nm		0.5 nm
-	600 - 770 nm		0.7 nm
-	900 - 1155 nm		1.0 nm
- Measurement range in altitude: 5 - 70 km
- Spatial resolution of 1 km in altitude
- Scanning rate: 1/s
- Range of sun tracker look angles:	$\pm 65°$ (in azimuth plane)
	-25° to 5° (in vertical plane)
- Data rate: 320 kbit/s

- **ALISSA**[284] = l'Atmosphere par LIdar Sur SAliout (French sensor. Note: Alissa did not fly on Salyut). Alissa is conceived as a simplified fixed frequency lidar, using Mie scattering to detect clouds and aerosols, the primary objective is to study the impact of altitude determination on the description of the cloud field provided by geostationary satellites. Operations are considered for night-time to limit the difficulty of a more accurate alignment of the system. The main characteristics of the sensor are as follows:

Emitter:

- Second harmonic of Nd-Yag lasers: $\lambda = 532$ mm
- Energy pulse: 40 mJ
- Repetition rate: 50 Hz
- Natural divergence: $10^{-3}$ rad
- Divergence after collimator: $10^{-4}$ rad

Receiver:

- Cassegrain telescope of area A = 0.12 m$^2$
- Field of view: $10^{-3}$ rad
- Bandwidth of the filter: 0.5 nm

Energy requirements: 3 kW

The analysis of the signal is made by a pulse counting system with a gate of 1 µs which provides a height resolution of: $\Delta_z = 150$ m.

The integration of 6 consecutive pulses (to increase the SNR) gives a horizontal resolution along the satellite track of: $\Delta_{x,y} = 1$ km

---

284) M. L. Chanin, M. Desbois, A. Hauchecorne, 'ALISSA a French Russian cooperation in the PRIRODA mission. Paper of CNRS - Service d'Aeronomie

The pulse counting system has 512 channels of 1 µs which can be adjusted at 2, 4, 8 µs.

| Energy $E_o$<br>$T^2$ (window) | 24 mJ<br>83% | 40 mJ<br>83% | 24 mJ<br>40% | 40 mJ<br>40% |
|---|---|---|---|---|
| Cumulus | 200 | 340 | 50 | 80 |
| Cirrus ($\tau = 0.2$) | 20 | 30 | 4 | 7 |
| Boundary Layer | 5 | 8 | 1 | 2 |
| Stratospheric aerosol ($R = 2$) | 1 | 2 | 0.2 | 0.4 |

**Table 61:**  **Nr. of Photoelectrons expected for 6 pulses**

Alissa science objectives:
-   The cloud radiation problem; improvement of large scale cloud cover and cloud properties retrieval.
-   The cloud radiation problem; interpretation of small scale structures of the cloud tops
-   Measurement of tropospheric aerosols and the boundary layer.
-   Stratospheric studies

## 4.  PRIRODA  future Sensor (MOMS-02P)

**MOMS-02**[285],[286]= Modular Optoelectronic Multispectral Scanner (see A.70 for definition of MOMS-02).

Current (1993) plans call for the MOMS-02 payload to be flown on PRIRODA-1 in 1995 (after its assignment on the D-2 Shuttle mission in the spring of 1993). The idea is to take the MOMS-02 payload along with a regular MIR re-supply flight (in 1995) to be added to the PRIRODA-1 module (which is at that time already docked to MIR and operational). MOMS-02 on Priroda will in addition be outfitted with an integrated GPS subsystem (for more accurate orbit determination, hence ground positioning of the data).
-   Swath width (H = 400 km) = 105 km for MS and ST channels; = 50 km for HR channels.
-   Ground pixel size = 18 m for MS and ST channels; = 6 m for HR channels
-   Data rate = 84 Mbit/s

| Sensor<br>Type | IR - Spec-<br>trometer<br>ISTOK-1 | Imaging Spectrometer | | VIS Scanners | | TV<br>Camera | Lidar<br>ALISSA |
|---|---|---|---|---|---|---|---|
| | | MOS-A | MOS-B | MUS-SK | MSU-E | | |
| Wave-<br>length | 4-16 µm | 755-768<br>nm | 408-1010<br>nm | 0.5-1.1 µm<br>8-12.5 µm | 0.5-0.9 µm | 0.4 - 0.7<br>µm | 527 nm |
| Nr. of<br>channels | 64 | 4 | 13 | 5 | 3 | 1 | 1 |
| Geo. Reso-<br>lution (km) | 1 x 6 | 2.8x2.8 | 0.7x0.65 | 0.12x0.12<br>IR:0.3x0.3 | 23x25 m | 0.3x0.3 | 1.0 vertical<br>0.15 horiz. |
| Swath (km) | 6 | 83 | 83 | 350 | 2x27 | 90 | 3' |
| Spectral<br>Resolution | .125 (4-8µ )<br>.25 (8-16µ | 1.4 nm | 10nm | 0.1µm | 0.1µm | | |
| Look<br>Angle | 0-90° | Nadir | Nadir | 39° | Nadir | | |

**Table 62:**  **Overview of PRIRODA Optical Instruments**

285)  German User Requirements to PRIRODA Mission, Annex 1 of Protocol to MOMS-02 for the PRIRODA Mission, DLR paper of PRIRODA Workshop, May 1991
286)  Protocol of the Meeting of Specialists of USSR and Germany on MOMS-02 for the PRIRODA Mission. DLR paper, May 1991

| Sensor Type | Passive MW-Radiometers | | | | Active MW Sensors | |
|---|---|---|---|---|---|---|
| | IKAR-N | IKAR-D | IKAR-P | R-400 | SAR Travers | Altimeter Greben |
| Wavelength (cm) Frequency (GHz) | cm - GHz 0.3 - 100 0.8 - 37.5 1.35 - 22.22 2.25 - 13.76 6.0 - 5.0 | cm - GHz 0.3 - 100 0.8 - 37.5 1.35 - 22.22 | cm - GHz 2.25 - 13.76 6.0 - 5.0 | cm - GHz 4.0 - 7.5 | cm - GHz 9.2 - 3.28 23.0 - 1.30 | cm - GHz 2.25 - 13.76 |
| Sensitivity (K) | 0.15 | 0.4-1.5 | 0.15 | | | |
| FOV (km) | 60 | 400 | 750 | 400 | 70-100 | 2 |
| Geo. Resolution (km) | 60 | 5-15 | 50-75 | 50 | 0.1-0.15 | ± 10 cm |
| Look Angle | 0° | 40° | 40° | 40° | 30-40° | |

**Table 63:**     **Overview of PRIRODA Microwave Instruments**

## Data Distribution:

PRIRODA sensor data are transmitted from MIR to CIS ground stations NPO ENERGIA and/or to IRE at Moscow (IRE is the Space Management Organization of the USSR Academy of Sciences). IRE is designated as the prime archive of all **PRIRODA data** (note: the IRE archives are not regarded as long-term). IRE performs preprocessing of all sensor data and provides all required ancillary data. The time delay for preprocessing is currently projected for about 1 week after data reception.

Agreements between IRE and its partners (the sensor providers and participating countries of the PRIRODA mission) consider data sharing, they have primary access to all science data and ancillary information. The sensor data of MOS-OBSOR and of 'SAR Travers' is being processed and archived by DLR-IKF Berlin as the responsible data center for all German user access. All thematic processing may be done at the user sites (SAR Travers data calibration is being planned at DLR in Oberpfaffenhofen). The data transfer from Moscow to Germany may be provided via suitable transportable storage media (HDDT and/or optical disks) and/or via satellite communication.

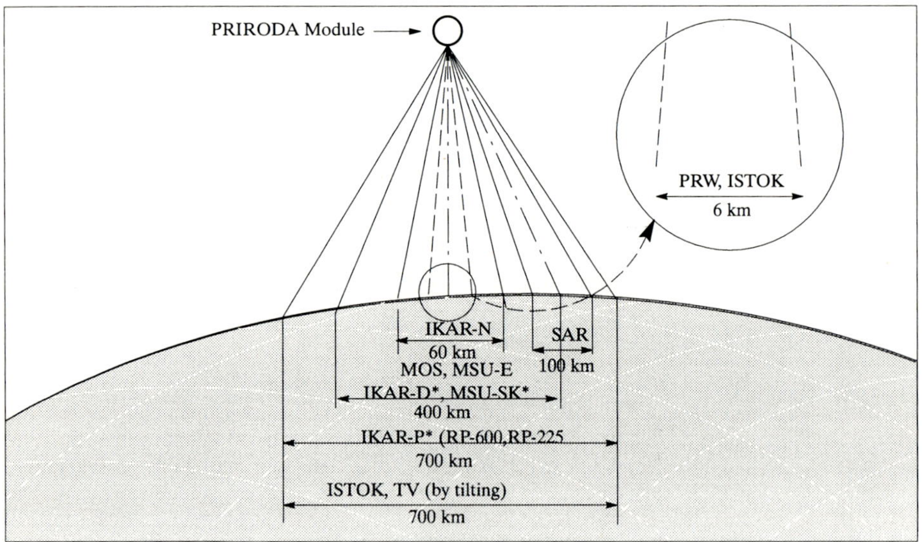


**Figure 79:**     **Scheme of Overlapping FOV's of PRIRODA Sensors[287]**

287)  * IKAR-D, -P and MSU-SK with forward look angle (in flight direction) of 40° against nadir

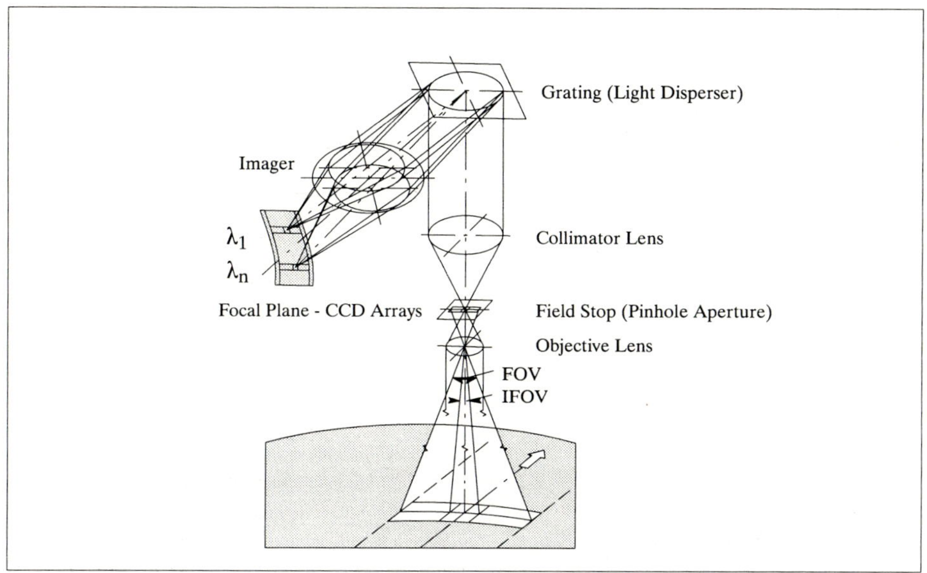


**Figure 80:     Principle of the MOS-Obsor Electromechanical Imaging Spectrometer**

## A.83    RADARSAT

RADARSAT is a Canadian (CSA/CCRS = Canadian Space Agency / Canada Center for Remote Sensing) Earth observation satellite with SAR equipment. Planned launch: January 1995 (NASA is a major partner in the mission providing launch services, Delta II vehicle from Vandenberg AFB, NASA will receive RADARSAT data at ASF, Fairbanks Alaska)[288]. Five year design life of satellite.

Application: Observation of the arctic ice movement (navigation), agriculture and forestry, topography, oceanography, management of fisheries, etc. .

Orbit: Near circular sun-synchronous dawn-dusk orbit at an altitude of 798 km; inclination = 98.6°, ascending node at 18:00 (6 PM equatorial crossing); period = 101 min; repeat cycle = 24 days (subcycles of 7 and 17 days); the dawn-dusk Orbit conveys the following advantages:

-    the platform will be in near continuous solar illumination, providing equivalent opportunities for image acquisition during ascending and descending passes (energy consumption).
-    a late equatorial crossing time will reduce reception conflicts with other remote sensing satellites

**Sensor:**

**SAR** (Multi-Mode instrument; observation in C-Band (5.3 GHz, 5.6 cm wavelength), transmitting and receiving horizontally polarized radiation. Choice of 3 transmit pulses and selection of numerous beams permits a wide range of swath widths, incidence angles and image resolutions. The SAR can operate at a 'high duty cycle' capable of imaging for up to 28 minutes per orbit (orbital period of 101 minutes).

288)  R. K. Raney, A.P. Luscombe, E.J. Langham, S. Ahmed "Radarsat', reprint from Proceedings of the IEEE, Vol. 79, Nr. 6, June 1991

Special SAR design features: calibration, rapid data processing, the phased array antenna to provide controlled beam steering, and the first satellite implementation of a radar technique known as ScanSAR.

ScanSAR is a technique permitting extended observation coverage (wider swath) on command. In this mode, rapid steering of the elevation beam pattern of the antenna is essential. Extended range coverage can be obtained by using a set of contiguous beams, enabling images to swath widths of up to 500 km. This is accomplished at no increase in mean data rate from the sensor, but at the cost of degraded resolution of the resulting image.
The principle of ScanSAR is to share radar operational time between two or more separate subswaths in such a way, as to obtain full image coverage of each.

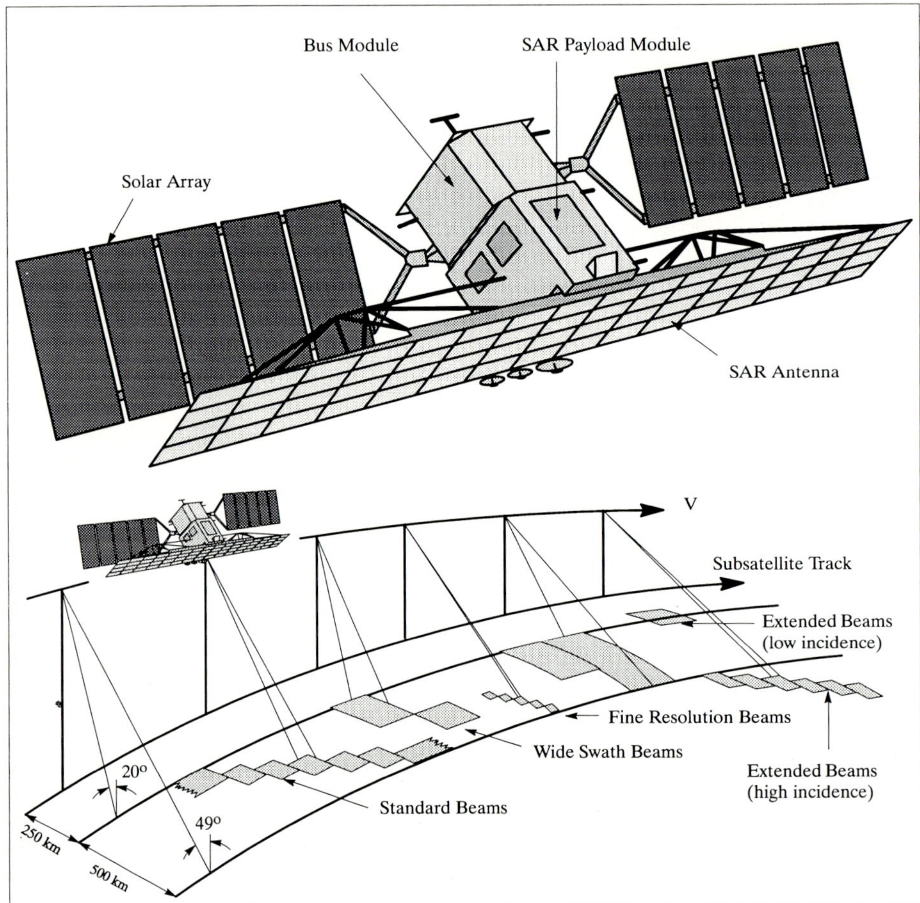


**Figure 81:     The Radarsat S/C Model and Illustration of Observation Geometries**

## Data:

Onboard recording capability of the total stream for up to 10 minutes (satellite outside of receiving stations).
Ground stations: Prince Albert (Saskatchewan), Gatineau (Quebec), Fairbanks (Alaska).
Data distribution by Radarsat International Ltd. (RSI) of Ottawa.

| Mode | Resolution (Range x Azimuth, m)[*] | Looks[**] | Swath Width (km) | Incidence Angle (Degree) |
|---|---|---|---|---|
| Standard | 25 x 28 | 4 | 100 | 20 - 49 |
| Wide (1) | 48-30 x 28 | 4 | 165 | 20 - 31 |
| Wide (2) | 32-45 x 28 | 4 | 150 | 31 - 39 |
| Fine Resolution | 11-9 x 9 | 1 | 45 | 37 - 48 |
| ScanSAR (N) | 50 x 50 | 2-4 | 305 | 20 - 40 |
| ScanSAR (W) | 100 x 100 | 4-8 | 510 | 20 - 49 |
| Extended (H) | 22-19 x 28 | 4 | 75 | 50 - 60 |
| Extended (L) | 63-28 x 28 | 4 | 170 | 10 - 23 |

**Table 64:      RADARSAT Imaging Modes[289]**


# A.84    RESURS-F

The Russian/CIS Resurs-F[290] (Resource-F, F1 and F2) satellite series is a photoreconnaissance S/C with short mission life times, in the order of 14 to 21 days. The onboard sensors are film cameras whose data (namely the films) are recaptured after the end of each mission (Vostok-based film-return craft, the film camera systems are returned in small spherical descent capsules, which are reused an average of three times)[291]. The flight tests for the F2 series were completed in 1989. The altitude of the orbit is changed during the mission from 400 km to 170 km. A 'Resurs-F' Satellite was launched for instance on May 21, 1991 by a Soyuz launch vehicle.[292]

The Resurs F1 S/C was space-tested in 1975-80, the Resurs F2 S/C in the period of 1989-90. As of 1993 there are hundreds of flown F1 and F2 missions. The F1 flies missions typically of two weeks duration (up to 11 days in standby mode) with 3 KFA-200 and 2 KFA-1000 cameras onboard. F2 undertakes 30 day missions in lower orbits using the MK-4 camera apparatus.

The data products of the Resource-F series are commercially distributed by Sojuzkarta since Oct. 1990 (in Germany by GAF = Gesellschaft für Angewandte Fernerkundung, in the US through Central Trading Systems, Fort Worth, Texas).

Objective: Provision of high-resolution photographic images of the Earth's surface. Monitoring of natural resources, ecology studies, etc.

Orbit:
F1: Non-Sun-synchronous polar orbit. Altitude = 275 km, inclination between 82.3°, coverage between adjacent orbits: once between the latitudes of 30 - 80°.
F2: Non-Sun-synchronous polar orbit. Altitude = 240 km, inclination between 82.3°, coverage between adjacent orbits: once between the latitudes of 0 -83° with 2/3 overlapping.

**Sensors:**

- **KFA-200** Camera System. Focal length = 200 mm, swath width = 0.9 altitude (245 km), duty cycle = 12%. KFA-200 provides images in three ranges: 0.5 - 0.6 µm, 0.6 - 0.7 µm, and 0.7 - 0.9 µm. The cameras provide synchronous images of the Earth surface in an image format size of 180 x 180 mm to a scale of $1 : 10^6$. The spatial resolution is between 25 and 30 m. These images provide very suitable correlation data for color synthesis in thematic data products. The geometric properties of the images permit a photogrammetric analysis. The precision of the altitude determination (method of image triangulation) is in the order of 50 m.

---


[*]     Nominal; range and processor dependent
[**]    Nominal; ground range resolution varies with range
290)    'Sowjetisches kosmisches System zum Studium der Naturschätze der Erde und zur Umweltkontrolle - der heutige Stand und die Perspektiven für den Zeitraum 1991 -1995', the paper is a translation of a presentation given by L. Dessinow from the USSR Academy of Sciences in 1989.
291)    Interavia Space Directory 1990-91, p. 436
292)    'Soviets Launch New Resurs-F Satellite', Space News, May 27 - June 2, 1991, p. 26

- **KFA-1000** Camera System. Focal length = 1000 mm, swath width = 0.3 altitude (82 km for one camera), duty cycle = 12%. KFA-1000 provides spectro-zonal and panchromatic images. The concept employs a fan-like mounting configuration of two cameras (with a large focal length and wide format) in order to achieve an extra wide swath wide of 120 to 165 km. The film width is 300 mm. The scale of the images (focal length of 1000 mm) is 1 : 200 000 to 1 : 270 000, the resolution on ground (terrain) is 5 to 10 m. The total number of images of one film is about 1800. Spatial resolution = 8 - 10 m.

- **MK-4** = 4-Channel Camera System. Apparatus with four photographic lenses. Image frame = 180 x 180 mm, focal length = 300 mm, swath width = 0.6 altitude (144 km), duty cycle up to 19%, spatial resolution = 12-14 m.
  Provision of multispectral data in 3 channels and spectro-zonal image data (colored) in 1 channel. The operational imaging program allows the selection of 4 zones (channels) out of six available.

The apparatus operation is based on photometric concepts. This assures images with high-quality geometric references. The sum average distortion of coordinate points in an image is max. ± 9 µm.

| Parameter | Nr. of channels | Spectral Bands | Filter | Film | Film Length |
|---|---|---|---|---|---|
| KFA-200 (3) | Camera 1 | 510 - 600 nm | ZHS-18+SZS-23 | T-42 | 230-250 m |
| | Camera 2 | 700 - 840 nm | KS-19 | I-840K | 230-240 m |
| | Camera 3 | 600 - 700 nm | KS-10 | T-38 | 230-250 m |
| KFA-1000 (2) | 1 (right side) | 570 - 800 nm | OS-14 | SN-10 | 580 m |
| | 2 (left side) | 570 - 800 nm | OS-14 | SN-10 | 580 m |
| MK-4 (4 chan- | Channel 1 | 640 - 690 nm | KS-13 | T-30M | 500 m |
| nels are used | Channel 2 | 810 - 860 nm | KS-17 | I-840K | 500 m |
| of 6 possible | Channel 3 | 515 - 565 nm | interference filter | T-30M | 500 m |
| channels) | Channel 4 | 460 - 510 nm | interference filter | T-30M | 500 m |
| | Channel 5 | 610 - 750 nm | OS-14 | CN-10 | 500 m |
| | Channel 6 | 435 - 680 nm | ZHS-11 | TsN-4 | 250 m |

**Table 65:**    **Photocamera Characteristics of the Resurs-F S/C Series**[293]

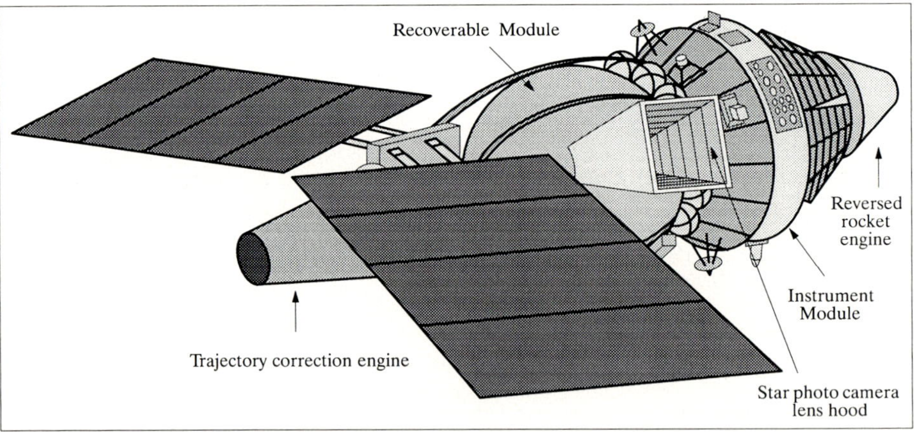


**Figure 82:**    **The Resurs-F2 S/C Model**

Current plans call for space-testing of a modernized Resurs-F1 series in 1994-95, and for a Resurs-F2M series in 1996.

---

293) Information provided by the State Center "PRIRODA", Moscow

| Parameter | Resurs-F1M | | Resurs F2M |
|---|---|---|---|
| Lifetime | 19 days of operation, up to 6 days in standby | | 30 days of operation |
| Orbit | altitude = 235 km, inclination =82.3° | | altitude = 240 km |
| Coverage | once/orbit (lat. 0 - 83°, 1/2 overlapping) | | once/orbit (lat. 0 - 83°, 2/3 overlapping) |
| Photocameras | KFA-200 (1) | KFA-1000 (3) | MK-4 (4-channel camera) |
| Image size | 180 mm x 180 mm | 300 mm x 300 mm | 180 mm x 180 mm |
| Focal length | 200 mm | 1000 mm | 300 mm |
| Swath width | 0.9 altitude | 0.3 x altitude (1 camera) | 0.6 x altitude |
| Duty cycle | up to 12% | up to 10% | up to 19% |
| Type of data | B&W | Spectro-zonal B&W | Multispectral in 3 channels, spectro-zonal in 1 channel |
| Resolution | 23-25 m | 6-8 m | 7 -10 m          8 - 11 m |

**Table 66:**     **Main Characteristics of the modernized Resurs-F S/C**


# A.85    RESURS-O

Resurs-O1[294] is an operational Russian/CIS satellite series (operator: NPO Planeta; S/C integrator: VNIIEM, Moscow). The objective is the observation and monitoring of natural resources (similar in function and objectives to the Landsat series). Operation of the Resource-O1 series was started in 1985 and is planned to continue at least until 1995/6. Applications: observation of the state of agricultural crops, assessment of hydrological conditions, forest and tundra fires, pollution monitoring.[295]

Orbit: Sun-synchronous orbit, altitude = 650 km, inclination = 98°. Orientation accuracy = 6 arc/min along the velocity vector; service life =2 (3) years, payload mass = 420 kg (Resurs-O1), 900 kg (Resurs-O2); A satellite orbit correction engine provides the capability for day-and-night observations in regions of special interest (Resurs-O2 series). Note: the Resurs and Meteor series S/C have the same platforms, they are being designed and built by VNIIEM, Moscow.

| Satellite Series | Launch Date | Sensor Complement | Remarks |
|---|---|---|---|
| Resurs-O1-1 | 3.10.1985 - 26.12.1986 | 2 MSU-E, 2 MSU-SK MSU-S, Travers | analog sensor output and digital transmission |
| Resurs-O1-2 | 20. April 1988 (operational as of 1993) | 2 MSU-E, 2 MSU-SK | analog sensor output and digital transmission |
| Resurs-O1-3 | built, scheduled after end of previous mission (Feb.-March 1994) | 2 MSU-E, 2 MSU-SK | analog sensor output and digital transmission |
| Resurs-O2-1 | 1995, advanced series | 2 MSU-E, 2 MSU-SK, MSU-E resol.=27 m, SAR | digital sensor output and digital transmission |
| Resurs-O2-2 | | | |

**Table 67:**     **Overview of Resurs-O Mission Program**

294) T.M. Wasjuchina, A.M. Wolkow, ”Zustand und Perspektiven der Entwicklung Kosmischer Systeme zur Erforschung natürlicher Ressourcen der Erde und der Hydrometeorologie”, Moscow 1988, translated into German by R. Müller, 1989 (IKF)
295) COSPAR-90-Paper by A. Karpov, USSR State Committee for Hydrometeorology, Moscow. Title of paper: ”Hydrometeorological, Oceanographic and Earth-Resources Satellite Systems operated by the USSR”.

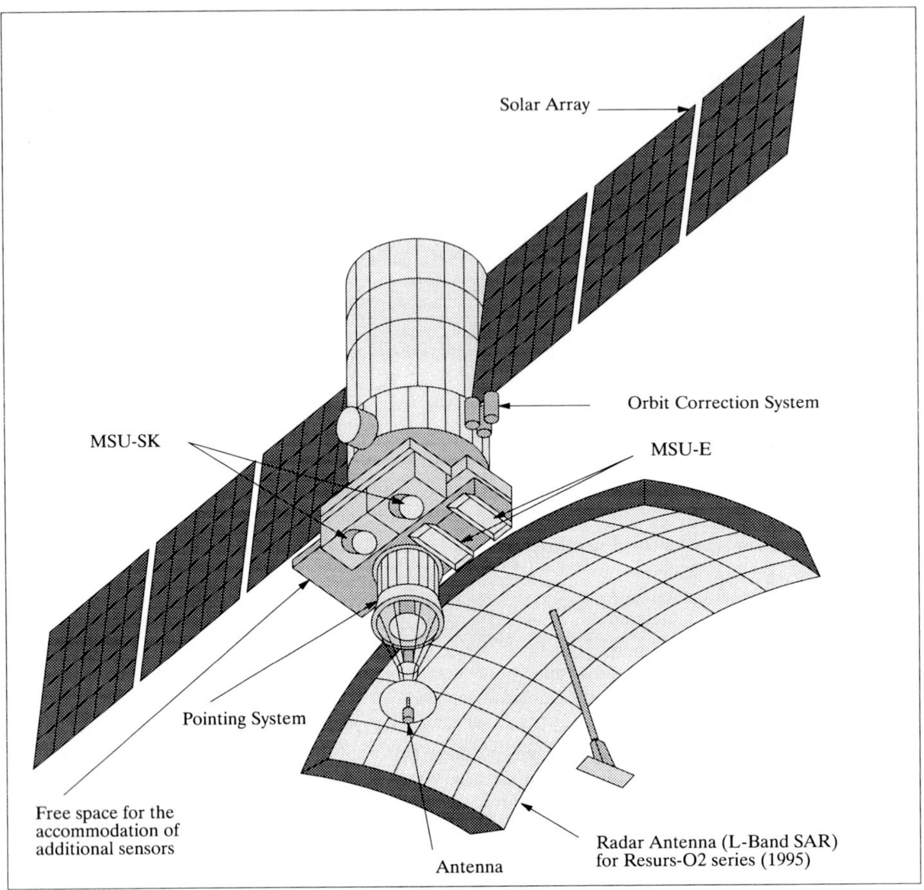


**Figure 83:    The Resurs-O S/C Model**

Sensors:

- **MSU-E** = High-Resolution Multispectral Scanner with Electronic Scanning. Spectral ranges (3) of 0.5 - 0.6 µm, 0.6 - 0.7 µm, 0.8 - 0.9 µm. Spatial resolution = 45 x 33 m; swath = 45 km (for one device) , and 80 km for both devices (some overlapping), repeat cycle = 18 days. MSU-E provides look angles: ± 32° (the instrument may be pointed in the cross-range direction thereby extending the potential coverage to a 600 km wide strip ( ± 300 km to each side of the S/C). The observation direction of the instrument can be set by ground commands in steps of 2°.

- **MSU-SK** = Multispectral Scanner of moderate resolution with conical Scanning. (see PRIRODA), Spectral ranges: 0.5-0.6, 0.6-0.7, 0.7-0.8, 0.8-1.1, and 10.4 - 12.6 µm. Spatial resolution = 170m (in VIS) and 600 m (in the IR). Swath = 600 km , repeat cycle = 3-5 days.

- **Travers** = SAR with wavelength 9.2 cm (3.28 GHz, S-Band), resolution = 200 m

Data transmission:
Frequency at 460 MHz, data rate at 7.86 Mbit/s. Beginning with the Resurs-O2 series the frequency band will be X-Band (8.2 GHz) with data rates of 16 or 45 Mbit/s.

MSU-E and SK have 6 Mbit/s nominally (but a selectable data rate of any one device is possible). Data is being received at the ground stations in Moscow, main ), Novosibirsk (regional), and Khabarovsk (regional). The received data is communicated among all processing centers in real-time and/or store-and-forward modes. Provision of Resurs-O data is available to any customer on a commercial basis.[296],[297]

- **SAR** L-Band (1.28 GHz = 23 cm); new sensor on Resurs-O2 series only
    - Swath width:                     50 km
    - Resolution in azimuth:           150 m (for satellite processing)
    - Resolution in azimuth:           50 m (for ground processing)
    - Range resolution:                100 m
    - Look angle                       35°

| | Parameter | Resurs-O1 Series | Resurs-O2 Series |
|---|---|---|---|
| S/C mass | including payload mass | 1900 kg | 2400 kg |
| | payload mass | 500 kg | 900 kg |
| Earth pointing accuracy | | 10 arc min | 6 arc min |
| | along the velocity vector | 30 arc min | 6 arc min |
| Power supply | average per day | 500 W | 800 W |
| | peak value | 1200 W | 2000 W |
| Power supply | voltage range | 23-34 V | 24-34 V |
| | for additional payload | - | 27 ± 1 V |
| Orbit correction engine | | NA | Available |
| Lifetime | | 2 | 3 |

Table 68:     Resurs-O Series Satellite Characteristics

## A.86    SAFIR

SAFIR (**SA**tellite **F**or **I**nformation **R**elay) is a commercially operated data collection system which is being developed and built by OHB-System GmbH of Bremen, Germany, and sponsored in part by DARA. The system is intended to offer data collection and positioning services on a global scale for an international user community. SAFIR systems operations are intended to be provided by OHB-Teledata (a daughter of OHB-System).[298]

The space segment of SAFIR consists in phase 1 (1994-95) of an independent and autonomous SAFIR payload on the Russian RESURS-01-3 spacecraft (referred to as SAFIR-R), to be launched in Feb./March 1994, for technological verification of the system concept. Phase 2 considers an autonomous polar-orbiting micro-satellite, called SAFIR-1, in 1995 (also being launched by a Russian launch vehicle), to be followed later on by SAFIR-2, etc.. The nominal life of SAFIR-1 is 5 years.

The SAFIR satellite is physically a cube of 0.45 m side length with a total mass of 55 kg, its outer skin is made up of solar cells. SAFIR is equipped with an electronic box, containing two on-board computers and two data storage units (5 MByte) for redundancy, a TM/TC transmitter/receiver unit plus antennas, a GPS receiver and a clock. The satellite is capable of measuring the frequency difference of any ground station based on the Doppler effect which serves in turn as input into the on-board computer in the form of integrated Doppler counts.

The ground segment consists of a main ground station at OHB-Teledata (for central scheduling of all customer traffic, distribution of all customer data, and for satellite control), and two types of small portable ground stations that collect the data at the customer-selected sites.

---

296)  Y. V. Trifonov, "Resurs-O Space System for Ecological Monitoring", VNIIEM, Moscow, Dec. 1990
297)  Y. V. Trifonov, "VNIIEM Proposals on Cooperation in the Field of Space Engineering", Moscow, 1991
298)  Brochures and documentation provided by OHB-System

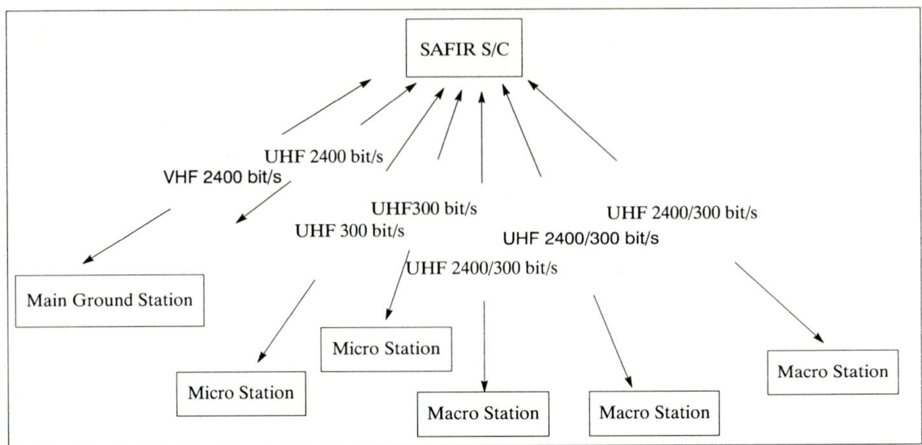


**Figure 84:**   **Concept Overview of the SAFIR System**

Orbit: (SAFIR-1) polar orbit, altitude = 700 km, inclination = 98°, period = 98 minutes.

**User Ground Stations:**

The data collection concept considers two types of user ground stations: a macro station and a micro station.

- Macro stations (in the order of 0.5 kg) provide a storage capacity of 190 kByte for user data, they can collect and store data from their own monitoring instruments or from the satellite. Macro stations may be able to transmit their data with 300 bit/s or with 2400 bit/s to the satellite, depending on communication requirements. A standard macro station consists of a CPU board, a HF board, and an interface board. The I/F board provides the following external interfaces: RS 232, digital input channels, and digital output channels. The macro station may optionally be equipped with a GPS receiver board (for position monitoring), a battery pack, and a telephone modem.

  The ground concept considers a number remote macro stations as well as a 'home station' for each user. The home station consists of a macro station with a PC interface, a battery pack, and a loading unit.

- Micro stations are miniaturized light weight communication units (in the order of 0.1 kg) providing basically the function of position-fixing (animal tracking, etc. via Doppler effect). They transmit their data at a rate of 300 bit/s to the satellite. A micro station has a data storage capacity of 8 kByte.

Each ground station knows the next nominally scheduled contact time. In case of failure (no nominal contact between the ground station and SAFIR) the ground station uses the next possible contact time that known in advance and stored in its memory. A minimum power level of 1 W EIRP is required at the ground station to communicate with the satellite. A ground station may be configured go into a 'sleep mode' during non-contact periods (in order to save power, in this case only real-time measurements at contact time are possible), it goes back into listening mode according to calculated schedule. It is also possible to configure a station to remain in operational (listening) mode and to record measurements during non-contact periods if called for by the requirements.

**Communication Concept:**

The principal task of the satellite is scheduled data collection when in view of user ground stations, intermediate on-board storage of the data, and relay of the data to the main ground station (store-and-forward mode with the use of a mail box). The bidirectional communication principle permits also data transmission from the main ground station or from the user home station via SAFIR to the individual user ground stations (for configuration control, or the capability to reprogram the remote station, etc.). A user is normally equipped with a home station which permits him to communicate with his remote stations in store-and-forward or relay mode, depending on the visibility conditions within the satellite link.

SAFIR operates with two communication frequencies:

- VHF-Band (137.175 MHz) link between the satellite and the main ground station at a transmission rate of 2.4 kbit/s.
- UHF-Band (401.00 MHz uplink, and 400.75 MHz downlink) between the ground stations and the satellite, and also between the main ground station and the satellite, at transmission rates of 300 or 2400 bit/s respectively.

The digital transmission of messages between the satellite and the ground stations is based on an HDLC protocol for error-free communication (error-free in the sense that the SAFIR communication concept provides either correct data or no data). The protocol permits any system user to select his individual message length for a ground station. The transfer frame length depends on the total amount of user data. Shorter messages are transmitted with short frames. The maximum frame length is 1 kByte.

**Access Method:**

SAFIR is equipped with one receive and one transmit channel for servicing the user ground segment. Each ground station has a scheduled time slot within the time of visibility in which the communication with SAFIR is performed. A communication link is established by addressing the user station by its individual identification number. The addressed user station must respond to this addressing prior to a message transfer. The link is closed by SAFIR after the data transfer is completed. The overall efficiency of this access method depends on synchronization delays and the error rate of the channel. The access method of SAFIR is TDMA.

With an average contact time of 6 seconds per remote link (5 seconds of data transfer on the average plus 1 second of synchronization time) SAFIR is capable of servicing up to 100 remote stations within a 10 minute circle of view (of 5000 km diameter). A Doppler location service requires 3 to 6 consecutive contacts for each remote station (depending on accuracy requirements), which amounts to a SAFIR serviceability of 34 or 17 moving stations respectively within a circle of view (instead of the 100 normal stations).

The swath of SAFIR provides the following coverage conditions (changing with latitude):

- 3 times of daily coverage at equatorial latitudes (min)
- 4 times of daily coverage at 40-45° latitudes (min)
- 5 times of daily coverage at 50° latitude (min)
- > 10 times of daily coverage in polar regions

| Communication between Main Ground Station and Satellite | | Communication between Macro Station and Satellite | |
|---|---|---|---|
| | | Frequency uplink | 401.00 - 401.10 MHz |
| Frequency | 137.175 - 137.275 MHz | Frequency downlink | 400.75 - 400.85 MHz |
| Modulation | BPSK, $\Phi = 0, 180°$ | Modulation | FSK, $\Delta f = 9$ kHz |
| Transmission rate | 2400 bit/s | Transmission rate | 300/2400 selectable |
| Coding | Manchester code | Coding | Manchester code |
| On-board computer | | Transputer-based multiprocessor | |
| SAFIR-1 mass | | 55 kg | |
| SAFIR-1 stabilization | | Earth pointing by gravity gradient | |
| SAFIR-1 power consumption | | 20 W average, 50 W maximum | |

**Table 69:**     **Some SAFIR Specification Parameters**

## A.87    Salyut Space Station

USSR space station program. Salyut development began in the 1960s. The first launch occurred in April 1971. The different Salyut space stations were equipped with orbital maneuvering engines that permitted a controllable reentry capability (S/C debris could fall harmlessly into the oceans). The different space stations were raised periodically into higher orbits by the Soyuz visiting spacecrafts.[299]

Earth observation sensors (civil program) on Salyut-6 and -7 consisting of a floor-mounted photographic complex.
Note: there was also an astronomical and a material science instrument package onboard.

**MKF-6** = Multispectral Camera (Carl Zeiss, Jena), first operation of this IKF sensor flown on Soyus-22 in 1976, afterwards in operation on Salyut-6 and -7. MKF-6 operated in 6 spectral bands and could cover 10 million $km^2$ of the Earth's surface with each film cassette. Spectral bands: 0.46-0.50 µm, 0.52-0.56 µm, 0.58-0.62 µm, 0.64-0.68 µm, 0.7-0.74 µm, 0.79-0.89 µm. Number of lenses = 6; focal length = 125 mm; weight of camera = 73 kg; power = 100 W; FOV=40.5°,

**KATE-140**  = Camera System. Spatial resolution of 50 - 70 m.

| Name | Launch Date | Apogee (km) | Perigee (km) | Inclination (°) | Remarks |
|---|---|---|---|---|---|
| Salyut-1 | April 19. 1971 | 222 | 200 | 51.6 | reentry Oct. 11. 1971 |
| Salyut-2 | April 3. 1973 | 260 296 | 215 261 | 51.6 | reentry May 28. 1973 |
| Salyut-3 | June 25. 1974 | 270 275 | 219 253 | 51.6 | reentry Jan. 24. 1975 |
| Salyut-4 | Dec. 26. 1974 | 270 355 369 | 219 343 349 | 51.6 | reentry Feb. 3. 1977 |
| Salyut-5 | June 22. 1976 | 260 275 | 219 256 | 51.6 | reentry Aug. 8 1977 |
| Salyut-6 | Sept. 29. 1977 | 275 377 521 | 219 335 470 | 51.6  83 | crew occupation for a total of 676 days reentry July 28 1982 |
| Salyut-7 | April 19. 1982 | 260 | 212 | 51.6 | crew occupation for a total of 812 days, reentry 1989 |

**Table 70:**    **Overview of Salyut Space Stations**


## A.88    SAMPEX (Solar Anomalous and Magnetospheric Particle Explorer)

SAMPEX is a NASA/GSFC mission in the Small Explorer Program (SMEX) with the objective to measure energetic electrons as well as ion composition of particle populations from ~0.4 MeV/nucleon to hundreds of MeV/ nucleon from a zenith-oriented satellite in near polar orbit. Study the energy, composition, and charge states of particles from supernova explosions, from the heart of Solar flares, and from the depths of nearby interstellar space; also monitoring of the magnetospheric particle populations which plunge occasionally into the middle atmosphere of the Earth. SAMPEX uses the magnetic field of the Earth as an essential component of the measurement strategy. Mission design life = 1 year with a goal of 3 or more years of operation.

SAMPEX[300] was launched on July 3, 1992 from NASA's Western Test Range (Lompoc, Ca.), Scout launch vehicle.

Orbit: Polar orbit, perigee = 520 km, apogee = 670 km, inclination = 82°

---

299)  Interavia Space Directory 1990-91 (previously Jane's Space Flight Directory), pp. 122-124

Sensors:

**LEICA** = Low Energy Ion Composition Analyzer (University of Maryland under NASA contract).
LEICA is a time-of-flight spectrometer that monitors the incident ion mass and energy by simultaneously measuring the time-of-flight and the kinetic residual energy of particles that enter the telescope and stop in an array of 4 solid state detectors. The time-of-flight is determined by START and Stop pulses from chevron microchannel plate (MCP) assemblies that detect secondary electrons emitted from the entrance foil and a foil in front of the solid state detector, respectively, when the ion passes through them. These secondary electrons are accelerated to approximately 1 kV and deflected onto the MCPs by electrostatic mirrors. The measured energy and velocity are combined to yield the mass of the ion and the energy per nucleon.

**HILT** = Heavy Ion Large Telescope (MPE Garching/NASA).
HILT monitors heavy ions from He to Fe in the energy range from 8 - 220 MeV/nucleon for oxygen, covering the medium-energy solar energetic ions, galactic cosmic rays, and the range of maximum intensity of the anomalous cosmic ray component. The sensor consists of a three-element ion drift chamber with two thin multi-layer entrance windows followed by an array of 16 solid state detectors and a scintillation counter with photodiodes. HILT uses a flow-through isobutane system for the drift chamber.

**MAST** = Mass Spectrometer Telescope (Caltech/GSFC).
MAST monitors the isotopic composition of elements from Li (Z=3) to Ni (Z=28) in the range $\sim 10$ MeV/nucleon  several hundred MeV/nucleon. MAST consists of a combination of surface barrier and lithium-drifted solid state detectors (11 total). Combined matrix detector positions determine the particle trajectories, allowing accurate corrections to be made for the pathlength variation with angle and detector response for non-uniformities. MAST also performs measurements of stopping He isotopes from $\sim 7$ - 20 MeV/nucleon.

| Energy Range for: | LEICA | HILT | MAST | PET |
|---|---|---|---|---|
| Electrons | - | - | - | 0.4-30 MeV |
| H | 0.76 - 6.1 | - | - | 18 - 250 MeV |
| He | 0.45 - 6.1 | 4.3 - 38 | 7 - 20 | 18 - 350 MeV/n |
| C | 0.44 - 11.4 | 7.2 - 160 | 14 - 210 | 34 - 120 MeV/n |
| Si | 0.33 - 5.5 | 9.6 - 177 | 21 - 330 | 54 - 195 MeV/n |
| Fe | 0.21 - 3.1 | 11 - 90 | 27 - 450 | 70 - 270 MeV/n |
| Charge range for elements | 1 - 26 | 2 - 28 | 2 - 28 | 1 - 2 (1 -28*) |
| Charge range for isotopes | 2 - 16 | 2 | 2 - 28 | 1 - 2 (1 - 10*) |
| Geometry factor ($cm^2$ sr) | 0.8 | 60 | 7 - 14 | 0.3 - 1.6 |
| FOV ($^o$, full angle) | 24 x 20 | 68 x 68 | 101 | 58 |
| Mass (kg) | 7.4 | 22.8 | 8.8 | (incl. with MAST) |
| Power (W) | 4.9 | 5.6 | 5.3 | (incl. with MAST) |
| Telemetry (kbit/s) | 1.3 | 0.9 | 1.4 | 0.5 |
| * commandable low-gain mode | | | | |

**Table 71:**     **Performance Overview of SAMPEX Instruments**

**PET** = Proton/Electron Telescope (Caltech/GSFC).
PET complements MAST by monitoring the energy spectra and relative composition of protons (18 - 250 MeV) and helium nuclei (18 - 350 MeV/nuclei) of solar, interplanetary, and galactic origins, and the energy spectra of solar flare and precipitating electrons from $\sim 0.4$ to $\sim 30$ MeV. The instrument measures both trapped and precipitating energetic particles in different parts of the SAMPEX orbit.

---

300)  D. N. Baker, G. M. Mason, O. Figueroa, G. Colon, J. G. Watzin, R. M. Aleman, " The Solar, Anomalous, and Magnetospheric Particle Explorer (SAMPEX) Mission", Preprint 93-128, Uni. of Maryland - see also (by the same authors): IEEE Transactions on Geoscience and Remote Sensing, Vol. 31, Nr. 3, May 1993, pp. 531-541

Data: Onboard data storage is provided by solid-state memory (26.5 Mbyte of data storage). Data transmitted to the ground is formatted as a "Packet Telemetry" stream in accordance with CCSDS standards. S/C operation by GSFC on a continuous basis. Data are transmitted to the ground twice per day via the Wallops Flight Facility.

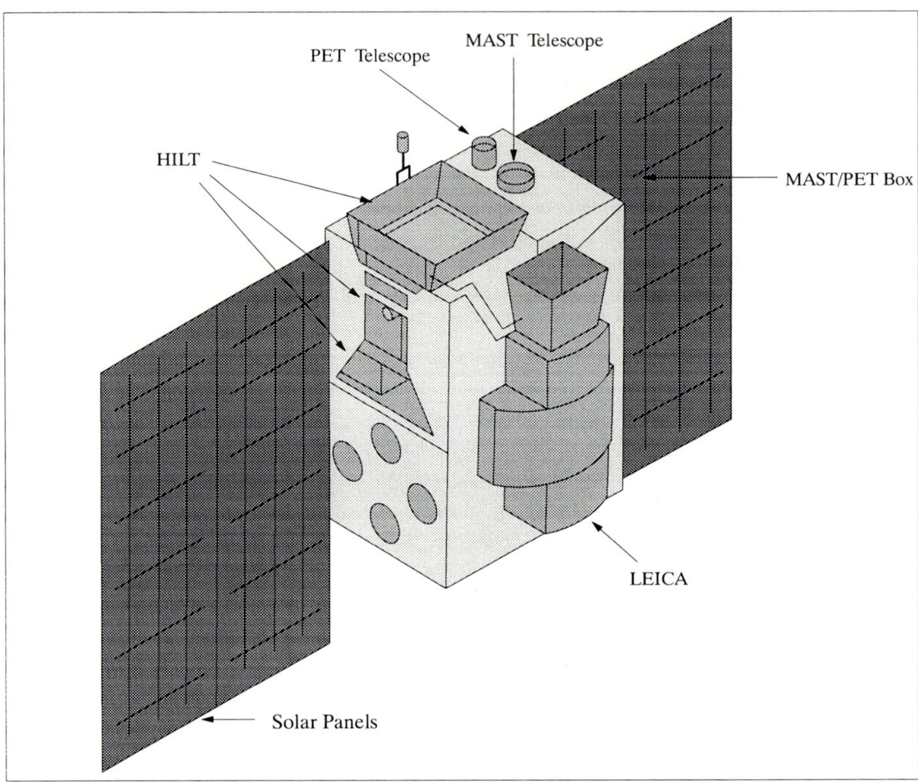


**Figure 85:**    **The SAMPEX S/C Model**

## A.89    SAN MARCO D/L

San Marco D/L[301),302)] is a cooperative mission of CRA (University of Rome/Italy) and NASA/GSFC with sensor contributions from several agencies. NASA Scout launch vehicle, spin-stabilized satellite (axis parallel to Earth's axis), payload mass = 234 kg, satellite diameter = 0.940 m. Launch site: San Marco Range, Kenya; Mission duration: 25 March 1988 until 6. Dec. 1988.

Objective: the primary objective is to explore the possible relationship between solar activity and meteorological phenomena. Investigation of the structure, dynamics, and aeronomy of the equatorial thermosphere, in particular:

-    The relationship between density, composition, temperature, and airglow as a function of solar flux
-    The interhemispheric transport of ions and neutral particles

301)  Payload Definition Document for SAN MARCO D/L Satellite, CRA, Oct. 1987
302)  G. Schmidtke, H. Doll, C. Wita, and S. Chakrabarti, "Solar EUV/UV and equatorial airglow measurements from San Marco-5", Journal of Atmospheric and Terrestrial Physics, Vol. 53, Nr. 8, pp. 781-785, 1991

- The dependency of the neutral wind, ion drag, temperature, and particle densities on local time
- The relationship between neutral wind and the electric field
- Ionosphere irregularities and special events, and other parameters keyed to longitude
- The equatorial distribution of ozone
- Magnetic storm induced variations in the measured parameters
- The dependency of the correlation between the meridional wind and ion drift on the magnetic field
- The variation of the midnight temperature maximum

Orbit: Near-equatorial orbit, apogee = 619 km, perigee = 262 km, inclination = 2.9°.

Sensors:

- **DBI** = Drag Balance Instrument (PI: L. Broglio, CRA)
  Measurement of the total aerodynamic drag force on the S/C.

- **ASSI** = Airglow Solar Spectrometer Instrument (PI: G. Schmidtke, Institut für physikalische Meßtechnik, Freiburg, Germany)
  Measurement of airglow, solar radiation, reflected solar radiation, and radiation of interplanetary and intergalactic sources. Four scanning spectrometers with solar pointing control cover a broad spectral range [from 20 - 700 nm (EUV - VIS)] using channels with 18 overlapping wavelength ranges with a spectral resolution of 0.7 - 4 nm (each spectrometer contains 4 or 5 detectors for a total of 18 detectors, there are two units of ASSI, namely ASSI-A and ASSI-B). Large dynamic ranges up to $10^{11}$ permit the measurement of very faint airglow or interplanetary radiation as well as intense solar emissions.
  The units ASSI-A and ASSI-B are mounted on separate solar pointing devices, viewing 180° away from each other from the equatorial plane of the spacecraft.

- **WATI** = Wind and Temperature Spectrometer (PI: N. Spencer, GSFC)
  Measurement of two wind components (horizontal and normal to orbit plane), along with the kinetic temperature.

- **EFI** = Electric Field Instrument (N. Maynard, PHG; T. Aggson, GSFC)
  Plasma transport measurements. EFI consists of two orthogonal 40 m wire antenna pairs that are extended in the spin plane of the satellite, and a shorter antenna pair in the S/C spin axis. DC and electric field measurements.

- **IVI** = Ion Velocity Instrument (PI: W. Hanson, University of Texas)
  IVI measures the 3-D bulk velocity of the ambient ions in the S/C velocity frame. In addition IVI measures the ambient plasma concentration and the ion temperature.

Data: Onboard recorder, VHF band data link, data rate = 6 kbit/s. Sensor data at CRA and at World Data Center (GSFC).

The mission San Marco D/L is the fifth mission in Italy's San Marco program series.[303]
- San Marco 1 - Launch: Dec. 15 1964 from Wallops. Orbit 194 x 697 km, incl. = 38°. Measured atmospheric density until its decay on Sept. 13 1965
- San Marco 2 - Launch: April 26 1967, Scout vehicle from San Marco (Floating platform base in the Indian Ocean off the coast of Kenya). Orbit 185 x 211 km, incl. = 3°. Observation of atmospheric density. S/C re-entry: Oct. 14 1967.
- San Marco 3 - Launch: April 24 1971 from San Marco. Orbit: 222 x 718 km, incl. = 3°.
- San Marco 4 - Launch: Fe. 18 1974 from San Marco. Orbit: 231 x 910 km, incl. = 3°. Observation of auroral zone over the equator. S/C re-entry on May 4, 1976.

---

303) Jane's Spaceflight Directory 1988-89, pp. 35-36

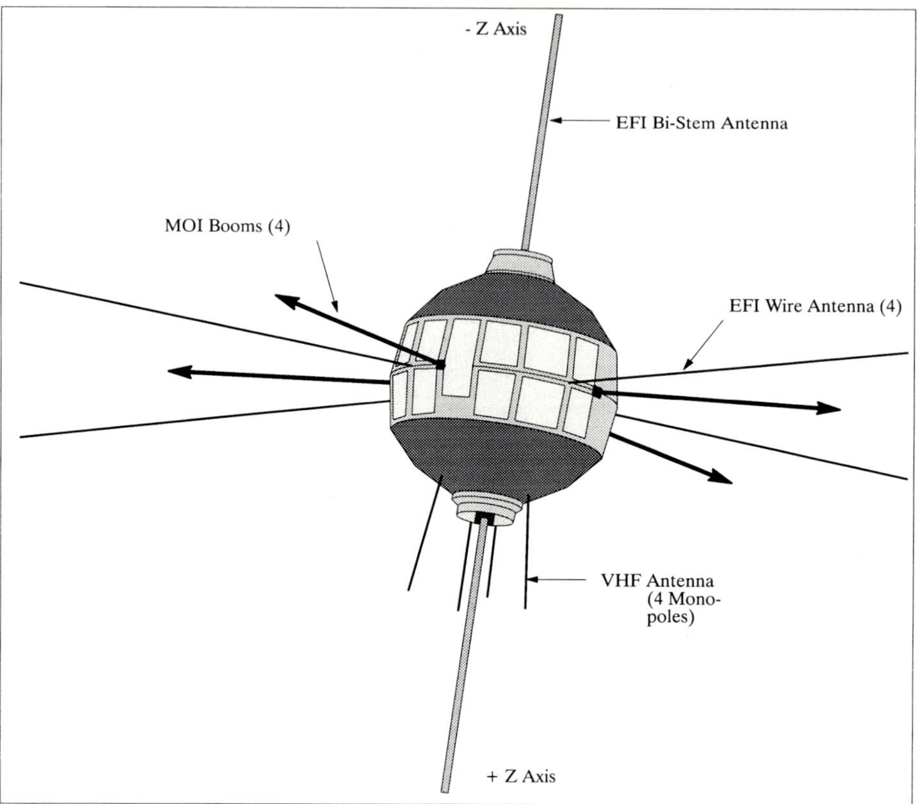


**Figure 86:     San Marco D/L Spacecraft Model**

## A.90     SCD1 (Satélite de Coleta de Dados)

SCD1[304],[305] is Brazil's first experimental Data Collection Satellite, as portion of MECB (Missao Espacial Completa Brasileira). INPE's responsibility in MECB is the space segment and associated ground segment of the program [four satellites are planned within this program: two for environmental data collection (SCD1 and SCD2) and two for remote sensing of the Earth (SSR1 and SSR2)]

A pilot DCP network in Brazil is equipped with sensors suitable in the fields of meteorology, oceanography and atmospheric chemistry. Spin-stabilized S/C; payload weight = 115 kg; power = 90 W; design life = 1 year (projected life is now 2.5 years). Launch of SCD1: Feb. 9, 1993 from NASA's Kennedy Space Center by OSC of Fairfax Va. (Pegasus launch vehicle). SCD1 is performing nominally. SCD2 is planned for a launch in late 1995.

The shape of SCD1 is an octagonal prism, 80 cm tall, whose base fits into a 1 m diameter circle. The outer surface of the S/C is cover with solar cells. S-Band TT&C communication is provided by a TM/TC subsystem for housekeeping telemetry transmission, telecommand and ranging (S/C mission control functions). Onboard storage capability of TT&C data.

304)  INPE brochure 'SCD1 Data Collection Satellite', and Fax information from Prof. P. M. Fagundes, Rio De Janeiro
305)  "SCD1 Satellite Description", and "The Brazilian Data Collecting System", papers provided by C. E. Santana of INPE

Orbit: Near-circular orbit; altitude = 750 km; inclination = 25°; period = 98.8 minutes.

The SCD1 satellite is a real-time repeater of environmental data gathered on the ground by automatic data collection platforms (DCPs) random access service for DCPs. The real-time repeater concept implies that there is no on-board processing and storage capability but simply a relay function of data by the satellite. The data reception on the ground is performed by a single station in Cuiabá, Brazil, every time SCD1 comes into view. This concept permits a data collection capability from all sites in Brazil and well beyond (coverage region of 3000 km radius). It is estimated that the system setup is capable of servicing up to 500 DCPs randomly located in the coverage area with a probability of 85% of acquiring the data of any DCP at least once per day. The DCP data received by the Cuiabá station are processed by the MCC in Cachoeira Paulista and subsequently stored in a database for user access.

The random access method employed by SCD1 for data collection is similar and compatible with the ARGOS system (2 uplink channels: ARGOS uplink frequency at 401.650 MHz plus a private frequency 30 kHz away from the ARGOS frequency). This means:

- Access scheme is pure (i.e. unslotted) ALOHA, without a return channel
- Message format, data rates, and transmission frequencies are ARGOS compatible (message length of 32-256 bits, 32 bit/s uplink rate, 401.650 MHz uplink frequency).
- Brazilian DCPs in the ground segment can also be serviced by other satellites carrying ARGOS.

As of 6/1993 there are 35 DCPs operational in Brazil. SCD1 is receiving about 120 DCPs in its region of coverage. About 200 new users are seeking access to the system (including those from other South American countries).

Note with regard to the SCD1 access scheme:[306]
SCD1 uses a "random time-division multiple access" scheme which is similar to "unslotted ALOHA". However, SCD1 and also ARGOS access are simpler than ALOHA due to the missing return channel from the satellite to the DCP. This means in fact loss of data due to interference (interference occurs when the demand for service exceeds the system's capability, the result is loss of data from system 'blockage'). However, the system works satisfactorily in spite of this disadvantage due to the following reasons:

- interference is mitigated by the Doppler shifts, tending to spread the DCP transmissions
- the system operates at a low input rate (400 bit/s) so that a data packet has at least a probability of 80-90% of getting through without interference
- there is a lot of redundancy in the data of consecutive transmissions from the same DCP, so that successful reception of one out of two or even three is still satisfactory
- there is no catastrophic effect for subsequent communications if some data are lost.

### A.90.1   SSR1

SSR1 (Satelite de Sensoriamento Remoto) is Brazil's Remote Sensing Satellite (INPE). Objective: repetitive monitoring of Brazil through spaceborne imagery. SSR1 has a total mass of 170 kg. Nominal design life of 2 years. A launch is planned for the end of 1996.[307],[308] A follow-up SSR2 satellite is projected for launch in 1997.

Orbit: Sun-synchronous circular orbit; altitude = 640 km; equator crossing time = 9:30 AM; repeat cycle = 4 days.

Sensor: (same sensor as on CBERS)

- **WFI** = Wide-field Imager (CCD Linear Array Camera)
  Objective: medium resolution VIS/NIR remote sensing (in particular over areas of ex-

306) Information provided by C. E. Santana of INPE
307) "The first Brazilian Earth Observation Satellite (SRR)", paper by C. E. Santana and J. Kono of INPE
308) 'Satellite Launch to Advance Brazilian Space Program', Space News Aug. 31-Sept. 6, 1992, p. 43

tensive vegetation in the Amazon region). Operation of the camera is only considered over Brazil (direct sensor data transmission to the ground).

- Spectral bands:      0.63 - 0.69 µm, and 0.77 - 0.87 µm
- Swath width          755 km (FOV = 60°)
- Spatial resolution   212 m (IFOV = 0.331 mrad)
- Data quantization:   8 bit
- Data rate:           1.35 Mbit/s
- Band-to-band registration: better than 0.3 pixel
- Modulation transfer function: better than 0.35

The MCC function will be performed over a single station located in Cuiabá (Brazil). Sensor data and TT&C communications via S-Band.


## A.91   SEASAT

NASA (JPL) Earth observation mission. Launch: 27. June 1978 (Atlas-Agena from Vandenberg), End of mission (abrupt power system failure in its 1,502 orbit) October 9. 1978. Mission duration: 70 days (data generation) of 106 operational days.

Application: Ice and ocean monitoring, land use, Geology, Forestry, Mapping

Orbit: Polar orbit, 108° inclination, altitude = 800 km, Swath width = 100 km, period = 101 minutes

Sensors: (all Seasat sensors are microwave sensors)

- L-Band **SAR** (Synthetic Aperture Radar) in HH Polarization, Look Angle = 20°; Pixel size = 25 x 25 m (spatial resolution on the surface at 4 looks); Radiometric resolution = 5 Bit raw data. Spectral range: 1275 MHz (frequency, L-Band); wave length ($\lambda$)= 23 cm; swath width=100 km. Antenna: 1024-element phased array antenna 10.7 m x 2.4 m. The Seasat SAR sensor is regarded the first imaging SAR system used in Earth orbit.

- **SMMR** (Scanning Multichannel Microwave Radiometer); Objectives: Monitoring of sea surface temperatures, wind speeds, rain rate, atmospheric water content and ice conditions.
  Spectral ranges:      6.6 GHZ (45 mm)
                        10.7 GHz (28 mm)
                        18.0 GHz (17 mm)
                        21.0 GHz (14 mm)
                        37.0 GHz (8 mm)

- **Radar Altimeter** (first attempt to achieve 10 cm altitude precision from orbit)
  Spectral range:       13.499 GHz (Ku-Band)
  Ground resolution:    1.6 km
  Objective: Determination of sea surface profiles, currents, wind speeds and wave heights

- **Radar Scatterometer** (SASS = Seasat-A Scatterometer System);
  Objective: mapping of the surface windfield on a global basis over the oceans)
  Spectral range:       14.599 GHz
  Ground resolution:    280 km (nadir)
  Swath width:          750 km

**SAR Data:**

Image size: 100 x 100 km (JPL); 100 x 75 km (DLR); 4 Looks full digital Image Processing with 12.5 m pixel spacing; 8000 Lines x 8000 Pixels (JPL)
Transmission: Frequency= 11.25 MHz; Data rate = 110 Mbit/s

Seasat provided global statistics on ocean mesoscale variability, detected strong currents, and showed the potential to detect the longer wavelengths of the circulation.

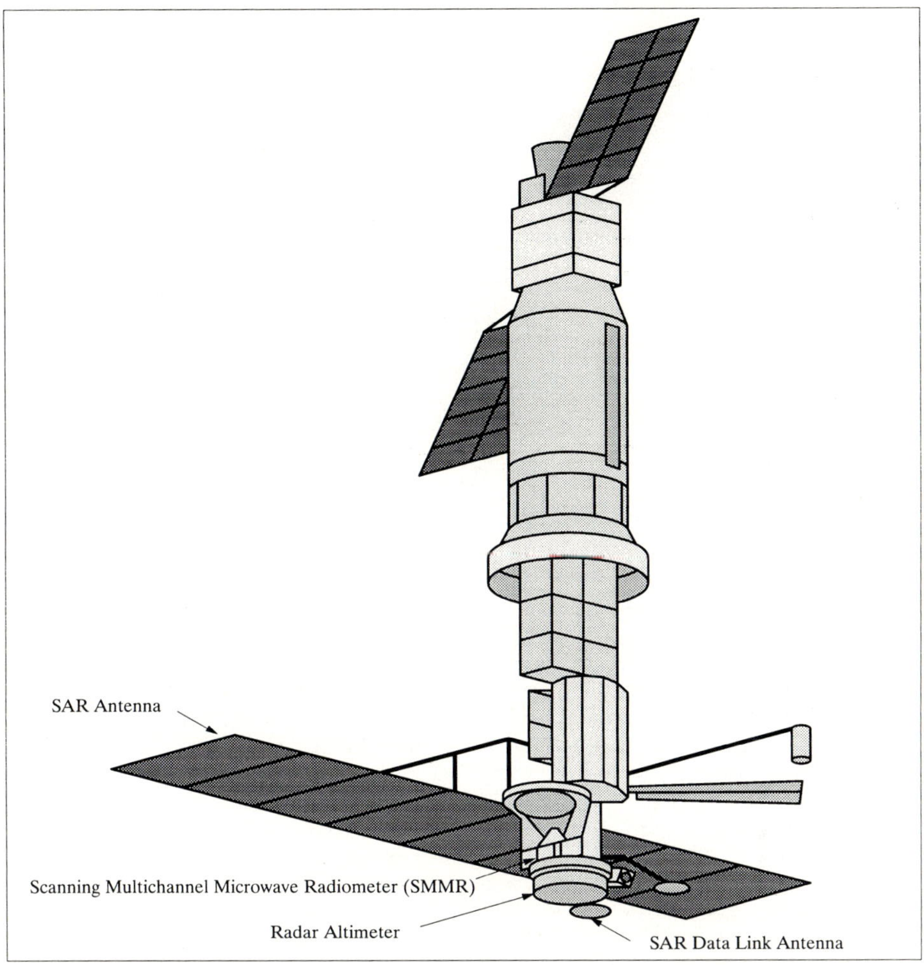


**Figure 87:** **The Seasat S/C Model**

## A.92   SeaStar

A NASA (GSFC) Earth observation mission, defined around the sensor **SeaWiFS**, scheduled for launch in July 1994. Nominal life = 5 years. The SeaStar S/C will be launched by a Pegasus rocket from Vandenberg (of Orbital Sciences Corporation = OSC)[309],[310]. The operation of the SeaStar satellite will also be contracted to OSC, as well as the science data processing and distribution.

Application: Ocean color data, Ocean biology and ecology, Phytoplankton concentrations, etc.

---

309)  "Orbital Sciences Captures $120 Million in Business, Pegasus Launches Ocean Satellite Ordered", Space News, March 11-17, 1991, p. 7
310)  "OSC Reviews Seastar Design", Space News, Oct.28-Nov.3, 1991, p. 22

Orbit: Sun-synchronous polar orbit, altitude = 705 km, inclination =98.2°, equator crossing time at local noon (descending node), successive orbit equatorial crossing longitude = -24.721°, period = 98.2 minutes, orbital repeat time = 16 days (233 orbits)

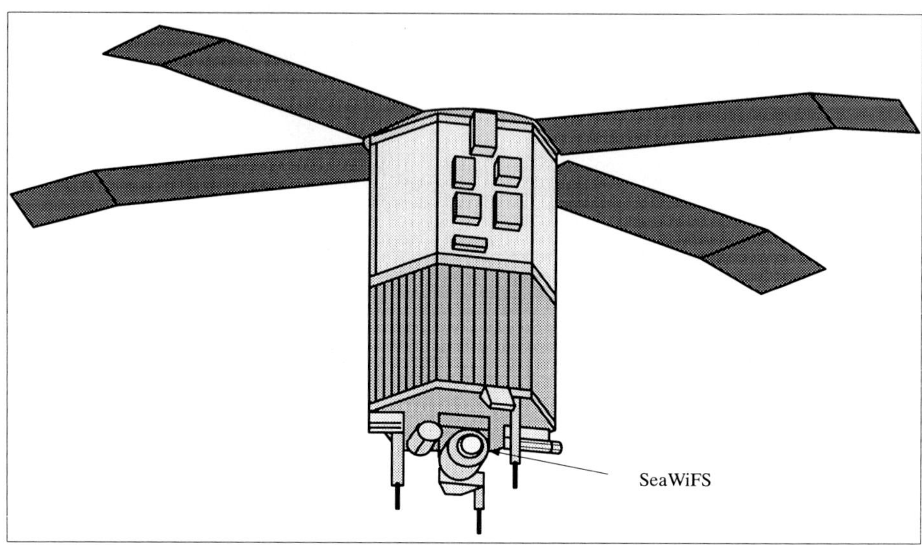


**Figure 88:**      **The SeaStar S/C Model**

Sensor:

**SeaWiFS** (Sea-Viewing Wide Field-of-View Sensor)[311],[312]. The SeaWiFS sensor is of CZCS heritage (Nimbus-7 sensor), and was originally planned to fly on Landsat-6. It is scheduled now for its own mission with the name of SeaStar. The SeaWIFS sensor is being built by Hughes Space Systems.

-    Radiometric accuracy:        < 5% absolute each band
-    Calibration/Stability Monitor: solar diffuser, lunar view
-    Relative Precision:           < 1% linearity of signal output to radiance
-    Between Band Precision:      < 5% relative band/band over 50 - 90% of saturation
-    Polarization Sensitivity:     < 2% worst case, all scan and tilt angles
-    Dynamic Range:            10 bit quantiz.; 4 gains (each channel, 0.75:1:1.5:2.5)
-    Bright Target Recovery:      < 10 samples
-    Location knowledge:        0.5 km, at 1-sigma level for instrument

| Band | Wavelength (nm) | Saturation Radiance (milliwatt/cm²) | Input Radiance (milliwatt/cm²) | SNR (Measured at Input Radiance) |
|---|---|---|---|---|
| 1 | 402 - 422 | 13.63 | 9.10 | 499 |
| 2 | 433 - 453 | 13.25 | 8.41 | 674 |
| 3 | 480 - 500 | 10.50 | 6.56 | 667 |
| 4 | 500 - 520 | 9.08 | 5.64 | 640 |
| 5 | 545 - 565 | 7.44 | 4.57 | 596 |
| 6 | 660 - 680 | 4.20 | 2.46 | 442 |
| 7 | 745 - 785 | 3.00 | 1.61 | 455 |
| 8 | 845 - 885 | 2.13 | 1.09 | 467 |

**Table 72:**      **SeaWiFS Performance Summary**

---


311) 'System Concept for Wide-Field-Of View Observations of Ocean Phenomena from Space', NASA-NOAA-Eosat publication, 1987
312) H. v.d. Piepen, V. Amman, R. Doerffer, 'Remote Sensing of Substances in Water', GeoJournal 24.1, 24-27, 1991 (May) by Kluwer Academic Publishers

| Sensor Parameter | GAC Mode (Global Area Coverage) | LAC Mode (Local Area Coverage) |
|---|---|---|
| Scan width (degree) | 45° | 58.3° |
| IFOV at sensor (mrad) | 1.5835 | 1.5835 |
| Ground IFOV at nadir (km) - spatial resolution | 1.12 | 1.12 |
| Pixels along scan | 248 | 1285 |
| Pixel numbers | (147 - 1135, by 4) | (1-1285) |
| Scans/second | 1.5 | 6 |
| Scan plane tilt | 20° | 20° |
| Ground swath width, km | 1502 | 2802 |

**Table 73:     SeaWiFS Sensor Characteristics**

| | |
|---|---|
| Satellite Transmitting Power | 5 Watt |
| HRPT S-Band Frequencies: | 1702.5 MHz |
| Frequency Stability | ± 20 ppm |
| Modulation | Split-Phase |
| Modulation Rate | 0.6654 Mbit/s |
| Satellite Transmitter Antenna Loss | 2.0 dB |
| Satellite Antenna Gain | 1.8 - 2.1 dBiC |
| EIRP | 31 - 37 dBm |
| Slant Range | 2900 km |
| Free Space Loss | 166.1 dB |
| Fading and Rain Margin | 5 dB |

**Table 74:     Requirements for a SeaWiFS LAC Downlink Ground Station HRPT**

**Data:**[313]

Onboard storage of sensor data is limited to 125 MByte. With two dumps/day, this allows global observations at reduced spatial resolution to be recorded, and a limited limited amount (about 20 min/day) of high resolution Local Area Coverage (LAC) data. To increase the coverage of LAC data, NASA encourages the operation of HRPT (High Resolution Picture Transmission) stations by the user community throughout the world. These HRPT stations can collect real time LAC data via direct broadcast whenever the spacecraft is in view.

However, to protect the commercial interests of the builder and operator of the mission, OSC, SeaWiFS real time LAC data will be encrypted. Commercial stations requesting SeaWiFS data must purchase a decryption capability directly from OSC. HRPT stations for research data use must obtain prior approval from NASA (SeaWiFS Project Office at GSFC). The research licences will be issued at no cost, but will be subject to certain restrictions. See also Figure 67 for NOAA HRPT stations.

## A.93     Shuttle EO Imaging Cameras

A special camera system was developed for Shuttle flights with the objective to obtain high-resolution image observations of special areas of interest from piloted space systems.

Applications: spot target observations.

- **ESC - Electronic Still Camera**[314]
  The system consists of the following components:
  - a hand-held battery-operated high-resolution camera (fully digital and programmable)

---

313) 'Roles and Responsibilities of HRPT Stations for SeaWiFS', SeaWiFS Project Office, GSFC, Dec. 19, 1991

- a lap-top computer-based playback/downlink unit for onboard image processing
- a ground station capable of receiving Shuttle data, processing images, producing hardcopies, and distributing the data and hardcopies to end users.

The camera is designed and built with astronomical-grade CCDs which resulted in near-film-quality imaging. The downlink capability provides the means of investigator ground inspection and interaction during Shuttle flights. This interaction capability optimizes the yield of experiments and observations. The Orbiter's Ku-Band downlink scheme permits an image transmission within 15 seconds, however, ESC images may also be downlinked through any channel, regardless of bandwidth and bit rate. As of 1992 the Naval Research Laboratory is developing a latitude-longitude locating system, which when used with the ESC, will identify the location of the Earth target imaged within 1 nautical mile. The camera flown on STS-48 was set up for a 1 k x 1 k CCD array.

ESC specifications:

- ESC spectral range: 0.4 - 1.1 $\mu$m, the spectral response allows for multispectral imaging (human interaction onboard permits quick decisions: for instance with the best way to photograph a scene, lenses may be exchanged, filters may be selected for multispectral imaging, a new data storage cartridge may be loaded, etc.).
- Ground resolution of about 40 m (at 470 km orbit with 180 mm lens); = 15 m (at 290 km orbit with 300 mm lens, theoretical limit)
- ESC is fully programmable to support a variety of sensors: 1k x 1k, 2k x 2k, and 1k x 1k color (note: a 2k x 2k array is planned)
- dynamic range = 60 dB; sensor dynamic range = 80 dB
- removable image storage media for unlimited storage capability
- adaptable to all Nikon lenses or any instrument that uses a Nikon mount
- retains the functions of a Nikon F4, such as autofocus, autoexposure, and manual override

The ESC was first used on mission STS-48 in September 1991. Since then, ESC has flown on STS-42 and is planned to fly on STS-45, STS-49, and STS-53.

Playback/Downlink capabilities:

- employs virtual imaging, which allows the user to display ESC images on a variety of monitors with a variety of solutions
- provides standard and non-standard image processing
- allows the user to select and deselect desired images for downlinking
- allows notes to be included with each downlinked image

Ground station capabilities:

- compatible with Orbiter digital downlink
- primary image processing performed with Sun 3/470 and PIXAR imaging computers
- capability of processing 4k x 4k true color images
- data can be networked anywhere in the world
- hardcopy capability for ground crew

## A.93.1   Shuttle Film Camera Systems

Earth photographs are collected using 35 mm, 70 mm, and 5 inch format still cameras, equipped with a variety of lenses and film types.

- Modified Hasselblad single-lens reflex camera, model 500 ELX (prior to Jan. 1990

---

314) S. D. Holland, 'The NASA Electronic Still Camera System', IEEE IGARSS '92 Volume I, pp. 149-151

ELM)[315),316)]. A 70 mm- format camera for hand-held astronaut photography. An important accessory is the data recording module for the film magazine, which records the exact time when the photograph was taken (for post-flight identification of the scene). Data Recording Modules (DRMs) were installed on Hasselblad cameras in April 1985 with STS-51-B. Before DRMs were installed, determining the location of ocean features required identifying geographic references in the photograph, e.g. coastlines or islands.

- Rolleiflex single-lens reflex camera, model 6008 (was flown several times during 1990-92, first flight on STS-32 in Jan. 1990). A 60-frame, 70 mm film magazine was devised. The camera features automatic exposure control.

- Linhof Aero Technika 5-inch format camera system. The camera is flown on a space-available basis. First flight on STS-41G in Oct. 1984. Lens focal lengths: 90 mm or 250 mm. Time and date recording of Linhof photographs began in Dec. 1988 with STS-27.

- Nikon camera systems F3 and F4. These cameras are being used primarily for interior documentary. They have also been used for Earth observation photography with tele-photo lenses of 180 mm and 300 mm focal lengths. Cataloging of these 35 mm Earth-looking images has been instituted since STS-37 (April 1991).

Postflight Support:
A catalog is published after each Shuttle flight giving the geographic location of each frame of the Earth-looking photography. Each identified frame is listed with its geographical name, flight ephemeris data, and some descriptive feature information. A typical Shuttle flight produces 3000-6000 pictures.

Electronic Data Bases:
The Space Shuttle Earth Observations Project (SSEOP) Photography Database[317)], containing geographical references to more than 90000 space photographs (over 26000 from Shuttle flights until 1991) is operational at the VAX computer of JSC, Houston. Access to this database is free of charge and available 24 hours a day.


## A.94    SIR-A (Shuttle Imaging Radar)

SIR-A = Shuttle Imaging Radar with Payload A. A JPL payload on Shuttle. Launch: Nov. 12, 1981 (duration: 1 week, second orbital test flight of the Shuttle Columbia). The overall science payload aboard this Shuttle flight was referred to as OSTA-1.

SAR Application: Land use, Geology, Hydrology, Forestry, Mapping

Orbit: Shuttle orbit, 38° inclination, 260 km altitude, SAR swath width: 50 km

Sensor:
**SAR** = Synthetic Aperture Radar in HH Polarization; L-Band (freq. = 1.28 GHz); Look Angle = 47 degrees; Pixel size = 40 x 40 m; optical data only

**Data:**
Image size: 50 km x length of active orbit
Optical image correlation, transparencies and photos available
Transmission: no transmission, optical recording onboard

315) D. L. Amsbury, J. M. Bremer, ' Recent Developments in Space Shuttle Remote Sensing, using hand-held Film Cameras', IGARSS '92, Volume I, pp. 152-154

316) S. G. Ackleson, D. E. Pitts, 'Global Distribution of hand-held Photographs of Ocean and Coastal Regions Taken during Space Shuttle Missions, 1981-1991', IEEE IGARSS '92 Volume II, pp. 1550-1552

317) R. M. Nelson, K. J. Willis, W. J. Daley, F. R. Brumbaugh, J. M. Bremer, ' Cataloging and Indexing - The Development of the Space Shuttle Mission Data Base and Catalogs from Earth Observations hand-held Photography', IEEE IGARSS '92 Volume I, pp. 155-157

**Other remote sensing instruments of the OSTA-1 payload were:** (see also LFC in A.57)

**SMIRR**[318] = Shuttle Multispectral Infrared Reflectance Radiometer (NASA/JPL experimental instrument). SMIRR was also part of the science payload (OSTA-1)

Objectives of SMIRR: a) obtain radiometric measurements from orbit of land surfaces in 10 wavelength channels (rock and mineral identification). b) determine the spectral response of known rock types under different climatic conditions worldwide; c) establish the utility of orbital narrow-band radiometry in the 2.0 to 2.5 μm region; d) establish the requirements for future narrow-band imaging systems.

**OCE** = Ocean Color Experiment[319].
The experiment consists of a multispectral image scanner of high sensitivity, dedicated to the measurements of water color and its interpretation in terms of major water constituents (chlorophyll, sediments, pollutants, etc.).

Sensor (OCE):
The instrument is an 8-channel opto-mechanical image scanner. Total wavelength range from 464 - 773 nm. FOV = ±45°. IFOV = 3.5 milliradian.

Prior to and during the Shuttle mission with the OSTA-1 payload, various OCE-related experiments took place. The main objective of these activities was to perform comparative measurements from aircraft, ship and ground for correlation and verification of the Shuttle OCE data.

A duplicate instrument of the OCE, referred to as **OCS** (=Ocean Color Scanner) was shipped from GSFC to DLR and installed in aircraft for pre-Shuttle test flights in various regions of Europe.

The following organizations participated in the OSTA-1/OCE related activities: GSFC (USA), DLR (Germany), University of Lisbon (Portugal), Hydrologic Institute of the Navy (Portugal), University of Lille (France), GKSS (Germany), INTA Madrid (Spain), IIP Barcelona (Spain).


# A.95   SIR-B

SIR-B = Shuttle Imaging Radar with Payload B. A JPL payload on Shuttle as part of the OSTA-3 payload. Launch: Oct. 5, 1984 (duration: 1 week).

Application: Land use, Geology, Hydrology, Forestry, Mapping

Orbit: Shuttle orbit, 57° inclination, 225 km altitude, Swath width: 30 - 60 km

Sensor:**SAR** = Synthetic Aperture Radar in HH Polarization; L-Band (freq. = 1.28 GHz); Look Angle = variable from 15 - 60°; Pixel size = 25m azimuth; Radiometric resolution = 3-6 Bit raw data

**Data:**
Image size: 30 - 60 km x length of active radar
Data transmission: Shuttle Downlink via TDRSS; Data rate: 40 Mbit/s

318) Manual of Remote Sensing, Second Edition, American Society of Photogrammetry, 1983, pp. 1707-1710
319) H. v.d. Piepen, V. Amann, H. Helbig, HH. Kim, W. Hart, et al. "The Promise of Remote Sensing, IEEE paper presented at IGARSS '82, June 1-4, Munich

## A.96   SIR-C/X-SAR

Shuttle Imaging Radar with Payload C (SIR-C)[320],[321]. A NASA/JPL, DARA/DLR, and ASI payload on Shuttle. This is also known under the name SRL-01 = Space Radar Laboratory. Total payload mass = 5600 kg, power consumption of payload sensors = 3 - 8.5 kW. A total of 3 Shuttle missions are planned, each of 10 days duration:

1. SRL-01: April 1994

2. SRL-02: Aug. 1994

3. SRL-3: 1996 [as of 1993 concepts are worked on to convert this Shuttle mission into a 'free flyer' mission (launched by the Shuttle) to extent the mission life to about one year for a much better coverage and data return of a multifrequency SAR payload].

Application: Land use, geology, hydrology, oceanography, snow and ice, vegetation, calibration, and technological experiments.

Objectives:
* Conduct geoscience investigations that require the observational capabilities of orbiting radar sensors alone, or in conjunction with other sensors, that will lead to a better understanding of the surface conditions and processes on the Earth.
* Explore regions of the Earth's surface that are not well characterized because of vegetation, cloud, or sediment cover in order to better understand land and ocean surface conditions and processes on a global scale.
* Incorporate this new knowledge into global models of surface and subsurface processes.

Orbit: Shuttle orbit, 57° inclination, 225 km altitude, swath width: 15 - 90 km

Sensors: All 3 sensors are flown on each mission. The SAR sensors are independent of the day/night cycle and mostly independent of the weather.

1. **L-Band SAR** (1.28 GHz), JPL

2. **C-Band SAR** (5.3 GHz), JPL. The L-Band and C-Band SARs allow multi-frequency and multi-polarization measurements. Parallel image generation in L-Band and C-Band with HH,VV,HV, or VH polarization. Look angle = variable from 20 - 55°. Pixel size = 25 m in azimuth and 25 m in slant range. Data Rate = 90 Mbit/s for L-Band and 90 Mbit/s for C-Band.

3. **X-SAR** (SAR for X-Band Measurement (9.6 GHz), provided by DARA/DLR and ASI) X-SAR uses only vertical polarization (VV). Swath width = 15 - 45 km. Azimuth resolution = 30 m; Slant range resolution = 10-20 m. Look angle (off nadir) = 20 - 55°. Data rate = 45 Mbit/s.

**Data:**
- The source data are digitally coded and onboard recorded.
- Transmission: only 1 SAR data stream at a time is possible. Ku-Band via TDRSS to White Sands, etc.
- Image size: 100 km (flight direction) x 50 km
- An X-SAR POCC is being built by DLR at JSC (Houston)

Nominally, 50 hours of SIR-C data (on each of the four channels) and 50 hours of X-SAR data will be recorded by onboard tape recorders. In addition, a limited amount of SIR-C

320) "X-Band Synthetic Aperture Radar (X-SAR) and its Shuttle-Borne Application for Experiments", Paper by Herwig Öttl and Francesco Valdoni
321) R.L. Jordan, B. L. Huneycutt, M. Werner, 'The SIR-C/X-SAR Synthetic Aperture Radar System', Proceedings of the IEEE, Vol. 79, No. 6, June 1991, pp. 827-838

data may be transmitted to the ground (via TDRSS Ku-Band 50 Mbit/s data link). for near-real-time digital processing during the mission.

The goal is to provide data calibrated in such a way as to allow comparisons with other spaceborne SAR data (e.g. ERS-1, JERS-1, Radarsat, etc.) so that a time-series view of key geophysical parameters may be realized.

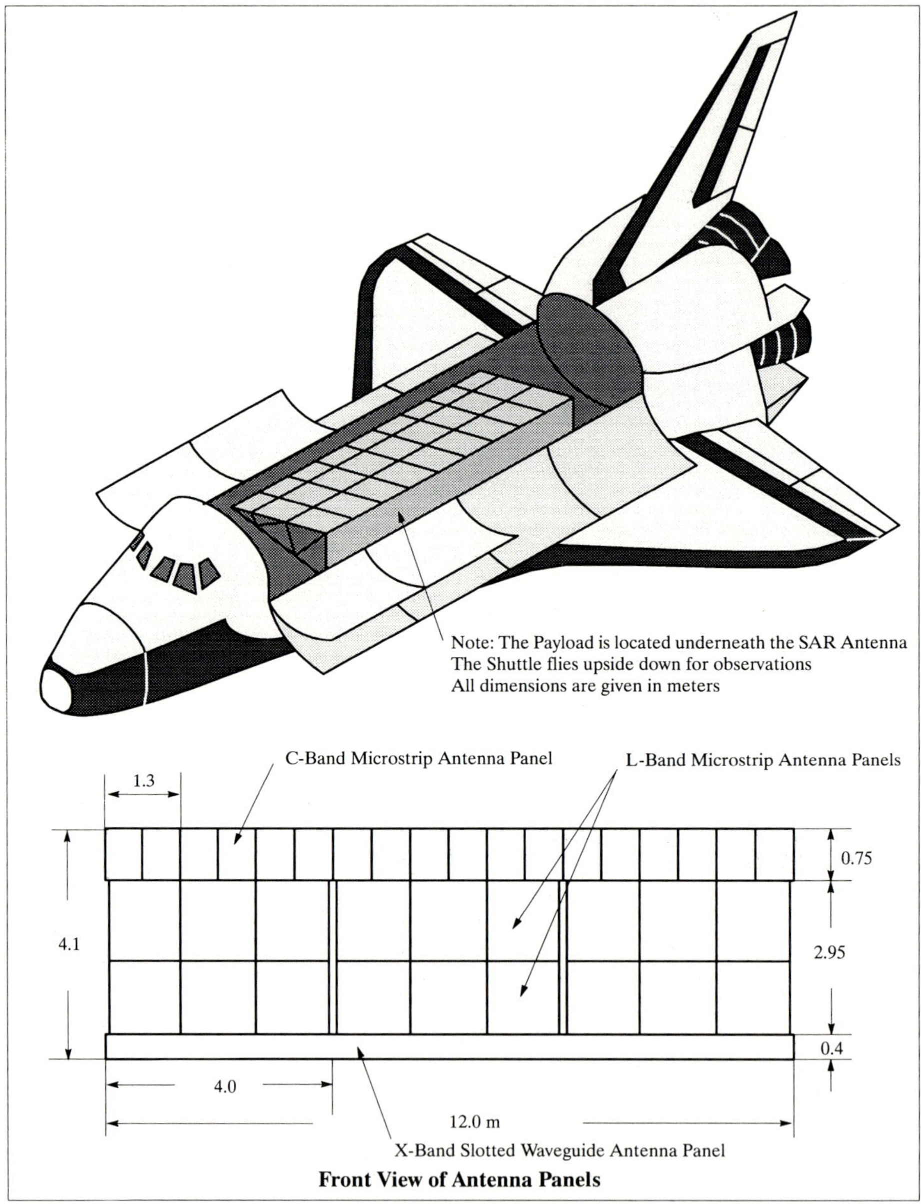


**Figure 89:      Configuration of the SIR-C/X-SAR Payload in the Shuttle Bay**

## A.97     SKYLAB Space Station

NASA Program. Skylab was launched on 14. May 1973 (Saturn 5) and subsequently visited by three Apollo astronaut crews, who lived and worked in it for periods of 28, 59, and 84 days respectively (1. crew launch: 25. May 1973 to 8. Feb. 1974 (last splash down of a crew).[322]

- Skylab-1   Launch: 14. May 73 (unmanned)      34,981 orbits      Re-entry: 11.7.79

- Skylab-2   Launch: 25. May 1973 (manned)      404 orbits         Re-entry: 22.6.73

- Skylab-3   Launch: 28. July 1973 (manned)     858 orbits         Re-entry: 25.9.73

- Skylab-4   Launch 16. Nov. 1973 (manned)      1,214 orbits       Re-entry: 3.2.74

Orbit: Circular orbit, altitude = 435 km, inclination = 108°, period = 101 minutes.

Sensors: (Earth Resources Experimental Package, there were a total of six sensors)

- **S190A** = Multispectral Photographic Camera System (used six identical cameras with different film/filter combinations in order to view the same ground area simultaneously in the visible region)
  Spectral range:        400 - 900 nm
  Focal length:          152 mm
  Image format:          5.7 x 5.7 cm
  Image scale:           1 : 2,850,000
  Image overlapping:     60%
  FOV                    21.2°
  Image size             163 x 163 km
  Ground resolution      100 - 260 m

- **S190B** = Earth Terrain Camera
  Spectral range         400 - 880 nm
  Focal length           457 mm
  Image format           11.4 x 11.4 cm
  Image scale            1 : 950,000
  Image overlapping:     60%
  FOV                    14.24°
  Image size             109 x 109 km
  Ground resolution      55 - 100 m

- **S 191** = Infrared Spectrometer (camera system)[323]. The S191 experiment was basically a visible and infrared spectroradiometer (hand-held instrument). Measurement of radiation flux in the bands from 0.4 to 2.4 µm, and from 6.2 to 15.5 µm. FOV = 1 mrad. Note: The S191 and S192 experiments were operated from May 1973 to Feb. 1974. Both systems provided useful data. The data return was limited by the amount of magnetic tape that could be transported in resupply flights.

- **S 192** = Multispectral Opto-mechanical Scanner. Spectral range: 0.4-12.5 µm; number of spectral bands = 13; radius of circular scan = 42 km; swath width = 72.4 km; IFOV = 79 m (0.182 mrad.); instrument weight = 57 kg; power = 266 W (peak).

- **S 193** = Passive Microwave Radiometer/Active Scatterometer and radar altimeter. First implementation of a combined passive and active microwave sensor. Objective: simultaneous measurements of radar backscatter and radiometric brightness temperature in a number of scanning modes, primarily for the purpose of studying of surface winds, precipitation over the oceans.
  Frequency = 13.9 GHz; swath width = 180 km; spatial resolution = 16 km; mechanically scanning parabolic antenna

322) 'Skylab', Jane's Spaceflight Directory 1988-89, 4. Edition, pp. 117-122
323) P. Slater, 'Remote Sensing', Optics and Electronics Systems, Addison-Wesley Publishing Co., 1980, pp.456-462

- **S 194** = Passive Microwave Radiometer. Objective: measurement of soil moisture. Frequency = 1.4 GHz (21 cm wavelength, L-Band); spatial resolution = 115 km; nadir viewing array antenna.


## A.98    SME (Solar Mesosphere Explorer)

Atmospheric mission operated by the University of Colorado at Boulder (JPL management of SME project).[324],[325],[326],[327] NASA launch: 6 Oct. 1981 with Delta vehicle from Vandenberg. The satellite was operational until April 13, 1989. Objective: Study of atmospheric (mesosphere) ozone and the processes that form and destroy it.
SME is a spin-stabilized S/C at 5 rpm, the spin axis is maintained nearly perpendicular to the orbit velocity vector by means of a magnetic torquing attitude control system. The payload (instrument package) was developed at the Laboratory for Atmospheric and Space Physics (LASP) of the University of Colorado.

Orbit: Sun-synchronous near-circular orbit; altitude = 534 km, inclination = 97.7°,

**Sensors:** (total of 5 instruments with 4 limb-scanning optical instruments)

- UV Ozone Experiment. Objective: observation of ozone concentrations (ozone absorption of Rayleigh scattered sunlight at the limb) in the altitude region of 50 - 70 km. The instrument is dual-channel Ebert-Fastie spectrometer operating in the spectral range of 2460 - 3350 Å (channel A) and 2710 - 3350 µm (channel B); 64 grating steps per scan. The Cassegrain telescope focal length = 250 mm. FOV = 0.1° x 1.5°.

- Four-Channel Infrared Radiometer (passive device). Objective: radiance measurements in the infrared region.
  The instrument makes temperature measurements by viewing carbon dioxide emission with narrow and wide spectral bandwidth channels centered at 15.5 µm. Ozone and water emissions are measured at 9.6 µm and 6.3 µm, respectively. IFOV sweeps through a 4° angle that is equivalent to a limb scan from 20 - 150 km in altitude. Each channel is sampled 20 times per S/C revolution.

- Airglow Instrument. Objective: measurement of the ozone altitude profile by limb scanning at 1.27 µm.
  The instrument consists of a grating spectrometer (Ebert Fastie) and an uncooled detector system with a narrow FOV.

- Visible Nitrogen Dioxide Experiment. Objective: measurement of nitrogen dioxide concentrations in the altitude region from 20 - 40 km ($NO_2$ absorption of Rayleigh scattered sunlight at the limb).
  The instrument is an Ebert-Fastie spectrometer (of same design as before). Spectral range = 3250 - 4500 Å (2nd order, channel A), = 5200 - 7700 Å (1st order, channel B); 256 grating steps per scan.

- Solar UV Monitor. Objective: monitoring of solar flux scattered from a diffusing screen. The instrument is an Ebert-Fastie spectrometer (of same design as before). Operation of instrument in two modes: 1. scanning of the spectral range of 1600 Å to 3100 Å by rotation of the grating; 2. setting of the grating to a required wavelength with the instrument monitoring the solar flux.
  The instrument is mounted to the side of the S/C scanning once per S/C revolution through the sun.


324)  J. R. Cowley, G. M. Lawrence, "Earth Limb Altitude Determination for the Solar Mesosphere Explorer", AIAA-83-0429
325)  Ch. Barth, "Solar Mesosphere Explorer to study ozone", Nature, Volume 293, 24 Sept. 1981
326)  J. R. Stuart, K. A. Gause, "Solar Mesosphere Explorer Mission", AIAA paper, 17th Aerospace Sciences Meeting, Jan. 15-17, 1979,
327)  "Solar Mesosphere Explorer - Experiment Description", paper of LASP, University of Colorado at Boulder

Data: The SME data have been archived in the NSSDC (National Space Science Data Center) at GSFC and in the SME database at the University of Colorado.[328]

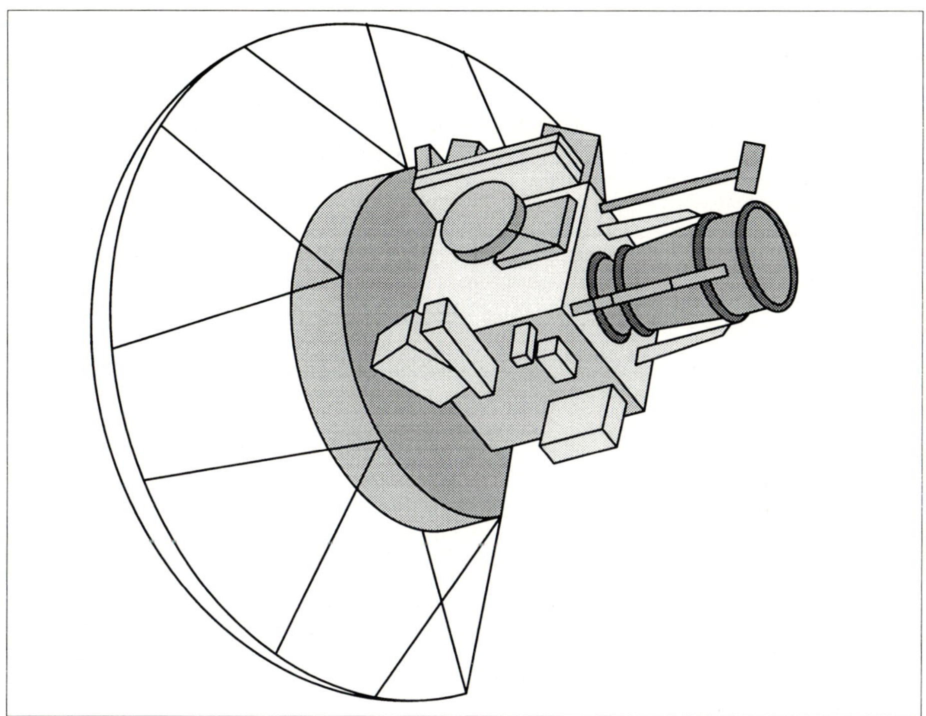

**Figure 90:    The SME S/C Model**

## A.99    SMM (Solar Maximum Mission)

NASA/GSFC mission. Objectives: Observation of solar flares (trigger mechanisms) over a wide band of wavelengths in the UV, X-ray and gamma-ray regions of the spectrum; coverage of the maximum period of a sunspot cycle, .... for a better understanding of the violent nature of the Sun and its effects on the Earth. Measurement of radiation fluxes.
Launch Feb. 14, 1980 by Delta vehicle from Cape Canaveral. Satellite mass = 2315 kg. End of mission: 1989.

SMM performed flawlessly until an attitude-control system failure (actually a failure of a fuse) occurred on Nov. 23, 1980 which disabled the S/C to point precisely at the Sun. In April 1984 SMM became the first satellite to be retrieved, repaired, and redeployed in orbit by a Shuttle crew. [329], [330] The repair was also used to replace in orbit a dish antenna, enabling SMM to transmit its data via TDRS-1.

Orbit: circular Earth orbit, apogee = 569 km, perigee = 566 km, inclination = 28.5°, period = 96 min (the S/C spends about 60 min of each orbit in full view of the Sun).

328)  "Solar Mesosphere Explorer - Scientific Data & Publications", Final Report, LASP, December 1989
329)  S. P. Maran, B. E. Woodgate, "A Second Chance for Solar Max", Sky & Telescope, June 1984, pp. 498-500
330)  A. Chaikin, "Solar Max: Back from the Edge", Sky & Telescope, June 1984, pp. 494-497

**Sensors:**[331), 332)]

**ACRIM** = Active Cavity Radiometer Irradiance Monitor (PI: R. C. Willson, JPL).[333)]
Objective: regular observations of the total solar irradiance. The goals of the experiment
are:

-    i) to begin a long-term climatological data base on solar irradiance variability
-    ii) to provide a shorter-term data base for solar physics investigations.

ACRIM has three active cavity radiometer (ACR) type IV sensors. The sensors view the
Sun through a 5° (full angle) field of view. ACRIM uses black-body cavity to measure the
total irradiance of the visible hemisphere of the sun.

ACRIM is a cavity pyrheliometer, with solar input during half of each 131.072 s measurement interval and servo-adjusted heat input during the rest of each cycle, when the shutter
blocks the solar input. The solar irradiance at the Earth is therefore proportional to the difference between the heating rates with shutter open and closed. The accuracy of each irradiance measurement obtained in this way is $\approx 0.1\%$.

**C-P** = Coronagraph - Polarimeter (PI: R. M. MacQueen, NCAR Boulder, Co.).
Objective: corona change monitoring over time, observation of the corona's large-scale
magnetic structures and their evolution.

The C/P instrument is an externally occulted Lyot coronagraph with a vidicon detector.
Images of the corona with 10" spatial resolution are built up in four quadrants by means of
sector mirrors (spectral range: 4465 - 6385 Å in 7 bands). In addition, with seven filters and
three polaroids with principal axes 60° apart, the C/P is able to obtain measurements of coronal electron densities and distinguish ejected material at chromospheric temperatures
from coronal features.

**GRS** = Gamma-Ray Spectrometer (PI: E. L. Chupp, Uni. of New Hampshire).
The instrument includes seven integral-line calibrated NaI(Tl) detectors for nuclear lines
between 0.3 and 0.9 MeV. The energy resolution (full width at half maximum) of these detectors is > 7% at .662 MeV. In addition, a thick CsI(Na) crystal, along with the NaI(Tl)
spectrometers, is sensitive to $\gamma$-rays in the range 10-140 MeV, and neutrons with energies
above 20 MeV. An auxiliary system of two NaI detectors yields hard X-ray fluxes from 10 to
140 keV. Time resolutions are typically about 2 s and 16 s, but can be as fine as 64 ms for the
300-350 keV energy band.

Among other things, GRS made important contributions to the international study of Supernova 87A, which in Feb. 1987 provided astronomers with their first local opportunity to
study such an explosion since 1604.

**HXRBS** = Hard X-Ray Burst Spectrometer (PI: B. R. Dennis, GSFC). Objectives:

-    to determine the nature of the mechanisms which accelerate electrons to 20-200 keV in
  the first stage of the solar flare, and to >1 MeV in the second stage of many flares
-    to characterize the spatial and temporal relation between electron acceleration, storage and energy loss throughout the solar flare

The instrument is a collimated X-ray spectrometer. The detector consists of two primary
scintillation crystal components. Scintillation events in its actively collimated CsI(Na) detector are read out every 128 ms in fifteen energy channels between $\sim 25$ to $\sim 500$ keV. A
circulating memory is able to accumulate relatively brief periods of data during the more
intense flares with time resolution down to 1 ms. The full width at half maximum of the field
of view is $\sim 40°$.

**HXIS** = Hard X-Ray Imaging Spectrometer (PI: C. de Jager, Astronomical Institute,
Utrecht, NL, and Uni. of Birmingham, UK).

---


331) 'NASA's Solar Maximum Mission: A Look at the New Sun', June 1987, NASA brochure, edited by J. B. Gurman
332) "The Solar Maximum Mission Experiments", Solar Physics, Volume 65, pp. 5-109
333) R. C. Willson, S. Gulkis, M. Janssen, S. H. Hudson, G. A. Chapman, "Observations of Solar Irradiance Variability", Science, Volume 211, Feb. 1981, pp. 700-702

Objective: Study of the energy problem associated with hard X-ray generating electrons. Study of flare models and mechanisms.

The instrument consists of an imaging collimator of ten grid plates, each divided into 576 sections, and a position-sensitive detector system consisting of 900 mini-proportional counters. The grids form a coarse field of view 6' 24" in extent and a fine field of view 2' 40" across (with 8" square pixels). Each pixel is sampled in six energy bands ranging from 3.5 to 30 keV, with time resolution down to 1.25 s.

HXIS served as "flare alarm" to alert the other instruments electronically. Note: HXIS malfunctioned in June 1981 and could not be repaired in the SMM rescue mission in April 1984. The 1980 HXIS data set, however, forms a most impressive record of the spatial and energetic distribution of X-rays in solar flares, with a spatial resolution never before achieved at such high energies.

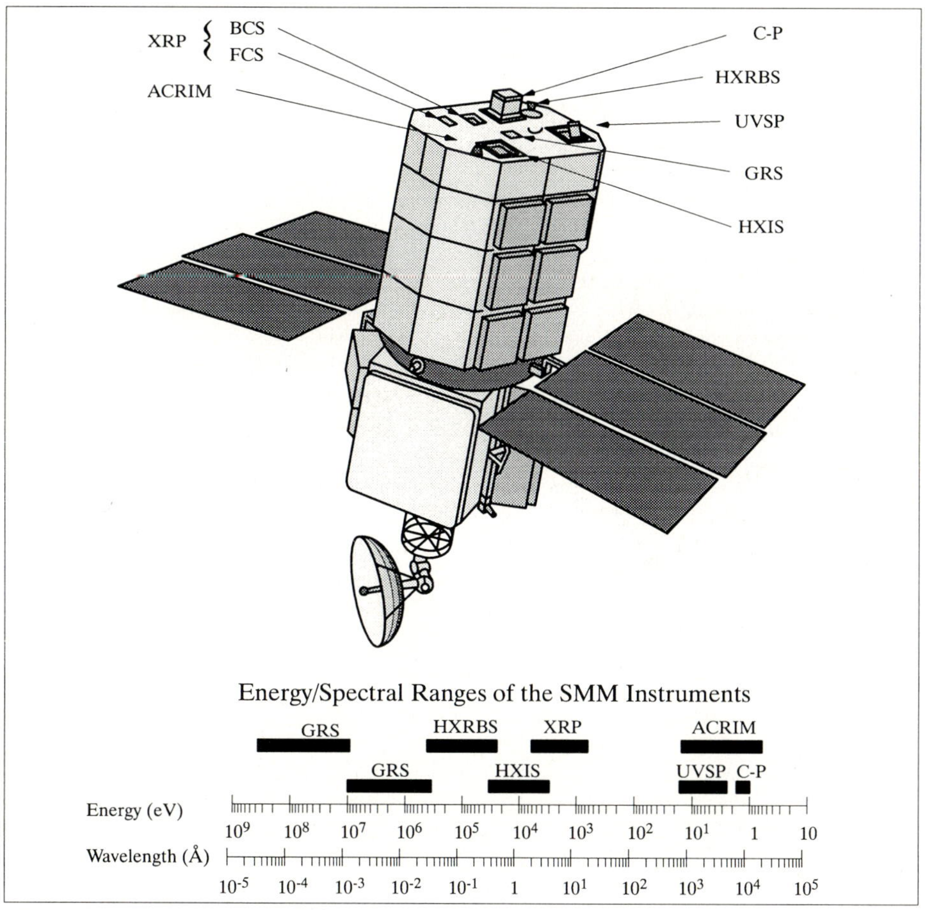


**Figure 91:    The SMM S/C Model**

**UVSP** = Ultraviolet Spectrometer and Polarimeter (PI: E. Tandberg-Hanssen, MSFC). Objective: Study of the UV radiation from the solar atmosphere, in particular from the active regions, flares, prominences, and the active corona.

UVSP consists of an aplanatic Gregorian telescope of ~2" resolution, an Ebert-Fastie spectrometer, and five photomultiplier detectors. The telescope secondary is mounted on gimbals and may be rastered to build up an image of an area on the Sun of up to 256" x 256", in steps of as little as 1". The entrance apertures range in size from 1" x 1" to 15" x 286", and the exit slits range from 0.1 to 3.0 Å in second order.

A polarimeter may be inserted behind the exit aperture; it consists of two systems: a three-quarter wave retarder for those wavelengths, where the spectrograph grating is an effective linear polarizer, and a quarter-wave retarder and linear polarizer for use at wavelengths at which the grating polarization is very low. The instrument is sensitive to wavelengths of 1170 - 1800 Å in second order and up to 3600 Å in first order.

In April 1985, the UVSP grating drive mechanism became stuck, the grating appeared to be fixed at a wavelength of $\approx 1380$ Å (second order; $\approx 2760$ Å in first order). During periods of solar activity, UVSP is used in second order to provide pointing and timing information for other SMM instruments; **at other times, first order measurements of ozone concentration in the Earth's atmosphere are carried out at S/C sunrise and sunset (occultation measurements). These unique observations are making possible a mapping of ozone concentrations at latitudes of $\pm 50°$ at altitudes of 50 - 75 km. Approximately 20000 separate ozone altitude profiles have been obtained from late 1985 through March 1989.**[334])

**XRP** = (Soft) X-Ray Polychromator (PIs: J. L. Culhane, Uni. College, London; K. J. H. Phillips, RAL, UK; K. T. Strong, Lockheed, Ca.).
Objective: Investigation of those aspects of solar activity that lead to the production of plasma temperatures in the 1.5 to 50 million degree range.
The XRP consists of two distinct instruments, the **BCS** = Bent-Crystal Spectrometer (PI: J. L. Culhane) and **FCS** = Flat-Crystal Spectrometer. The FCS is able to rotate its crystals to provide its seven detectors with access to the spectral range 1.40-22.43 Å. The BCS, with a collimator field of view of 6', is able to obtain spectra in the range 2-3 Å, with a resolution $\lambda/\Delta\lambda$ of $\approx 10^4$, with eight position-sensitive proportional counters.

Data: The downlink of science data was considerably improved with the availability of TDRS in April 1983. Data from ground-based Sun observations (collation of telescopic data and magnetograms with Solar Max's maps) served as forecasts for likely event predictions on the Sun for instrument precision pointing setup of the flare watch on the actual flare site.


## A.100   SOHO (Solar and Heliospheric Observatory)

ESA/NASA collaborative mission within the ESA's 'Solar Terrestrial Science Programme' (STSP), part of ISTP (International Solar Terrestrial Physics Programme). The following overall mission objectives are pursued:

- solar spectroscopy at soft X-ray and EUV wavelengths (study of the composition of the solar corona, of the structure and dynamics of the magnetic structures making up the corona, and of the coronals holes, etc.)
- study of the structure and dynamics of the solar interior through the observation of minute oscillations on the Sun's surface (helio-seismology).
- study of the solar wind and solar energetic particles, interaction with the Earth, plasma processes in both the solar and magnetospheric context.

A NASA[335),336)] launch is projected for July 1995 aboard an ATLAS-II AS vehicle from Cape Canaveral. The spacecraft is spin-stabilized (3 axis), power = 1350 W (solar cells) and 950 W from NiCd batteries, scientific payload mass = 600 kg, nominal lifetime = 2 years,

---

334) A. C. Aikin, W. Henze, D. J. Kendig, R. Nakatsuka, H. J. P. Smith, " Variations of Mesospheric Equatorial Ozone as observed by the Solar Maximum Mission", Geophysical Research Letters, Vol. 17, No. 3, March 1990, pp. 299-300,
335) P. Lo Galbo, M. Bouffard, "SOHO - A Cooperative Scientific Mission to the Sun", esa bulletin, Aug. 1992, pp. 21-25
336) "The Solar-Terrestrial Science Project of the Inter-Agency Consultative Group for Space Science", esa SP-1107, November 1990, pp. 21-24

mission operations are performed from GSFC, the S/C is provided by ESA. The C/D phase started in May 1991.

Orbit: Halo orbit around the Sun-Earth Lagrangian Point ($L_1$), about 1.5 million km from Earth (see Figure 153 and Table 238). The pointing accuracy of the S/C in the direction of the Sun $\leq$ 1 arcsec.

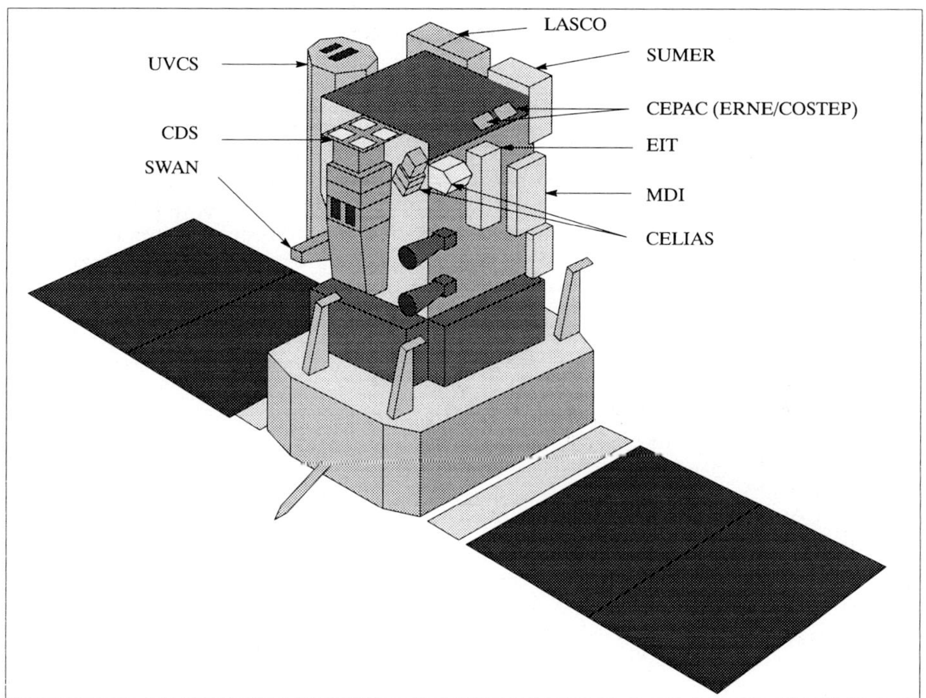


**Figure 92:    The SOHO S/C Model**

**Sensors:** (8 from Europe, 3 from USA)[337]

### 1.    Solar Atmosphere Remote Sensing Instruments

**SUMER** = Solar Ultraviolet Emitted Radiation (PI: K. Wilhelm, MPI, Germany)
Study of plasma flow characteristics (turbulent motions, waves. temperatures, and densities of the plasma in the upper atmosphere of the Sun).
The instrument takes images of the Sun in EUV light with high resolution in space, wavelength and time. The spatial resolution is 1.2 arcsec, the spectral resolving power is characterized by a $\lambda/\delta\lambda$ of up to $4.0 \times 10^4$ ($\delta\lambda$ corresponds to the pixel size). Spectral shifts can be determined with sub-pixel accuracy. Wavelength range = 500 - 1600 Å. Integration can be as short as 1 second. Data: on line profiles, shifts and broadenings, ratios of temperature- and density-sensitive EUV emission lines in the range between $10^4$ and $2 \times 10^6$ K.

**CDS** = Coronal Diagnostic Spectrometer (B. E. R. Harrison, RAL, UK)
Study of the transition region and corona temperature and density (intensity ratios of selected EVU line pairs, with spatial and temporal scales appropriate to the fine-scale features of the solar atmosphere). There will be close collaboration and correlation with SUMER.

---

337) The SOHO Mission - Scientific and Technical Aspects of the Instruments, esa SP-1104, ISSN 0379-6566, Nov. 1988

The instrument consists of a telescope [outer f-number = 9.48, FOV = 4 arcmin, on axis resolution = 2 arcsec (FWHM)], a normal incidence spectrometer (NIS) with a wavelength = 2 bands in 240 - 800 Å range, and a grazing incidence spectrometer (GIS) with a wavelength = 4 bands in 150 - 800 Å range.

**EIT** = Extreme-Ultraviolet Imaging Telescope (J. P. Delaboudinière, IAS/CNRS, France)
Study of the evolution of chromospheric and coronal structures (identification and interpretation of the spatial and temperature fine structures of the solar atmosphere). EIT provides full disk images in emission lines formed at temperatures that map the solar structures. Images in four narrow band passes are obtained using normal incidence multi-layered optics deposited on quadrants of a Ritchey-Chretien telescope. Temperature range of solar structures to be measured: $6 \times 10^4$ - $3 \times 10^6$ K. Resolution = 1 arcsec (uniform over FOV = 50 x 50 arcmin).

**UVCS** = Ultra-Violet Coronograph Spectrometer (J. L. Kohl, SAO, USA)
Study of electron and ion temperatures, densities and velocities in the corona (for interpretation in coronal heating, solar wind acceleration, and solar wind composition). Spectroscopic measurements of the solar corona out to 10 solar radii from the sun-center.
The UVCS instrument consists of an occulted telescope and a high resolution spectrometer assembly. Three off-axis paraboloidal mirrors focus co-registered images of the extended corona onto the three entrance slits of the spectrometer assembly. Three sections in spectrometer. One section is optimized for line profile measurements of H I Lyman-$\alpha$, its wavelength range is 1148 - 1248 Å. Another section is optimized for line intensity measurements of O VI $\lambda$ 1032, and is also used to observe Si XII $\lambda$ 499 and 521 in second order. The first order range is 932 - 1068 Å. The Mg X doublet at $\lambda$ 610 and $\lambda$ 625 may be observed in first order. The third section is used to measure polarized radiance of the visible corona.
Telescope: off-axis parabolic mirror, focal length = 750 mm, image scale = 0.218 mm/arcmin, FOV = 42' (tangential) x 141'.
Spectrometer: gratings = 105 (disp) x 70 mm; nominal radius = 750 mm; radius ratio = 1.0259 (1216 Å), = 1.0215 (1032 Å, 625Å), = 1.0040 (visible); ruling frequency = 2400 mm$^{-1}$; reciprocal dispersion = 5.63 Å/mm (1$^{st}$ order), = 2.82 Å/mm (2$^{nd}$ order).

| Spectral Line | Observed Quantity | Spectral Resolution FWHM (Å) | Spatial Resolution FWHM |
|---|---|---|---|
| H I 1216 | Profile | 0.2 | 10" x 10" |
| H I 1216 | e⁻ Profile | 2.0 | 1.6' x 1' |
| Fe XII 1242 | Intensity | 1.23 | 1' x 1' |
| Fe XII 1242 | Profile | 0.14 | 10" x 5' |
| O VI 1032/37 | Intensity | 1.23 | 1' x 1' |
| O VI 1032 | Profile | 0.14 | 10" x 5' |
| Si XII 499/521, 2nd order | Intensity | 0.61 | 1' x 1' |
| Si XII 499, 2nd order | Profile | 0.07 | 7" x 5' |
| Mg X 610/625, 2nd order | Intensity | 0.61 | 1' x 2.5' |
| MG X 610, 2nd, order | Profile | 0.07 | 7" x 5' |

**Table 75:**     **Observational Parameter Specifications for the UVCS Instrument**

**LASCO** = Large-Angle and Spectrometric Coronograph (Pi: G. E. Brueckner, NRL, USA)
Study of structure evolution, mass, momentum and energy transport in the corona. Measurement of electron column densities from just above the limb, at 1.1 $R_{Sun}$, out into deep heliospheric space, at 30 $R_{Sun}$. Spectral analysis of the inner corona with a high resolution scanning. imaging interferometer. Measurement of spectral profiles of three emission lines and one Fraunhofer line for each picture point (providing temperatures, velocities, turbulent motions and volume densities). Direction of coronal magnetic fields via polarization analysis.

The LASCO instrument consists of the coronograph optical system (with three optical systems and CCD cameras) and the instrument control system.

**SWAN** = Solar-Wind Anisotropies (J. L. Bertaux, CNRS, France)
Study of the solar-wind mass flux anisotropies and temporal variations. The SWAN instrument is a Lyman photometer which maps the sky interplanetary hydrogen emission almost entirely every other day. From these sky maps, the latitude distribution of the solar wind mass flux from the equator to the pole can be measured, as well as the variation of this distribution (time scale $\approx$ one month).
The instrument consists of two identical sensor units placed on opposite sides of the S/C and driven by a common electronic box. Each sensor consists of a two-mirror scanning system, a light baffle (to keep out stray light), a hydrogen cell in which a tungsten filament is alternately heated to produce Lyman-$\alpha$ absorbing hydrogen atoms, and cooled to produce the reference signal. Resolution of Lyman-$\alpha$ light distribution is $1^\circ$.

## 2.    Solar Wind 'In Situ" Instruments

**CELIAS** = Charge, Element and Isotope Analysis (D. Hovestadt, MPI, Germany)
Study of ionic energy distribution and composition. Measurement of the mass, ionic charge and energy of the low and high speed solar wind, of suprathermal ions, and of low energy flare particles.
CELIAS includes 3 mass- and charge-discriminating sensors based on the time-of-flight technique: CTOF (Charge Time-of-Flight) for the elemental, charge and velocity distribution of the solar wind; MTOF (Mass Time-of-Flight) for the elemental and isotopic composition of the solar wind, MTOF also includes a proton monitor (PM); and STOF (Suprathermal Time-of-Flight) for the mass, charge and energy distribution of suprathermal ions. Mass resolution: $M/\Delta M > 100$ (MTOF; charge resolution: $\Delta Q \approx 0.3 - 1, 4 < M < 60$ (typical for CTOF and STOF).

| Sensor | Area, Geometric Factor | $\Delta E/E$ | Efficiency of the TOF unit |
|---|---|---|---|
| CTOF | $0.08\ cm^2$ | 0.04 | 0.25 - 0.65 |
| MTOF | $0.013\ cm^2$ | 0.03-5 | 0.03 - 0.1 |
| PM | $7 \times 10^{-5}\ cm^2$ | 0.05 | - |
| STOF | $0.1 - 0.2\ cm^2\ sr$ | 0.1 | 0.2 - 0.8 |

**Table 76:**    **Measurement Capabilities of the CELIAS Instrument**

**CEPAC** = COSTEP/ERNE Particle Analyzer. Collaboration (Pi: J. Torsti, ERNE Finland, H. Kunow, U. of Kiel, Germany). Note: COSTEP = Comprehensive Suprathermal and Energetic Particle Analyzer. Three sensors are furnished by the COSTEP consortium: LION (Low Energy Ions and Electrons), MEICA (Medium Energy Ion Composition Analyzer), and EPHIN (Electron Proton Helium Instrument).
Study of energy distribution of particles, energy spectrum and composition. Measurement of particle emissions from the Sun over a wide range of species (electrons through iron) and energies (60 keV/particle to 500 MeV/nucleon).

LION instrument consists of two sensor heads each containing a double telescope to measure energy spectra of ions and electrons in the range of 60 keV - 5 MeV for protons, and 60 keV - 300 keV for electrons.

EPHIN is a multi-element array of solid state detectors with a plastic scintillator guard counter to measure energy spectra of electrons in the range of 150 keV to > 5 MeV, and hydrogen and helium isotopes in the range of 4 MeV/n to > 53 MeV/n.

## 3.    Helio-seismology Instruments

**GOLF** = Global Oscillations at Low Frequencies (Pi: A. Gabriel, IAS/CNRS, France)
Study of the internal structure of the Sun by measuring the spectrum of free global oscilla-

tions [global velocity and magnetic field oscillations (low degree modes)]. GOLF measures both p and g mode oscillations, with the emphasis on the low order long period waves which penetrate the solar core (frequencies between $10^{-7}$ and 6 x $10^{-3}$ Hz, with a sensitivity of 1 mm/s). The method involves an extension to space of the ground-based technique for measuring the mean line-of-sight velocity of the viewed solar surface. A sodium vapor resonance scattering filter is used in a longitudinal magnetic field to sample the two wings of the solar absorption line.

**VIRGO** = Variability of Solar Irradiance and Gravity Oscillations (Pi: C. Froehlich, PMOD/WRC, Switzerland)
Study of irradiance oscillations (low degree modes) and solar constant.
The VIRGO instrument contains two types of active cavity radiometers for monitoring the solar "constant", two three-channel sun photometers (SPM) for the measurement of spectral irradiance at 335, 500, and 865 nm, and a low resolution imager (LOI) with 12 pixels. The prime scientific objective is probing the solar interior by helioseismology with p- and g-mode solar oscillations determined from spectral irradiance (SPM) and radiance (LOI) variations on time scales of minutes to the mission time.

**MDI** = Michelson Doppler Imager (P. H. Scherer, Stanford University, Ca., USA)
Study of velocity oscillations (high degree modes). Measurement of line-of-sight velocity by Doppler shift, transverse velocity by local correlation tracking, line and continuum intensity, and line-of-sight magnetic fields with both 4 and 1.4 arcsec resolution (2 and 0.7 arcsec pixels respectively).
MDI provides velocity maps by sampling the solar Ni I 6768 Å line profile at four points and finding the Doppler shift by determining the phase of the first Fourier coefficient of the line shape with respect to a reference wavelength.

SOHO Data: Onboard recording capability of 1 Gbit of data (requirement of 48 h autonomous operation without ground contact). Science data transmission via S-Band. Downlink date rate = 220 kbit/s.


## A.101   SOLAR-A (Yohkoh)

Japanese (ISAS) Solar-Terrestrial mission. The primary objective is the study of high energy phenomena on the Sun through X-ray and Gamma-ray observations.[338]

The S/C has dimensions of 1 m x 1 m x 2 m (height), mass = 420 kg (100 kg science payload), power supply = 450 W (solar cells). Launch on Aug. 30, 1991 by an ISAS M-3SII-6 vehicle; nominal life = 2 years; S/C is 3-axis stabilized pointing toward the Sun.

Orbit: apogee = 792 km, perigee = 525 km, inclination = 31.3°, period = 97 minutes

**Sensors:**

- **HXT** = Hard X-Ray Telescope. Multi-pitch bi-grid modulation collimators with Na I scintillators. HXT is a Fourier telescope with 64 collimators covering an energy range from 10 - 100 keV at 7 arcsec resolution for the whole solar disk. HXT mass = 48 kg.

- **SXT** = Soft X-Ray Telescope[339] (US-Japan collaboration between Lockheed Palo Alto Lab and Tokyo Astronomical Observatory). A grazing incidence soft x-ray mirror with an x-ray CCD detector. SXT uses a 23 cm diameter Wolter-Nariai optics of 155 cm focal length, it covers the energy range from 0.25 - 3 keV at 2 arcsec resolution. Imaging is provided by a 1024 x 1024 pixel CCD behind two six-position filter wheels. Mass = 27 kg. - SXT has produced more than 1 million images of the sun since the S/C was launched.

---

338) 'The SOLAR-A Mission', The Solar-Terrestrial Science Project of the Inter-Agency Consultative Group for Space Science, esa SP-1107, November 1990, pp. 74-76
339) "Yohkoh's Prodigious Output helps Scientists Study Sun", Space News, June 7-13, 1993, p. 12

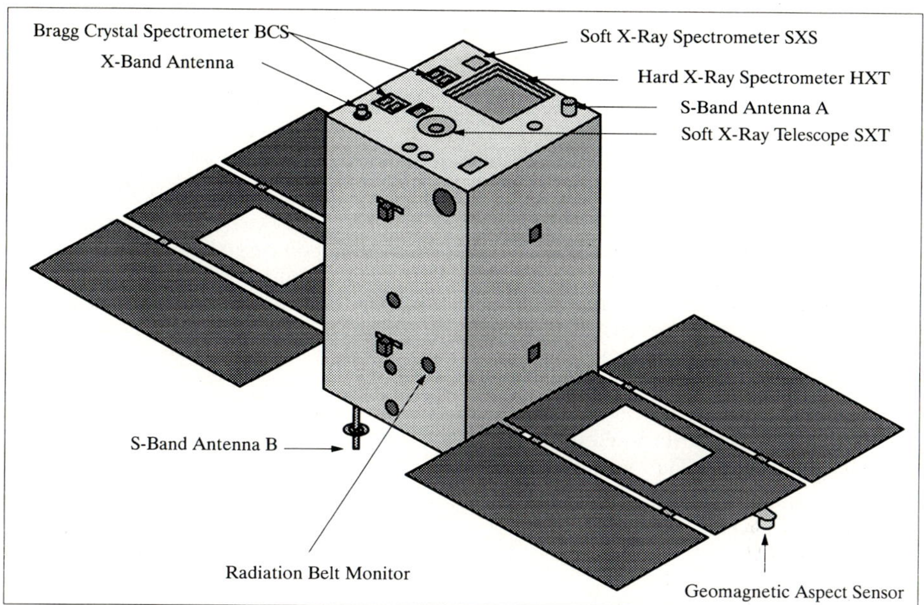


**Figure 93:** **The Solar-A S/C Model**

- **BCS** = Bragg Crystal Spectrometer (UK-Japan collaboration, Mullard Space Science Lab and RAL, and Tokyo Astronomical Observatory). Bent crystals with position sensitive proportional counters. Mass = 13 kg

- **WBS** = Wide-Band Spectrometer. A set of proportional counters, an Na I scintillator and two BGO scintillators. Measurement ranges: 2 - 30 keV, 20 - 400 keV, and 0.2 - 100 MeV. Mass = 16 kg

Data: Onboard recorder (80 Mbit capacity, magnetic bubble). Real-time data transmission of 32 kbit/s, or 262 kbit/s playback data, both to Kagoshima. Downlink communication in S-Band and X-Band.

# A.102 Spacelab-1

Shuttle Mission (STS-9, Columbia) 28. Nov. 1983 - 8. Dec. 1983 (10 day flight). Orbit: 240 - 257 km, inclination = 57°. NASA-ESA mission.

Sensors:
**Metric Camera** (German/ESA Experiment; the camera was a slightly modified Aerial Survey Camera of the type Zeiss RMK A 30/23)[340]. Objective: to test the mapping capabilities of high-resolution space photography on a large film format.
Application: Topographic and thematic mapping. The metric camera provided high-resolution photographs and experimental results on planimetric and topographic mapping from outer space.

| | |
|---|---|
| Spectral range | 535 - 900 nm |
| Focal length | 305 mm |

---

340) 'Spacelab-1 Metric Camera, User Handbook and Data Catalogue', compiled by M. Schroeder, E Suckfüll, G. Todd, and P. Lohmann of DLR, Oberpfaffenhofen, Dec. 1986

| | |
|---|---|
| Image size | 23 cm x 23 cm |
| Image Scale | 1 : 820 000 |
| Image overlapping | 60 - 80 % |
| FOV | 41.2°. |
| Swath width | 189 km |
| B/W film | Kodak Double-X Aerographic film 2045; pixel size = 12 x 12 m |
| Color film | Kodak Aerochrome infrared film 2443; pixel size = 12 x 12 m |
| Data rate | 550 frames per film |

For Metric Camera operations, the Shuttle flew with the open cargo bay oriented towards the Earth. In this flight attitude, the camera's optical axis looked vertically down to the Earth's surface. A total of 21 camera operations (exposure series) were taken which varied in duration from 2 to 18 minutes.

**MRSE** = Microwave Remote Sensing Experiment

MRSE operates in two modes, the scatterometer mode or the thermal radiometer mode. The two frequency scatterometer measured backscatter from the the ocean surface at two adjacent frequencies. The thermal radiometer measured surface temperature.

**GRILLE** = Infrared spectrometer (French/Belgian sensor). GRILLE has a high resolution in the infrared region (2 μm), it measures vertical profiles of: CO, $CO_2$, NO, $H_2O$, $CH_4$, $H_2O$ and HCl.

## A.103    Spacelab-3

Shuttle (Challenger, STS-17) mission between April 29 and May 6, 1985.[341] A total of 15 experiments onboard, most of them in the area of microgravity (crystal growth).
Orbit: 352 km altitude, 57° inclination

Sensor:

**ATMOS** = Atmospheric Trace Molecule Spectroscopy (JPL sensor). Objective: Investigation of the distribution of neutral constituents in the Earth's atmosphere.
The ATMOS sensor is a Michelson interferometer, with a response to radiation in the near- to infrared regions, designed to make observations from on board the Shuttle in the solar occultation mode.

ATMOS gathered data during 20 occultation events; 13 sunset occultations located between 26° - 34° N latitudes and 7 sunsets around 48° S.
ATMOS employs a high-resolution FT (Fourier Transform) interferometry and limb viewing in the Bands of 2.2 μm to 16 μm.

## A.104    SPOT

Spot[342],[343],[344],[345] = Système Probatoire d'Observation de la Terre (also: Satellite Pour l'Observation de la Terre). The Spot Earth observation program was started by France in 1977. Spot (owned jointly by France, Sweden, and Belgium) is built by Matra and operated by CNES. Spot-1 is operational since 1986 (launch: Feb. 22, 1986 by Ariane V16 from Kou-


341) "Overview of ATMOS Results from Spacelab-3', in Optical Remote Sensing of the Atmosphere, 1990 Technical Digest Series of the Optical Society of America, Volume 4, pp. 64-66
342) CNES viewgraphs of 1991
343) Jane's Spaceflight Directory 1988-89, Fourth Edition, pp. 22-23
344) Note: Spot-1 was retired from normal operations in Sept. 1990. Both of its recorders are defect. Spot Image wants to reactivated Spot-1 to meet increased demand for satellite imagery. See Space News Dec. 4 1991 p. 4

rou ). Spot-2 is operational since Jan. 22, 1990 (almost identical S/C to Spot-1). Spot-3 was launched on September 26, 1993. Spot-3 is basically identical to its predecessors. Spot-4 has been authorized. Nominal life time of Spot-1,2,3 = 3 years.

Application: Land use, Agriculture, Forestry, Geology, Cartography, Regional Planning, etc.

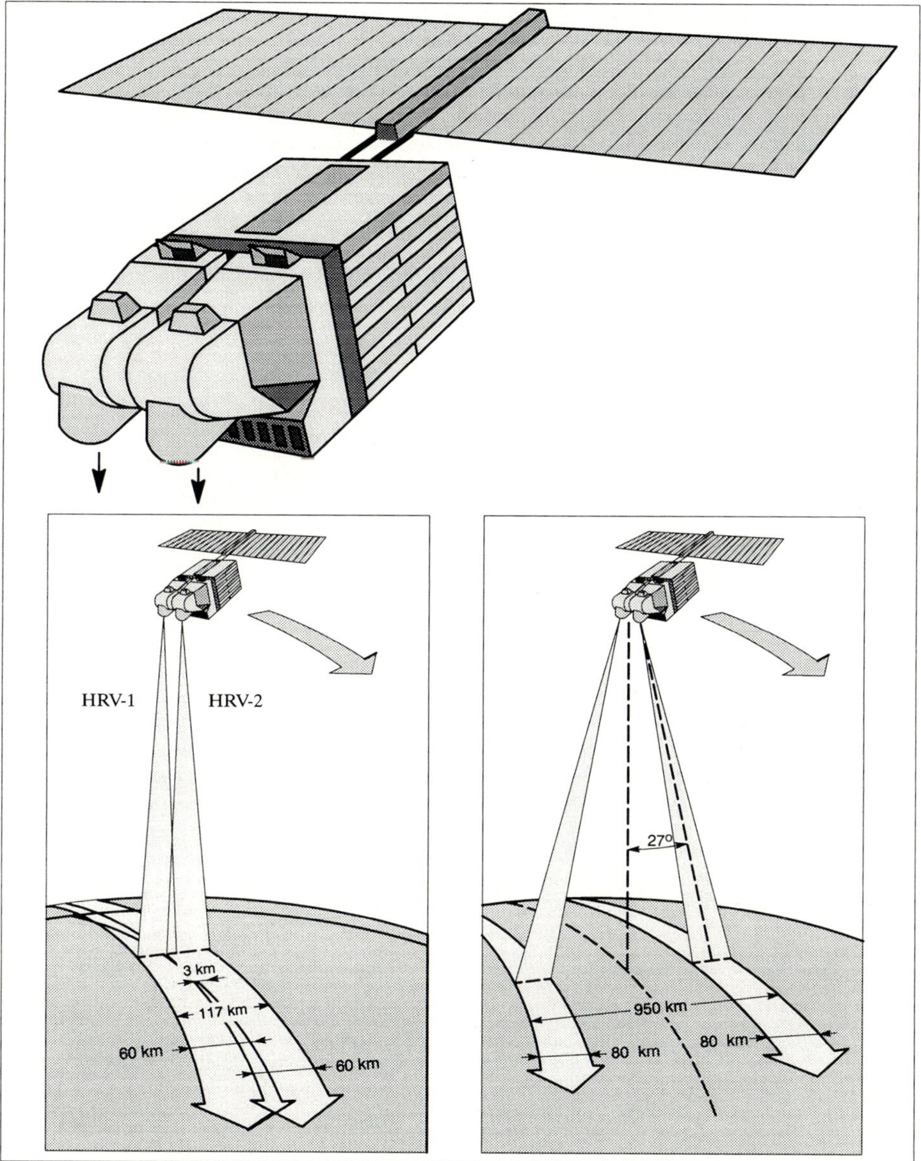


**Figure 94:**     **The Spot S/C Model and its Observation Geometry Bounds**

345) Note: Spot-1 was retired from normal operations in Sept. 1990. Both of its recorders are defect. Spot Image wants to reactivated Spot-1 to meet increased demand for satellite imagery. See Space News Dec. 4 1991 p. 4

Orbit: Sun-synchronous polar orbit, altitude= 832 km, Inclination = 98.7° , Period = 101 minutes, repeat coverage = 26 days, 3-axis stabilized platform, 10:30 AM local time (descending) equator crossing. Duty cycle: daylight coverage only.

Sensors:

- **HRV = Haute Resolution Visible** (= High-Resolution Visible Sensor); Push-Broom Charge-Coupled Device (CCD), 6000 Detectors, each with 1728 elements. The HRV sensor operates in two modes, namely the **Multispectral Mode (MS)** and the **Panchromatic Mode (P)**. Each Spot payload consists of two identical HRVs imaging instruments that are pointable in the cross-track direction up to 27° from nadir. Each instrument covers a swath width of 60 km when vertically positioned (both instruments have a TFOV of 117 km, 3 km overlap, nadir pointing). The swath width is 80 km (each) when the instrument is angled 27° to the vertical. Measurement angle: +/- 27° from the vertical (in 45 steps)

    - **HRV Multispectral mode** (MS) 3 Spectral ranges :   500 - 590 nm
                                                            610 - 680 nm
                                                            790 - 890 nm
      Spatial resolution in MS = 20 m.

    - **HRV panchromatic mode** (black and white images), spatial resolution = 10 m
      Spectral range          510 nm - 730 nm

With the side-viewing feature on Spot, there is the capability to plot contours by taking stereoscopic images, as well as enabling a particular location to be imaged more frequently, without waiting for an overhead pass.

**Data:** (commercial data distribution by Spot-Image of Toulouse and of Reston Va.)

- Image size: 60 x (60 - 85) km
- Image size: MS Mode: (3000 - 4900 Lines) x (3000 - 5200 Pixels)
- Image size: P Mode: (6000 - 9800 Lines) x (6000 - 10400 Pixels)

Transmission: Frequency = 8253 MHz (X-Band downlink); Data rate = 25 Mbit/s

**DORIS** (on Spot-2 as an experiment and prototype, on Spot-3, etc.). DORIS is a CNES-developed tracking system for the determination of precise orbits (see DORIS description under 'TOPEX/POSEIDON", A.110.1 ).

| Station Location | Country | Commissioning Date | Organization |
|---|---|---|---|
| Toulouse/Aussaguel | France | February 1986 | CNES |
| Kiruna | Sweden | February 1986 | ESA |
| Prince Albert | Canada | June 1986 | CCRS |
| Gatineau | Canada | June 1986 | CCRS |
| Hyderabad | India | May 1987 | NRSA |
| Maspalomas | Spain, Canary Islands | Nov. 1988 | ESA |
| Cuiaba | Brazil | April 1988 | INPE |
| Lad Krabang | Thailand | May 1988 | NRCT |
| Hatoyama | Japan | Oct. 1988 | NASDA |
| Islamabad | Pakistan | June 1989 | SUPARCO |
| Hartebeetshoek | South Africa | August 1989 | CSIR |
| Riyadh | Saudi Arabia | Oct. 1989 | KACST |
| Alice Springs | Australia | May 1990 | NATMAT |
| Tel Aviv | Israel | Feb. 1991 | IAI |

**Table 77:       Some of the Spot Series direct receiving Stations around the World**

## A.104.1  Spot-3

Operational spacecraft (as of Nov. 29, 1993), launched September 26, 1993. Spot-3 is equipped with 2 HRV imaging sensors + DORIS + POAM-II.

**POAM-II** = Polar Ozone and Aerosol Measurement (atmospheric chemistry experiment). The instrument was provided by the Naval Research Laboratory (NRL) of Washington D. C. it is sponsored by the Innovative Science and Technology Office of the Ballistic Missile Defense Organization (BMDO) and the Space Test Program of DOD. The objective is to measure vertical profiles of polar ozone, aerosols, water vapor, nitrogen dioxide, atmospheric density and temperature in the stratosphere and upper troposphere. The measurement technique is by solar occultation through the Earth's atmospheric limb at nine wavelengths from the UV to NIR. There are separate filtered optics for each of the nine channels, which are co-aligned with each other and the sun tracker. FOV = 0.01° by 0.75° (slit). Observation of the full width of the solar disk with a vertical resolution of 0.6 km (at the tangent point in the Earth's atmosphere).[346]

| Channel | Primary Measurement | Center Wavelength (nm) | Bandwidth at FWHM (nm) |
|---------|--------------------|-----------------------|-----------------------|
| 1 | Aerosols | 353.0 | 5 |
| 2 | $NO_2$ off | 442.0 | 2 |
| 3 | $NO_2$ on | 448.3 | 2 |
| 4 | $O_3$ | 600.0 | 15 |
| 5 | $O_2$ on | 760.8 | 2 |
| 6 | $O_2$ off | 780.0 | 15 |
| 7 | $H_2O$ off | 920.0 | 2 |
| 8 | $H_2O$ on | 935.5 | 2 |
| 9 | Aerosols | 1059.0 | 10 |

**Table 78:**     **Spectral Coverage of the POAM-II Instrument**

POAM-II observes 14 sunrise and 14 sunset events per day. The primary measurements are through the Earth's limb for each of the 9 wavelengths as a function of tangent altitude. These measurements are inverted to yield the vertical profiles of ozone, aerosols, nitrogen dioxide, water vapor, and oxygen.

| Parameter | Height Range (km) | Accuracy (%) |
|-----------|-------------------|--------------|
| Aerosols | 10 - 40 | 5 - 15 |
| $O_3$ | 10 - 60 | 5 |
| $H_2O$ | 15 - 40 | 5 - 7 |
| $N_2O$ | 20 - 40 | 5 - 10 |
| Temperature | 10 - 60 | 1 - 2 |

**Table 79:**     **POAM-II Measurement Capabilities**

The POAM data rate is 32.768 kbit/s. Downlinks are via a dedicated POAM antenna. The downlink is received by a network of ground stations operated by the US Air Force. Uplinks are part of the general SPOT uplinks. Once validated, the reduced data (transmittances) are archived in the BMDO's Background Data Center at NRL, and will be publicly available.

## A.104.2  Spot-4

The Spot-4 S/C is approved and being built for a launch in 1997. Same orbit and geometrical constellations as before. Spot 4 and 5 (Spot 5 is planned for 2000/1) are considered 2nd gen-

346) R. M. Bevilacqua, et al., "Polar Stratospheric Studies with the Polar Ozone and Aerosol Measurement Experiment (POAM-II)", Proceedings of the American Meteorological Society, Eighth Conference on Atmospheric Radiation, 23-28 January 1994, Nashville TN

eration satellites. The most important advance will be the addition of the "Vegetation"[347] payload, with 5 spectral bands to allow continuous, world-wide monitoring of crops and spontaneous vegetation for crop forecasts and environmental studies.

The Spot-4 design differs from the earlier Spot series (Spot-1,2,3) in the following aspects: five years design lifetime; a new extended platform design; increased onboard storage capacity; the 10 m sampling will be implemented in the 610 - 680 nm band enabling the 10 and 20 meter resolution data to be co-registered onboard the satellite instead of on the ground; the addition of a short wave infrared band (1.58-1.75 µ) for improved vegetation analysis capability.

| Channel | Spectral Range | optimized to detect |
|---|---|---|
| $B_0$ | 0.43 - 0.47 µm | chlorophyll |
| $B_1$ | 0.50 - 0.59 µm | vegetation, turbidity |
| $B_2$ | 0.61 - 0.68 µm | vegetation |
| $B_3$ | 0.78 - 0.89 µm | vegetation, atmospheric correction |
| SWIR | 1.58 - 1.75 µm | vegetation, atmospheric correction |

**Table 80:     Spectral Ranges of the Vegetation Monitoring Instrument**

Sensors:

- **HRVIR** (2) = High-Resolution Visible and Infrared sensor (improved version of HRV). There will be a supplementary band in the 20 m multispectral mode at 1.5-1.7 µm. Resolution: 20 m (multispectral) and 10 m (panchromatic)

- **VEGETATION**. An additional payload for Spot-4, called VEGETATION, with a ground swath width of 2200 km, and a resolution close to 1 km is under consideration in order to provide the capability of large scale monitoring of the Earth's vegetation. Calibration accuracy: interband =3%, multidate = 5%, absolute = 10%; TFOV = 101°; Pixel size = 1.1 km; Location accuracy = 2.5 km.

  The combination of high and low spatial resolutions (of HRVIR and VEGETATION) provide a major contribution to the measurement of temporal changes over a few points or at certain time periods for a proper determination of the influence of the various kinds of ground cover.

- PASTEL. Spot-4 will fly in addition a joint ESA/CNES experiment called PASTEL. PASTEL is a prototype high data rate intersatellite transmission system based on laser technology.

## A.104.3  Spot-Radar

Proposed CNES mission for launch in the year 2000. An all-weather viewing capability with a high-resolution radar is considered. Spot-Radar is seen as a complement to the Spot series for inter-tropical regions with an all-weather day and night imaging capability. A multipolarization SAR sensor in X-Band is under consideration.

## A.105   Starlette

Starlette[348] is a CNES 'Solid Earth' mission, a passive satellite dedicated for geodetic and geophysical studies with SLR (Satellite Laser Ranging) observations support. Starlette was launched Feb. 6 1975 from Kourou with Diamant B. Starlette is the world's first passive laser satellite for Solid Earth research.

---

347)  F. Achard, J. P. Malingreau, T. Phulpin, G. Saint, B. Saugier, B. Segun, D. Vidal-Madjar, "The Vegetation Instrument on Board SPOT-4 - A Mission for Global Monitoring of the Continental Biosphere", LERTS brochure, Toulouse, 1990

Orbit: Altitude varies between 810 (perigee) and 1105 (apogee) km, inclination = 50°, projected orbital life = 2000 years.

Objectives: Studies of the gravity field of the Earth (both long term average values and temporal variations due to the tides within the Earth).

Satellite: sphere with a radius of 12 cm; weight = 47.29 kg; coefficient of reflectivity = 1.1 (approx.). The core parameters are:

- shape:    isocaedron
- metal:    alloy of uranium 238 and 0.2% of vanadium
- density:  18.7
- weight:   35.5 kg

The skin features are:

- shape:    twenty spherical caps with triangular bases, fixed on the faces of the isocaedron
- metal:    alloy of aluminum and 5% magnesium

Measurement campaigns require the support of the world-wide SLR network.


## A.106   Stella

Stella is a CNES satellite mission launched on Ariane as a passenger experiment (piggyback) along with the Spot-3 launch, launch date. 26.9. 1993. Stella was put on top of the third stage (Ariane) with a special device, including a spring to give the required increment of velocity and a spin of 5 to 8 rev/minutes (spin axis perpendicular to the ecliptic).

Orbit: Sun-synchronous quasi circular polar orbit, altitude = 780-800 km, inclination = 98.2°, orbital life: several centuries.

Objectives: Studies of the gravity field of the Earth. Contribution to the modelling of non-gravitational forces. Contribution to the modelling of Earth and ocean tides in conjunction with other passive laser satellites (LAGEOS-I and -II, ETALON-1,-2, Starlette, AJISAI). Contribution to temporal variations of the gravity field due to geophysical causes like post glacial rebound, allowing determinations of mantle viscosity, and possible geographical variations, again in conjunction with other laser satellites.

Satellite: Stella is an exact twin of Starlette. Sphere with a diameter of 24 cm, weight of 47.29 kg, etc. Tracking requirement of the world-wide SLR network (international cooperation and active participation is desired).


## A.107   TEMISAT (Telespazio Micro Satellite)

TEMISAT[349),350),351)] is a data collection and distribution satellite for environmental monitoring of autonomous ground stations. The satellite and a portion of the corresponding ground segment is owned and operated by Telespazio of Rome Italy, and was built by Kayser Threde of Munich, Germany. TEMISAT-1 was launched (piggyback on Meteor-2-24) on August 31, 1993 on a Russian Cyclone launcher from Plesetsk. The spacecraft is essentially a cube of 0.35 m side length with a mass of about 40 kg, and passive magnetic field stabilization.


348)  M. Lefebvre, "Stella", CSTG Bulletin No. 11, Title: New Satellite Missions for Solid Earth Studies, 1989, pp. 25-32
349)  "Telespazio Readies Temisat Satellite for Summer Launch", Space News, April 19-25, p.24
350)  "Temisat", Kayser Threde paper
351)  "Blackbird: A Family of Microsatellites for Communications and Remote Sensing", Kayser Threde brochure

Application: world-wide commercial data collection, in particular for the science community in remote areas.

Orbit: Polar circular orbit, altitude = 950 km, inclination = 82.5°, period = 112 minutes.

The ground segment consists of a Mission Control Center (MCC, .... one of its major tasks is the central scheduling and configuration control of all customer traffic), a number of User Control Stations (UCS, .... for those users who operate under their own control a sizable number of remote terminals), Remote Terminals (RTs, ... which control the monitoring instrument(s), receive and store data from the monitoring instrument, and function as communication nodes with the satellite), and finally the monitoring instruments themselves.

The TEMISAT polar orbit provides about a 10-12 minute contact period for all the monitoring platforms in the instantaneous area of visibility, or the footprint, of its orbit. TEMISAT is constantly polling the ground segment for possible customers (RTs, UCSs, or MCC) to initiate communication opportunities. It is foreseen that all the platforms with their RTs in this footprint will have a chance to communicate their data to the satellite in due time.

All instrument location knowledge is provided with a built-in GPS receiver in the monitoring instrument configuration (this applies to moving instrumentation in the ground segment such as drifting buoys, or other moving objects). The GPS concept offers a position accuracy of about 30-100 m.

**Communication Concept:**

TEMISAT covers the VHF and UHF frequency spectrum allocated by WARC 92 for micro satellite LEO communication applications. A total of 8 parallel receiving channels and 2 simultaneous transmitting channels are provided for the monitoring task of the ground segment. With this capability TEMISAT operates as an intelligent communications node in orbit supporting two different operational support modes:

- Transparent regenerative transponder mode (data relay function when data sender and receiver are in the same coverage area of the satellite)
  In this mode, data from the receiving channels (e.g. from the ground segment platforms) are decoded, forward corrected - then again recoded and formatted for retransmission to the ground in real time (either to the MCC or to the UCS or to both).

- Store-and Forward Mode (operational function when data sender and receiver are in different coverage areas of the satellite)
  The received data are decoded, forward corrected, and stored on board (total storage capability of 30 MByte RAM) until the intended receiving station is in satellite visibility. Then the specific data file is recoded, formatted, and transmitted to the ground.

Data can generally be collected by the MCC and be provided/distributed to the user community via electronic interface.

**Access Methods:**

TEMISAT uses a TDMA/SCPC (Time Division Multiplexing Access/Single Channel per Carrier) access scheme for the monitoring task of the ground segment. TDMA/SCPC is centrally controlled by MCC (computer scheduling of customer traffic up to 4 times per day) this avoids data collision in the uplink and maximizes the channel throughput rate.

The TDMA/SCPC access scheme permits a data rate of 1200 bit/s (2400 bit/s viterbi-encoded) for every one of the 8 channels. The MCC has configuration control over the channel assignment and time slot allocation for each customer.

Note: The TDMA/SCPC access scheme allows a system throughput capacity of over 80% which is practically independent of the number of simultaneous users (30 or 300); on the other hand, the access scheme of ARGOS is 'unslotted ALOHA' which permits under 0.5 normalized offered channel traffic a channel throughput rate of about 18% (max).

| Nr. of 'Receive' cha. for data collection & telecommand (TC) Operational Receive rate Data rate Coding Modulation Frequency | 10 (8 + 2) 8 data/1 TC (all uplink ground - S/C) 2400 bit/s 1200 bit/s Viterbi 1/2 MSK 9 cha. in VHF, 1 experimental (UHF) |
|---|---|
| Nr. of 'Transmit' channels Operation Transmission rate (TM/polling) Data rate (TM/polling) Coding (TM/polling) Modulation (TM/polling) | 3 + 3 cold redundant (3 cha. downlink) 1 TM (S/C housekeeping) / polling, 2 data 2400 bit/s 1200 bit/s Viterbi 1/2 MSK |
| Transmission rate Data rate Coding Modulation RF-Output power (per channel) Frequency | 9600 bit/s (downlink) 4800 bit/s Viterbi 1/2 GMSK >5W / 2.5 W VHF |
| Number of antennas Antenna type Polarization | 5 quarter-wave monopole linear |
| On-board computer Mass memory | transputer-based multiprocessor 2 x 12.5/8.5 MB SECDED protected |
| S/C dimensions S/C total mass S/C stabilization Power consumption operational (max) Power consumption standby Average available power | 350 x 350 x 350 (mm) 40 kg passive, magnetic field 60 W 2 W 18-20 W |

**Table 81:    Summary of TEMISAT Specification**

All RTs in the ground segment are always in listening mode ready to establish contact with the TEMISAT polling channels at the scheduled time slot. If an average contact time of 5 seconds per RT is considered, in which the communication link (about 200 ms of sync time) is established and the data transmitted (uplinked), then TEMISAT is capable of servicing up to 960 RTs within a 600 second period with 8 parallel channels [10 minute pass period over any ground point in the TEMISAT swath width of 5000 km (10° elevation above the horizon)].

The wide swath of TEMISAT visibility provides the following coverage conditions:

- 3 times of daily coverage at equatorial latitudes (min)
- 4 times of daily coverage at 40-45° latitudes (min)
- 5 times of daily coverage at 50° latitude (min)
- > 10 times of daily coverage in polar regions

TEMISAT recognizes every remote terminal (RT) in the ground segment by its unique address in the message header. The message length for an uplink transmission can be variable, depending on the individual RT requirements. With an available net data rate of 1200 bit/s and an actual transmission time of 4 seconds in a contact, this amounts to 600 bytes of data in a single transmission. All user data is packetized with 600 bits per packet (2 packets per second in the transmission link). The user has the freedom to insert his data into the frame according to his specific requirements (this can be sensor data along with a time stamp plus GPS data, there can be super- or subcommutations, etc.). A transmission can consist of some real-time data and some stored data. In general every RT stores its collected sensor data and provides it for the next available uplink opportunity. Data can be sent twice (i.e. on the next possible pass) in case of an unsuccessful transmission connection.

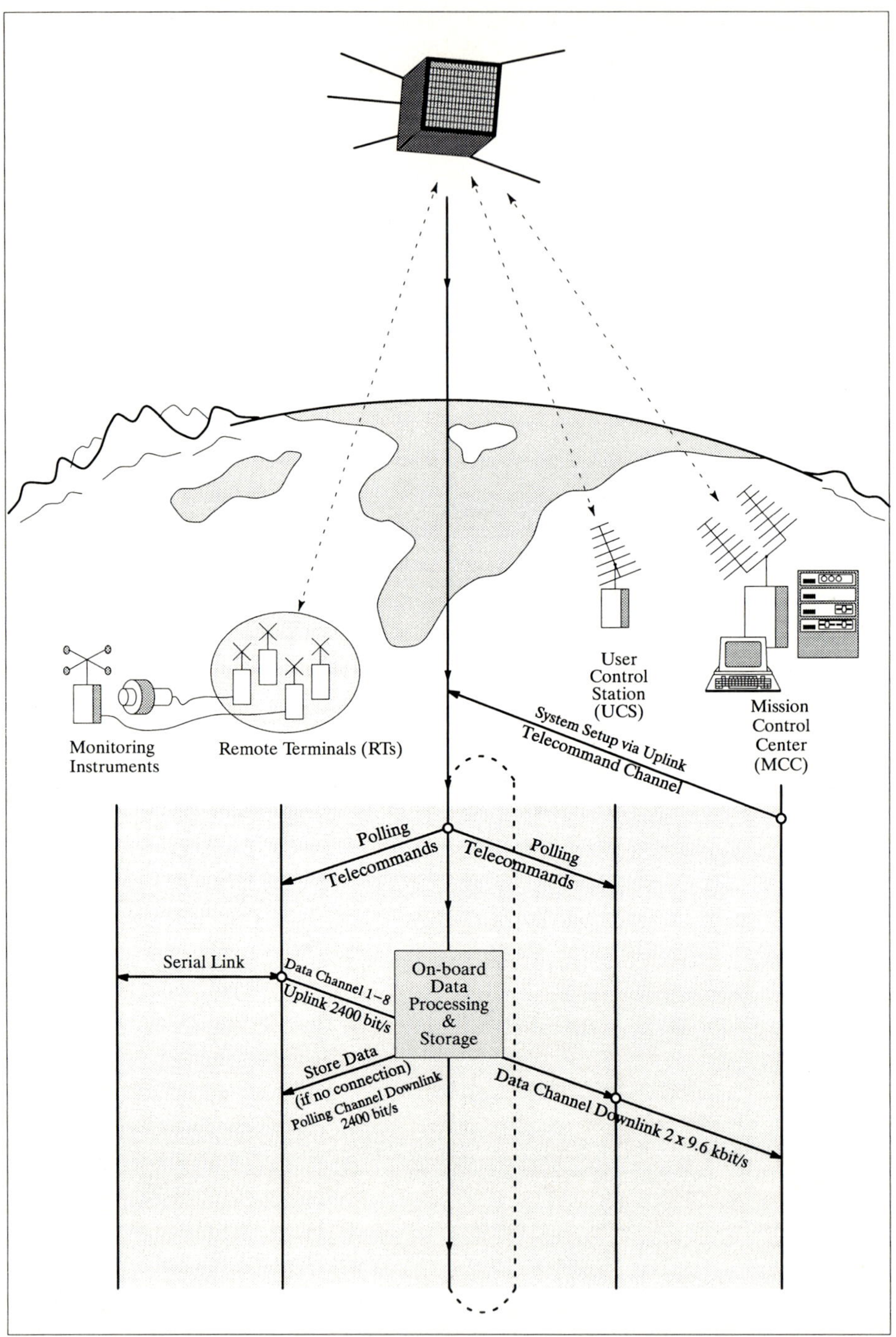


**Figure 95:**     **Scenario/Communication Logic of the TEMISAT Data Collection System**

| VHF Uplink Frequencies (MHz) | UHF Uplink Frequencies (MHz) | VHF Downlink Frequencies (MHz) |
| --- | --- | --- |
| 149.6625 | 402.3375 | 137.645 |
| 149.6875 | 402.3625 | 137.670 |
| 149.7125 | 402.3875 | 137.695 |
| 149.7375 | 402.4125 | 137.720 |
| 149.7625 | 402.5875 | 137.745 |
| 149.7875 | 402.6125 | |
| 149.8125 | 402.6375 | |
| 149.8375 | 402.6625 | |
| 149.8625 | | |
| 149.8875 | | |

**Table 82:     Operational Frequencies of TEMISAT (allocated by WARC)**

The concept and technology of TEMISAT operation represent a new dimension - in fact a new generation - in data collection services for the user community. This applies in particular with the introduction of the following features:

- Selection of the most efficient access scheme technology (TDMA/SCPC)
- The availability of sufficient data rates for the data collection function along with 8 communication channels for instantaneous parallel service assignments
- The bi-directional store-and-forward communication concept offers a remote control capability of every remote sensing instrument by the user (or by the MCC on user request).
- Provision of real-time/stored data transmission which offers full-time (round-the clock) coverage (as compared to spot coverage) of the measured parameters.
- Position data can be included into every packet if needed. There are no constraints of several contacts per RT within the same footprint for Doppler position fixing (as is the case for ARGOS PTTs).


# A.108   TIMED

TIMED (Thermosphere, Ionosphere, Mesosphere Energetics and Dynamics) is a planned NASA/GSFC exploratory mission, designed to carry out a comprehensive investigation of physical and chemical processes acting within, and upon, the Mesosphere and Lower-Thermosphere/Ionosphere (MLTI), the region in the Earth's atmosphere from about 60 to 180 km in altitude. The mission requires two spacecraft: one in a high inclination orbit to give broad latitudinal coverage, and one in a lower inclination orbit for rapid local time coverage. The high inclination spacecraft is called TIMED-H, the low inclination spacecraft is called TIMED-L. At the end of 1993 the TIMED project is in Phase B.[352]

Launch: the launch of TIMED-H is planned for the end of 1998 from the Western Test Range. The TIMED-L launch is planned for early 1999 from the Eastern Test Range. Launch vehicle candidates are the Pegasus XL, the Conestoga class, and Taurus.

Orbit: polar circular orbit of TIMED-H, inclination = 95°, altitude = 400 km; circular orbit of TIMED-L, inclination = 49°, altitude = 400 km. A mission design life of two years.

Objectives: The primary science objective of the TIMED mission is to investigate and to understand quantitatively the processes responsible for the energy and momentum budgets of the MLTI. The following objectives are considered:

- Determination of the temperature, density and wind structure of the MLTI, including the seasonal and latitudinal variations (energetics).
- Estimation of the relative importance of the various radiative, chemical, electrodynamical, and dynamical sources or sinks of energy for an understanding of the thermal structure of the MLTI (energetics).

---

352)  Information provided by H. G. McCain of NASA/GSFC

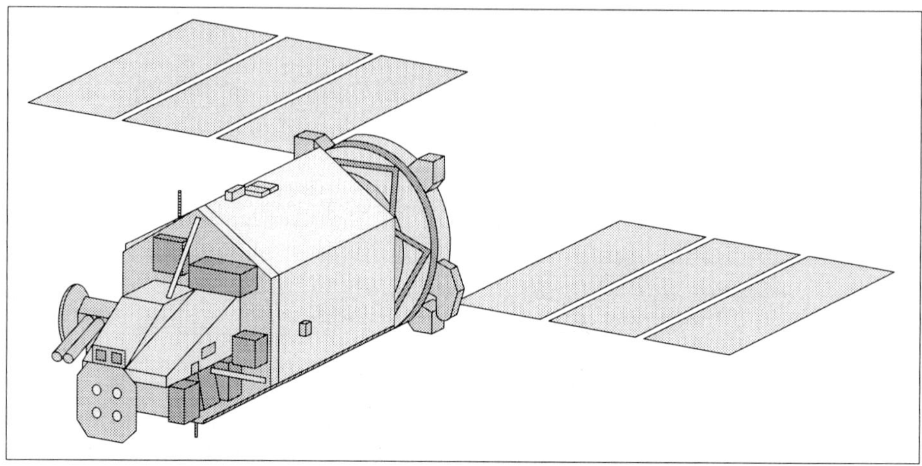

**Figure 96:**     **The TIMED S/C Model**

- Estimation of the sources and magnitudes of gravity waves and planetary waves (extra-tropical and equatorial) in the MLTI region, their spatial and temporal variability and their forcing of the large-scale circulation. Thermal and constituent structures (dynamics).
- Estimation of the mean tidal structures, their annual and height variations, and their effects on the large-scale circulation in the MLTI (dynamics).
- Estimation of the relative importance of the interactions occurring among the different wave motions (tidal, gravity, and planetary) and the extent, causes and consequences of their variability (dynamics).
- Estimation of the spatial and temporal variability of the radiatively significant minor species and the role that atomic oxygen plays in influencing the radiative balance through collisional energy transfer (chemistry).
- Estimation of the relative importance of the disturbance dynamo, tidal dynamos and magnetospheric penetration fields in determining the global electrodynamics at mid- and at equatorial latitudes (ionosphere).
- Estimation of the spatial and temporal variability of the odd-oxygen, odd-hydrogen and odd-nitrogen compounds in the MLTI (chemistry).
- Estimation of the roles of neutral winds and electric fields in controlling large- and small-scale variability in ionospheric composition and density structures (ionosphere).

Areas of applications: The MLTI is a very poorly understood region of the atmosphere. The results of the TIMED mission will enable the scientific community to establish the first quantitative MLTI baseline and will serve as a basis for future investigations.

**Sensors:**

**SEE** = Solar EUV Experiment. The instrument measures the absolute fluxes of solar UV, EUV, and XUV radiation to determine the rates of energy deposition, dissociation and ionization.

**SABER** = Sounding of the Atmosphere using Broadband Emission Radiometry. The sensor measures the IR emissions emitted at altitudes generally below 100 km by limb scanning to drive vertical distributions of temperature and concentrations of energetically important species ($O_3$, $H_2O$, $NO$, $NO_2$ $CO$, $CO_2$) as well as radiative energy loss.

**TIDI** = TIMED Doppler Interferometer. TIDI measures the VIS/NIR emissions emitted at altitudes between 60 and 300 km by limb scanning techniques to determine the temperature and horizontal winds with the use of the Doppler effect. The instrument makes also density measurements, mostly on the day side of the orbit.

**TIPE** = TIMED Imaging Photometer Equipment. The sensor measures the VIS/NIR emissions emitted at altitudes between 80 and 105 km by imaging to determine the atmospheric wave structure and rotational temperature.

**GUVI** = Global Ultraviolet Imager. GUVI measures the UV radiation emitted at altitudes generally above 150 km by limb scanning and imaging to determine during daytime conditions the concentrations of $N_2$, $O_2$, O, and the temperature; infers the fluxes of precipitating auroral particles.

**TONE** = Temperature, Ozone and Nitric-Oxide Experiment. The objective is to measure UV radiation emitted at altitudes between 50 and 160 km by limb scanning to determine during daytime the vertical distributions of temperature, total density, NO and $O_3$ concentrations. The instrument obtains also data about the distribution of noctilucent clouds.

**DATES** = Density and Temperature Spectrometer. The objective is to measure the $O_2$ band emissions at altitudes below about 160 km by limb scanning to determine during daytime the vertical distributions of temperature and density, plus concentrations of O (1D) and $O_3$ (below 100 km). Also measures the solar energy input.

| Sensor | Detector Type | Measurement Type | Spacecraft |
|---|---|---|---|
| SEE | MCP (Codocon) | Remote | H and L |
| SABER | HgCdTe | Remote | L |
| TIDI | CCD | Remote | H and L |
| TIPE | CCD | Remote | H and L |
| GUVI | MCP (Sealed) | Remote | H |
| TONE<br>    Medium Resolution Spectrometer<br>    Low Resolution Spectrometer<br>    Infrared Spectrometer | <br>MCP (Codocon)<br>Reticon<br>Ge Photodiode | Remote | H |
| DATES | CCD | Remote | L |
| TWINNIE<br>    Mass Spectrometers<br>    Retarding Potential Analyzer | <br>Channeltrons<br>Solid State | In-situ | H and L |
| IDI<br>    3 Mass Spectrometers<br>    2 Drift Meter Sensors<br>    Ion Trap Sensor | <br>Channeltrons<br>Solid State<br>Solid State | In-situ | H and L |

**Table 83:     TIMED Sensors and Detectors**

**TWINNIE** = Temperature, Wind, Neutrals, and Ions Experiment. The objective is to make in-situ measurements to determine the concentrations of neutrals and ions and the neutral temperature and winds (horizontal and vertical profiles).

**IDI** = Ionospheric Dynamics Experiment. The instrument makes in-situ measurements to determine the ion drift velocities (and related electric fields) as well as the ion densities and temperature.

Average data rates: = 10.9 kbit/s for TIMED-H and 6.9 kbit/s for TIMED-L. Science data are transmitted in S-Band.


## A.109   TOMS Missions

Within NASA's Earth Probes Program several TOMS (Total Ozone Mapping Spectrometer) or TOMS/NSCAT missions are planned. The objectives are:

- Continuation of the global ozone data set that began in 1978 with the flight of TOMS on NIMBUS-7
- Continue observation of environmentally important areas such as the Antarctic ozone hole
- Observation of sulfur dioxide clouds resulting from volcanic eruptions

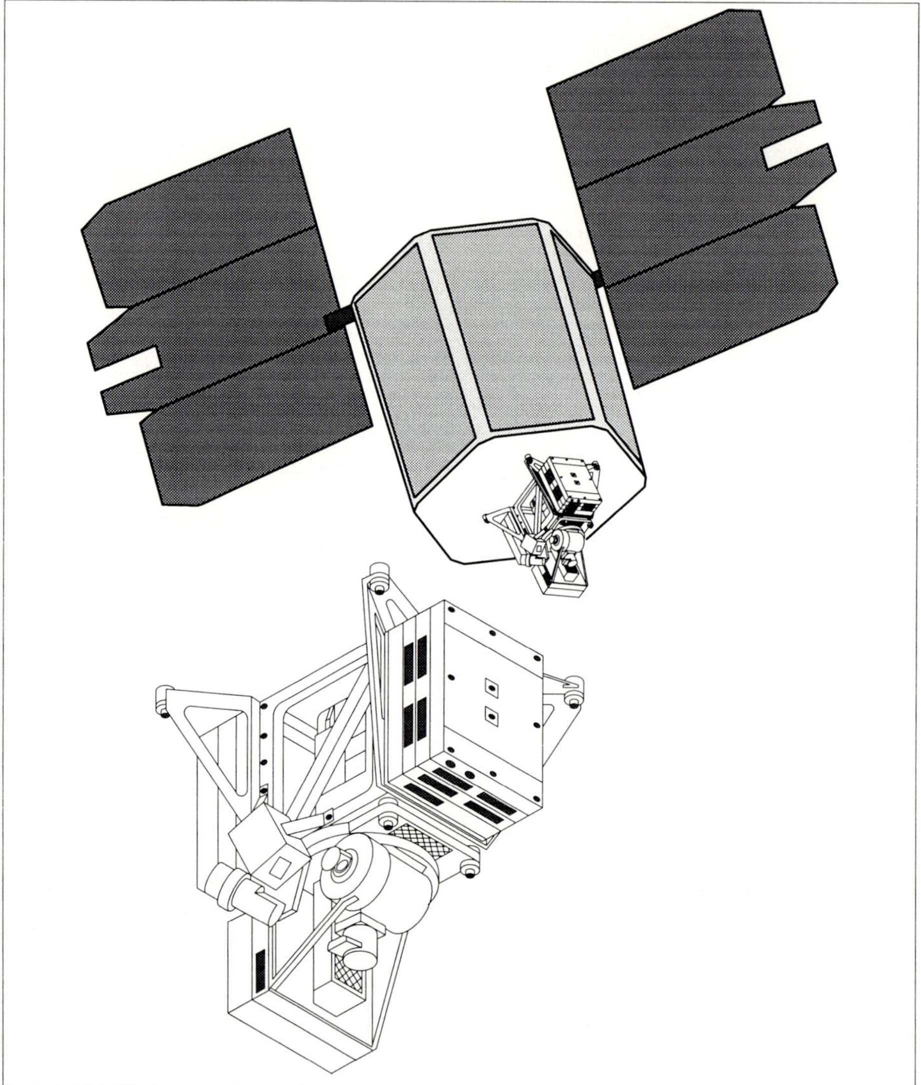

**Figure 97:    The TOMS-EP S/C and Sensor Models**

## A.109.1  TOMS-EP

TOMS-EP (Earth Probes) is a free-flying spacecraft with a payload of the TOMS instrument. It is nadir pointing and provides a contiguous survey of the Earth every day. Life: 2 year mission. Launch vehicle: Pegasus. Launch in July 1994.

Orbit: circular, polar, sun-synchronous, altitude = 955 km, inclination = 99.3°

Sensor: TOMS

- 105° scan of 3x3° FOV, TOMS scanner is not pointed at sun.
- Fastie-Ebert monochromator
- Wavelength range: 308.6 - 360.0 nm
- Ozone trend measurement capacity: 0.1% yearly (goal)
- Swath width: 2722 km
- Spatial resolution: Nadir - 50 km x 50 km; average: 62 km x 62 km
- Dimensions: 15 x 30 x 27 cm
- Mass: 33 kg; Power: 21 W
- Data rate: 600 - 700 bit/s (average). Data will be available to worldwide community of global change researches through the National Space Science Center (NSSDC).

### A.109.2  TOMS/NSCAT on ADEOS

TOMS and NSCAT instruments will be flown on the Japanese ADEOS mission in 1996 (see ADEOS mission).

Orbit: Sun-synchronous polar and circular orbit, altitude = 797 km, inclination = 98.6°

TOMS and NSCAT data will be available to the worldwide community of global change researchers through the Earth Science Data and Information System (ESDIS)

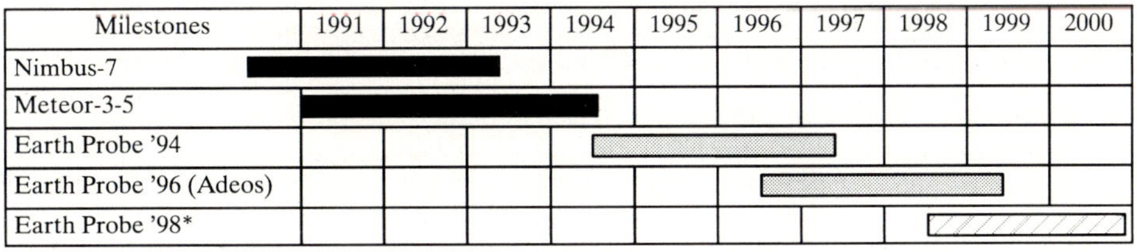

| Milestones | 1991 | 1992 | 1993 | 1994 | 1995 | 1996 | 1997 | 1998 | 1999 | 2000 |
|---|---|---|---|---|---|---|---|---|---|---|
| Nimbus-7 | | | | | | | | | | |
| Meteor-3-5 | | | | | | | | | | |
| Earth Probe '94 | | | | | | | | | | |
| Earth Probe '96 (Adeos) | | | | | | | | | | |
| Earth Probe '98* | | | | | | | | | | |

**Figure 98:**    **Overview of TOMS Missions**[353]

### A.110  TOPEX/POSEIDON

TOPEX/POSEIDON[354],[355],[356],[357] = Topography Experiment for Ocean Circulation. A NASA (JPL) / CNES joint Earth observation mission. Launch: Aug. 10, 1992 (Ariane 4 launch vehicle from Kourou).[358] POSEIDON was originally a separate CNES Mission, but later combined with TOPEX (1985). TOPEX/POSEIDON is the heart of the WOCE (World Ocean Circulation Experiment) Program, it is also considered for TOGA. Nominal Life: 3-5 years. TOPEX is regarded as the SEASAT successor mission. Satellite weight = 2402 kg. The S/C has a variety of communication antennas to link the mission with TDRSS, with the DORIS tracking system and with GPS.

Objectives: Dedicated altimetry mission. Combination of high altimetric precision and high orbital accuracy for the purpose of ocean topographic mapping. Measurements of sea surfaces for the modelling of global changes in ocean circulation and sea level (global panoramic maps of sea-surface topography). Development of climate models for long-term

353)  * TOMS instrument available for a mission of opportunity (possibly on a Russian spacecraft)
354)  "Predicted Topex Positioning Accuracy with Differential GPS Techniques", presented at, and published in the 'Proceedings of the first International Symposium on Precise Orbit Positioning with GPS' April 15, 1985
355)  Lee-Lueng Fu, M. Lefebvre, "TOPEX/POSEIDON: Precise Measurement of Sea Level From Space", CSTG Bulletin No. 11, Title: New Satellite Missions for Solid Earth Missions, June 1989, pp. 51-54
356)  'Currents' - the JPL Topex/Poseidon Newsletter, March 1990, Issue 1
357)  Topex/Poseidon Science Investigation Plan, NASA (Document Resource Facility), Sept. 1 1991
358)  'Topex-Poseidon Partners Discuss Sequel', Space News, Aug. 17-23, 1992, p. 3

forecasts (in the order of a season or longer). Geoid model improvements. Requirements: Precision orbit.

Orbit: Circular non-sun-synchronous orbit; 1334 km altitude (2 hour period), inclination = 66°, 10-day repeat orbits.
Note: The satellite orbit tracking coverage provided by Laser and DORIS is not continuous in time; hence, orbit computation based on dynamical equations is required to produce precise orbit for the mission. Expected orbit accuracy: 3 cm (rms for a single pass).

**Sensors:**

- **ALT** = Radar Altimeter (of GEOS-3, Seasat, and Geosat heritage). ALT uses a linear FM chirp pulse centered at 13.6 GHz and at 5.3 GHz (2-Frequency, to correct for ionospheric path delays), built by APL, managed by GSFC. Prime sensor for the measurement of sea surface heights, wave heights, and surface wind speed. ALT mass = 206kg, power = 237 W, altitude measurement accuracy of 2.4 cm. Simultaneous measurements at both frequencies so that ionospheric range delay can be directly estimated from the two measurements.
  Note: The altimeter antenna is shared between ALT and SSALT - with ALT using 90% of the time during the first 6 months. After that time will be allocated on assessment based on the first 6 months.

- **TMR** = Topex Microwave Radiometer (JPL). Operation at 18, 21, and 37 GHz to measure the total water vapor content along the altimeter pulse path to correct for the water vapor induced range delay. The uncertainty in the altimeter range measurement made by such a system under normal ocean conditions is expected to be less than 5 cm at 7 km spatial resolution (3 cm at 100 km resolution). Mass = 50 kg, power = 25 W.

- **GPSDR** = GPS Demonstration Receiver system for direct position measurement, JPL; Demonstration of GPS differential ranging as an experiment. The GPS receiver measures incoming signals from the GPS satellites and uses in addition ITRF (International Terrestrial Reference Frame, i.e. a set of reference ground stations) measurements for DGPS results. GPSDR operates at 1227.6 MHz and at 1575.4 MHz. Mass = 28 kg, power = 29 W, altitude accuracy < 10 cm.
  PS: The TRANET orbit determination (with the ground network) was dropped in favor of GPSDR.

- **SSALT** =Single-Frequency Solid-State Altimeter. (1-Frequency of 13.65 GHz). Experimental sensor (CNES) to demonstrate the concept of low-power, low-weight, low data rate (1/7 the rate of ALT due to extensive onboard processing) and low-cost altimeter for future earth observing missions. The ionospheric range correction is provided by a model the makes use of simultaneous DORIS measurements. A measurement accuracy of 2.5 cm is expected. SSALT mass = 24 kg, power = 49W.

- **LRA** = Laser Reflector Array, JPL. This sensor will be used by a ground laser network (of 10-15 SLR stations) to track the position of the satellite for precision orbit determination (verification of altitude measurements). Mass = 29 kg, accuracy = 2 cm.

- **DORIS** = Doppler Orbitography and Radiopositioning Integrated by Satellite (CNES-GRGS-IGN development)[359],[360]

## A.110.1 DORIS

DORIS is a one-way microwave tracking system for the determination of precise orbits (goal of 5-10 cm radial distance and 0.3 mm/s in range rate accuracy). The concept is based


359)  "Other Satellite-Based Microwave Systems', Lecture Notes in Earth Sciences - The Interdisciplinary Role of Space Geodesy, Springer Verlag I. Mueller, S. Zerbini, chap. 5, pp. 161
360)  DORIS - Precision Satellite-Based Orbit Determination, CNES brochure

on a ground segment (of globally positioned tracking stations) and a space segment (i.e. DORIS as a passenger payload in a satellite). There is also a control center as part of the ground segment, located at CNES.

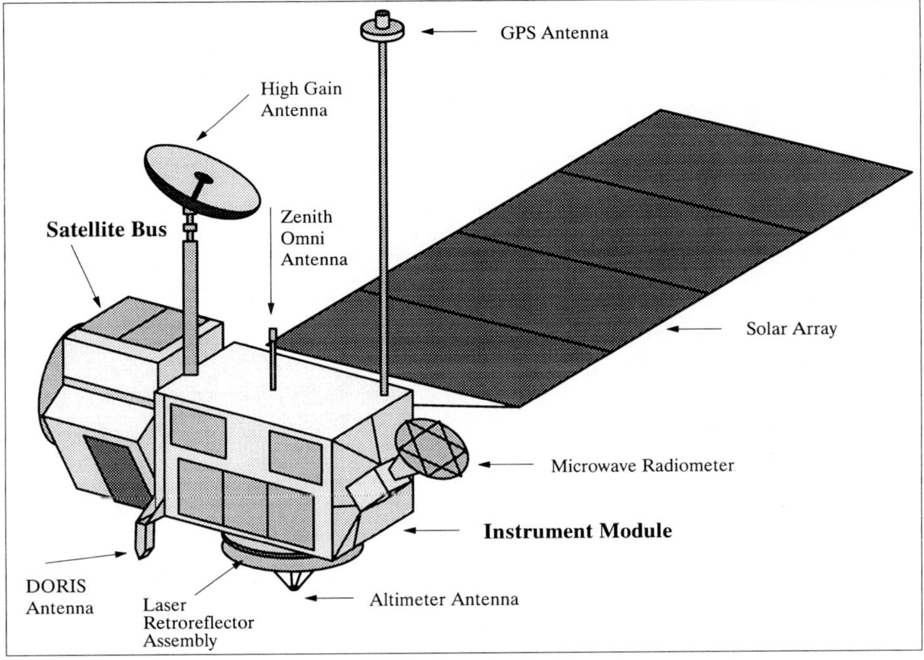


**Figure 99:    The Topex/Poseidon S/C Model (Zenith View)**

The onboard receiver measures the Doppler shift of 'Uplink Beacons' in two frequencies ($f_1$ = 2036.25 MHz, $f_2$ = 401.25 MHz), which are transmitted continuously by the DORIS ground network of stations (30 to 50). One measurement is used to determine the radial velocity between spacecraft and beacon, the other to eliminate errors due to ionospheric propagation delays. Only one beacon can be received by the space segment at any time. Orbit determinations with a precision of 10 to 20 cm are supposed to be possible.

The DORIS onboard package comprises a receiver, [or radial velocity measurement unit, consisting essentially of two receiving chains; total mass = 17 kg, power consumption = 20 W; size = 385 x 280 x 210 mm], an ultrastable crystal oscillator, and an omnidirectional antenna.

The DORIS ground segment comprises:
- the DORIS Control Center (DCC) at CNES
- a beacon installation and management center, managed by IGN. A network of Orbit-Determination Beacons (ODBs) is positioned throughout the world.
- Precision orbit determination computations performed by CNES (Earth's gravitational field computation on the basis of DORIS data by GRGS).

An ODB comprises two transmitters (one operating at 401.25 MHz, the other at 2036.25 MHz), an ultrastable oscillator, and a microprocessor performing the necessary control and management functions, transmission of timing, housekeeping, and failure diagnosis. An ODB also includes an antenna and 3 meteorological sensors (atmospheric pressure, air temperature, and relative humidity), these parameters are needed for atmospheric propagation delays. An ODB message carries meteorological data, the beacon ID, and information concerning the beacon operating status. The complete message lasts 0.8 seconds and is repeated once every 10 seconds.

A second class of beacons is termed Ground Location Beacons (GLBs). These are at positions that are either unknown or not known to sufficient accuracy. GLBs use the results of high-precision orbit determination as input for the precise determination of ground positions. GLBs are functionally identical to ODBs. Each GLB transmits independently of all others for 10 seconds, once, twice, or tree times every minute, but only while the satellite is in range.

The master beacon (MB) is the link between DCC and the onboard package. On each pass the DCC transmits data and instructions for onboard programming.

A possible disadvantage of this one-way DORIS system might be in the synchronization of all clocks involved (onboard and on ground at each station).

DORIS application: All-weather global tracking of Topex/Poseidon, estimate of the total content of ionospheric free electrons. DORIS will be a supporting instrument to LRA.

**Data:** The geophysical data produced by the Topex/Poseidon mission will be accessible to the international scientific community through US and French national data centers. The French data center for oceanography is called AVISO (CNES). The data products include:

- Sea surface topography
- Significant wave height
- Surface wind speed
- Ocean tides
- Vertically integrated atmospheric water vapor
- Vertically integrated ionospheric electron content

S/C status 1/1994: all instruments operational.


## A.111   TRMM (Tropical Rainfall Measuring Mission)

TRMM = Tropical Rainfall Measuring Mission, a joint NASA-NASDA mission with a low-inclination (equatorial) orbit. NASA/GSFC provides the satellite, 4 passive sensors, and mission operations, NASDA the launch vehicle (H-II rocket) and the precipitation radar instrument. Planned launch: Aug. 1997, design life = 3 years, satellite mass = 3500 kg.

Spacecraft: mass = 3620 kg (including 725 kg of fuel); 3-axis stabilization, power = 1000 W; data rate = 170 kbit/s average and 2 Mbit/s on playback; TDRSS S-Band communications (8.5 minutes/orbit playback time)

Orbit: Non Sun-synchronous circular orbit. Altitude: 350 km (approx.), inclination: 35°.

Objectives: Global change studies, especially in developing an interdisciplinary understanding of atmospheric circulation, ocean-atmospheric coupling, and tropical biology. General circulation models require detailed data on the latent heating of equatorial air masses, and the forcing and propagation speed of waves involved in the 30- to 60-day tropical oscillations[361],[362].

- Measurement of diurnal variation of precipitation and evaporation in the tropics to provide an increased understanding of how substantial rainfall affects the global climate patterns.

361) 'The Early Observing System Reference Handbook, ESAD Missions 1990-1997, NASA-GSFC, pp. 62-64
362) T. Keating, T. Ryan, 'Tropical Rainfall Measuring Mission (TRMM): US/Japan Science Operations',
     AIAA-92-0594

- Obtain a minimum of 3 years of climatologically significant observations of rainfall in the tropics.
- In tandem with cloud models, provide accurate estimates of the vertical distributions of latent heating in the atmosphere.

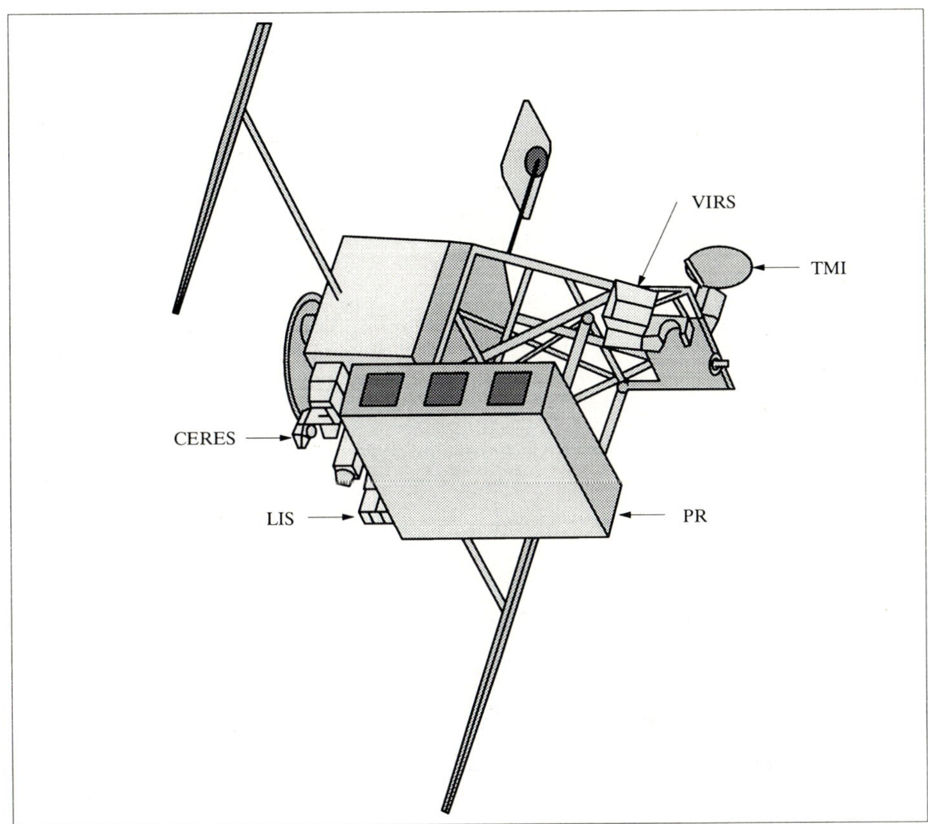


**Figure 100:   The TRMM S/C Model**

**Sensor Complement:**[363]

Prime Sensors:

- **PR** = Precipitation Radar (NASDA instrument)[364].
  Active MW radar operating at 13.796 GHz and at 13.802 GHz with horizontal polarization (orbit permits monthly sampling over the complete diurnal cycle). Resolution: range resolution 250 m, horizontal resolution = 4.3 km at nadir, swath width: 220 km, minimum measurable rain rate of 0.5 mm/h. Objective: 3-D rainfall distribution over land and oceans (combined with TMI sensor).
  Antenna type: Planar array - a 128-element active phase array system is adopted to achieve a contiguous scanning within the swath. Beam width: 0.71° x 0.71°; aperture: 2.1 m x 2.1 m; scan angle: ± 17°; peak power = 600 W; data rate: 93.5 kbit/s.

---

363)  NASA paper provided by ESAD and OSSA.
364)  T. Kozu, M. Kojima, K. Oikawa, K. Okamoto, T. Ihara, T. Manabe, 'Development Status of Rain Radar for Tropical Rainfall Measuring Mission', IEEE IGARSS '92, Volume II, pp. 1722-1724

The PR has 3 basic operational modes: observation mode, internal calibration mode, and external calibration mode.

- **VIRS** = Visible Infrared Scanner (NASA instrument)
  Five spectral ranges are considered: 0.63 µm, 1.6 µm, 3.75 µm, 10.8 µm, and 12 µm. Resolution = 2 km at nadir. Swath width = 720 km. Applications: Data will be used in conjunction with data from CERES to determine cloud radiation . VIRS will enable "calibration" of precipitation indexes derived from data of other sources (rain estimation from brightness temperature).

- **TMI** = TRMM Microwave Imager (NASA instrument)
  TMI is a passive MW radiometer with frequencies in the multichannel dual polarization range (5 discrete channels at 10.7, 19.4, 22, 37, and 85-91 GHz). Resolution: 4.4 km to 45 km depending on the frequency used. Swath width = 680 km. Applications: Data is related to rainfall rates over oceans (vertically integrated rainfall distribution).

Secondary Sensors:

- **CERES** = Clouds and the Earth's Radiant Energy System (NASA instrument). The production instrument is based on ERBE technology. CERES is a broadband scanning radiometer with the capability of operating in either an elevation scan mode (primary mode) or in a biaxial scan mode. Applications: Earth's radiation budget and atmospheric radiation from the top of the atmosphere to the surface.
  CERES measures shortwave and longwave infrared radiation using thermistor bolometers for the determination of the radiation budget. FOV = 156° (limb to limb). Resolution = 21 km at nadir.
  Three spectral channels:    0.3 - 5 µm (shortwave)
                              8.0 - 12.0 µm (longwave)
                              0.3 - 50 µm (total)

- **LIS**[365] = Lightning Imaging Sensor (US instrument).
  Measurement of lightning distribution and variability over the Earth.
  Measurement approach: LIS is a staring imager that detects the rate, position, and radiant energy of lightning flashes. LIS will detect intra-cloud and cloud-to-ground lighting. Applications: cloud characterization, hydrological cycle, cerial cycles.
  - special filter to image at 777.4 nm onto a 128 x 128 CCD array detector
  - event processor to subtract out the bright background during daylight (sensor takes data during day and night)
  - swath: 600 km
  - spatial resolution: 8.5 km (4 km at nadir)

  Accommodation issues:
  - mass: 23 kg
  - power: 42 W
  - Data rate: 6 kbit/s
  - FOV: 75 x 75°, IFOV: 0.7°

TRMM Mission Data:

Primary Products:

- Average monthly rainfall over the tropics and subtropics for at least 3 years

Secondary Products:

- Cloud cover (VIRS, CERES)
- Rain rates (TMI)
- Rain rate vertical profile (PR)

---

365) EOS Reference Handbook, NASA/GSFC, 1991

- Path-averaged rain rate and liquid water content (PR)
- Lightning distribution and variability (LIS)

Data Validation Program:

- Rain rate spatial distribution (surface radars)
- Rain rate point measurements (in situ measurements)

| Application | Total Precipitation Rate | | |
|---|---|---|---|
| | Spatial Average | Time Average | Accuracy |
| Climate Models | 500 x 500 km | monthly mean | 1 mm/day (10% in heavy rain) |
| Diurnal cycle over ocean | 20° Longitude | bimonthly | 10% first harmonic amplitude 20% second harmonic amplitude |
| General circulation model vertical distribution | 500 m | N/A | N/A |
| Tropical rain systems structure and evolution | 20 km | N/A | 30 - 50% |

**Table 84:     TRMM Scientific Accuracy Requirements**

| Sensor | Observation Objectives | Frequency | Horizontal Resolution | Swath Width |
|---|---|---|---|---|
| Precipitation Radar PR | 3-D rainfall distribution | 13.8 GHz | 4.3 km (nadir) | 220 km |
| TMI | Vertically integrated rainfall distribution | 10.7, 19.4, 21.3, 37, and 85.5 GHz | 5-41 km | 790 km |
| VIRS | Cloud distribution and height, rain est. from brightness temp. | 0.63, 1.6, 3.75, 10.7, and 12 μm | 2 km | 1500 km |
| CERES | Radiation from top of clouds and Earth, energy budget | 0.3 - 3.5 μm<br>8.0 - 12.0 μm<br>0.3 - 50 μm | 25 km (nadir) | Scan angle: ±80° |
| LIS | Lightning distribution | 0.7774 μm | 4 km (nadir) | 600 km |

**Table 85:     Overview of TRMM Sensor Complement and Objectives[366]**


# A.112   UARS  (Upper Atmosphere Research Satellite) 

UARS = Upper Atmosphere Research Satellite. The first NASA mission in the series: 'Mission to Planet Earth'. Launch date: Sept. 13. 1991 with Space Shuttle (Discovery), UARS is a free-flying laboratory; 3 years nominal life time; GSFC = POCC. UARS weight = 6480 kg, 3-axis stabilized S/C.[367]

Application: Measurement of the energy flux (input and loss) in the upper atmosphere. Global photochemistry in the upper atmosphere, in particular in the stratosphere and the mesosphere (trace gases and temperature profiles); Dynamics of the upper atmosphere; Transport phenomena of the different processes; Correlations between the upper and lower atmosphere and their changes.

- Atmospheric chemistry and temperature
- Atmospheric winds

366)  Courtesy of K. Maeda, NASDA
367)  'UARS Seen as Earth Observing System's Dress Rehearsal', Space News September 9-15, 1991, p. 24

- Solar energy
- Energetic particles

Orbit: Circular orbit, 57° inclination; Altitude = 585 km, period = 97 minutes,

**Sensors:**[368],[369]

**CLAES** = Cryogenic Limb Array Etalon Spectrometer (Solid-hydrogen cooled spectrometer sensing atmospheric infrared emissions); NASA sensor, (A.E Roche, Lockheed Palo Alto Research Lab)
Measurement with 4 Etalons and 8 filters the following wave spectrum: 3.5 µm, 6 µm, 8 µm and 12.7 µm. Because the detectors and optics generate their own thermal emissions, they must be cooled to temperatures which suppress this emission. The coolant inside CLAES is expected to be depleted after about 18 months of service.
CLAES measures concentrations of members of the nitrogen and chlorine families, as well as ozone, water vapor, methane, and carbon dioxide. To obtain a vertical profile of species concentration, CLAES utilizes a telescope, a spectrometer, and a linear array of 20 detectors to make simultaneous measurements at 20 altitudes ranging from 10 to 60 km.
Observables: $N_2O$, $NO$, $NO_2$, $HNO_3$, $CF_4$, $CF_2Cl_2$, $CFCl_3$, $HCl$, $O_3$, $ClONO_2$, $CO_2$, $H_2O$, $ClO$, $CH_4$ and temperature.
Ops status: CLAES ran out of oxygen in May 1993, as planned.

**ISAMS** = Improved Stratospheric and Mesospheric Sounder (Mechanically cooled spectrometer sensing atmospheric emissions); (F.W. Taylor, Oxford University)
This is a SAMS successor using filter radiometry and pressure modulation techniques in the following ranges:: 4.6 - 16.6 µm (medium infrared Band). Instrument is provided by UK (improved version of SAMS which operated aboard Nimbus-7 from 1978 to 1983).
ISAMS is a filter radiometer employing 8 detectors. It observes infrared molecular emissions by means of a movable off-axis reflecting telescope. In addition to scanning the atmosphere vertically, the telescope can also be commanded to view regions to either side of the UARS observatory, thus providing increased geographic coverage. One feature of ISAMS is that it carries samples of some of the gases to be measured in cells within the instrument. Atmospheric radiation collected by the telescope passes through these cells on its way to the detectors (spectra matching).
ISAMS measures the concentrations of nitrogen chemical species, as well as ozone, water vapor, methane, and carbon monoxide. Data rate: 1.25 kbit/s.
Observables: $CO$, $H_2O$, $CH_4$, $N_2O_5$, $NO$, $N_2O$, $O_3$, $HNO_3$ and aerosols.

Status: the ISAMS instrument experienced a chopper motor failure in late July 1992, preventing its collection of atmospheric chemistry data.

**HALOE** = Halogen Occultation Experiment (J.M. Russell, LaRC)
Gas filter Radiometer Correlation in sun occultations. Infrared range = 2.43 - 10.25 µm. Spatial resolution: vertical = 1.6 km at limb; horizontal = 6.2 km at limb. Measurement of the vertical distribution of hydrofluoric and hydrochloric acids as well as those of methane, carbon dioxide, ozone, water vapor, and members of the nitrogen family. The HALOE experiment uses samples of the gases to be observed as absorbing filters in front of the detectors to obtain a high degree of spectral resolution.
During every UARS orbit, at times of S/C sunrise and sunset, HALOE will be pointed toward the Sun to measure the absorption of energy along this line of sight. There are 28 solar occultation opportunities per day, providing data for 14 different longitudes in each of the Northern and Southern Hemispheres. Data rate = 4 kbit/s.
Observables: $HF$, $HCl$, $CH_4$, $NO$, $H_2O$, $O_3$, $NO_2$ and pressure.

**MLS** = Microwave Limb Sounder (J. W. Waters, JPL). MLS is a microwave radiometer which measures atmosphere thermal emission from selected molecular lines at mm wave-

---

368) Portion of a UARS publication put out by NASA (provided by B. Needham of NOAA)
369) 'Upper Atmosphere Research Satellite', Summaries of papers presented at the Optical Remote Sensing of the Atmosphere Topical Meeting, Feb. 12-15 1990, Optical Society of America, Volume 4, pp. 1-22

lengths.
3-channel Heterodyne Limb Sounder. Measurement frequencies: 63 GHz (1 band), 183
GHz (2 bands) and 205 GHz (3 bands, corresponding to wavelengths of 4.8, 1.64, and 1.46
mm respectively). Each band is 500 MHz wide.

MLS is a microwave radiometer providing global measurements of chlorine monoxide (key
reactant that destroys ozone), hydrogen peroxide, water vapor, and ozone. The MLS observations will provide, for the first time, a global data set on chlorine monoxide in the upper
atmosphere. MLS will also determine the altitudes of atmospheric pressure levels. Spatial
resolution in each band: ~400 km horizontal and 4 km vertical; measurement are made
along the tangent track of the limb view, with no cross-track scanning, swath width: 5 - 85 km
(vertical limb coverage). Data rate: 1.25 kbit/s.
Observables: $O_3$, $ClO$, $H_2O_2$, $H_2O$ and pressure.

Calibration: radiometric views of cold space, and an on-board ambient-temperature blackbody target. Calibration stability: better than 1% over mission lifetime.

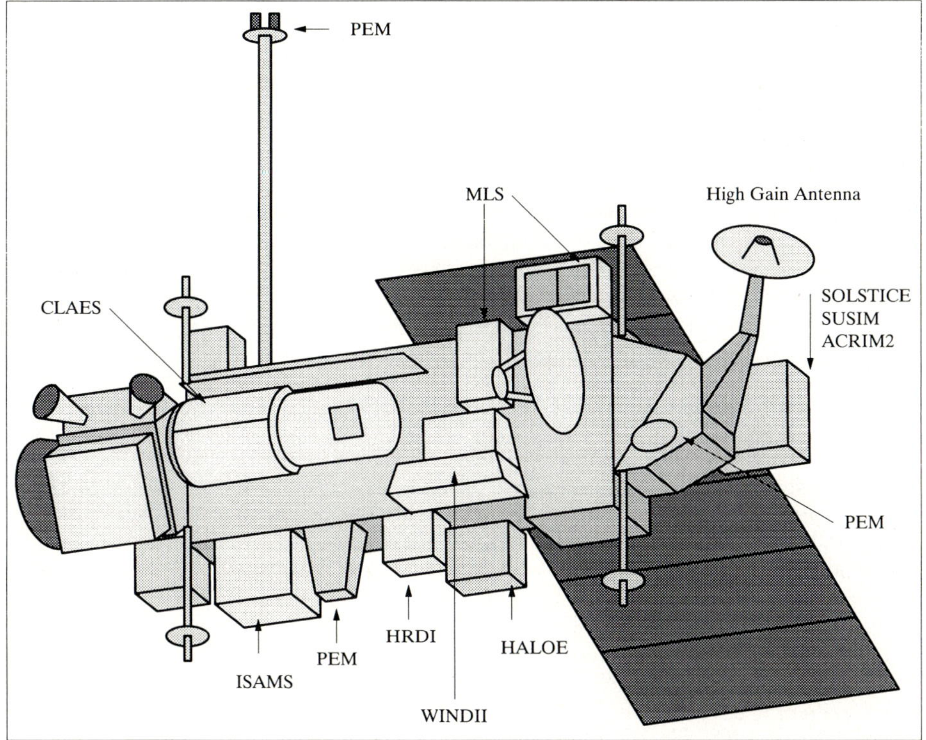


**Figure 101:    The UARS S/C Model**

**SOLSTICE** = Solar/Stellar Irradiance Comparison Experiment (G.J. Rottman, University
of Colorado at Boulder). Three-channel grating spectrometer for the measurement of solar
and stellar irradiation (UV radiation in the wavelength range from 115 to 430 nm with a
resolution of 0.12 - 0.25 nm). The device compares the ultraviolet output of the sun with
similar radiation produced by 30 stable, bright blue stars. The stars constitute the standards
against which the solar irradiance is measured.

The experiment consists of a spectrometer with three spectral channels, each with a separate grating and photomultiplier tube. SOLSTICE will be pointed toward the Sun during the daylight portion of the orbit, and toward one of the calibration stars during most of the nighttime orbit. Data rate: 250 bit/s.

**SUSIM** = Solar Ultraviolet Spectral Irradiance Monitor (G.E. Brueckner, Naval Research Lab). SUSIM is a double-dispersion scanning spectrometer.
Spectrometer for the measurement of solar irradiance (UV radiation in the spectral range from 120 - 400 nm with a resolution of 0.1 nm). SUSIM incorporates 2 spectrometers, 7 detectors, and a set of 4 deuterium UV calibration lamps. One spectrometer observes the Sun and measures the variation in solar UV flux as a function of time. The second spectrometer monitors independently the calibration lamps. Data rate: 2 kbit/s.

**PEM** = Particle Environment Monitor (J.D. Winningham, Southwest Research Institute). PEM represents a collection of 4 instruments called AXIS, HEPS, MEPS and VMAG. PEM is mounted at three separate locations on UARS, including the boom.

- AXIS (X-Ray, Proton and electron spectrometer). Measurement of Bremsstrahlung x-rays from Earth, solid-state detection. Energy range from 3 keV to 100 keV. Spectral band: 0.012 to 0.41 nm
- HEPS (Charged particle spectrometer). Measurement of in situ electron energies from 0.04 MeV to 5 MeV, proton energies from 0.07 MeV to 139 MeV. Measurement technique: solid state identification.
- MAPS (Charged particle spectrometer). Measurement of electrons and protons. Energy range of electrons: 1 eV - 32 keV, energy range of protons: 1 eV to 32 keV; measurement technique with electric deflection with mechanical separation.
- VMAG (3-axis fluxgate magnetometer). VMAG DC: center at 0 nT, width: ± 65500 nT. VMAG DC: center at 0 nT with width dependent on AXIS. Measurement of the Earth's magnetic field and the type, number and distribution of charged particles flowing into the upper atmosphere from space.

**HRDI** = High Resolution Doppler Imager (P.B. Hays, University of Michigan).
Mapping of atmospheric winds at altitudes below 45 km. HRDI observes the Doppler shifts of spectral lines within the atmospheric band system of molecular oxygen to determine the wind field. There are no sharp emissions lines in the radiance of the Earth's limb at such altitudes, but the oxygen bands contain many lines that appear as deep absorption features in the brilliant spectrum of scattered sunlight.
A triple-etalon Fabry-Perot interferometer, serving as a high-resolution spectral filter, will ensure efficient rejection of the intense emission continuum outside the absorption lines. HRDI will exploit these daytime absorption features to provide wind data for the stratosphere and upper troposphere to an accuracy of 5 m/s or better.
At altitudes above about 60 km, HRDI observes emission lines of neutral and ionized atomic oxygen in the visible and near-infrared regions. Unlike the molecular absorption lines, however, the emission lines are observable both day and night. Mesosphere and Thermosphere wind profile to an accuracy of 15 m/s or better.

Measurement in 13 spectral bands with center wavelength and bandwidth in brackets, all in nm: 557.5 (0.84), 630.0 (.74), 630.5 (1.48), 686.8 (0.88), 687.6 (0.88), 692.3 (1.06), 692.9 (1.36), 723.5 (0.85), 760.7 (0.92), 763.5 (1.45), 764.9 (0.86), 766.6 (1.01), 775.6 (0.93).

The interferometer is tunable within the spectral width and can simultaneously measure 31 samples in a 0.5 cm$^{-1}$ spectral region at a resolution of 0.05 cm$^{-1}$. - Data rate = 4.75 kbit/s.

**WINDII**[370] = Wind Doppler Imaging Interferometer (G.G. Shepherd, York University, Canada; Canadian/French Sensor, 145 kg weight). WINDII is a field-widened Michelson interferometer measuring the Doppler shift and line broadening of atmospheric emission in

---

370) 'Wind Imaging Interferometer (WINDII) for the UARS Mission', Optical Remote Sensing of the Atmosphere, 1990 Technical Digest Series of the Optical Society of America, Volume 4, pp. PD3-1 to 4

the visible and near-infrared.
Measurement of temperatures and wind conditions in the stratosphere[371].
WINDII utilizes emission lines for the basic Doppler-shift measurements. In addition to lines of neutral and ionized oxygen, these include two lines of the OH molecule and a molecular-oxygen line. WINDII obtains measurements both day and night at altitudes above 80 km.

The WINDII spectral filter is a high-resolution Michelson interferometer. The instrument consists of a telescope, the interferometer, and a detector array. The telescope views 45° and 135° from the S/C velocity vector simultaneously. In normal operation, the detector provides a vertical resolution of 20 km. Wind velocity accuracy is within 10 m/s in the altitude range between 80 and 300 km. Data rate: 2 kbit/s.

**ACRIM2** = Active Cavity Radiometer Irradiance Monitor (R.C. Willson, JPL; see ACRIM in EOS)

**Data:**

There are two onboard recorders available (each capable of recording two orbits of data at 32 kbit/s). Tape recorder playback = 512 kbit/s.
Data transmission via TDRSS with 32 kbit/s (R/T) and 512 kbit/s (P/B).

Correlative data sets are provided in parallel, in particular NOAA SUBV data sets, as well as ground truth measurements (Balloons, Sounding Rockets, Shuttle flights) and meteorological data.


## A.113  Viking

The Viking project is a Swedish mission, funded by the Swedish Board of Space Activities, and managed by SSC (Swedish Space Corporation) with international cooperation on sensors. Objectives: Study of the physical processes associated with the interaction of the solar wind with the Earth's magnetosphere (establish the region for acceleration of Auroral particles and the mechanisms involved, behavior of the resulting Aurora Borealis) in the high latitude medium height (1-2 $R_e$) regions.[372), 373),374)]

Launch Feb. 22 1986 by Ariane V16 from Kourou (piggyback payload with Spot-1). The satellite was attitude controlled (spin stabilized at 3 rpm, cartwheel mode, magnetic control), size approximately 1.9 x 0.5 m, satellite mass in orbit = 286 kg, payload mass = 44.7 kg (68 kg including booms and antennas). The design life of Viking was 8 months, the satellite provided data for 15 months (May 12, 1987 end of mission).

Orbit: Polar orbit, apogee = 13530 km, perigee = 817 km, inclination = 98.8°, orbital period = 262 min, drift of rising node = 0.17°/day, rotation of line of apsides = - 0.50°/day. It took 8 months for the apogee to move from a starting point with a latitude of about 40° N up over the polar region and down to the equator. After these 8 months the apogee was in the southern hemisphere. Note: there was no onboard storage capability and there was no receiving station in the Antarctic region; hence, measurements could only be made at low altitudes from the Esrange/Kiruna ground station.

Sensors:

**V1: Electric Field Experiment** (PI: Lars Block of RIT, also involved: ESTEC and UC Berkeley). Measurement of the dc electric field by 3 orthogonal pairs of spherical probes.


371) 'Windii To Read Upper Atmosphere In Depth', Space News September 16-22, 1991, p. 8
372) B. Hultqvist, "The Swedish Satellite Project Viking", Journal of Geophysical Research, Vol. 95, Nr. A5,, May 1, 1990, pp. 5749-5752
373) B. Hultqvist, "The Viking Project", Geophysical Research Letters, Volume 14, Nr. 4, April 1987, pp. 379-382
374) "The Viking Program", EOS Transactions, American Geophysical Union, Volume 67, Nr. 42, Oct. 21, 1986, pp. 793-795

- 4 wire booms, each 40 m long
- 2 rigid booms, 4 m each
- 53 or 106 samples per second
- data rate = 4.3 kbit/s
- mass = 4.7 kg (without booms)
- power = 7.3 W

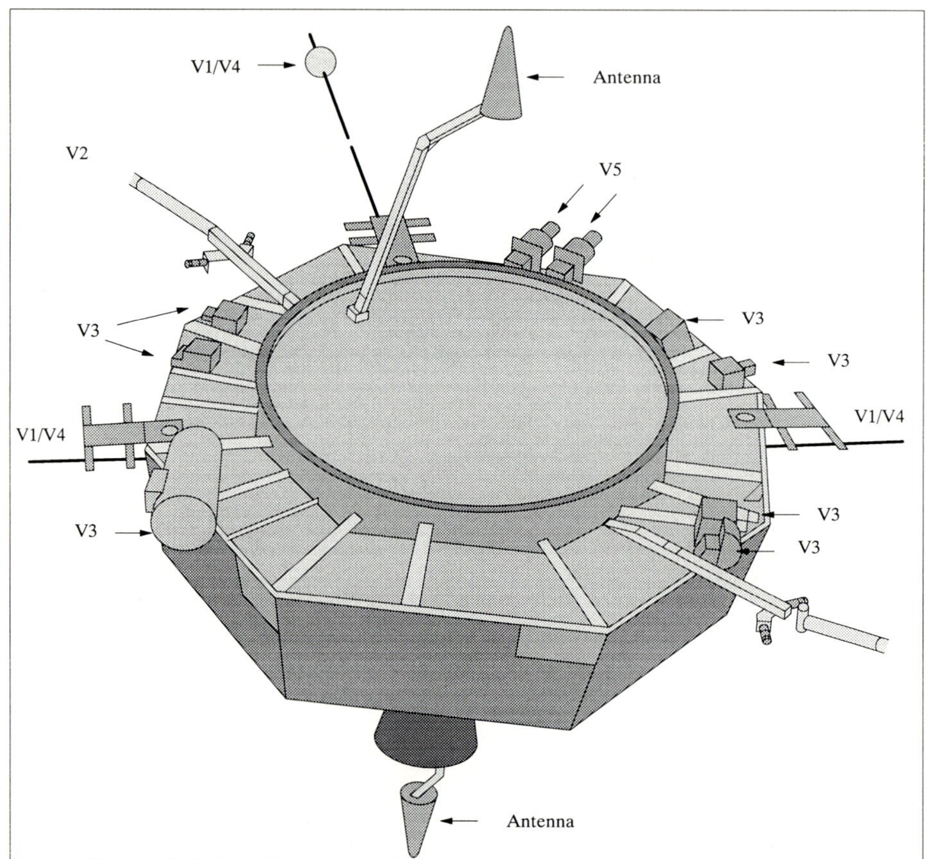


**Figure 102:**    **The Viking S/C Model**

**V2: Magnetic Field Experiment** (PI Tom Potemra of APL, also involved: IRF-U of Sweden, GSFC/NASA, and U. of Kyoto). The experiments V1 and V2 have the same sampling rates to determine whether the variations in the electric and magnetic fields are due to fine-scale currents or waves.

- three-axial fluxgate magnetometer system on a 2 m long radial boom
- 53 samples per second
- four ranges of measurement: from $\pm 65536$ ($\pm 8$) nT to $\pm 1024$ ($\pm$o.125) nT
- data rate = 2.2 kbit/s
- mass = 2.1 kg
- Power = 1.1 W

**V3: Hot Plasma/Energetic Particle Experiment** (PI R. Lundin of IRF, also involved: U. of Bergen, MPI Lindau, Los Alamos National Laboratory, Aerospace, APL, RAL). Measurement objectives:

- electrons with spectral resolution $\Delta E/E \sim 0.05$, from 10eV to 40 keV
- electrons with angular resolution $\Delta\alpha \leq 2^o$ from 0.1 keV to 300 keV
- energy and pitch angle distributions of ions $\Delta E/E \sim 0.08$, $\Delta\alpha \leq 6$ from 40 eV to 40 keV
- three-dimensional distribution functions of ions from the satellite potential to 10 keV
- composition, energy, and pitch angle distribution of ions up to 10 MeV/nucleon.

All these measurements can be made simultaneously. The instrument consists of 7 units covering different kinds of particles and different energy ranges.

- data rate = 12.4 kbit/s
- mass = 18.2 kg
- Power = 13.7 W

**V4L: Low-Frequency Wave Experiment** (PI: G. Gustavsson of IRF-U, also involved IRF of Sweden, DSRI of Denmark, Cornell University, N. Y.). The wave experiment is divided into two parts, depending on the frequency. V4L measures waves up to about 10 kHz. Two wave signals can be processed simultaneously that may come from any of the electric field antennas, from a magnetic loop antenna measuring fluctuations in the direction of the spin axis, or from the V1 experiment.
The experiment uses 2 wire booms of the Electric Field Experiment and in addition a magnetic coil antenna for the measurement of one component of the magnetic wave field (mounted on a 2 m long rigid boom).

- frequency range: 0 - 15 kHz
- experiment measures also $\Delta N_e/N_e$ and the wave phase velocity

**V4H: High-Frequency Wave Experiment** (PI: A. Bahnsen of DSRI, also involved: CRPE of France, IRF-U of Sweden). V4H has two sets of eight filters each, with center frequencies from 4 kHz to 512 kHz, and a pair of stepped frequency analyzers from 10 kHz to 500 kHz. Analog signals are provided by one of the electric field antennas or from the magnetic loop antenna.

- the experiment uses the same sensors as the low-frequency experiment
- frequency range: 10 kHz - 500 kHz
- the experiment contains also an active resonance experiment which among other things gives the electron density from the plasma frequency.
- data rate = 25.6 kbit/s (for low and high frequency experiments together)
- mass = 11.4 kg (for low and high frequency experiments together)
- power = 21.2 W (for low and high frequency experiments together)

**V5: Auroral Imaging Experiment** (PI: C. Anger/S. Murphree of U. of Calgary, also involved: NRC of Can., U. of York, Can., U. of Saskatchewan, Can., U. of Alberta, Can., IRF of Sweden, MISU of Sweden). Study of the dynamic behavior of auroral forms.

- The instrument determines the auroral luminosity distribution over the polar cap and the auroral regions.
- field of view: $29^o$ x $24^o$
- wavelength ranges: 1240-1500 Å and 1330-1900 Å (two different cameras)
- at apogee one pixel covers 20 x 20 $km^2$.
- an image can be produced every spin (in 20 seconds)
- the aurora can be seen with good contrast also on the sunlit side of the Earth
- data rate = 5.1 kbit/s (a higher bit rate of 30.7 kbit/s can be used at the expense of the wave experiments)
- mass = 8.5 kg
- power = 5.3 W

## A.114  WIND

NASA/GSFC Solar-Terrestrial mission within the US GGS (Global Geospace Science) ISTP initiatives/programs with the objective to study sources, acceleration mechanisms and propagation processes of energetic particles and solar wind. Investigation of the solar wind input first from the day-side double lunar swingby orbit, and later from a small halo orbit at $L_1$.

Projected launch: April 1994 with a Delta II vehicle from Cape Canaveral. S/C mass = 1250 kg (300 kg propellant, 195 kg science payload). Nominal lifetime = 3 years. Spin-stabilized S/C at 20 rpm, 2.4 m $\oslash$ and 1.8 m height. Solar arrays provide 370 W, including 144 W for payload instruments.

Orbit: lunar swingby to a 250 $R_E$ apogee and a perigee of at least 5 $R_E$ during first 2 years; thereafter $L_1$ halo orbit (figure-eight orbit around the Earth and the moon on the Sun side of the Earth).

**Sensors:**

**MFI** = Magnetic Field Investigation (PI: R. Lepping, GSFC). Objectives: investigation of the structure and fluctuations of the interplanetary magnetic field (transport of energy and acceleration of particles in the solar wind).
Instrument: Magnetometer measures the intensity and direction of magnetic field vector. Measurement rate of 44 vectors per second.

**WAVES** = Radio and Plasma Wave Experiment (PI: J. Bouqeret, Observatoire de Meudon, France). Objectives: measurement of the plasma over a very wide frequency range.

**SWE** = Solar Wind Experiment (PI: K. Ogilvie, GSFC). Objectives: measurement ions and electrons in the solar wind and the foreshock regions. Rates of once per minute for ions and 20 times per minute for electrons. Deduction of solar wind velocity, density, temperature, and heat flux.

**EPACT** = Energetic Particles Acceleration, Composition, Transport (PI: T. von Rosenvinge, GSFC). Objectives: investigation of the elemental and isotopic abundances of the minor ions making up the solar wind with energies in excess of 20 keV. Measurements at a rate of once per minute for ions and 20 times per minute for electrons. Deduction of solar-wind velocity, density, temperature, and heat flux.

**TGRS** = Transient Gamma Ray Spectrometer (PI: B. Teegarden, GSFC). Objectives: observation of the transient gamma-ray events, spectroscopic survey of cosmic gamma-ray transients, measurements of gamma-ray lines in solar flares. Measurement ranges: 15 keV - 8.2 MeV.

**SMS** experiment, consisting of: **SWICS** (Solar Wind Ion Composition Study) and **STICS** (Mass Sensor, and Suprathermal Ion Composition Study); (PI: G. Gloeckler, U. of Maryland). Objectives: determine the abundance, velocity, spectra, temperature, and thermal speeds of solar-wind ions (plasma investigations in conjunction with EPACT).

**PLASMA** = 3-D Plasma and Energetic Particles Experiment (PI: R. Lin, U. of Ca. Berkeley). Objectives: measurement of ions and electrons with energies above that of the solar wind and into the energy particle range. Energy range: 0.03 - 30 keV, sampling rate: 20 times per minute; wide angular coverage, good directional sensitivity. Study of particle in the bow shock and in the foreshock regions.

**KONUS** = Gamma Ray Burst Investigation (PIs: E. Mazets/T. Clinc, Ioffe Physical Technical Institute, St. Petersburg, GSFC, instrument sponsored by Russia). Objectives: gamma-ray burst studies similar to the TGRS studies. KONUS has a lower resolution than TGRS but a broader coverage to complement TGRS. KONUS performs also event detection and measures time history.

Data: Onboard recording capability (1.3 Gbit digital tape recorder). Transmission via DSN (Deep Space Network) for 2 hr nominal daily contact. Science data rates: 5.6 kbit/s realtime and 128 kbit/s playback data. The WIND S/C provides onboard interconnection of instrumentation for data communication. Data sharing among the instruments can be triggered by pattern recognition schemes in onboard computers.

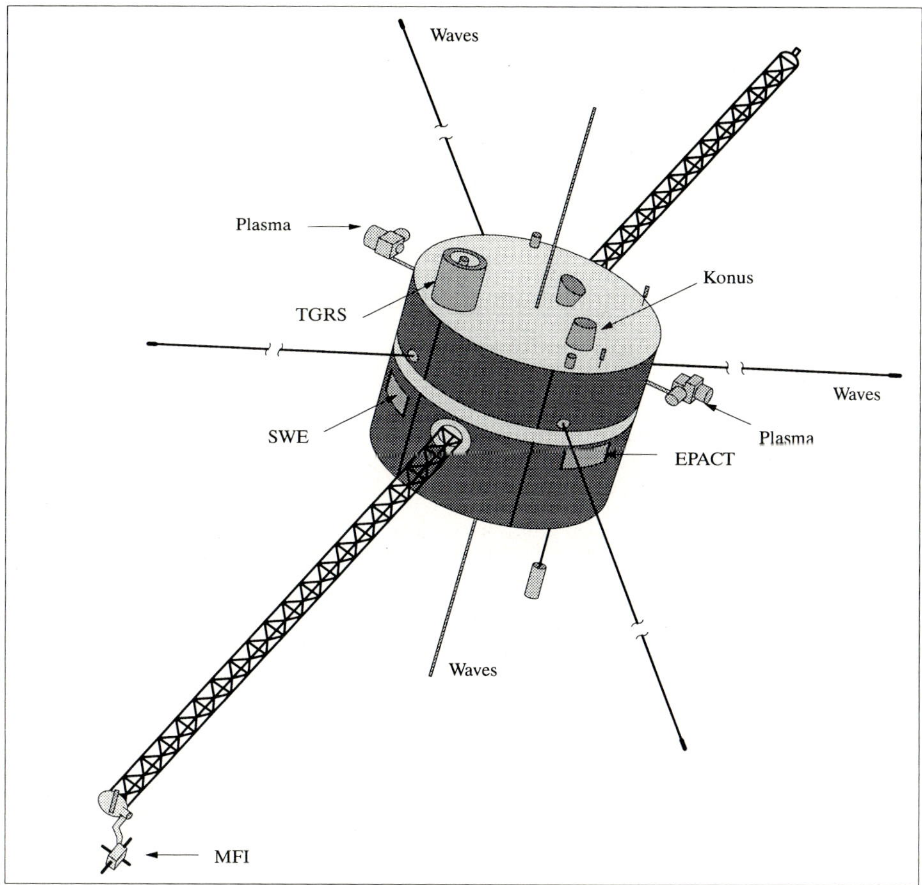


**Figure 103:    The WIND S/C Model**

# Part B    Survey of Airborne Sensors

Airborne observations with dedicated instruments have their own history in contributing to advances in the Geosciences and to extend our general knowledge in many fields, including the numerous surveys for resources. These observations simply cannot be regarded an appendix to spaceborne observations, as one might assume with virtually all the publicity on the spaceborne front; rather, airborne observations have a wide field of applications of their own, they are also the most efficient means by which laboratory- and ground-based observations can be extended to observations from space. The science community requires data on the local, regional, and global scales for an understanding of the overall picture.[375]

The logistical and operational complexity in the field of airborne sensors easily matches that of spaceborne sensors. Spaceborne sensors are usually identified with a unique spacecraft and mission of considerable length, the mission and its goals are well publicized, there is usually a long lead time prior to any data processing (complex space missions require in general long planning periods).

On the other hand, airborne missions are characterized by their relative short durations, by their regional confinement for a particular survey, by their adaptability to change to suit new requirements, by the relatively rapid availability of airborne sensor data. Airborne instruments may be upgraded, when a new technology becomes available, or when a better storage concept is affordable. Airborne sensors appear on and disappear from the scene of Earth observation, sometimes without much notice. It is indeed a tough job for anyone to keep track of the sensor zoo. The following list presents only a few scenarios of the degrees of freedom and the general environment encountered with airborne sensors.

1. Aircraft sensor complement.

- Dedicated instrumentation (fixed sensor complement) in support of a particular program (like atmospheric research, oceanography, land processes, cloud and meteorological research, photo-mapping surveys. etc., some support may also be multidisciplinary in nature).

- Changing sensor complement. A mission is planned for a certain time period with instruments brought in and out of the aircraft for each mission by the investigators.

- Support of instrument development.
  The introduction of a new or advanced (complex) spaceborne sensor technology requires generally a considerable degree of maturity in technology development and instrument testing before it can be flown successfully on a satellite. There are normally a number of steps (programs) prior to the commitment of satellite sensor experimentation/operation, that validate the required instrument performance of critical subsystems, the overall measurement performance, data processing algorithms, and other system parameters to provide a sound basis for the intended applications.

  The field of airborne sensors can be regarded as the proving ground for future spaceborne sensors. The airborne concept permits an environment for repeated instrument access on all levels, ideally suited for experimentation and validation; test flights can get along with a lot of provisional gadgets and arrangements, limitations on instrument weight and power consumption are not as restrictive as on the intended satellite mission. Complexity can be handled in an orderly way, this reduces considerably the risk of failure. The cost of airborne instrument experimentation along with its infrastructure is relatively small compared to spaceborne experimentation.

---

375) W. B. Johnson, S. H. Melfi, "Airborne Geoscience: The Next Decade", The Report of the Special Interagency Task Group on Airborne Geoscience, February 1989, NASA, NOAA, NSF

2. Applications
- Airborne data collection is being provided for a number of research projects or simply for surveys, seemingly independent of spaceborne data.
- Airborne in-situ sensors provide complementary data for use with other remotely collected data.
- In-situ data gathered from airborne sensors is being used as a calibration standard for surface-based or for spaceborne sensors.

3. Data resolution requirements on mesoscale.
   There are a number of remote sensing applications on a regional level that require a certain degree of repeat coverage and/or very high spatial resolutions that cannot be provided with the instrument from a satellite orbit (high-quality topographic maps). Other airborne applications require a very fine quantization of spectral information (very many and very narrow bands, which so far could not be provided in spaceborne instruments due to severe data constraints; HIRIS may be the first attempt on the spaceborne scene, see A.21) to enable the investigative data user to discriminate the individual constituents in an area much more effectively.

4. International programs and campaigns
   Recent years have seen worldwide coordinated campaigns with data collected over a predefined region from similar spaceborne and airborne sensors for the analysis of better models.

5. Types of aircraft for mission requirements (and combinations thereof):
- Small aircraft
- Mid-sized aircraft
- Long-range aircraft
- High-altitude jet aircraft
- Ultra-high subsonic aircraft

6. Peripheral equipment and services

- Aircraft are generally equipped with computer-controlled data-logging and display systems that record data and provide real-time control and graphic output during flight. Some future applications may also request a communication link to a ground station in order to provide the experimenter with real-time access to his data.

- Aircraft provide also a number of online supportive services for its payload such as: instrument power supply, positional and timing data, environmental data, etc.. Some research aircraft offer several navigation systems as well as GPS for navigation and instrument pointing. For an experimenter the accurate knowledge of platform and target position, of velocity and of time are fundamental parameters for the interpretation of his data.

7. Users of airborne data:

- Investigators and Co-investigators of a particular instrument
- Researches who require access to specific databases of airborne instrument data
- Users who query directories and catalogs for relevant data

Some countries maintain a strong engagement in the field of airborne observations. It is an effective and usually also an affordable field of activities for government-sponsored programs as well as for commercial service providers to contribute with their observational data to advances in science and to survey the country in its multi-faceted aspects (resources, mapping, etc.).

This survey of airborne sensors does not spell out the observational needs for particular national programs; rather, it provides technical sensor descriptions so that the reader may use this information for his own interest. There are certainly many opportunities for resource sharing on various levels, this applies for aircraft, for instruments, as well as for airborne observational data. Some context information is provided whenever available and suitable [this applies to data (such as chapter B.45) as well as to special aircraft (such as chapters B.31, B.91 and B.109)]. In view of the widespread use of airborne sensors in all fields of Geoscience and the broad spectrum of commercial applications, there is no claim to completeness of this survey. The sensors are listed in alphabetical order.

The reader will notice a more detailed description of airborne sensors in comparison with the spaceborne sensors. This is intended for a better presentation of concepts. Readers who are looking for more detailed information are referred to the references or to the particular agencies.

## B.1     AES (Airborne Emission Spectrometer)

AES is a JPL (sponsored by NASA, NOAA and EPA) infrared Fourier Transform Spectrometer intended for the investigation of the chemistry and physics of the Earth's lower atmosphere (the troposphere) from airborne platforms such as the NASA/JPL DC-8 and P-3 research aircraft. AES is complementary to, and a test-bed for the spaceborne TES (Tropospheric Emission Spectrometer) on EOS platforms. AES itself is scheduled for flight in the spring of 1994, JPL is the lead center for development and operations.

AES is a Michelson interferometer with the plane mirrors replaced by cube-corners. It has 4 detector systems (one for each wavelength region) each containing 4 elements. This permits AES to generate 16 spectra simultaneously over the same target area. As a prototype of TES, AES will provide critical data on both, the acquisition methodology and on regional atmospheric chemistry. After TES is launched, AES will continue to play its role in correlative measurements through underflights of the EOS spacecraft.

Application: tropospheric chemistry.
The key issue is to understand how the lower atmosphere cleanses itself of both natural and anthropogenic trace gases (commonly referred to as pollutants). These gases can be injurious to the health of all living things and in some cases, through transport into the stratosphere, participate in depletion of stratospheric ozone. The sources, sinks and distributions of these gases is nor well understood - no global inventory exists (PI: R. Beer).[376]

| Item | TES | AES |
|---|---|---|
| Sponsors | NASA/GSFC | IC, NASA-HQ, NOAA, EPA |
| Platform | EOS (AM-2) | Aircraft, ground |
| Spectral range | 600 - 4350 $cm^{-1}$ (2.3-16.7 µm) | 650-4250 $cm^{-1}$ (2.4-15.4 µm) |
| Spectral coverage | 105% | 96% |
| Spectral resolution | 0.1 $cm^{-1}$ downlooking<br>0.025 $cm^{-1}$ limb-viewing | 0.1 $cm^{-1}$ |
| Operating temperature | 150 K (radiative) | ambient |
| Focal plane arrays | 4 x 1 x 32 (MCT PC & PV) | 4 x 1 x 4 (MCT PV & PC) |
| Focal plane temperature | 65 k (Stirling cycle cooler) | 65 K (pumped $LN_2$) |
| Pointing | automatic | interactive |
| Field of regard | Limb-to-limb (45° forward) | 30° cone about nadir |
| Spatial resolution | 2.3 x 23 km limb viewing<br>0.5 x 5 or 5 x 50 km downlooking | 7 x 70 m |

**Table 86:     A Comparison of TES and AES Instruments**[377]

---

376)  Paper provided by R. Beer of JPL
377)  R. Beer, "6th TES/AES Science Team Meeting", The Earth Observer, Vol. 4, Nr. 6, Nov/Dec 1992, pp. 8-10

## B.2 AIMR (Airborne Imaging Microwave Radiometer)

AIMR is a passive microwave radiometer developed and built by MPB Technologies Inc. of Claire, Quebec, for the Ice Branch of the Atmospheric Environment Service (AES) of Canada. The instrument was flown on ice reconnaissance missions off the coast of Labrador on an Electra aircraft starting in February 1989.[378]

AIMR consists of two major units. The External Payload Assembly (EPA) is mounted below the aircraft and collects data. The Cabin Mounted Equipment (CME) is an AIMR workstation located inside the aircraft providing control, display and data recording for the system. EPA provides a four-channel beam, a scanning system (mirror type) and the calibration loads. AIMR images are produced by a narrow beam antenna which scans across-track. The images are brightness temperature maps which are displayed in color and in real-time. Two frequency bands are collected, 37 GHz and 90 GHz, each at two orthogonal polarizations.

Applications: ice reconnaissance, soil moisture, oil spill mapping, snow depth mapping, crop and forestry monitoring, ocean temperature and salinity, water pollution studies.

MPB Technologies is in the process of building a new AIMR designed for more operational than research use. The new system will fly in a DASH-7 aircraft.

| Four channels at 2 frequencies and 2 polarizations | 37 and 90 GHz |
|---|---|
| Resolution (3 dB beamwidth, 90 GHz)<br>Resolution (3 dB beamwidth, 37 GHz) | $1^\circ$ corresponds to 50 m (altitude of 3 km)<br>$2.7^\circ$ |
| Radiometric sensitivity (37 GHz) | $< 1$ K |
| IF bandwidth | 0.5 - 2.0 GHz for 37 GHz channel<br>2.0. - 4.0 GHz for 90 GHz channel |
| FOV | $\pm 60^\circ$ |
| Quantization | 12 bit |

**Table 87:** **AIMR System Parameters**

## B.3 AIR-93 (Airborne Imaging 93 GHz Radiometer)

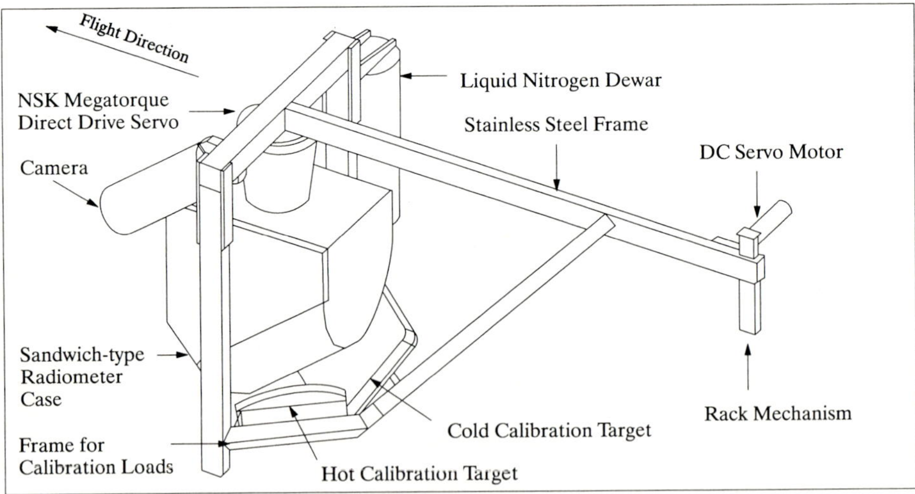


**Figure 104:** **Schematic Illustration of the AIR-93 Instrument**

378) J. L. Paul, B. W. Gibbs, P. Nguyen, F. G. R. Warren, "Design of an Airborne Imaging Microwave Radiometer", μrad 92, Proceedings of the Specialist Meeting on Microwave Radiometry and Remote Sensing, pp. 454-459

AIR-93 is a compact two-channel (H and V) imaging radiometer for a helicopter (Bell 206 Jet-Ranger) platform. The instrument was designed and built at the Helsinki University of Technology and the engineering office Ylinen Oy (RF section). AIR-93 is a mechanically scanned sensor employing a conical scan. The conical scan is realized by rotating the entire RF-unit - to preserve polarizations - by a direct drive servo, whose motion is fully programmable. Due to programmability, the scan pattern can be altered and tuned by software as required.[379] The instrument features also a pitch-servo, which enables a two-axis active aircraft attitude stabilization, which is feed by the aircraft attitude data source.[380] The radiometer is calibrated by the ambient temperature (hot load) and cooled liquid nitrogen (cold load) reference loads.

| Center frequency | 93 GHz |
|---|---|
| Incidence angle | $50^\circ$ |
| Polarization | Vertical and horizontal |
| Imaging method | Parallel scan (left to right) |
| Look direction | Along-track (forward) |
| Scan angle | $35^\circ$ |
| Swath width | 662 m (at a flight altitude of 500 m) |
| 3 dB beam width | $1.6^\circ$ |
| Footprint size | 22 x 34 m (cross-track x along-track) |
| Pixel overlap | 29% in cross-track and 26-39% in the along-track direction |
| In-flight calibration | Hot load (ambient temperature absorber)<br>Cold load (liquid nitrogen cooled absorber) |
| Data storage | Hard disk |
| Additional equipment | Video camera |
| Screen display | Microwave image and video image of target area |
| Instrument mass | Sensor = 18 kg; signal processing and battery = 35 kg + 50 kg |
| Motion compensation | Attitude GPS and active servo stabilization |
| Platform | Helicopter or aircraft |

**Table 88:     Technical Specification of the AIR-93 Instrument**

The prototype instrument is to be flown in early 1994. Its applications are to sense snow and ice cover extent and derivation, as well as soil moisture distribution, and, under ideal conditions, forest biomass estimates.


## B.4     AIRSAR (Airborne SAR)

AIRSAR is a NASA/JPL multi-frequency instrument package aboard a DC-8 aircraft and operated by NASA's Ames Research Center in Moffett Field, Ca. (operational since late 1987) in support of assigned measurement campaigns.[381]

The AIRSAR system has a single STALO clock and a single digital chirp generator which generates the chirp waveform at L-Band. The L-Band chirp signal is up-converted to C-Band and down-converted to P-Band and provided to TWT amplifiers in the case of the L-Band and C-Band radars, and to a class C solid state amplifier in the P-Band case. After amplification, these signals are transmitted to the AIRSAR antennas, all of which are dual polarized microstrip patch arrays. The AIRSAR antennas are all mounted body-fixed to the fuselage of the DC-8 aft of the left wing.

Transmit polarization diversity is achieved by alternately transmitting the signals using horizontal or vertical polarizations. Receive polarization diversity is accomplished by measur-

---


379)  M. Kemppinen, M. Hallikainen, "The Theory and Mechanical Realization of an Ideal Scanning Method for a Single-Channel Imaging Microwave Radiometer", IEEE Transactions on Geoscience and Remote Sensing, Vol. 30, July 1992, pp. 743-749

380)  E. Lantto, M. Kemppinen, "Kuvaavan radiometrin aktiivistabilointi", Helsinki University of Technology, Laboratory of Space Technology, Report A13, April 1993

381)  J. van Zyl, R. Carande, Y. Lou, T. Miller, K. Wheeler, "The NASA/JPL Three-Frequency Polarimetric AIRSAR System", JPL paper

ing six channels of raw data simultaneously, both H and V polarizations at all three frequencies. The video data are digitized using 8-bit ADCs, providing a dynamic range in excess of 45 dB. The digital data from the six radar channels are then multiplexed to form an extended 'range line', combined with navigation data, and stored on tape using one or more of three high density recorders.

In addition to the polarimetric mode, the AIRSAR system can also be operated in the along-track (ATI) and cross-track interferometry (XTI) modes. These modes are implemented by the addition of three more antennas; one L-Band antenna located just forward of the left wing of the DC-8, and two C-Band vertically polarized antennas arranged one above the other and placed just forward of the three dual polarized antennas used in the polarimetric mode.

The ATI mode, used to measure wind surface velocities, utilizes the aft polarimetric L-Band and C-Band antennas, the front L-Band antenna, and the bottom XTI antenna. Signals are transmitted alternatively out of the aft and front L-Band and C-Band antennas, and signals are received through the front and aft antennas simultaneously. The P-Band radar is operated in the standard polarimetric mode.

| Parameter | P-Band | L-Band | C-Band |
|---|---|---|---|
| Wavelength (cm) | 67-70 | 24 | 5.65 |
| Look angle (°) | 0-70 | | |
| Chirp bandwidth (MHz) | 20 [or 40] | | |
| Chirp RF frequency (MHz) | 448.75-428.75 [447.50-407.50] | 1258.75-1238.75 [1257.50-1217.50] | 5308.75-5288.75 [5307.50-5267.50] |
| Calibration tone (MHz) | 428.5546875 | 1238.5546875 | 5288.5546875 |
| Video frequency (MHz) | $450.00 - f_{rf}$ | $1260.00 - f_{rf}$ | $5310.00 - f_{rf}$ |
| Chirp duration (μs) | 10 or 5 | | |
| Peak transmit power (dBm) | 60 | 67 | 59 |
| Antenna polarization | H and V dual microstrip patch array | | |
| Bit/sample | 8 | | |
| Data recording | 3 HDDR's, full multiplex redundancy | | |
| Data rate (Mbit/s) | 10 | | |
| Data acquisition modes | 3 frequency quad-polarization, dual frequency ATI, XTI | | |
| Image Characteristics (Polarimetry Products) | | | |
| Multi-look (standard) | 16-look compressed Stokes matrix | | |
| Image size (range x azimuth, pixels) | 1280 x 1024 per frequency | | |
| Bytes/pixel | 10 | | |
| Pixel size (range x azimuth, m) | 6.66 x 12.1 [3.33 x 6] | | |
| Single-look (special) | 1-look compressed scattering matrix | | |
| Image size (range x azimuth, pixels) | 4 files of 1280 x 1024 per frequency | | |
| Bytes/pixel | 10 for 4 polarizations | | |
| Pixel size (range x azimuth, m) | 6.66 x 12.1 [3.33 x 6] | | |
| | | | |
| Nominal aircraft altitude | 26,000 ft (~7300 m) | | |
| Nominal aircraft velocity | 420 knots (~216 m/s) | | |

**Table 89:    Summary of AIRSAR System Parameters**[382]

In the XTI mode, used to measure topography (see TOPSAR), two vertically polarized antennas are used which are mounted above each other at C-Band. The C-Band signals are transmitted out of the bottom antenna and the returned signals are received simultaneously through the top and bottom antennas. The L-Band and P-Band radars are operated in the standard polarimetric mode.

AIRSAR data processing: A variety of processors and processing techniques is used to process AIRSAR data to imagery. A real-time correlator is part of the AIRSAR radar equip-

---

382)  Note: the values in the square brackets [ ] are for the 40 MHz Mode

ment and produces low-resolution ($\approx 25$ m) two-look survey imagery. The same on-board equipment can generate a slightly higher resolution (15 m) 16-look image of a smaller area (12 km x 7 km) within 10 minutes of acquisition using the quicklook processor. These on-board processors are useful for assessing the health of the radar.

Final processing of selected portions of the AIRSAR data to high-quality image products is done offline after the campaign. Products are provided by two different systems: the frame processor and the synoptic processor.

### B.4.1  TOPSAR (Interferometric Radar Topographic Mapping Instrument)

The TOPSAR project at JPL is an experimental interferometric radar experiment for topographic mapping applications on a DC-8 aircraft  along with a modified AIRSAR instrument. For this purpose the AIRSAR on the DC-8 aircraft was augmented with a pair of C-Band antennas displaced across track to form an interferometer sensitive for topographic variations of the Earth's surface. The antennas were sponsored by ASI, while the AIRSAR modifications were sponsored by NASA. A new data processor was developed at JPL for the production of topographic maps. Data were acquired over several sites in the US and Europe in the summer of 1991 in a number of flight campaigns.[383]

Concept: A radar interferometer is formed by relating the signals from two spatially separated antennas. The separation of the two antennas is called the baseline. The spatial extent of the baseline is one of the major performance drivers in an interferometric radar system - if the baseline is too short, the sensitivity to signal phase differences may be too small for detection, if the baseline is too long, additional noise due to spatial decorrelation may corrupt the signal.

The TOPSAR implementation uses much of the AIRSAR hardware (several modifications were made to optimize the performance in the topographic mapping mode. When in use TOPSAR effectively replaces the C-Band polarimeter instrument, but the L- and P-Band systems are undisturbed and operate together with the topographic mapper. The combined instrument produces simultaneous L- and P-Band fully polarimetric plus C-Band VV polarized backscatter images in addition to the topographic product.

The objective was to provide an operational instrument capable of delivering digital elevation models (DEMs) at a height accuracy of 2 m and a spatial resolution of 10 m. This goal was not met - topographical maps with a height accuracy of 3-40 m were produced. The discrepancy is due to a lack of proper aircraft motion compensation algorithms in the data processor.

As a consequence, a new processing scheme was developed featuring aircraft motion compensation, absolute phase retrieval, and three-dimensional location. The new processor has been tested using data which were acquired with extreme aircraft motion. This technique permits the generation of rectified height maps without any use of ground reference points. The topographic maps generated by the radar were compared to DEMs derived using conventional optical stereo techniques. The rms errors typically varied from 2.2 m in relatively flat terrain up to 5.0 m in mountainous regions.[384]

The TOPSAR experiments and results have led to the definition of a TOPSAT mission at JPL which still requires program approval as of Fall 1993.


383)  H. A. Zebker , S. N. Madsen, J. Martin, K. B. Wheeler, T. Miller, Y. Lou, G. Alberti, S. Vetrella, A Cucci, "The TOPSAR Interferometric Radar Topographic Mapping Instrument", IEEE Transactions on Geoscience and Remote Sensing, Vol. 30, Nr. 5, Sept. 1992, pp 933-940

384)  S. N. Madsen, H. A. Zebker, J. Martin, "Topographic Mapping Using Radar Interferometry: Processing Techniques", IEEE Transactions on Geoscience and Remote Sensing, Vol. 31, Nr. 1, January 1993, pp. 246-256

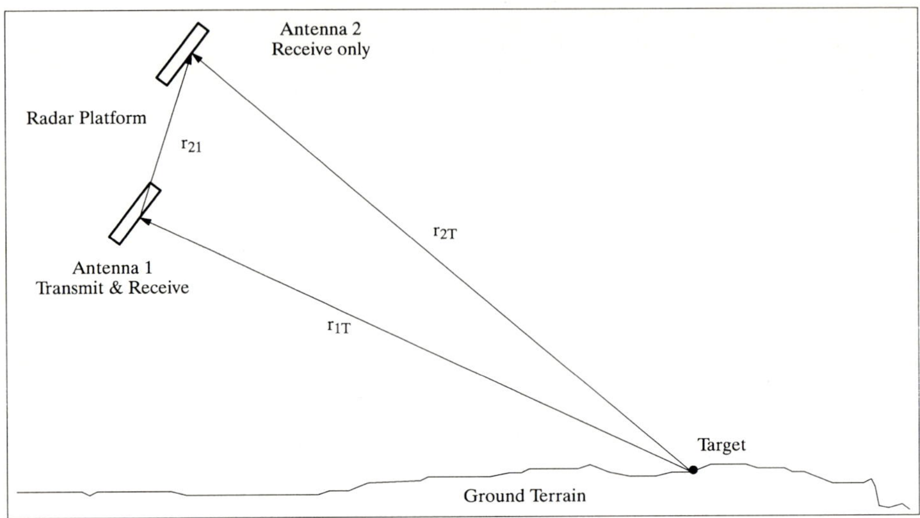


**Figure 105:**   **General Two-Antenna Interferometer Geometry**

| Parameter | Value |
|---|---|
| RF frequency | 5.288 GHz (C-Band) |
| Bandwidth | 40 MHz |
| Peak power | 1 kW |
| Pulse repetition rate | 2.64 m$^{-1}$ |
| Pulse length | 5 µs |
| Antenna length<br>Antenna width<br>Baseline length | 1.5 m<br>0.1 m<br>2.58 m |
| Antenna Azimuth beamwidth<br>Antenna elevation beamwidth | 2°<br>30° |
| Look angle | 30-55° |
| Flight altitude | ≈9 km |
| Slant range (near/far) | 10/15 km |
| Data rate | 80 Mbit/s (onboard recording) |

**Table 90:**   **TOPSAR Radar System Parameters**

## B.5    AIS (Airborne Imaging Spectrometer)

NASA instruments AIS-1 and AIS-2 (AIS-1 was flown from 1982-1985, AIS-2 from 1986-1987). AIS is one of the first types airborne imaging spectrometers ever built, it is a pushbroom scanner developed at JPL.[385]

- AIS-1 sensor: CCD area array detectors
  Spectral range:   900 - 2100 nm ("tree" mode)
                   1200 - 2400 nm ("rock" mode)
  Spectral resolution: 128 bands, 9.3 nm bandwidth
  IFOV: 1.91 mrad
  Spatial resolution: 12 m (at 6 km altitude). Swath width: 365 m (at 6 km altitude).


385)  Imaging Spectroscopy: Fundamentals and Prospective Applications", F. Toselli and J. Bodechtel (Editors), Kluver Academic Publishers, Remote Sensing Volume 2, ISBN 0-7923-1535-9, 1992, pp. 97-102, Annex - Airborne Imaging Spectrometers (J. Bodechtel , S. Sommer)

- AIS-2 sensor: CCD area array detectors
  Spectral range:   800 - 1600 nm ("tree" mode)
                    1200 - 2400 nm ("rock" mode)
  Spectral resolution: 128 bands, 10.6 nm bandwidth
  IFOV: 2.05 mrad
  Spatial resolution: 12 m (at 6 km altitude)
  Swath width: 787 m (at 6 km altitude)

Applications: Geo-chemistry, mineral identification, identification of altered rocks, geo-botany, vegetation stress, etc.

## B.6   AISA (Airborne Imaging Spectrometer for different Applications)

AISA is a commercially available instrument (operational since beginning of 1993) of Karelsilva Oy of Finland, with the objective to provide users with wide variety of application configurations. The instrument has been developed in cooperation with the Finnish Technical Research Institute and with contractual expertise from NASA. AISA is a programmable imaging spectrometer based on CCD (pushbroom) technology. One of the dimensions of the 2-D CCD elements is aligned to the image spectral content of the light dispersed by a reflection grating. The other dimension is used for imaging the spatial content of the light passing through a slit. Applications: hydrology, geology, forestry, agriculture, etc. .

AISA provides the following selectable features:[386],[387]

- Number of channels [1-286]
- Frequency spectrum [450 - 900 nm]
- Bandwidth [1.6 - 9.4 nm]
- Spectral, spatial, or combined spectral / spatial modes
- Data rate for data storage [max. 0.5 MByte/s]
- Integration time [ >4 ms]

The AISA instrument is made up of two modules - the 'front-end' and the 'data acquisition and control unit' (DACU). The front-end consists of the optics, spectrograph, CCD elements, the CCD detector electronics, and the control unit. A PC (486), a tape recorder (8 mm tape drive), a data acquisition board, and a quicklook color monitor make up the DACU. The DACU can be loaded with configuration files containing information on all parameters of the system. The CCD data is synchronized and digitized with 12 bit quantization. The recorded data are annotated with auxiliary flight data and GPS data (RS - 232 port).

AISA provides four different operational modes in order to solve the data rate problem at the storage device. The data rate constraint for the tape recorder is 500 kByte/s. This may lead to the concept of 'intermittent frames', a discontinuous measurement technique in the path of flight. A proper interval is chosen between individual frames to allow for data recording of the previous frame. The following modes are defined:

- Mode A: The complete CCD information is stored. The minimum frame interval for data storage is 350 ms.
- Mode B: A number of spectral channels is selected for storage. The width of each spectral channel is selected between 1.6-9.4 nm. The maximum number of channels which can be used within the tape data constraint of 500 kByte/s, is a function of integration time, channel spacing and total sum of channel bandwidth.
- Mode C: Full spectral information is stored for 47 equally spaced spatial lines. The minimum frame interval for data storage is 55 ms.

---

386) Information brochure of AISA provided by Karelsilva Oy, Finland
387) B. Braam "Design and first results of the Finnish Airborne Imaging Spectrometer for different Applications, AISA", presented at SPIE Symposium on Optical Engineering and Remote Sensing, Orlando Fla., 12-16 April, 1993

- Mode D: The user may select a set of spatial channels and a set of looks with full spectral information within the limits of the data recorder.

The AISA instrument can be flown on light aircraft (Cessna 152 or 182). A nominal flight altitude is 1000 m, this configuration provides spatial resolutions of 1 m across track.

| Optical Parameters | | | |
|---|---|---|---|
| Front optics | 24 mm, f/2.8 | Slit | 23 µm width |
| FOV per pixel | 1 mrad | FOV across track | 21° |
| Tilt angle | dependent on used mount | FOV along track | 0.055° |
| Spectral range | 450 - 900 nm | Dispersive element | Prism-Grating-Prism |
| Dispersion | 1.56 nm/pixel | | |
| Sensor Parameters | | | |
| CCD element | Thomson TH7863 | CCD size | 288x384 (image & memory) |
| Pixel size | 23 µm x 23 µm | Spatial elements used | 384 video pixels+8 dark ref. |
| Nr. of channels | 286 | Sensitivity | 30 dB SNR at 30 millilux |
| Saturation exposure | 0.18 µJ/cm$^2$ | Responsivity | 11 V/µJ/cm$^2$ (550 nm) |
| Radiometric Parameters | | | |
| Dynamic range | 12 bit | | |
| Integration time | programmable, minimal ~4 ms in mode B with one spectral channel | | |
| Sensor sensitivity | 0.9 DN/(µW ms cm$^{-2}$ sr$^{-1}$); 1.4 DN/(µW ms cm$^{-2}$ sr$^{-1}$ nm$^{-1}$) per 1.56 nm cha. | | |
| Data Recording Parameters | | | |
| Recorder type | 8 mm, helical scan tape | Nr. of recorded cha. | 286 max |
| Resolution | 12 bits | Data packing | selectable |
| Data rate (max) | 500 kByte/s sustained | Storage capacity, max | 5 GB/tape |
| Performance Parameters | | | |
| Ground resolution across track | | ~0.001 x altitude | |
| Ground resolution along track | | integration time x velocity | |
| Spectral resolution | | 1.56 nm over 450 - 900 nm | |
| Minimal bandwidth | | 1.56 nm | |
| Maximal bandwidth | | 9.36 nm | |
| Power and Weight Parameters | | | |
| Power front-end | 220V, 15W | Power DACU | 110 or 220 V, 185W |
| Weight front end | 12 kg | Weight DACU | 26 kg |

**Table 91:**     **Technical Specification Parameters of AISA**


# B.7     ALAS (Airborne Laser Altimeter System)

ALAS is a NASA/GSFC developed research instrument which measures the laser pulse time-of-flight and the distortion of the pulse waveform for reflection form Earth surface terrain features. The objective is to provide a capability for ranging and waveform studies of Earth surface topography (precursor instrument for a spacecraft sensor) with high resolutions and accuracies. The instrument was developed in the period 1986-90. The platforms for instrument operations are: T-39 Sabreliner and P-3-B aircraft, both stationed at WFF (Wallops Flight Facility, Va.).[388]

ALAS is a nadir-pointing lidar operated in a repetitively pulsed mode. The resulting laser ranging echo information (series of pulse time-of-flight values) is analyzed for the derivation of the surface elevation profiles. This is also referred to as laser altimetry. In addition, the shape of the reflected waveform in the laser footprint contains information whose interpretation (three axis position knowledge) provides such parameters as: surface slope, roughness, and reflectivity.

The source laser of ALAS is a diode-pumped, Q-switched Nd:YAG operating at its fundamental wavelength of 1064 nm. A reflector telescope and silicon avalanche photodiode (Si APD) are the basis of the optical receiver. A high-speed time-interval unit and a separate high-bandwidth waveform digitizer (computer-controlled) are used to process the back-

388)  J. L. Bufton, J. B. Garvin, J. F. Cavanaugh, L. Ramos-Izquierdo, T. D. Clem, W. B. Krabill, "Airborne Lidar for Profiling of Surface Topography", Optical Engineering, January 1991, Vol. 30, Nr. 1, pp. 72-78

scattered pulse information. The platform is equipped with proper navigation equipment, roll and pitch gyros, and a DGPS receiver for two-axis pointing and for a time reference.

| Laser transmitter | | Lidar Receiver | |
| --- | --- | --- | --- |
| Laser type | diode-pumped Nd:YAG | Telescope type | f/0.8 diamond-turned aluminum parabola |
| Wavelength | 1064 nm | Telescope diameter | 0.38 m |
| Pulse energy | 0.3 mJ | Optical filter | 2 nm bandpass |
| Pulse width | 4 ns FWHM | Detector type | Si APD, 1.5 mm dia. |
| Repetition rate | 1- 100 Hz (55 Hz for contiguous measurements) | FOV = IFOV | 2.5 mrad, 10 mrad |
| Divergence | 2 mrad | Bandwidth | 120 MHz |
| Cooling | conduction to telescope | Time-interval counter | 156 ps resolution |
| Total instrument mass | 35 kg | Waveform digitizer | 100 channels, 742 ps res. 11-bit amplitude resolution 400 MHz bandwidth |

**Table 92:     Parameter Specification of the ALAS Instrument**

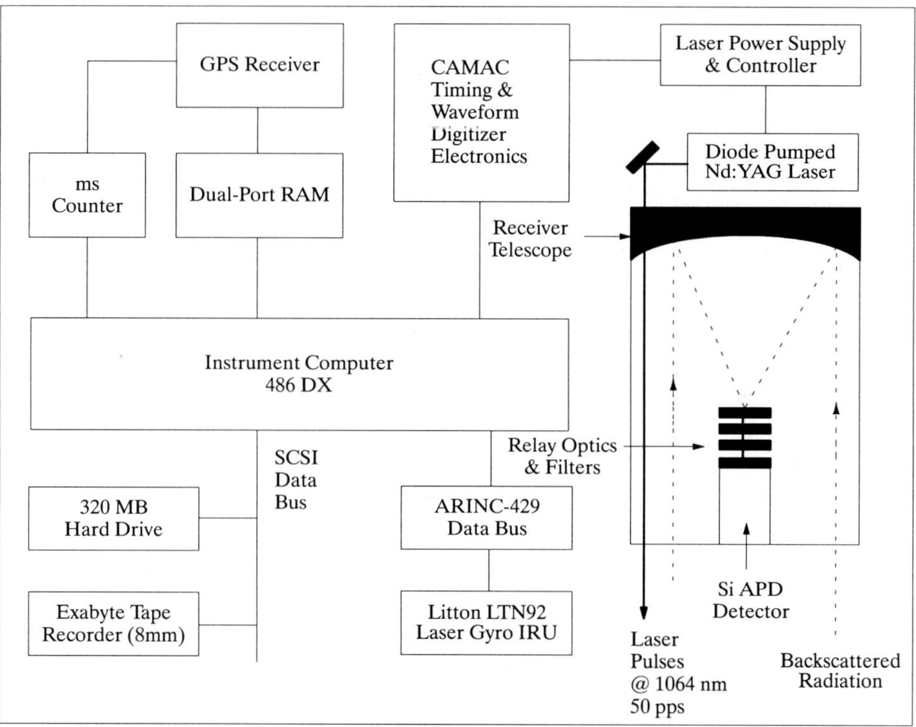


**Figure 106:     Functional Block Diagram of ALAS**

The ALAS telescope is a narrow-field device used for high resolution profiling of terrain in low altitude ( < 1 km) missions and lidar measurements of vegetation in higher altitude ( > 1 km) profiling missions. For simulation of space flight laser altimeter operation with large sensor footprints in the 30-70 m class, a wide-field-of view (Sasquatch) telescope is used in place of the actual ALAS telescope. Sasquatch has a 10 mrad FOV. The Sasquatch configuration is typically flown at high altitudes in order to maximize the footprint size.

The ALAS pulsed laser transmitter is a new device designed and built at GSFC in the Spring of 1993. It incorporates a diode-pumped, Q-switched, cavity-dumped Nd:YAG laser oscillator design that delivers a single spatial mode output pattern. In operation two in-line, Nd:YAG oscillator crystals are optically pumped with pulsed diode laser radiation at 808 nm wavelength for a duration of ~200 $\mu$s. The polarization-coupled oscillator pulse output is directed into a double pass diode-pumped amplifier. There it encounters a further rotation of polarization that prevents the laser pulse from returning to the oscillator and produces the observed output pulse with a factor of 2 or 3 more pulse energy. This new ALAS laser is capable of producing 300 $\mu$J pulses at a repetition rate of 50 pulses per second ~4 ns FWHM pulsewidth. IFOV = 2 mrad.

Airborne operation of ALAS is based on a high SNR environment (> 40 dB) in which each laser pulse can be used for a unique range measurement and for waveform data interpretation. No pulse-to-pulse averaging is used for terrain profiles. The laser altimeter signal strength depends on the laser pulse power backscattered from the target surface and collected by the receiver telescope. With an IFOV of 2.0 mrad the laser footprint on the Earth surface turns out to be 2.0 m in diameter per km of altitude.

## B.8 ALF (Airborne Laser Fluorosensor)

ALF is a proprietary survey exploration instrument of World Geoscience Corp. LTD of Perth, Australia (ALF was purchased from BP Exploration in 1992, version MK3). The instrument is used to detect petroleum on the sea surface which has leaked or seeped from undiscovered oil fields. ALF is installed in a twin engine aircraft and flown at altitudes of 100 m above sea level or less.

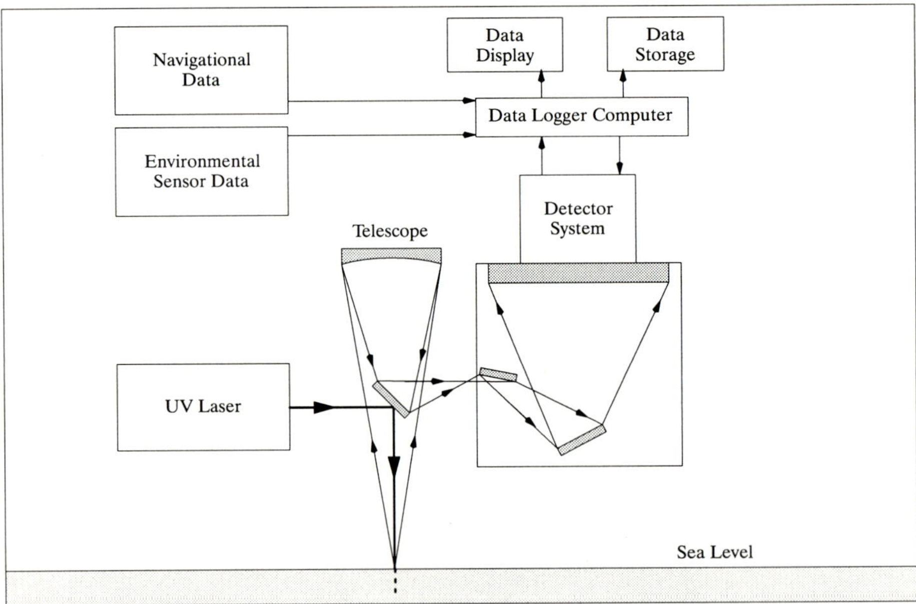


**Figure 107:  Basic System Components of the ALF Instrument**

The ALF instrument consists of a pulsed (50 Hz) UV laser, an imaging telescope, an electro-optical detector, and a data logger computer. The laser beam is transmitted vertically downward to illuminate the sea surface. A characteristic fluorescence response is produced by the presence of surface oil. The returning light is collected by a 16 inch Newtonian reflecting telescope. This beam is split into its constituent spectral colors and converted into digital signals by a diode array detector. The data are stored digitally by the data logging computer system, which simultaneously records GPS navigational information. Oil films can be detected down to a thickness of about 0.05 µm. ALF offers the following surveying capabilities:[389]

- Pollution: ALF can distinguish fresh petroleum films on the sea-surface from other oil slicks of non-seep origin (which have negligible fluorescence). It cannot distinguish seeped oil from freshly spilled crude oil. However, repeat flying eliminates pollution problems, as pollution slicks lose their fluorescent component quickly by evaporation (and also change their position, whereas seeped oil reappears at or near its point of origin).
- Algae: Some algae such as trichodesmium produce a very weak blue fluorescence (but concentrations are orders of magnitude lower than very thin oil films).
- Flat water slicks: ALF can distinguish flat water slicks due to wind and current action. These are non-fluorescent and can confuse other passive slick mapping sensors.
- Sun glint: Sun glint occurs when the detectors are swamped by sunlight reflecting directly of the sea-surface into the telescope. Sun glint and oil spectra are easily distinguished by their spectral signature.

| |
|---|
| Miniaturized solid state laser with an emission wave length of 255 nm. The wavelength of the older MK2 instrument was 308 nm. The new wavelength (255 nm) provides a wider spectral window and better discrimination of slick types. |
| Installed system mass of 300 kg |
| Maximum current draw of 90 A from aircraft 28 V DC supply |
| Sampled area of 20 m² per laser shot |
| Laser firing and fluorescence spectra data collected synchronously at 50 Hz |
| 176 channels (minimum) of spectral data acquired over a wavelength range of 250 - 700 nm, with channels positioned to concentrate on features of interest, i.e. the laser backscattered light, the seawater Raman signal, and the 300 - 700 nm fluorescence signal. |
| In-flight quicklook display capability of a number of system parameters |
| An integral part of the system is a UV/VIS camera and video recorder system targeted at the sea surface being surveyed by ALF. This arrangement is very useful for post flight checking. |

**Table 93:     Technical Features/Specifications of the ALF MK3 Instrument**

In a typical survey, ALF data are acquired over uniformly spaced flight lines aligned across (and/or perpendicular to the dominant current direction) the site of interest. The ALF operator monitors data as it is collected and records environmental parameters such as sea state, plus any unusual features on the sea surface. After the flight the data are processed to identify the fluorescence anomalies (fuors), the video flight logs are screened to eliminate anomalies due to non-oily features such as algal mats. Finally, a survey map showing the locations of the fluorescence anomalies is generated for exploration interpretation.


## B.9     ALIAS (Aircraft Laser Infrared Absorption Spectrometer)

ALIAS[390] is a JPL instrument, developed for NASA's 'Upper Atmospheric Research Program'(UARP), which started its operation in 1990 (instrument mass = 75 kg). The instrument is of BLISS (Balloon-borne Laser In-Situ Sensor) heritage which had 11 successful flights during the period 1983-92 (instrument mass of 1000 kg). ALIAS is flown on ER-2 aircraft with over 50 flights by the end of 1993. Most of the flights resulted from two major campaigns, namely AASE-II (at Fairbanks Alaska, Bangor, Maine) and SPADE (Strato-

---

389)  Company brochure of World Geoscience Corporation LTD, Perth, Australia
390)  Information provided by C. R. Webster of NASA/JPL

spheric Photochemistry, Aerosols and Dynamics Experiment). Since 1992 a further instrument, ALIAS-II, is being developed and built for PERSEUS A with an instrument mass of about 25 kg.

## B.9.1   ALIAS-I on ER-2 Aircraft

ALIAS (or ALIAS-I) is a high-resolution (0.0003 $cm^{-1}$) scanning TDL (Tunable-Diode-Laser) spectrometer designed to make direct and simultaneous measurements of HCl, $NO_2$, $HNO_3$, CO, $CH_4$, and $N_2O$ (including vertical profiles of $CH_4$ and $N_2O$) in the polar stratosphere at sensitivities of tens of parts per $10^{12}$ over a 30 second integration time.[391]

The measurement technique is based on using tunable lead-salt diode lasers operating from 3.4 - 8 μm, and scanning over absorption lines recorded with second-harmonic absorption spectroscopy over an 80 m path length in a 1 m multipass optical Herriott cell. A single liquid-nitrogen dewar contains all four TDLs and four HgCdTe IR detectors on a single cold finger. The light from the laser is collected by half-inch Zn-Se lenses and directed through a wedged Zn-Se window through one of four coupling holes in a 15.24 cm diameter Au-coated Al spherical mirror forming the Herriott cell. After traversing 80 passes of the Herriott cell, the beams exit the same entrance holes and are directed back to the dewar, where they are directed onto Zn-Se collection lenses in front of the appropriate HgCdTe detector. Second-harmonic detection frequencies in the range 20-40 kHz are used in conjunction with a 10 Hz fast-sweep ramp for spectral scanning. Data collection is based on averaging spectral scans for recording spectra from 0.3-60 seconds.

The ALIAS-I instrument has four optical subsystems: a liquid-nitrogen-cooled dewar containing lasers and detectors, a moveable lens assembly, an optical bench supporting steering of mirrors and reference wavelength cells, and the multipass Herriott cell.

## B.9.2   ALIAS-II on Perseus Aircraft

ALIAS-II objectives for NASA's AESA/HSRP Program:

1.  To provide measurements (high spatial resolution in latitude and longitude) of $N_2O$, $CH_4$, and $H_2O$ as part of a reference chemical climatology data base for the lower stratosphere and upper troposphere regions. This data base can be used for calibration and testing of global models, and as an indicator of change for atmospheric composition.
2.  To quantitatively characterize the emission of $H_2O$ into the emission plumes, and its mixing by shear winds and diffusion from injection through the 5-20 day formation of zonally-uniform distributions. This occurs in conjunction with the additional ALIAS-II measurements of the tracer $N_2O$ for identifying transport and redistribution of emissions near and above the tropopause, and of $CH_4$ to assess perturbations to the initial air parcel photochemistry.
3.  To investigate the mechanisms responsible for the stratosphere/troposphere exchange of water vapor, through measurements in the middle/upper tropopause and the tropopause region. ALIAS-II will make extensive measurements of $H_2O$ as a function of both altitude and latitude, as input for model predictions of global warming.
4.  To investigate plume chemistry and physics with regard to the formation and growth of ice particles within the plume, and their possible fall out. ALIAS-II measurements of gas-phase $H_2O$ (and $N_2O$ as tracer) will identify the extent of dehydration locally within regions of particle formation from an aircraft platform entering either the plume region or a similar environment.

---

391)  C. R. Webster, R. D. May, C. A. Trimble, R. G. Chave, J. Kendall, "Aircraft (ER-2) laser infrared absorption (ALIAS) for in-situ Stratospheric measurements of HCl, $N_2O$, $CH_4$, $NO_2$ and $HNO_3$", Applied Optics, January 20,1994

| Channel Nr. | Trace Gas | Wavenumber | Expected Mixing-Ratio | Minimum-detectable Mixing-ratio | |
|---|---|---|---|---|---|
| 1 | $N_2O$<br>$N_2O$ | 1277<br>1277 | 305 ppbv<br>80 ppbv | at 500 mbar<br>at 50 mbar | 4 ppbv |
| 1 | $CH_4$<br>$CH_4$ | 1277<br>1277 | 1.7 ppmv<br>0.8-1.4 ppmv | at 500 mbar<br>at 50 mbar | 20 ppbv |
| 2 | $H_2O$<br>$H_2O$ | 1636<br>1636 | 50-700 ppmv<br>1.5-5.5 ppmv | at 500 mbar<br>at 50 mbar | 20 ppbv |

**Table 94:**   **ALIAS-II Measurement Capability of Trace Gases**

| Instrument | Species measured simultaneously | Measurement time |
|---|---|---|
| ALIAS-I (ER-2)<br>(5 gases) | $N_2O$, $CH_4$, CO,<br>HCl,<br>$NO_2$ | 0.3 s<br>30 s<br>60 s |
| ALIAS-II (Perseus)<br>(2 or 3 gases) | i) CO plus $NO_2$ or HCl<br>ii) $N_2O$, $CH_4$, $H_2O$ | 0.3 s, 60/30 s<br>0.3 s |

**Table 95:**   **Overview of measured Species of the ALIAS Instruments**

The ALIAS-II instrument is a 2-channel version of the ALIAS (ER-2) instrument, with identical electronics, dewar configuration, and instrument- and data processing- software. In this configuration ALIAS-II measures the trace gases in a very clean, open path (important for $H_2O$) defined as the 5 m round-trip from the nose-to-the wing. This is achieved by using a miniaturized multi-reflector assembly on the wing.

In an alternative configuration, a Herriott-cell design copied from ALIAS-I spans the 2.5 m distance, and achieves a pathlength of about 120 m in an optical system designed to accommodate wing motions. With a change of diode laser sources, this alternative configuration will allow ALIAS-II to make measurements of CO, and $NO_2$ (or HCl) with higher sensitivities than the ER-2 experiment.

# B.10   ALPS (Airborne Laser Polarization Sensor)

ALPS is an active NASA/GSFC sensor featuring multispectral radiometric and polarization measurements using a polarized laser light source. ALPS provides a source of optical scattering measurements in the field to characterize terrestrial surfaces in their natural state (information that depends on the surface texture and dielectric properties, its transmittance and reflectance, and its absorption). ALPS is a prototype, applications-oriented instrument for active optical remote sensing. The instrument is a single spot, non-imaging sensor, measuring the degree of depolarization by a surface.[392]

The ALPS instrument consists of a pulsed, polarized laser source, an optical receiver system, a video camera and recorder, and data acquisition and analysis hardware and software. The choice of laser wavelengths is limited to two frequencies from the UV to the NIR spectral range by the photo-cathode response of the Photo Multiplier Tube (PMT) detectors. A pulsed (7ns) Nd:YAG laser is employed. It operates in the NIR at 1064 nm and in the VIS at 532 nm. ALPS employs twelve PMT detectors configurable to measure desired parameters such as the total backscatter and the polarization state , including the azimuthal angle and ellipticity, at different UV and NIR wavelengths simultaneously.

ALPS has been flown in several NASA campaigns on a HU-1B helicopter. The instrument is also known under its old name PALIS (Polarized Airborne Laser Imaging Sensor) prior to the designation ALPS.

---

392)   J. E. Kalshoven, P. W. Dabney, "Remote Sensing of the Earth's Surface with an Airborne Polarized Laser", IEEE Transactions on Geoscience and Remote Sensing, Vol. 31 Nr. 2, March 1993, pp. 438-446

## B.11 AMMR (Airborne Multichannel Microwave Radiometer)

AMMR is an operational passive radiometer system since 1980, owned by NASA/GSFC. AMMR consists of an array of single-beam radiometers at the frequencies of 10, 18.7, 21, 37, and 92 GHz. These radiometers are installed in the windows on the left-hand side of the DC-8 aircraft. The 18.7, 37, and 92 GHz units are dual polarized. All radiometers have a beam width of about 6°, all have a temperature sensitivity of about 0.5 K and a calibration accuracy of ± 4 K. The major application of AMMR is for precipitation measurements, other surface parameters like sea ice, snow and vegetation covers can also be measured. AMMR took part in the TOGA/COARE mission in January/February 1993.

## B.12 AMMS (Airborne Microwave Moisture Sounder)

AMMS is a 4-channel mechanically scanned imaging radiometer with frequencies at 92, 183.3 ± 2, 183.3 ± 5, and 183 ± 9 GHz. The sensor is owned and operated by NASA/GSFC. The major applications of AMMS are to measure precipitation, clouds and water vapor profiles. It has a 15 cm aperture giving an angular resolution of about 2° at 92 GHz and 1° at 183 GHz. AMMS is a cross-track scanner having an angular swath of ± 45°. It first became operational in 1981.

In recent operation the temperature sensitivity of the sensor was found to deteriorate appreciably. Based on the data acquired in a previous mission in September 1990, the temperature sensitivities at the four frequencies (92 GHz, 183.3 ± 2, 183.3 ± 5, and 183 ± 9 GHz) were found to be about 4.1 K, 4.5 K, 4.3 K, and 7.5 K respectively. The calibration accuracy was in the order 1 K in the brightness temperature range of 250-300 K and could be as much as 2-3 K at brightness temperature range of < 200 K. Because of the poor temperature sensitivity, the sensor is no longer being used for water vapor profiling; it can provide very useful microwave signatures from precipitation associated with convective systems.[393]

AMMS is normally installed in the NASA DC-8 aircraft, flying in the altitude range of 10-12 km. The sensor took part in the TOGA/COARE mission in the Pacific during January/February 1993.

## B.13 AMPR (Advanced Microwave Precipitation Radiometer)

AMPR is a NASA/MSFC-sponsored instrument [of Georgia Technical Research Institute (GTRI) heritage] with the objective to support the development and validation of current and future satellite precipitation retrieval algorithms of oceanic and land-based systems. The instrument is a low-noise passive system providing multifrequency microwave imagery with high spatial resolution.[394]

AMPR is a cross-track scanning total power microwave radiometer with four channels centered at 10.7, 19.35, 37.1 and 85.5 GHz. It has a dual lens antenna to accommodate two separate feedhorns. One horn is a copy of the feedhorn used for SSM/I (on DMSP, see A.19). This horn feeds the 19.35, 37.1 and 85.5 GHz channels with a 13.5 cm diameter lens. The other AMPR feedhorn was designed by GTRI to accommodate the 10.7 GHz frequency. The lens for this horn has a diameter of 24.7 cm.

AMPR is flown in the Q-bay of the ER-2 aircraft. The dual antenna and scanning mirror extend below the aircraft body into a hatch opening. The instrument has a 90° total scan centered at nadir. The observations are contiguous at 85.5 GHz and coincident at all four channels leading to oversampling at the lower frequencies. A rotating polarization scheme

393) Information provided by J. R. Wang of GSFC
394) Information provided by R. Hood and R. W. Spencer of NASA/MSFC

for all four channels is being used during scanning. The polarization varies from vertical (V) at 45° to the left of nadir to horizontal (H) at 45° to the right of nadir. At nadir the polarizations are equally mixed. The scanning strategy allows dual-polarized information to be collected at the extreme scan angles with opposite aircraft passes over a given target.

AMPR has been flown on numerous missions (within TOGA/COARE) over oceanic convection systems in the vicinity of Florida measuring the brightness temperature at the four frequencies of the instrument. The analysis and interpretation of its data is considered valuable design input for the future spaceborne TRMM sensors.

| Parameter | 85.5 GHz | 37.1 GHz | 19.35 GHz | 10.7 GHz |
|---|---|---|---|---|
| Bandwidth (MHz) | 1400 | 900 | 240 | 100 |
| Temperature resolution (°C) | 0.3 | 0.2 | 0.5 | 0.2 |
| Integration time (ms) | 50 | 50 | 50 | 50 |
| Horn type | SSM/I | SSM/I | SSM/I | GTRI |
| Lens diameter (cm) | 13.5 | 13.5 | 13.5 | 24.7 |
| Beamwidth | 1.8° | 4.2° | 8.0° | 8.0° |
| Footprint @ 20 km altitude | 0.6 km | 1.5 km | 2.8 km | 2.8 km |
| Beam efficiency | TBD | 98.8% | 98.7% | 97.8% |
| Cross-polarization | TBD | 0.4% | 1.6% | 0.2% |
| Swath width | | 40 km | | |
| FOV | | ± 45° | | |
| Polarization (all channels) | | H @ - 45° scan angle, V @ + 45° scan angle | | |
| Instrument mass | | 152 kg (radiometer + power supply + data system) | | |
| Contiguous footprints @ 85.5 GHz; lower frequencies are oversampled by the beamwidth ratio 4 SSM/I ports are available for additional channels External warm (heated) and cold (ambient) calibration targets enclosed in Styrofoam | | | | |

**Table 96:     Characteristics of the AMPR Instrument**


## B.14    AMSS MK-II (Airborne Multi-Spectral Scanner)

AMSS is a commercially operated sensor of Geoscan PTY Ltd., West Perth, Australia aboard an aircraft with a stabilized (roll, pitch and yaw, drift compensation) optical sensing platform. Specifications:[395]

- Pixel size = 2 to 20 m (8 - 10 m typical)
- Swath width = 1.5 - 19 km
- Pixels/line of swath: = 768 pixels (acquired data), = 1024 pixels (resampled data)
- Spectral bands: up to 24 recorded bands are selected from 46 detector channels
- FOV = 92°
- IFOV at nadir: = 2.1 x 3.0 mrad (acquired data), = 2.1 x 2.1 mrad (resampled data)
- Data storage: Optical (WORM) disks, 5.25 inch, 325 MByte per side
- Data is normally provided corrected for atmospheric backscatter and geometric distortion.

Application: mineral exploration, agriculture and forestry mapping, environmental monitoring.

| Spectral Range | Number of Channels | Wavelength Range | Band Pass Range |
|---|---|---|---|
| VNIR | 32 | 0.49 - 1.09 µm | 0.17 - 0.24 µm |
| SWIR | 8 | 2.02 - 2.37 µm | 0.43 - 0.44 µm |
| TIR | 6 | 8.50 - 12.0 µm | 0.55 - 0.59 µm |

**Table 97:     AMSS Spectral Parameters**

---

395) "Airborne Multi-Spectral Scanner MK2", brochure of GEOSCAN PTY, Ltd, Australia

## B.15    AOL (Airborne Oceanographic Lidar)

NASA experimental instrument flown on NASA P-3A aircraft of GSFC Wallops Flight Facility. AOL has been in use since 1977, it was upgraded many times to suit certain support requirements.[396] AOL was and is also used to test and support satellite sensor and algorithm development (such as CZCS, SeaWiFS, OCTS, EOS-Color, MERIS, MODIS).

AOL is a scanning laser radar system having a multispectral time-gated receiving capability. The system is designed to allow adjustments in most transmitter and receiver settings. The flexibility gives the AOL system a potential application in many oceanographic areas.

AOL can be operated in either of two modes: temporal- or fluorosensing mode. In the temporal mode, the sensor resolves the timing of the reflected laser pulse in 0.7 ns channels over a period of 250 ns.[397]

In the fluorosensing mode, laser stimulated florescence from water ice, or ground targets is spectrally resolved between 380 nm and 740 nm in 32 contiguous channels of 11.25 nm width.

In either mode, the AOL also functions as a high precision laser altimeter measuring the range between the sensor and the surface. This range-measurement capability permits the AOL to conduct airborne surveys of topographic surfaces including water and ice.

| | |
|---|---|
| Transmitter: | Nitrogen Laser with energy of 1 mJ |
| Wavelength | 337.1 nm (oil film thickness measurements) |
| Bandwidth | 0.1 nm |
| Pulse width | 10 ns |
| Pulse rate | $\leq$ 100 Hz |
| Peak output power (max) | 100 kW |
| Beam divergence | 4 mrad |
| Receiver: | |
| Bandwidth | 3500 - 8000 Å |
| Spectral resolution (min) | 11.25 nm |
| FOV | 1-20 mrad, variable, vertical, and horizontal |
| Temporal resolution | 8 - 150 ns, variable |
| Aircraft altitude | 150 m |
| Aircraft velocity | 75 m/s |

**Table 98:**    **Typical AOL Operating Parameters for Oil Fluorosensing Mode**

Applications: chlorophyll mapping experiments[398], oil spill detection, tracer dye, oceanic turbidity cell structure, water depth, laser backscatter investigations, etc.

Past AOL programs: 1984 DOE/SEEP; 1985 DOE/SPREX; 1987 DOE/FLEX and DOD/BIOWATT; 1988/89 DOE SEEP; 1989 NSF/JGOFS/NABE; 1992 NSF/JGOFS/EQPAC.

Future AOL programs: 1993 NSF/JGOFS; 1994 SeaWiFS validation; 1995 NSF/JGOFS Arabian Sea; 1997 NSF/JGOFS South Ocean; 1998 EOS-Color validation; 1999 NSF/JGOFS/NABE.

396)  F. E. Hoge, R. N. Swift, "Oil film thickness measurement using airborne laser-induced water Raman backscatter", Applied Optics, 1 October 1980, Vol 19, Nr. 19, pp. 3269-3281

397)  F. E. Hoge, "Oceanic and Terrestrial Lidar Measurements", Chapter 6 of 'Laser Remote Chemical Analysis', R. M. Measures (Editor), John Wiley & Sons, 1988, pp. 409-503

398)  F. E. Hoge, R. N. Swift, "Photosynthetic Accessory Pigments: Evidence for the Influence of Phycoerythrin on the Submarine Light Field", Remote Sensing Environment, 34, pp. 19-35, 1990

## B.16    ARGUS (Two-Channel Atmospheric Tracer Instrument)

Argus, a 'two-eyed' sensor for upper tropospheric and stratospheric tracer measurements, is named after the hundred-eyed Greek giant, Argus Panoptes. The instrument was developed at NASA's Ames Research Center and is considered for the Perseus aircraft platform at altitudes up to 30 km. Argus is of ATLAS (Airborne Tunable Laser Absorption Spectrometer, see chapter B.19) heritage.[399]

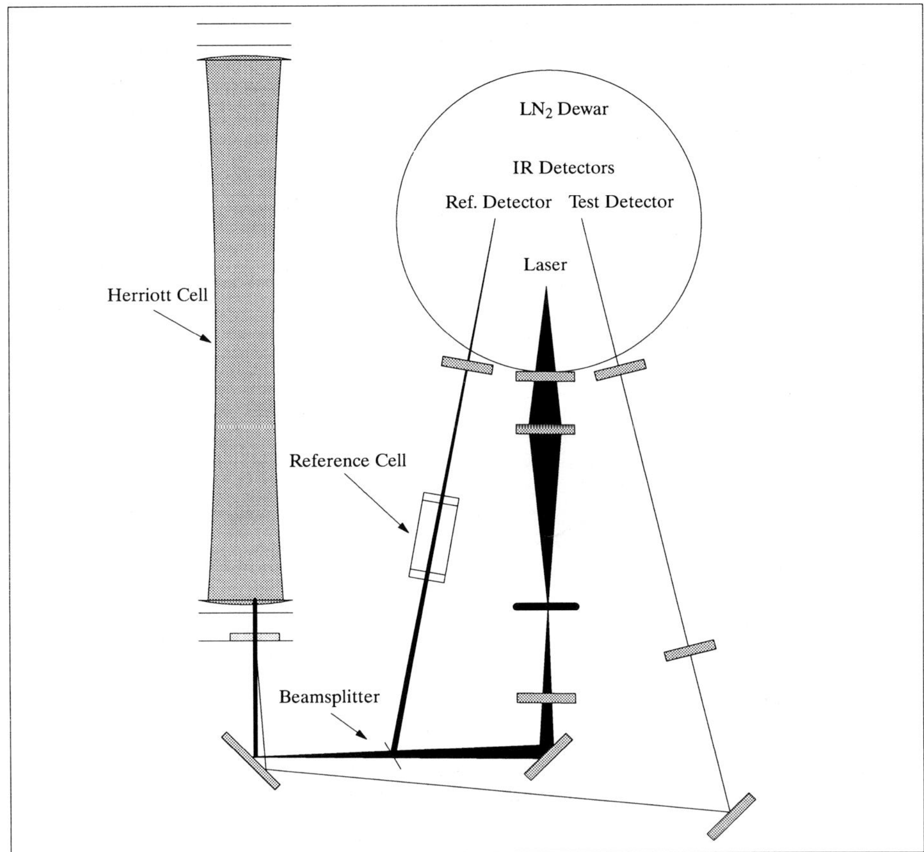


**Figure 108:    Schematic Diagram of the Argus Measurement Concept**

The tracer elements currently (1993) measured by Argus are NO$_2$ (at 4.5 μm) and CH$_4$ (at 3.3 μm) at a sampled rate of 0.1 Hz. Argus is a two-channel TDL (Tunable-Diode-Laser) spectrometer operating in the second harmonic mode and employing sweep integration at a 10 Hz sweep rate. The combination of second harmonic spectroscopy and sweep integration provides great stability and excellent signal to noise ratio, and therefore high precision in the retrieval of stratospheric tracer fields. Direct fits (Marquardt-Levenberg algorithm) of the data to the second harmonic spectra are used to retrieve the tracer molecule number densities using the known spectral line parameters.

The instrument employs 3 processors, 2 for control of the individual channel lasers and one to manage overall data acquisition and storage. Each laser is current- and temperature-con-


399) M. Loewenstein, "ARGUS: A New Instrument for PERSEUS A", The Perseus Data Link, Issue #3, Third Quarter 1993

trolled on separate laser mounts inside the dewar. The lasers are sine-modulated at about 40 kHz. The 80 kHz second harmonic data are detected with phase-sensitive detection/integration electronics. The four InSb detectors, as well as two lasers, are mounted in the 1.8 liter capacity liquid nitrogen dewar.

A diagram of one optical channel of Argus is shown in Figure 108. Transfer optics convert an f/2 laser beam to a quasi-collimated f/40 beam which is injected into a 26.1 cm base path Herriott cell. The beam traverses the cell 72 times for a total path of 18.8 m. A beamsplitter provides a second beam that passes through a frequency marker cell to provide wave-number calibration for the second harmonic line fitting procedure. The marker cell contains a gas at low pressure providing several sharp, Doppler-broadened, spectral lines of accurate known frequency.

| | |
|---|---|
| Measurement accuracy: | 3% |
| Measurement precision: | 1% |
| Response time | 10 s |
| Instrument mass: | 23 kg |
| Instrument power: | 65 W |

## B.17    ARMAR (Airborne Rain Mapping Radar)

ARMAR is an active microwave radar sensor operating at a frequency of 13.8 GHz. The instrument has been developed by NASA/JPL for the purpose of supporting future spaceborne rain radar systems, in particular the PR (Precipitation Radar) instrument of TRMM (Tropical Rainfall Measuring Mission, see A.111). ARMAR is installed on the AMES DC-8 aircraft and is operated by JPL. The sensor was completed in late 1991, the first airborne testing was in May 1992. Additional tests were completed in December 1992, and the system was successfully deployed during TOGA/COARE campaign in the Western Pacific in early 1993.[400)]

The primary design goal for ARMAR was to develop a system which matches the PR sensor of TRMM in both frequency and scanning geometry. ARMAR therefore operates at 13.8 GHz and has a cross-track scanning geometry as illustrated in Figure 109(for spaceborne sensor and for algorithm development as well as for post-launch calibration of PR). A number of capabilities have been included on ARMAR which improve its ability to support PR and will allow it to serve as a testbed for future spaceborne systems.

For example, ARMAR has been designed to have a finer spatial resolution than PR so that specific topics such as the effect of partial beam filling can be studied. ARMAR is capable to make like-polarization, cross-polarization, or alternating dual-polarization measurements. ARMAR can also obtain a greater number of independent samples than PR by using frequency diversity, transmitting up to four slightly different frequencies. When using a single transmit frequency, ARMAR is coherent, providing Doppler information. For situations in which a high accuracy in Doppler measurement is required, the antenna can be pointed at nadir rather scanned, allowing a substantially larger dwell time and improved Doppler resolution. While operating as a radar, a small fraction of time is spent measuring the brightness temperature in a radiometer mode at the same frequency and viewing geometry as the radar mode. Finally, ARMAR uses pulse compression to achieve the required range resolution.

The ARMAR instrument is equipped with a mechanically scanned antenna (diameter = 0.41 m). Measurements are performed by looking in the nadir direction with a FOV of $\pm 20°$.

---

400)  Information provided by S. Durden of JPL

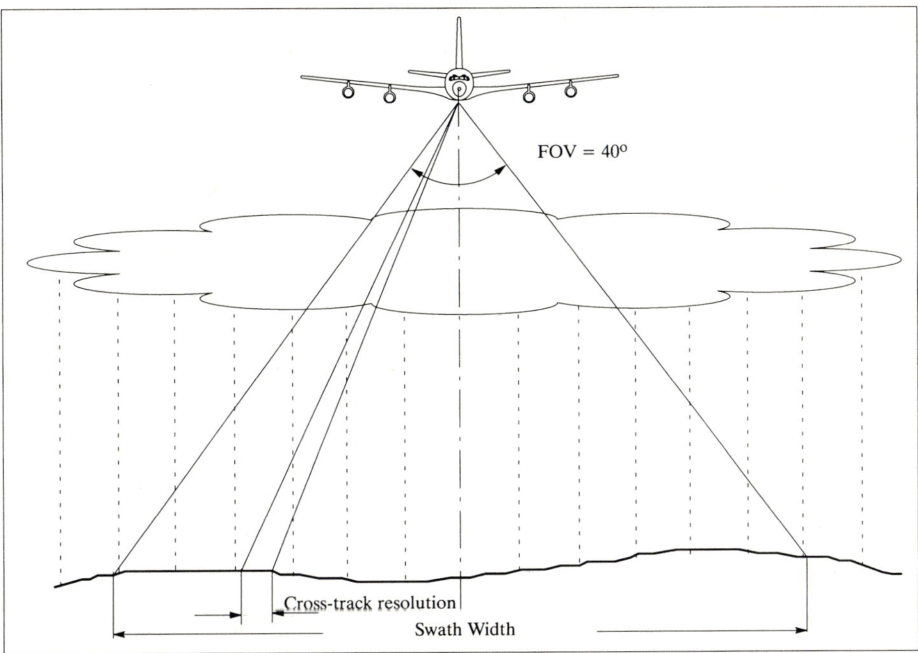


**Figure 109:    ARMAR Scanning Geometry**

| Performance Characteristics: | |
| --- | --- |
|     Range resolution (6 dB width) | 80 m |
|     Surface cross-track resolution (12 km altitude) | 800 m |
|     Swath width | 9 km |
|     Frequency | 13.8 GHz |
|     Polarizations | HH, VV, HV, VH |
| Antenna Characteristics: | |
|     Aperture diameter | 0.4 m |
|     Gain | 34 dB |
|     3 dB beamwidth | 3.8° |
|     Sidelobe level | -32 dB |
|     Polarization isolation | -28 dB |
| Transmitter Characteristics: | |
|     Peak power | 200 W |
|     PRF (Pulse Repetition Frequency) | 1-8 kHz |
|     Number of transmit frequencies | 1-4 |
|     Pulse duration | 5-45 µs |
|     Chirp bandwidth | 4 MHz |
| Receiver Characteristics: | |
|     System noise temperature | 650 K |
|     Sample frequency | 10 MHz |
|     ADC resolution | 12 bits |
| Radiometer Characteristics: | |
|     Bandwidth | 40 MHz |
|     $\Delta T$ per pixel | 1 K |

**Table 99:    ARMAR System Parameters**

## B.18    ASAS (Advanced Solid-State Array Spectroradiometer)

ASAS is a NASA airborne, off-nadir-pointing imaging spectroradiometer with the objective to acquire bidirectional radiance data from terrestrial targets. The original instrument has been modified for off-nadir pointing by GSFC in order to study the directional anisotropy of solar radiance reflected from terrestrial surfaces. As a consequence, ASAS is able to track and image a target site through a discrete sequence of fore-to-aft view directions from 45° forward to 45° aftward. The pointing capability was first utilized in 1987.[401]

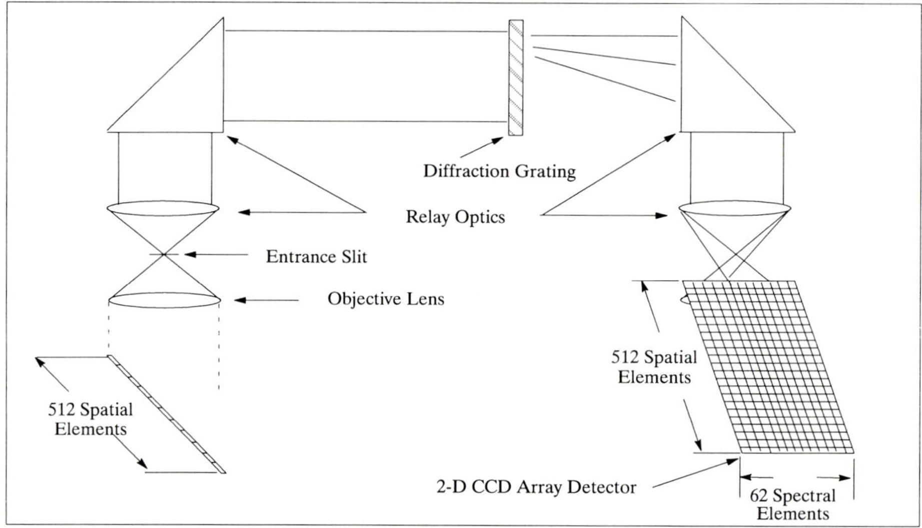


**Figure 110:    Schematic of the upgraded ASAS Optical System**

ASAS acquires data from 29 spectral bands in the range 455 - 871 nm with a resolution of approximately 15 nm. The ASAS optics features a f/1.4 objective lens with a 57.2 mm focal length for a 25° FOV. A diffraction grating, located between two prisms, disperses the received radiant energy into its wavelength spectrum (see Figure 110) . The second prism directs the dispersed energy onto the focal plane.

A 512 x 32 element silicon CID (Charge Injection Device) detector array is located at the focal plane to generate digital image data. The first two rows of the array are blacked-out., with the remaining 29 rows intended for digital image data acquisition. The long dimension of the array is reserved for the spatial resolution of 512 cross-track ground elements. The cross-track spatial resolution is 4.25 m from an altitude of 5000 m.

As the aircraft flies forward, each row of the 512 elements is electronically scanned to generate 29 channels of digital data in the "pushbroom' fashion. The channels are spatially registered and each channel corresponds to the spectral band of radiant energy incident on the element row (only 29 channels of the 32 were operable). The scan rate is selectable (3, 6, 12, 24, 48, or 64 frames/s. The quantization is 12 bit.

Multi-angle data.
The sensor FOV is tilted forward as the aircraft approaches the target site. The optical head is then rotated through a discrete sequence of fore-to-aft tilt angles as the aircraft flies over


401) J. R. Irons, K. J. Ranson, D. L. Williams, R. R. Irish, F. G. Huegel, "An Off-Nadir-Pointing Imaging Spectroradiometer for Terrestrial Ecosystem Studies", IEEE Transactions on Geoscience and Remote Sensing, Vol 29, Nr. 1, January 1991, pp. 66-74

and then past the site. A typical sequence consists of seven angles from 45° forward to 45° aft in 15° increments.

ASAS upgrades in 1991/92.[402]

*   A new tilting system for the optical head was installed to allow tilting angles up to 75° forward and up to 60° aft.

*   A replacement of the ASAS detector array subsystem. A CCD array was installed along with a new data acquisition system to accommodate the new array. The new array provides acquisition of data in 62 spectral bands ranging from 400 to 1060 nm with a spectral resolution of 11.5 nm.

The data of the ASAS sensor are being used for the development, testing, and validation of algorithms requiring multi-angle data. Such algorithms are required for the EOS era.

History: The ASAS optics were originally part of the Scanning Imaging Spectroradiometer (SIS) of NASA/JSC in the early 1970's. SIS employed a vidicon detector for imaging. ASAS was created from SIS in 1981, when a CID detector array was incorporated with the optical system (NASA/JSC and the Naval Ocean System Center). ASAS was transferred to NASA/ GSFC in 1984, the mounting bracket of the instrument was modified (gimbal mount) to permit off-nadir tilting (first utilized in 1987).

## B.19   ATLAS (Airborne Tunable Laser Absorption Spectrometer)

ATLAS is a second-harmonic absorption spectrometer using a tunable diode laser of PbSnTe composition. The instrument is NASA-sponsored and flown on ER-2 aircraft at Ames Research Center since 1987 for applications (in-situ measurements) in atmospheric chemistry.[403]

Principle of operation: The instrument detects an infrared active target gas (e.g. $N_2O$, $CH_4$, CO, or $O_3$) by measuring the fractional absorption of the infrared beam from a tunable diode laser as it traverses a multipass White cell, containing an atmospheric sample at ambient pressure. The laser source is tuned to an individual revibrational line in an infrared absorption band of the target gas, and is frequency modulated at 2 kHz. Synchronous detection of the resultant amplitude modulation at 2 kHz and 4 kHz yields the first and second harmonics of the generally weak absorption feature with high sensitivity ($\Delta I/I < 10^{-5}$).
Part of the main beam is split off through a short cell containing a known amount of the target gas to a reference detector. The reference first harmonic signal is used to lock the laser frequency to the absorption line center, while the second harmonic signal is used to derive the calibration factor needed to convert the measurement beam second harmonic amplitude into absolute gas concentration. A zero beam is included to correct for background gas absorption occurring outside the multipass cell. The response time of the instrument is set by the gas flow rate through the White cell, which is normally adjusted to give a new sample every second. Periodic standard additions of the target gas are injected into the sample stream as a second method to calibrate the measurement technique and as an overall instrument diagnostic.

402)   J. Irons, "The Advanced Solid-State Array Spectroradiometer (ASAS)", The Earth Observer, Vol. 3. Nr. 7, 1991, pp. 31-35
403)   M. Loewenstein, J. R. Podolske, K. R. Chan, S. E. Strahan, "Nitrous Oxide as a Dynamical Tracer in the 1987 Airborne Antarctic Ozone Experiment", J. of Geophysical Research, Vol. 94, Nr. D9, 1989, pp.11.589-11.598

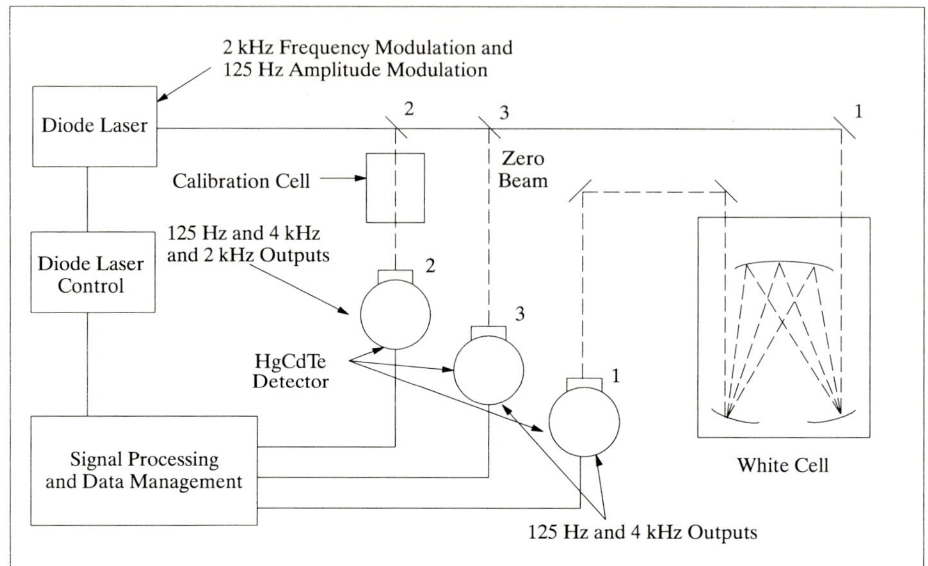


**Figure 111:    Schematic Diagram of ATLAS[404)]**


## B.20    ATLAS (Airborne TerrestriaL Applications Scanner)

ATLAS is a NASA-sponsored instrument designed and developed at Stennis Space Center (SSC, Mississippi). The instrument is in its final testing stages (as of Fall 1993) and expected to be operational in early 1994 (on LearJet 23). ATLAS employs an electromechanical rotating mirror scanner system.[405)]

Objectives: ATLAS was designed specifically to combine the CAMS and TIMS capabilities into one instrument. Correlation or supplementing of ATLAS airborne data with TM (Landsat), ASTER or Spot spaceborne data. ATLAS observation altitudes may range from 1 - 12 km.

ATLAS duplicates CAMS (see B.24) channels 1-8 (ATLAS 1-8) and TIMS channels 1-6 (ATLAS 10-15) with the addition of a MWIR (Mid Wavelength Infrared) channel at 3.35 - 4.2 µm (ATLAS 9) for a total of 15 spectral channels. An integrating sphere and high- and low-temperature blackbodies provide calibration data. No calibration is provided for channel 9.

NASA/SSC also operates a sensor calibration lab. This lab is used to perform spectral, spatial, and radiometric calibrations on the instruments: CAMS, TIMS and ATLAS. Numerous calibrations are performed on sensors provided by outside customers.

---

404)  The seven output signals shown are stored at a 2-Hz rate. The 125-Hz data carry laser power information; 2-kHz data carry first-harmonic feedback information for laser line locking; 4-kHz data carry second-harmonic information proportional to $N_2O$ mixing ratios.
405)  Information provided by B. A. Spiering of NASA/SSC

| Channel | Bandwidth (μm) | Cooling | NER Sensitivity mW/(cm$^2$ str μm) |
|---|---|---|---|
| 1 | 0.45 - 0.52 | ambient | 1.04 10$^{-7}$ |
| 2 | 0.52 - 0.60 | | (NER = Noise Equiva- |
| 3 | 0.60 - 0.63 | | lent Radiance) |
| 4 | 0.63 - 0.69 | | |
| 5 | 0.69 - 0.76 | | **Preliminary values** |
| 6 | 0.76 - 0.90 | | |
| 7 | 1.55 - 1.75 | 77 K | 4.79 10$^{-8}$ (NER) |
| 8 | 2.08 - 2.35 | 77 K | 2.46 10$^{-6}$ (NER) |
| 9 | 3.35 - 4.20 | 77 K | 3.38 10$^{-5}$ (NER) |
| | | | (NEΔT) °C |
| 10 | 8.20 - 8.60 | 77 K | 0.072 |
| 11 | 8.60 - 9.00 | | 0.070 |
| 12 | 9.00 - 9.40 | | 0.068 |
| 13 | 9.60 - 10.2 | | 0.045 |
| 14 | 10.2 - 11.2 | | 0.028 |
| 15 | 11.2 - 12.2 | | 0.029 |

| | |
|---|---|
| Optical parameters | Entrance aperture: 180.5 mm<br>Aperture: f/8 |
| IFOV | 2.0 mrad |
| FOV | 73.34° |
| Scan rate | 6 - 50 rps (revolutions per second) |
| Scan rate increment | 1 rps |
| Data quantization<br>Video words per scan line<br>Analog bandwidth<br>Housekeeping words per scan line | 12 bit<br>640<br>9.425 - 78.54 kHz<br>200 |
| Recorder data rate<br>Recorder capacity | 0.8 MByte/s<br>10 GByte per tape |
| Instrument mass | 197 kg |
| Calibration Parameters | |
| Thermal reference sources<br>    Emissivity<br>    Aperture field-filling<br>    Uniformity<br>    Stability<br>    Temperature control<br>    Adjustability<br>    Accuracy | 2 blackbody units<br>0.99<br>100% (205.9 mm)<br>> 95%<br>> 95%<br>-15 °C to 60 °C<br>0.1 °C<br>0.1 °C |
| Visible reference source<br>    Aperture field filling<br>    Uniformity<br>    Stability | 1 integrating sphere<br>100% (205.9 mm)<br>> 95%<br>> 95% |
| GPS accuracy (lat, long, alt)<br>INS accuracy (lat, long)<br>Gyroscope accuracy<br>Roll correction | 25 m<br>177 m (at equator)<br>0.206"<br>± 15° |

**Table 100:** **Specification of the ATLAS Instrument**

## B.21 AVIRIS (Airborne Visible/Infrared Imaging Spectrometer)

AVIRIS (of AIS heritage) is a NASA/JPL-developed/owned instrument operated by NASA/AMES Research Center (ARC) aboard an ER-2 aircraft. The instrument measures transmitted, reflected, and scattered solar energy from the Earth's surface and atmosphere in 224 channels over extended regions at high spatial resolutions. AVIRIS is regarded as the first operational hyperspectral instrument.

Science objectives: AVIRIS radiance spectra are used to identify, measure and monitor constituents of the Earth's surface and atmosphere based on molecular absorption and particle scattering signatures. Research areas include: ecology, oceanography, geology, snow hydrology, cloud and atmospheric studies. AVIRIS data are also used for satellite calibration, modeling, algorithm development and validation. Research with AVIRIS is dominantly directed towards understanding processes related to the global environment and climate change.[406],[407]

AVIRIS was flown for the first time in 1986 (first airborne images), 1987 first science data, fully operational since 1989 [in June/July 1991 the instrument was flown over numerous European test sites in the framework of EMAC (European Multi-Sensor Airborne Campaign)]. AVIRIS uses scanning optics and a group of four spectrometers to image a 614 pixel swath width simultaneously in 224 contiguous spectral bands. A spatial image is built up through the scanner motion, which defines an image line 614 pixels wide perpendicular to the aircraft direction, and through the aircraft motion, which defines the length of the image frame. Sensor: CCD line array detector technology. Spectral range: 380 - 2500 nm with a total of 224 channels. - All AVIRIS data is decommutated and archived at JPL. Since 1989 over 4000 scenes were acquired (a scene consists of 614 pixels x 512 lines x 224 bands).

AVIRIS is of a modular construction, consisting of 6 optical subsystems and 5 electrical subsystems. The optical subsystems (a scanner, 4 spectrometers, and a calibration source) are coupled together through optical fibers.

Data: The recorded data set forms an image cube of which two axis represent spatial dimensions, and the third represents a spectral dimension (see Figure 144). The data recorder of AVIRIS has been upgraded in 1992 (by a Metrum VLDS, 10 GByte storage capacity).

The general system performance of the instrument is continually being upgraded.[408] In addition the ground data system at JPL was modernized and its functionality expanded to handle the large data volumes of the sensor expediently and to provide data retrieval.

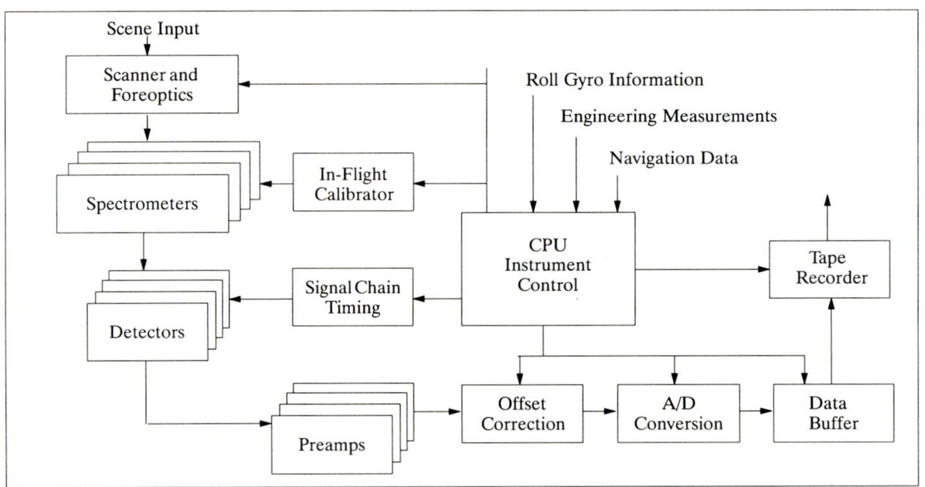



**Figure 112:  AVIRIS Functional Block Diagram**


406)  Information provided by R. O. Green of NASA/JPL

407)  W. M. Porter, H. T. Enmark, "A System of the Airborne Visible/Infrared Imaging Spectrometer (AVIRIS)", SPIE, Vol. 834 Imaging Spectroscopy II, 1987

408)  W. M. Porter, T. G. Chrien, E. G. Hansen, Ch. M. Sature, "Evolution of the Airborne Visible/Infrared Imaging Spectrometer (AVIRIS) Flight and Ground Data Processing System", SPIE, Vol. 1298, 1990, pp. 11-17

| Parameter | Value | Parameter | Value |
|---|---|---|---|
| Spectrometer A<br>　Spectral range<br>　Number of bands<br>　Bandwidth<br>　Grating | <br>380 - 690 nm<br>32<br>9.7 nm<br>117.65 lines/mm | Detectors of Spect. A<br>　Type<br>　Number of elements<br>　Material<br>　Integration time | <br>Line array<br>32<br>Silicon<br>87 μs |
| Spectrometer B<br>　Spectral range<br>　Number of bands<br>　Bandwidth<br>　Grating | <br>670 - 1270 nm<br>64<br>9.5 nm<br>128.2 lines/mm | Detectors of Spect. B<br>　Type<br>　Number of elements<br>　Material<br>　Integration time | <br>Line array<br>64<br>Indium Antimonide<br>87 μs |
| Spectrometer C<br>　Spectral range<br>　Number of bands<br>　Bandwidth<br>　Grating | <br>1260 - 1880 nm<br>64<br>10.0 nm<br>124.2 lines/mm | Detectors of Spect. C<br>　Type<br>　Number of elements<br>　Material<br>　Integration time | <br>Line array<br>64<br>Indium Antimonide<br>87 μs |
| Spectrometer D<br>　Spectral range<br>　Number of bands<br>　Bandwidth<br>　Grating | <br>1880 - 2500 nm<br>64<br>12.0 nm<br>128.6 lines/mm | Detectors of Spect. D<br>　Type<br>　Number of elements<br>　Material<br>　Integration time | <br>Line array<br>64<br>Indium Antimonide<br>87 μs |
| Foreoptics<br>　FOV<br>　IFOV<br>　Eff. focal length | <br>30°<br>1 mrad<br>19.76 cm | Fiberoptics<br>　Material<br><br>　Diameter | <br>Silica (A,B); Fluoride<br>glass (C and D)<br>200 μm |
| Swath width | 11 km (20 km altitude) | Spatial resolution | 20 m x 20 m |
| Scanner rate | 12 scans/s | Pixels/scan line | 614 |
| Quantization | 12 bit | Data rate | 20.4 Mbit/s |
| Tape recorder | Metrum VLDS | SNR ((@ 490 nm) | 100:1 |
| Instrument mass | 300 kg | Instrument power | 28 VDC, 41 A |
| Flight altitude | 20 km | Ground-track velocity | 735 km/h |
| Velocity/height | 20 knots/km | | |

**Table 101:　AVIRIS Instrument Parameters**


# B.22　CAESAR

CAESAR = CCD Airborne Experimental Scanner for Applications in Remote Sensing.
CAESAR is a sensor developed and built by NLR (the Netherlands) and the Institute of
Applied Physics (TU Delft). It is operated from NLR's Metro II laboratory aircraft.[409]

Sensor: CAESAR is a single-lens triple CCD camera with filter sets in front of the CCD
sensor arrays. A set of three cameras form the 9-channel down-looking module. A fourth
camera is mounted as a 3-channel forward-looking module and is pointing up to 52° forward. Sensor type: Thomson CSF TH 7801; elements/channel: 1728; element size: 13 x 13
μm. CAESAR is operational since 1988.

Applications: Measurement of water quality, coastal research, agricultural and forestry resource mapping, geo-botanical research (discrimination between chlorophyll, gelbstoff and
other organic matter; shallow water depth mapping, bottom topography mapping), bidirectional reflectance measurements.

CAESAR can be operated in different basic modes:
- 　Land mode, using the 3 central channels of the down-looking module
- 　Special Land mode, using the 3 central channels of the down-looking module with
  a higher geometric resolution
- 　Sea mode, using all 9 channels of the down-looking module
- 　Forward-looking mode, using all 3 channels of the forward-looking module

---

409)　"CAESAR CCD Airborne Experimental Scanner for Applications in Remote Sensing", an NLR brochure

Spectral range: 400 - 1050 nm
Spectral resolution:

- 535-895 nm, land observation mode, 3 bands, 30-50 nm bandwidth
- 400-1050 nm, water observation mode, 9 bands, 20-60 nm bandwidth.

Data: all channels are handled in parallel. Recorded raw CCD data and aircraft data are used to correct the CAESAR data radiometrically and geometrically. A quick-look facility is available.

| Advised Configurations | Land mode | Special Land mode | Sea Mode | Forward-Looking mode |
|---|---|---|---|---|
| Altitude | 3 km | 2 km | 6 km | 3 km |
| Integration time | 7.5 ms | 5 ms | 40 ms | 7.5 ms |
| Ground Speed | 100 m/s | 100 m/s | 105 m/s | 100 m/s |
| Number of pixels | 1728 | 1280 | 1728 | 1728 |
| Swath width | 1296 m | 640 m | 2592 m | 1200 m |
| Spatial resolution | 0.75 x 0.75 m | 0.5 x 0.5 m | 4.0 x 4.0 m | 2.0 x 2.0 m |
| Radiometric resolution | 10 bit | 9 bit | 12 bit | 10 bit |

| Optical Parameters | Down-Looking | Forward-Looking | | |
|---|---|---|---|---|
| Lens | 52.1 mm | 84.9 mm | | |
| IFOV | 0.25 mrad | 0.15 mrad | | |
| FOV | 24.3° | 14.7° | | |
| View angle of<br>ahead channels<br>central channels<br>backwards channels<br>tilt angle | 11.5°<br>0°<br>-11.5°<br>0-20° | 7°<br>0°<br>-7°<br>0-52° | | |

Table 102:     **Technical Specifications of CAESAR**

| Mode | Sea | Inland Water | Land | Special Land | Forest | Stereo | Forward |
|---|---|---|---|---|---|---|---|
| Nr. of Channels | 9 | 9 | 3 | 3 | 3 | 3 | 3 |
| Spectral filters (nm)<br>center + (width) | 410 (20)<br>445 (20)<br>520 (20)<br>565 (20)<br>630 (20)<br>685 (20)<br>785 (30)<br>1020 (60)<br>1020 (60) | 520 (20)<br>565 (20)<br>600 (20)<br>630 (20)<br>650 (10)<br>665 (10)<br>677 (15)<br>705 (18)<br>785 (30) | 550 (30)<br>670 (30)<br>870 (50) | 550 (30)<br>670 (30)<br>870 (50) | 650 (10)<br>685 (20)<br>785 (30) | t.b.s. | 550 (30)<br>670 (30)<br>870 (50) |
| Signal to Noise Ratio<br>Quantization (bit)<br>IFOV (mrad)<br>Swath width (m)<br>Pixels<br>Pixel width<br>Pixel length<br>Integration time (ms)<br>Flight altitude (km)<br>Tilt (°) | 2000<br>12<br>0.25<br>2600<br>1728<br>1.5<br>4.0<br>40<br>6<br>0-20 | 2000<br>12<br>0.25<br>2600<br>1728<br>1.5<br>4.0<br>40<br>6<br>0-20 | 200<br>10<br>0.25<br>1300<br>1728<br>0.75<br>0.75<br>7.5<br>3 | 200<br>9<br>0.25<br>640<br>1280<br>0.5<br>0.5<br>5<br>2 | 200<br>9<br>0.25<br>640<br>1280<br>0.5<br>0.5<br>5<br>2 | 200<br>10<br>0.25<br>1300<br>1728<br>0.75<br>0.75<br>7.5<br>3 | 200<br>10<br>0.15<br>ca. 1400<br>1728<br>ca. 0.8<br>ca. 1.5<br>7.5<br>4<br>0-52 |

Table 103:     **Specification of the CAESAR CCD Pushbroom Scanner**

## B.23    CALS (Cloud and Aerosol Lidar System)

CALS is a NASA/GSFC autonomous lidar system that has been flown on high-altitude aircraft (the WB-57F and ER-2) since 1979. Objectives/applications: cloud and radiation studies, development of cloud remote sensing, provision of satellite ground truth measurements. The instrument is also referred to as CLS (Cloud Lidar System).

The transmitter of the lidar system is a Nd:YAG laser with a nominal pulse energy of 50 mJ at the doubled wavelength of 532 nm. Backscattered light is collected by an 18 cm diameter Questar telescope and is split by a prism type polarization beamsplitter into parallel and perpendicular components. The instrument formerly employed Photomultiplier Tube (PMT) detectors but now uses Avalanche Photo Diodes (APD). Two polarizations of the 532 nm signal and a single 1064 nm channel are detected. Data compression is being done by 4 decade logarithmic amplifiers. Data acquisition and storage is fully automatic. As part of the operational and data analysis procedures, the instrument is calibrated for the observed backscatter cross-section of the atmosphere and the signal depolarization.[410]

The instrument was upgraded in the period of 1991-93, it currently features a new diode-pumped Nd:YAG laser with an energy of 200 mJ and a pulse repetition rate of 20 Hz. The instrument may be flown for periods of 9 hours with a 5 GByte data storage system. CALS provides the following measurement characteristics:[411],[412]

- Nadir viewing from 20 km altitude (on ER-2 aircraft)
- Profiling capability of all cloud and aerosol structures
- Resolutions: 7.5 m vertical and 20 m horizontal
- 532 nm and 1064 nm frequencies

Recent field experiments with CALS participation:

- FIRE Cirrus and STRATUS I (1986-87)
- COHMEX STORM Experiment (1986)
- FIRE Cirrus II (1991)
- ASTEX (1992)
- TOGA/COARE (1993)
- CEPEX (1993)

## B.24    CAMS (Calibrated Airborne Multispectral Scanner)

A NASA-sponsored instrument developed and operated at Stennis Space Center (Mississippi). CAMS is flown on a LearJet 23 (equipped with INS and GPS) at altitudes up to 12 km, the instrument became operational in 1987.[413]

The CAMS instrument provides spectral coverage in 6 contiguous channels from 0.45 - 0.90 μm (VIS/NIR), plus 2 channels in the SWIR and 1 channel in the TIR range. An on-scanner integrating sphere provides calibration data for channels 1-8. A high-temperature blackbody ($< 50$ °C) and a low-temperature blackbody ($> -10$°C) provide calibration data for channel 9.


410)  J. D. Spinhirne, M. Z. Hansen, J. Simpson, "The Structure and Phase of Cloud Tops as Observed by Polarization Lidar", Journal of Climate and Applied Meteorology, Vol. 22, Nr. 8, August 1983, pp. 1319-1331
411)  Information provided by J. D. Spinhirne of NASA/GSFC
412)  J. D. Spinhirne, "Cirrus Structure and Radiative Parameters from Airborne Lidar and Spectral Radiometer Observations: The 28 October 1986 FIRE Study", Monthly Weather Review, Vol. 118, Nr. 11, November 1990, pp. 2329-2343
413)  Information provided by B. A. Spiering of NASA/SSC

| Channel | Detector | Bandwidth (μm) | NER (Sensitivity) mW/(cm² str μm) |
|---|---|---|---|
| 1 | Silicon Array | 0.45 - 0.52 | ≤ 0.06 |
| 2 | | 0.52 - 0.60 | ≤ 0.06 |
| 3 | | 0.60 - 0.63 | ≤ 0.11 |
| 4 | | 0.63 - 0.69 | ≤ 0.06 |
| 5 | | 0.69 - 0.76 | ≤ 0.23 |
| 6 | | 0.76 - 0.90 | ≤ 0.22 |
| 7 | Ge (Germanium) | 1.55 - 1.75 | ≤ 0.28 |
| 8 | InSb | 2.08 - 2.35 | ≤ 0.07 |
| 9 | HgCdTe | 10.2 - 12.5 | ≤ 0.2 K (NE$\Delta$T) |
| IFOV | | 2.5 mrad | |
| FOV | | 100° | |
| Scan rate | | 6 - 60 rps | |
| Data quantization | | 8 bit video | |
| Pixels/scan line | | 700 | |
| Instrument mass | | 106.5 kg | |
| Cooling | | Ch. 7 (Thermoelectric) Ch. 8 (LN$_2$) Ch. 9 (LN$_2$) | |
| Optical parameters | | Telescope: Dall Kirkham Aperture: f/2.8 Focal length: 267 mm NIR: dichroic bandpass SWIR: dichroic grating TIR: dichroic bandpass | |

**Table 104:     Specification of the CAMS Instrument**


## B.25     CAR (Cloud Absorption Radiometer)

CAR[414] is a NASA/GSFC developed multi-wavelength scanning radiometer for measuring angular distributions of scattered radiation deep within a cloud layer. The objective is to provide measurements from which the single scattering albedo of clouds can be derived as a function of wavelength. The instrument was built in the early 1980's.

CAR is a 13 channel scanning radiometer providing a 1° field of view with a scan geometry in the vertical plane (on the right-hand side of the aircraft) from 5° before zenith to 5° past nadir (190° aperture). This arrangement permits observations of both the zenith and nadir intensities.

| | |
|---|---|
| CAR optical system | non-dispersive type - consisting of a complex configuration of dichroic beam splitters and narrow-band interference filters |
| Dall-Kirkham (Cassegrain) telescope | 12.4 cm diameter |
| IFOV | 1° (17.5 mrad) |
| FOV | 190° along the scan line  and 1° along the plane's velocity vector |
| Detectors channels 1-5 channels 6-7 channels 8-18 (filter wheel selected) | hybrid silicon photodiodes operating at 308 K Germanium detectors operating at 255 K InSb cold-filtered detector cryogenically cooled |
| Scanner | Electro-mechanical scanning system |
| Scan cycle | 600 ms |
| Calibration | Laboratory integrating sphere measurements (1.83 m diameter) from time to time |

**Table 105:     CAR Instrument Parameters**

414)  M. D. King, L. F. Radke, P. V. Hobbs, "Determination of Spectral Absorption of Solar Radiation by Marine Stratocumulus Clouds from Airborne Measurements within Clouds", Journal of the Atmospheric Sciences, Vol. 47, Nr. 7, 1 April 1990, pp. 894-907

| Channel Number | Center Wavelength ($\mu$m) | Spectral Resolution FWHM ($\mu$m) | Minimum Intensity (mW cm$^{-2}$ $\mu$m$^{-1}$ sr$^{-1}$) | SNR |
|---|---|---|---|---|
| 1 | 0.503 | 0.016 | $1.085 \times 10^1$ | 3622 |
| 2 | 0.673 | 0.020 | $8.684 \times 10^0$ | 1903 |
| 3 | 0.744 | 0.019 | $7.268 \times 10^0$ | 2785 |
| 4 | 0.866 | 0.020 | $5.404 \times 10^0$ | 3022 |
| 5 | 1.031 | 0.020 | $3.884 \times 10^0$ | 3052 |
| 6 | 1.198 | 0.022 | $2.382 \times 10^0$ | 624 |
| 7 | 1.247 | 0.046 | $2.244 \times 10^0$ | 2446 |
| 8 | 1.547 | 0.030 | $4.502 \times 10^{-1}$ | 436 |
| 9 | 1.640 | 0.041 | $6.091 \times 10^{-1}$ | 685 |
| 10 | 1.722 | 0.038 | $4.470 \times 10^{-1}$ | 340 |
| 11 | 1.996 | 0.039 | $7.494 \times 10^{-3}$ | 29 |
| 12 | 2.200 | 0.040 | $7.708 \times 10^{-2}$ | 191 |
| 13 | 2.289 | 0.023 | $5.135 \times 10^{-2}$ | 33 |

**Table 106:   Spectral Characteristics of the CAR Instrument[415]**

The first seven channels of CAR are continuously and simultaneously sampled, while the eighth registered channel is selected from among the six channels on a filter wheel. With automatic sequencing the filter wheel rotates to a new filter position every fourth scan. Since the scan rate of the radiometer is 1.67 Hz, each minute of flight duration results in 100 measurements of the angular intensity field for each of the first seven channels, and typically 12 measurements for each of the six filter wheel channels. At a nominal aircraft speed of 80 m/s, it follows that the zenith and nadir intensity measurements are obtained within a distance of approximately 24 m for each scan of the radiometer.

The CAR sensor was tail-mounted in a B-23 aircraft of the University of Washington until 1984. After May 1985 CAR was mounted into the nose of a C-131-A aircraft, also of the University of Washington. In 1987 CAR participated in the FIRE campaign (comprehensive measurements of marine stratocumulus clouds, coordination of multiple-aircraft and multiple-satellite observations).

Applications: Use of the diffusion domain method for the determination of the spectral similarity parameter, i.e. the single scattering albedo of clouds. Determination of the optical thickness and effective particle radius of stratiform cloud layers from reflected solar radiation measurements. CAR was also utilized in May/June 1992 (coordinated campaign with data from Landsat TM) for the measurement of directional and spectral reflectances of the Kuwait oil-fire smoke.[416),417),418),419)]

# B.26   CARABAS (Coherent All RAdio BAnd Sensing)

CARABAS is a Swedish airborne experimental SAR instrument designed and built at FOA (National Defense Research Establishment, Linköping, Sweden), which operates in the lower part of the VHF-Band. The objective is good penetration of vegetation/foliage and to some extent of the ground surface. The frequency band was chosen: 1) to reduce the image


415)  M D. King, M. G. Strange, P. Leone, L. R. Blaine, "Multiwavelength Scanning Radiometer for Airborne Measurements of Scattered Radiation within Clouds", Journal of Atmospheric and Oceanic Technology, Vol. 13, Nr. 3, September 1986, pp.513-522

416)  T. Nakajima, M. D. King, "Determination of the Optical Thickness and Effective Particle Radius of Clouds from Reflected Solar Radiation Measurements. Part I: Theory", Journal of Atmospheric Sciences, Vol. 47, Nr. 15, 1 August 1990, pp. 1878-1893

417)  T. Nakajima, M. D. King, J. D. Spinhirne, L. F. Radke, "Determination of the Optical Thickness and Effective Particle Radius of Clouds from Reflected Solar Radiation Measurements. Part II: Marine Stratocumulus Observations", Journal of Atmospheric Sciences, Vol. 48, Nr. 5, 1 March 1991, pp. 728-750

418)  M. D. King, L. F. Radke, P. V. Hobbs, "Optical Properties of Marine Stratocumulus Clouds Modified by Ships", Journal of Geophysical Research, Vol. 98, Nr. D2, February 20, 1993, pp. 2729-2739

419)  M. D. King, "Directional and Spectral Reflectance of the Kuwait Oil-Fire Smoke", Journal of Geophysical Research, Vol. 97, Nr. D13, Sept. 1992, pp.14,545-14,549

speckle level without sacrificing resolution, and 2) to obtain a system with diffraction-limited resolution (i.e. a system with a dimension of the resolution cell comparable with those of the wavelengths employed, and in this way minimize the influence of speckle). A feasibility study started the project in 1985, the instrument was built and integrated during 1988-90, test flights on a Rockwell Sabreliner have been carried out during 1992, with the first major SAR campaign in October 1992.[420])

| Aircraft | Rockwell Sabreliner |
|---|---|
| Nominal altitude | 1500 - 6500 m |
| Nominal speed | 100 m/s |
| Antenna | 2 wideband dipoles |
| Polarization | Horizontal |
| Frequency | 20 - 90 MHz |
| Number of frequencies, n | $\leq 57$ |
| Frequency stepping factor | 1.25 MHz |
| Pulse length | 0.5 μs |
| Receiver bandwidth | 2.5 MHz |
| Peak power | 1 kW |
| System PRF, $PRF_s$ | 10 kHz |
| Effective PRF, $PRF_e$ | 10/2/n kHz |
| Intermediate frequency (IF) | 2.5 MHz |
| Digital sampling rate | 10 MHz |
| Maximum slant range, $R_{max}$ | 7.5 km |
| Number of bits/real sample | 12 |
| Data rate | 80 Mbit/s |
| Tape recorder capacity | 107 Mbit/s |
| Cassette capacity | 60 minutes |

**Table 107:**      **System Parameters of the CARABAS VHF SAR Instrument**

The antenna system consists of inflatable and flexible canvas sleeves (individually mounted on an air inlet in each end of the T-bracket; the antenna system arrangement has a two-pronged shape) which is trailed behind the aircraft in flight. The antenna works as a wide-band dipole over the frequency range 20 - 90 MHz and has a total length of 5.5 m with a diameter of 0.3 m. The antennas must be used in order to resolve the backscatter from from the left-hand and right-hand side of the aircraft since the antenna directivity is low. Dipole elements of various length are sewn into the envelope surface of the sleeve. Analog filters tune the dipole elements for each frequency.

The external modification to the aircraft consists of a T-bracket attached on the rear of the body, just below the tail, and includes two air inlets. The antennas are folded to the body when the aircraft moves on the ground and unlocked manually just before takeoff.

The SAR measurement concept is based on a stepped-frequency technique - instantaneous narrow-band signals are used while stepping through the required bandwidth. The scattering model used assumes that the reflectivity from a ground surface element is proportional to the scalar product between the ground surface normal and the aspect vector from the aircraft to that element. The inverse scattering problem consists of restructuring the reflectivity function with knowledge of all circular averages measured along the synthetic aperture.

Derived signal processing algorithms have been applied on the acquired radar data to form images. The capability of foliage penetration has been demonstrated. Further methods of testing and analysis are in progress.

A typical spatial resolution of CARABAS is 5 m in slant range and 7 m in azimuth. So far, no motion compensation has been carried out in the processing. Another parameter that has a great impact on the resolution is the available signal bandwidth, not severely disturbed by strong external radio sources. As of 1993, a lot of effort is devoted to signal processing with

420)   A. Gustavsson, P. O. Frölind, H. Hellsten, T. Jonsson, B. Larsson, G. Stenström, "The Airborne VHF SAR System CARABAS", IGARSS'93, Vol. II, Kogakuin University, Tokyo, Japan, Aug. 18-21, 1993, pp. 558-562

respect to motion compensation and filtering techniques of interfering radio sources, to improve the resolution further.

A second improved CARABAS system (CARABAS-2) is planned to be assembled in 1994, based on the experiences gained from CARABAS-1.

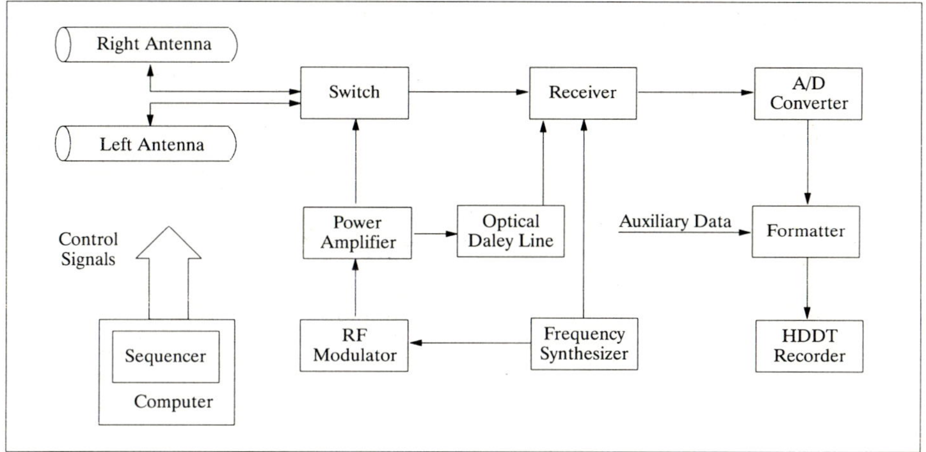


**Figure 113:    Functional Block Diagram of the CARABAS Radar Electronics**

## B.27    CASI (Compact Airborne Spectrographic Imager)

CASI is a commercially available instrument (since 1990) of ITRES Research Limited of Calgary, Alberta, Canada. CASI is a lightweight spectrometer/multispectral pushbroom (CCD) imager system (of FLI heritage) for airborne remote sensing applications. It covers the visible and near infrared spectrum (0.430 - 0.870 µm) with 1.8 nm sample interval (288 spectral bands) and 512 spatial pixels. IFOV = 0.3 - 2.4 mrad. FOV = 35°. The frame transfer CCD has a 578 x 288 (lines x bands) pixel image area. The CCD is oriented in such a way to provide 512 spatially resolved points per line. Quantization level = 12 bits.[421]

CASI consists of the sensor head, instrument control unit, power supply, power inverter, monitor and keyboard. The system is equipped with a data recorder (standard 8 mm video cassette tape recorder with a volume of about 1 GByte per cassette). Data are recorder at a maximum rate of 230 kByte/s. Operator selectable band sets based on the requirements of the application. Quicklook capability in either grey-scale or pseudo color display. CASI can operate either as a multispectral imager or as a high-speed multi-point spectrometer with co-registered monochromatic imagery. In imaging mode, the spectral band configurations are defined interactively with a graphical user interface. In multi-spectrometer mode, high resolution spectra are recorded for the full spectral range of the instrument for up to 39 regularly spaced points in every data frame. CASI can be provided with optional roll correction and calibration systems.

CASI owners/operators:
G. A. Borstad & Associates, Sidney, B. C., Canada;
Aerospace Image Production, Herrenburg, Germany;
Geomatic Technologies Inc., St. John's, Newfoundland, Canada;

---

421)  Information brochures provided by ITRES Research Limited, Calgary, Alberta, Canada

Defense Research Establishment Suffield, Alberta, Canada;
Freie Universität Berlin, Weltraum Institut, Berlin, Germany;
Institute for Space & Terrestrial Science, York University, North York, Ontario, Canada;
EG&G Energy Measurements Inc., Las Vegas, Nevada, USA, SM Systems and Research
Corporation.

| | |
|---|---|
| Spectral coverage | 418 - 926 nm sampled by 288 detector elements, sampling interval = 1.8 nm, spectral resolution = 2.9 nm |
| Spatial coverage | 35.5° swath width, with standard lens; single camera gives 612 pixels, other lenses can be used to vary swath width , sampling interval = 1.2 mrad, spatial resolution = 1.6 mrad |
| Spectral mode | 39 spectra of the full range (418-926 nm) are recorded with 2.9 nm resolution, from 39 different directions across the swath. A full-resolution image at a pre-determined wavelength is also recorded to assist in track recovery. |
| Spatial mode | Spectral pixels are grouped to form up to 15 bands (15 push-broom images each 512 pixels wide). The bandwidth and spectral position are under software control. The number of bands governs the integration time. |
| Sensitivity | Depends on signal level |
| Integration times | 50 ms, typical in spatial mode, 100 ms, typical in spectral mode |
| Digitization | 12 bit |
| Detectors | 612 by 576 element EEV UT104 array, 512 of the 612 elements are recorded, 288 of the 576 elements are used for storage |
| Optics | Reflection grating with f/2.0 optics |
| Data recording | Exabyte digital recording onto 8 mm tapes, 1.1 GByte/tape, 280 kbit/s recorded |
| Power | 3.9 A at 110 VAC (400W) or 20 A at 28 V (560W) |
| Weights | Total = 55 kg, head = 6 kg |
| Aircraft mountings | Designed for light aircraft |

**Table 108:     Specifications of the Compact Airborne Spectrographic Imager (CASI)**


## B.28     Chinese Airborne Instruments


### B.28.1     CIS (Chinese Imaging Spectrometer)

CIS is an imaging spectrometer being developed and built by the Shanghai Institute of Technical Physics (SITP), Chinese Academy of Sciences (CAS), with the following objectives:[422]

- to get familiarized with the key technologies in imaging spectrometry
- to develop and test an airborne sensor (along with the algorithms for data processing and interpretation); the airborne sensor serves as a prototype and testbed for the spaceborne sensor (research tool)
- to develop an engineering model of a spaceborne sensor
- to build a flight unit for spaceborne testing and operation on a polar orbiting platform.

The CIS airborne instrument employs the following techniques:

- line scanning technique (45° scan mirror)
- a Cassegrain nonspherical telescope as primary optics
- parallel co-axis coupling of the dispersion system and the telescope
- multi-band separation by co-axis splitter
- image transformation coupling with the use of bunched optical fibers
- VNIR dispersion with type III holographic concave grating
- SWIR dispersion by planar blaze grating

422) Zheng Qinbo, Zhang Zhimin, Zhang Baolong, Xu Xuerong, Feng Qi, Gu Gong, "Spaceborne Chinese Imaging Spectrometer", Proc. of the 5th ISCOPS, Shanghai, 7-9 June, 1993

- 3-element mosaic filter/detection combination
- 64 element Si detector line array
- 24-element HgCdTe line array
- PC-based information processing system
- Digital tape recording system with SCSI interface

The airborne engineering model (prototype) of CIS is in the test phase as of 1993. A space-borne sensor CIS is considered for the turn of the century.

| Spectral/Optical Parameters | | | | |
|---|---|---|---|---|
| Spectral Ranges | VNIR | SWIR | MWIR | TIR |
| Wavelength (µm) | 0.4 - 1.04 | 2.0 - 2.48 | 3.55 - 3.95 | 10.5 - 12.5 |
| Nr. of channels | 64 | 24 | 1 | 2 |
| Bandwidth (nm) | 10 | 20 | 400 | 1000 |
| Sensitivity | NEΔR ~1% | NEΔR ~2% | NEΔT ~0.5 K | NEΔT ~ 0.5 K |
| IFOV (mrad) | 1.2 x 3.6 | 1.2 x 1.8 | 1.2 x 1.2 | 1.2 x 1.2 |
| Effective focal length (mm) | 500 | 400 | 200 | 200 |
| Detector element size (mm²) | 0.23 x 4 | 0.38 x 0.72 | 0.24 x 0.24 | 0.24 x 0.24 |
| FOV | | 80° | | |
| Quantization level | | 10 bit | | |
| Aperture | | 200 mm | | |
| Scan speed | | 6 Rev/s | | |
| Pixels per line | | 1226 | | |
| Sample rate | | 1.6 MHz | | |
| Data rate | | 10 Mbit/s | | |

Table 109:    Instrument Specification of CIS (Airborne Prototype Version)

## B.28.2   AMS (Airborne Multispectral Scanner)

AMS is a 19-channel instrument designed and built by SITP (no relation to Daedalus AMS).[423]  The scanner equally divides the spectrum from 0.46 - 1.1 µm into 16 bands. If required, such bands may be grouped to form some channels compatible with TM, MSS (of Landsat) or HRV (of Spot). The AMS optics features a Kennedy scanner for high scanning efficiency of the ground. Applications of the instrument are mainly in the field of resource surveys.

## B.28.3   TIMS (Thermal Imaging Multispectral Scanner)

TIMS is a 7-band thermal infrared multispectral scanner developed and built by SITP. It's remote sensing applications are in the area of geological surveys. TIMS uses a slant 45° rotative mirror for scanning. The main optical unit consists of a primary mirror and a collimating mirror. An infrared blaze grating with blaze wavelength of 9.0 µm is used as the dispersive element. The convergence unit is a specially designed Ge-lens with a FOV of 4.91° and f/0.78.

## B.28.4   MAIS (Modular Airborne Imaging Spectrometer)

The MAIS[424] instrument is designed, built and operated by the Shanghai Institute of Technical Physics (SITP) of the Academy of Sciences of China. The instrument is installed on a Citation S/II aircraft, the first flying tests with MAIS were conducted in November 1990. During the period September-October 1991, MAIS was also flown successfully in a joint Sino-Australian remote sensing campaign near Darwin and at several other test sites in Western Australia.

---


423)  Xue Yongqi, et al. "New Progress of Airborne Scanners at SITP from 1986 to 1990", paper presented at the 11th Asian Remote Sensing Conference in 1990

424)  Y. Xue, M. Shen, C. Yang, J. Wang, W. Yu, "Modular Airborne Imaging Spectrometer (MAIS)", paper provided by Z. Zhang of SITP, Shanghai

MAIS is a 71 channel imaging spectrometer with a spectral coverage from 0.44 µm to 11.8 µm. MAIS features a modular design, the optical system has four independent modules: the scanning unit, and three spectrometer modules for the different spectral ranges (separate calibration of each module is possible).

Applications: The instrument is being used for a number remote sensing applications by SITP for geological and environmental (pollution) surveys, some campaigns were conducted in Xinjiang and in the Quilian mountain area of the Gansu province, China.

As of 1993 the 64-channel prototype scanner is being built into the MAIS instrument.

| Scanner type | 45 degree rotating mirror (opto-mechanical scanner) | | |
|---|---|---|---|
| | Parameters: | Aperture of the primary telescope | 180 mm |
| | | Focal length of primary telescope | 180 mm |
| | | Focal length of collimator | 60 mm |
| | | TFOV (total field of view) | 90° |
| | Modular structure: | compact arrangement, flexible functions | |
| | | Focal length of primary telescope | 180 mm |
| | | Focal length of collimator | 60 mm |
| | | TFOV (total field of view) | 90° |
| | Scan rate: | 10 lines/s; 512/1024 pixels per line | |
| | Quantization: | 12 bit | |
| | Recording data rate | 640 kByte/s | |
| Spectrometer A | Wavelength range: | VNIR (0.44 - 1.08 µm) | |
| | Number of channels: | 32 | |
| | Spectral resolution: | 20 nm | |
| | IFOV: | 1.5 or 3 mrad | |
| | Detector type: | Silicon linear array | |
| | Dispersive element: | blaze wavelength 0.64 µm, 210 grooves/mm | |
| Spectrometer B | Wavelength range: | SWIR (1.5 - 2.5 µm) | |
| | Number of channels: | 32 | |
| | Spectral resolution: | 30 nm | |
| | IFOV: | 4.5 mrad | |
| | Detector type: | PbS linear array | |
| | Dispersive element: | blaze wavelength 1.8 µm, 110 grooves/mm | |
| Spectrometer C | Wavelength range: | TIR (7.8 - 11.8 µm) | |
| | Number of channels: | 7 | |
| | Spectral resolution: | 0.4 or 0.8 µm | |
| | IFOV: | 3 mrad | |
| | Detector type: | MCT linear array | |
| | Dispersive element: | planar blaze grating, blaze wavelength 9.2 µm, 20 grooves/mm | |
| Onboard Electronics | Multi-channel analog tape recorder<br>Preprocessor with spectral and spatial program<br>Real-time monitor displaying pseudo-color images<br>Auxiliary parameter recording (e.g. status of plane (position) and sensor, date, etc. | | |
| Flight altitude | The normal operational altitudes are between 4-6 km | | |

**Table 110: Specification of the MAIS Instrument**

## B.28.5 Prototype Scanner

This research instrument (image spectrometer development) co-uses the main optical unit of TIMS. A long wave pass filter separates the VIS range from the NIR range. In spectrometer I (0.46-1.1 µm), a concave holographic grating performs dispersing and converging. A special filter is used on the surface of the 32-element Si detector array to remove the higher order spectra. In spectrometer II (1.4-2.5 µm), the dispersive element is a plane blaze grating having a blaze wavelength of 1.6 µm. The grating sways back and forth between two positions synchronously with the scan line and stays at each position for a scan line. In this way the 16 detector elements cover 32 bands.

| Sensor | DGS | AMS | TIMS | Prototype Scanner | UV/IR Scanner | VNIR/TIR Scanner |
|---|---|---|---|---|---|---|
| Application | remote sensing | remote sensing | remote sensing | remote sensing | sea pollution monitoring | forest fire detection |
| FOV | 100° | 90° | 90° | 90° | 100° | 100° |
| IFOV | 3 mrad | 3 mrad | 3 mrad | 3-4.5 mrad | 3 mrad | 3 mrad |
| Scan rate/s | 25-100 | 20-50 | 10-30 | 10-20 | 100 | 100 |
| Optical area | 52 cm$^2$ | 52 cm$^2$ | 200 cm$^2$ | 200 cm$^2$ | 64 cm$^2$ | 52 cm$^2$ |
| Focal length | 666 mm | 666, 217 mm | 180 mm | 180 mm | 800 mm | 666, 217 mm |
| Scan mirror | 4-sided | 4-sided | 45° mirror | 45° mirror | 45° mirror | 4-sided |
| Spectral Bands (µm) | 0.40-0.43<br>0.43-0.48<br>0.48-0.54<br>0.53-0.62<br>0.60-0.70<br>0.68-0.90<br>3.0-5.0<br>or TM<br>1,2,3,4,5,7<br>8 cha. total | 16 bands in<br>0.46-1.1<br>1.55-1.75<br>2.08-2.35<br>8.0-12.5<br><br><br><br><br>19 cha. total | 8.2-8.6<br>8.6-9.0<br>9.0-9.4<br>9.4-9.8<br>9.8-10.6<br>10.6-11.4<br>11.4-12.2<br><br><br>7 cha. total | 32 bands in<br>0.46-1.1<br><br>32 bands in<br>1.4-2.5<br><br><br><br><br>64 cha. total | 0.28-0.38<br>8.0-12.5 | 0.4-0.8<br>3-5<br>8-12.5 |
| Detector & working temperature | PMT Insb (77 K) | Si line array HgCdTe (77 K) | HgCdTe (77 K) line array | Si+ HgCdTe (77 K) line array | PMT HgCdTe (77 K) | Si HgCdTe (77 K) |
| Record & Display | Multi-channel analog tape, CRT display, film producing | Analog tape, laser-desk, Multi-color R/T display | Analog tape, laser-desk, Multi-color R/T display | Analog tape, laser-desk, Multi-color R/T display | Analog tape, Multi-color R/T display, film producing | Multi-color R/T display, Transmission in TV format |
| Flying parameter | | display & record | display & record | display & record | display | display |
| On-board processing | | programmable in bands | programmable in bands | programmable in bands | | |

**Table 111:    Overview of Chinese Scanners in the 1986-1990 Period**


## B.28.6   CAS-SAR (Chinese Academy of Sciences SAR)

CAS-SAR was developed by the Electronics Institute of CAS and is operated by IRSA-CAS (Institute of Remote Sensing Applications of CAS), Beijing, China. The SAR instrument is flown on a Cessna Citation S2 aircraft along with other instruments (cameras: Wild RC-10 and RC-10A, an imaging spectrometer (MAIS), etc.).[425]

Application: flood control (disaster) monitoring, geological exploration, a possibility of vegetation discrimination is considered.

The aircraft is equipped with a right- and a left-looking radar antenna mounted on platforms under the fuselage. The SAR instrument inside the plane may be switched in-flight to either one of these antennas, permitting a corresponding swath switch. The PRF is governed by the ground speed of the aircraft ($V_g$).

---

425)  Information provided by Tong Qingxi of ISRA, Chinese Academy of Sciences (CAS)

| RF center frequency | 9.375 GHz (X-Band) | |
|---|---|---|
| Wavelength | 3.3 cm | |
| IF bandwidth | 30 MHz | |
| Pulse repetition frequency (PRF) | $6.4\,V_g$ | |
| Polarization | HH, VV, HV, VH, in-flight configurable | |
| Looking modes | left-side or right-side of track, in-flight changeable | |
| Antenna gain | $\geq 27$ dB | |
| Receiver noise | $\leq 5$ dB | |
| Spatial resolution | 10 m x 10 m (range x azimuth), independent of altitude | |
| Data recording | on film | |
| Dynamic range of imaging film | $\geq 15$ dB | |
| Gray level of imaging film | 32 | |
| Geometric distortion | $\leq \pm 3\%$ | |
| Ground speed of aircraft $V_g$ | 130 - 200 m/s | |
| Platform | Cessna Citation S/II | |
| Data transmission capability: real-time air-ground in S-Band (for flood disaster monitoring) | | |
| Data rate | 90 Mbit/s | |
| Nominal flight altitude | 5 - 10 km | |
| Survey swath width | 35 km | |
| Working mode | Flight altitude (km) | Swath width (km) |
| A | < 6 | 15 - 50 |
| B | 6 - 10 | 25 - 60 |
| C | 6 - 10 | 35 - 70 |
| D | 6 - 10 | 45 - 80 |

**Table 112:**     **Specification Parameters of CAS-SAR**

## B.29    CHRISS (Compact High Resolution Imaging Spectrograph Sensor)

CHRISS is a commercially-developed-and-built-to-order high resolution hyperspectral imaging instrument by SAIC (Science Applications International Corporation of San Diego, Ca.) for SETS Technology Inc., Hawaii. The system was developed during 1991-92 and is scheduled for customer delivery in April 1994. There is one flight model of CHRISS with two different configurations, one is referred to as the IR&D (Internal Research and Development) prototype configuration, the second one is called the SETS configuration. Both instrument configurations are flown (1993) on a light twin-engine aircraft (Piper Aztec).[426]

CHRISS consists of the following system components:
- a fixed focal length camera lens front end
- an all-reflective imaging spectrograph to disperse and re-image the radiation ($\sim 10^\circ$ FOV).
- a CCD digital camera in the output plane of the spectrograph
- a video frame storage system
- a CCD TV color camera and SVHS tape recorder (for data correlation, processing)
- a GPS navigation system (aircraft)

CHRISS is a nadir-pointing CCD Pushbroom 2-D imaging spectrometer with an all-reflective design, capable of imaging contiguously a spectral range from the ultraviolet to the infrared region with a conventional blazed diffraction grating. Images are acquired by a strip of n spatial elements (e.g., pixels in the cross-track direction) in many (simultaneous and contiguous) spectral bands, inherently registered - thereby forming an image frame - m image frames (in the flight direction) make up an image cube.

The instrument is configurable with regard to the spatial and spectral resolutions. Within the spatial dimension between 192 and 385 pixels may be chosen, in the spectral dimension

---

426) Courtesy of B. A. Speer of SAIC

the number of bands may range from 79 to 144. (Note: The active chip area of the CCD camera contains 385 spatial pixels x 288 spectral pixels, the spectral dimension is bisected with a quantization of 12 bit).

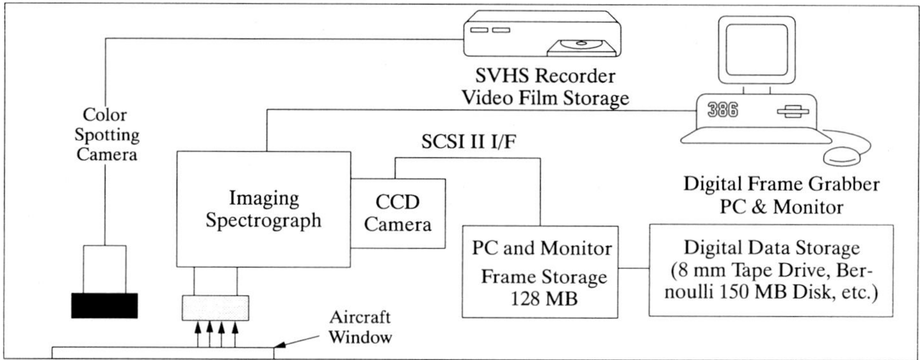


**Figure 114:** **Flight Data Recording System of CHRISS**

The spatial/spectral sampling varies according to the CCD camera format selected for a particular application. The CCD camera frame rate and noise characteristics can also be matched to the mission and platform to provide the optimum resolution and SNR. Frame rates are typically ≥ 30 frames per second. The spectrograph has a limiting blur of approximately 40 to 80 μm in the visible range, providing up to 385 channels (pixels) in the 20 mm slit plane. The spectrograph may be tuned to a range in the VIS/NIR from approximately 400 to 900 nm, consistent with blaze wavelength and a single octave spectral range. The instantaneous spectral range is tailored to the 7 mm spectral format of the spectrograph; grating dispersion is currently set to provide approximately 430 nm over the 7 mm format. Figure 114 depicts the systems general arrangement. Table 113 summarizes the system parameters for the CHRISS SETS configuration.

Possible instrument applications: surveying for petroleum seepage, vegetation identification, forestry inventory surveys, ocean color monitoring, environmental monitoring.

| Optical Parameters | | Spectral Format | |
|---|---|---|---|
| First focal length | 100 mm | Spectral range | 430 - 860 nm |
| First focal ratio | f/4 | Spectral channels[428] | 40 |
| Spectrograph magnification | 1.0 | Image plane dispersion | 63 nm/mm |
| | | Channel bandwidth | 11 nm |
| | | CCD Array[429] | |
| Along-Track FOV | | Operational Parameters | |
| Slit width | 50 μm | Altitude | 1000 m |
| Geometric IFOV | 500 μrad | Ground speed (0 wind) | 40.24 m/s |
| | | Geometric IFOV | 0.5 m |
| Cross-Track FOV | | Frame rate | 45/s |
| Image plane format | 16.9 mm | Along-track sample | 0.89 m |
| FOV | 169 mrad | Lateral GSD (Ground Sample Distance) | 0.88 m |
| Spatial Channels[427] | 192 | Swath width | 169 m |
| Spatial Channel IFOV | 44 μm | Integration time | 21.32 ms |

**Table 113:** **CHRISS System Parameter Setup for the SETS Configuration**

427) Two, 44 μm pixels are binned on-chip during readout to produce a single spatial pixel.
428) Two, 44 μm spectral pixels are binned to produce a single 11 nm channel FWHM.
429) The CCD has 288 x 385 active area binned 2 spatial x 4 spectral (192 x 72). Forty (40) spectral channels are within the 7 nm spectral format and are used read out, the remaining 32 are dumped. Quantization is 12 bits per pixel; the data rate is 1.37 MByte/s.

## B.30    CNC (Condensation Nucleus Counter)

CNR is an airborne instrument of the University of Denver, Co. (PI: J. C. Wilson). The instrument became operational in the early 1980s, it is flown on ER-2 aircraft and has been utilized in the following campaigns/studies: Aerosol Climatic Effects Study, Stratosphere-Troposphere Exchange Project, Airborne Antarctic Ozone Experiment, and the Airborne Arctic Stratospheric Expedition (AASE).

CNC measures the number concentration of aerosol particles having diameters in the range of 0.02 to about 1.0 μm. The instrument operates at altitudes from 8 to 21.5 km.[430]

The instrument functions by saturating the aerosol sample with warm alcohol vapor and then cooling the sample so that the alcohol vapor condenses on particles in the sample, causing them to grow to the sizes which can easily be detected by a simple optical particle counter.

Data quality: Laboratory studies have been made of the response of the instrument as a function of size and pressure. The precision of CNC has been checked in the lab and was found to be within a few percent of that implied by counting statistics.


## B.31    Condor-APV (Autonomously Piloted Vehicle)

Condor is a high-altitude research aircraft built by Boeing for DOD and acquired by the Lawrence Livermore National Laboratory (LLNL) in April 1993.[431]

Condor is a HALE (High Altitude Long Endurance) type unmanned aircraft designed, built and tested by Boeing during the late 1980's. It demonstrated its capabilities during a 141 hr test flight program (between October 1988 and November 1989) that included a flight to an altitude of 67,028 ft (20 km) and endurance times over 58 hours, carrying payloads in excess of 700 kg. Its design envelope includes a 70,000 ft ceiling altitude and mission durations in excess of 150 hrs. Condor's wings, with a total span of 200 ft, use graphite/Kevlar, epoxy sandwich, and Nomex honeycomb construction resulting in a weight distribution of 2 lb/ft$^2$. The wing design is a Liebeck airfoil with a lift-to-drag ratio exceeding 40. The APV is powered by two six-cylinder opposed 175 hp liquid-cooled engines (Teledyne Continental). Two stages of turbo-charging maintain engine efficiency at high altitudes resulting in extremely low fuel consumption (0.38 lb/hr/hp) and hence in long-endurance flights. The engines drive three-blade composite propellers of 16 ft diameter, with a high aspect ratio and thrust efficiencies approaching 0.9. - The autonomous operation of the vehicle, including take-offs, landings, and emergency maneuvers, was successfully demonstrated utilizing two Delco Magic 3 flight control computers and a microwave landing system (MLS). As of 1993 investigations are underway to upgrade the navigation system with GPS and to utilize DGPS for take-offs and landings.

Condor will be utilized in the future by the Department of Defense (DOD) as well as by the civilian sector (NASA, NOAA, NCAR, and others) for Earth observing applications. The Condor owner/operator and program coordinator is in all cases LLNL. The potential high-altitude research applications for Condor are numerous. It may be used as a HALE platform to test ballistic missile defense weapons including lasers and interceptors; it may also be used for various surveillance needs, including radar ocean imaging and wake detection. The fields of atmospheric science, environmental- and weather monitoring are considered prime candidates for a science support program (in situ measurements and remote observations made over many diurnal cycles to study the chemistry and dynamics of the upper troposphere and stratosphere). As of 1993 a science program for Condor is in the planning phase along with the corresponding payload candidates.

---

430)  Information provided by R. F. Pueschel of NASA/ARC
431)  Courtesy of T. W. Lawrence of LLNL

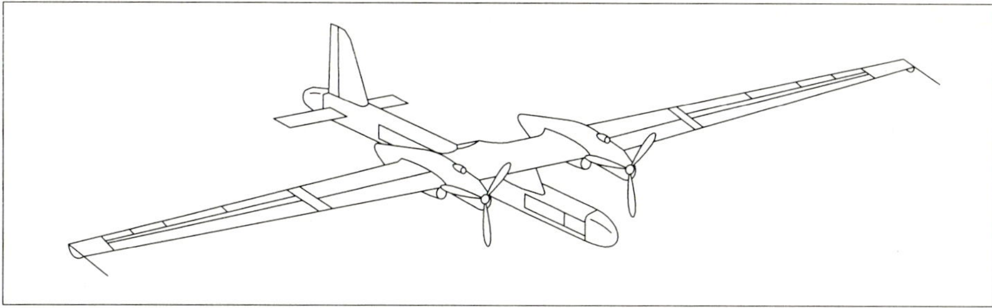

**Figure 115:    Illustration of the Condor-APV**

## B.32    C/X-SAR

C/X-SAR[432],[433] is a CCRS (Ottawa, Canada) airborne sensor (built by MacDonald Dett-wiler, and by Canadian Astronautics Ltd) aboard a Convair 580 aircraft. C/X-SAR is used as a research tool, there are also contractual flight assignments for individual customers with specific needs. The aircraft and instrument system is being operated for CCRS under contract by industry (Innotech Aviation Ltd. with Intera Technologies Ltd.). The C/X-SAR system is also commercially available.

Applications: imaging of terrain, ocean or ice scenes, monitoring of resources (renewable such as agriculture and forestry, or non-renewable such as geological resources), stereo imaging (part of special applications, different incidence angles are used to create parallax for targets, however, the stereo imaging capability is not being used by CCRS). Preparatory support for the spaceborne Radarsat sensor.

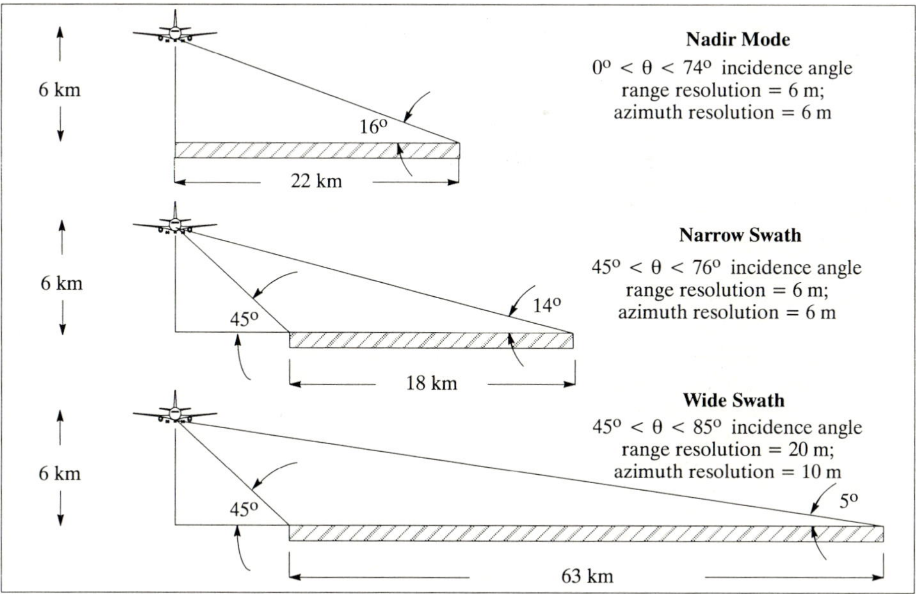


**Figure 116:    Geometries of C/X-SAR Standard Operating Modes**

---

432)  "CCRS Airborne C/X-SAR", brochure of CCRS, 1988
433)  Information brochure provided by Innotech Aviation

| Parameter | C-Band System | | X-Band System | |
|---|---|---|---|---|
| **Transmitter** | | | | |
|   Frequency | 5.30 GHz | | 9.25 GHz | |
|   Wavelength | 5.66 cm | | 3.24 cm | |
|   Transmitter peak power | 64, 16, 1 kW | | 6 , 1 kW | |
|   Polarization | VV, HH, VH, HV | | VV, HH, VH, HV | |
|   Polarization cross-coupling | < -40 dB | | < -40 dB | |
|   Pulse repetition frequency | 2.32 or 2.57 Hz/m/s | | 2.32 or 2.57 Hz/m/s | |
|   Estimated noise equivalent backscatter coefficient | -40 dB | | -33 dB | |
| Resolution modes | High | Low | High | Low |
|   chirp length | 7 μs | 8 μs | 15 μs | 30 μs |
|   chirp coding | nonlinear FM | linear FM | linear FM | linear FM |
| **Receiver** | | | | |
|   A/D converter dynamic range | 30 dB | | 30 dB | |
|   Resolution modes | High | Low | High | Low |
|     compressed pulse width | 40 ns | 120 ns | 32 ns | 134 ns |
|     noise figure | 5.2 dB | 3.7 dB | 5.3 dB | 5.2 dB |
| **Antenna** | | | | |
|   Polarization | Horizontal | Vertical | Horizontal | Vertical |
|     azimuth beam width | 3° | 3.3° | 1.4° | 1.4° |
|     elevation beam width | 28° | 25° | 26° | 26° |
|     peak gain | 26 dB | 24.8 dB | 28 dB | 28.5 dB |

| Recording | |
|---|---|
| **Signal** <br>   4096 range cells <br>   8-bit, in-phase and quadrature components <br>   1 channel full swath, 2 channel half swath <br>   not azimuth processed <br>   range compressed, motion compensated <br>   annotation in record header | **Image** <br>   4096 or 2048 range cells <br>   8 bits (magnitude) per range cell <br>   1 channel <br>   annotation in auxiliary data block <br>   output from same cha. as dry-silver paper <br> **Video** <br>   512 range cells per range line <br>   fully annotated <br>   NTS-C (VHS) format |

| Real-Time Processing | | | |
|---|---|---|---|
| **Hard-copy Display** <br>   dry-silver paper <br>   2000 pixels per range line <br>   maximum line rate 39/s <br>   number of grey levels 16 <br>   full header annotation, time and position blocks | Resolution ($\tau$) <br> high <br> low | Azimuth ($\tau_a$) <br> 6 m <br> 10 m | Range ($\tau_r$) <br> 6 m <br> 20 m |
| slant range or ground range corrected <br> 4096 8-bit pixels per range line | azimuth-compressed 7 looks <br> 1 channel per radar | | |

**Table 114:    Technical Specifications of C/X-SAR**

C/X-SAR Features:
- Mapping to the left or right of the aircraft
- Slant range or ground range correction
- Video monitoring and recording of real-time imagery
- C- and X-Band receivers each with dual channel receivers
- Digital motion compensation across the swath
- Dynamic antenna stabilization
- Variable range-dependent gain for different applications (e.g. ocean, land)
- Radar and navigation parameter logging
- Built-in test equipment for receiver calibration
- Operation in three selectable modes: nadir, narrow swath, and wide swath
- As of 1993 the C/X SAR instruments can be configured in 3 different ways to perform special SAR imaging functions. The original configuration uses the two radars as synchronized, C- and X-Band polarization diversity SARs. Two additional experimental configurations have been added: C-Band polarimetry and C-Band interferometry.

The C/X-SAR instruments are computer-controlled and use identical algorithms to generate the specified configurations. These specifications include: the desired imaging function (dual frequency, polarization diversity, polarimetry, interferometry, etc.), the desired analysis objective, the scene contents, and the predetermined targets of interest within a scene.

The two radars (of C/X-SAR) are constrained to operate in the same imaging geometry by the commonality of the antenna drive system and are constrained to operate at the same pulse repetition frequency (synchronized) by the common image data recording system. All other parameters can be adjusted independently.

## B.33 Daedalus Instruments (Digital Multispectral Scanner)

Commercially available instrument package of Daedalus Enterprises of Ann Arbor Michigan, USA. As of 1993 Daedalus has delivered over 85 instruments/systems to its customers into 24 countries.[434]

### B.33.1 ATM (Airborne Thematic Mapper)

ATM and ATMX (extended) systems are the most widely used airborne line-scanners in the world.

Note: the ATMX (AADS1278) is designed for higher altitude applications and special spectrometer configurations that require additional collecting aperture to achieve a higher sensitivity than is provided by the standard ATM (AADS1268) system. The electronics include all the the ATM features, the scan head is designed to interface to the ATM, TIMS, or AOCI.

History: Daedalus built the AADS1260 sensor model from 1976 to 1982. This instrument preceded the era of the TM sensor on Landsat. The AADS1260 instrument enables users to emulate the spectral bands of the Landsat-4 MSS sensor from an airborne platform. The 4 channels of the MSS satellite are obtained by combining 7 of the 11 channels contained in the AADS1260 airborne multispectral system.

| Owners/Operators of Daedalus AADS1260 Systems | Aircraft |
|---|---|
| Argentine Air Force; Parana, Argentina | Lear 35 |
| Canada Centre for Remote Sensing (CCRS), Ottawa, Canada | Falcon 20-C |
| Dept. of Conservation and Environment, Melbourne, Australia | King Air |
| Energy & Resources Labs (ERL)/Industrial Technology Research Institute (ITRI), Chutung, Hsinchu Taiwan | King Air |
| Environmental Protection Agency (EPA), Las Vegas, Nevada, USA | Aero Commander |
| Eurosense - Belfotop N.V., Wemmel, Belgium | DO228 / Cessna 404 |
| NOAA, Rockville, Md, USA | |
| | |
| Owners/Operators of Daedalus ATM & ATMX Systems | Aircraft |
| Alenia of Rome, Italy | Partenavia |
| Centre for Remote Sensing in Geology, Beijing, China | Twin Otter |
| CSIRO Office of Space Science & Applications, Canberra, Australia | King Air |
| DLR, Institute of Optoelectronics, Oberpfaffenhofen, Germany | DO 228 |
| EG&G Energy Measurements, Inc., Las Vegas, Nevada, USA | Cessna Citation II |
| Natural Environmental Research Council (NERC), Swindon, UK | Beech Queen Air/Aero Com. |
| Instituto Nacional de Technica Aeroespacial (INTA), Madrid, Spain | Casa 212 |
| Instituto Geografico Militare (IGM), Florence, Italy | Lear 25 |
| NASA/AMES Research Center, Moffett Field, Ca., USA (ATMX) | ER-2 |
| National Remote Sensing Agency (NRSA), Hyderabad, India | King Air 200 |
| Italian Merchant Marine Ministry, Rome, Italy | Piaggio 166 |
| Telespazio S. p. A., Rome, Italy (ATMX) | Partenavia |

**Table 115:   Owners/Operators of Daedalus AADS1260, ATM and ATMX Systems**

---

434) Information brochures provided by Daedalus Enterprises, Inc., Ann Arbor, Michigan 48106-1869, USA

The AADS1268 was given the acronym "ATM" (Airborne Thematic Mapper), it was developed to match the Landsat TM sensor bands. This instrument (AADS1268) includes spectral channels that match 6 of the 7 bands contained in the Landsat TM, and 4 additional visible and NIR channels. The thermal band (channel 6 of TM) was broadened for airborne operations. The spectral match of bands enables users to compare and complement data from both systems. The ATM now covers channels of Landsat MSS, TM and of Spot (HRV).

ATM (AADS1268) is an opto-mechanical linescanner with 8 bit data quantization. The ATMX (AADS1278) is similar but has a large collecting aperture and can be configured for 10 bit data quantization on selected channels. The total mass of of the system is about 200 kg.

| Channel Nr. ATM | Channel Nr. Landsat TM | Spectral Band ( μm) |
|---|---|---|
| 1 |  | 0.420 - 0.450 |
| 2 | 1 | 0.450 - 0.520 |
| 3 | 2 | 0.520 - 0.600 |
| 4 |  | 0.605 - 0.630 |
| 5 | 3 | 0.630 - 0.690 |
| 6 |  | 0.695 - 0.750 |
| 7 | 4 | 0.760 - 0.900 |
| 8 |  | 0.910 - 1.050 |
| 9 | 5 | 1.550 - 1.850 |
| 10 | 7 | 2.080 - 2.350 |
| 11 | 6 | 8.500 - 12.50 |

**Table 116:     Spectral Channels of the ATM and TM Sensors**

| Measurement Mode | Wide Angle | Normal Angle |
|---|---|---|
| FOV | 85.92° | 42.96° |
| IFOV | 2.5 mrad | 1.25 mrad |
| Swath width (at 1000 m altitude)<br>Swath width (at 20 km altitude of ER-2) | 1860 m | 787 m<br>15.4 km |
| Spatial resolution at 1000 m altitude (nadir)<br>Spatial resolution at 20 km altitude of ER-2 | 2.5 m x 2.5 m | 1.25 m x 1.25 m<br>25 m x 25 m |
| Nr. of pixels per scan line | 716 | 716 |
| Scan rates (scan/s or lines/s) | 12.5, 25, 50, 100 | 12.5, 25, 50 |
| Roll correction | ± 15° | ± 15° |

**Table 117:     Geometric Characteristics of the ATM**

The AADS1280 scanner is a one-of-a-kind instrument built in 1978 for NASA for the purpose of crop classification. It then became the property of DGGTN (Direction General de Geografica del Territorio Nacional) of Mexico to detect poppy crops. The instrument is flown on a Lear 35 aircraft.

| Channel | Wavelength (μm) | Channel Width (μm) |
|---|---|---|
| 1 | 0.54 - 0.59 | 0.05 |
| 2 | 0.62 - 0.67 | 0.05 |
| 3 | 0.74 - 0.80 | 0.06 |
| 4 | 0.84 - 0.92 | 0.08 |
| 5 | 0.95 - 1.06 | 0.11 |

**Table 118:     Spectral Channels of the AADS1280 Scanner**

## B.33.2   Analog Bispectral Instruments

The following instruments are in this category: AADS1220, AADS1221, and AADS1230.

- **AADS1220 (Dual Channel Terrain Surveillance Scanner)**

This passive sensor uses TIR (Thermal Infrared) line scan techniques for wide-area day/night monitoring of land and water surfaces of interest. The system senses and records relative temperature differentials of objects in the target area and translates the resultant signal differences into a near photo-quality black and white image (on-board film recorder).

The two-channel design offers "hot spot" and fire detection capabilities. By combining the two detector signals, it is possible to create a high resolution image (MCT detector) with a very sensitive hot spot or fire detection capability (InSb detector).

Applications: Environmental monitoring, energy loss detection, geological explorations, forest fire monitoring, volcanology, etc.

- Operating wavelengths     3.0 - 5.5 μm, and 8.0 - 13.0 μm
- Aperture     12.7 cm
- Focal length     15.2 cm
- Optical aperture (effective)     f/2
- Scan rate     160 scans/s
- TFOV     87° 20'
- Gated FOV     77° 20'
- IFOV     1.7 mrad
- Temperature resolution     0.2 °C
- Velocity/Height ratio     0.26 radians/s
- Roll correction     total: ±10°, unvignetted ±5°

- **AADS1221 (Maritime Surveillance Scanner)**

The AADS1221 IR/UV instrument is a proven and operational sensor [being offered by SSC (Swedish Space Corporation) as a licensee of Daedalus, see B.78] for monitoring of oil slicks (spreading of oil and measurement of the relative thickness of the oil film), in support of cleanup operations.

The IR/UV instrument is used at close range to obtain high-resolution imagery. The passive sensor measures reflected radiation (in IR and UV bands) from the sea surface. A mirror scanning normal to the aircraft flight direction allows for mapping line by line, yielding images 400 m wide from a 300 m flight altitude, with a ground resolution of a few meters.

The IR/UV signals are presented in real-time on a split-screen format color TV monitor. There is support for quicklook documentation, data recording, air-to-ground image transfer, a provision of SLAR instrumentation (Ericsson, see B.78.1), and camera instrumentation.

- Operating wavelengths     0.32 - 0.38 μm (UV), and 8.5 - 14 μm (IR)
- System sensitivity     NEΔR ≤0.06% (UV), NEΔT ≤0.2 °C (IR)
- IFOV     5 mrad
- TFOV     87°
- Scan rate     160 scans/s
- Velocity/Height ratio     0.4 radians/s
- Roll correction     output signal for: ±10°

- AA2000 is identical to AADS1221 except that it includes an image printer to allow it to be operated as a "stand-alone" system.

- **AADS1230 (Dual Channel Quantitative Infrared Scanner)**

- Operating wavelengths     0.35 - 13.0 μm
- Aperture     12.7 cm
- Focal length     15.2 cm
- Optical aperture (effective)     f/2
- Scan rate     80 scans/s
- TFOV     87° 20'

| - | Gated FOV | 77° 20' |
| - | IFOV | 2.5 mrad |
| - | Temperature resolution | 0.2 °C |
| - | Velocity/Height ratio | 0.2 radians/s |
| - | Roll correction | total: ± 10°, unvignetted ± 5° |

With regard to the AADS1230, 1220, 1221, and AA2000 systems, Daedalus offers optional detectors to fill the two ports. The customer can thus acquire combinations of 3.0-5.5 µm, 0.4-1.1 µm, and 0.32-0.38 µm (UV). Normally these systems are operated with one thermal detector plus one UV detector (for oil spill monitoring).

| Owners/Operators | System | Aircraft |
|---|---|---|
| Canada Centre for Remote Sensing (CCRS), Ottawa, Canada | AADS1230 | Falcon 20-C |
| Centre for Remote Sensing in Geology, Beijing, China | AADS1230 | Citation II/ Twin Otter |
| Clyde Surveys Ltd.; Berkshire, UK | AADS1230 | Beechcraft Queen Air |
| Committee for Scientific Research, Portugal | AADS1220 | CASA 212 |
| German Federal Ministry of Transportation (BMV) | AADS1221 | DO 228 (German Navy) |
| Impressa Luigi Rossi, Brescia, Italy | AADS1230 | Cessna 402 |
| Indian Coast Guard | AADS1221 | DO 228 |
| Industroprojekt; Zagreb, Yugoslavia | AADS1230 | DO 228 |
| Institute Geographique National (IGN), Creil, France | AADS1230 | B-17 |
| Intera-Kenting, Ontario, Canada | AADS1230 | Cessna 402 |
| LAPAN, Jakarta, Indonesia | AADS1230 | Beech D18 |
| Ministry of Defense, Middlesex, UK | AADS1220 | Britton-Norman Islander |
| National Land Survey of Sweden, Gaevle, Sweden | AADS1230 | |
| Norwegian State Pollution Control Authority, Oslo, Norway | AADS1221 | |
| Ontario Centre for Remote Sensing, Ontario, Canada | AADS1230 | Piper Navajo Chieftain |
| Portuguese Air Force, Amadora, Portugal | AADS1221 | CASA 212 |
| Rijkswaterstaat, Netherlands | AADS1220 | Cessna 212 |
| Rinaldo Piaggio / Italian Merchant Marine, Rome, Italy | AA2000 | Piaggio P-166 |
| Royal Thai Air Force, Bangkok, Thailand | AADS1220 | Merlin IV |
| Swedish Coast Guard, | AADS1221 | Cessna Skymaster / CASA 212 |
| Swedish Space Corporation, Solna, Sweden | AADS1220 | Cessna 337 |
| Tennessee Valley Authority, Chattanooga, TN, USA | AADS1230 | Rockwell Grand Commander 680 |
| Texas Aerial Surveys, Houston, TX, USA | AADS1230 | |
| United States Geological Survey (USGS), Denver Co., USA | AADS1230 | |

**Table 119:    Daedalus Analog Bispectral Instrument Owners**

## B.33.3    Analog and Digital Bispectral/Multispectral Instruments

The following instruments are in this category: AADS1250,  and AA3500.

- AADS1250 (Eleven Channel Analog Multispectral System)

The system consists of a dual optical path single aperture scanner permitting simultaneous acquisition of 10 channels of calibrated reflected data in the spectral region of 0.38 - 1.10 µm, and one channel of calibrated infrared data in the 3.0-14.0 µm region. The system specifications correspond to those of the AADS1220 instrument.

- AA3500  = Airborne Bispectral Scanner (ABS)

ABS is a dual optical port, bispectral scanner consisting of a compact scan head with one or two detectors, and an electronics package. ABS provides 12-bit digital precision and a broad dynamic range with interchangeable detectors for a wide range of applications plus a continuous video monitoring and operator control. Standard 8 mm digital tape recorder.

The system performs simultaneous scanning for TIR and a second spectral region: either SWIR, VNIR, or UV. Dual blackbody reference sources permit calibration of the TIR channel.

Applications: ABS collects data for applications as diverse as geological mapping, vegetation studies, pollution monitoring, maritime surveillance, heat loss detection, etc.

| Spectral ranges (2 available) | TIR + (8.5-12.5 μm), | MWIR or (3.0-5.5μm) | VNIR or (0.4-1.1μm) | UV (0.32-0.38 μm) |
|---|---|---|---|---|
| IFOV<br>FOV<br>Scan rates | 2.5 (1.25 or 5.0 optional) mrad<br>86°<br>100, 50, 25, 12.5, 6.25 scans/s (operator selectable) | | | |
| Velocity/Height ratio (V/H)<br>Roll correction<br>Navigation data interface | 0.25 radians/s at 100 scans/s<br>± 15° of roll correction (automatic)<br>Data: date, time, ground speed, latitude, longitude | | | |
| Power requirements<br>Mass of instrument | 28 ± 3 VDC, 40 A continuous<br>81 kg | | | |

**Table 120:   Specifications of the ABS Sensor**

| Owners/Operators | System | Aircraft |
|---|---|---|
| Asia Air Survey Co., Tokyo, Japan | AA3500 | Aero Commander 685 |
| JPL/US Forest Service, Pasadena, CA, USA | AA3500 | Beechcraft King Air / Merlin IV |
| Rinaldo Piaggio/Guardia di Finanza, Finale Ligure, Italy | AA3500 | Piaggio P-166 |
| TERMA, Lystrup, Denmark | AA3500 | DO 228 |

**Table 121:   Daedalus Digital Bispectral Instrument Owners**

| Owners/Operators | System | Aircraft |
|---|---|---|
| Asia Air Survey Co., Tokyo, Japan | AADS1250 | Aero Commander 685 |
| Centre National d'Etudes Spatiales (CNES),Toulouse, France | AADS1250 | B-17 |
| Japan Weather Association, Tokyo, Japan | AADS1250 | Cessna 402-B |
| NASA/Kennedy Space Center, Florida, USA | AADS1250 | Beech Model 18 |
| Royal Thai Air Force, Bangkok, Thailand | AADS1250 | Merlin IV |
| Spacetec Datengewinnung GmbH; Freiburg, Germany | AADS1250 | |

**Table 122:   Daedalus Analog Multispectral Instrument Owners**

## B.33.4   AOCI (Airborne Ocean Color Imager Spectrometer)

AOCI[435] is a sensor development under NASA contract in response to user requests to emulate the spectral and radiometric characteristics of the spaceborne CZCS (Coastal Zone Color Scanner) on the Nimbus-7 satellite (launched in 1978) to expand the possibilities of biological oceanography. The CZCS instrument failed in 1986 on Nimbus-7, Sea-WiFS (see A.92) is the next NASA sensor to continue ocean color monitoring in the summer of 1994 (other spaceborne sensors in this category are OCTS on ADEOS with a launch in late 1995 and MERIS on ENVISAT-1 (see A.78.1) with a launch in 1998).

Application: AOCI is a prime tool in oceanographic research for the study of biomass, chlorophyll, fishery management, pollution detection, investigation of aquatic ecosystems for a wide variety of uses.

Owners/operators of AOCI instruments: NASA/AMES Research Center, Moffett Field, Ca., on ER-2 aircraft; Telespazio S. P .A., Rome, Italy on Partenavia aircraft.

---

435)  R. C. Wrigley, R. E. Slye, S. A. Klooster, R. S. Freedman, M. Carle, L. F. McGregor, "The Airborne Ocean Color Imager: System Description and Image Processing", Journal of Imaging Science and Technology, Vol 36, Nr. 5, Sept./Oct. 1992, pp. 423-430

Note: The NASA/AMES AOCI is actually the reconfigured AADS1278 (ATMX) instrument depending on the support/application requirements.

**CZCS (Nimbus-7)**

| Band Nr. | Band center (nm) | Band width (nm) | SNR |
|---|---|---|---|
| 1 | 443 | 20 | 158 |
| 2 | 520 | 20 | 200 |
| 3 | 550 | 20 | 280 |
| 4 | 670 | 20 | 176 |
| 5 | 750 | 100 | 118 |
| 6 | 11500 | 2000 | - |

**AOCI**

| Band Nr. | Band center (nm) | Band width (nm) | SNR |
|---|---|---|---|
| 1 | 444 | 23 | 450 |
| 2 | 490 | 20 | 1010 |
| 3 | 520 | 21 | 915 |
| 4 | 565 | 20 | 615 |
| 5 | 619 | 21 | 440 |
| 6 | 665 | 21 | 350 |
| 7 | 772 | 60 | 360 |
| 8 | 862 | 60 | 250 |
| 9 | 1012 | 60 | 120 |
| 10 | 10395 | 3900 | - |

**SeaWiFS (Seastar)**

| Band Nr. | Band center (nm) | Band width (nm) | SNR |
|---|---|---|---|
| 1 | 412 | 20 | 500 |
| 2 | 443 | 20 | 675 |
| 3 | 490 | 20 | 665 |
| 4 | 510 | 20 | 640 |
| 5 | 555 | 20 | 595 |
| 6 | 665 | 20 | 440 |
| 7 | 765 | 40 | 455 |
| 8 | 865 | 40 | 465 |

**Table 123:**     Spectral/Radiometric Characteristics of CZCS, AOCI and SeaWiFS Sensors[436]

The AOCI system was developed from the AADS1278 system. The AADS1278 optics system consists of a rotating scan mirror, a telescope, folding optics, and a field stop. A new AOCI spectrometer section replaces the original TM spectrometer section of the AADS1268. Note: Band number 9 and 10 of the AOCI provide 8-bit quantization, the rest 10-bit quantization.

| | |
|---|---|
| FOV | 1.49 rad (85°) |
| FOV (=swath width) at 20 km altitude | 37 km |
| FOV (=swath width) at 12.5 km altitude | 23 km |
| IFOV | 2.5 mrad |
| IFOV (=spatial resolution) at 20 km altitude | 50 m |
| IFOV (spatial resolution) at 12.5 km altitude | 31 m |
| Nr. of pixels per scan line | 716 |
| Scan rate | 6.25, 12.5, 25 scan/s |
| Quantization level | 10 bit (for band 1-8), 8-bit (for band 9-10) |

**Table 124:**     Geometric Characteristics of the AOCI Instrument

## B.33.5   AMS (Airborne Multispectral Scanner)

The AMS (AA3600) is a dual-port system which records up to six spectral channels simultaneously directly onto digital tape. The standard configuration offers a dual element thermal-infrared detector and an 8-channel VNIR spectrometer for choice from 10 spectral bands. Applications: Water mapping of chlorophyll, geologic mapping, suspended sediment and temperature conditions, forest inventory and crop vigor studies.

| AMS Systems | Aircraft |
|---|---|
| EG&G Energy Measurements Inc. (EG&G); Las Vegas, NV, USA | Cessna Citation II |
| Geoscan GmbH, Hildesheim, Germany | Cessna 404 |
| Korea Institute of Geology Mining and Minerals (KIGAM), Taejon, Korea | Aero Commander 690 |

**Table 125:**     Owners/Operators of AMS Systems

---

436) SeaWiFS band centers and band widths (at full width, half maximum) are given in nanometers, the SNRs use radiances expected at the top of the atmosphere.

| Spectral Channels | Channel Width (µm) |
|---|---|
| UV Channel (Optional with 5.0 mrad IFOV) | 0.32 - 0.38 (oil spill detection) |
| VNIR Spectrometer | 0.42 - 0.45<br>0.45 - 0.52<br>0.52 - 0.60<br>0.60 - 0.63<br>0.63 - 0.69<br>0.69 - 0.75<br>0.76 - 0.90<br>0.91 - 1.05 |
| IR Channels | 3.0 - 5.50<br>8.50 - 12.5 |
| Geometric Parameters | |
| IFOV | 2.5 mrad (1.25 optional) |
| FOV | 86° |
| Scan rates (scans/s) | 6.25, 12.5, 25, 50, 100 |
| Roll correction | ± 15° |
| Quantization level | 8 or 12 bit |

**Table 126:**   **Spectral and Geometric Characteristics of the AMS Instrument**


## B.33.6   TIMS (Thermal Infrared Multispectral Scanner)

TIMS is a digital airborne scanner (specified by JPL, developed by Daedalus) and an optional imaging spectrometer for the AADS 1278 system (ATMX) since 1981. TIMS provides 6 spectral bands in the thermal infrared region for geological applications.

A unique version of the TIMS with a larger collecting area was built for NASA/SSC (on Lear 23 aircraft) and is designated: AADS1285. The TIMS instrument is occasionally installed on the AMES C-130 or ER-2 aircraft for specific project support.

Applications: Monitoring of spectral emittance data on geologic materials (rock type discrimination, geological mapping, oil/mineral surveys, etc.). Since TIMS has on-board blackbody calibration sources, its use has been extended to surface temperature mapping and for thermal inertia studies.

| Channel | Detector | Spectral Region (µm) | NEΔT (Sensitivity) |
|---|---|---|---|
| 1 | HgCdTe | 8.2 - 8.6 | ≤ 0.3°C |
| 2 | HgCdTe | 8.6 - 9.0 | ≤ 0.3°C |
| 3 | HgCdTe | 9.0 - 9.4 | ≤ 0.3°C |
| 4 | HgCdTe | 9.40 - 10.2 | ≤ 0.3°C |
| 5 | HgCdTe | 10.2 - 11.2 | ≤ 0.3°C |
| 6 | HgCdTe | 11.2 - 12.6 | ≤ 0.3°C |
| Geometric Parameters | | | |
| FOV | 86° (80° for SSC TIMS) | | |
| IFOV | 2.5 mrad | | |
| Ground resolution | 7.6 m (at 3000 m altitude) | | |
| Pixels/scan line | 714 (638 for SSC TIMS) | | |
| Swath width | 4.8 km (at 3000 m altitude) | | |
| Scan rates (selectable) | 6.25, 12.5 and 25 scans/s (7.3, 8.7, 12, 25 for SSC TIMS) | | |

**Table 127:**   **Spectral Coverage and Geometric Parameters of the TIMS Instrument**


| TIMS Systems | Aircraft |
|---|---|
| Instituto Geografico Militare (IGM), Florence, Italy (AADS1278) | Lear 25 |
| NASA/ Stennis Space Center (SSC), Mississippi, USA (AADS1285) | Lear 23 |
| Telespazio S. P. A., Rome, Italy (AADS1278) | Partenavia |

**Table 128:**   **Owners/Operators of TIMS Systems**

## B.33.7   Wildfire

Wildfire is a Daedalus development for NASA (1987) with the objective to measure fires, as well as terrestrial and atmospheric parameters. The instrument is an optical spectrometer for the ATMX system. Wildfire provides 50 channels in the infrared region. It has the capabilities for infrared studies of geological and volcanic features as well as for environmental applications (PI: J. Brass and J. Arvesen at AMES).

| Spectrometer | Spectral Range ( µm) | Nr. of Channels | Band Width (µm) |
|---|---|---|---|
| 1 | 1.15 - 1.55 | 8 | 0.05 |
| 2 | 1.55 - 2.35 | 16 | 0.05 |
| 3 | 3.00 - 5.40 | 16 | 0.15 |
| 4 | 8.20 - 12.7 | 10 | 0.40 - 0.50 |

**Table 129:    Spectral Coverage of the Wildfire Instrument**

Owners/operators of Wildfire: NASA/AMES Research Center, Moffett Field Ca., on ER-2 aircraft.
The Wildfire instrument no longer exists (as of 1993), according to J. Myers of NASA/AMES, it was converted into the MAS [MODIS Airborne Simulator].

## B.33.8   MAS (MODIS Airborne Simulator)

MAS is a 50 channel modified spectrometer based on the AADS1278 scanner. The instrument was built by Daedalus for NASA/AMES Research Center and GSFC for an airborne simulator of the MODIS instrument (initial flights of MAS in 1992).[437] The system has successfully flown over 15 flights on the NASA ER-2 aircraft. MAS utilizes four detector arrays to simultaneously acquire data in the spectral range from 0.53 to 14.5 µm, with bands from 0.04 µm to 0.5 µm.

| Spectral channels | 50 (17 terrestrial, 23 Fire) |
|---|---|
| Output channels | 12, 8-bit (optional 8 10-bit plus 2 8-bit), Future modification: 50 channels at 16-bit digitization in 1994 |
| IFOV<br>Spatial resolution (ground)<br>Total scan angle<br>Swath width<br>Pixel/scan line<br>Scan rate<br>Roll correction | 2.5 mrad<br>50 m (at 20 km spacecraft altitude)<br>85.92°<br>37 km at 20 km altitude<br>716<br>6.25, 12.5, 25 scan/s<br>± 15° |

**Table 130:    MAS Sensor Parameters**

## B.33.9   MIVIS (Multispectral Infrared and Visible Spectrometer)

MIVIS is a new Daedalus development for geologic and environmental investigations. The first instrument delivery with a 102-channel configuration is scheduled for the Fall of 1993 for operational use. MIVIS provides the following features:

* a large area scan mirror for increased energy gathering power
* a common field stop design to ensure co-registration of all channels in the described configuration (providing spatially and spectrally coregistered images)
* a four-port spectrometer assembly with spectral coverage in the VIS,NIR, IR, and TIR regions
* 12 bit quantization (touch panel for operator interaction)

437) M. King, D. Herring, "The MODIS Airborne Simulator (MAS)", The Earth Observer, Vol 4, Nr. 6, 1992, pp. 15-19, the paper provides a historical background of MAS

- VLDS tape recorder system
- a companion digital image processing system known as MIDAS (Multispectral Interactive Data Analysis System)

| Spectrometer | Spectral Range ( μm) | Nr. of Channels | Band Width (μm) |
|---|---|---|---|
| 1 | 0.433 - 0.833 | 20 | 0.02 |
| 2 | 1.15 - 1.55 | 8 | 0.05 |
| 3 | 2.00 - 2.50 | 64 | 0.008 |
| 4 | 8.200 - 12.70 | 10 | 0.40 - 0.50 |
| Total = 102 channels, expandable to 128 channels | | | |

**Table 131:    Example of a MIVIS Spectral Coverage**

Owners/operators of MIVIS: Consiglio Nazionale Delle Richerche (CNR), Rome, Italy; on Casa 212 aircraft.

| Record band (bits) | Band Nr. | Band Center (μm) | Band Width (μm) | Spectral Range (μm) | SNR |
|---|---|---|---|---|---|
| 8 | 1 | 0.547 | 0.043 | 0.529 - 0.572 | 130 |
|  | 2 | 0.664 | 0.055 | 0.635 - 0.688 | 260 |
|  | 3 | 0.707 | 0.042 | 0.688 - 0.729 |  |
|  | 4 | 0.745 | 0.040 | 0.729 - 0.769 |  |
|  | 5 | 0.786 | 0.040 | 0.770 - 0.810 |  |
|  | 6 | 0.834 | 0.042 | 0.810 - 0.852 |  |
| 8 | 7 | 0.875 | 0.041 | 0.852 - 0.893 | 270 |
|  | 8 | 0.910 | 0.031 | 0.896  0.927 | 80 |
| 8 | 9 | 0.945 | 0.043 | 0.926 - 0.969 | 180 |
| 8 | 10 | 1.623 | 0.057 | 1.595 - 1.652 | 200 |
|  | 11 | 1.680 | 0.050 | 1.655 - 1.705 |  |
|  | 12 | 1.730 | 0.050 | 1.705 - 1.755 |  |
|  | 13 | 1.780 | 0.050 | 1.755 - 1.805 |  |
|  | 14 | 1.830 | 0.050 | 1.805 - 1.855 |  |
|  | 15 | 1.880 | 0.050 | 1.855 - 1.905 |  |
|  | 16 | 1.930 | 0.050 | 1.905 - 1.955 |  |
|  | 17 | 1.980 | 0.050 | 1.955 - 2.005 |  |
|  | 18 | 2.030 | 0.050 | 2.005 - 2.055 |  |
|  | 19 | 2.080 | 0.050 | 2.055 - 2.105 |  |
| 8 | 20 | 2.142 | 0.047 | 2.126 - 2.173 | 187 |
|  | 21 | 2.180 | 0.050 | 2.155 - 2.205 |  |
|  | 22 | 2.230 | 0.050 | 2.205 - 2.255 |  |
|  | 23 | 2.280 | 0.050 | 2.255 - 2.305 |  |
|  | 24 | 2.330 | 0.050 | 2.305 - 2.355 |  |
|  | 25 | 2.380 | 0.050 | 2.355 - 2.405 |  |
|  | 26 | 3.000 | 0.150 | 2.925 - 3.075 |  |
|  | 27 | 3.150 | 0.150 | 3.075 - 3.255 |  |
|  | 28 | 3.300 | 0.150 | 3.255 - 3.375 |  |
|  | 29 | 3.450 | 0.150 | 3.375 - 3.525 |  |
|  | 30 | 3.600 | 0.150 | 3.525 - 3.675 |  |
| 10 | 31 | 3.725 | 0.151 | 3.659 - 3.810 | 115 |
|  | 32 | 3.900 | 0.150 | 3.825 - 3.975 |  |
|  | 33 | 4.050 | 0.150 | 3.975 - 4.125 |  |
|  | 34 | 4.200 | 0.150 | 4.125 - 4.275 |  |
|  | 35 | 4.350 | 0.150 | 4.275 - 4.325 |  |
|  | 36 | 4.503 | 0.142 | 4.436 - 4.578 |  |
|  | 37 | 4.651 | 0.150 | 4.582 - 4.732 |  |
|  | 38 | 4.800 | 0.150 | 4.725 - 4.875 |  |
|  | 39 | 4.950 | 0.150 | 4.875 - 5.025 |  |
|  | 40 | 5.100 | 0.150 | 5.025 - 5.175 |  |
|  | 41 | 5.250 | 0.150 | 5.175 - 5.325 |  |
| 10 | 42 | 8.563 | 0.396 | 8.342 - 8.738 | 71 |
|  | 43 | 9.642 | 0.426 | 9.451 - 9.877 | 65 |
|  | 44 | 10.438 | 0.466 | 10.259 - 10.275 | 73 |
| 10 | 45 | 11.002 | 0.448 | 10.791 - 11.239 | 95 |
| 10 | 46 | 12.032 | 0.447 | 11.799 - 12.246 | 79 |
|  | 47 | 12.775 | 0.447 | 12.539 - 12.986 | 79 |
| 10 | 48 | 13.186 | 0.352 | 13.023 - 13.375 | 78 |
| 8 | 49 | 13.952 | 0.517 | 13.630 - 14.147 | 77 |
|  | 50 | 14.302 | 0.358 | 14.163 - 14.521 | 63 |

**Table 132:    Spectral Coverage of the MAS Instrument**

## B.34   DLR Lidar Instruments

As of 1993, DLR has four operational airborne lidar systems that are used intermittently for a number of research applications, including studies/demonstrations for the usefulness of spaceborne backscatter lidars. The sensor is considered as part of an airborne test program for the ESA spaceborne ATLID (ATmospheric LIDar) sensor.

- **ALEX** = Aerosol Lidar Experiment (see table 133). ALEX is a nadir-looking, three-wavelength Nd:YAG sensor. The instrument participated in the ELAC 1990 campaign.

- **OLEX** = Ozone-Aerosol Lidar Experiment (see table 134).[438],[439]

- **H$_2$O-DIAL** = Water Vapor Differential Absorption Lidar (see table 135). The system is based on a frequency-doubled Nd:YAG pumped narrow-band dye laser for online and offline measurements.

- **Microlidar** (see table 136). Microlidar is a diode-pumped Nd:YAG laser with a high repetition rate. The sensor is considered as part of an airborne test program for the ESA spaceborne ATLID sensor.

| Lidar instrument | **ALEX = Aerosol Lidar Experiment**, operational at DLR (Institute: Physics of the Atmosphere) since 1979 |
|---|---|
| Measurement technique | Incoherent backscatter |
| Parameters or constituents measured | Aerosols and clouds |
| Measurement range | 0 - 30 km |
| Vertical resolution | 3 m - 12 m |
| Frequency of measurement (typically) | 4 - 5 periods of 2 weeks per year |
| Measurement times (typically) | 1 hr - 10 hr, day and night |
| Laser type and transmitter wavelength | Nd:YAG: 1064 nm, 532 nm, and 354 nm |
| Pulse repetition frequency | 5 - 10 Hz |
| Laser energy/pulse | 500 mJ at 1064 nm; 150 mJ at 532 nm; 70 mJ at 354 nm |
| Airborne platform of instrument | Falcon 20, DO 228, and Transall |
| Receiver size and configuration | 35 mm diameter, Cassegrain telescope |
| Receiver FOV | 1.5 mrad (full angle) |
| Receiver bandwidth | 10 nm for 1064 nm frequency, 3 nm for 532 nm and 354 nm frequency |
| Detectors used | PMT (R928 Hamamatsu) |
| A/D converter | LeCroy 8-bit, 100 Hz, 4-channel, polarization capability |
| Research objectives/applications | 2-D aerosol and boundary layer structures, clouds, cirrus, volcanic aerosols, PSCs (Polar Stratospheric Clouds) |

**Table 133:**    **Specification of the ALEX Instrument**[440][441]


438)  W. Renger, G. Ehret, P. Mörl, "Airborne Lidar for Atmospheric Research in the Arctic", 15. ILRC Conference Abstracts, Tomsk, Russia, 1990, pp. 74-76

439)  M. Wirth, W. Renger, G. Ehret, "Airborne DIAL Remote Sensing of the Arctic Ozone Layer", 16th International Laser Radar Conference, Cambridge, Ma. USA, Conference Abstracts, pp. 107-108, 1992

440)  J. Fischer, W. Cordes, A. Schmitz-Pfeiffer, W. Renger, P. Mörl, "Detection of Cloud Top Height from Backscattered Radiances within the Oxygen A Band - Part 2: Measurements", JAS, Vol. 30, 1991, pp. 1260-1267

441)  M. Kästner, K. T. Kriebel, R. Meerkötter, W. Renger, G. H. Ruppersberg, P. Wendling, "Comparison of Cirrus Height and Optical Depth Derived from Satellite and Aircraft Measurements", submitted to Monthly Weather Review, Florida State University, Tallahassee, 1992

| Lidar instrument | **OLEX = Ozone-Aerosol Lidar Experiment**; operational at DLR (Institute: Physics of the Atmosphere) since 1990 |
|---|---|
| Measurement technique | Incoherent backscatter |
| Parameters or constituents measured | Ozone, aerosols and clouds |
| Measurement range | 0 - 30 km |
| Vertical resolution | 3 - 12 m rangebins, ozone: 500 - 1000 m |
| Frequency of measurement (typically) | 3 - 4 periods of 2 week campaigns per year |
| Measurement times (typically) | 1 hr - 10 hr, day and night |
| Laser type and transmitter wavelength | Nd:YAG: 532 nm, and 354 nm; XeCl: 308 nm |
| Pulse repetition frequency | 5 - 10 Hz |
| Laser energy/pulse | 150 mJ at 532 nm; 70 mJ at 354 nm; 150 mJ at 308 nm |
| Airborne platform of instrument | Transall (BMVg) |
| Receiver size and configuration | 35 mm diameter, Cassegrain telescope |
| Receiver FOV | 1.5 mrad (full angle) |
| Receiver bandwidth | 3 nm for 532 nm frequency, 10nm for 354 nm and 308 nm frequency |
| Detectors used | PMT (R928 Hamamatsu) |
| A/D converter | LeCroy 8-bit, 100 Hz, 4-channel |
| Research objectives/applications | 2-D distributions of volcanic aerosols, PSCs (Polar Stratospheric Clouds) and ozone in the stratosphere |

**Table 134:     Specification of the OLEX Instrument**

| Lidar instrument | **$H_2O$-DIAL = Water Vapor Differential Absorption Lidar**; operational at DLR (Institute: Physics of the Atmosphere), 1988 |
|---|---|
| Measurement technique | Incoherent backscatter |
| Parameters or constituents measured | Water vapor, aerosols and clouds |
| Measurement range | 0 - 10 km |
| Vertical resolution | 10 - 15 m for aerosols, 300 - 1000 m for water vapor |
| Frequency of measurement (typically) | one week periods for about 5 times per year |
| Measurement times (typically) | 1 hr - 3 hr, day and night |
| Laser type and transmitter wavelength | Tunable dye laser around 720 nm pumped by a frequency-doubled Nd:YAG laser |
| Pulse repetition frequency | 10 Hz |
| Laser energy/pulse | 30-40 mJ at 724 nm |
| Airborne platform of instrument | Falcon 20, NCAR Electra |
| Receiver size and configuration | 35 mm diameter, Cassegrain telescope |
| Receiver FOV | 1 - 3 mrad adjustable |
| Receiver bandwidth | 0.6 nm |
| Detectors used | PMT (R928 Hamamatsu) |
| A/D converter | Transiac 20, 10 MHz |
| Total instrument mass | 240 kg |
| Research objectives/applications | 2-D distributions of water vapor and aerosols, boundary layer and cloud top heights, cirrus clouds |

**Table 135:     Specification of the $H_2O$-DIAL Instrument**[442],[443],[444]


442) G. Ehret, C.. Kiemle, W. Renger, G. Simmet, "Airborne Remote Sensing of Tropospheric Water Vapor Using a Near Infrared DIAL System, submitted to Appl. Opt., 1992

443) G. Ehret, W. Renger, "Atmospheric Aerosol and Humidity Profiling Using an Airborne DIAL System in the Near IR, OSA Tech. Digest, Optical Remote Sensing of the Atmosphere, paper ThAG, 1990, pp. 586-589

444) G. Ehret, C. Kiemle,W. Renger, G. Simmet, "Airborne  remote sensing of tropospheric water vapor with a near-infrared differential absorption system", Applied Optics, Vol. 32, Nr. 24, 20 August, 1993, pp. 4534-4551

| Lidar instrument | **Microlidar**; operational at DLR (Institute of Opto-Electronics) since 1988 |
|---|---|
| Measurement technique | Incoherent backscatter |
| Parameters or constituents measured | Dense clouds, especially under multiple scattering conditions |
| Measurement range | 0 - 1 km |
| Vertical resolution | 1.5 m |
| Frequency of measurement (typically) | 2 - 3 periods of 1 week each per year |
| Measurement times (typically) | 1 hr - 4 hr, day and night |
| Laser type and transmitter wavelength | Diode pumped Nd:YAG: 1064 nm |
| Pulse repetition frequency | 100 Hz |
| Laser energy/pulse | 1.5 - 4 mJ (tunable) |
| Airborne platform of instrument | Falcon 20 |
| Receiver size and configuration | 35 mm diameter, Cassegrain telescope |
| Receiver FOV | 6 mrad (full angle); 18 mrad (separate detection of multiple scattering) |
| Detectors used | 4 PIN photodiodes |
| A/D converter | 2 x Philips 10-bit, 100 MHz, 2-channel |
| Research objectives/applications | distributions of dense clouds, cirrus; dense fog (ground based) |

**Table 136:    Specification of the Microlidar Instrument**[445],[446],[447]


## B.35    DOE Airborne Instruments in ARM Program

The US Department of Energy (DOE) and the Strategic Environmental R&D Program (SERDP) are sponsoring and developing four airborne instruments for use on Unmanned Aerospace Vehicles (UAVs) such as PERSEUS. The research activities are considered as part of the ARM (Atmospheric Radiation Measurement) program for atmospheric modeling. Major measurement objectives are: radiation balance determination at the tropopause, flux profiles through the troposphere, cloud top properties, and upper tropospheric water vapor profiles.

All instruments will be integrated into the UAV payload by Sandia National Laboratories, Livermore, Ca. with the instrument developers serving as PIs and mentors. The scientific data will be made available both to the ARM Science Team as well as to the general scientific community.


### B.35.1    MPIR (Multispectral Pushbroom Imaging Radiometer)

As of 1993 MPIR is being designed and built at the Sandia National Laboratories in Albuquerque NM (PI: G. Phipps). Objective: investigation of radiation/cloud interactions. First test flights with MPIR are planned for 1995.[448]

The instrument offers nine spectral bands between 0.58 and 11.5 μm with wavelengths chosen for the retrieval of cloud reflectivity, cloud droplet phase (ice/water) and size. An objective is also to provide radiometric calibration for spaceborne sensors, such as AVHRR. MPIR generates multispectral images by using linear CCD detector arrays in a pushbroom imaging arrangement. Each channel (band) of MPIR is independent, with its own detector array, filter, all reflective optics, and electronics. The detector and optics package for each

445)  W. Krichbaumer, "Airborne Cloud Measurements with the DLR-Microlidar during the CLEOPATRA Campaign", submitted to JTech, 1992
446)  Ch. Werner, J. Streicher, H. Herman, H. G. Dahn, "Multiple-Scattering Lidar Experiments, Opt. Eng. Vol. 31, 1992, pp. 1731-1745
447)  W. Krichbaumer, et al., "A diode-pumped Nd:YAG lidar for airborne cloud measurements", Optics & Laser Technology, Vol. 25, Nr. 5 , 1993, pp. 283-287
448)  Information provided by J. Vitko of Sandia National Laboratories, Livermore Ca.

of the nine spectral channels is being designed as a modular unit that can be interchanged with any of the other detector units. To change wavelengths one simply exchanges the detector modules. This design method thus allows not only cloud radiometry, but also such applications as crop assessment, detection of environmental contaminations, etc.

Instrument calibration: a mix of pre-, post-flight and on-board calibration will be used to attain calibration goals of 1% in the thermal- and 3% in the visible spectrum.

| | |
|---|---|
| Spectral bands (μm) | 0.58 - 0.68; 0.74 - 0.76; 0.93 - 0.95; 1.12 - 1.16; 1.35 - 1.41; 2.11 - 2.22; 3.5 - 4.0; 8.7 - 9.3; 10.0 - 11.5 |
| Linear detector arrays (9) with LN$_2$ cooling | 256 elements per array |
| Detector material | Si < 1 μm<br>InGaAs (or PbS), up to 2.2 μm<br>PbSe (or InSb or HgCdTe) at 3.5 μm<br>HgCdTe from 8.7 to 12 μm |
| FOV | 80° |
| IFOV | 6 mrad at FOV center, 8 mrad at FOV edge |
| Output | 2 samples per second |
| Quantization | 12 bit |
| Data rate | 55 kbit/s |

**Table 137:    Specification Parameters of the MPIR Instrument**


## B.35.2    CDL (Cloud Detection Lidar)

CDL is an active instrument under development (1993) at the Lawrence Livermore National Laboratories (LLNL) in Livermore, Ca. CDL is a small backscatter lidar for detecting and profiling thin cirrus clouds and for accurate determination of cloud top altitudes. The instrument is intended to fly on small UAV's, such as PERSEUS. Test flights are planned for Spring 1995 (PI: A. Ledebuhr).

Energy source: Raman-shifted, flash-pumped Nd:YAG laser (the Raman cell is a pressurized methane cell, which shifts the laser wavelength from the 1.06 μm characteristic of an Nd:YAG to the 1.54 μm operating wavelength). This 1.54 μm operating wavelength qualifies CDL as an eye-safe lidar, where eye safety limits are 4 to 5 orders of magnitude higher than at a wavelength of 1.06 μm. The laser provides a pulse energy of 10 - 20 mJ with a repetition rate of 1 Hz. The aperture of the receiver is between 20-25 cm in diameter for the detection of thin clouds with an optical density of 0.03 and a minimum thickness of 100 m. The telescope is of a simple parabolic reflector design with an f/number close to 1. CDL has a full aperture narrowband filter and a detector mounted at the prime focus of the primary optics. Instrument requirements call for an in-flight control capability to operate CDL in either nadir- or zenith-viewing mode.

- Detector type: ADP (InGaAs)
- Vertical resolution: 100 m
- Receiver FOV: 0.8 mrad
- Instrument mass: < 20 kg
- No polarization capability is currently planned


## B.35.3    HONER (Hemispherical Optimized Net-flux Radiometer)

HONER is under development at the Los Alamos National Laboratories (LANL), Los Alamos, NM (PI: P. La Delfe). The objective is to make high-accuracy net radiative flux measurements for climate modeling. Test flights are planned for 1995.

The net flux is the difference between the up- and down-welling radiation. The knowledge of the net flux between two altitudes represents the heating rate of that layer. However, the quantity (net flux) is difficult to measure since it is the difference of two large numbers. HONER uses a novel optical differencing technique to measure this net flux directly.

The HONER measurement technique is based on an optical integrating sphere with two apertures: one looking downward and the other upward. The radiation through each aperture is chopped at the same frequency, but out of phase, and then is incident on filtered pyroelectric detectors. Pyroelectric detectors have the intrinsic property of being sensitive only to the AC (Alternating Current) component of the signal. Hence, if the phase difference between the chopped upward and downward radiation is 180°, then HONER gives a direct measure of the net flux. If the phase difference is selected to be 90°, then HONER measures the absolute fluxes (in quadrature). The HONER measurement principle should be able to obtain net fluxes with an accuracy of 3%.

| Spectral ranges (total broadband) | < 0.3 - 4 µm, and 4 - > 50 µm |
|---|---|
| FOV | > 170° (upward and downward) |
| Net flux accuracy | 3% |

- Optical differencing for common mode rejection
- Direct difference measurements between upwelling and downwelling fluxes
- Frequency and phase multiplexed optical signals for continuous data acquisition and calibration, simultaneously
- Material absorption filtering for stability. Additional spectral channels can easily be added.
- Insensitive to drift
- In-flight conversion from differential mode to simultaneous measurement of upwelling and downwelling fluxes
- Low production cost

**Table 138:    Specification Parameters and Features of HONER**

In practice, HONER consists of two integrating spheres. The instrument provides broadband, hemispherical flux and net flux measurements of both solar and thermal fluxes. HONER offers in-flight calibration with two calibration sources (blackbodies in IR and lamp/small integrating sphere combinations in the solar portions of the spectrum).

## B.35.4   UAV-AERI (UAV Atmospheric Emitted Radiance Interferometer)

The instrument is a miniaturized version of existing ground-based AERIs that are utilized for DOE climate programs [one is currently operational at the U. of Wisconsin, the other at DOE's Cloud and Radiation Testbed (CART) site in north/central Oklahoma]. As of 1993, the UAV-AERI sensor is being developed and built at CIMSS of the University of Wisconsin, Madison (PI: H. Revercomb). The objective is to provide contributions for the study of radiative properties in various fields:

- Atmospheric spectroscopy
- Cloud microphysics and radiative transfer modeling
- Atmospheric chemistry (direct observations of radiatively active trace gases)
- Spectral radiative forcing of dynamic processes

UAV-AERI measures the upwelling and downwelling radiances over a spectral range from 3 to 20 µm with a spectral resolution of 0.5 cm$^{-1}$ or better and with a spatial footprint of 1 km diameter from 10 km altitudes.

The instrument is in its early stages of design and development (end of 1993) and will be available for flight in 1996. UAV-AERI uses a modified commercial Michelson interferometer (repackaged for small UAV), calibration sources are provided to ensure the required radiometric accuracy.

- Spectral range: 3 - 20 µm (3333 cm$^{-1}$ - 500 cm$^{-1}$)
- Spectral resolution: 20 nm (0.5 cm$^{-1}$)

## B.36    DO-SAR (Dornier SAR)

The DO-SAR instrument was developed and built by Dornier (DASA) during the period 1986-89 and became operational in 1989. The objective is gain SAR sensor experience with respect to high-resolution imaging technology. The instrument has been flown since then on DO 228 and Transall aircraft.[449]

The main features of the instrument are:

- high resolution and MTI (Moved Target Indication) capability
- multifrequency and multipolarization capability (simultaneous operation)
- interferometric SAR modes in C-Band (azimuth and elevation)
- programmable chirp signals and pseudo-noise coding

The operational frequencies are 5.3 and 9.6 GHz, representing centimeter wavelengths, and 35 GHz, representing millimeter wavelengths. An inertial measurement unit within the antenna mount is used for motion compensation. The measured radar echo data is augmented by parallel measurements of a GPS receiver, a barometer, and by autofocus techniques. The correction of the flight path end velocity is done in realtime by the on-board computer. The swath width depends on the number of selected modes (parallel operation) and the desired resolution. The swath width is limited to 10 km by the tape recorder data rate of 100 Mbit/s.

| Parameter | C-Band | X-Band | Ka-Band |
|---|---|---|---|
| Center Frequency | 5.3 GHz | 9.6 GHz | 35 GHz |
| Wavelength | 5.6 cm | 3.1 cm | 8.5 mm |
| Polarization | VV, VH, HH, HV | VV, VH, HH, HV | VV |
| Transmitter peak power | 1 kW | 2 kW | 1 kW |
| Interferometric modes | azimuth basis =0.6 m; elevation basis = 1 m | | |
| Bandwidth | 50, 100, 200, or 400 MHz | | |
| Pulse Repetition Rate | 1.5, 3.0, or 6.0 kHz | | |
| Resolution | < 1 m for 4 looks | | |
| Quantization | 8 bit (I and Q) | | |
| Number of samples | 1024, 2048, or 4096 | | |
| Data rate | 100 Mbit/s | | |
| Swath width | 300 m - 10 km (depending on altitude) | | |
| Operating altitude | 100 m - 4 km | | |
| Incidence angle range | 30° - 89° | | |
| Pulse generation | programmable (digital signal generator) | | |
| Operating modes | a maximum of 4 different radar modes can be supported (data limited) example: C-Band (VV+HH)+ Ka-Band (VV+VV) or: C-Band (VV+VH+HH+HV) | | |
| Positioning | integrated GPS | | |
| Motion compensation | integrated INS, GPS and barometer, supported by autofocus | | |
| Image processing | integrated online multilook SAR processor programmable SAR ground processor SAR processing software for UNIX | | |
| Storage media | Ampex DCRSi cassette for raw radar data Video-8-cassette recorder for video data DAT-cassette for online-processed SAR images CCT for ground processed SAR images Diskettes for INS- and GPS-data Diskettes for log- and system files | | |

**Table 139:    Specification of the DO-SAR Instrument**

Three antenna platforms/configrations are available for DO-SAR:

- An X-Band reflector antenna together with a C-Band reflector antenna (only for Transall aircraft operation)

---

449)  Information provided by B. Fritsch and R. Röber of Dornier

- Three slotted waveguide array antennas for stereo- and MTI-SAR applications, they can only be mounted on a DO-228 aircraft. One of these antennas is mounted in the first door, and two in the second door. The Ka-Band antenna is mounted in the second door as well. Furthermore, a C-Band reflector antenna is mounted in the tailboard of the aircraft.
- The third antenna configuration can be mounted onto either aircraft type. It consists of the Ka-Band antenna and two slotted waveguide array antennas (C-Band) for vertical and horizontal polarization measurements.

The equipment of the C-, X-, and Ka-Band radars, the data unit/recorders, and the SAR processor are installed in the aircraft cabin.

## B.37    DRA-SAR (Defense Research Agency SAR)

Two SAR instruments (one X-Band and one S-Band) developed by DRA (Defense Research Agency) at Malvern, UK. The two SARs are installed on different aircraft. The X-Band system has been operational since 1983, the C-Band system became operational in 1993. Both radar systems are used for research into information extraction from images, targets and background. They are both available for civil use, the principal field in the civil context is the development of tools for remote sensing.[450],[451]

| Parameter | C-Band SAR | X-Band SAR |
|---|---|---|
| RF center frequency | 5.7 GHz | 9.65 GHz |
| Wavelength | 5.3 cm | 3.1 cm |
| IF center frequency | 300 MHz | 300 MHz |
| System bandwidth | 90 MHz | 100 MHz |
| SAW chirp signal bandwidth | 90 MHz | 100 MHz |
| Expanded pulse length | 5-12 µs | 5 µs |
| Compressed (analog) pulse length | 10 or 100 ns | 10 ns |
| Digital chirp signal bandwidth | 90 MHz | 100 MHz |
| Operating range | < 10 km | 10 to > 50 km |
| Swath width | 6 km | 6 km |
| Number of range samples | 4096 | 4096 |
| Range sampling interval | 1.5 m | 1.5 m |
| Azimuth sampling interval | > 3/16 | > 3/16 |
| Azimuth sampling rate | adjustable | adjustable |
| Flight altitude | < 3.5 km | < 14 km |
| Angle of incidence (from grazing) | 0 - 70° | 0 - 30° |
| Antenna gain | 19.9 dB | 35 dB |
| Antenna 3 dB beamwidth,    azimuth<br>elevation | 8°<br>24° | 1°<br>10° |
| Transmit peak power | 9.4 W | 50 kW |
| Receive noise | 8.5 dB | 8.5 dB |
| Nominal PRF (Pulse Repetition Frequency) | 10 kHz | 1 kHz |
| Quantization (bit/sample, I & Q) | 16 | 16 |
| Sampling rate (MHz) | adjustable | adjustable |
| Nominal data rate on high density DCRS | 80 Mbit/s | 80 Mbit/s |
| Maximum recording time per tape | 90 min | 90 min |
| Spatial resolution, range x azimuth | 2.2 m x 2 m | 2.5 m x 1 m |
| Number of statistically independent looks | 8 | 1 |
| Radiometric distortion | 1 in 10000 | 1 in 10000 |
| Polarization | full | HH |
| Along-track velocity | 62 m/s | 200 m/s |

**Table 140:    Specification of the DRA-SAR Instruments**

---

450)  Information received from C. J. Oliver of DRA, Malvern
451)  C. J. Baker, A. M. Horne, R. G. White, "Grazing Angle Dependency of SAR Imagery", RADAR 92, pp. 359-362

**SAR Image processing**

SAR data processing of the DRA-SAR instruments is performed on the ground. The facilities at DRA include a real-time rate programmable processor based on the parallel Meiko CS-1 computer (2.5 Gflop/s) providing in effect real-time focused, linear, image formation.

The image formation algorithms developed at DRA include features for SAR image auto-focussing and phase correction. There are also a number of tools for image interpretation such as (note: the algorithms are available under license from DRA for implementation on neural networks or on single- and multi-processors):

- Despeckling
- Intensity segmentation
- Target change detection
- Target super-resolution
- Clutter simulation
- Texture segmentation and classification

## B.38 Dual Polarized 37 GHz Radiometer

A passive NASA/GSFC instrument flown on a C-130 aircraft at the Wallops Flight Facility and operational since 1992. Objective: measurements of canopy characteristics (fractional ground cover, water content, and structural attributes, e. g., foliage and woody stems).[452]

The instrument employs noise injection with null balance for measuring the antenna energy. Dual-polarized (H and V) coincident observations are performed; the center frequency is 37 GHz; the bandwidth 800 MHz; the 3 dB beamwidth is 6°, with a beam efficiency of 90%. The temperature sensitivity is 1 K. The instrument may be also be mounted on a truck.

## B.39 DUTSCAT (DUT Airborne Radar Scatterometer)

The airborne radar scatterometer DUTSCAT[453] is an instrument for measuring radar backscatter from the Earth's surface. It operates in six frequencies between 1 and 18 GHz. DUTSCAT was developed by the Laboratory for Telecommunications and Remote Sensing Technology of the Electrical Engineering Department of the Delft University of Technology (DUT) in cooperation with NLR, it is owned by BCRS. DUTSCAT is installed in a Beech-craft Queen Air research aircraft of NLR.

| Parameter | Specification |
| --- | --- |
| Radar Type | coherent pulse radar |
| Frequencies | 1.2, 3.3, 5.3, 9.65, 13.7, and 17.25 GHz |
| Pulse repetition frequency | 78.125 kHz |
| Pulse width | 100 ns |
| Peak power | 250 mW |
| Operating range | 50 - 1920 m |
| Polarization | VV and HH |
| Incidence angle | 0 - 80 ° |
| Data acquisition | 8 bits 40 M-samples/s |
| Output | A-scan 5 scans/frequency |
| Aircraft speed | 110 knots |

**Table 141:   DUTSCAT Specifications**

The DUTSCAT system uses a 0.9 m diameter parabolic dish antenna mounted on a support structure and protected by a radome. The high-frequency parts of the radar (transmitters/

---

452) Information provided by B. J. Choudhury of NASA/GSFC
453) "DUTSCAT - Airborne Radar Scatterometer", NLR/TU Delft brochure

receivers) are mounted in the support structure. DUTSCAT can be tilted between 0-80° from vertical, looking to the left of the aircraft flight direction.

The facility has been used since November 1983 for investigations of land and sea in a single-frequency mode. Since August 1986 the instrument is operational in its six-frequency configuration.

The scatterometer data are recorded in a digital format on magnetic tape. Aircraft position, attitude, speed and altitude data are recorded on the same tape simultaneously.


## B.40    EDOP (ER-2 Doppler Radar)

EDOP is a NASA-sponsored X-Band Doppler instrument designed and developed by GSFC and nose-mounted into the ARC ER-2 aircraft, it is operational since September 1993. The instrument has two antennas: one is nadir-pointing with pitch stabilization, the other antenna is forward pointing. The EDOP system took its first reflectivity data in the CAMEX campaign during September/October 1993; Doppler measurements will be taken during subsequent flights. Objectives: measurement of the vertical structure of air motions and precipitation in mesoscale convective systems; development of combined passive microwave/radar algorithms for precipitation estimation (related to the TRMM mission (A.111) PR (Precipitation Radar) sensor development and GEWEX).[454]

EDOP measures high resolution time-height sections of reflectivity and vertical hydrometeor velocity (and vertical air motion when the hydrometeor fall speed and aircraft motions are removed). The forward beam measures in addition the linear depolarization ratio (LDR), which provides orientation information of the hydrometeors (i.e. the canting angle), hydrometeor phase, size, etc. The dual beam geometry has advantages over a single beam. Horizontal air motions along the aircraft track can be calculated by using the displacement of the aircraft to provide dual Doppler velocities (forward and nadir looks) at a particular altitude.

EDOP has been designed as a turn-key system with real-time processing on-board the aircraft. The R/F system consists of a coherent frequency synthesizer which generates the transmitted and local-oscillator frequencies used in the system, a pulse modulator (PIN diode) which converts the transmitted frequency into a 0.25 - 2.0 µs pulse signal, a high gain 20 kW Traveling Wave Tube Amplifier (TWTA) which is coupled through the duplexer to the antenna, and the receiver which is comprised of a low-noise ($\sim$ 2 dB) GaAs preamplifier, mixer, IF amplifier and synchronous detectors for each of the antennas. The composite system generates a nadir oriented beam with a co-polarized receiver and a 35° forward-directed beam with co- and cross-polarized receivers. The antenna gain consists of two separate offset-fed parabolic antennas with tri-mode feed horns mounted in the nose radome of the ER-2. The two beams operate simultaneously from a single transmitter.

The system obtains high vertical resolution profiles (37.5 m spacing) of radar reflectivity, vertical velocity, Doppler spectral width, and linear depolarization ratio (forward beam only). The antenna footprint on the ground is about 1.2 km in diameter (with a 2.9° beamwidth) for flight altitudes of 20 km.

EDOP is furnished with a real-time processor to accommodate the very high data and processing rates required by the system's 4400 Hz pulse repetition frequency, the 4 linear and 3 logarithmic analog output channels, and the 0.25 µs range resolution. The input data rate is 56 MByte/s, the required processing rate is 380 MFLOPS to calculate reflectivity, Doppler velocity, and spectral width at 0.1 s intervals (440 pulses) and 900 range gates. The data system also provides automatic gain control to achieve > 100 dB dynamic range.

---

454)  G. Heymsfield, NASA/GSFC, "TOGA/COARE Mission Plan", NASA paper, November 1992, pp. 51-54

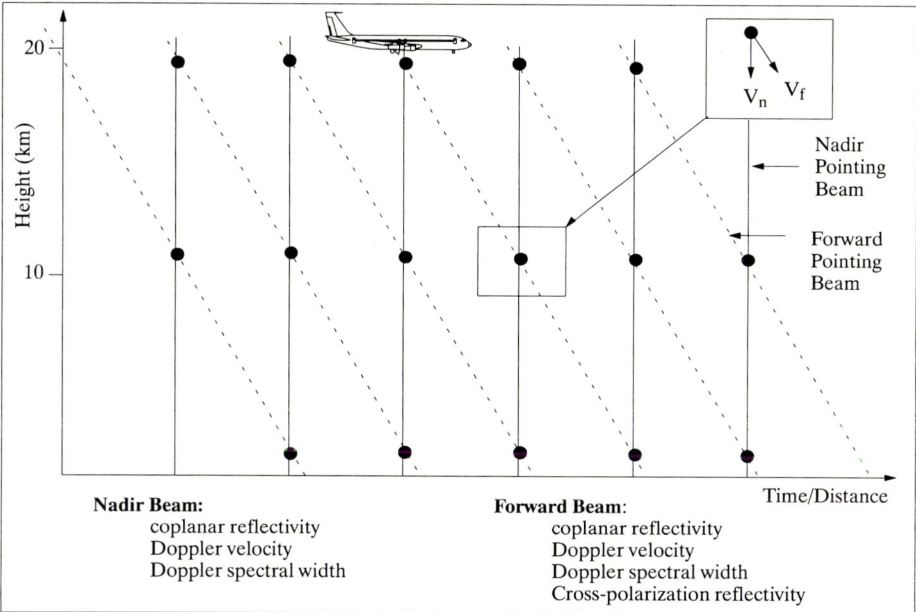



**Figure 117:**    **Measurement Geometries of EDOP**

| Center frequency | 9.6 GHz (X-Band) | Antenna diameter | 0.76 m |
|---|---|---|---|
| Peak power | 20 kW | Antenna beamwidth | $2.9^o$ |
| Duty cycle | 0.01 max. | First sidelobe level | $\leq 30$ dB |
| Pulse length | 0.25, 1.0 ms | Cross-polarization level | $\leq 38$ dB |
| PRF | 2200, 4400 | Receiver dynamic range | 110 dB |
| Nr. of Doppler channels | 2 | Nr. of logarithmic re-flectivity channels | 3 |
| Nadir beam:<br>  transmit polarization<br>  received polarization | horizontal<br>co-polarized | Forward beam:<br>  transmit polarization<br>  received polarization | vertical<br>co-polarization and<br>cross-polarization |

**Table 142:**    **Specification of the EDOP Instrument**

# B.41   ELDORA/ASTRAIA

The ELDORA[455],[456] (ELectra DOppler Radar) / ASTRAIA (Analyse Stereoscopique par Radar Aeroporte sur Electra) is an X-Band, dual beam, airborne Doppler radar that has been jointly built by the NCAR (Boulder Co.) and by CRPE, Paris (France). Background: The initial designs of ELDORA by NCAR and of ASTRAIA by CRPE began in 1984. The efforts of both parties were combined in 1988 for a joint development of a single airborne radar system. NCAR provided the transmitter receiver, signal processor, data system, and aircraft, CRPE provided the antenna/rotodome and the intermediate frequency signal processor. Development of the instrument was largely completed in 1991. A first test of a complete, but simplified radar occurred in 1991. In June 1992, two complete radars were tested in the lab, each using a three frequency, stepped waveform. Flight testing occurred until the end of 1992, the first TOGA/COARE experiment was supported in January/February 1993.

---

455) P. H. Hildebrand, J. Testud, "ELDORA/ASTRAIA Capabilities for TOGA COARE", paper provided by P. H. Hildebrand

456) P. H. Hildebrand, C. A. Walther, Ch. Frush, J. Testud, "The ELDORA/ASTRAIA Airborne Doppler Weather Radar: Goals Design, and First Field Tests", submitted Nov. 4, 1993 to Proceedings of IEEE

Scientific objectives: Measurement of storm structures and kinematics (radial velocity and reflectivity) with an accuracy level to be able to derive thermodynamic properties of the storm (buoyancy, heat transport, water transport). These measurements are required over large domains, over remote domains, and at high enough sampling density to resolve the important scales of motion and structure.

Measurement requirements:
- The temporal and spatial data sampling needs range greatly for different types of clouds. For a large storm system, e.g. hurricanes and meso-scale convective systems, the spatial resolution requires 0.5 - 0.8 km in the horizontal and 0.3 - 0.4 km in the vertical direction with measurements completed in a few seconds.
- Observation of isolated cumulonimbi require measurements each 0.3 km in the horizontal and vertical plane with measurements completed in about 3 minutes.
- For small cumulus requirements call for a spatial resolution of 0.1 km in both planes in a time frame of about 1 minute.
- Stratiform cloud systems require resolutions slightly lower than those of cumulonimbus.
- Measurement accuracy for radial velocities: $\pm 0.3$ - 0.5 m/s and for reflectivities within $\pm 1$ - 1.5 dBZ (note: Z is the radar reflectivity factor, a meteorological radar intensity scale).

| Parameter | TOGA/COARE | Design Goal |
|---|---|---|
| Wavelength | 3.2 cm | 3.2 cm |
| Transmit frequency | 9.35 and 9.45 GHz | 9.2 - 9.8 GHz |
| Beamwidth | 1.8° circular | 1.8° circular |
| Antenna gain | 38.75 dB, 38.7 dB | 40 dB |
| Polarization (antenna) | horizontal | horizontal |
| Sidelobes | -23 to - 30 dB | -40 dB |
| Beam tilt angle (fore or aft) | $\pm 18.5°$ | $\pm 15$ - 19° |
| Antenna spin axis parallel to | heading | heading |
| Dwell time | 15 ms | 7-50 ms |
| Rotational sampling rate | 1° | 1° |
| Peak transmit power | 35 - 50 kW | 35 - 50 kW |
| Minimum detectable signal at 10 km: | -12 dBZ | -12 dBZ |
| Receiver bandwidth | 0.5 - 8 MHz | 0.5 - 8 MHz |
| Receiver temperature at antenna | < 600 K | < 600 K |
| Pulse repetition frequency | 2000 pps | 2000 - 5000 pps |
| Unambiguous range | 55 km | 20 - 90 km |
| Unambiguous velocity | 16 m/s | 13 - 20 m/s |
| Number of radars | 1 (switched) | 2 (one fore, one aft) |
| Number of transmit frequencies per radar | 2 | 3 |
| Pulse chip length | 1 µs | 0.1 - 3.0 µs |
| Range averaging | 1 chip | 1 - 4 chips |
| Total gate length | 0.15 km | 0.015 - 1.2 km |
| Elevation step | 1° | 1° |
| Along-track beam spacing | 0.72 and 1.44 km | 0.3 - 1.0 km |

**Table 143:    Radar Characteristics in TOGA/COARE versus Design Goals**

The ELDORA/ASTRAIA design features a dual-beam, fore-aft scanning technique permitting the radar (two complete radar systems) to collect dual Doppler information in flight. The scan pattern enables collection of dual fore- and aft-pointing Doppler velocity data, the analysis of this data provides 2-D wind fields on observation surfaces passing through the storm. The instrument makes use of a complex transmitted waveform which consists of 3 - 5 different frequencies. In addition the radar transmits a shorter pulse than is needed in the output data sample. These data are averaged in the digital signal processor to provide the required additional samples.

The radar consists of 5 major functional blocks: the RF signal generator/receiver unit, the high power amplifiers, the signal processor, the antenna/rotodome system, and the radar control equipment. The two 35 kW peak power traveling wave tube amplifiers are driven by a common synthesizer (a high-precision 10 MHz crystal oscillator). The synthesizer generates all the needed reference frequencies for the fore and aft radar transmitters. The 60 MHz intermediate frequency (IF) is generated by the RF signal generation hardware and is used by the receivers, the IF signal processors, the master timing module, and the digitizers. The transmitted signal consists of pulses which contain sub-pulses, or chips for each of the three to five separate transmitted frequencies. The transmitted pulses can be transmitted at staggered intervals in order to assist with range-velocity ambiguity reduction. - The transmitted signal passes through a dual channel rotary joint to separate fore and aft antennas which are mounted in the rotodome. The rotodome spins about the aircraft longitudinal axis at a rate of 20 - 40 rpm to provide the required scanning motion.

Calibration of the received signals must include calibration of the high and low channel A/D converters for each frequency and for each radar. Thus for TOGA/COARE, with three transmitted frequencies per radar, a total of 12 calibrations must be monitored.

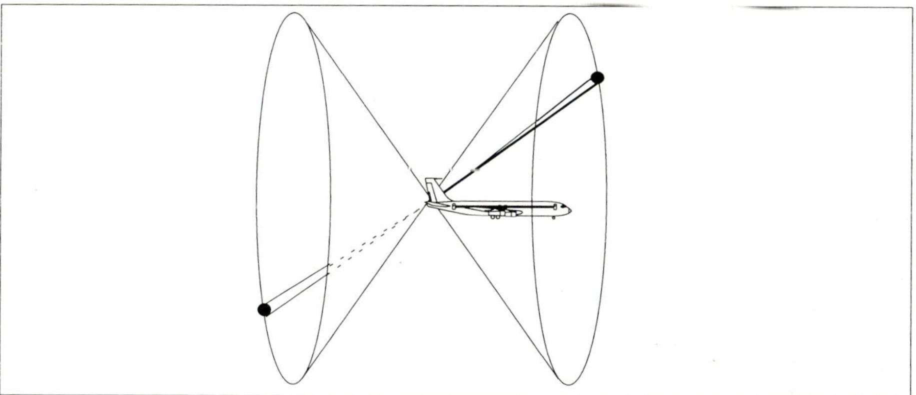

**Figure 118:**   **Schematic Illustration of the Fore-Aft Scan Technique**

## B.42    EMIRAD (Electromagnetics Institute Radiometer)

EMIRAD is an airborne, multifrequency passive radiometer system designed and built by the Electromagnetics Institute (EMI) of the Technical University of Denmark. EMIRAD became operational in 1980. Applications: EMIRAD has been used for several sea -ice missions in the high Arctic and for measuring oil pollution on sea surfaces in different European waters. EMIRAD is comprised of the following systems:[457]

- an imaging multifrequency antenna
- receivers at 5, 17, and 34 GHz
- digital processing and control units
- digital tape recorders
- real-time false color display

The antenna is based on an offset 1 m aperture parabolic reflector scanning sinusoidally around the vertical axis, and illuminated by 3 microwave horns pointing upwards along that very axis. The antenna can be configured for horizontal and vertical polarization - individually for each channel. The beam efficiency is in access of 95%, the sidelobe level below -30 dB, and the cross-polarization level is below -23 dB in the maximum scan range of ± 25°.

---

457)  Information provided by N. Skou of the Technical University of Denmark, Lyngby, Denmark

The antenna and the receivers are mounted on a cargo pallet, which is positioned on the
loading ramp of a Hercules C-130 aircraft. The ramp is closed during take-off, landing and
transit, but lowered to its horizontal position when measurements are to be carried out.
Thus, the antenna has an unobstructed view of the scene below the aft of the aircraft. The
ground speed of the aircraft is 140 knots during operation when the ramp is open.

| Radiometer type | Noise-injection with microwave front-end temperature stabilized to 40°C | | |
| --- | --- | --- | --- |
| Frequency | 5 GHz | 17 GHz | 34 GHz |
| RF bandwidth | 500 MHz | 1 GHz | 1 GHz |
| Noise figure | 4.5 dB | 5 dB | 5 dB |
| Integration time | 8 ms | | |
| Sensitivity | 1 K | | |
| Polarization | H or V | H and V | H or V |
| Angle of incidence | 41° | 54° | 50° |
| Footprint (1000 m) | 160 x 248 m | 47 x 73 m | 24 x 37 m |
| Antenna type | Off-set parabolic reflector viewed by 3 vertical horns | | |
| Scan angle | ± 12.5° for 1000 m altitude; ± 25° for 2000 m altitude | | |
| Scan speed | 1.0 Hz | | |
| Scan pattern | sinusoidal | | |
| Swath width | 500 m for 1000 m altitude, 2000 m for 2000 m altitude | | |
| Aperture | 1 m | | |

**Table 144:    EMIRAD Characteristics**

The radiometer system is designed to give contiguous ground coverage down to a flight altitude of 1000 m. In the past the radiometer system has been used in two imaging modes: 2000
m altitude with ± 25° scan giving a swath of 2000 m; and 1000 m altitude with ± 12.5° scan
giving a 500 m swath (and a better ground resolution than in the 2000 m mode).

EMIRAD is provided with an operator's console, including a digital processing and control
unit and the CCT recorders. Navigational data from the aircraft INS is recorded onto the
CCT along with the measured radiometric data. The console also provides a real-time false
color display system for in-flight monitoring of the radiometer data, and a TV monitor and
video cassette recorder both connected to a TV camera mounted on the antenna pallet. The
TV camera has a FOV comparable to that of the antenna, it gives valuable visible support to
the task of data interpretation.


# B.43    EMISAR (Electromagnetics Institute SAR)

EMISAR is an airborne SAR instrument designed and built by the Electromagnetics Institute of the University of Denmark at Lyngby, Denmark. A single polarization C-Band system has been flown since 1989, including a full-resolution, full swath on board real-time
processor and display. As of 1993 the instrument is being upgraded for dual-frequency polarimetric operation. The C-Band polarimetric SAR is scheduled for engineering tests in the
Fall of 1993, the L-Band polarimetric SAR is planned to be completed by the end of 1994. In
addition, the system will be upgraded in 1994 to include interferometric modes. The C-
Band antenna is mounted in a pod, it is 3-axis stabilized.

EMISAR is installed and flown on a Gulfstream G-3 aircraft of the Danish Air Force (maximum altitude of about 14 km). The cruising speed is 480 knots ($\sim$ 250 m/s). The platform is
equipped with GPS and INS navigation systems. EMISAR is expected to fly also on other
aircraft as part of the JRC EARSEC initiative.

The spatial resolution of the system is in-flight programmable, the finest resolution being
1.5 m (2 m with side lobe suppression). Single polarization data can be processed off-line to
multi-look data with a resolution of 2 m by 2 m. The system records 8192 complex range

samples, hence, the swath width is approximately equal to the pixel spacing defined times 8192. The recording geometry can be programmed in-flight, it is possible to cover incidence angles in the range of 20° to 80°. The system includes motion compensation correction.

The system[458] has been financed by the Thomas B. Thriges Foundation, the Technical University of Denmark, the Danish Technical Research Council, and by the European Economic Community.

| System Parameter | C-Band (single pol. 1989) | C-Band (polarimetric 1993) | L-Band (polarimetric 1994) |
|---|---|---|---|
| RF center frequency | 5.3 GHz | 5.3 GHz | 1.25 GHz |
| System bandwidth | 100 MHz | 100 MHz | TBD |
| Polarization | VV | Quad-polarization | Quad-polarization |
| Max. pulse length | 20 μs | 20 μs | 20 μs |
| Antenna gain | 27 dB | 25 dB | 15 dB (estimated) |
| Azimuth 3 dB beamwidth | 2.7° | 2.5° | 12° |
| Elevation pattern width | 30° | 30° | 55° |
| Transmitted peak power | 2 kW | 2 kW | 6 kW |
| Received noise | 2.5 dB | 1.5 dB | 1.5 dB |
| System loss | 4 dB (estimated) | 4 dB (estimated) | 4 dB (estimated) |
| Quantization (bits/channel I+Q) | 8+8 | 8+8 | 8+8 |
| PSLR (Peak Side Lobe Ratio) | 30 dB | 30 dB | TBD |
| ISLR (Integrated Side Lobe Ratio) | 25 dB | 25 dB | TBD |
| Spatial resolution | 2 m x 2 m | 2 m x 2 m | TBD |
| Cross polarization ratio | 25 dB | 25 dB | 25 dB |
| Range pixel spacing (m) | 1.5 x N | 1.5 x N | 1.5 x N |
| Azimuth pixel spacing (m) | programmable | programmable | programmable |
| Azimuth presummer | programmable | programmable | programmable |
| Tape recorder rate | 107 Mbit/s | 240 Mbit/s (shared) | |
| Range 10 dB SNR, σ=-10dB(m²) | 80 km | 55 km | TBD |
| Flight altitude | > 12 km | | |

**Table 145: Technical Specifications of EMISAR**

## B.44 ER-2 High-Altitude Aircraft Program

Two ER-2 (Earth Resources) survey aircraft are operated by NASA's Ames Research Center (ARC) in Moffett Field, California. The ER-2's are flown from various deployment sites in support of scientific airborne missions (national and international campaigns).[459]

The ER-2 is the modern successor of the 1950's vintage U-2 aircraft. ARC has operated two U-2's since 1971, they were phased out of operation and replaced by two ER-2's. The ER-2 is a versatile aircraft for multiple mission work. It is approximately 30% larger than the U-2 for an increased payload capacity [four large pressurized experiment compartments and a high-capacity AC/DC electrical system). Flying at altitudes of 70,000 feet (21 km), the ER-2 operates above 95% of the Earth's atmosphere and yields an effective horizon of almost 500 km.

As of 1993 more than one hundred different sensors and experiments have been flown aboard the ER-2. Some of these sensors are part of the regular ARC inventory and are used by various investigators from all NASA centers, other government agencies and universities; other sensors are flown on a particular mission(s) in support of experimental work (test of concept, etc.) for an investigator. - The ER-2 high-altitude missions involve collecting data in three principal areas: atmospheric data within the stratosphere, Earth and celestial observations using electronic sensors, and thirdly, photographic data acquisition. Some examples:

---

458) Information provided by S. N. Madsen and N. Skou of the Electromagnetics Institute of the Technical University of Denmark
459) Information provided by J. Myers of ARC

1. Atmospheric investigations:

- Measuring the levels of fluorocarbons, hydrocarbons and other atmospheric constituents capable of interacting with and degrading the Earth's ozone layer. The ER-2 was deployed to Punta Arenas, Chile during August/September 1987 in order to conduct atmospheric experiments in the ozone hole discovered over the Antarctic.

- Studying the dynamics of pollutants moving from the troposphere to the stratosphere and the resulting changes in atmospheric chemistry. Measuring these phenomena was accomplished from an operational base in Darwin, Australia.

- Direct observations of severe thunderstorms using lasers, infrared and microwave scanners, spectrometers and electric field antennas.

- Sampling and identifying volcanic dust plumes injected into the upper atmosphere from the volcanic eruptions of El Cichon in Mexico and Mount St. Helens in the Washington State.

2. Earth and celestial observations:

- Simulating, prior to launch, the spatial and spectral characteristics of the scanners aboard Landsat 1 through 5.

- Measuring background microwave radiation in intergalactic space predicted by the "Big Bang" theory. The magnitude variations in the measurements were used to determine the velocity and direction of our own galaxy.

- Defining the spectral bands for measuring food productivity and pollution in the world's oceans through data flights with the Ocean Color Scanner. These flights were important in developing the CZCS instrument on Nimbus 7. Currently (1990's) the Airborne Ocean Color Imager (AOCI, see B.33.4) is collecting data on board the ER-2 and will provide design information for the next generation instruments.

- Defining the concept, and proving the feasibility of registering day and night thermal imagery to measure the heat capacity of the Earth's surface. Input for HCMM (see A.46).

- Directly observing forest fires and controlled burns using a multispectral scanner to image thermal emissions from the fire. The resulting imagery is transmitted in real-time to firefighter personnel for fire management and containment. Also, smoke plume constituents are collected and evaluated for their role in changing the radiation balance of the Earth's atmosphere.


## B.45   EROS Digital Imagery and Photographic Products

The USGS (US Geological Survey's) EROS Data Center at Sioux Falls, South Dakota serves as the archive and product distribution facility for NASA-AMES aircraft-acquired digital imagery and photography. Digital imagery is available to users on CCTs.[460]

The Airborne Science and Applications Program (ASAP) has acquired photographic data over much of the coterminous US, Alaska, and Hawaii. Since 1971 NASA U-2 and ER-2 high altitude aircraft have employed sophisticated aerial mapping cameras of different formats and focal lengths for obtaining high-resolution photography for Earth observation and science applications. Additionally the C-130-B aircraft acquires mapping camera photography at medium altitudes during its digital data acquisition missions. This NASA acquired photography comprises an archive of over 500,000 frames. These data are accessible

---

460) Paper provided by Jeff Myers of NASA/AMES Research Center

through a geographically referenced data base that allows users to specify a study area and select frames of photography by film emulsion, format, and scale. Photographic products available through the EROS Data Center include 1:1 transparency reproductions and 1x, 2x, 3x, and 4x paper print enlargements of any frame selected. Film formats include 9 x 9 inch photography taken with six and twelve inch focal length lenses and 9 x 18 inch photography taken with 24 inch focal length lenses.

### B.45.1   Airborne Science and Applications Program (ASAP)

The NASA ASAP is supported by three ER-2 high altitude Earth Resources Survey aircraft. These aircraft are operated by the High Altitude Missions Branch at NASA/Ames Research Center, Moffett Field, Ca.. The ERS-2s are used as readily deployable high altitude sensor platforms to collect remote sensing and in situ data on Earth resources, celestial phenomena, atmospheric dynamics, and oceanic processes. Additionally, these aircraft are used for electronic sensor research and development and satellite investigative support.

The ERS-2s are flown from various deployment sites in support of scientific research sponsored by NASA and other federal, state, university, and industry investigators. Data are collected from deployment sites in Kansas, Texas, Virginia, Florida, and Alaska. Cooperative international scientific projects have deployed the aircraft to sites in Great Britain, Australia, Chile, and Norway.

### B.46   E-SAR (Experimental SAR)

DLR airborne SAR system designed and built by the Institute for RF Technology at Oberpfaffenhofen, Germany. The instrument is installed aboard a DO 228 aircraft, it is used as a research tool to elaborate SAR-related problems in particular in the areas of system performance and data analysis. E-SAR has been flown in many campaigns since the beginning of 1989 in preparatory support of spaceborne sensors such as AMI (of ERS-1), X-SAR (on SIR-C/X-SAR).[461]

E-SAR is a high-resolution SAR operating in single mode either at L-, C- and X-Band with either horizontal or vertical polarization. A fourth band, P-Band, will be added in 1994. The sensor is versatile with many support options for flight and radar configurations. Applications: Imaging of all surface types, flat or mountainous terrain, ocean or ice surfaces; monitoring of resources (agriculture, forestry, etc.) or urban growth. There are also contractual flight assignments for  individual customers with specific needs.

The system platform is equipped with IRS (Inertial Reference System) and GPS receivers. The maximum operating altitude is 8000 m. The maximum cruising speed is 440 km/h.

Operating geometry: The nominal ground speed for SAR operation is 70 m/s ($\sim$ 252 km/h). The radar is capable of mapping in three swath modes: narrow, wide, and super wide mode (as of 1993 the super wide mode is in the process of being added). The optimum altitude for SAR operation is 3500 m. The depression angles of the antennas can be varied either mechanically (C- and X-Band) or electronically (L-Band) within a range of 20-50°. The off-nadir (look) angles typically range from 15-70°. The aircraft motion is compensated for good image quality.

A single digital conversion and recording system is used to record the SAR raw data on HDDT's formatted in the SAR 580 format. In addition there is the feature of onboard real time SAR quicklook processing.

The ground system consists of the following units:

461)  "The DLR Airborne Experimental SAR System, E-SAR", DLR brochure

- Radar Raw Data Transcription
- Standard SAR Processing.

Image sizes are: 4 x 4 km for narrow swath width and 5 x 6.5 km for wide swath width.

| Parameter | L-Band | C-Band | X-Band | P-Band |
|---|---|---|---|---|
| RF center frequency (GHz) | 1.3 | 5.3 | 9.6 | 0.450 |
| IF center frequency (MHz) | | 300 | | |
| System bandwidth (MHz) | | 120 | | |
| SAW chirp signal bandwidth (MHz) | 98 | - | - | - |
| Expanded pulse length (µs) | | 4.98 | | |
| Compressed (analog) pulse length (ns) | | 17 | | |
| Digital chirp signal bandwidth (MHz):<br>    narrow swath mode<br>    wide swath mode<br>    super wide swath mode | | 90<br>50<br>18 | | |
| Antenna gain (dBi) | 14 | 17 | 17.5 | ca. 12 |
| Antenna 3 dB beamwidth,  azimuth (°)<br>                      elevation (°) | 18<br>35 | 19<br>33 | 17<br>30 | ca. 30<br>ca. 60 |
| Transmit peak power (W) | 500 | 90 | 2500 | 200 |
| Receiver noise (dB) | 8.5 | 4.0 | 5.0 | ca. 5 |
| Receiver dynamic range with AGC/STC (dB) | | ≥ 40 | | ≥ 40 |
| Nominal pulse repetition frequency (PRF, Hz) | | 952.38 | | |
| Quantization (Bit/sample, I or Q) | | 6 | | |
| Sampling rate (MHz)    narrow swath mode<br>                    wide swath mode<br>                    super wide swath mode | | 100<br>60<br>20 | | |
| Echo buffer memory capacity (I or Q; words) | | 2560 | | |
| Nominal date rate on high density tape (Mbit/s) | | 28 | | |
| Maximum recording time per tape (14 inch; min) | | 15 | | |
| | | | | |
| Spatial resolution, range x azimuth (m)<br>    narrow swath width<br>    wide swath width<br>    super wide swath width | | 2.5 x 2.5<br>4.5 x 4.5<br>11.5 x 11.5 | | |
| Number of statistically independent looks | | 8 | | |
| Radiometric resolution (8 looks) | | < 2 dB | | |
| Geometric distortion | less than one resolution cell | | | |

**Table 146:    Technical Specifications of the E-SAR**

## B.47    E-SLAR (Experimental Side-Looking Airborne Radar)

E-SLAR[462] is an experimental DLR system (of the type: real aperture radar) designed and built by the Institute for RF Technology at Oberpfaffenhofen, Germany. The initial version was built in 1976 and redesigned and improved as a research tool ever since. The original instrument provided a single-frequency X-Band SLAR capability, the current (1993) 2$^{nd}$ generation E-SLAR version offers two front-ends, a $K_a$-Band for the study of radar back-scatter at different frequencies, and an X-Band system. The current instrument is flown on a DO 228 aircraft.

The X-Band antenna is a slotted waveguide antenna of 3.6 m length. It provides vertical polarization and is fed from the one end. The $K_a$-Band antenna is a reflector type and verti-cally polarized. The signal is detected by an incoherent rectifier in both transmitter/receiver units (X-Band/$K_a$-Band), the output from each receiver is a unipolar video signal.

E-SLAR is provided with an on-board data recording system, operator controls and moni-toring along with a quicklook capability.

---

462)  F. Witte, "The Archimedes IIa experiment: remote sensing of oil spills in the North Sea", International Journal of Remote Sensing, 1991, Vol. 12, Nr. 4, pp. 809-821

Applications: detection of oil slicks on water surfaces, land use monitoring (tropical rain forest, flood surveys, etc.).

| Parameter | X-Band | K$_a$-Band |
|---|---|---|
| Transmitter/Receiver | | |
| Frequency | 9.39 GHz | 35 GHz |
| Peak power (short pulse) | 23.0 kW | 20 kW |
| Peak power (long pulse) | 19.5 kW | |
| PRF (recorded) | 43 Hz | 43 Hz |
| Pulse width | 60 ns/280 ns | 60 ns |
| IF center frequency | 60 MHz | 60 MHz |
| IF amplifier bandwidth | 16 MHz | 15 MHz |
| IF dynamic range | 70 dB | 80 dB |
| Noise figure | 8.6 dB | < 10 dB |
| Antenna | | |
| Antenna gain | 29.5 dB | 33 dB |
| 3 dB beam width | Az. = 0.53°, El. = 45° | Az. = 0.5°, El. = 20° |
| Polarization | linear, VV | linear, VV |
| Antenna length | 3.6 m | 1.2 m |
| Data Recording | | |
| Digital | Videocassette recorder (VCR) | |
| Word length | 8 bit | |
| Recording time | 60 min (on one cassette) | |
| Quicklook | | |
| Rolling map | 512 x 512 pixels | |
| Display | b/w TV monitor | |
| Operating Conditions | | |
| Altitude | 1000 - 3000 m | |
| Airspeed | about 150 knots | |
| Swath width | 3 - 15 km | |
| Depression angle | 22° | |
| Resolution (short pulse, altitude = 1000 m) | Incidence angle: 10° 40° 65° | |
| | Azimuth (m): 53.3° 14.4° 10.2° | |
| | Elevation (m): 9.1° 11.8° 21.4° | |

**Table 147:     Technical Parameters of the E-SLAR**


## B.48     ESMR (Electronically Scanned Microwave Radiometer)

ESMR is a NASA/GSFC passive sensor. The instrument measures the upwelling thermal microwave radiation emitted by the Earth's surface and atmosphere at a frequency of 19.35 GHz (1.55 cm wavelength). It is a cross-track imager scanning an angular swath of ± 50° in 39 steps. ESMR is usually flown on a DC-8 aircraft at altitudes of 10-12 km. Its beam width is 3°, the polarization is horizontal with the electric vector parallel to the direction of flight. The temperature sensitivity of this sensor is about 2 K, the calibration accuracy is about ± 5 K. The main application of ESMR is to provide microwave images for precipitation and sea ice studies. ESMR took part in the TOGA/COARE mission in early 1993.[463]


## B.49     ESTAR (Electronically Steered Thinned Array Radiometer)

A prototype airborne synthetic aperture microwave radiometer[464],[465],[466] (a passive instrument), developed by NASA/GSFC and the University of Massachusetts at Amherst, with the objective to improve the spatial resolution of spaceborne passive microwave imagers; in other words: to apply microwave interferometry to an Earth remote sensing problem.


463)  Information provided by J. R. Wang of GSFC

464)  C. S. Ruf, C. T. Swift, A. B. Tanner, D. M. Le Vine, "Interferometric Synthetic Aperture Microwave Radiometry for the Remote Sensing of the Earth", IEEE Transactions on Geoscience and Remote Sensing, Vol. 26, Nr. 5, September 1988, pp. 597-611

465)  D. M. Le Vine, M. Kao, A. B. Tanner, C. T. Swift, A. Griffis, "Initial Results in the Development of a Synthetic Aperture Microwave Radiometer", IEEE Transactions on Geoscience and Remote Sensing, Vol. 28, Nr. 4, July 1990, pp. 614-619

The instrument is a 1.4 GHz (L-Band) radiometer suitable for the remote sensing of soil moisture and ocean salinity. ESTAR uses Fourier synthesis to derive high-resolution images from a minimum number of antenna elements. The concept is borrowed from radio astronomy where it is commonly used to obtain high-resolution stellar images by synthesizing very large apertures from a few relatively small antennas.

The ESTAR prototype has been flown aboard NASA Wallops aircraft (P-3) in the summer of 1988. The instrument consists of five antenna elements and a bank of 10 correlators that cross-correlate all possible pairs of antennas. The antennas are deployed along a line and are spaced at certain integer multiples of half the RF wavelengths so that the set of all cross-correlations generates every half-wavelength spacing up to a maximum spacing, while sampling each one as few times as possible. The configuration is referred to as a minimum redundancy linear array.

| Center Frequency | 1.4 GHz (L-Band) |
|---|---|
| Bandwidth | 25 MHz |
| Polarization | horizontal |
| Resolution along-track | $\pm 4^\circ$ |
| Resolution cross-track | $\pm 3^\circ$ |
| Swath width | $\pm 45^\circ$ |
| Integration time | 0.25 s |
| Platform | NASA/ARC C-130 and NASA WFF P-3 |
| Data | 15 correlator outputs + housekeeping data |

**Table 148:    Specification of the ESTAR Instrument**

The ESTAR prototype is a hybrid instrument which uses real aperture antennas to obtain resolution along-track (with stick antennas) and employs aperture synthesis to obtain resolution across-track. The array of stick antennas is oriented with their axis pointed in the direction of motion. The stick antennas produce a fan beam which is narrow in the along-track direction and broad in the cross-track direction. The fan beam is swept along the surface as the sensor moves and provides resolution along-track. Resolution across-track is achieved using aperture synthesis (which divides the swath into small cigar-shaped resolution cells) by detecting the signal from pairs of the stick antennas with a correlation receiver. All correlation pairs are measured simultaneously.

> An imaging radiometer maps the brightness temperature distribution over a given field of view (FOV). Real aperture radiometers do this by scanning their antenna - either mechanically or electrically - across the FOV. The resolution of the image is consequently determined by the beamwidth of the antenna.
> Interferometric imaging radiometers, on the other hand, generate an image indirectly, by measuring the Fourier transform of the brightness temperature distribution over the FOV. This measurement, which is referred to as the visibility function, is inverse Fourier-transformed by the user to form an image.

At each antenna an RF front end amplifies and mixes the received signal down to an intermediate frequency (IF). At the IF a 10-way complex cross-correlator then generates the visibility samples by correlating all antenna pair combinations. Each correlator consists of two multipliers - one produces the in-phase component by multiplying the IF signals directly, the other produces the quadrature component by multiplying one IF with a $90^\circ$ shifted version of the other IF. In all there are 21 analog signals which are low pass filtered, sampled by an A/D converter every quarter second and sent to the computer.

The raw data produced by ESTAR is a set of complex cross-correlation voltages as taken between various pairs of antennas in the array. Each data point is a sample of the visibility

466) D. M. Le Vine, T. T. Wilheit, R. E. Murphy, C. T. Swift, "A Multifrequency Microwave Radiometer of the Future", IEEE Transactions on Geoscience and Remote Sensing, Vol. 27, Nr. 2, March 1989, pp. 193-199

function and is a measure of a spatial harmonic in the brightness temperature scene. The spatial harmonic depends on the spacing between a given antenna pair. If the elements are spaced at every integer multiple of one half of an RF wavelength, $\lambda/2$, then the resulting visibility samples are sufficient to reconstruct an image through a 180° field of view. The largest spacing, or baseline, determines the spatial resolution of the imager. The key feature of a thinned array is that all the required baselines are obtained with fewer antennas than there are spacings. The ESTAR prototype in Figure 119, for example, simultaneously measures visibility at eight contiguous cross-track spacing with just five antennas.

The ESTAR instrument has also been calibrated by several methods on the ground and in further test flights in the summer of 1989. Models have been developed that are able in relating temperatures to the color in the images. The demonstration of the aperture synthesis (microwave interferometry) technique for remote sensing is regarded as promising and may be utilized in future spaceborne systems.[467]

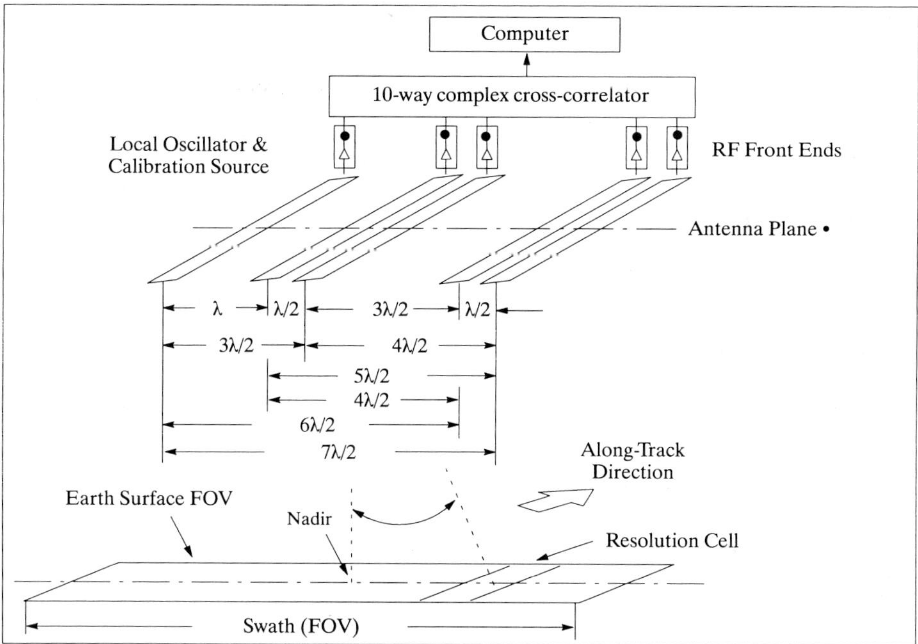

Figure 119: Schematic Elements of the ESTAR Prototype Sensor

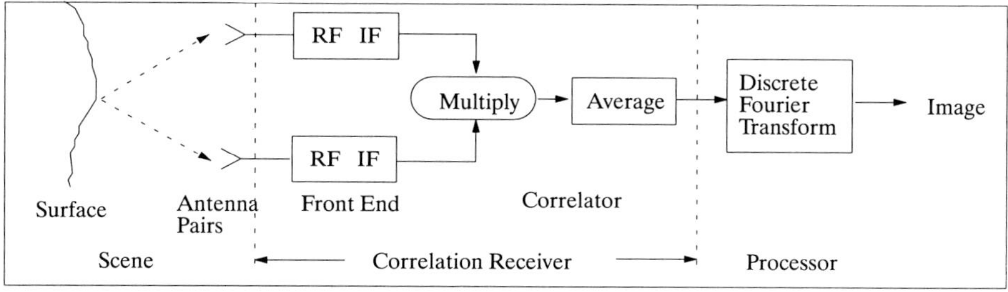

Figure 120: Schematic Illustration of the ESTAR Signal Processing

467) A. B. Tanner, C T. Swift, "Calibration of a Synthetic Aperture Radiometer", IEEE Transactions on Geoscience and Remote Sensing, Vol. 31, Nr. 1, January 1993, pp. 257-267

## B.50  FLASH (FOA Laser Airborne Sounder for Hydrography)

FLASH is a Swedish airborne depth-sounding lidar system, owned and operated by FOA, providing data from water surfaces for charting purposes. The instrument is flown on a helicopter (Boeing Vertol), first trials of a prototype and demonstrator were conducted in 1989. Two operational systems are assembled in 1993 based on the experiences with FLASH, they will be ready for use in 1994. The two new systems are called Hawk Eye, they are manufactured by SAAB Combitech (Sweden) in cooperation with FOA.[468],[469]

The FLASH system consists of a sensor platform (transceiver/scanner), a signal handling module, a data processing/storage system (computer), a navigation module, and a power supply. The transceiver unit is made up of a laser head, the optical receiver, and the receiver electronics unit. The scanner is programmable enabling the beam to search the water surface in different patterns. The FOV can be altered from 5 to 50 mrad. The signal extraction unit converts the received signal into digital form, and of extracting the depth value in real-time for on-board display. The computer controls the system and handles the data. The lidar data are stored on magnetic tape along with a number of system parameters and navigation data. The navigation module employs a GPS receiver for differential positioning.

The laser is a frequency-doubled Nd:YAG laser with emission wavelengths of 532 and 1064 nm. The pulse length is approximately 7 ns. The laser beam divergence is automatically controlled via a beam expander to values between 2 - 10 mrad to allow studies of the horizontal resolution of bottom features close to the surface. The longer infrared wavelength (1064 nm) is used to measure the slant range to the water surface (the wavelength of 1064 cannot penetrate the surface, only a few centimeters). The resolution in slant range measurements is ~8 cm. The IR beam is also used for discrimination between reflected signals. The short wavelength (green) beam of 532 nm is used for water penetration measurements.

The IR and green beams are swept by the scanner. The programmable scanner offers a wide range of system performance optimization and utility with regard to selectable scan patterns and to spot densities. - Typical altitudes for operating the system are in the range of 200 - 500 m above ground. The water depth that typically can be reached depends on the water quality, it is given by the formula:

$$D_{max} \approx 4 \, K, \text{ where } K = \text{diffuse attenuation coefficient (1/m)}$$

The footprint on the water surface can be altered by changing the laser beamwidth within the range: $\Phi_{Laser} \approx 1 - 10$ mrad.

| | |
|---|---|
| Mirror diameter<br>Receiver aperture | 280 mm<br>200 mm |
| Beam nadir angles | ± 25° (lateral), -15" (backward), 25" (forward) |
| Sensor angle encoders | 0.005° resolution |
| Vertical gyro accuracy | ± 0.3° |
| Rate gyro resolution | 0.025°/s |
| Digital servo-loops sampling frequency | 1 kHz |
| Scanner mass | 24.6 kg |

**Table 149:    FLASH Scanner Characteristics**

468)  K. O. Steinvall, K. R. Koppari, U. C. M. Karlsson, "Experimental evaluation of an airborne depth-sounding lidar", Optical Engineering, June 1993, Vol. 32, Nr. 6, pp. 1307-1321
469)  R. Axelsson, O. Steinvall, P. Sundberg, "Programmable Scanner for Laser Bathymetry", reprint from the International Hydrographic Review, Monaco, LXVII(1), January 1990, pp. 161-170

## B.51    FLI (Fluorescence Line Imager)

The FLI (also known as Programmable Multispectral Imager, PMI)[470],[471]is an airborne imaging spectrometer developed for the Canadian Department of Fisheries and Oceans (built by Moniteq Ltd and Itres Ltd.). The instrument was in operation by Moniteq, of Concorde, Ontario, from 1984 - 1990. Objectives: imaging of ocean chlorophyll fluorescence and spectral reflectance changes in water caused by phytoplankton in the sea. Note: the instrument has also been operated in missions for applications related to water quality and aquatic vegetation monitoring as well as over land. Support of a campaign in Germany in 1986 to monitor spectral reflectance from forests as part of a forest damage survey.

The imaging spectrometer makes use of a 2-D multi-element array of detectors (CCD push-broom technology) in the focal plane of a dispersive optical system. The system can operate in two modes.

*   Spatial mode: high spatial resolution mapping in 8 selectable spectral bands.
*   Spectral mode: high spectral resolution mapping in 288 spectral bands (spatial resolution is reduced to 40 look directions).

The system comprises five separate optical camera modules, aligned to provide a field-of-view (FOV) of about 70°, with 1925 detector elements and 1.3 mrad resolution (IFOV). Each module has a silicon diode array with 385 elements across track by 288 elements. The spectral response is from 430 - 805 nm, all wavelengths are accessible in steps of 1.3 nm, the spectral resolution is 2.5 nm.

| | |
|---|---|
| Spectral coverage | 430 - 800 nm using 288 detector elements, pixel size = 1.3 nm, spectral resolution = 2.5 nm |
| Spatial coverage | 70° swath with five cameras having a total of 1925 detectors, pixel size = 0.65 mrad, spatial resolution = 1.3 mrad |
| Spectral mode | Spectra are recorded from 40 different directions across the swath (8 per camera). The band width and look direction are under software control. |
| Spatial mode | Spectral pixels are grouped into a push-broom image about 1900 pixels wide in each of eight spectral bands. The bandwidth and spectral position are under software control. |
| Sensitivity | Maximum 1800:1 SNR for a band of 16 detector elements at full signal |
| Integration time | 40 ms minimum, typically 90 ms in spatial mode, 150 ms in spectral mode |
| Digitization | 12 bits on chip, summation to 16 bit readout |
| Detectors | Five 385 by 576 element EEV P8600 arrays. Half of the area is used for frame storage |
| Optics | Transmission grating with f/1.4 Nikon lenses |
| Data recording | Bell and Howell 14-track high-density tape drive 8 by 875 kbit/s recorded |
| Weights | Head = 70 kg, data acquisition = 26 kg, recorder extra, cooling unit = 41 kg |
| Power | Head and acquisition = 450 W, cooling unit up to 1 kW |
| Aircraft mountings | DC-3, Falcon Fan-jet, Cessna 402, Piper Navajo, DO 228, Twin Pioneer |

**Table 150:    Specifications of the Fluorescence Line Imager**


## B.52    FOLPEN (Foliage Penetration VHF Impulse SAR)

SRI International of Menlo Park Ca.[472] operates two fixed-wing VHF impulse SAR systems with foliage penetration capability, they are designated: FOLPEN-I and FOLPEN-II. The instruments are operational since December 1990, they are currently used for target detection hidden under foliage.


470) J. F. R. Gower, G. A. Borstad, C. D. Anger, H. R. Edel, " CCD-Based Imaging Spectroscopy for Remote Sensing: The FLI and CASI Programs", Canadian Journal of Remote Sensing, Vol. 18, Nr. 4, 1992, pp. 199-208
471) S. M. Till, "Airborne Electro-Optical Sensors for Resource Management", Geocarto International, Vol 3, 1987, pp. 13-23, the article contains also the LARSEN instrument
472) Information provided by R. S. Vickers of SRI International

The system provides on-board real-time image display (1024 x 1024 pixels) and data recording onto optical disk (1.6 MByte/s data rate), the aircraft is equipped with GPS.

The FOLPEN antenna consists of two phased arrays, one on each wing for the transmit and receive functions. The array elements are loaded dipoles designed to suppress and reflections and to be relatively flat from 100 - 500 MHz. The sampling rate in I&Q on FOLPEN-I is 500 MHz, on FOLPEN-II it is 250 MHz. The swath width is 3 km maximum, dependent on altitude.

| | |
|---|---|
| Frequency | 100 - 500 MHz in two bands (VHF/UHF-Band SAR) |
| Bandwidth | 200 MHz per band |
| Pulse Repetition Frequency (PRF) | 200 Hz |
| Pulse width | 5 ns |
| Polarization | HH |
| Resolution | 1 m x 1 m |
| Coverage | 500 km$^2$/hr at a flight altitude of 3 km |
| Surveying altitudes | 300 m to 3.5 km |
| Average Power | 5 W |
| Instrument mass | 115 kg |

**Table 151:    FOLPEN Parameter Specification**


## B.52.1   GPR (Ground Penetrating Radar)

GPR is a helicopter-borne ground penetrating radar profiler of SRI International. The current system (1993) is an upgrade of previous systems built for special-purpose client applications over the past 20 years. The first system was flown for ice penetration studies in 1974. As of 1993 SRI has two operating systems installed in helicopters.

GPR is capable of detection of buried targets up to a few meters depth in favorable soils, up to 1 m in most soils. A deeper penetration requires a change in antennas. At mid-range frequencies the ground swath is approximately equal to the altitude - nominally 15 m. As in all GPR systems, performance is severely limited in high-loss soils, such as wet clay or silt.

| | |
|---|---|
| Frequency range | 200 - 1000 MHz (VHF/UHF Profiler) |
| PRF | 200 MHz |
| Pulse width | 5 ns |
| Polarization | not applicable |
| Vertical resolution | 0.5 m |
| Average power | 1 W |
| Survey speed | up to 60 kts |
| Survey altitude | 15 m (above ground) |

**Table 152:    Specification Parameters of SRI-GPR**

## B.53    GER Corporation Instruments

GER (Geophysical & Environmental Research Corporation of Millbrook New York) is a commercial builder of remote sensing imaging spectrometers [ASTER Simulator (custom-built), DAIS-2815, DAIS-7915, DAIS-16115, GERIS, etc.]. Recording data rates vary from 300 kByte/s to 32 MByte/s depending on instrument and recording device.

### B.53.1    ASTER Simulator

The ASTER simulator is an airborne instrument developed (by GER Corporation of Milbrook, N. Y. for the 'JAPEX Geoscience Institute in Tokyo') for algorithm verification of ASTER (Advanced Spaceborne Thermal Emission and Reflection Radiometer) on the EOS/AM-1 payload. The ASTER simulator is in operation since 1991.[473]

The ASTER simulator is basically a 20-band TIR system (200 nm wide, contiguous bands) with one additional band in the VNIR and three bands in the MWIR spectrum. The MWIR bands were chosen to investigate the utility in this region for geological and environmental remote sensing.

A 3-axis gyroscope is attached to the scanner housing to measure the aircraft's attitude. The aircraft position is determined by a GPS receiver. The on-board computer has a display monitor for quicklooks. Recording media: IBM 3480 compatible cartridge recorder (12 inch, 18 track parallel), 200 MB capacity/tape, up to 10 tapes for auto-loading.

|  | VNIR (μm) | MWIR (3-5 μm) | TIR (8-12 μm) |
|---|---|---|---|
| Channel(s) | 0.76-0.85 | 3.0-3.6;3.6-4.2;4.4-5.0 | 20 channels each 0.2 μm |
| NEΔT |  | 0.2-0.3 K (300 K) | 0.2-0.3 K (300 K) |
| Detector | 2 Si adjoining | 3-element InSb array | 4 6-element HgCdTe arrays |
| Detector size | 2 x 4 mm each | 2.5 x 7.3 mm total | 2.0 x 4.182-5.532 mm total |
| SNR |  | 75-200 | 75-200 |
| IFOV (3 apertures)<br>Ground IFOV<br>Sampling interval in scan direction<br>Flight speed range<br><br>Unvignetted scanning field of regard (FOV)<br>Scanning speed<br>Data encoding (radiometric resolution)<br>Pixels per scan line<br>No inter-band registration error<br>Liquid nitrogen cooling of detectors | selectable: 1, 2.5, or 5 mrad for all bands<br>5, 12.5, or 25 m at 5000 m flight altitude<br>i (2.5/7) mrad where i=1-7<br>85-190 Knots on Piper Aztec<br>0-3000 m (0 - 6000 m when electronics pressurized)<br>approximately 80° full angle<br>5-15 scans/s using optical disk recorder<br>15 bit<br>512-1024 | |
| Collecting Optics | | | |
| Kennedy-type reflecting scanner<br>Parabolic focussing mirror<br>Focal length<br>Effective clear aperture size | 9 x 15 inches<br>13.5 inches; approx. f/1<br>approximately 540 cm² | | |

**Table 153:    Specification Parameters of the Airborne ASTER Simulator**

The spectrometer optics separate the incident radiation into 24 distinct spectral bands using dichoric beamsplitters and gratings to eliminate inter-band registration errors. The VNIR detector plane has two silicon detectors mounted adjoining each other. Their outputs are electronically combined and recorded as a single data channel. A 3-element InSb array is mounted in the MWIR detector plane. The output from each element is recorded separately. A 12-element HgCdTe array is mounted in each of the TIR detector planes (each 12-element HgCdTe array is composed of two 6-element arrays positioned adjoining each other).

473)  H. Watanabe, M. Sano, F. Mills, S. H. Chang, S. Masuda, "Airborne and Spaceborne Thermal Multispectral Remote Sensing", 1992

One set measures radiation between 8.0-10.4 μm; the other set measures radiation between 9.6-12.0 μm. A total of 24 channels of data are produced and recorded. Four spectral bands (9.6-10.4 μm) are recorded twice. The two outputs for each of these four bands may be digitally combined during ground processing.

| VNIR (0.52-0.88 μm) | | | | | |
|---|---|---|---|---|---|
| (VNIR uses a 4000 element linear CCD array for each spectral image) | | | | | |
| Band (3) | 1 | 2 | 3 Nadir | 3 Forward | |
| Channel | 0.52-0.60 | 0.63-0.69 | 0.76-0.86 | 0.76-0.86 | |
| NEΔR | 0.5% | 0.5% | 0.5% | 0.5% | |
| MTF | 0.25 | 0.25 | 0.25 | 0.25 | |
| Detector | Si CCD | Si CCD | Si CCD | Si CCD | |
| IFOV (μrad) | 21.3 | 21.3 | 21.3 | 18.1 | |
| Ground IFOV | 15 m | 15 m | 15 m | | |
| Quantization | 8 bit | 8 bit | 8 bit | 8 bit | |

| SWIR (1-3 μm) | | | | | | |
|---|---|---|---|---|---|---|
| (SWIR uses a 2048 element staggered IR-CCD array for each spectral channel) | | | | | | |
| Band (6) | 4 | 5 | 6 | 7 | 8 | 9 |
| Channel | 1.6-1.7 | 2.145-2.185 | 2.185-2.225 | 2.235-2.285 | 2.295-2.365 | 2.360-2.430 |
| NEΔR | 0.50% | 0.60% | 0.80% | 1.00% | 1.00% | 1.30% |
| MTF | 0.25 | 0.25 | 0.25 | 0.25 | 0.25 | 0.25 |
| Detector | PtSi-Si | PtSi-Si | PtSi-Si | PtSi-Si | PtSi-Si | PtSi-Si |
| IFOV (μrad) | 42.6 | 42.6 | 42.6 | 42.6 | 42.6 | 42.6 |
| Ground IFOV | 30 m | 30 m | 30 m | 30 m | 30 m | 30 m |
| Quantization | 8 bit | 8 bit | 8 bit | 8 bit | 8 bit | 8 bit |

| TIR (8-12 μm) | | | | | | |
|---|---|---|---|---|---|---|
| (TIR uses a 10 element staggered array with mechanical scanning for each channel) | | | | | | |
| Band (5) | 10 | 11 | 12 | 13 | 14 | |
| Channel | 8.125-8.475 | 8.475-8.825 | 8.925-9.275 | 10.25-10.95 | 10.95-11.65 | |
| NEΔT | 0.3 K | 0.3 K | 0.3 K | 0.3 K | 0.3 K | |
| MTF | >0.25 | >0.25 | >0.25 | >0.25 | >0.25 | |
| Detector | HgCdTe | HgCdTe | HgCdTe | HgCdTe | HgCdTe | |
| IFOV (μrad) | 127.6 | 127.6 | 127.6 | 127.6 | 127.6 | |
| Ground IFOV | 90 m | 90 m | 90 m | 90 m | 90 m | |
| Quantization | 12 bit | 12 bit | 12 bit | 12 bit | 12 bit | |

**Table 154:  Characteristics of the ASTER Satellite System (EOS/AM1)**

## B.53.2   DAIS-2815  (Digital Airborne Imaging Spectrometer)

| Spectrometer | Spectral Range (μm) | Nr. of Bands | Bandwidth (μm) |
|---|---|---|---|
| 1 (VNIR) | ≈0.7 - ≈1.0 | 1 | 0.3 |
| 2 (MWIR) | 3 - 5 | 3 | ≈0.6 |
| 3 (TIR) | 8 - 12 | 20 | 0.2 |
| Radiometric resolution | 15 bit | | |
| IFOV (3 apertures) | selectable: 1, 2.5, or 5 mrad | | |
| Swath width | 82° | | |
| Pixels per line | 512-2048 | | |
| Dispersion element | grating | | |
| Maximum mirror scan frequency | 50 Hz | | |
| Recording media | IBM 3480 cartridge (1/2 inch, 18 track parallel) | | |
| Detectors | Si (VNIR) | | |
| | InSb: liquid nitrogen cooling (MWIR) | | |
| | MCT: liquid nitrogen cooling (TIR) | | |

**Table 155:   Specification of the DAIS-2815 Instrument**

The DAIS-2815 instrument is the commercial version of the ASTER Simulator. First operational model in September 1991. Owners/operators of DAIS-2815 instruments: JGI and JAROS.

### B.53.3 DAIS-7915 (Digital Airborne Imaging Spectrometer)

DAIS[474] is 79-channel/15 bit quantization opto-mechanical scanner instrument built by GER (Geophysical Environmental Research Corporation) of Milbrook, N.Y. and funded by the European Community (Joint Research Centre, Ispra, Italy) and DLR. DAIS covers the spectral range from the visible to the thermal infrared wavelengths at variable spatial resolutions from 2-30 m. The system is integrated and operated by DLR (Institute of Opto-Electronics) and scheduled to be flown at the beginning of 1994. DAIS will be accessible through DLR to serve EARSEC (European Airborne Remote Sensing Capabilities) as well as other contractual applications worldwide.

Applications: environmental monitoring of land and marine ecosystems, vegetation stress research, agriculture and forestry resource mapping, geological mapping, mineral exploration and provision of data for geographic information systems.

| Spectral Parameters | | |
|---|---|---|
| Instrument | Nr. Channels | Spectral Range |
| Spectrometer 1 | 32 | VNIR   400 - 1010 nm (silicon detectors) |
| configuration a: | | VNIR   498 - 1010 nm, bandwidth. 16 nm |
| configuration b: | | VNIR   400 - 720 nm, bandwidth 10 nm (water applications) |
| configuration c: | | VNIR   530 - 850 nm, bandwidth 10 nm (vegetation) |
| Spectrometer 2 | 8 | SWIR   1500 - 1788 nm, bandwidth 36 nm (InSb detectors) |
| Spectrometer 3 | 32 | SWIR   1970 - 2450 nm, bandwidth 36 nm (InSb detectors) |
| Mid-Wavelength IR band: | 1 | MWIR   3000 - 5000 nm (InSb) |
| Spectrometer 4 | 6 | TIR   8700 - 12700 $\mu$m, bandwidth 600 nm (HgCdTe array) |
| General System Parameters | | |
| IFOV | 3.3 mrad | |
| Swath width (FOV) | $\pm 39°$ | |
| Tilt angle | $\pm 20°$ longitudinal direction | |
| Radiometric resolution | 15 bits | |
| Radiometric sensitivity | VNIR   $< 0.02$ Wcm$^{-2}$sr$^{-1}$ $\mu$m$^{-1}$<br>SWIR   $< 0.02$ Wcm$^{-2}$sr$^{-1}$ $\mu$m$^{-1}$<br>MWIR/TIR   $= 0.1$ K | |
| Signal-to-noise ratio (SNR) | VNIR   $> 150$<br>SWIR   $> 80$<br>MWIR/TIR   $> 80$ | |

**Table 156:**    **DAIS-7915 Instrument Spectral Ranges and System Parameters**

The DAIS system consists of the optoelectronic module (OM) and the electronic module (EM). The OM houses the Kennedy scanner with scan motor, encoder, two blackbody radiation sources, a folding mirror assembly, telescope and gyros as well as beam splitters, grating spectrometers, detector assemblies and preamplifiers. The EM includes the data acquisition module and control electronics, the blackbody controller and the power supply/distribution units. The mass of OM is 172 kg; the mass of EM is 100 kg.

Image data are measured with a radiometric resolution of 15 bits/pixel and coregistered bands. Housekeeping data are recorded as channel 80. Selectable IFOV: 3.3 (2.5, or 5.0 optional) mradians. The TFOV = 64-78° depending on the size of the aircraft hatch. Pixels per line = 512.

Expected system performance: SNR = 100-200 (VNIR), and SNR = 20-50 (SWIR), for a 30% ground albedo and a solar zenith angle of 45°.

Data: onboard cassette recorder; instrument data rates = 1.58 Mbit/s (for a 3 Hz scan rate) to 15.8 Mbit/s (for a 30 Hz scan rate). The selected scan rate depends on the aircraft altitude.

---

474)   S. H. Chang, M. J. Westfield, F. Lehmann, D. Oertel, R. Richter, "79-channel Airborne Imaging Spectrometer", GER/DLR paper

## B.53.4 DAIS-16115 (Digital Airborne Imaging Spectrometer)

DAIS-16115 is a GER imaging spectrometer of 161 channels at 15 bit quantization. All spectrometer channels are spatially registered (image cube). Applications: geological, environmental, ecological and marine life monitoring.

| Spectrometer | Spectral Range ($\mu$m) | Nr. of Bands | Bandwidth (nm) | SNR |
|---|---|---|---|---|
| 1 (VNIR) | 0.4 - 1.0 | 76 | 8 | > 200 |
| 2 (SWIR1) | 1.0 - 1.8 | 32 | 25 | > 100 |
| 3 (SWIR2) | 2.0 - 2.5 | 32 | 16 | > 100 |
| 4 (MWIR) | 3.0 - 5.0 | 6 | 333 | > 80 |
| 5 (TIR) | 8.0 - 12.0 | 12 | 333 | > 80 |
| 6 (Stereo) | 0.4 - 1.0 | 2 bands stereo fore and aft pointing | | > 500 |
| 7 | 1 general data channel (gyro data, housekeeping, etc.) | | | |
| IFOV | 3 mrad | | | |
| Swath width (FOV) | $\pm 39^\circ$ (8 km swath at 5 km altitude) | | | |
| Scan speed | up to 50 Hz | | | |
| Detectors | Si, InSb, MCT | | | |
| MTBF | 600 hrs | | | |
| Power | 28V, 50 A | | | |

**Table 157:**  **Specification of the DAIS-16115 Imaging Spectrometer**

## B.53.5 GER-63 Channel Scanner

GER-63 is an airborne 63-channel imaging spectrometer designed for environmental studies and acquisition of spectral information pertinent for geological studies. A Kennedy-type scanner is used to acquire the images, which are formed at the entrance slit to the spectrometer. Note: The instrument was sometimes also referred to by the name of GERIS or as AIS).

| Spectrometer | Spectral Range ($\mu$m) | Nr. of Bands | Bandwidth (nm) | Detector |
|---|---|---|---|---|
| 1 (VNIR) | 0.4 - 1.0 | 24 | 25 | Si |
| 2 (SWIR1) | 1.5 - 2.0 | 4 | 125 | PbS |
| 3 (SWIR2) | 2.0 - 2.5 | 29 | 17.2 | PbS |
| 4 (TIR) | 8.0 - 12.5 | 6 | 750 | HgCdTe |

**Table 158:**  **Specification of the GERIS Imaging Spectrometer**

## B.54 Harvard Atmospheric Chemistry Instruments

During the past 15 years Harvard University (J. G. Anderson) has been developing instrumentation for the measurement of trace species in the stratosphere. One aspect of this research has been to improve man's understanding of stratospheric ozone depletion.

### B.54.1 OH/HO$_2$-Instrument

The objective is to measure OH and HO$_2$ in the altitude region of 8 to 25 km (PI: J. G. Anderson). The instrument consists of a high repetition rate pulsed dye laser pumped with a pair of Q-switched diode pumped YLF (Yttrium Lithium Fluoride) solid-state lasers, a detection axis to detect laser-induced fluorescence at 309 nm, an NO gas addition system to convert HO$_2$ to OH. The OH/HO$_2$ instrument was test-flown on ER-2 aircraft in the Fall Of 1992 and participated in SPADE in the Spring of 1993.

Principle of operation: OH is detected by direct laser-induced fluorescence in the (O-1) band of the $^2\Sigma$-$^2\Pi$ electronic transition. The instrument produces a frequency-tunable laser light at a wavelength of 282 nm. An on-board frequency reference cell is used by a computer to lock the laser to the appropriate wavelength. Measurement of the signal is then made by tuning the laser on and off resonance with the OH transition. Stratospheric air is channeled into the instrument using a double-ducted system that both maintains laminar flow through the detection region and slows the flow from the free-stream velocity (at about 200 m/s) to 40 m/s. The laser light is beam-split and directed to two detection axes where it passes through the stratospheric air into a multipass White cell. Fluorescence from OH, centered at 309 nm, is detected orthogonal to both, the flow and the laser propagation using a filtered PMT (Photo Multiplier Tube) assembly. Optical stability is checked periodically by exchanging the 309 nm filter with a filter centered at 302 nm where Raman scattering of $N_2$ is observed. $HO_2$ is measured as OH after chemical titration with nitric oxide: $HO_2$ + $NO{\rightarrow}OH$+ $NO_2$. Variation of added NO density and flow velocity as well as the use of two detection axes aid in the diagnostics of the kinetics in this tritation. Ancillary measurements of ozone and water vapor are made as diagnostics of potential photochemical interfaces.

### B.54.2    ClO/BrO Instrument

A dual-axis resonance fluorescence instrument (PI: R. C. Cohen). Objective: to measure ClO and BrO from 10 to 25 km in altitude. The instrument has been flown in the 1987-1992 AASE and AAOE campaigns as well as in SPADE.

The instrument consists of an NO gas addition system and a resonance fluorescence detection system of Cl and Br atoms respectively. A 5 cm x 5 cm inlet samples air from the free stream. Laminar flow in the sampling duct is maintained at velocities of 20-80 m/s, insuring that the walls of the instrument have a negligible effect on the measurement. The Vacuum Ultraviolet (VUV) radiation produced in a low pressure plasma discharge lamp is used to induce resonance scattering in Cl and Br atoms within the flow sample. The radicals ClO and BrO are chemically converted to Cl and Br respectively by addition of NO and the rapid reaction $XO+NO{\rightarrow}X+NO_2$.

### B.54.3    H$_2$O Instrument

A photofragment fluorescence hygrometer instrument for the measurement of water vapor in the stratosphere (PI: J. G. Anderson). Water vapor (and ozone) molecules are critical for an integrated understanding of deep convection. Water vapor is the primary drive of convection, and ozone, in addition to being radiatively important, can be used to identify regions of deep convection. This is due to a steep gradient in the lower stratosphere. The instrument was test flown on the ER-2 aircraft (nose) in 1992 and and in 1993 in the CEPEX and SPADE campaigns, it is considered for future flights on PERSEUS A. Note: The combination of the $HO_2$ and $O_3$ instruments flew on the ER-2 aircraft under the designation WOX (Water Ozone Experiment) in CEPEX.[475)]

The water vapor instrument uses the technique of photofragment fluorescence combined with dual-path absorption to measure water vapor concentrations ranging from $10^{13}$ to $10^{16}$ molecules/cm$^3$ at pressures from 50 to 500 mb. The detection scheme utilizes Lyman-alpha photons at a wavelength of 121.6 nm to photo-dissociate water vapor and produce excited OH molecules, which emit photons that are detected by a PMT (Photo Multiplier Tube) near 314 nm. Dual-path absorption measurements (path length is 9.9 cm), which provide a self-consistent check in the laboratory, are carried out during the ascent and descent part of each flight of the aircraft to verify the fluorescence calibration. In-flight diagnostics, such as periodically changing the air flow velocity, confirm that the water vapor measurements are not contaminated by the walls of the instrument.

---

475)  Information provided by E. Weinstock of Harvard University

Laboratory calibration: A stable water vapor concentration is provided by an air flow established by a 5 SLM (Standard Liters per Minute) flow controller bubbled through water and premixed with a flow of air from a larger flow controller. The mixture is then fed into the fast flow system through a loop injector. The water vapor concentration in the flow tube can be measured by absorption down the center of the flow tube, and for water vapor concentration $\geq 2\,e^{14}$ by the dual-path absorption measurement that is part of the flight instrument.

In-flight calibration: A vacuum photodiode, positioned across the duct from the lamp, serves as a beam flux monitor of Lyman-alpha emission. A doughnut-shaped VUV (Vacuum UV) spherical mirror surrounds the photodiode and focuses the Lyman-alpha radiation back across the duct to a second photodiode. In the presence of sufficient water vapor, the two diode measurements provide a dual-path absorption measurement, independent of the lamp intensity. This absorption measurement made simultaneously with fluorescence provides an in-flight calibration.

## B.54.4  $O_3$ Instrument

An absorption instrument with the objective to measure ozone in the altitude region between 10 and 25 km (PIs: E. Hintsa, E. Weinstock). The instrument consists of an absorption cell, an ozone scrubber, and a stabilized 254 nm light source.

The instrument determines ozone concentrations by measuring the absorption of 253.7 nm radiation. Ambient air is alternately drawn through a scrubber that chemically removes ozone and through a Teflon inlet tube. With ozone scrubbed air flowing through the detection cell, the reference signal ($I_o$) is determined. The ozone signal ($I$) is determined with ambient air flowing through the cell. The ozone concentration is then determined with ambient air flowing through the cell. The ozone concentration is the determined by Beer's law, where $\sigma$ is the cross section and $\lambda$ is the path length.

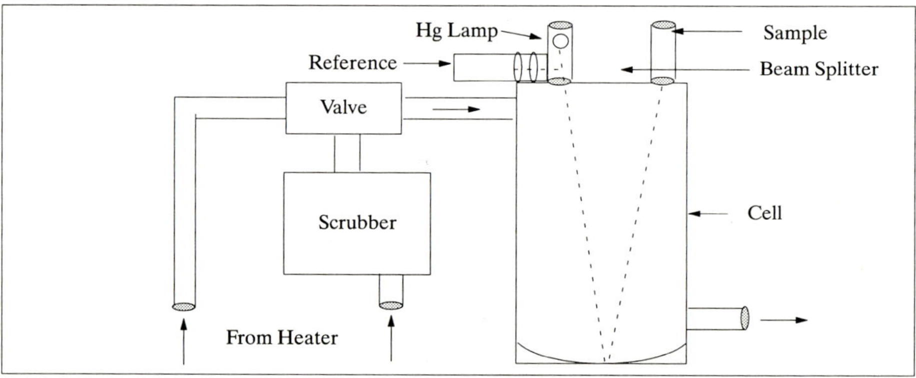


**Figure 121:**　**Schematic Arrangement of the Ozone Instrument**

## B.54.5  $ClONO_2$-Instrument

The instrument consists of a heater to thermally dissociate $ClONO_2$ followed by titration of the ClO fragment with NO gas addition and resonance fluorescence of the Cl atom produced in the reaction with NO. The objective is to measure $ClONO_2$ from 15 to 25 km to help understand the partitioning of inorganic chlorine in the stratosphere.

Plans are underway to develop and build a $ClONO_2$-instrument for the Perseus aircraft. The method of detection will focus on the selective thermal dissociation of $ClONO_2$ into ClO and $NO_2$ fragments. The ClO fragment will then be quantitatively detected along with ClO.

## B.54.6   $NO_y$-Instrument

A Harvard-developed instrument scheduled for test flights on PERSEUS A in April 1994 and for mission flights in July and October 1994. The same instrument will also be flown on PERSEUS B (PI: S. C. Wofsy).[476]

The objective is to measure $NO/NO_y$ in the region between 8 and 25 km (20 - 200 mb) using a lightweight design suitable for use on remote pilotless aircraft. The detection scheme utilized is chemiluminescence of NO with $O_3$, where $NO_y$ is reduced to NO by $H_2$ over gold at 300 °C, with photon counting at wavelengths longer than 625 nm. The measurement period is 1 second with ± 5 ppt. The lightweight design includes operation at ambient pressure, allowing a small roots pump, replacing a large vacuum pump, to move a large volume of sample (2l/s). A new cell design has been implemented that rapidly mixes reagent $O_3$ with the sample just upstream of the cell, to eliminate artifacts associated with incomplete mixing. The instrument is calibrated by adding known flows of NO, $NO_2$, or other species to the sample, the null signal is determined by frequently switching the reagent ozone flow to a reaction volume where the sample NO is tritrated to $NO_2$, out of view of the photomultiplier.

## B.54.7   $CO_2$-Instrument

A Harvard-developed instrument for measuring $CO_2$ in the stratosphere with high precision ( ± 0.05 ppm) and accuracy ( ± 0.1 ppm). (PI: S. C. Wofsy). The measurement period is 2 s. The instrument measures absorption of light at 4.26 μm wavelength (IR) in a 12 cm cell maintained at 340 torr, and compares to absorption in an identical cell containing a reference gas with known concentration of $CO_2$. Calibration is maintained by filling the sample cell with three different standard mixtures at frequent intervals during the flight. Stability and high precision are obtained by carefully regulating temperatures and pressures throughout the instrument. Accuracy is determined by comparing primary and secondary standards to archival standards at the Scripps Institution for Oceanography.

## B.55   HIS (High-Resolution Interferometer Sounder)

The HIS[477],[478] instrument is an airborne Michelson-type interferometer and a calibrated Fourier Transform Spectrometer that was developed (from 1983-85) by the Cooperative Institute for Meteorological Satellite Studies (CIMSS) of the University of Wisconsin at Madison (sponsored by NASA/GSFC and NOAA). Objectives: atmospheric temperature and humidity sounding; demonstrate the capability of an interferometer to measure precisely the thermal emission spectrum; explore the use of high-resolution IR spectra over a variety of weather conditions. The primary focus is on the retrieval of temperature and water vapor profiles with a high spectral resolution ($\lambda/\Delta\lambda = 2000$) and high radiometric precision (0.1-0.2 °C RMS noise equivalent temperature).

HIS is a nadir-looking instrument that is flown on ER-2 aircraft. The three spectral bands, covering most of the region from 3.6 to 16.4 μm, are split inside a single liquid helium dewar, which contains three sets of bandpass cold filters, focussing optics, and arsenic-doped silicon detectors. The preamplifiers are external and operate near the ambient pod temperature of about 260 K. The gain of each channel is fixed, the signals are digitized with a 16 bit A/D converter. Onboard numerical filtering is used to reduce the sample rate from the HeNe Laser rate by factors of 14, 8, and 8 in bands I, II, and III.

476)   Information provided by S. Wofsy of Harvard University
477)   "High-Resolution Interferometer Sounder (HIS) Phase II", A Report from the Cooperative Institute for Meteorological Satellite Studies, University of Wisconsin-Madison, October 1988
478)   H. E. Revercomb, et. al, "Radiometric calibration of IR Fourier transform spectrometers: solution to a problem with the High-Resolution Interferometer Sounder", Applied Optics, Vol. 27, Nr. 15, 1 August 1988, pp. 3210-3218

Absolute instrument calibration at each wave number is provided by viewing two high-emissive blackbodies, that are temperature-controlled to 300 K and to about 240 K. The noise-equivalent temperature and calibration accuracy are approximately 0.1-0.2°C and 0.5-1.0°C over much of the spectrum. HIS calibration observations of the two on-board reference blackbodies are made every two minutes. Each group of HIS interferograms consists of two cold blackbody views, two hot blackbody views, and six Earth views from each scan direction of the mirror.

The Bomem Michelson interferometer (developed by Bomem Inc. of Quebec, Canada) provides double-sided interferograms from both scan directions. Its auto-alignment system makes it possible to operate in the ambient thermal environment of the pod and in very close proximity to the aircraft jet engine. The optical bench is shock-mounted to damp high frequency vibration, the interferometer is evacuated to protect the beamsplitter during descent.

HIS has been flown on many flights and participated in the following campaigns/projects: Kitt Peak (April 1986), COHMEX (June-July 1986), FIRE-I (October-November 1986), Pacific Ocean (May 1991), CAPE (July-August 1991, SERON (August 1991), FIRE-II (November -December 1991), STORM-FEST (February-March 1992), CAMEX (September-October 1993). The HIS instrument has also been adapted to function as a ground-based temperature and water vapor profiler.

| Parameter | Value | |
|---|---|---|
| Spectral ranges: | | |
|     Band I | 590 - 1070 (cm$^{-1}$) | 9.3 - 16.4 ($\mu$m) |
|     Band II | 1040 - 1930 (cm$^{-1}$) | 5.1 - 9.6 ($\mu$m) |
|     Band III | 2070 - 2750 (cm$^{-1}$) | 3.6 - 4.63 ($\mu$m) |
| Spectral resolution | | |
|     Band I | 0.5 cm$^{-1}$ | (2 nm) |
|     Band II | 1 cm$^{-1}$ | (1 nm) |
|     Band III | 1 cm$^{-1}$ | (1 nm) |
| FOV (Telescope) | 100 mrad | |
| FOV (Interferometer) | 30 mrad | |
| Blackbody reference sources | | |
|     Emissivity | > 0.998 | |
|     Aperture diameter | 1.5 cm | |
|     Temperature stability | ± 0.1 K | |
| Auto-aligned interferometer | modified Bomem BBDA2.1 | |
| Beamsplitter | | |
|     Substrate | KCl | |
|     Coatings (1/4$\lambda$ at 3.3 $\mu$m) | Ge + Sb$_2$S$_3$ | |
| Maximum delay (double-sided, (cm) | | |
|     Band I (hardware limit = ± 2.0) | ± 1.8 | |
|     Bands II and III (limited by data system) | + 1.2, - 0.8 | |
| Michelson mirror optical scan rate | 0.6 - 1.0 cm/s | |
| Aperture stop (at interferometer exit window) | | |
|     Diameter | 4.1 cm | |
|     Central obscuration area fraction | 0.17 | |
|     Area | 10.8 cm$^2$ | |
| Area-solid angle product | 0.0076 cm$^2$ sr | |
| Detectors | | |
|     Type | Arsenic-doped Silicon | |
|     Diameter | 0.16 cm | |
|     Temperature | 6 K | |
| Ground resolution (for 20 km altitude) | 2 to 4 km (approximately) | |

**Table 159:    Characteristics of the HIS Instrument**

HIS[479] is being used as a research tool for a number of applications, such as: estimation of cloud radiative properties in the infrared region (spectral emissivity and reflectivity), cross-validation of the spectral correction algorithm applied by other sensors (such as ERBE).[480] Observations with HIS have already made significant contributions to validating and improving line-by-line radiative transfer models. A long-term conceptual study program is underway at CIMSS (funded by NOAA, NASA, and EUMETSAT) called GHIS (Geostationary HIS). The objective is to investigate on the utility of an interferometer sounder for future weather satellites, based on the experiences gained with HIS.[481]


## B.56    HUTSCAT (Helsinki University of Technology Scatterometer)

HUTSCAT is an active dual-frequency radar sensor designed and developed at the Helsinki University of Technology. The instrument measures the backscattering properties of a target with a range resolution of 0.65 m. The real-time ranging capability is obtained by performing the Fast Fourier Transform (FFT) to the received time-domain signal (the scatterometer can identify the backscattering sources within a distributed target, like forests).

| Parameter | Value |
|---|---|
| Center frequency | 5.4 GHz (C-Band) and 9.8 GHz (X-Band) |
| Modulation type | FM-CW |
| Modulation bandwidth | 300 MHz |
| Polarization modes | HH, VV, HV, and VH |
| Antenna type | Parabolic with ring-loaded dipole-disk feed |
| Antenna size (diameter) | 75 cm (5.4 GHz); 36 cm (9.8 GHz) |
| Incidence angle | 1. Mechanically adjustable support: 0-45° off nadir<br>2. Electrically adjustable support: 0-60° off nadir |
| Antenna look direction | Across flight track |
| Pitch angle compensation (slow) | max. 10° (preset prior to take-off) |
| Radar control | PC |
| Data storage | Cartridge tape unit (60 MByte)<br>Bernoulli box disk drive (44 MByte)<br>Floppy disk drive (1.2 Mbyte) |
| Stored data | Radar return versus range<br>Fourier spectra<br>Time domain signal |
| Calibration methods | Internal (delay time)<br>External (active radar calibrator) |
| Additional equipment | video camera (synchronized to radar) |
| Signal Processing Parameters | |
| Modulation frequency | 60 Hz |
| Number of samples in each FFT | 1024 |
| Sampling frequency (one channel) | 160 kHz |
| Range resolution | 0.65 m |
| A/D converter | 12 bit |
| Data rate (data recording) | 80 kByte/s |

**Table 160:    Technical Parameters of HUTSCAT**

Measurements[482],[483],[484] are made simultaneously at eight channels (VV, HH, HV, and VH polarization modes at 5.4 GHz and at 9.8 GHz). HUTSCAT is operational since 1988 and flown as a research tool on a helicopter (Bell 206 Jet Ranger) with the objective to mon-


479) W. L. Smith, et al., "Remote Sensing Cloud Properties from High Spectral Resolution Infrared Observation", Journal of the Atmospheric Sciences, Vol. 50, Nr. 12, 15 June 1993, pp. 1708-1720

480) S. A. Ackerman, W. L. Smith, H. E. Revercomb, "Comparison of broadband and high-spectral resolution infrared observations", International Journal of Remote Sensing, Vol. 14. Nr. 15, 1993, pp. 2875-2882

481) W. L. Smith, R. E. Revercomb, et al., "GHIS - The GOES High Resolution Interferometer Sounder", Journal of Applied Meteorology, Vol. 29, Nr. 12, December 1990, pp. 1189-1204

482) M. Hallikainen et al., "A Helicopter-Borne Eight-Channel Ranging Scatterometer for Remote Sensing: Part I: System Description", IEEE Transactions on Geoscience and Remote Sensing, Vol. 31 Nr. 1, January 1993, pp. 161-169

483) J. Hyyppä, M. Hallikainen, "Development of a Helicopter-Borne 8-Channel Ranging Scatterometer", HUT, Laboratory of Space Technology Report 4, July 1991

itor forests, sea ice, and snow; HUTSCAT is also being used for the development of algorithms for data interpretation from spaceborne radars. The most potential application in Finland is a forest survey for an estimation of the forest inventory and characteristics.

HUTSCAT provides two microwave configurations - one for ranging and the other for backscattering measurements. Two antennas are used for ranging to achieve a better isolation between the transmitter and receiver. For backscattering measurements the scatterometer employs a single antenna for each frequency. The system measures the radar return spectrum for eight channels in 16.6 ms which corresponds to an along-track distance of 0.33 m (helicopter speed 20 m/s).

## B.57    HUTSLAR (HUT Side-Looking Airborne Radar)

The HUTSLAR real aperture radar was designed and built at the Helsinki University of Technology (HUT) and became operational in 1988. The initial on-board signal processing of HUTSLAR was partially performed in analog form, the data storage system was equipped with an analog video tape recorder. This system was replaced in 1993 by an all digital system.[485]

Applications: Mapping of sea ice, forests and oil spills. In Finland the monitoring of sea ice conditions is useful maritime information. The data of HUTSLAR are used in combination with HUTSCAT and with satellite data.

| Operational frequency | 9.445 GHz (X-Band), $\lambda = 3.174$ cm |
|---|---|
| Polarization | Vertical |
| Antenna beam width | 0.5° (horizontal) |
| Peak power output | 4 kW |
| Pulse repetition frequency | 45 - 450 Hz (depending on flight speed) |
| Pulse width | 0.5 µs |
| Resolution | 75 x 75 m (approximately) at 8.6 km |

**Table 161:    Technical Specification of the HUTSLAR Instrument**

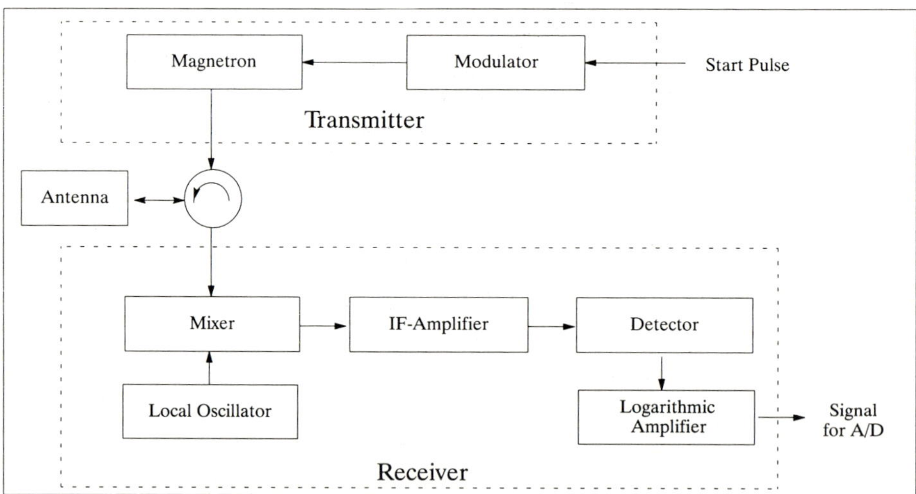


**Figure 122:    Schematic Blockdiagram of the HUTSLAR Instrument**

484) J. Hyyppä, "Development and feasibility of airborne ranging radar to forest assessment", doctoral dissertation at HUT, Finland, November 1993
485) Information provided by P. Ahola of HUT

The HUTSLAR transmitter/receiver signals are incoherent. The transmitter sends 256 pulses/s (constant count, PRF is dependent on air speed), the received echo pulse is digitized at a sample rate of 2 MHz providing a spatial resolution of 75 m. The slant-range resolution is achieved by a narrow horizontal beam of the antenna (beam width = 0.86° x 80°). The slant-range resolution is 87 m at a range of 10 km (167 m at a range of 19 km. HUTSLAR uses a slotted waveguide antenna of 2m length in the flight path direction.

HUTSLAR was initially flown on a helicopter (flight altitudes of 1 km). Current plans (1993) call for an aircraft mounting in order to eliminate the yaw-axis vibration of the helicopter. A GPS system is also considered (as input for better data processing).

## B.58    HYDICE (Hyperspectral Digital Imagery Collection Experiment)

The HYDICE instrument is under development at the Naval Research Laboratory (NRL) in Washington D. C. (built by Hughes Danbury Optical Systems Inc., Danbury, CT, USA). Project phases: contract awarded in Sept 1992, preliminary design review: December 1992, critical design review: June 1993, aircraft integration: April 1994, sensor acceptance: October 1994. It is expected to be operational at the end of 1994 for military and civilian use.[486]

HYDICE is a 206 channel imaging spectrometer in the spectral range from 0.4 - 2.5 μm (VNIR/SWIR/MWIR), that employs the CCD pushbroom technique with a single detector array for coverage of the entire spectral region (full spectral range contiguous sampling). The instrument requirements call for a high SNR and high resolutions (spatial and spectral) of the imaging data.[487]

The on-board system provides tape recording of all data (data rate of 45 Mbit/s), and a quicklook display. The aircraft is equipped with INS and GPS providing navigation data for the engineering telemetry record.

| | |
|---|---|
| Optics system<br>Aperture diameter<br>Objective lens | Paul Baker fore-optics, Schmidt prism spectrometer<br>27 mm<br>f/3.0 |
| IFOV, (FOV) | 0.5 mrad, (8.94°) |
| Spectral coverage<br>Nr. of spectral channels<br>Bandwidth (FWHM)<br>Integration mode<br>Integration time<br>Frame time<br>Readout time<br>Quantization | 0.4 - 2.5 μm, contiguous<br>206<br>7.6 - 14.9 nm<br>simultaneous<br>1.0 -42.5 ms<br>8.3 - 50 ms<br>7.3 ms<br>12 bit |
| Polarization<br>Pixel size<br>Array size<br>Number of pixels per line (swath width)<br>Detector array | <3%<br>40 x 40 μm<br>320 x 210 pixels<br>312 pixels<br>320 x 210 element InSb |
| Mass of instrument<br>Power of instrument | 176 kg<br>1700 W |
| Aircraft<br>Velocity/Height (V/H) ratio<br>Altitude<br>Pointing<br>Orientation | CV-580 (the sensor is mounted outside the pressure hull)<br>0.028 radians/s (0.01 - 0.06, selectable)<br>8 - 10 km<br>± 1° of nadir<br>± 1° of ground track |
| SNR (at 5% albedo) | 250-280 in VNIR (0.4-1.0 μm)<br>100-107 in SWIR (1.0-1.9 μm)<br>50-58 in MWIR (1.9-2.5 μm) |

**Table 162:    HYDICE Instrument Specifications**

---

486) R. Basedow, P. Silverglate, W. Rappoport, R. Rockwell, D. Rosenberg, K. Shu, R. Whittlesey, E. Zalewski, "The HYDICE Instrument Design", International Symposium on Spectral Sensing Research (ISSSR), 15-20 Nov. 1992, Maui, Hawaii
487) Information provided by D. Pope of NRL

Program objective: demonstrate the utility of hyperspectral imaging through the acquisition and analysis of high-quality hyperspectral data. - Civil applications are in the following fields:

- Agriculture: crop analysis, pest control, stress analysis
- Forestry: monitoring of inventory, habitat mapping, pest control, reforestation
- Environment: monitoring of toxic waste, acid rain, air pollution, eutrophication, soil conservation, water pollution
- Resource management: land use, mineral identification
- Mapping: area classification, bathymetry, wetlands, critical habitats
- Disaster management: damage assessment, search and rescue support
- Law enforcement: a host of applications

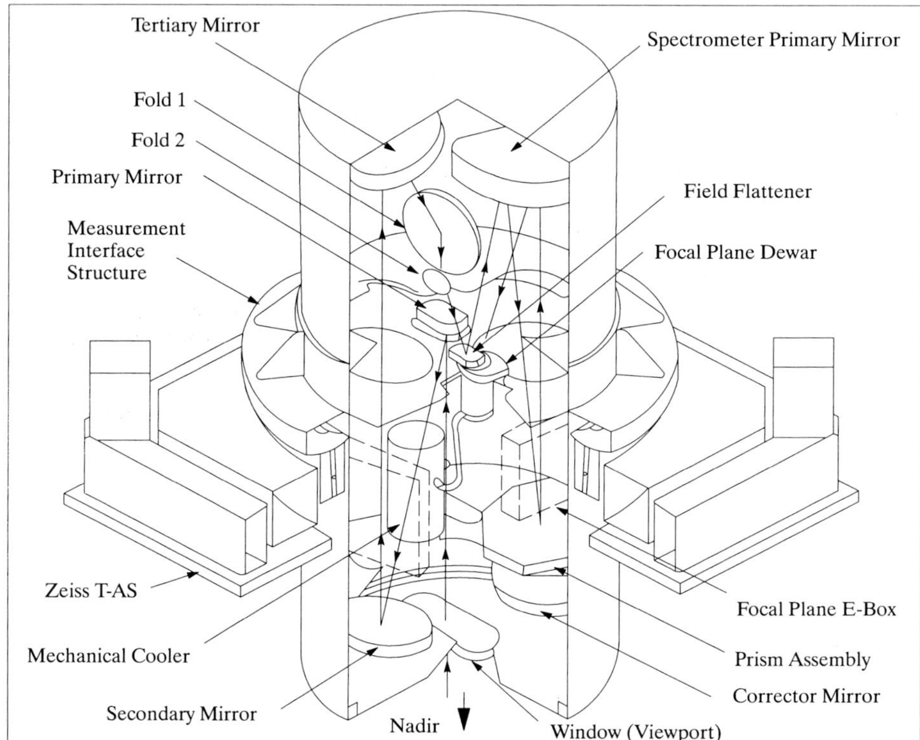


**Figure 123:    Schematic View of the HYDICE Instrument**

## B.59    ISM (Infrared Imaging Spectrometer)

ISM is a French airborne spectrometer developed by DESPA of the Paris/Meudon observatory, as part of a planetary mission, in collaboration with IAS. CESR (Centre d'Etude Spatiale des Rayonnements) realized an airborne version of the instrument on the ARAT (Atmospheric Research and Remote Sensing Aircraft). The first test flight occurred in 1991.

ISM operates in the NIR and MWIR portion of the spectrum, it provides 128 spectral channels between 0.8 and 3.2 µm. Objectives: geology, clouds, ice and snow, vegetation (i.e. forest and agricultural), with special emphasis on the Mid-Wavelength Infrared (MWIR) region for canopy chemistry (lignin, nitrogen, cellulose) analysis.[488),489)]

ISM employs an electro-mechanical scanner design [scan mirror located in front of the collecting aperture, (whiskbroom principle)] that projects the entire spectrum of a resolved element onto two CCD arrays. The spectrometer provides a fixed grating system which measures the radiation wavelengths in three ranges. The optical system features a dichroic blade and beam splitters for wavelength selection. ISM has two 64 PbS detectors located in parallel rows providing acquisitions with 128 contiguous spectral bands. The detection system is cooled by passive cryogenics.

| Spectral coverage | 0.80 - 1.6 µm (64 bands of 12.5 nm bandwidth)<br>1.6 - 3.2 µm (64 bands of 25 nm bandwidth) |
|---|---|
| IFOV (across track) x (along track)<br>Ground resolution at 3 km altitude<br>FOV (selectable) | 3.3 mrad x 11.7 mrad<br>9 m (across track) x 35 m (along track)<br>40° |
| Swath width (at 3 km altitude)<br>Scan rate (selectable)<br>Quantization<br>Integration time<br>SNR<br>Detector | <2 km<br><30 steps/s<br>12 bit<br>30 ms<br>~100 ($\lambda$ <2.7 µm)<br>PbS cooled |
| Sun calibration<br>Instrument internal calibration | with optical fiber<br>Blackbody |
| Telescope characteristics | |
| Aperture<br>Diameter<br>Focal distance<br>Magnification<br>System temperature | 25 mm<br>25 mm<br>100 mm<br>8<br>-70°C to -30°C |

**Table 163:     Specification of the ISM Instrument**


# B.60    LAC (Large Area Collector)

LAC is a NASA/JSC designed and developed instrument flown on NASA/Ames ER-2 aircraft in support of the JSC Cosmic Dust Program. The original objective was to collect extraterrestrial dust samples from the stratosphere; the goals have been broadened since then to encompass terrestrial and space debris particulates as well.[490)]

Background: JSC began the Cosmic Dust Program in May 1981 with the aim of systematic collection and curation of cosmic dust for scientific investigation. Initial collections were made with instruments mounted underneath a wing using NASA WB-57F aircraft from JSC, the program was extended in 1988 to include Ames U-2 aircraft and since 1989 ER-2 aircraft. A new generation of collection system, namely LAC, was introduced along with the ER-2 aircraft dust collection support of the program. Catalogs of dust samples are kept at JSC. As of 1993 the program continues to fly both small collectors and LACs.

The dust collectors have surface areas of 300 $cm^2$ which are coated with silicone oil (dimethyl siloxane) and flown ER-2 aircraft at high altitudes (20 km). The collectors are installed in a specially-constructed wing pylon, which ensures the required level of cleanliness during active sampling periods. LAC instrumentation is exposed in the stratosphere by barometric controls and then retracted into sealed storage containers prior to aircraft descent. In the laboratory at JSC, particles are individually removed from the collectors using glass-needle micro-manipulators under a binocular stereo-microscope. All processing and storage of

488)  F. Zagolski, et al., "Preliminary Results of the ISM Campaign - The Landes, South West Frances-", IGARSS '92,
       Vol. I, Houston Texas, May 26-29, 1992, pp. 6-8
489)  Information provided by J. P. Gasstellu-Etchegorry of CNRS, Toulouse, France
490)  'Cosmic Dust Courier', Nr. 1, March 1982, NASA/JSC

each particle is performed in a Class-100 clean room. Particles are cataloged with descriptions to size, shape, transparency, color, luster, type, etc. All samples collected are available to scientists worldwide.[491),492)]

Sufficient quantities of extraterrestrial materials are collected to allow chemical and mineralogical compositions of individual particles to be determined. Study of these materials, whose sources may be comets, dust of asteroid collisions or of planetary impacts, and meteorite ablation, provide valuable information about our solar system. Most collected particles have terrestrial origins - they are also available for study from NASA/JSC.

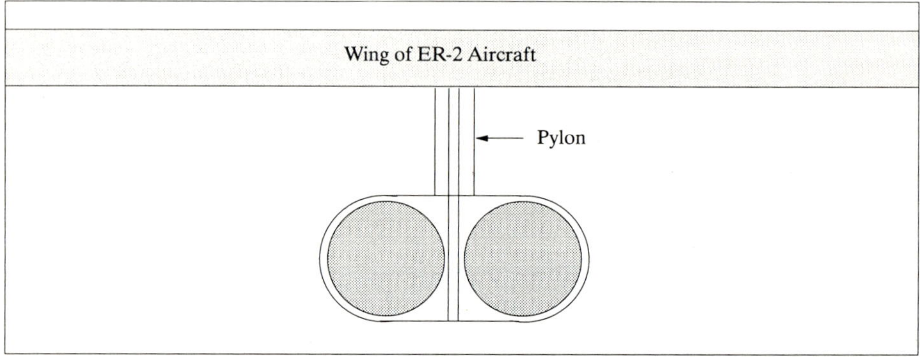


**Figure 124:    Frontview Illustration of opened Large Area Collectors on ER-2 Aircraft**

## B.61    LARSEN (Airborne Scanning Lidar)

Larsen is a CCRS sensor (developed and built by Optech Inc., CCRS, and the Canadian Hydrographic Service) operational since 1985 for the survey of coastal areas. Larsen measures water depths of up to 40 m in clear coastal water with an accuracy of 0.3 m. The system has a swath width of 270 m from an altitude of 500 m and generates a uniform sounding grid pattern, with the soundings spaced about 30 m apart and a position accuracy of about 15 m.

Operating principle: The system uses an optical radar (lidar) technique to measure water depth from an aircraft. A short pulse of infrared radiation and a short pulse of green radiation are simultaneously emitted from the laser toward the water surface. The infrared pulse is scattered from the water surface, while the green pulse penetrates the surface and is scattered from the bottom. Both scattered pulses are detected by a receiver in the aircraft, the elapsed time between the pulses is used to measure the water depth.

Larsen consists of the following major subsystems: transmitter, receiver, electronics, guidance system for positioning, a video camera/video disk system to provide along-track imagery, and a data logger for digital data storage.

The lidar uses a frequency-doubled Nd:YAG laser, operating simultaneously at 1064 and at 532 nm. The average repetition rate is 20 Hz. The laser output beam is directed towards the water by a scanning mirror which with the laser firing control allows a specific and uniform grid pattern to be generated on the water. The location of each laser sounding is determined using aircraft position, altitude and attitude information. Aircraft position is provided by a microwave ranging system, using ground-based transponders of known location, and by GPS. Aircraft height above water is measured by an infrared laser altimeter, and attitude data by an inertial reference system mounted to the sensor.

491)  R. A. Barrett, A. L. Dodson, K. L. Thomas, J. L. Warren, L. A. Watts, M. E. Zolensky, "Cosmic Dust Catalog",
       Vol. 13, NASA/JSC, September 1992
492)  'Cosmic Dust Courier', Nr. 8, Feb. 1989, NASA/JSC

# B.62 LASE (Lidar Atmospheric Sensing Experiment)

A NASA-sponsored instrument[493] developed at Langley Research Center (Hampton Va.) with the objective to measure water vapor and aerosol in the atmosphere with a fine vertical resolution in the altitude range from 0 to 10 km during daytime and nighttime surveys. LASE is based on the DIAL (Differential Absorption Lidar) technique, and uses a tunable Ti:Sapphire laser transmitter. The instrument is regarded the first step in a NASA effort to develop and demonstrate the DIAL technology for airborne flight experiments.

| Atmospheric Parameter | Investigation Region | Altitude Range (km) | Spatial Resolution Horizontal (km) | Vertical (m) | Measurement Uncertainty (%) Night | Day (ocean albedo) |
|---|---|---|---|---|---|---|
| $H_2O$ Profile | Mesoscale | 0 - 10 | 10 | 200 | 10 | - |
| | | 0 - 10 | 20 | 300 | - | 10 |
| $H_2O$ Column Content | Sub-cumulus scale | - | 40 m | - | 8 | 10 |
| Atmospheric Backscatter | Aerosol and Clouds (sub-cumulus clouds) | 0 - 14 | 40 m | 50 | 3 | 5 |

**Table 164: LASE Measurement Objectives**

The tunable laser subsystem (TLS) output laser beam is from a Ti:Sapphire laser pumped with a doubled Nd:YAG laser. The wavelength of the Ti:Sapphire laser is tunable from 813 - 818 nm and controlled by injection seeding with a tunable laser diode. The ND:YAG is a flash-lamp pumped and the doubling crystal is CD*A with an output of 660 mJ of energy at 532 nm.

The telescope is a Dall-Kirkham design. The FOV is adjustable remotely by an electronic field stop at the focal point of the telescope. A 300 pm FWHM bandpass filter is used in the aft optics to minimize background noise during daytime missions. To accommodate the in-flight selection of two different pre-selected $H_2O$ absorption lines, the narrow-band filter can be tilted up to $10°$ for center-band shifts of approximately 2.5 nm (accuracy of tilt mechanism = $\pm 0.1°$). - As of 11/1993 LASE is in the test phase and scheduled for integration on the AMES ER-2 aircraft in the winter of 1993/94.

| Transmitter: (Ti:Sapphire laser) | |
|---|---|
| energy | $\geq$ 150 mJ average |
| line width | 1.1 pm |
| pulse repetition rate | 5 Hz |
| wavelength (tuning range) | 813 - 818 nm |
| beam divergence | 0.73 mrad |
| pulse width | 300 ns |
| Receiver: effective area | 0.11 m$^2$ |
| FOV | 1.23 mrad |
| Filter bandwidth | 0.4 nm (day) |

**Table 165: LASE Sensor Specifications**

# B.63 LEAF (Laser Environmental Airborne Fluorosensor)

LEAF[494] is a Canadian/US airborne sensor [supported since 1987 by: Canadian Panel on Energy Research, US Minerals Management Service, Fisheries and Oceans Canada, CCRS (Canada Centre for Remote Sensing), American Petroleum Institute, and the US Coast Guard; built by Barringer Research Ltd. of Rexdale, Ontario and BP] for the detection and

493) W. R. Vaughan, E. V. Browell, "A Lidar Instrument to measure $H_2O$ and Aerosol Profiles from the NASA ER-2 Aircraft", paper presented at the Specialty Meeting on Airborne Radars and Lidars, Meteo-France, Toulouse, July 7-10, 1992

494) R. Dick, M. Fruhwirth, M. Fingas, C. Brown, "Laser Fluorosensor Work in Canada", presented at the First Thematic Conference on Remote Sensing for Marine and Coastal Environments, New Orleans, La, Ju. 15-17, 1992

mapping of oil spills on water, ice and shorelines, and of chlorophyll and other environmental variables in near surface waters. As of 1992 the LEAF system is installed aboard a twin engine Aero Commander aircraft and operated in test programs by Barringer Research Ltd.

LEAF is an active sensor (a short laser light pulse is being emitted to the surface, the reflected echo is analyzed) capable of day and night operation. It uses a XeCl Excimer laser, operating at a wavelength of 308 nm (UV range), measuring fluorescence at 64 (channels) colors between 322 nm and 665 nm.

The excitation wavelength of 308 nm seems to be well suited to the stimulation of fluorescence in oils. But laser fluorescence can also be applied to monitor other objects/targets in the field of oceanography and fishery, requiring other (longer) excitation wavelengths.

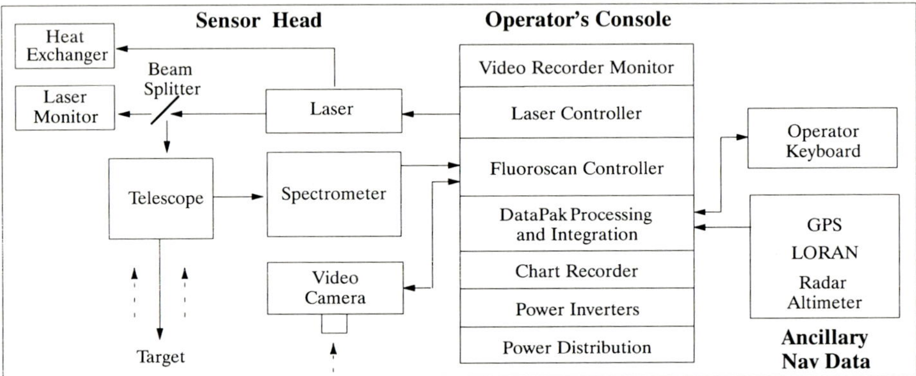


**Figure 125:**    **LEAF Functional Block Diagram**

| Laser Transmitter | | Spectrometer | |
|---|---|---|---|
| Type | Excimer, XeCl | Optics type | Czemy-Tuner |
| Wavelength | 308 nm | F - Ratio | Entrance, f/3.1; exit, f/2.2 |
| Output energy | 20 mJ/pulse (typical) | Grating | 600g/mm; blazed @ 400 nm |
| Average power | 2 W | Detector | Gatable, intensified 1024 diode |
| Peak power | 3 MW | Gate width | > 20 ns |
| Pulse width | 7 ns | Spectral range | 320-635 nm or 525-696 nm |
| Beam divergence | 2.5 x 7.0 mrad | Channels | 32 to 64 user selectable |
| Pulse rate | 100 Hz | Resolution | 4.7 nm or 2.8 nm |
| Telescope | | Lidar Altimeter | |
| Type | Dahl-Kirkham | Range | 30 - 300 m (100 - 1000 ft) |
| Diameter | 203 mm | Cal constant | 10 mv/ft |
| Focal ratio | f/3.1 | Resolution | + 0.5 ft RMS |
| Focus | 30 m to infinity | Temp. coefficient | 0.013 ft/C |
| Field of View (FOV) | 1 x 3 mrad | Accuracy | +0.5 ft @ 25C (calibrated) |
| | | Gate offset range | surface + 30 ft in 5 ft steps |
| | | Gate width | 20 - 180 ns in 10 ns steps |
| Controller | | DataPak & Peripherals | |
| Function | triggers laser; digitizes and pre-processes data; communicates with DataPak | DataPak | applies background correction calculates spectral averages laser power, altitude, nav data formated for tape archive |
| System | STD-Bus with 7909A termination | Chart recorder | 5 spectrally averaged banbs laser backscatter |
| Master CPU | RLC SBC-188 | | peak fluorescence |
| Slave CPU | ZT-8830 | | ancillary data |
| DataPak I/O | SBX-20 GPIB I/F | Aircraft navigation | GPS |
| Panel I/O | PL7605-1 32-bit p. | | LORAN |
| GDAS I/O | 2x RS422 I/F boards | | Radar altimeter |
| A/D | 2xRTI 1260 | | |

**Table 166:**    **LEAF System Specifications**

## B.64   LEANDRE

LEANDRE (Lidar Embarqué Aerosols Nuages Dynamique Rayonnement Environnement; or:Airborne Lidar for dynamic radiation studies of the environment in particular of aerosols and clouds) is an airborne lidar of CNRS/CNES, France for the study of the lower atmosphere with emphasis on mesoscale boundary layers and free troposphere meteorology. Observation of the following atmospheric profile variables: aerosols, water vapor, pressure, temperature, and wind. The instrument can be used in a nadir or zenith pointing direction depending on mission requirements. LEANDRE serves also as a test tool for the ESA ATLID sensor. LEANDRE is flown on a Fokker F27 aircraft.

| Lidar instrument | LEANDRE (CNES/CNRS, France), operational since 1990 |
|---|---|
| Measurement technique | Incoherent backscatter |
| Laser, transmitter wavelength | Nd:YAG: 532 nm, and 1064 nm, |
| Detection channels | one at wavelength 1064 nm for polarized backscatter<br>two at wavelength 532 nm for depolarized backscatter |
| Pulse repetition frequency | 10 Hz |
| Laser energy/pulse | 150 mJ at 532 nm; 250 mJ at 1064 nm; |
| Resolution | 15 m vertical; 10 m horizontal |
| Lidar beam pointing configurations | nadir pointing in two modes: fixed or scanning at $\pm 15^\circ$, or fixed zenith pointing |
| Telescope | 30 cm diameter, Ritchey Chretien |
| A/D converter sampling frequency | 10 MHz |
| Research objectives/applications | Clouds and boundary layer distributions. The depolarized channel (532 nm) is the most suitable for low cloud and BL objectives. |

**Table 167:    Specifications of the LEANDRE Instrument**

## B.65   LFS (Laser Fluorosensor)

The LFS is an active sensor (of OLS heritage) with the objective to analyze the upper layers of the sea surface (in the nadir direction) from airborne altitudes in the 100 - 300 m range. In particular, the following applications are pursued:

- identification and classification of pollutants on the sea surface
- determination of the film thickness of oil spills (0.1 - 10 μm range) and estimation of the oil volume from a 2-D mapping of the spill
- detection of swimming chemicals and of substances drifting underneath the sea surface
- measurement of gelbstoff concentrations for hydrographic interpretation
- measurement of algae for biological productivity estimations

LFS[495] was financed by the German Ministry of Research and Technology (BMFT), it was developed by the University of Oldenburg jointly with Krupp MaK, Kiel. Regular operation within maritime surveillance is done by Optimare, Wilhelmshaven.

The instrument was installed in 1990 on a DLR aircraft (DO 228) for initial test flights. It is now (1993) being installed on a German Navy aircraft along with the instrument MERES (Do 228, equipped with OMEGA and GPS navigation) for regular operational maritime pollution surveillance of the North Sea and Baltic Sea, starting in October 1993.

The LFS profiling instrument consists of a high power pulse laser and of a telescope receiver and spectrograph for the detection of laser-induced fluorescence and scattering. The laser wavelength is chosen in the UV or blue/green portion of the spectrum where a penetration depth of a few meters into the water can be achieved. At these wavelengths, fluorescence of suspended and dissolved organic matter of natural origin can be detected, also chlorophyll

---

495)  EARSeL Advances in Remote Sensing, Vol. 1, Nr. 2, Feb. 1992, p. 85-90

contained in algae, and gelbstoff. In addition to gelbstoff and chlorophyll fluorescence, Raman scattering of water molecules is observed.[496]

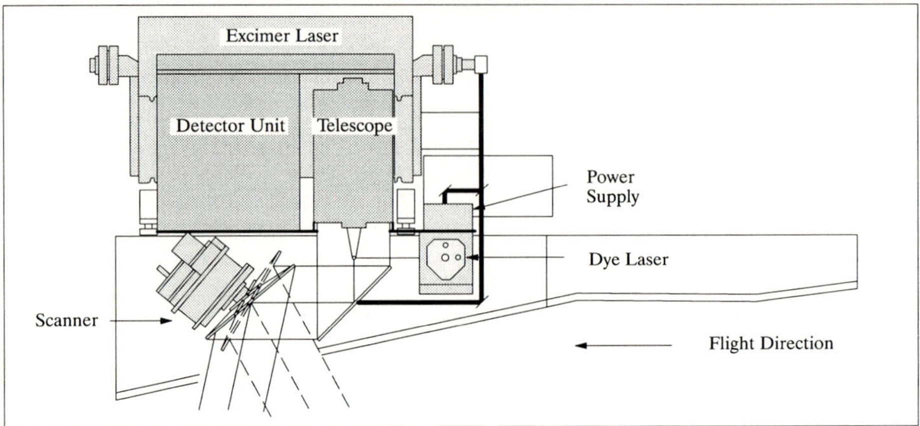


**Figure 126:     Schematic of the Laser Fluorosensor LFS**

| | |
|---|---|
| **Operating parameters:** | |
| Size (L x W x H) | 1270 x 355 x 958 mm excimer laser; 961 x 460 x 944 mm detector unit |
| Weight | 317 kg |
| Flight altitude | 300 m (typical for daylight); 1000 m (typical for night flights) |
| Aircraft ground speed | 100-200 knots |
| Electrical power | 1.0 kVA (in standby), 3.4 kVA (operation, at 110 Hz pulse rate) |
| Data interpretation | Real-time |

| | | |
|---|---|---|
| **Lasers:** | XeCl excimer | Dye (excimer laser pumped dye laser) |
| Emission wavelength | 308 nm | 382 nm |
| Pulse energy | 150 mJ | 20 mJ |
| Pulse length | 20 ns | 15 ns |
| Beam divergence | 5 mrad | 3 mrad |
| Repetition rate | 200 Hz max, 110 Hz average | 20 Hz |

| | |
|---|---|
| **Receiver Telescope** | reflective, Schmidt-Cassegrain |
| Entrance aperture | 20 cm (f/10) |

| | |
|---|---|
| **Scanner** | conical type |
| Full scan angle | 28° across-flight direction, 35° in-flight direction |
| Scan frequency | ≤ 20 Hz, selectable |
| Swath width | 150 m at 300 m flight altitude |
| Pixel-to-pixel distance | 10 m typical at 300 m flight altitude (200 Hz max laser rep. rate) |

| | |
|---|---|
| **Spectrograph** | 12 discrete channels, modular |
| Wavelengths (nm) | 344 $\qquad$ Raman scatter, $\lambda_{las}$ = 308 nm |
| | 330/365/380 $\qquad$ Raman baseline, oil and gelbstoff fluorescence |
| | 440 $\qquad$ Raman scatter, $\lambda_{las}$ = 382 nm |
| | 410/470 $\qquad$ Raman baseline, oil and gelbstoff fluorescence |
| | 500/550/600/650 oil and gelbstoff fluorescence |
| | 685 $\qquad$ chlorophyll fluorescence |
| Wavelength selection | dichroic splitters, interference and blocking filters |
| Optical bandwidth | 10 nm typical |
| Detectors | head-on PMT, range gated |
| A/D conversion | 12 channel gated integrator, 11 bit |

| | |
|---|---|
| **Computers** | All VME-Bus, 68020 processor, 16 MHz |
| Sensor control | Selection of lasers & detectors, control of cycle, film thickness estimation |
| Classification | Identification of substances |
| Image processing | Central operator console; images of LFS ground pixel distribution, evaluation of multisensor imagery, data storage on streamer or optical disk |

**Table 168:     Overview of LFS Parameters**

496) K. Grüner, R. Reuter, "A new Sensor System for Airborne Measurements of Maritime Pollution and of Hydrographic Parameters", GeoJournal 24.1, 1991, pp. 103-117

# B.66    LIP (Lightning Instrument Package)

LIP is a MSFC sensor complement used to investigate relationships between lightning and storm electrification along with a number of underlying and interrelated phenomena - including structure, dynamics, and evolution of thunderstorms and thunderstorm systems, precipitation distribution and amounts, atmospheric chemistry processes, the global electric circuit, and ionospheric/magnetospheric coupling. In these investigations, an emphasis is placed on developing algorithms that employ lightning data (and coordinating access to that data), that are observed by optical satellite sensors and that will be observed by future satellite sensors for lightning detection such as OTD (Optical Transient Detector), LIS on TRMM (A.111) and the Optical System on FORTE (A.33).[497]

LIP has a heritage that goes back to 1980 and included sensors to measure the optical and electrical waveforms associated with lightning and lightning spectra in the VIS and NIR ranges. There are upgraded LIP versions as of 1990 that are being flown on ER-2 and DC-8 aircraft in various campaigns (like TOGA/COARE in 1993, CAPE in 1992, StORM-FEST in 1992) along with a number of other sensors (e.g. infrared and microwave sensors) for the acquisition and analysis of multi-parameter datasets. The LIP instrumentation can detect total storm lightning and differentiate between intra-cloud and cloud-to-ground discharges. LIP is considered a technical input provider for satellite sensor development as well as a ground truth data collector for the satellite instruments.

The program's objective is also to 'quantify' the lightning relationships such as: occurrence and location of embedded convection, the strengths of updrafts and downdrafts, thermodynamic and electrical energy budgets, precipitation amounts and distribution, storm type, dimensions, life cycle, lightning rates, -distribution, -characteristics (i.e. number of strokes per flash, ratio of intra-cloud to cloud-to-ground lightning, discharge energy, etc.).

**ER-2 LIP Instrumentation (since 1986)**
LIP consists of two electric field mills, a conductivity probe, and a data system. One of the mills is installed on the upper Q-bay hatch cover, the second one is mounted the aft AMPR faring or on the lower E-bay hatch cover. The conductivity probe is integrated on the right-hand superpod nose cone.

The electric field mills measure the vertical component of the electric field ($E_z$) over a dynamic range exceeding 3 orders of magnitude (i.e. there are 3 gain channels x1, x46, and x2116) for the measurement of fair weather electric fields as well as large storm systems (e.g. 10-20 kV/m). Also measurement of the electric charge on the aircraft.
The conductivity probe provides a measure of the air conductivity at the aircraft altitude. Conductivity contributions due to positive and negative ions are measured simultaneously.

All data are recorded on-board. The data products are: a) electric field components and aircraft self-charge (continuous record throughout the flight), resolution = 10 Hz; b) air conductivity (continuous record throughout the flight), resolution = 10 Hz;

**DC-8 LIP Instrumentation**
LIP consists of 4 electric field mills, a lightning location detector (i.e. a 3 M Stormscope), and a data system. The field mills measure the electric field vector components over a dynamic range from $E \leq 1$ V/m to over 106 V/m. Abrupt electric field charges are used to identify lightning discharges (flash type and characteristics, storm lightning activity, etc.). The 3M Stormscope is a lightning magnetic direction finder that detects lightning sferics, obtains a bearing to the discharge, and estimates the distance (detection capability up to a radius of 200 nautical miles).

Absolute calibration of the field mills on the aircraft requires the addition of a "stinger" to the plane. This is a device that allows the aircraft to be charged to a high known voltage in

---

497)  Information provided by R. J. Blakeslee of NASA/MSFC

order to more accurately determine the field mill response to electric fields, the geometric enhancement of the aircraft, and the aircraft charge.

## B.67    M-7 (Mapper Multispectral Testbed)

The M-7 multispectral sensor was developed by ERIM (Ann Arbor, Michigan) - for NASA and DOD in 1971/72 to provide airborne imagery and to support research in the application of remote sensing. [Note: the 'M' in the sensor name stands for 'Michigan', when ERIM was still part of the University of Michigan - the term 'testbed' refers to the fact that instrument configuration is being changed to meet customer needs, and to keep the technology to state-of-the-art]. The sensor and aircraft instrumentation experienced various upgrades during the 70's and 80's including: digital recording of data, increase of the spectral bands from 12 to 16, utilization of a radar altimeter, installation of an IMU for sensor attitude information, and the use of LORAN-C and GPS positioning systems. - The following upgrades of the system were completed and demonstrated in the summer of 1993:[498]

- Increase of the wavelength bands from the original 15 to a total of 31 bands in the spectral range from 0.35 to 12.5 μm.
- Installation of a Laser Inertial Navigation System (LINS, from Honeywell, H-770) and dual Trimble 4000 GPS receivers for differential mode operation (DGPS).

These upgrades were the result of user needs for high-quality, calibrated and spatially registered multispectral imagery in map coordinates. Both navigation systems (LINS and DGPS) are integrated into a MMSS (Motion Measurement Subsystem); both LINS data and the multispectral imagery are precision time-tagged using the GPS clock reference. Modified software in the LINS allows access not only to the real-time navigation solution, but also to all raw accelerometer and gyro signals and compensation coefficients for subsequent post-processing on the ground.

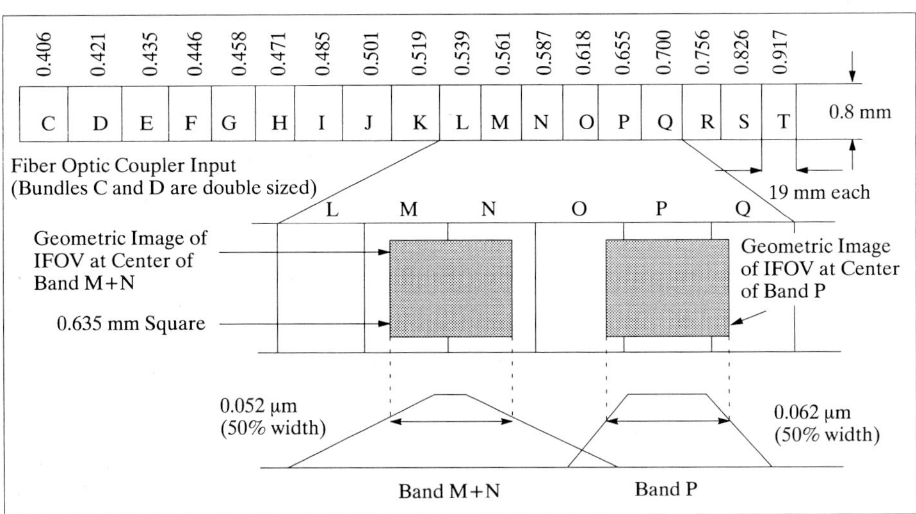


Figure 127:    Typical Spectrometer Band Responses and Geometrical Relations

[Post-processing consists of first developing the differential GPS (DGPS) solution, and then iteratively processing the LINS/DGPS data through a 61-state Kalman filter. This combines


498) D. C. Carmer, R. Horvath, D. P. Rice, J. W. Sisak, "M-7 Mapper Multispectral Testbed", paper presented at the Second Thematic Conference on Remote Sensing for Marine and Coastal Environments, New Orleans, La., 31 January - 2. February, 1994

the best qualities of the LINS navigation data (short-term precision) with those of the DGPS data (long-term accuracy) to produce an optimized total (six degrees of freedom) navigation solution for use by the M-7 Mapper camera model for image reconstruction.]

M-7 applications include terrain and topographic mapping, multipass stereo imaging and simulation of imagery from broad-area, fine-spatial resolution space-based sensors [Landsat (TM), Spot (HRV), etc.]. The overall M-7 system provides a capability to accomplish precision image correction, orthorectification, geocoding and multi-pass mosaics. ERIM provides its services of data collection and processing on a contractual basis to any customer (past customers have been: NASA, DOD, USGS, Department of Agriculture, state and local governments, and some private companies (oil, geological surveys).

The M-7 mapper is an opto-mechanical imaging spectrometer, the design features four clearly-defined detector regions. A common aperture and a 12.7 cm diameter 45° scan mirror is used to collect radiation for all four detector positions.[Note: the term 'position' means 'optical port' for a particular detector type covering a particular spectral range]. Two-dimensional detector arrays (CCDs) are used in positions 1 and 3, this requires special geometrical preprocessing to de-rotate the array pixels and to bring all spectral bands to the location of a common reference band, usually to the one of the spectrometer bands in position 4.

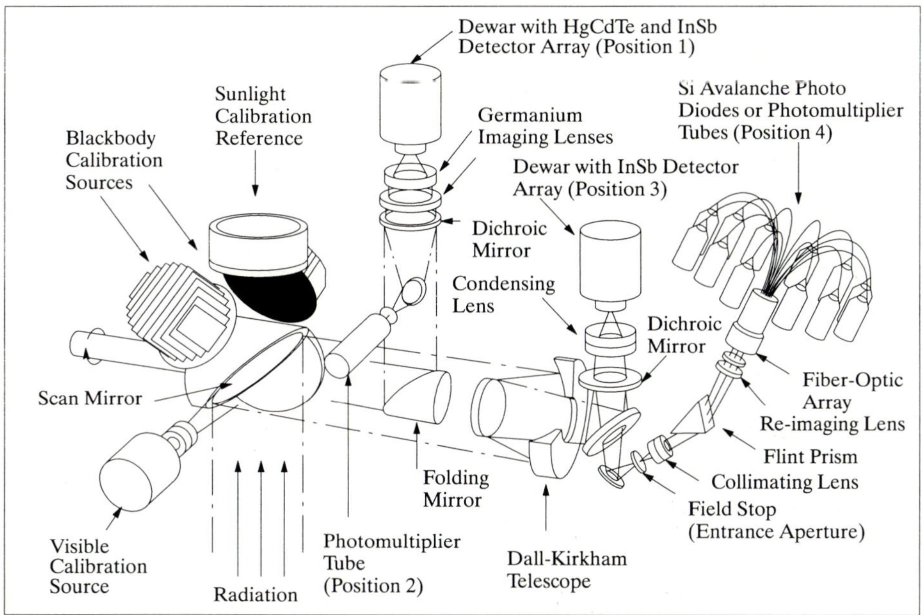


**Figure 128:    M-7 Optical-Mechanical Layout**

| Position | Area (cm²) | f/number | Effective Focal Length |
|---|---|---|---|
| 1 | 40 | 1 | 7.62 cm |
| 2 | 40 | 2.8 | 21.6 cm |
| 3 | 81 | 1.25 | 12.7 cm |
| 4 | 81 | 2.0 | 25.4 cm |

**Table 169:    M-7 Mapper Optical Parameters**

The four detector positions are achieved by: folding out the obscured center region of the Dall-Kirkham telescope (position 1); a dichroic reflector in this folded-out path (position 2); the output of the telescope (position 3), and a dichroic reflector in the telescope output (position 4).

The spectrometer in position 4 is comprised of an entrance aperture, a collimating lens, the prism, a re-imaging lens and a fiber-optic coupler that distributes incoming radiation to the detectors. Typically each individual segment of the fiber optic coupler is terminated in a cap mounted over a photomultiplier tube (PMT) or Si APD detector. - Six-element detector arrays (dewar-mounted) are in positions 1 and 3. The arrays are comprised of two rows of 3 elements each with individual filters supported in a small frame mounted to the dewar's cold finger. Because of the refractive optics used to focus radiation onto the detectors, the spacing of each element from the dewar's window was specified to assure optimum simultaneous focussing.

A dichroic reflector at position 2 reflects radiation from 0.35 to 1.0 μm and transmits from 2.8 to 14 μm. This, in combination with a choice of UV, VIS or NIR band limiting filters offers considerable configuration flexibility for narrow or broad spectral bands. Instead of adding blue, green and red bands in the data processing to achieve a panchromatic image, direct collection capability of panchromatic imagery is provided by the choice of a properly specified broadband filter.

| Position | Detector Arrangement | Spectral Bands at FWHM (μm) | |
|---|---|---|---|
| 1) Folded from central obscuration in primary telescope (6 bands in MWIR and TIR) | Dewar with detector array (2 x 3) MWIR detectors: InSb TIR detectors: MCT (HgCdTe) | 3.3 - 4.4<br>4.6 - 5.3<br>8.2 - 8.7 | 8.7 - 9.2<br>10.4 - 11.4<br>10.4 - 12.5 |
| 2) Dichroic beam splitter from path (1) (1 band option: UV/VIS/NIR) | PMT or Si APD and filter | 0.35 - xx, UV,or or option of a panchromatic band | 0.50 - 0.90 VIS<br><br>0.35 - 1.0 |
| 3) From primary telescope<br><br>(6 bands in NIR/SWIR) | Dewar with detector array (2 x 3)<br><br>InSb detector elements | 1.0 - 1.1<br>1.2 - 1.3<br>1.55 - 1.65 | 1.55 - 1.75<br>2.08 - 2.21<br>2.08 - 2.35 |
| 4) Dichroic beam splitter from path (3)<br><br><br><br><br><br><br><br><br><br>(18 bands in VIS/NIR) | Prism spectrometer with fiber optic coupler and PMT or Si APD detectors (the letter behind the band is the fiber designation) | 0.399 - 0.413  C<br>0.413 - 0.431  D<br>0.428 - 0.443  E<br>0.438 - 0.455  F<br>0.449 - 0.468  G<br>0.461 - 0.482  H<br>0.474 - 0.497  I<br>0.489 - 0.514  J<br>0.505 - 0.534  K | 0.523 - 0.556  L<br>0.544 - 0.581  M<br>0.567 - 0.610  N<br>0.594 - 0.645  O<br>0.626 - 0.688  P<br>0.665 - 0.741  Q<br>0.712 - 0.808  R<br>0.771 - 0.893  S<br>0.846 - 0.950  T |
| Scan angle | | ± 45° | |
| IFOV | | 2.5 mrad | |
| Line scan rates | | variable: 10 - 67 per second | |
| In-flight relative calibration  (UV, VIS/NIR,SWIR)<br>TIR | | Filtered lamps<br>ambient BB and 2 adjustable BBs | |
| Electrical bandwidth | | DC to 120 kHz | |
| Digital record bits | | 8 | |
| Simultaneous bands recorded | | 16 | |
| Image stabilization and rectification | | post-processing to ground coordinates via recorded roll, pitch and yaw angles, altitude, velocity and position data | |
| NEΔR<br>NEΔT | | ≤ 0.5% for reflective bands<br>≈ 0.1 K for thermal bands | |
| Platform | | ERIM Caribou | |
| Observation altitudes<br>Spatial resolutions | | 0.4, 0.8, 1.2, 2.0, and 4.0 km<br>1, 2, 3, 5, and 10 m | |
| Mass of instrument (total) | | 270 kg | |

**Table 170:    M-7 Mapper Specification of Spectral Bands and Sensor Parameters**

## B.68    MAMS (Multispectral Atmospheric Mapping Sensor)

A NASA-sponsored instrument. MAMS is a modified Daedalus scanner flown aboard an ER-2 aircraft at ARC. Objective: study of weather-related phenomena including storm system structure, cloud-top temperatures, and upper atmosphere water vapor. The scanner retains 8 silicon-detector channels in the VNIR region as provided on the Daedalus Thematic Mapper Simulator, with the addition of four channels in the SWIR and TIR regions relating to specific water vapor features.[499]

MAMS is one of several spectrometer arrays at ARC that mount on a standard Daedalus AADS1268 scanner system [along with TMS (NS001), AOCI, and MAS]. ARC has two basic scanner systems, the spectrometer sections are interchangeable. MAMS has been flown many times since 1985, as of 1993 the system is more and more being replaced by MAS.

Note: MAMS data are not archived at the EROS Data Center since MAMS is an experimental system with low spatial resolution (there is little terrestrial application for the data as all scenes will be predominantly cloud-covered.

| Channel Nr. | Wavelength of Channel (µm) |
|---|---|
| 1 | LSBs for bands 9-12 |
| 2 | 0.45 - 0.52 |
| 3 | 0.52 - 0.60 |
| 4 | 0.57 - 0.67 |
| 5 | 0.60 - 0.73 |
| 6 | 0.65 - 0.83 |
| 7 | 0.72 - 0.99 |
| 8 | 0.83 - 1.05 |
| 9 | 3.55 - 3.93 low range |
| 10 | 3.55 - 3.93 high range |
| 11 | 10.3 - 12.1 |
| 12 | 12.5 - 12.8 |
| Geometric Parameters | |
| FOV | 86° |
| IFOV | 5 mrad |
| Ground resolution | 99 m (at 20 km altitude) |
| Swath width | 37 km (at 20 km altitude) |
| Pixels/scan line | 716 |
| Scan rate | 6.25 scans/s |
| Ground speed | 200 m/s |
| Quantization | 8-bit for bands 2-8 <br> 10 bit for bands 9-12 |

**Table 171:    Spectral Coverage and Geometric Parameters of the MAMS Instrument[500]**

## B.69    MARA (Multimode Airborne Radar Altimeter)

MARA is a beam-limited 36 GHz radar altimeter prototype designed, built and operated by NASA/GSFC. The system was developed in 1988 to support spaceborne multibeam altimetry investigations and to provide high-precision airborne measurements for ocean, land and ice topographies.[501] Two distinct sensor configurations are available: The first is a fixed, multibeam altimeter mode; the second mode is a Scanning Radar Altimeter (SRA), which rapidly scans a single beam across-track to achieve a 64 pixel raster-type scan. Both systems utilize large aperture antennas to generate a small surface footprint which provides a higher spatial resolution in comparison to a pulse-limited altimeter. For this reason a beam-limited altimeter is better suited for surfaces where topology is varying. MARA has been oper-

---

499)  Information provided by NASA-HQ (D. Dokken) and by ARC (J. Myers)

500)  Note: the 3.55-3.93 µm channels can be operated at 6.20 - 6.90 µm as well. The LSBs refer to the Least Significant Bits, which give 10-bit resolution to channels 9-12 on an otherwise 8-bit system.

501)  C. L. Parsons, E. J. Walsh, "Off-nadir radar altimetry", IEEE Transactions on Geoscience and Remote Sensing, Vol. 27, 1989, pp. 215-224

ated on several aircraft since 1990, among them NASA and NOAA P-3s and a Fokker F-27. The maximum unpressurized operating altitude is 3 km.[502]

Multibeam Mode:
The nominal multibeam geometry consists of five separate beams, one pointed at nadir and the others pointed 12° off-vertical with azimuth directions 90° apart and centered on the along and cross-track axes. The surface footprint is 2.9 m for a 400 m altitude (slightly wider off-nadir). The high-resolution footprint and 6 ns transmit pulse translate to nominal rms deviations on MARA surface elevation measurements of ± 6 cm at nadir and ± 15 cm off-nadir (12° pointing angle). The multiple beam capability is exploited to provide surface slope information and study schemes for improved satellite altimeter range tracking over rough surfaces. A key feature of the system is the ability to digitize and store each return waveform. Return waveform characteristics are used to improve range-to-surface tracking and to derive geophysical information related to the surface and/or volume backscatter from the target. This mode was used on a recent ice sheet mapping mission over Greenland (1991). The data is being used to study continental ice sheet penetration at microwave frequencies.
Multibeam mode applications: topographic measurements, mm-wave surface/volume backscatter studies.

Scanning Radar Altimeter Mode (SRA):
The SRA is a successor to the now-retired SCR (Surface Contour Radar), which was flown in numerous ocean remote sensing experiments.[503] SRA participated in the following campaigns: TOGA/COARE (1993), SWADE (1991), and Grand Banks ERS-1 SAR Wave Experiment (1991). It scans a pencil beam (1° two-way) perpendicular to the aircraft ground track and measures the range to 64 points evenly spaced across a swath whose width is 0.8 times the aircraft height. SRA corrects for the off-nadir geometry and generates a false-color-coded elevation map of the sea surface elevation below the aircraft in real-time. Post-flight data processing routinely produces ocean directional wave height-variance spectra that have shown the SRA's ability to produce high-quality topographic maps for terrain in remote regions.
SRA mode applications: topographic mapping, ocean directional wave spectra measurements, terrain and ocean backscatter studies.

| Transmitter | | Receiver(s) | |
|---|---|---|---|
| Center frequency | 36 GHz | Noise figure | 6 dB DSB |
| Peak power | 1.7 kW | IF frequency | 600 MHz |
| PRF (Multibeam) | 200 Hz maximum | Bandwidth | 220 MHz |
| PRF (SRA) | 10 cross-track scan/s 64 pulses/scan | Detector | square law |
| Maximum duty cycle | 0.01 | Dynamic range | 80 dB |
| Multibeam Mode Antenna | | SRA Mode Antenna | |
| Lens aperture | 85 cm diameter | Lens aperture | 45.7 cm diameter |
| 3 dB beamwidth | 0.62° one-way | 3 dB beamwidth | 1.4° one-way |
| Gain | 47 dB | Gain | 41 dB |
| Feed configuration | 5 separate feeds, pointing angles adjustable to any point within 15° from vertical | Feed configuration | single horn feed, scanning pattern generated using reflector |
| | | Scan swath | ± 22° from vertical, cross-track (64 separate angles) |

**Table 172:    MARA System Characteristics**

502)  Information provided by D. Vandemark of NASA/GSFC
503)  E. J. Walsh, "Surface contour radar directional wave spectra measurements during LEWEX", Directional Ocean Wave Spectra, R. Beal (ed.), Johns Hopkins University Press, pp. 86-90, 1991

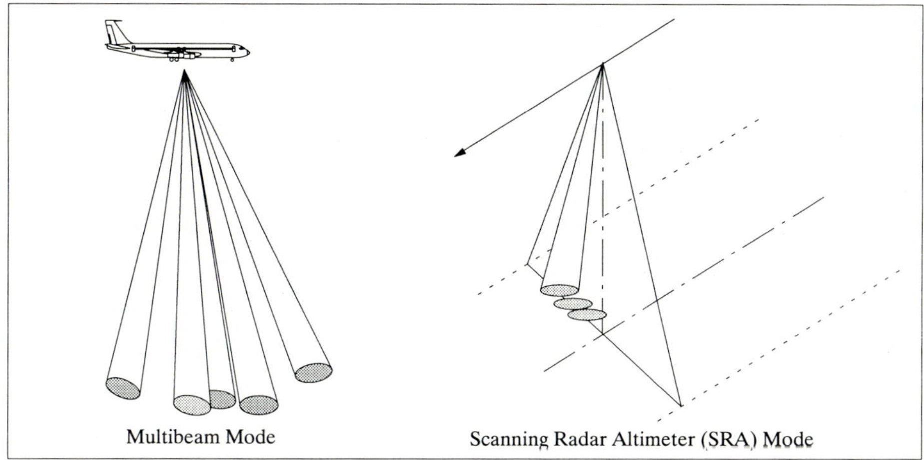


**Figure 129:   MARA Observation Configurations**

## B.70    MCR (Multispectral Cloud Radiometer)

MCR is a NASA/GSFC seven-channel scanning radiometer, operational since 1976, with the objective to measure cloud properties (optical thickness, infrared reflection of clouds, cloud top heights, cloud top temperature, etc. The instrument has been flown on various missions and aircraft (among them CV-990, WB-57F, ER-2, etc. ) at altitudes between 13 and 20 km.[504],[505]

| Channel Number | Center Wavelength ($\mu$m) | Resolution FWHM ($\mu$m) | Remarks, Principal Measurement |
|---|---|---|---|
| 1 | 0.7539 | 0.0010 | Clear air thickness |
| 2 | 0.7605 | 0.0013 | $O_2$ A-Band, altimetry, volume scattering coefficient |
| 3 | 0.7635 | 0.0012 | $O_2$ A-Band, altimetry, volume scattering coefficient |
| 4 | 1.367 | 0.002 | $H_2O$ vapor, $H_2O$ vapor amount |
| 5 | 1.644 | 0.089 | Clear, cloud phase, cloud/snow discriminator |
| 6 | 2.164 | 0.094 | Clear, cloud phase, particle size |
| 7 | 10.84 | 0.81 | Clear, temperature |

**Table 173:    Spectral Characteristics of MCR**

MCR is a mechanical cross-track scanning radiometer employing 45° rotating mirrors. FOV = ± 42°, IFOV = 7 mrad. Nadir pointing. All channels are sampled simultaneously. The optical system of MCR is of non-dispersive, filter-dichroic design (Dall-Kirkham telescope). The thermal channels are internally calibrated with two blackbody sources, which are viewed in the upward part of the mirror scan. The visible and NIR channels are calibrated from integrating sphere measurements, made before and after each mission. Calibration accuracies are in the order of 5%.

As of 1993 MCR has been transferred to NCAR (Boulder Co.). An upgrade of the data system is planned. The sensor will be flown on a WB-57F aircraft.

504)  R. J. Curran, H. L. Kyle, L. R. Blaine, J. Smith, T. D. Clem, "Multichannel scanning radiometer for remote sensing cloud physical parameters", Revision of Science Instruments, Vol. 52, Nr. 10, Oct. 1981, pp. 1546-1555
505)  J. D. Spinhirne, "Cirrus Structure and Radiative Parameters from Airborne Lidar and Spectral Radiometer Observations: The 28 October 1986 FIRE Study, Monthly Weather Review, Vol. 118, Nr. 11, November 1990, pp. 2329-2343

## B.71    MEIS (Multi-detector Electro-optical Imaging Sensor)

MEIS is a commercially available multispectral pushbroom scanner of MacDonald Dett-
wiler and Associates (MDA) in Richmond B. C., Canada.[506),507),508)]

MEIS was originally developed by MDA under contract to CCRS. MEIS I prototype development in 1978, the MEIS II instrument is in operation since 1983 by CCRS (more than 300
missions with MEIS II as of 1993). The data acquired by MEIS II are characterized by high
radiometric sensitivity and high spatial resolution ($>0.5$ m). Image data are acquired in
narrow spectral bands. The integrated charges from each element of the array are sampled
and digitized to produce a line image of the scene below. The aircraft motion provides the
scanning in the forward direction.

| Scanner type | Pushbroom CCD scanner, 8 spectral channels |
|---|---|
| Spectral response | 390 - 1100 nm spectral range |
| FOV<br>IFOV | 40°<br>0.7 mrad |
| Detector type | Fairchild CCD 122 silicon, buried channel<br>1728 element linear array, 13 μm x 13 μm photo-elements |
| Lens type | Angenieux, type R2, format 35 mm; focal length = 24.61 mm; f/2.2-f/22 |
| Line sample rates | 25, 50, 100, 200 Hz (operator selectable to match aircraft velocity) |
| Gains | 1, 2, 4, 8, 16, 32 (operator selectable) |
| Exposure times | 5 ms per line sample rate, by powers of 2 (operator selectable) |
| Data Format | 1024 pixels recorded per channel and per scan line<br>8 bit digitization<br>ancillary data recorded (sensor parameters, navigation parameters, etc.) |
| Signal/Data Processing | • correlated double sampling of CCD signal<br>• real-time dark current subtraction<br>• real-time relative gain correction (i.e. radiometric uniformity correction)<br>• real-time image data resampling (provides pixel registration, geometric corrections for optical distortions, and roll correction)<br>• storage in programmable memory of radiometric and geometric coefficients for ten sets of filters<br>• real-time aircraft roll-correction<br>• data rate: 1.75 Mbit/s per channel |
| Interfaces | Alice real-time display (quicklook capability)<br>High-density digital tape recorder<br>Inertial navigation system |

**Table 174:    Technical Parameters of the MEIS II Sensor**

In addition to the nadir-mode of operation, the MEIS system is used to acquire continuous
fore-aft stereo imagery by the addition of a precision stereo mirror module. In stereo mode,
two of the 8 channels look fore and aft at angles of ± 35°, while 6 channels look to nadir. The
data may be combined with the aircraft navigation and attitude data to provide digital terrain information and standard geometrically corrected products.

Applications: vegetation (crop and forest mapping), environmental monitoring (water
quality and bathymetry), geology (exploration and mapping), topographic mapping, etc.

Owners/operators of MEIS instruments: CCRS/Innotech Aviation (Canada) aboard a Fal-
con-20C aircraft (owned by Innotech Aviation). MEIS II is the principal component of the

506)) MEIS Information brochure provided by MacDonald Dettweiler and Associates (MDA) in Richmond B. C.,
       Canada
507)) S. M. Till, "Airborne Electro-Optical Sensors for Resource Management", Geocarto International, Vol 3, 1987,
       pp. 13-23
508)) S. M. Till, R. A. Neville, W. D. Mc Coll, R. P. Gauthier, "The MEIS II Pushbroom Imager - Four Years of Opera-
       tion", Progress in Imaging Sensors, Proc. ISPRS Symposium, Stuttgart, 1-5 September 1986,, ESA SP-252, No-
       vember 1986, pp. 247-253

electro-optical sensor package, the other imaging instruments on the Falcon aircraft are an AADS-1260 multispectral scanner from Daedalus, a 230 mm format mapping camera, and in the future, SFSI (see B.103).

Since 1983 the MEIS instrument has been flown on close to 60 remote sensing missions per year. A data acquisition service is also provided on a commercial basis. The Falcon 20-C aircraft can operate at maximum altitudes of 11 km at maximum speeds of 800 km/h.


## B.72    MERES (Multifrequency Radiometer for Remote Sensing of the Sea Surface)

MERES[509],[510] is an airborne multifrequency passive microwave radiometer instrument of DLR (Institute for RF Technology, Oberpfaffenhofen) with the prime objective to monitor maritime oil pollution (coastal waters of Germany) aboard a DO228 aircraft. The system consists of two offset rotating parabolic mirrors and two radiometer sets, each set of which contains in turn three radiometers at the center frequencies of 18.7, 36.5, and 89 GHz. In addition, another 89 GHz radiometer is used for the measurement of the average radiometric sky temperature (sky radiometer).

The system is continuously calibrated with the use of a "hot load" at ambient temperature and a Peltier cooled "cold load". A computer system controls, configures, and calibrates the instrument, records the data, and allows onboard data reduction and quick-looks for 'amount of oil estimations' on the sea surface. The system was tested and qualified between Nov. 1991 and 1993, it is scheduled to be in a pre-operational phase by the end of 1993.

| Sensor type | Linescanner | | Sky Radiometer | |
|---|---|---|---|---|
| Antenna type | Rotating offset parabolic reflector | | Fixed sector-horn antenna | |
| IFOV | $0.8°$ / $2.0°$ / $4.0°$ | | $8° \times 80°$ | |
| Swath (FOV) | 76° | | | |
| Scan frequency | 10 revolutions/s = 20 scan lines/s | | | |
| Radiometric resolution | 1.0 K (89), 2.3 K (36.5); 3.2 K (18.7) $(T = 300\ K, \tau = 360\ \mu s)$ | | | |
| Scanning principle | mechanical whiskbroom (2 parab. mirrors in a continuously rotating cylinder) | | | |
| Receiver Front-End | | | | |
| Channel | Receiver Type | Center Frequency (GHz) | Bandwidth (GHz) | Noise Temperature K |
| 1 | heterodyne DSB (double side band) | 89.0 | 2 | 590 |
| 2 | heterodyne DSB | 36.5 | 0.4 | 565 |
| 3 | heterodyne SSB | 18.7 | 0.2 | 550 |
| Sky Radiometer | heterodyne DSB | 89.0 | 0.9 | 590 |

**Table 175:    Summary of MERES Characteristics**

509) K. Grüner, G. Kahlisch, H. Schreiber, P. Sliwinski, "A new Passive Microwave Linescanner for Airborne Measurements of Maritime Oil Pollutions", paper presented at IEEE-MTT Conference 1992, Albuquerque, N. M.
510) H. Süß, K. Grüner, "Present Activities of DLR in Microwave Radiometry", µrad 92, Proc. of Specialist Meeting on Microwave Radiometry and Remote Sensing Applications, Boulder Co., June 1992, pp. 408-415

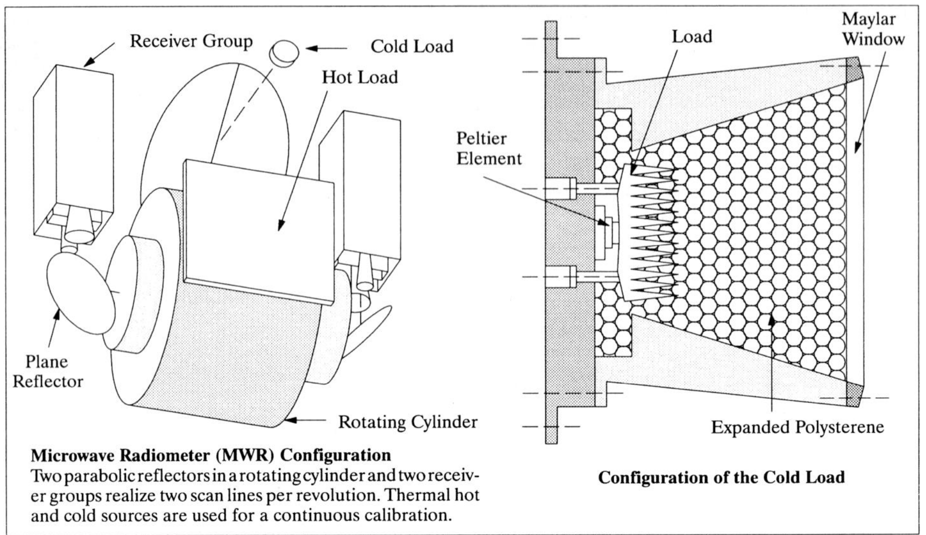



**Microwave Radiometer (MWR) Configuration**
Two parabolic reflectors in a rotating cylinder and two receiver groups realize two scan lines per revolution. Thermal hot and cold sources are used for a continuous calibration.

**Configuration of the Cold Load**

**Figure 130:    The MERES Instrument Model**

## B.73    MINISCAT

MINISCAT is an active single-frequency HUT scatterometer (a reduced version of HUT-SCAT, see B.56) which participated in the Arctic-91 campaign.[511] The sensor is a non-imaging FM-CW radar using a digital signal processor and Fast Fourier Transform processing. The instrument is flown on a helicopter.

The principle of operation is comparable to HUTSCAT. The main differences are:

- MINISCAT is a single-frequency (5.3 GHz) scatterometer
- MINISCAT has two ranging modes with resolutions of 0.31 m and 0.62 m
- MINISCAT is a light-weight version of HUTSCAT

| | |
|---|---|
| Center frequency | 5.3 GHz |
| Sweep bandwidth | 300/600 MHz |
| Modulation type | FM-CW |
| Range resolution | 31/62 cm |
| Distance range | 10 - 75 m / 10 - 150 m |
| Polarization modes | HH, VV, HV, VH |
| Antenna type | 75 cm parabolic antenna with ring-loaded dipole disk feed |
| Calibration capability | external and internal |

**Table 176:    Parameters of the MINISCAT Instrument**

Applications: MINISCAT is a research tool at HUT, it is also used for the measurement of stand profiles of forest canopies to produce stem volume estimates.

---

511)  J. Pallonen, "Scatterometer for arctic measurements", Thesis, HUT, Finland, May 1992

## B.74    MIR (Millimeter-Wave Imaging Radiometer)

MIR is a newly developed cross-track imager (passive radiometer) owned and operated by NASA/GSFC. The system measures radiation at the following frequencies: 89, 150, $183.3 \pm 1$, $183.3 \pm 3$, $183.3 \pm 7$, and 220 GHz. Three additional channels at $325 \pm 1$, $325 \pm 3$, and $325 \pm 8$ GHz will be installed in the future (after 1993). The beam width of MIR is 3.5° for all frequencies, the scan range is $\pm 45°$. The system has a temperature sensitivity of 1 K or less and a calibration accuracy of $\pm 1$ K in the brightness temperature range of 250-300 K. Below 100 K brightness temperature, the calibration accuracy is in the order of $\pm 2$ K. The sensor is flown on NASA ER-2 aircraft at altitudes of 20 km. MIR became operational in May 1992, it took part in the TOGA/COARE mission in early 1993. MIR measures water vapor profiles, clouds and precipitation.[512]

## B.75    MkIV (Mark-IV Interferometer)

MkIV is a NASA-sponsored FTIR (Fourier Transform Infrared) spectrometer developed and built by JPL for remotely monitoring the composition of the atmosphere.[513] It makes observations from ground-based, airborne and balloon platforms using the solar absorption technique. The instrument derives from the ATMOS sensor, which flew on Shuttle mission in 1985, 1991, and 1992. The main improvement incorporated into the MkIV is the use of two detectors in parallel to cover the entire 600 - 5500 $cm^{-1}$ region simultaneously. Over 30 different atmospheric gases can be measured including: $H_2O$, $CO_2$, $O_3$, $N_2O$, CO, $CH_4$, $N_2$, $O_2$, $NH_3$, NO, $NO_2$, $IINO_3$, $IINO_4$, $N_2O_5$, $ClNO_3$, HOCl, HCl, HF, $SF_6$, $COF_2$, $CF_4$, $CH_3Cl$, $CHF_2Cl$, $CFCl_3$, $CF_2Cl_2$, $CCl_4$, OCS, $SO_2$, HCN, $C_2H_2$, $C_2H_6$, and many isotopic variants.

| Configuration | Double-passed Michelson Interferometer with passive tilt and shear compensation |
|---|---|
| Spectral coverage | 600 - 5500 $cm^{-1}$ (1.8 - 16 μm) in solar absorption mode |
| Spectral resolution | 0.008 $cm^{-1}$ (1 million points per spectrum) |
| Sampling rate | 10 kHz (controlled by 633 nm He:Ne reference laser) |
| Scan time | 105 s for 66 cm OPD; 210 s for 133 cm OPD (Optical Path Difference) |
| Detectors (77 K) | HgCdTe photoconductor (600 - 2000 $cm^{-1}$); InSb photodiode (1850 - 5000 $cm^{-1}$) |
| Dynamic range | 19 bits (12 bit ADC+7 bit programmable gain) |
| SNR | HgCdTe > 400:1; InSb > 500:1 |
| Ops temperature | 20-30°C |
| Beamsplitter | KBr (5.5 inch diameter; 1.0 inch thick; 21 minute wedge) |
| Beam diameter | 25 mm |
| FOV diameter | HgCdTe = 3.6 mrad; InSb = 3.0 mrad |
| Throughput | HgCdTe = 50 x $10^{-6}$ $cm^2$ x sr; InSb = 35 x $10^{-6}$ $cm^2$ x sr |
| Sun tracker | Two-axis, controlled by quadrant detector |
| Spatial resolution | Limb viewing mode: 400 km horizontal, 2 km vertical<br>Zenith viewing mode: 10 km horizontal, 10 km vertical |
| Size of instrument | 1.3 m x  0.8 m x 0.7 m |
| Mass | 250 kg |
| Power | 350 W |
| Data rate | 360 kbit/s (162 MByte/h) |

**Table 177:    Specification Parameters of the MARK-IV Instrument**

The main components of the MkIV are: an active 2-axis suntracker which feeds direct solar radiation to the beamsplitter; a double-passed Michelson interferometer employing cube-corner retro-reflectors (one moving and one stationary); a leadscrew and drive motor for translating the moving retro-reflector, InSb and HgCdTe detectors, preamplifiers, filters

---

512)  Information provided by J. R. Wang of GSFC
513)  G. C. Toon, "The JPL MkIV interferometer", Optics and Photonics News, October 1991

and analogue-to-digital converters (ADCs); a single-frequency He:Ne laser and detection system to trigger the sampling of the infrared signals chains; and a computer to record the digital data and Fourier transform it.

| April 1985 | JPL, first ground-based spectra |
|---|---|
| September/October 1986 | McMurdo, Antarctica; ground-based measurement campaign |
| September/October 1987 | Punta Arenas, Chile; Antarctic aircraft campaign (DC-8) |
| January/February 1989 | Stavanger, Norway; Arctic aircraft campaign (DC-8) |
| October '89, September '90, May '91 | Fort Sumner, New Mexico, Balloon Flight |
| January - March 1992 | Alaska/Norway/Maine,Second Arctic Aircraft Campaign (DC-8) |
| September 1992, September 1993 | Fort Sumner, New Mexico, UARS correlative balloon flight |
| April 1993 | Daggett, California, UARS correlative balloon flight |

**Table 178:    MARK-IV Instrument Utilization History**


## B.76    MMS (Meteorological Measurement System)

MMS is a NASA/ARC-developed instrument for ER-2 aircraft. The objective is to collect in-situ meteorological data at high altitudes with high resolutions and accuracies. MMS is operational since 1986 and has been modified in 1992-93.[514]

The instrument consists of the following major components:

- An air motion sensing system to measure the velocity of the air with respect to the aircraft.
- A high-resolution inertial navigation system to measure the velocity of the aircraft with respect to the Earth.
- A data acquisition system to sample, process and record the measured quantities.

The instrumentation for the air motion system is located in the ER-2 nose and the lower fuselage, the inertial navigation system and data acquisition systems are located in the aircraft equipment bay. MMS took part (or will take part) in the following campaigns:

- STEP (Stratosphere Troposphere Exchange Project), 1987
- AAOE (Airborne Antarctic Ozone Experiment), 1987
- AASE (Airborne Arctic Stratospheric Expedition), 1989
- AASE-II (Airborne Arctic Stratospheric Expedition), 1991-92
- SPADE (Stratospheric Photochemistry, Aerosol, and Dynamics Experiment), 1992-93
- MAESA (Measurement for Assessing the Effects of Stratospheric Aircraft), 1994
- ASHOE (Airborne Southern Hemisphere Ozone Experiment), 1994

Performance: MMS provides in-situ measurements of pressure (p), temperature (T), potential temperature ($\Theta$), wind vector (u,v,w), position (altitude, longitude, latitude), pitch ($\theta$), roll ($\phi$), heading ($\psi$), angle of attack ($\alpha$), angle of sideslip ($\beta$), true airspeed, aircraft eastward velocity ($\dot{x}$), northward velocity ($\dot{y}$), vertical acceleration ($\dot{z}$), and time (t) at a sample rate of 5 times per second. MMS data products are presented in the form of either 1-Hz or 5 Hz time series. The accuracies of the primary products are:

| Typical value at 65000 ft | | Accuracy |
|---|---|---|
| Pressure | 60 mb | ± 0.3 mb or 0.5% |
| Temperature | 180 K | ± 0.3 K or 0.2% |
| Horizontal wind | 25 m/s | ± 1 m/s or 4.0% |
| Vertical wind | < 1 m/s | Resolution: 0.1 m/s |

**Table 179:    Accuracies of Primary MMS Data Products**

---

514)  Information provided by K. R. Chan of NASA/ARC, Moffett Field, Ca.

## B.77    MMW-SAR (Millimeter Wave SAR)

The MMW-SAR is an ARPA-funded instrument, designed at the MIT Lincoln Laboratory and built by Loral Defense Systems (Phoenix, Az.) with the objective to a) investigate the detection and classification of stationary targets using ultra-high resolution data (fully polarimetric SAR mode and Real Aperture Radar (RAR) mode capability), and b) advance the understanding of millimeter-wave imaging technology. The system is maintained and operated at MIT since 1988, it is flown at altitudes from 500 feet above ground level up to 16500 feet, depending on the mode of depression angle selected for the mission. Typically, SAR data is collected at depression angles of 20° and 45° (at altitudes of 8500 ft and 16500 ft).[515]

The system consists of an airborne radar that operates at 33.56 GHz (Ka-Band), plus a ground processing and archiving system at Lincoln Laboratory. MMW-SAR is regarded a research tool and designed to collect large data sets of distributed clutter, target signatures, and targets in clutter. The data collected is used to investigate the detection and classification of stationary targets using ultra-high fully polarimetric SAR and RAR data.

| Frequency | 33.56 GHz (Ka-Band) |
|---|---|
| RF bandwidth | 600 MHz |
| Waveform | Linear FM (deramp on receive) |
| Transmit polarization<br>Receive polarization | V then H, or L then R<br>V and H, or L and R |
| Transmitter power | 84 W peak |
| Receiver noise figure | 6.8 dB |
| Gimbal order | Azimuth roll-elevation-scan |
| Antenna beamwidth | 1.4° |
| Antenna peak sidelobe | -27 dB |
| Platform | Grumman Gulfstream G-1 aircraft |
| Recording media | 28-channel high-density tape, digital disk |
| Operational modes (4) | SAR stripmap, SAR spotlight, RAR, and Calibration |
| Spatial resolution | 0.3 m in range and in cross-range (SAR mode)<br>0.3 m in range direction (RAR mode) |

**Table 180:    MMW-SAR System Parameters**

MMW-SAR has four operational modes:

1.    SAR stripmap mode: - a side-looking mode used to collect long strips of data at a given depression angle. The strips have a cross-range extent of about 440 m. Strip lengths are usually between 15 and 25 km. This mode is used to collect large amounts of clutter data needed to support the statistical analysis that forms the basis of accurate clutter modeling.

2.    SAR spotlight mode: - the antenna is aimed at a fixed reference point in a specific target area. The area imaged in this mode is 150 m x 150 m. The spotlight mode is used to collect data on specific targets, either in the clear or in clutter, from all aspect angles.

3.    RAR mode: - a real-beam radar mode. The cross-range resolution is a function of the antenna beamwidth and of the radar slant range to the target. There are two RAR mode which correspond to the SAR modes in that one scans a strip and the other tracks a spot; they are called 'RAR search' and 'RAR track', respectively. The RAR modes use the innermost axis of the antenna gimbal to scan the antenna over a wide azimuth swath. The RAR modes emulate the searching and tracking of a radar seeker.

515)  J. C. Henry, T. J. Murphy, K. M. Carusone, "The Lincoln Laboratory Millimeter-Wave Synthetic Aperture Radar (SAR) and Imaging System", Proceedings of SPIE, Vol. 1630 Synthetic Aperture Radar, 1992, pp. 35-52

4.  Calibration mode: - prior to each data collection pass, the radar is configured in calibration mode. In this mode a replica of the transmitted waveform, very carefully controlled in amplitude and phase, is injected into the radar front end ahead of the low-noise amplifiers. This calibration pulse is processed through the system in the same manner as a normal radar pulse. The purpose is to verify that the receiver gain and phase do not change during mission operations. If any changes are detected by processing the calibration pulses, then these changes can be compensated by appropriate modification to the calibration values used for each pass.

Polarimetric properties. The MMW-SAR instrument is fully polarimetric. The measurement of the complete polarization scattering matrix for each pixel is achieved by transmitting, on alternating pulses, two orthogonal senses of polarization (e.g. horizontal and vertical). There are two orthogonal receive channels, which operate simultaneously following each transmitted pulse. One pair of transmitted pulses generates all four components of the scattering matrix (HH, HV, VH, VV).

The MMW-SAR consists of the following systems: Aircraft Position Location System, Antenna Motion Compensation System, Gimbal Control System, Data Recording System, and Control Computer System.

- Aircraft Location System: INS, Ground-based beacon location system, GPS, and Barometric altimeter.
- Antenna Motion Compensation System: The short-term (high-frequency) motion of the antenna is measured by IMU (Inertial Measurement Unit). This data is used as input to compensate the pulse-to-pulse phase of the radar to maintain coherence during synthetic aperture.
- Gimbal Control System: a four-axis (azimuth, roll, elevation, and scan) gimbal stabilizes and points the antenna in the desired direction.
- Receiver and Data Recording System: the received signals (H and V) are amplified by GaAs Low Noise Amplifiers. The preprocessor includes: the A/D converters, the PRF buffer, the cross-plane filter, the bus multiplexer, and the high-density tape interface. The tape recording system consists of an Ampex 28-channel HDT recorder.
- Control Computer System: VAX 4300 computer plus peripherals.

Lincoln Laboratory maintains an 'image formation and data analysis facility'. The Image Formation Processor (IFP) consists of an AMPEX HBR 3000 high density tape playback unit, a Numerix 432 Processor, a very large disk-storage unit, a high-speed Visi-net data bus and a control computer. - Some selected data sets are available to U.S. researchers upon prior written approval of the ARPA sponsor. Instrument upgrades planned for the future include higher resolution capabilities.


## B.78   MSS (Maritime Surveillance System)

MSS[516)] is a commercially available integrated airborne system of SSC, Sweden, consisting of several instruments, with the objective to offer a cost-effective set of tools for the surveillance of sea traffic, pollution monitoring, fishery protection, and border patrol. The system is composed of the following sensors and support systems:

- SLAR (Side Looking Airborne Radar)
- IR/UV (Infrared/Ultraviolet Line Scanner) IR/UV is licensed from Daedalus (see AADS1221 description under B.33.2)
- MWR (Scanning Microwave Radiometer)
- Camera (Photographic Camera System)
- Video (Video Camera System)
- MP (Maritime Central Processor)

---

516)  Information provided by O. Fäst of SSC, Sweden

- Image Link (X-Band transmitter/receiver for air-ground communication and vs versa)
- DET (Data Evaluation Terminal, ground-based)

| Operation-al since | Aircraft Type | Country | Organization | Application | MSS Sensor Complement |
|---|---|---|---|---|---|
| 1991 | CASA 212-200 | Sweden | Swedish Coast Guard | Pollution, fishery, ship traffic control | SLAR, IR/UV, MWR, Camera, Video |
| 1991 | CASA 212-100 | Portugal (3 systems) | Portuguese Air Force | Fishery control | SLAR, IR/UV, MWR, Camera, Video |
| 1988 | Cessna 404 | The Netherlands | Rijkswaterstaat | Pollution & ship traffic control | SLAR, IR/UV, Camera |
| 1988 | Cessna 402 B | UK | Department of Transport | Pollution control | SLAR, IR/UV, Camera |
| 1987 | CASA 212-200 | Sweden | Swedish Coast Guard | Pollution, fishery, ship traffic control | SLAR, IR/UV, MWR, Camera, Video |
| 1987 | CASA 212-200 | Sweden | Swedish Coast Guard | Pollution, fishery, ship traffic control | SLAR, IR/UV, MWR, SLOS, Camera, Video |
| 1987 | Y-12 | China | Oceanic Administration | Pollution control | SLAR, IR/UV, MWR, Camera |
| 1987 | Fairchild Merlin | Norway | Statens Forurensningstilsyn | Pollution and fishery control | SLAR, IR/UV, Camera |
| 1986 | DO 228 | India | Coast Guard | Pollution control | IR/UV |
| 1985 | DO 228 | Germany (2 systems) | Marineflieger-geschwader 5 | Pollution control | SLAR, IR/UV, MWR, Camera |

**Table 181:    Overview of Operational Airborne MSS Systems**


## B.78.1    SLAR (Side-Looking Airborne Radar)

The SLAR (Ericsson model) is used for high resolution surveillance of large areas of sea surface. The swath width is 80 km (40 km to each side of the aircraft or 80 km to one side, as selected by the operator). The vertically polarized antenna gives an enhanced sea clutter response, which brings out oil films on the sea surface with high contrast.

| | | |
|---|---|---|
| Frequency | | 9.3 GHz (X-Band) |
| Transceiver | transmitted peak power pulse length pulse repetition rate | 10 kW<br>0.5 µs (0.125 µs option)<br>1.0 kHz |
| IFOV | | 9 mrad  horizontal (along track)<br>650 mrad vertical (across track) |
| Antennas | Horizontal beam width<br>Vertical beam width<br>Gain<br>Polarization | 0.5°<br>$\geq 37°$<br>31 dBi<br>Vertical |
| SLAR mass (including two antennas) | | 90 kg |

**Table 182:    System Parameters of SLAR**

The system solution also gives the SLAR an outstanding performance in small target detection, making it an efficient instrument for fishery surveillance and for search and rescue missions. SLAR features a real-time display with several display modes and a number of selectable image enhancement capabilities, in addition it provides automatic target positioning supported by TV screen displays.


## B.78.2    IR/UV (Infrared/Ultraviolet System)

The instrument is described under B.33.2. The instrument is capable of observing minute temperature differences on the water surface, it is ideal for mapping oil spills and other types of pollution. Instrument mass = 30 kg.

### B.78.3  MWR (Scanning Microwave Radiometer)

MWR is used for detailed mapping of an oil spill, it measures the oil film thickness and provides an estimate of the total oil volume on the surface. The oil film thickness image is real-time color-coded, the data is presented together with the corresponding data from IR/UV.

MWR has four antennas mounted on a rotating platform. This concept has been chosen to give a continuous scanning capability. An important advantage of this concept is that the incidence angle and the polarization angle are both kept constant during a sweep reducing the influence from other variables to a minimum.

| | |
|---|---|
| MWR frequency: | 35 GHz |
| Angle of incidence: | 25° (constant) |
| Polarization: | Horizontal |
| Swath width | 150 m, 300 m, 600 m |
| Ground resolution: | 7 m, 15 m, 30 m |
| MWR mass | 32 kg |

### B.78.4  Camera (Photographic Camera System)

The camera is a modified Nikon professional handheld camera for standard 35 mm film. Each frame is automatically annotated with date, time, position, heading, connection number, mission ID and frame number.

| | |
|---|---|
| Camera type | 35 mm single-lens reflex |
| Picture format | 24 mm x 36 mm (standard 35 mm film) |
| Lens mount | Nikon bayonet mount |
| Shutter speed | 1/8000 s to 30 s |
| Stutter release | electromagnetic |
| Annotation imprint | special modification for exposure of annotation on each frame |
| Camera mass | 6 kg |

### B.78.5  Video (Video Camera System)

Video allows close-up documentation of activities on the sea surface beneath the aircraft. The video signal is complemented with annotation information from the computer. The information is added to each image frame (in same fashion as for the Camera) prior to VHS recording.

| | |
|---|---|
| Imager | 2/3" integrated color mosaic filter signal chip CCD (556 x 581 pixels) |
| Resolution | ≥ 460 lines horizontal |
| Min. illumination | 7 lux |
| SNR | ≥ 46 dB |
| Shutter speed | strobe effect shutter available for in-flight operation; 1/2000 s |
| Format | VHS/S-VHS |
| Annotation data | Video signal integrated with annotation data + navigation data |
| Video mass | 6 kg |

### B.79  MTP (Microwave Temperature Profiler)

MTP[517] is an airborne passive microwave radiometer designed and built by NASA/JPL. The radiometer measures brightness temperature versus elevation angle, which is converted by statistical retrieval coefficients to profiles of air temperature versus altitude above and below the aircraft. Several MTP's have been flown on four different NASA aircraft during the past 15 years.

---

517)  Courtesy of B. L. Gary, NASA/JPL

An MPT was first flown on a NASA CV-990 aircraft in 1978, and provided tropopause altitude information during the Clear Air Turbulence Flight Test Program. An improved MTP was flown for several years on the C-141 Kuiper Airborne Observatory, which produced data revealing that tropopause shapes having inversion layers are twice as likely to produce turbulence than other tropopause shapes.

A smaller MTP was built for NASA's ER-2 aircraft for the Stratospheric/Tropospheric Exchange Project. A project series of flights based in Darwin, Australia in 1986 for the study of exchange processes that account for the dehydration of the lower stratosphere. The same MTP was used during the 1987 Airborne Antarctic Ozone Experiment (AAOE), the 1989 Airborne Arctic Stratospheric Expedition (AASE), and the 1991/1992 Airborne Arctic Stratospheric Expedition II (AASE II). Each of these missions studied processes producing stratospheric ozone depletion. An MTP was built for NASA's DC-8 aircraft for use in the AASE II.

The MTP/DC-8 instrument profiles temperature every 14 seconds, providing an accuracy exceeding radiosonde and satellite analysis of the temperature field throughout the altitude region 7 and 17 km. RMS accuracy, as measured against nearby radiosondes, is better than 1 K over a 7 km region, and better than 2 K over a 12 km region. Performance is unaffected by cirrus clouds or polar stratospheric clouds (PSC's).

The MTP/DC-8 and MTP/ER-2 instruments are total power radiometers that employ a horn antenna, with a 7.5° half-power, full-width beam pattern, which scans from -60° to +60° in 10 elevation angle steps. The MTP/DC-8 operates at 55.51, 56.66 and 58.79 GHz, and completes a scan every 14 seconds. The MTP/ER-2 instrument operates at 56.67 and 58.80 GHz. The MTP/ER-2 records data on a removable memory card. The electronics weigh 10 kg, and the fairing and other hardware weigh another 5 kg. A lighter MTP is planned to fly on the unmanned aircraft series Perseus (see Table 194).

## B.80    MTS (Millimeter-Wave Temperature Sounder)

The NASA-sponsored MIT (Massachusets Institute of Technology) MTS instrument is an improved version of a fixed-beam 118 GHz temperature sounder built in 1976 and operated aboard NASA's Convair 990 aircraft in 1977 and 1978. [518),519)] - MTS (operational since 1985) is capable of ground-based zenith-viewing at remote sites or of nadir-viewing operation aboard NASA/ARC's ER-2 aircraft. Since 1991 it has also been capable of zenith-scanning operation. Objective: passive observations for temperature sounding and precipitation parameter estimation.

MTS collected data in 33 ER-2 flights in 1986 during the GALE (Genesis of Atlantic Lows Experiment) and COHMEX (Cooperative Huntsville Meteorological Experiment) campaigns, yielding high spatial resolution microwave images of atmospheric $O_2$ brightness temperatures over clear air, stratiform precipitation, and convective precipitation. MTS has further observed clear air, precipitation, hurricanes, and zenith opacity during: 8 CAPE flights in 1991, 22 STORM-FEST flights in 1992, 3 TOGA/COARE flights in 1993, and 8 CAMEX (Convection and Atmospheric Moisture Experiment) flights in 1993.

MTS is a dual-band microwave radiometer system used for the measurement of atmospheric temperature as well as other phenomena affecting the transmission in the microwave absorption bands of molecular oxygen. The instrument consists of a scanhead housing, two radiometers and a video camera. Supporting equipment includes detectors, integrators, power supplies, temperature controllers, a video recorder, and a microcomputer. One radi-

518)  A. J. Gasiewski, J. W. Barrett, P. G. Bonanni, D. H. Staelin, "Aircraft-based Radiometric Imaging of Tropospheric Temperature and Precipitation Using the 118.75 GHz Oxygen Resonance", Journal of Applied Meteorology, Vol. 29, Nr. 7, 1990, pp. 620-632
519)  Update information provided by D. H. Staelin, MIT

ometer is a 53.65 GHz fixed-beam single-channel sensor; the other instrument is a 118 GHz cross-track scanning spectrometer. Since 1993 the 53-GHz system has been able to tune rapidly between 52.8 - 55.5 GHz under digital control. Both radiometers employ symmetric Dicke-switching to minimize drift-induced errors in the measured antenna temperature.

A rotating mirror allows for a planar scan of the antenna beam over a $\pm 47^\circ$ FOV. The MTS scanning spectrometer yields about 0.5 K brightness temperature resolution per spot for the probing channels. The corresponding horizontal spatial resolution is 1.6 - 2.3 km for features at 8 km altitude and 2.6 - 3.7 km at the Earth's surface (flight altitude of ER-2 aircraft at 19.5 km, flight speed of 200 m/s). Hot and ambient temperature calibration loads with about 1% reflectivity are viewed once during each scan.

The single-channel fixed beam radiometer senses oxygen emissions near 53.65 GHz. The nadir-viewing 53 GHz antenna has a beamwidth of $7.5^\circ$. The 53 GHz radiometer is calibrated by connecting the input port to hot and ambient waveguide terminations via a mechanical waveguide switch. The CCD camera was expanded to a $107^\circ$ FOV in 1993.

Data products: Data have been used to produce images of temperature and precipitation structure, to infer precipitation cell top altitudes, to detect atmospheric waves, and to measure atmospheric microwave transmittances.

| MTS Channel | Sideband Frequencies (GHz) | Sideband Bandwidth (MHz) |
|:---:|:---:|:---:|
| 0 | 53.65 $\pm 0.8$ (pre CAMEX) | 160 |
| 1 | 118.75 $\pm 0.66$ | 170 |
| 2 | 118.75 $\pm 0.84$ | 210 |
| 3 | 118.75 $\pm 1.04$ | 240 |
| 4 | 118.75 $\pm 1.26$ | 220 |
| 5 | 118.75 $\pm 1.47$ | 240 |
| 6 | 118.75 $\pm 1.67$ | 220 |
| 7 | 118.75 $\pm 1.90$ | 270 |
| 8 | 118.75 $\pm 0.50$ (post GALE) | 125 (post GALE) |

**Table 183:  MTS Channel Specifications**


# B.81    MUSIC (MUlti-Spectral Infrared Camera)

The MUSIC airborne hyperspectral instrument was developed and built at the Lockheed Palo Alto Research Laboratory. The current version is operational since 1989 (earlier versions since 1977) and flown on NASA research aircraft at altitudes up to 20 km, a ground-based configuration of the instrument has recently been developed.[520]
Applications: chemical vapor sensing, plume diagnostics, spectral signatures.

MUSIC collects infrared image data simultaneously in the MWIR and TIR spectrum. The sensor consists of two parallel optical telescopes, each with its own set of optical filters and detector arrays. Si:In detectors in MWIR and Si:Ga detectors are employed in TIR region. Each detector array is a 45 x 90 pixel matrix for a total of 4050 detectors in each array. All detectors collect data in parallel at up to 80 frames/s. The optical system provides a pixel angular spacing of 0.5 mrad, yielding a total of 22.5 x 45 mrad ($1.3^\circ$ x $2.6^\circ$). An indexing filter in each telescope allows for selection of any of 7 bandpass filters. Since the emphasis in the MUSIC flight program has been on multispectral data, typical modes of operation call for rapid interchange of filters, under computer control, to obtain near simultaneous data in several spectral bands.

---

520) Information provided by W. P. Rudolf of Lockheed

| Spectral Parameters | | |
|---|---|---|
| Spectral range | 2500 - 7000 nm | 6000 - 14500 nm |
| Instantaneous range (single CVF position) | 350 nm | 600 nm |
| Bandwidth | 25-70 nm | 60 - 1400 nm |
| Channels | 90 | 90 |
| Spatial Parameters | | |
| IFOV | 0.5 mrad | |
| FOV | 1.3° (across-track) x 2.6° (along-track) | |
| Pixel/line | 45 | |
| Swath | 68 m @ 3 km; 450 m @ 20 km altitude | |
| Electronic Parameters | | |
| Digitization | 12 bit | |
| Data Rate | 80 frames/s digital PCM | |

**Table 184:    Characteristics of the MUSIC Spectrometer**

The sensor is a cryogenic instrument, with all elements of the optical train behind the entrance window operating at the temperature of liquid nitrogen (78K). The detector arrays are cooled with liquid helium. Because of this cooling, the internal background is very low and the detectors operate close to the BLIP limit as defined by the external signal through the cooled spectral filters. Single frame sensitivity is about 1 $\mu W/cm^2/sr/m^{-6}$, varying slightly with the specific band.

In addition to the 7 spectral bandpass filters in each telescope, a circular variable filter (CVF) is installed which provides a continuously variable narrow bandpass ($\lambda/\Delta\lambda \sim 100$) which can be continuously tuned by rotation of the filter to cover the full spectral range in each telescope. Either the CVF or the discrete filter wheel can be used (pre-programmable control).

In flight, data is recorded in digital PCM format on wideband data tape for post flight analysis. For use in a special mountain top data collection exercise, an alternate gimbal pointing system and an alternate data recording system (direct high-speed disk recording) were developed. The system offers quicklook capability.

## B.82    NAILS (NCAR Airborne Infrared Lidar System)

NAILS[521] is a heterodyne Doppler lidar instrument of NCAR in Boulder Co. for the use in atmospheric science research (upgrade of instrument in 1993). NAILS was sponsored by NSF with the intention of supporting the university research community. NAILS measures range-resolved backscatter from the atmospheric aerosol using a transmitted wavelength of 10.6 $\mu$m (other $CO_2$ laser wavelengths are possible). The instrument has been used with direct (not heterodyne) detection for measuring profiles of infrared backscatter and for monitoring the structure of marine stratocumulus. NAILS is now being upgraded to include the capability of measuring the atmospheric velocity component along the lidar line of sight by means of the Doppler shift of the radiation scattered by the atmospheric aerosol.

Applications: NAILS can determine the horizontal wind field in quasi-stationary situations by flying a box pattern while pointing horizontally. This is useful for convergence beneath convective clouds and can be used to study entrainment into the side of cumulus clouds. When pointing vertically, the measured vertical velocity can be combined with lidar measurements of constituent concentration to infer vertical fluxes by an eddy-correlation method. Remote measurements of profiles of vertical velocity contribute as well to studies of entrainment in stratus clouds. When used with appropriate conical scanning (not yet added to the system), profiles of the horizontal wind are available to determine the transport of pol-

---

521)  Information provided by R. Schwiesow of NCAR

lutants. In combination with other lidars operating at shorter wavelengths, backscatter measurements from NAILS provide data on the aerosol particle size. The middle infrared operating wavelength also provides strong aerosol-to-molecular contrast to define the top of the surface mixed layer.

Operating parameters:

| | |
|---|---|
| operating range (nominal) | 0.6 to 10 km |
| range resolution | $\pm 50$ m |
| velocity accuracy | $\pm 0.5$ m/s |
| pulse repetition frequency | up to 64 Hz |

The lidar system is carried on the NCAR Electra aircraft and consists of an optical package of approximately 200 kg, electronics rack of approximately 150 kg, and auxiliary equipment (turning mirror, gas supplies, etc.) of about 50 kg. Data are recorded as raw time series of digitized signal return on 8 mm digital tape. The system provides a real-time display capability (range-time plot) of velocity component and backscatter intensity, the raw time series data can be processed (on the ground) to provide any desired type of analysis.

The system uses a pulsed $CO_2$ TEA laser as a transmitter and a continuous-wave $CO_2$ laser as a local oscillator for heterodyne detection and to seed the TEA laser for single-frequency output. In addition to allowing Doppler measurements, heterodyne operation makes NAILS more sensitive for backscatter profile measurements than it would be with direct detection, due to the reduction of background noise.


## B.83   NASIC (NASA Aircraft-Satellite Instrument Calibrator)

NASIC is a NASA/GSFC spectroradiometer flown on ER-2 aircraft for the calibration of satellite sensors in the spectral range from 400 to 1040 nm. The instrument is of NOAA heritage acquired by GSFC in 1988. The primary objective is to provide input for quantitative calibration estimates on long-term operational sensors with insufficient calibration knowledge due to one or more of the following factors:

- imprecise knowledge of prelaunch calibration parameters
- absence of adequate on-board calibration capabilities
- occurrence of significant in-flight calibration changes over extended periods.

The goal is the compilation of accurate, long-term records of environmental parameters for use in climate and global studies. The system is in operation since 1980 and has been upgraded several times. Satellite sensors that were calibrated with NASIC include (some only once): AVHRR channels 1 and 2 on NOAA-9, -10, and -11; Landsat-5 TM bands 1-4; GOES-6 and -7 VAS (visible bands), and Nimbus-7 CZCS.[522]

The calibration method with NASIC underflights requires accurate prediction of the satellite-target viewing geometry, which is necessary to enable the aircraft spectroradiometer to be co-aligned with the satellite viewing vector during the satellite overpass. Small corrections must be applied to account for the effects of the atmospheric path between the aircraft and the satellite, and to account for the difference between the footprints of the two instruments on the target. These corrections, and knowledge of the spectral response function of a given channel of the satellite sensor (radiometer), allow the calculation of equivalent sets of radiance values (from the aircraft measurements) and count values (from the satellite measurements) that correspond to the altitude of the satellite radiometer and the field of view of the aircraft spectrometer.

A typical data collection period for underflights of a polar-orbiting satellite is in the order of 3 minutes. Data are collected of the same target area in parallel from the satellite sensor and

---

522)  P. Abel, B. Guenther, "Calibration Results for NOAA-11 AVHRR Channels 1 an 2 from Congruent Path Aircraft Observations", Journal of Atmospheric and Oceanic Technology, Volume 10, Aug. 1993, pp. 493-508

from NASIC. The method assumes that the two datasets correspond with to identical states of scene structure and illumination. The footprint of NASIC is chosen much larger than the footprint of the satellite sensor to allow for footprint corrections.

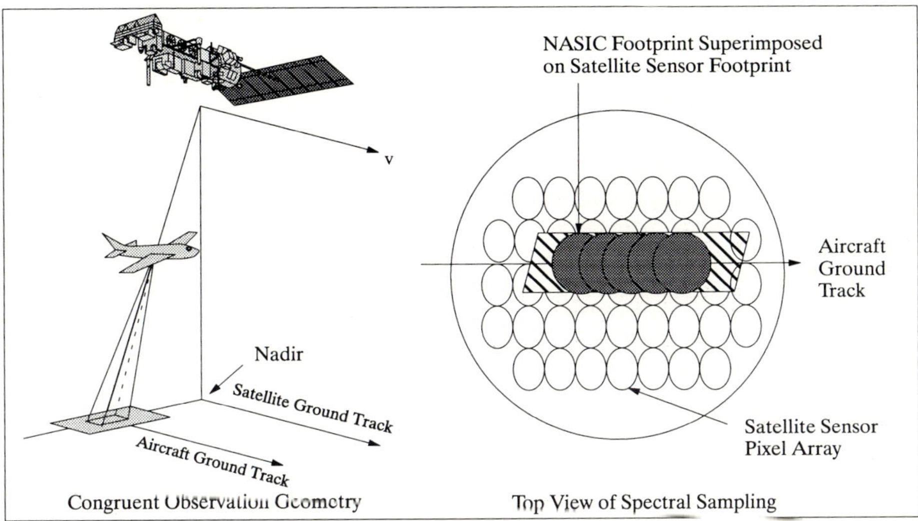


**Figure 131:   Arrangement of Viewing Geometries for NASIC Calibration**

The NASIC instrument is a spectroradiometer of Ebert design with zero-dispersion, a double monochromator[523] and a CCD silicon detector unit. The spectral scan takes approximately 5 seconds and covers 184 positions (25 ms per position or sample) equally spaced in wavelength from 400 - 1040 nm. The NASIC footprint on the ground is a single pixel (there is no spatial scan mechanism), its movement along the ground is controlled by the aircraft motion along the flight path. The spectral resolution is approximately 3.5 nm.

The spectral range of NASIC is configured for each particular application. The measurement of the AVHRR sensor channels 1 and 2 may serve as an example: A blocking filter on front of the detector prevents energy at wavelengths below 380 nm from entering the system. Spectral scans are performed sequentially in pairs, one scan in each pair having a second blocking filter in place to absorb energy at wavelengths below 700 nm thereby giving a valid radiance spectrum in the range from 700 - 1040 nm; the other with the second blocking filter removed provides a valid radiance spectrum from 400 - 760 nm.

NASIC is gimbal-mounted in the aircraft to allow its optical axis to be directed to a range of angles to the right (i.e. to starboard) and below the aircraft axis. These motions are controlled by an on-board PC through azimuth and elevation drive motors with a positioning accuracy of $\approx 1°$.

The instrument has been calibrated in the laboratory at GSFC on an irregular schedule usually before and after most flights.

In post-flight processing and analysis radiances observed by the satellite sensor are modelled from the measured NASIC radiance spectra by performing a convolution integral of the radiance spectra with the sensor's spectral response profiles.

---

523)  Note: "Zero dispersion double monochromator": two monochromators in series, arranged with aligned spectral scanning functions and with zero spectral dispersion at the final exit slit. This arrangement significantly reduces stray light effects relative to the performance of a single monochromator.

| Telescope focal length | 125 mm |
|---|---|
| Aperture | f/4.8 |
| Grating | 600 groove/mm, blazed at 500 nm |
| Spectral range | 400 - 1040 nm |
| IFOV = FOV | 105 - 160 mrad (footprint of 2 - 3 km) |
| Number of spectral pixels/scan | 184 |
| Detector | CCD (Silicon) actively temperature controlled |
| Quantization | 12 bit |
| Platform | ER-2 aircraft with INS, GPS |

**Table 185:    Specification of the NASIC Instrument**

# B.84    NCAR Electra Aircraft Instrumentation[524]

| Measured Variable | Instrument Type | Manufacturer and Model | Combined Performance of Transducer, Signal Conditioning, and Recorder | | |
|---|---|---|---|---|---|
| | | | Range | Accuracy | Resolution |
| Absolute Humidity (VLA) | Lyman-alpha Hygrometer | NCAR-developed LA-3 | 0.1-25 g/m$^3$ | +5% | 0.2% |
| Angle of attack (ADIFR) | Flow angle sensor radome | Rosemount Inc. 1221 | ± 10° | ± 0.134° | 0.002° |
| Angle of sideslip (BDIFR) | Flow angle sensor radome | Rosemount Inc. 1221 | ± 5° | ± 0.096° | 0.002° |
| Radiometric surface temperature (RSTB) | Bolometric Radiometer | Barnes Eng. Co. Optitherm | -29 to 75°C | ± 1.0 °C | 0.005°C |
| Infrared Radiation (IRT, IRB) | Pyrgeometer 4-45 µm (Silicon Dome) | Eppley, PIR | 0 - 600 W/m$^2$ | - | 0.40 W/m$^2$ |
| Visible Radiation (SWT, SWB) | Pyranometer 0.285-2.8 µm Clear Dome WG7 | Eppley, PSP | 0 - 1500 W/m$^2$ | - | 0.12 W/m$^2$ |
| UV Radiation (UVT, UVB) | Photometer 0.295-0.385 µm | Eppley, TUVR | 0-200 W/m$^2$ | - | 0.12 W/m$^2$ |
| Ozone Monitor (TEO3C) | UV Photometer | Thermo Electron, Model 49 | 0-1 ppm | 4 ppb | 1 ppb |
| Carbon Monoxide (CO) | Gas Correlation Filter | NCAR-developed 1989 | 0 - 5 ppm | 20 ppb | 10 ppb |
| Aerosol Concentra-tions (CONCNC) | Butanol Condensation Nuclei | TSI, Model 3760 | 0 - 10$^5$ count/cm$^3$ | ± 3% | 1 count/cm$^3$ |
| Geometric Altitude (HGM) | Radio Altimeter | Sperry Rand, RT-221 | 0-762 m | 0-152 m ± .6 152-762m ± 3% | 0.1 m |
| Geometric Altitude (HGME) | Radio Altimeter | Stewart Warner, APN-159 | 0-21,000 m | ± 9.7 m | 0.1 m |
| Cloud Liquid Water Content (XLWC) | Hot-wire | Johnson-Williams, LHW | 0-5 g/m$^3$ | - | 0.005 g/m$^3$ |
| Cloud Liquid Water Content (PLWC) | Heated-wire | PMS-CSIRO King Probe | 0-3 g/m$^3$ | - | 0.006 g/m$^3$ |
| Ice Detector (RICE) | Accretion of Cloud Drop-lets | Rosemount Inc., 871F | 0-0.5 mm increments | - | 0.0005 mm |
| Aerosol Spectrum (PCASP) | Laser Spectrometer | PMS Inc. | 0.12-3.12 µm | - | 0.025-.375 µm |
| Cloud Droplet Spec-trum (FSSP) | Laser Spectrometer | PMS Inc. | 0.5-47 µm | - | 0.5 or 1, 2, 3 µm |
| Cloud Droplet Spec-trum (260X) | Laser Spectrometer | PMS Inc. | 40-620 µm | - | 10 µm |
| Cloud Droplet Spec-trum (200X) | Laser Spectrometer | PMS Inc. | 40-280 µm | - | 20 µm |
| Cloud Droplet Spec-trum (200Y) | Laser Spectrometer | PMS Inc. | 300-4500 µm | - | 300 µm |
| Cloud Particle Spec-trum (2D) (2D-C) | Laser Spectrometer | PMS Inc. | 25-800 µm | - | 25 µm |
| Hydrometer Spec-trum (2D-P) | Laser Spectrometer | PMS Inc. | 200-6400 µm | - | 200 µm |
| Photography | VHS Video Cameras: Forward-looking, B&W, PULNIK: Model TM-34K Side side, color, GE camera: Model 1CVK 5032A Images recorded on GE Model 1CVK 5022X VCR | | Up to 6 hours of recording (with/without voice per cas-sette) | | |

524)  Information provided by P. Spyers-Duran of NCAR

| Measured Variable | Instrument Type | Manufacturer and Model | Combined Performance of Transducer, Signal Conditioning, and Recorder | | |
|---|---|---|---|---|---|
| | | | Range | Accuracy | Resolution |
| Navigation  Measurements | | | | | |
| Aircraft Latitude (ILAT) | Inertial Navigation System (INS) | Honeywell Laser ref. SM IRS | ±90° | 0.164° (6 hr) | 0.00017° |
| Aircraft Longitude (ILON) | Inertial Navigation System | Honeywell Laser ref. SM IRS | ±180° | 0.164° (6 hr) | 0.00017° |
| Aircraft Latitude (CLAT, CLON) | LORAN-C System | Advanced Navigation ANI-7000 | ±90° ±180° | 0.19 km (location dep'dent) | 0.0002° |
| Aircraft Position GLAT,GLON,GALT | GPS Receiver | Trimble Navigation Model 400 | 3-D pos. 2-D pos. | 25 m (horiz.) 35 m (vert.) 0.1 m/s (veloc) | ≤0.5 m ≤0.5 m ≤0.5 m/s |
| Aircraft G. Speed (IVNS, IVEW) | INS | Honeywell Laser ref. SM IRS | 0-400 m/s | 4.115 m/s (6 hr) | 0.0643 m/s |
| Aircraft Vert. Velocity (VZI) | INS | Honeywell Laser ref. SM IRS | ±200 m/s | ±0.1524 (6 hr) | 0.3048 m/s |
| Aircraft True Heading (ITHDG) | INS | Honeywell Laser ref. SM IRS | 0-360° | ±0.2° (6 hr) | 0.0055° |
| Aircraft Pitch Angle (IPITCH) | INS | Honeywell Laser ref. SM IRS | ±45° | ±0.05° (6 hr) | 0.0109° |
| Aircraft Roll Angle (IROLL) | INS | Honeywell Laser ref. SM IRS | ±90° | ±0.05° (6 hr) | 0.0109° |

## B.85   NEC-SAR (NEC Corporation SAR)

NEC-SAR is a prototype model of a small, light-weight, and high resolution SAR of NEC Corporation of Tokyo, Japan. The objective is to obtain high-resolution contour maps at a scale of 1:25,000. The test flight program of the original instrument started in July 1992, test flights for the interferometric SAR began in July 1993.[525]

NEC developed the instrument from its internal R&D funds. After prototype capability testing NEC will eventually offer the SAR instrument commercially for such applications as: topographic mapping, volcano surveillance, search and rescue operations at sea, reconnaissance, etc.

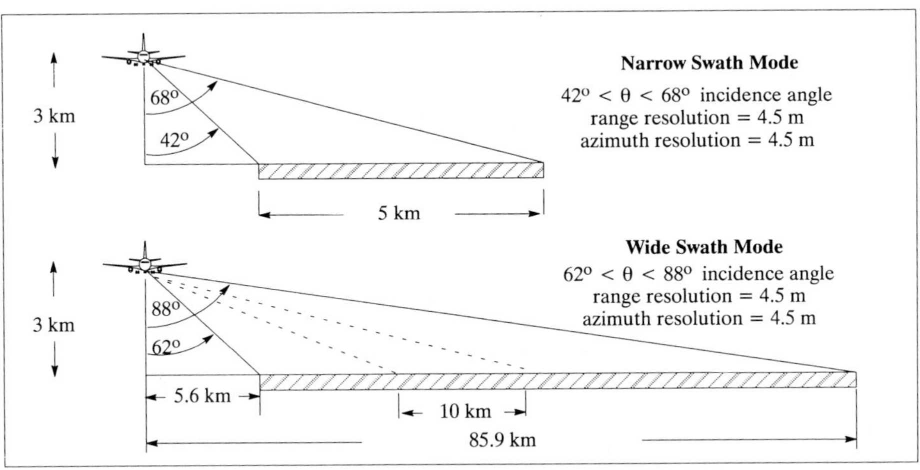


**Figure 132:   Radar Imaging Modes of the NEC-SAR Prototype Instrument**

525)  Information provided by H. Shinohara of NEC Corporation

| Parameter | Value |
|---|---|
| RF center frequency | 9.53 GHz (X-Band) |
| Wavelength | 3.1 cm |
| Expanded pulse length | 4 μs |
| Digital chirp signal bandwidth | 50 MHz |
| Polarization | HH |
| Antenna size | 340 mm x 100 mm |
| Baseline distance for interferometric mode | 600 mm |
| Antenna gain | 21.0 dBi |
| Antenna 3 dB beam width, azimuth | 7.5° |
| Antenna 3 dB beam width, elevation | 26.0° |
| Transmit peak power | 3000 W |
| Transmit average power | 12 W |
| Noise figure | $\leq 10$ dB |
| Receiver dynamic range with AGC/STC | $\geq 30$ dB |
| Nominal pulse repetition frequency (PRF) | $\leq 1200$ Hz |
| Nominal PRF for interferometric mode | 600 Hz |
| Quantization (bit/sample, I or Q), narrow swath mode | 6 |
| Quantization (bit/sample, I or Q), wide swath mode | 3 |
| Sampling rate | 63.5 MHz |
| Echo buffer memory capacity (I or Q) | 2176 words |
| Nominal data rate onto high density tape | 32 Mbit/s |
| Maximum recording time per tape (1/2 inch video) | 22 minutes |
| Spatial resolution, range x azimuth | 4.5 m x 4.5 m |
| Number of statistically independent looks | $\leq 6$ |
| Accuracy of calibration | $\leq 1$ dB |
| Aircraft platform | Cessna 208 |
| Ground speed | 280 km/h |
| Observation altitude (nominal) | 3000 m |
| Incidence angle,(near range, center range, far range) | 42°, 55°, 68° |
| Incidence angle,(near range, center range, far range) | 62°, 75°, 88° |
| Narrow mode swath width (nominal altitude of 3 km) | 5 km |
| Wide mode swath width (nominal altitude of 3 km) | 10 km |

**Table 186:    Specification Parameters of the NEC-SAR Prototype Model**

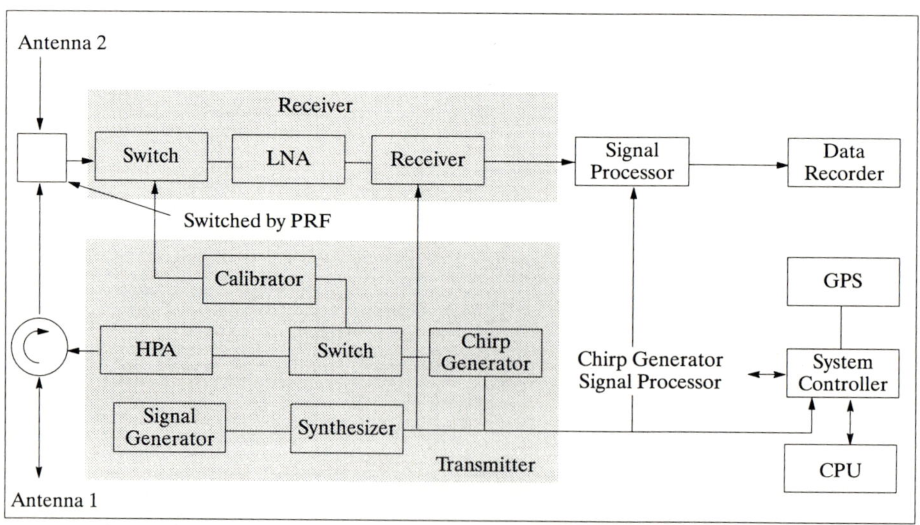


**Figure 133:    NEC-SAR Block Diagram (Interferometric SAR Mode)**

## B.86    NS001 (Thematic Mapper Simulator)

The NS001 multispectral scanner is a NASA-developed Thematic Mapper Simulator (TMS) flown onboard the C-130B aircraft based at Ames Research Center, Moffett Field Ca.. This scanner contains the seven Landsat 4 and 5 TM bands plus a band from 1.131-1.35 μm.

The NS001 is flown at low and medium altitudes. The format of the flight data consists of 838 8-bit words per frame (data for one wavelength band throughout a scan line). Of these, 699 are the video information and the remainder ancillary information (Greenwich time, scan line number, calibration lamp voltage and current, blackbody temperatures, etc.).

| Band Nr. | Wavelength of Band (μm) |
|---|---|
| 1 | 0.458 - 0.519 |
| 2 | 0.529 - 0.603 |
| 3 | 0.633 - 0.697 |
| 4 | 0.767 - 0.910 |
| 5 | 1.13 - 1.35 |
| 6 | 1.57 - 1.71 |
| 7 | 2.10 - 2.38 |
| 8 | 10.9 - 12.3 |
| Geometric Parameters | |
| FOV | 100° |
| IFOV | 2.5 mrad |
| Ground resolution | 7.6 m (at 3000 m altitude) |
| Swath width | 7.26 km (at 3000 m altitude) |
| Pixels/scan line | 699 |
| Scan rate | 10-100 scans/s |

**Table 187:    Spectral Coverage and Geometric Parameters of the NS001 Instrument**

## B.87    NUSCAT (Airborne $K_u$-Band Scatterometer)

NUSCAT[526] was developed by NASA/JPL for airborne experiments to improve geometrical model functions relating ocean backscatter to wind vectors, to serve as a test bed scatterometry technology demonstration, and to provide surface truth measurements for spaceborne scatterometers. NUSCAT was flown on the AMES C-130B aircraft and operated by JPL to take ocean backscatter data during SWADE (Surface Wave Dynamics Experiment) in February - March 1991. The experiment was conducted off the coast of Maryland and Virginia where several buoys are anchored. About 30 hours of data were taken in a total of 10 flights over a wide variety of oceanic conditions including wind speed as low as 2 m/s to more than 12 m/s and significant wave height from below 1 m to well above 5 m.

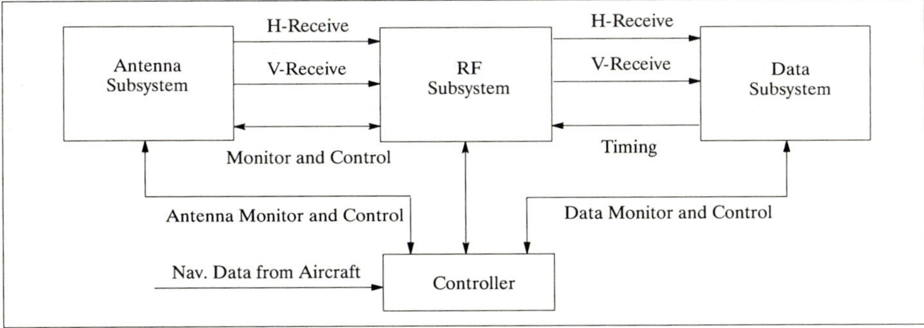


**Figure 134:    NUSCAT System Block Diagram**

---

526) Information received from S. Nghiem and S. Durden of JPL

The NUSCAT system is composed of an antenna subsystem, an RF subsystem, a data subsystem, and a controller. The antenna is a parabolic disk with a peak gain of 32 dB and a beam width of approximately 3°. The antenna is placed inside a radome on the tail of the C-130B aircraft, and is mounted on a gimbal which is used to rotate the antenna in complete azimuthal scans at selected incidence angles. The antenna subsystem is connected through a rotary joint to the RF subsystem. When the system is transmitting either H or V polarization, two receiver subsystems are used to collect simultaneously co- and cross-polarization returns. The radar echoes from each pulse are amplified, down-converted to I/Q samples and were digitally square-law detected. The returns from multiple echoes are integrated over 0.5 seconds and then recorded on computer-compatible tapes.

| | |
|---|---|
| Operational frequency | 13.9 to 13.995 GHz |
| Polarization | HH, VV, HV, VH |
| Incidence angle<br>Incidence angle slew rate | 0 to 70° (60 to 70° by using the aircraft roll)<br>10°/s |
| Azimuth angle<br>Azimuth angle slew rate | 0 to 360°<br>10°/s |
| Operational altitude | 0.5 km to 15 km |
| Pulse repetition frequency<br>Pulse length<br>Frequency diversity | 1.5 to 10 kHz<br>2 to 65 μs<br>pulse-pulse independence (1 MHz steps available in 100 MHz bandwidth) |
| $\sigma_0$ sampling rate | 2/second for quick-look |
| System dynamic range | observe winds from 2 - 50 m/s |
| Data collection | quick-look system, baseband digital recording |
| Calibration | transmit signal injection; 0.23 dB relative; 1.1 dB absolute |
| Transmit power | +51 dBm |
| Noise floor | -110 dBm |
| Frequency | 13.95 GHz |
| Frequency diversity | up to 32 with 100 MHz bandwidth |
| Sampling rate | 2/s |
| Data collection | 8 bit A/D converters |
| Instantaneous dynamic range | 30 dB |
| Antenna gain | 32.0 dB |
| Antenna beamwidth | ~3.1° |
| Timing | > >4 sets |
| Bandwidth | 4, 2, 1 and 0.2 MHz |
| Polarization | direct and cross-simultaneously |
| Gimbal RF path | rotary joints |

**Table 188:   NUSCAT System Parameters**

The internal system calibration is performed by injecting a transmit signal into the receiver through a calibration loop. The relative calibration accuracy involves the uncertainty in measurements of transmit power, receiver gain, position in antenna angles and altitude, sampling circuit, electromagnetic wavelength, rotary joint loss, radome loss, and attenuators and is estimated to be ±0.23 dB. The measured power accuracy depends on the number of independent samples and the signal-to-noise ratio. The operating frequency is dithered over 100 MHz to generate additional independent samples (N) which effectively reduce the communication noise $K_{pc} \propto N^{-1/2}$. The absolute accuracy is subjected to several sources of errors under laboratory test such as attenuators, calibration loop loss, antenna gain, beamwidth, and various losses from VSWR (Voltage Standing Wave Ratio), waveguide, and rotary joint. The antenna gain is determined by the three-horn measurement method at the JPL Antenna Range.

To evaluate the system stability and absolute calibration, data are taken over the ocean surface at 10° incidence where the backscatter is insensitive to surface roughness conditions. In experimental flights, the NUSCAT antenna is scanning in azimuth in 10° steps. During a 4-second period for each step, NUSCAT collects data and moves to the next step. The data are taken as functions of incident angle up to 60° in 10° steps, azimuthal angles in complete azimuthal circles, and horizontal and vertical polarizations.

## B.88   OLS (Oceanographic Lidar System)

OLS is an airborne instrument for fluorosensor detection of maritime oil spills. It was developed by the University of Oldenburg and utilized as an experimental sensor in the time frame of 1983-86 by the University of Oldenburg.

Application: synoptic mapping of hydrographic conditions in extended coastal regions.

OLS[527),528),529)] is a non-imaging sensor having excitation and detection wavelengths appropriate for the investigation of naturally occurring sea water compounds. The instrument is installed on a DO-228 aircraft and has been flown in many campaigns.

| **Lasers:** | Excimer | Dye (excimer laser pumped dye laser) |
|---|---|---|
| Emission wavelength | 308 nm | 450/533 nm |
| Pulse length | 12 ns | 6 ns |
| Peak power | 10 MW | 1 MW |
| Repetition rate | $\leq$ 10 Hz | $\leq$ 10 Hz |
| **Receiver Telescope** | f/10, Schmidt-Cassegrain (0.4 m diameter) | |
| **Spectrograph** | 8 discrete channels, modular | |
| Wavelength detection | 344,366, 380, 450, 500, 533, 650,,685 nm | |
| Wavelength selection | dichroic splitters, interference and blocking filters | |
| Detectors | PMT, EMI 9821/9818 | |
| Digitizer | Biomation 6500, 500 MHz, 11 bit | |
| Computer | LSI 11/23 with floppy disk, hard disk and magtape | |
| Total mass | 500 kg | |
| Resolution (Footprint) | 2.5 m at 245 m flight altitude | |

**Table 189:   Overview of the OLS Parameters**

## B.89   P-3/SAR

P-3/SAR[530),531),532)] is a high-resolution multifrequency polarimetric imaging SAR jointly developed for the US Government by the Naval Air Warfare Center, Aircraft Division (NAWC/AD) and the Environmental Research Institute of Michigan (ERIM). In addition, ERIM and NAWC/AD are continually upgrading the SAR system to meet new imaging research objectives. The P-3/SAR system is jointly owned and operated by ERIM and NAWC/AD, and is installed on a US Navy P-3A Orion aircraft. The sensor is a multimode SAR operating at X-, C-, and L-Band. In addition, ERIM is developing and integrating for the Advanced Research Projects Agency (ARPA) a polarimetric Ultra Wideband (UWB) frequency mode (200-900 MHz), to be operational in the summer of 1994. This new frequency mode is intended to support future foliage and ground penetration experiments/applications. Developed between 1985-87, the system has been operational since 1988 and has logged over 2000 hours of data collection (1993) worldwide, including flights in the USA, Canada, Caribbean, Europe, Africa, Greenland and the Pacific.

527)  T. Hengstermann, R. Reuter, "Lidar fluorosensing of mineral oil spills on the sea surface", Applied Optics, Vol. 29, Nr. 22, August 1990, pp. 3218-3227

528)  D. Diebel, T. Hengstermann, R. Reuter, R. Willkomm, "Laser Fluorosensing of Mineral Oil Spills", The Remote Sensing of Oil Slicks, edited by A. E. Lodge, 1989, Institute of Petroleum, published by John Wiley & Sons, pp. 127-142

529)  R. Reuter,, D. Diebel, T. Hengstermann, "Oceanic laser remote sensing: measurement of hydrographic fronts in the German Bight and in the Northern Adriatic Sea", International Journal of Remote Sensing, 1993, Vol. 14, Nr. 5, pp. 823-848

530)  Information provided by M. Dudzik and P. Wagner of ERIM

531)  A. R. Ochadlick, K. Birny, P. Cho, C. Duke, S. K. Krasznay, J. Evans-Morgis, J. S. Verdi, "A Description of the NADC SAR Facility and Examples of Observations and Measurements", CH2971-0/91/000-1785, © 1991 IEEE

532)  R. Sullivan, A. Nichols, R. Rawson, "Polarimetric X/L/C-Band SAR", Proc. IEEE Radar Conference, Ann Arbor, Michigan, 1998, pp. 9-14

Key operational features of the P-3/SAR system are:

- X-, C-, and L-Band modes of operation
  - Stripmap mode: a push broom mode with the antenna pointed 90° (broadside) to the line of sight (looking either left or right), imaging a continuous strip with a swath range of either 4.9 km or 9.8 km in the slant plane.
  - Spotlight mode: a mode in which the antenna is pointed 45° forward of broadside at a predetermined spot while the aircraft flies past until the antenna is pointing 45° aft of broadside. This mode allows data to be collected with a spot diameter of 1.5 km in L-Band (significantly smaller spots are collected at X- and C-Band) through a range of 90° in aspect angle variations.
  - Dragging spot mode: this is a variation of the spotlight mode in which the antenna rotation is slowed for a few seconds during it's rotation to increase the spot size of the area imaged in azimuth.
  - Circle mode: this is a variation of the stripmap mode in which the antenna is fixed broadside and aimed at a point on the ground, the aircraft flies in a circle around that point allowing the area to be imaged at all aspect angles.
  - Displaced Phase Center (DPC) mode: this is a variation of the stripmap mode in either X- or C-Band using two antennas to measure the radial position by observing a target from the same point in space at different times (lags).
- The UWB mode will operate in the stripmap mode with a fixed incidence angle imaging a continuous strip with a recorded swath range of approximately 1 km.
- Pass-to-pass single frequency, full polarimetric imaging in all frequencies.
- Simultaneous imaging at X-, C- and L-Band with selectable polarizations.
- Pass-to-pass UWB and X-, C-, L-Band imaging capability.

| Parameter | X-Band | C-Band | L-Band | UWB |
|---|---|---|---|---|
| Wavelength | 3.2 cm | 5.7 cm | 24.0 cm | 1.5 - 0.31 m |
| Center frequency | 9.35 GHz | 5.30 GHz | 1.28 GHz | 350 MHz |
| Polarization | VV, VH, HH, HV | VV, VH, HH, HV | VV, VH, HH, HV | VV, VH, HH, HV |
| Peak transmit power | 1.5 kW | 1.4 kW | 5.0 kW | 1.0 |
| Antenna<br>  Azimuth beamwidth<br>  Elevation beamwidth<br>  Gain<br>  Isolation | 1.8° (3.75 DPCA)<br>8.5°<br>27 dB<br>23 dB | 3.9° (8.0° DPCA)<br>15.0°<br>23 dB<br>23 dB | 10.0°<br>100°<br>16 dB<br>23 dB | 113.0° - 25.0°<br>83.0° - 18.0°<br>15.6° - 4 dBi<br>20 dB |
| Processed bandwidth | 120-60 MHz | 120-60 MHz | 120-60 MHz | 580 MHz |
| Impulse Response (IPR) Width (Range) | 1.5-3.0 m | 1.5-3.0 m | 1.5-3.0 m | 0.33 m |
| Sidelobes | 30 dB Taylor weighted | | | 10-15 dB |
| Pulse width | 4.0 µs | | | 26.58 µs |
| Peak duty cycle | 1.6% / 4 kHz | | | 35% |
| Samples | 4096 I&Q per channel | | | 4096 dechirped |
| Pulse Repetition Frequency (PRF) | 2000/channel max (proportional to velocity)<br>1000/Channel (DPCA) | | | TBD |

**Table 190:    Parameter Definition of the P-3/SAR Instrument**

There are three separate antenna or antenna combinations for the P-3/SAR system. The tri-band antenna is comprised of a dual horn feed dish for the X- and C-Band with a cross dipole array for the L-Band. This antenna is mounted in the belly radome just aft of the wings. The tri-band is capable of operating in all but the DPC and the UWB modes of operation. In addition, imaging incidence angles can be selected from 20-70° depending on the frequency, mode and altitude. The DPC antenna consists of two, dual frequency (X- and C-Band) quadridge horns mounted side-by-side. DPC operations require this antenna be mounted in place of the tri-band antenna in the belly radome. The UWB antenna is a one meter quadridge horn mounted in the magnetic anomaly detection (MAD) boom, located at the tail of the P-3 aircraft.

Applications include: sea ice and frozen terrain studies (Alaska, Greenland and Norway); seasonal forest studies (Maine), wave refraction, shelf break behavior, current-wave interactions, and sub-mesoscale features (North Sea, Gulf Stream and Continental Shelf); high wind studies (NASA windshear); buried object detection (29-Palms and Yuma Proving Grounds); foliage penetration for search and rescue (Duke Forest and Maine); environmental monitoring and disaster relief (Kenya, Mississippi River and Florida Everglades).

| Mode | Frequency Options | Polarization Options | Channel PRF (kHz) | Presum Factor | Recording Rate (Mbit/s) |
|---|---|---|---|---|---|
| Stripmap | | | | | |
|   Single swath | 1-3 | 1-4 | 2 | 6 | 65.5 |
|   Double swath | 1-2 | 1-2 | 2 | 6 | 65.5 |
|   Polarimetric | 1 | 4 | 1 | 3 | 65.5 |
|   No Presum | 1 | 1 | 2 | 1 | 98.3 |
| Spotlight | 1-3 | 1-4 | 2 | 6 | 65.5 |
| Dragging Spot | 1-3 | 1-4 | 2 | 6 | 65.5 |
| Circle | 1-3 | 1-4 | 2 | 6 | 65/5 |
| Displaced Phase Center | 1-2 | 1-4 | 1 | 1 | 98.3 |
| Ultra Wideband | 1 | 1-4 | TBD | TBD | TBD |

**Table 191:     Modes of Operation of the P-3/SAR Instrument[533]**


# B.90     PBMR (Pushbroom Microwave Radiometer)

PBMR is a multibeam L-Band (1.413 GHz) radiometer providing simultaneous cross-track radiometric measurements. The system is owned and developed by NASA's Langley Research Center (Hampton Va.) and operated by WFF of GSFC. Objectives: to provide an engineering prototype to investigate pushbroom technologies such as microwave integrated circuit receivers, digital signal processing within the radiometer, and local oscillator distribution. The instrument is considered as a research tool providing in addition some practical applications for soil moisture research.[534]

PBMR became operational in 1983 and has been flown in NASA C-130 or P-3 aircraft at low altitudes (< 1500 m), providing low-resolution images with a swath width of about 1.2 times the flight altitude. The instrument is mainly used for remote measurements of soil moisture and ocean salinity; it took part in the HAPEX campaigns in 1986 and 1992 as well as in the FIFE program in 1987-89.

| | |
|---|---|
| Center frequency | 1.413 GHz |
| RF bandwidth | 25 MHz |
| Sensitivity | 1.0 K |
| Calibration accuracy | 2.0 K |
| Antenna | 4 beam |
| Scan angles | $\pm 8^\circ$ and $\pm 24^\circ$ from nadir |
| Beam crossover | 3 dB |
| Sidelobe levels | -13 dB |

**Table 192:     PBMR System Parameters**

The radiometer is a Dicke Switching Noise Injection System. This technique provides a very stable radiometer by greatly reducing the effect of receiver noise and gain functions. The antenna system utilizes a Butler matrix feed to produce the required simultaneous beams at discrete cross-track angles. The system produces four beams, pointing at the cross-track direction of $\pm 8^\circ$ and $\pm 24^\circ$ from nadir, the beam width is about $16^\circ$.

---

533)  Note: 'PRESUM' refers to the number of pulses that are summed  together to allow increased data recording.
534)  Information provided by J. R. Wang of GSFC and by T. Schmugge of USDA Hydrology Lab

## B.91    PERSEUS (Unmanned High-Altitude Research Aircraft)

A NASA-supported (since 1991) commercially-developed high-technology aircraft and program for airborne high-altitude atmospheric research. Perseus is the platform to carry lightweight monitoring instruments - the objective is to offer adequate means for sampling the chemistry and dynamics of the upper troposphere and stratosphere, in particular in the polar regions. The idea is the provision of a low-cost platform (in equipment as well as in operation compared with spaceflight) to satisfy a multitude of requirements by the research community.[535),536),537),538)]

'Perseus' represents a family of remotely-controlled drones, designed and developed by 'Aurora Flight Sciences Corporation', Manassas, Virginia (aircraft can be bought or leased from Aurora, first monitoring flights with Perseus-A are scheduled for 1994). The family consists of the following models:

- Perseus-POC (Proof-of-Concept). This is a one-of-a-kind technology demonstrator. First flight of POC on 8 November, 1991.

- Perseus-A. Designed primarily for stratospheric research. Perseus-A uses a unique closed-cycle engine powered by liquid oxygen and gasoline to achieve altitudes of 25 - 30 km with payloads of 50 - 100 kg.

- Perseus-B. Designed for upper troposphere and lower stratosphere research. Perseus-B combines the airframe and flight control system from Perseus-A with a two-stage turbocharged engine. It is a an aircraft for long-range and long-endurance flights.

- Theseus. An extension of the Perseus program. Twin-engine aircraft with a fault-detecting, reconfigurable control system.

The remotely-controlled aircraft requires a ground station during its flight missions. Such a ground station consists of work stations for the aircraft pilot, an avionics engineer, a flight engineer, and the science payload operators.

The pilot has three video screens monitoring basic flight information (altitude, speed, heading, etc.). A navigation monitor traces the aircraft's location along its pre-programmed flight path on a map. The avionics engineer monitors both airborne and ground station computers, the integrity of RF links, and the system configuration. The flight engineer monitors engine performance and other aircraft systems.

| Parameter | POC | Perseus-A | Perseus-B | Theseus |
|---|---|---|---|---|
| Wing span | 17.9 m | 17.9 m | 17.9 m | 36 m |
| Wing area | 16.0 m$^2$ | 16.0 m$^2$ | 16.0 m$^2$ | 44 m$^2$ |
| Installed power | 35 kW | 50 kW | 60 kW | 120 kW |
| Empty mass | 348 kg | 500 kg | 450 kg | 1400 kg |
| Payload mass | 0-50 kg | 50-150 kg | 50-200 kg | 340 kg |
| Gross mass | 430 kg | 740 kg | 1000 kg | 2100 kg |
| Maximum altitude | 8 km | 25-31 km | 20 km | 27 km |
| Maximum range | 600 km | 1200 km | 19.000 km | 15.000 km |
| Maximum duration | 8 hr | 6 hr | 72 hr | ≈ 1 month |
| First Flight | November 1991 | November 1993 | March 1994 | 1995 |

**Table 193:    Performance Parameters of the Perseus Unmanned Research Aircraft**


535)  G. Taubes, "NASA Launches a 5-Year Plan to Clone Drones", Science, Vol. 260, 16 April 1993, p. 286
536)  Information provided by J. S. Langford of Aurora Flight Sciences Corporation
537)  S. Ashley, "Ozone Drone", Popular Science, July 1992, p. 60-64
538)  "The Perseus Data Link", Aurora quarterly

| Measurement (Constituent) | Technique/Instrument | Investigator | Sensor Mass (kg) |
|---|---|---|---|
| OH, HO$_2$, | Solid-state laser OH/HO$_2$-Instrument | J. G. Anderson (Harvard) | 30 |
| H$_2$O | Photofragment fluorescence hygrometer H$_2$O Instrument | | 30 |
| ClONO$_2$ | Thermal disassociation ClONO$_2$-Instrument | | 15 |
| ClO, BrO, | Resonance scattering / UV absorption ClO/BrO-Instrument | R. C. Cohen (Harvard) | 35 |
| O$_3$ | Absorption measurements of ozone O$_3$ Instrument | E. Hintsa, E. Weinstock (Harvard) | |
| NO, NO$_y$ | Catalysis/chemiluminescence NO$_y$-Instrument | S. Wofsy, B. Daube, J. Burley, D. Kliner | 45 |
| N$_2$O, CH$_4$ | TDL spectrometer (Argus) | M. Loewenstein, ARC | 25 |
| N$_2$O, CH$_4$, H$_2$O | TDL spectrometer (ALIAS II) | C. Webster, JPL | $\approx 30$ |
| CFC-11, 12, 113, CFC-22, N$_2$O, CH$_4$ | GC/EC | J. Elkins, NOAA | 15 / 15 |
| CO$_2$ | IR absorption, CO$_2$-Instrument | S. Wofsy, K. Boering, B. Daube, D. Toohey, R. Keeling, | 20 |
| CH$_4$, ClO/NO$_2$ | IR absorption | C. Kolb, Aerodyne | $\approx 40$ each |
| CFC-11 | Thermally tuned Ge interferometer | C. Kolb, Aerodyne | $\approx 50$ |
| T, P, RH | Drop-wind-sondes | Aurora | 0.25 each |
| T, P, RH, winds,clouds | Forward scattering nephelometer | Lockheed PRESSURS | 20 |
| Vertical temp. profiles | MTP (Microwave Temperature Profiler) | B. Gary, JPL | 10 |
| Rad./cloud interaction Flux/net-flux up/down Sub-visual cirrus cloud Radiative properties | MPIR (CCD Imaging Radiometer) HONER (Net-flux radiometer) CDL (Cloud Detection Lidar) UAV-AERI | Vitko et. al., Sandia LANL LLNL U. of Wisconsin | 85 |
| Aerosols | Condensation nucleus counter Optical particle spectrometer | Wilson, Denver | 5 / 6 |
| CH$_4$ | Laser absorption spectrometer | Deschler (Wyoming) | 6 |
| Optical depths Extinction profiles | HIRAASS (High-resolution Airborne Autotracking Sun Spectrometer) | P. B. Russell, (AMES) | 25 |

**Table 194:    Overview of prospective Investigations/Applications on Perseus Aircraft**

A ground station consists of a (mobile) container equipped with computers for all operations, a GPS receiver, and multi-channel narrow-band UHF (430-450 MHz band) radios for ground-air communications in both directions.

Airborne flight control relies on a central computer which supports communication, navigation, operation of the autopilots and control of the propulsion system. Sensors continuously monitor the aircraft state and pass the information to the pilot on the ground as well as to the airborne autopilot and navigation systems. Roll and pitch is sensed by a vertical gyro; a three-axis magnetometer reports heading; angular rates are sensed by rate transducers; while angle of attack and sideslip are measured with vanes. Indicated airspeed and barometric altitude are also measured by pressure transducers. Position data is determined by GPS.

The fully-digital autopilot can operate in several independent modes designed to improve flight safety. Perseus can be flown in 'autonav' mode. This is a computer flight with GPS input to follow waypoints along a preplanned flight path. In case of a loss of uplink command, autonav will automatically return the aircraft to a designated recovery area.

Applications: Large-area surveys due to long-endurance and long-range flights. Hurricane studies from above the storm; etc....

## B.92    PHARUS (PHased ARray Universal SAR)

PHARUS is a polarimetric C-Band airborne SAR, a Dutch project carried out by TNO Physics and Electronics Laboratory (TNO-FEL) in The Hague, together with NLR (National Aerospace Laboratory) in Amsterdam, and the Delft University of Technology. The program is sponsored in part by the Ministry of Defense and the Netherlands Research Sensing Board.

The objective is the operation of a full polarimetric C-Band SAR (wavelength = 5.66 cm) on an aircraft for Earth surface imaging. The system is scheduled to be completed in 1994 (the program started in 1988). The PHARUS system has an active phased-array patch antenna (image mode flexibility, selection of resolution). The antenna pointing on-board the aircraft is maintained through the phased array, rather than by mechanical stabilization.

A polarimetric radar is capable of measuring the complex scattering matrix of every resolution cell. In comparison with traditional radar systems which usually measure only a single fixed polarization, more information on the target surface is gathered by a polarimetric SAR. The information contained in the scattering matrix enables synthesis of every possible transmit and receive polarization by signal processing. Since the brightness of targets in a radar image depends on polarization, the contrast can be optimized, and targets may be classified or even identified.

The operating frequency of PHARUS is 5.3 GHz, the same as used by AMI on ERS-1 (the azimuth resolution of PHARUS will be approximately 1 m). This allows data comparison and additional aircraft data collection in projects demanding high temporal coverage of test sites.[539),540),541)]

| | |
|---|---|
| Radar type | Coherent pulse radar |
| RF frequency | 5.25 GHz (C-Band) |
| PRF | 3500 Hz |
| Waveform type | linear FM (no modulation) |
| Pulse length | 12.8 µs before pulse compression |
| Pulse compression ratio | 400 |
| Pulse bandwidth | 31 MHz (fixed) |
| Generation technique | I/Q memory read-out |
| DAC frequency | 87.5 MHz, 8 bit |
| Total memory capacity | 4096 bytes |
| Total peak transmit power | 160 W |
| Module peak transmit power | 20 W |
| Number of modules | 8 (transmit and receive) |
| Range | 3-14 km |
| Antenna | 8 elements antenna with 4 patches each uniform, no tapering |
| Polarization | vertical |
| Azimuth beamwidth | 9º |
| Azimuth presumming factor | 16 |
| Azimuth scan angle | -12º to +12º (1º step) |
| Elevation beamwidth | 24º |
| Elevation pointing angle | fixed |
| Mechanical elevation angle | 20, 30, and 40º (depression angle) |
| Resolution | 4.8 m in range |
| Sampling frequency in range | 87.5 MHz (A/D conversion) 4096 x 8 bit |
| Data storage rate | 8.2 Mbit/s |
| Airplane | Swaeringen Metro (NLR's laboratory aircraft) |
| Altitude | 3 - 6 km |
| Speed | 100 m/s |
| Position and motion registration | IRS, ARA |

**Table 195:    Parameters of the SAR Testbed PHARS**

539)  "Project PHARUS: Realization of a polarimetric C-Band airborne SAR", a TNO-FEL brochure

540)  P. Hoogeboom, P. Snoeij, P. J. Koomen, H. Pouwels, "The PHARUS Project, Results of the Definition Study Including the SAR Testbed PHARS", IEEE Transactions on Geoscience and Remote Sensing, Vol. 30, Nr. 4, July 1992

541)  P. Snoeij, P. Hoogeboom, P. J. Koomen, H. Pouwels, "A fully polarimetric airborne C-band SAR with an electronically steerable phased array, PHARUS", SEE & IEE Colloquium, SAR '93, ER93-391, pp. 48-52

| Mode | Resolu-tion (m) | Nr. of Looks | Altitude (km) | Noise equiv. $\gamma$ (dB) | Swath (km) | Range (km) | Incidence range ($^{\circ}$) |
|---|---|---|---|---|---|---|---|
| 1 polarization | 4 | 3 | 6.0 | -30 | 11.2 | 16 | 31-68 |
| 1 polarization | 8 | 8 | 6.0 | -30 | 14.6 | 20 | 31-73 |
| 1 polarization | 16 | 20 | 6.0 | -30 | 20.0 | 26 | 41-77 |
| 4 polarization | 4 | 4 | 4.5 | -40 | 4.4 | 8 | 26-56 |
| 4 polarization | 8 | 8 | 5.0 | -40 | 6.5 | 11 | 34-63 |
| 4 polarization | 16 | 20 | 6.0 | -40 | 7.9 | 13 | 31-62 |
| 2 polarization (ASAR) | 24 | 4 | 14.0 | -25 | 9.8 | 20 | 18-46 |

**Table 196:    Basic Pharus Modes**

### PHARS - Testbed/Prototype Phase of PHARUS

The objective of the PHARS testbed is to develop the technology for the PHARUS design. The first test flight of PHARS was performed on November 8, 1990. The PHARS records a swath width of 7 km up to a range of 13 km. The azimuth resolution is in the order of 6 m with 5 independent looks, or 1.2 m for single look. The aircraft motion is successfully compensated with off-line processing, using trajectory measurements from motion sensors both inside the aircraft and in the radar pod. The PHARS testbed results were introduced into the PHARUS design.

| | |
|---|---|
| Radar type | Coherent pulse radar |
| RF frequency | 5.3 GHz (C-Band) |
| PRF | 3500 Hz |
| Waveform type | linear FM (chirp) |
| Pulse length | 12.8 µs or 25.6 µs before pulse compression |
| Bandwidth | 40 MHz nominal, 100 MHz maximal |
| Total peak transmit power | 475 W |
| Range | 26 km |
| Antenna | 48 elements antenna with two rows of 24 patches |
| Polarization | transmit: H or V; receive: H and V |
| Azimuth beamwidth | 2.3$^{\circ}$ |
| Azimuth presumming factor | depends on the operating mode |
| Azimuth scan angle | -20$^{\circ}$ to + 20$^{\circ}$ (0.5$^{\circ}$ step) |
| Elevation beamwidth | 24$^{\circ}$ |
| Elevation pointing angle | fixed |
| Elevation scan angle | +15 to - 15$^{\circ}$ with respect to the pointing angle, 0.5$^{\circ}$ step |
| Resolution | 3.75 m in range |
| Sampling frequency in range | 100 MHz (A/D conversion) 16384 x 8 bit |
| Data storage rate | 100 Mbit/s |
| Airplane | Cessna Citation II |
| Altitude | 14 km max |
| Speed | 150 m/s |
| Position and motion registration | IRS, ARA, GPS |

**Table 197:    Nominal PHARUS Specifications**

## B.93    PMS (Particle Measuring Systems Inc.) Instruments

Commercially available meteorological instrument package of PMS of Boulder Co., USA. PMS has developed a complete line of aircraft-mountable instruments which allows real-time "in-situ" sizing of atmospheric particles (particle spectroscopy). The two-dimensional (2-D) probes provide image analysis, as well as size spectra. The 1-D probes provide only size spectra (see Table 198 for overview).[542]

All of the probes use a helium/neon laser as illumination source with light-scattering techniques used for sizing small particles down to 0.1 µm in diameter. Optical array imaging techniques are used for sizing large particles up to 12,400 µm in diameter.

---

542)  Several brochures were provided by PMS Inc. of Boulder Co.

Each optical array probe (OAP) uses a linear array of silicon photo-diodes as a sensor that is illuminated by a laser beam forming optics and mirrors. The beam is directed out from the probe enclosure through one hollow extension that has a mirror at its tip to reflect the beam across the sampling region of the probe to the tip of the other extension that also has a mirror to reflect the beam back into the enclosure and onto the array sensor.

According to PMS information (6/1993) over 350 instruments configured for aircraft operation have been sold worldwide. The FSSP-100 was delivered over 150 times, more than 160 OAPs are in operation (of these about 70 instruments are 1-D versions, 60 are 2-D versions, and the remainder are Grey Probes).

| | | Instrument Model Number | Sizing Range Diameter (μm) | Nr. of Size Channels | Resolution (μm) | |
|---|---|---|---|---|---|---|
| Light–Scattering Probes | | PCASP-100X | 0.10 - 3.00 | 15 plus oversize | 0.02 - 0.50 | One–Dimensional Probes |
| | | FSSP-300 | 0.3 - 20 | 31 | 0.05 - 1.00 | |
| | | FSSP-100 (Option A) | (R3) 0.5 - 8.0<br>(R2) 1.0 - 16<br>(R1) 2.0 - 32<br>(R0) 2.0 - 47 | 15<br>15<br>15<br>15 | 0.5<br>1.0<br>2.0<br>3.0 | |
| | | FSSP-100 (Option B) | (R3) 0.5 - 16<br>(R2) 1.0 - 32<br>(R1) 2.0 - 47<br>(R0) 5.0 - 95 | 15<br>15<br>15<br>15 | 1.0<br>2.0<br>3.0<br>6.0 | |
| Optical Array Probes | Cloud Droplet Probes | OAP-200X | 10 - 150 min<br>200 - 3000 max | 15<br>15 | 10 min<br>200 max | |
| | | OAP-230X | 10 - 300 min<br>200 - 6000 max | 30<br>30 | 10 min<br>200 max | |
| | | OAP-260X | 10 - 620 min<br>100 - 6200 max | 62<br>62 | 10 min<br>100 max | |
| | | OAP-2D2-C | 10 - 300 min<br>200 - 6000 max | 30<br>30 | 10 min<br>100 max | 2–D Probes |
| | | OAP-2D-GA2 Grey Probe | 10 - 620 min<br>100 - 6200 max | 62<br>62 | 10 min<br>200 max | |
| | Precipitation Probes | OAP-200Y | 300 - 4500 | 15 | 300 | 1–D Probes |
| | | OAP-230Y | 50 - 1500 min<br>200 - 6000 max | 30<br>30 | 50 min<br>200 max | |
| | | OAP-260Y | 50 - 3100 min<br>150 - 9300 max | 62<br>62 | 50 min<br>150 max | |
| | | OAP-2D2-P | 50 - 1500 min<br>200 - 6000 max | 30<br>30 | 50 min<br>200 max | 2–D Probes |
| | | OAP-2D-GB2 Grey Probe | 50 - 3100 min<br>150 - 9300 max | 62<br>62 | 50 min<br>150 max | |

**Table 198:    Overview of PMS Aircraft-Mountable Probes**

Sensors/Instruments:

**PCASP-100X (Passive Cavity Aerosol Spectrometer Probe)**

Measurement of the aerosol spectrum. The instrument selects 0.10 - 3.0 μm diameter particle sizing; resolution = 0.025-0.375 μm (progressively weighted). Owner/operator: NCAR Boulder Co. (mounted in Electra).

**FSSP-100 (Forward Scattering Spectrometer Probe)**

FSSP-100[543)] is an airborne instrument for particle size measurements with the following general fields of application: cloud physics, pre-cloud and haze measurements, general air pollution monitoring, stack plumes, cooling tower plumes, wind blown dusts, fuel plumes, traffic pollution assessment, etc. .

The FSSP instrument is an optical system using a laser beam in a high order multi-mode 5 mW He-Ne tube. The beam is focused to approximately 200 μm diameter using condensing optics with a 60 mm effective focal length. Particles passing through the laser beam in the sampling aperture scatter energy into the optics.

The FSSP has four overlapping size ranges with each size range divided into 15 linear size intervals providing up to 60 size channels in the 0.5 - 47 μm range. FSSP interfaces with a PMS data acquisition system.

- Option A:          0.50 - 47 μm diameter particle sizing
- Option B:          1.0 - 95 μm diameter particle sizing
- Size resolution:   One channel, 0.5 μm typical in most sensitive range
- Sampling:          Sample area: 0.3 mm$^2$, volume sampling rate: 30 cm$^3$s$^{-1}$ @ 100 ms$^{-1}$
- Environment:       -40 to +40°C; altitude: 0 - 12 km; humidity: 0 - 100% RH;
- Probe size:        101.6 cm length, 18 cm diameter
- System mass:       18.2 kg

Owners/operators of FSSP-100 instruments: DLR, Institute of Atmospheric Physics, Oberpfaffenhofen (mounted in Falcon); NCAR Boulder Co. (mounted in Electra), etc. ........

### OAP-230X (Optical Array Cloud Droplet Probe)

The instrument selects 10 - 200 μm per array element. 30 size channels. Note, the instrument designation with a n X suffix are designed for the 'cloud droplet configuration' with only 61 mm between the probe tips.

Owners/operators of instruments: DLR, Institute of Atmospheric Physics, Oberpfaffenhofen (mounted in Falcon); etc. ........

### OAP-200Y (Optical Array Precipitation Probe)

Instrument selects 300 μm per array element.; 15 size channels. The precipitation configuration is designated with a Y suffix, it measures larger particles and comes with a 261 mm distance between the probe tips.

### OAP-2D-GA2 (Greyscale Cloud Droplet Probe)

Instrument selects 10 - 100 μm per array element. The greyscale 2-D probes employ a 64-element array technology with 2 data bits stored for each element resulting in 128 bits per image size. The grey probes provide important depth-of-field information that is not available from standard 2-D probes (4 shadow levels for each element).

### OAP-2D-GB2 (Greyscale Precipitation Probe)

Instrument selects 50 - 150 μm per array element; 62 size channels.

---

543) "Forward Scattering Spectrometer Probe, Model FSSP-100", brochure of PMS, Boulder, Co. USA

## B.94   POLDER (Airborne Instrument)

POLDER (Polarization and Directionality of the Earth's Reflectances) is an airborne sensor of CNES/LOA of France (passive optical imaging radiometer). Objectives: validation of the spaceborne system concept of POLDER to be flown on ADEOS (see A.3); preparation and validation of operational algorithms for the spaceborne version; measurement of the polarization and directionality of the reflectances from natural surfaces and from the atmosphere.

POLDER makes sequential measurements at several wavelengths as well as at several polarizations. Each target area is measured at different wavelengths, polarizations, and view angles as the aircraft travels forward.

POLDER has been flown on a CNES balloon in 1989 and on the French research aircraft ARAT since 1990. Nominal flight altitudes are between 4 and 6 km. POLDER participated in campaigns over land surfaces in France, Germany, USA, UK, and Niger (HAPEX-SA-HEL), over the ocean (MEDIMAR), in the Antarctic (RACER), and in the clouds campaign ASTEX, EUCREX is scheduled for 1994.[544]

The airborne POLDER instrument design is very similar to that of the spaceborne version, consisting of three principal components: a CCD matrix detector, a rotating wheel carrying the polarizers and filters, and a wide FOV telecentric optics system. The main differences on the instrument itself are:

- The CCD matrix is 384 x 288 pixels in size on the airborne version, allowing a FOV of $\pm 51^\circ$ (along-track) x $\pm 43^\circ$ (across-track) - the spaceborne version has a CCD matrix of 274 x 242 effective elements.
- The filter wheel contains 10 filter slots (16 on the spaceborne version).
- The polarization filters are selected for each flight to suit the particular measurement objectives.

The spectral range of the instrument: 400 - 950 nm (same as spaceborne version), the spectral parameters are defined in Table 4. POLDER is calibrated in the lab with an integrating sphere.

POLDER observation principles: Owing to its two-dimensional characteristics, the CCD sensor matrix can observe any target within the instrument's swath under different viewing directions, allowing for the estimation of the target's bidirectional reflectance properties. The 10 channel sequence (10 filter slots) is repeated every 10 seconds, which corresponds to 4 rotations of the filter wheel. During this observation interval of 10 seconds, a given point on the Earth's surface remains within the instrument FOV, its bidirectional reflectance can be inferred. The data rate for a 10 second scene corresponds to: 384 x 288 x 10 x 2 bytes/10 seconds = 1 MByte/s. Note: the airborne pixels are not binned one by two, rather, the electronics encode each pixel value over two bytes. - As the aircraft passes over a target, radiance measurements are performed for each spectral band. These measurements provide a sampling of the target's BRDF (Bidirectional Reflectance and Distribution Function) and BPDF (Bidirectional Polarization Distribution Function).[545]

Measurement channels: 15 channels (3 channels for each polarized band, 1 channel for each unpolarized band). The ground spatial resolution is a linear function of altitude, at 5 km flight altitude the resolution is 35 m. The swath width is approximately twice the altitude dimension (with a cross-track FOV of $\pm 43^\circ$).

---

544)  Information provided by F. M. Bréon, Centre D'Etude de Saclay, Gif sur Yvette, France
545)  P. Y. Deschamps, F. M. Bréon, M. Herman, J. C. Buriez, J. L. Deuzé, A. Bricaud, M. Leroy, A. Podaire, G. Sèze, "The Polder Mission: Instrument Characteristics and Scientific Objectives", paper accepted by IEEE Transactions on Geoscience and Remote Sensing,

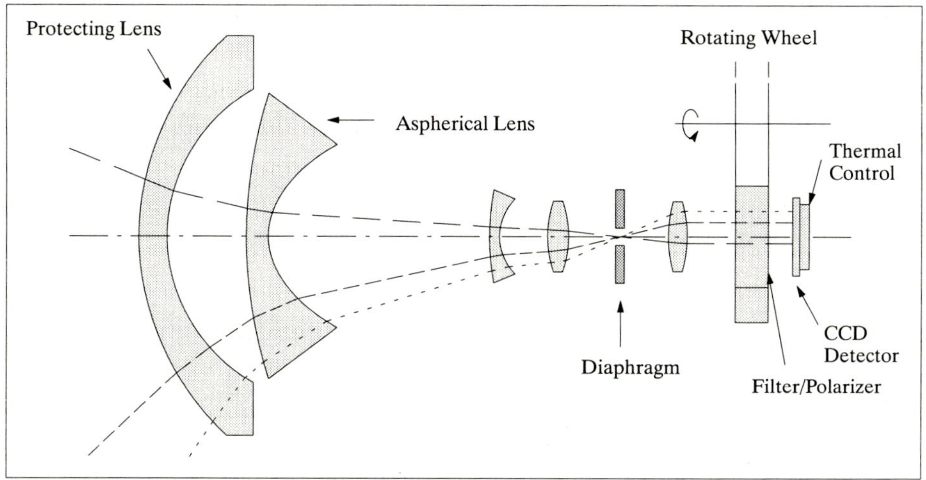


**Figure 135:    Optical Design Concept of the POLDER Instrument**

## B.95    PORTOS

PORTOS[546] is a French (CNES) passive microwave radiometer made by Matra Marconi Space. PORTOS has five, dual-polarized frequencies of 5, 10.7 23.8, 36.5, and 90 GHz. PORTOS became operational on the ground in 1991, in the airborne version in 1992. It is currently flown on the French ARAT (Fokker 27) aircraft at altitudes ranging from 300 m to 5000 m. PORTOS is a single-beam profiler (IFOV=FOV=11°). Since PORTOS is a profiler (i.e. measuring in the vertical plane), 'mapping' is performed by flying a series of parallel flight paths, separated by the length of a footprint (3 dB). The instrument resolution is linked to its mid-power beam of 11°. The incidence angle is variable from 0 to 50° in the along-track direction and from nadir toward the back of the plane.

The PORTOS range of frequencies corresponds to those of MIMR (Multifrequency Imaging Microwave Radiometer) to be flown on the spaceborne missions EOS-PM and on ME-TOP-1. Hence, PORTOS serves as a research tool for MIMR modeling with regard to instrument technology as well as with respect to algorithm development. PORTOS observation data allow the retrieval of the following parameters: soil moisture, biomass of vegetation, qualification of vegetation cover, seasonal vegetation cycles, surface roughness, atmospheric integrated water content, and surface-equivalent temperature.

PORTOS has been flown on a number of missions (including HAPEX-II Sahel) together with POLDER, also with other airborne instruments such as: PBMR, ASAS, NS001, TIMS on their respective aircraft platforms for reasons of observation data comparison and synergism. PORTOS measurements in conjunction with spaceborne missions were: DMSP (SSM/I), ERS-1 (ATSR/M, AMI, AMI-SCAT), NOAA (AVHRR), SPOT (HRV), Landsat (TM), etc.

Calibration methods employed with PORTOS:
- $N_{2L}$ and hot target on the ground
- use of water bodies in flight
- reference thermocouples and instrument model during acquisition

---

546)  Information provided by Y. H. Kerr of LERTS

| Frequency (GHz) | Polarization | Beam width at 3 dB | Radiometric Resolution | Bandwidth (MHz) | Efficiency at -20 dB |
|---|---|---|---|---|---|
| 4.9 - 5 | H, V | < 16° | 1 K | 100 | 70% |
| 10.6 - 10.7 | H, V | 13° | 0.9 K | 100 | 96% |
| 23.6 - 24 | H, V | 10° | 1.2 K | 400 | 97% |
| 36.3 - 36.7 | H, V | 11° | 1.1 K | 400 | 97.8% |
| 88.5 - 89.5 90.5 - 91.5 | H, V | 10.7° | 0.6 K | 2000 | 98% |

**Table 199:    Specification of PORTOS Observation Parameters**


## B.96    PRIRODA Airborne Instruments

The following PRIRODA/MIR sensors partake as airborne instruments (mostly with a reduced functional capability) aboard an Ilyshin-18 aircraft (see chapter A.82 for PRIRODA sensors) in a PRIRODA preparatory program, consisting of several campaigns. The program started in 1992. The nominal monitoring altitude is 6 km.

The Ilyshin-18 aircraft offers a range of speed from 330-700 km/h, a maximum altitude of 7 km, a maximum flight time of 6.5 hours, and a payload power source of 30 kW ( ± 27 V). The aircraft position is determined by the navigation system and by GPS.

- **SAR Travers** = Synthetic Aperture Radar (active microwave sensor). Objective: determination of reflection and scattering parameters of surface objects in the microwave region. SAR Travers operates to the left side of the aircraft (in flight direction).

    - Wavelength /frequency          10 cm / 3 GHz (S-Band)
    - Impulse frequency              3 kHz
    - Length of impulse              0.12 ns
    - Antenna size                   150 cm x 10 cm
    - Angle of incidence range       45-63°
    - Swath width                    20 km
    - Polarization                   VV
    - Spatial resolution             20 m

- **IKAR-D2** = Passive Microwave Radiometer (Delta), conical scanning 3-channel sensor. Objective: radiometric measurements of the Earth's surface. IKAR-D2 uses an antenna system with a rotating reflector for a conical scan with a beam look (scan) angle of 30° from nadir. The survey pattern is semi-cycloidal.

    - Wavelengths:          0.8 cm, 1.35 cm, and 2.25 cm
    - Spatial resolution    140 -370 m (depending on wavelength)
    - Swath width           6 km

- **IKAR-P** = Passive Trace Microwave Radiometers RP-225 and RP600. Antenna beam width of 9°; the look angle of the antenna beam from nadir is 40°; the radiometric resolution of RP-225 = 0.1 K and 0.15 K for RP-600.

    - Wavelengths:              2.25 cm and 6.0 cm respectively
    - Polarization              V and H
    - Beam diameter on ground   160 m

- **MKS-M** = Passive Multichannel Trace Radiometer (nadir pointing). The instrument provides 13 spectral bands with a band width of 15-20 nm. The pointing beam width of the telescope is 0.46°, providing a footprint of about 48 m x 48 m at 6 km flight altitude. The sampling rate is 31.25 Hz (this provides an over-sampling rate of about 6 at the flight speed of 110-125 m/s).

- **MSU-M** = Multispectral Scanner (electro-mechanical scanner, nadir-pointing).

- Wavelengths     0.5-0.6 µm, 0.6-0.7 µm, 0.7-0.8 µm, 0.8-1.1 µm
- FOV     90°
- IFOV     5.8' (spatial resolution = 18m)
- Scanning frequency     8 Hz
- Swath width     13 km

| Spectral Channel | Wavelength (nm) | Sensitivity $K_1$ (µW/cm nm sr) | Sensitivity $K_2$ (µW/cm nm sr) |
|---|---|---|---|
| 1 | 408.2 | 1.3392 | 6.4983 |
| 2 | 443 | 0.8858 | 4.2837 |
| 3 | 484.5 | 0.8986 | 4.1902 |
| 4 | 519.3 | 1.1401 | 5.1806 |
| 5 | 569.6 | 0.9555 | 4.3055 |
| 6 | 615.4 | 0.9070 | 4.1468 |
| 7 | 649.3 | 0.9632 | 4.0503 |
| 8 | 685.5 | 1.0142 | 4.7921 |
| 9 | 750 | 1.1437 | 5.3971 |
| 10 | 815 | 1.1715 | 5.5424 |
| 11 | 870.5 | 1.8492 | 8.8984 |
| 12 | 9.32.8 | 1.3909 | 7.4121 |
| 13 | 1026 | 1.5313 | 7.4559 |

**Table 200:    MKS-M Instrument Parameters**

- **ROSIS** (see B.101, scheduled for  October 1993 campaign)

- **AFA-41/20** = Aerial Foto Apparatus. The camera has an objective with a focal length of 200 mm.
  - Frame format     18 cm x 18 cm
  - Spatial resolution     47 lines per mm (at frame center)
  - Spatial resolution     10 lines per mm (at frame edge)

- **IRI** = Infrared Imager (IRE scanning instrument). Objective: surveying of oil spills on water surfaces and terrain subsurfaces fires (hot spots) and pipeline leaks. Spectral range = 8 - 14 µm; dynamic range = 100 K; swath width = 6 km; spatial resolution = 6-8 m; sensitivity = 0.05 K.

## B.97    Radius (Microwave Radiometer)

RADIUS is an airborne microwave radiometer owned and operated by NPO Vega of Moscow. The instrument was built by NPO Vega in 1986. RADIUS is utilized in agriculture, land reclamation, forestry, geology, oceanology, hydrometeorology, etc. Some of the measured parameters are:[547]

- Ground surface moisture content. Volume content of free water in the upper soil layer (20 - 100 cm thick)
- Biomass of the above-water part of vegetation (rice, reeds) 3-5 scale values
- Degree of mineralization and chemical pollution of interior water reservoirs with an accuracy of 1-3 g/l (depending on the concentration range).
- Water surface temperature variations in basins with an accuracy of 0.5-1.0°C.
- Depth of underground water level from 0 to 2.3 m (3-5 scale values).
- Thickness and packedness of ice on rivers and seas.

The scanning rate of the instrument can be adjusted for any speed or altitude of the aircraft. The instrument and science data are recorded on magnetic tape, a quicklook capability is provided. The instrument is offered for commercial use.

---

547) Information provided by Y. Krilov of NPO Vega, Moscow

| Frequency | Number and size of angle resolution cells in line | Survey Method |
|---|---|---|
| 15.2 GHz | 15 x 3.5° = 52.5° | Scanning |
| 5.475 GHz | 5 x 10° = 50° | Scanning |
| 1.425 GHz | 2 x 25° = 50° | Two commutated spatial beams |
| 0.700 GHz | 1 x 53° = 53° | One spatial beam |
| The swath width in each band is about equal to the flight altitude | | |
| Fluctuation sensitivity in channels | | 0.5 - 1.5 K |
| Power | | 800 W |
| DC Voltage | | 27 V |
| Total Instrument Mass | | 320 kg |

**Table 201:    Specification of the RADIUS Instrument**


## B.98    RAMS (RAdiation Measurement System)

A NASA-sponsored instrument being flown on since 1986 on a number of aircraft, such as: ER-2, DC-8, Electra, NCAR Electra, Sabreliner, NOAA P-3 and others. RAMS is an integrated system of several radiometers. The system provides airborne measurements to support analysis and theoretical calculations of cloud properties and radiation fields. RAMS data are usually compared and validated with satellite radiance measurements. RAMS consists of the following instruments:[548]

- BBHSR (Broad Bandpass Hemispheric Solar Radiometer) for total flux measurements. One BBHSR instrument is installed on top and one is installed on the bottom of the ER-2 fuselage. BBHSR features an electrically calibrated pyro-electric detector with optical chopping and null-balanced operation. Spectral region from 0.26 to 2.6 µm.

- BBHIR (Broad Bandpass Hemispheric Infrared Radiometer). The BBHIR system consists of two radiometers which flip for sequential up and down viewing (FOV: hemispheric nadir and zenith flipping). From the sequential measurements the IR net fluxes are found with very high accuracy. Spectral range from 5 to 40 µm.

- TDDR (Total Diffuse Direct Radiometer). The system consists of two instruments, one on the top and one on the bottom of the aircraft. TDDR is a multi-channel narrow spectral bandpass (5-10 nm) flux radiometer, narrow shadow arms sweep over the radiometer dome of the TDDR in orthogonal directions at regular intervals. This radiometer is used for optical depth determinations and direct/diffuse ratios (a measurement sequence permits the separation of the direct and diffuse components of the intensity field).

- NFOVR (Narrow FOV Radiometer). The instrument measures the IR irradiance in two spectral channels (5 to 40 µm region) in a 0.006 mrad FOV in the nadir direction. This radiometer uses a liquid nitrogen-cooled blackbody reference and provides upwelling infrared intensities above the clouds (high accuracy measurements of brightness temperature for cold cloud tops).

RAMS has been flown in the following projects/campaigns: STEP, FIRE, CEPEX, AGASP, Kuwait Oil Field Fires Experiment, Pinatubo Airborne Mission, AASE-II, TOGA/COARE, and others.

| Radiometer | Measurement | FOV |
|---|---|---|
| BBHSR | total flux | hemispheric nadir and zenith |
| BBHIR | IR flux | hemispheric nadir and zenith flipping |
| TDDR | 7 narrow band VIS/NIR filters | hemispheric with oscillating orthogonal shadow rings |
| NFOVR | 6.7 and 10.3 µm cryogenic BB | nadir viewing 15° FOV |

**Table 202:    Measurement Parameters of the RAMS Instrument**

---

548) "TOGA/COARE Mission Plan", NASA paper, November 1992, pp. 58-59

## B.99 RAMSES (Radar Aéroporté Multi-Spectral d'Etude des Signatures)

RAMSES is an experimental multi-band airborne radar of ONERA (French Aerospace Research Establishment, Fort de Palaiseau, Chatillon). The system is operational since 1990 and being flown on board a Transall C160 aircraft from the French Test Center (CEV), it has the capability of analyzing the effect of various instrument parameters such as carrier frequency, polarization or waveform.[549]

RAMSES is a high-resolution active radar (SAR) instrument operating in up to two simultaneous frequency bands, the choice is between L-, C-, X-, Ku- and W-Band. The system design is modular, providing a number of configuration options with regard to flight parameters (altitude, velocity, angle of incidence) and system (frequencies, polarization, waveform).

The aircraft (Transall C160) is equipped with INS (Inertial Navigation System), radio-altimeter, and GPS receiver. The maximum operating altitude is 3000 m; the ground speed for SAR operation is 75 m/s; the depression angle of the antennas can be varied within the range of 15° - 60°.

RAMSES is being used for civilian and military applications. Civilian uses include: remote sensing, wind-shear detection, and meteorological applications. Military uses are: air-to ground and air-to-air guidance, map-matching methods for navigation purposes, battlefield surveillance. RAMSES is also used as a research tool to investigate system-related problems with regard to system performance and data analysis.

| Frequency Band | L | S | C | X | Ku | Ka | W |
|---|---|---|---|---|---|---|---|
| Bandwidth (MHz) | 200 | | 300 | 300 | 300 | 600 | 500 |
| Power stage | SSA (Solid State Amplifier) SSA | | | TWT (Travelling Wave Tube) | | | EIA (Extended Inter-action Amplifier) |
| Transmit power (W) CW | 100 | | 20 | 200 | 200 | 100 | 100 |
| Antennas | array | array | horn | horn | horn | horn | horn |
| Site aperture (=elevation aperture) | 23° | | 13° | 15° | 15° | | 3°, 5°, 10°, 20° |
| Azimuth aperture | 16° | | 6.5° | 15° | 13° | | 3°, 5°, 10°, 20° |
| Transmit polarization (H and R only for pulse-to-pulse switching) | V or H | V or H | V | V or H L or R | V or H L or R | L or R | L or R |
| Receive Polarization | V, H | V, H | V | V, H L, R | V, H L, R | L, R | L, R |

**Table 203:** **System and Polarization Parameters of RAMSES**

Two waveforms are provided with the RAMSES instrument:

1. Stepped Frequency: - The frequency pilot of the radar is a fast-switching synthesizer; the frequency is switched pulse-to-pulse.

    - Synthesizer frequency range:   100 Hz - 4 GHz
    -           frequency step:    > 100 Hz
    -           switching time:    < 1μs

2. Digital Waveforms: - The frequency pilot is a digital generator including fast memory, high-speed digital-to-analog converter and low-pass filter. It enables any waveform such as linear-chirp, FMCW, non-linear modulations, etc. .

    - Digital generators:   maximum output clock   450 MHz
    -           memory depth    2 MByte
    -           resolution    8 bit

549) J. M. Boutry, D. Le Coz, "RAMSES: An Experimental Multi-Band Airborne Radar", Proceedings of the 'Speciality Meeting on Airborne Radars and Lidars', July 7-10, 1992, Toulouse France

**Data Acquisition:**
The recording system is capable to handle two frequency channels simultaneously (the recording system is the limiting factor to handle only two frequencies of the system).

- Digitizing of I and Q outputs under control of the system sampling clock (I and Q are the quadrature outputs of a balanced mixer; I = In-phase channel; Q = quadrature channel).
    - Maximum sampling rate     100 MHz
    - Number of bits     10 bit
    - Effective bits     6 bit

- Pre-filtering: Low-pass digital filtering by integration during a selectable time slot.
    - Number of range bits     1 to 8192
    - Number of integrated samples  1 to 64
    - Maximum input data rate     50 MHz
    - Input number of bits     10 bit
    - Output number of bits     16 bit

- Multiplexing of the radar, video, and navigation data. Storage on a digital rotating-head recorder.
    - Recorder model     AMPEX
    - Maximum data rate     13 MByte/s
    - Capacity     47 GByte

| Parameter | L-Band | C-Band | X-Band | Ku-Band | W-Band |
|---|---|---|---|---|---|
| RF center frequency (GHz) | 1.6 | 6.2 | 9.5 | 14.5 | 95 |
| IF center frequency (MHz) | | | 500 | | 500, 3500 |
| System bandwidth (MHz) | 200 | 300 | 300 | 300 | 500 |
| Digital chirp bandwidth (MHz) | | | 200 (max) | | |
| Antennas 3 dB beamwidth<br>azimuth (°)<br>elevation (°) | 16<br>23 | 6.5<br>13 | 15<br>15 | 13<br>13 | 3,5,10,20<br>3,5,10,20 |
| Transmit peak power (W) | 100 | 20 | 200 | 200 | 50 |
| Quantization (bit/sample,I&Q) | | | 6 (effective) | | |
| Sampling rate (MHz) | | | 100 (max) | | |
| Data rate on HDT (Mbit/s) | | | 107 (max) | | |
| Recording time/tape (min) | | | 60 (max) | | |
| Spatial resolution<br>range x azimuth (m) | | programmable, typical values:<br>2.5 x 2.5<br>5 x 5 | | | 0.5 x 0.5<br>1 x 1 |
| Radiometric resolution (8 looks) | | | < 1.5 dB | | |
| Geometric distortion | | | < 0.3% | | |

**Table 204:**   **Technical Specification of RAMSES**

Note: S-Band and Ka-Band operational capability of RAMSES is planned to be provided by the end of 1994.


## B.100   RMK (Reihenmeßkammer - Metric Camera)

RMK.[550] is a commercially available aerial survey camera series of Carl Zeiss, Oberkochen, Germany. Background: The first instrument series with the name of 'RMK' started in 1955, which was followed by the 'RMK A' series until the end of 1989. Over 800 'RMK A' series cameras have been sold worldwide. The current series, 'RMK Top', is being built since 1990. Each series represents a new generation in camera technology for a wide range

---

550) "Aerial Cameras and Accessories", and "RMK TOP - Survey Camera System for Aerial Photography", Carl Zeiss brochure

of photogrammetric applications. As of 8/1993 the RMK Top series has been sold into the following countries: Germany, United Kingdom, USA, Norway, Italy, France, Thailand, Indonesia, and India

DLR (Institute of Optoelectronics) operates 3 aerial survey cameras, type Zeiss RMK A (23 cm edge length image size) aboard a DO 228 aircraft with the primary objectives of forest mapping and aerial photography for topographic maps, etc. Cameras RMK A 8.5/23 and 15/23 are in operation since 1975, while RMK A 30/23 is in service since 1986.

The RMK modular system offers convenient interfaces for accessory and new component adaptation (identical for all cameras) such as: a suspension mount, an FK 24/120 film magazine, and an ICC Central Interval Computer. The system provides such features as:

- Automatic leveling control (HCON)
- Drift control (DCON)
- Automatic image sequencing and exposure mechanisms
- FMC (Forward Motion Compensation) up to 30 mm/s of forward motion is compensated without impairing the geometric quality of the camera.

| Designation | Camera Type | Lens Type Focal Length | Aperture (f-stops) | Angular field diagonal (lateral) | Max. nominal distortion | Principal uses/Comment |
|---|---|---|---|---|---|---|
| RMK A 8.5/23 | 125° super wide-angle | S-Pleogon A 85 mm | f/4, f/5.6, f/8 | 125° (107°) | 7 μm | Large-area photographic coverage for small-scale mapping |
| RMK A 15/23 | Standard wide-angle | Pleogon A 153 mm | f/4, f/5.6, f/8, f/11 | 93° (74°) | 2 μm | General work, i.e. aero-triangulation topographic and large-scale mapping |
| RMK A 30/23 | Standard normal-angle | Topar A 305 mm | f/5.6, f/8, f/11 | 56° (41°) | 3 μm | Aerial mosaics, orthophoto maps, first-order mapping and base maps for urban areas |
| RMK A 60/23 | Narrow-angle | Telikon A 610 mm | f/6.3, f/9, f/12.5 | 30° (21°) | 50 μm | Special purpose camera: a) high-altitude photography; b) city surveys; c) 1:250 or 1:500 scale flights; d) aerial mosaics |
| RMK Top 15 | Standard wide-angle | Pleogon A3 153 mm | f/4-f/22 | 93° | ± 3 μm | Successor to RMK A 15/23 |
| RMK Top 30 | Standard normal-angle | Topar A3 305 mm | f/5.6-f/22 | 56° | ± 3 μm | Successor to RMK A 30/23 |

**Table 205:**    **Specifications of the Zeiss Aerial Survey Camera Series RMK**

Note: due to the aircraft's motion, the image moves relative to the stationary film while the shutter is open. This forward motion causes object points to be represented as lines instead of points. so that the resolution is reduced. FMC counteracts this problem.

The RMK Top camera is in particular equipped for a functional combination with a GPS receiver subsystem. The shutter design features constant access time for mid-time of the exposure. The pulse emitted at the mid-point of exposure and, as a result, the instant of expose, are fed to the GPS with an accuracy of 0.1 ms. The software package T-Flight permits coordinate interpolation of the projection centers in relation to the position measurements, which are available at 1 second intervals. If the GPS serves also as a navigation system, the serial exposures can be triggered by GPS. GPS data can also be transferred onto the film exposure as auxiliary data.[551)]

For context information: the Metric Camera (German/ESA experiment on Spacelab-1 in Nov. 1983, see chapter A.102, was a slightly modified Zeiss camera of the type RMK A 30/23.

---

551) R. D. Becker, J. P. Barriere, "Airborne GPS for Photo Navigation and Photogrammetry (An Integrated Approach)", Paper Nr. 84 of 'Monitoring and Mapping Global Change', ASPRS/ACSM Convention, Washington, 1992

## B.101   ROSIS (Reflective Optics System Imaging Spectrometer)

ROSIS[552] is a compact airborne imaging spectrometer developed jointly by DASA(MBB), GKSS and DLR based on an original design by MBB for a flight on ESA's EURECA platform. The instrument has been modified slightly after some test flights in 1992, it is available again from the summer 1993. After completion of the first development phase, ROSIS belongs to both, GKSS (Institute of Physics) and DLR (Institute of Optoelectronics). Both organizations will test, maintain and operate the instrument and process the data. After the initial test phase of the modified instrument, it is planned to make the sensor available to a large user community inside and outside of Germany within a suitable environmental utilization program.

ROSIS will be operated onboard of a DLR Falcon Jet (max. flight altitude is $\approx 10$ km). Later flights on other carriers are foreseen as well.

The design driver of ROSIS was its application for the detection of sun-stimulated fluorescence of chlorophyll in coastal zones. This task determined the selection of the spectral range, bandwidth, number of channels, radiometric resolution and its tilt capability for sun glint avoidance. (Another spaceborne version of the instrument was designed according to MERIS requirements until the end of Phase B Study for the German ATMOS satellite).

The airborne ROSIS sensor can also be used for monitoring of features above land or within the atmosphere. The instrument can be operated either in the 'imaging mode' (selection of a maximum of 32 channels from a total of 84), or in a 'spectral made' (recording of all 84 spectral channels with a reduced pixel size). During data collection the gain, the tilt, the scan frequency and the channels to be recorded can be selected by the operator to match the requirements of the specific application. Data are recorded on a VLDS streamer tape. The entire system is controlled by means of a VME-bus computer. The specifications of the modified instrument are as follows:

- FOV                                  $= \pm 8^{\circ}$
- IFOV                                 $= 0.56$ mrad
- Max. scan frequency                  $= 60$
- Number of detector elements          $= 500$ (in each spectral channel)
- Spectral range                       $= 430 - 830$ nm
- Spectral sampling interval:
  - range: 430 - 550 nm                $= 12$ nm
  - range: 554 - 830 nm                $= 4$ nm
- Nr. of selectable channels           $= 84$
- Max. Nr. of recorded channels        $= 32$
- Radiometric encoding                 $= 12$ bit

ROSIS is calibrated spectrally during flight by means of a mercury lamp. After each flight it is calibrated radiometrically by means of a specially designed calibration sphere in the DLR laboratory in such a way, that the selected parameters (including the operational temperature) are reproduced. Based on this calibration, also dark currents and residual charges within the detector are corrected for, and a correction matrix is then produced for each detector element.

During flight navigation and attitude data from the aircraft are inserted into each frame of the image data. These data are used for the correction of aircraft instabilities, in particular roll motions. In addition ROSIS will be mounted on a stabilized platform in 1994.

After a flight campaign the data from the VLDS streamer tapes are processed (with radiometric and geometric corrections) either with the image processing system X-DIBIAS at DLR, or with GRIPS (GKSS Remote Sensing Image Processing System) at GKSS, into roll-corrected and calibrated radiance values for all detector elements in each selected channel.

---

552)  A handout was provided by H. van der Piepen of DLR

The following future developments of the present instruments are planned subject to available financial resources.

- Modification of the focal plane and the processing electronics so as to make use of the full field of view ($\pm 16°$) for which the instrument has been designed.
- Expansion of the instrument bandwidth into the SWIR spectral range (up to 2.4 μm).

## B.102   ROWS (Radar Ocean Wave Spectrometer)

ROWS is an active airborne sensor developed and operated by NASA/GSFC with the objective to measure the ocean's directional gravity-wave spectrum. ROWS is operational since 1978 and has been installed on several aircraft, the most recent aircraft were WFF's T-39 and P-3.[553],[554]

ROWS is a high-resolution radar with two distinct continuously-operating modes, namely the spectrometer mode and the pulse-limited altimeter mode.

- The spectrometer mode data are generated with a near-nadir-pointing, pencil-beam antenna which rotates in azimuth. The data are used to derive two-dimensional ocean wave spectral estimates and directional radar backscatter information.
- The pulse-limited altimeter mode radar returns are derived using a vertically-pointed horn antenna. Ocean significant wave height and surface wind speed are inferred from the altimeter return.

Applications: Measurements of the directional ocean gravity-wave spectrum, ocean wave height, and surface wind speed/stress. Recent field measurements with ROWS participation are:

- High Resolution Remote Sensing Program in 1993
- Surface Wave Dynamics Experiment (SWADE) in 1991
- Grand Banks ERS-1 SAR Wave-Mode Validation Experiment in 1991
- Labrador Extreme Wave Experiment (LEWEX) in 1987

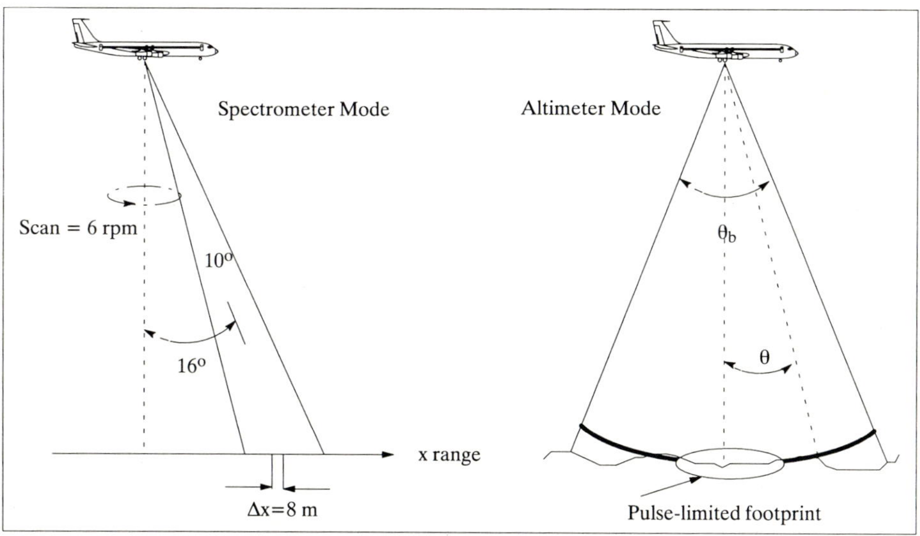


**Figure 136:   Illustration of ROWS Dual-Mode Measurements**

553) Information provided by D. Vandemark of NASA/GSFC/WFF
554) F. C. Jackson, W. T. Walton, P. L. Baker, "Aircraft and Satellite Measurements of Ocean Wave Directional Spectra using Scanning-Beam Radars", Journal of Geophysical Research, Vol. 90, Nr. C1, January 1985, pp. 987-1004

| Parameter | Value |
|---|---|
| Frequency | 13.9 GHz |
| Pulse type | linear FM, 100 MHz bandwidth, 1.2 µs chirp |
| Pulse length | 12.5 ns compressed |
| Peak power | 2 kW |
| Pulse Repetition Frequency (PRF) | 100 Hz |
| Dynamic range | 70 dB |
| Detection | noncoherent, square law |
| Antennas | a) 10 x 4° (elevation x azimuth) printed circuit array, vertical polarization, 16° boresight, 6 rpm rotation rate<br>b) vertically-pointing 29° pyramidal horn |
| Data system | PC-based with full waveform capture using a 100 or 500 MHz digitization rate |

**Table 206:   ROWS Instrument Characteristics**


## B.103   SFSI (SWIR Full Spectrographic Imager)

The SFSI instrument is a development by CCRS of Canada (in cooperation with the National Research Council of Canada) with the focus on the SWIR (ShortWave Infrared) region of the electromagnetic spectrum. First test flights with SFSI are planned for the beginning of 1994.[555]

| Detector | |
|---|---|
| Orientation in focal plane | V = spectral dimension; H = spatial dimension |
| Number of elements | V = 488 (244 after on-chip summing), H = 512 |
| Center spacing | V = 25 µm (50 µm after on-chip summing), H=31.5 µm |
| Cell area A | $7.88 \times 10^{-6} \ cm^2$ |
| Fill factor | 0.367 |
| Full well capacity | $5 \times 10^5$ el |
| Array size | V = 12.2 mm, H = 16.1 mm |
| CCD transfer mode | Interline |
| Quantization (digitization) | 12 bit |
| **Optics** | |
| Effective focal length | 78.75 mm |
| Pixel angular IFOV | 0.4 mrad |
| Total FOV | 11.7° |
| Pitch-scan range | ± 50° (maximum) |
| Focal ratio | f/2 |
| Mean linear dispersion | $10.17 \ \mu m \ nm^{-1}$ |
| Element band width | 2.46 nm |
| Slit-image-to-detector ratio | 1.26 |
| Number of detector rows summed | 4 |
| Effective band width | 9.84 nm |
| **Data Recording (Burst mode)** | |
| Image frame size | 512 lines x 512 pixels |
| Number of spectral bands | 122 |
| CCD frame rate / image line rate | 50 Hz |
| Pixel rate (spectral x spatial) | 6.25 MHz (after on-chip summing), 3.12 MHz after processor summing (post digitization) |
| Selectively recorded dynamic range | 8 bit |
| Data rate into VRAM | 25.0 Mbit/s |
| Buffer size (VRAM) | 32 MByte |
| Data rate onto MO (magneto-optical) disk | 335 kByte/s |

**Table 207:   SFSI System Parameter Specification**

SFSI covers a wavelength range from 1.2 µm to 2.4 µm in 122 contiguous 0.01 µm wide spectral bands. A 2-D detector array technology is used providing a full image cube with an area coverage of 512 x 512 pixels. The special feature of this sensor is its ability to achieve a spa-

---

555)  R. A. Neville, I. Powell, "Design of SFSI: An Imaging Spectrometer in the SWIR", Canadian Journal of Remote Sensing, Vol. 18, Nr. 4, October 1992, pp. 210-222

tial ground resolution of 4 m to a minimum of 0.5 m using the pitch-scan technique proposed for HIRIS.

SFSI employs a Loral-Fairchild CAM6003 camera, which has a 488 x 512 PtSi Schottky barrier CCD array. The control electronics, data processing, and data recording system have been developed to store 'image cubes' (see Figure 144) consisting of 512 lines by 512 pixels by 122 spectral bands by 8 bits in RAM (they are subsequently downloaded onto MO disk). The slowest operation in the chain, namely the MO disk recording rates, limit the operation to the acquisition of discrete image cubes in the flight path. A future upgrade to a high data rate tape recorder (helical scan type) will permit continuous image acquisition and recording at full 12-bit digitizer resolution at a data rate of 75 Mbit/s.

SFSI will be able to measure absorption band feature depths, relative to out-of-band radiance levels (i.e. to background), from 5% to 10% in the 1.20 - 1.32 $\mu$m and 1.50 - 1.79 $\mu$m windows for most target types (trees and soils).

## B.104   SILVACAM (Real-time False Color CCD Video Camera)

SILVACAM is a commercially available instrument of Karelsilva Oy of Finland for environmental monitoring (operational since 1992). SILVACAM produces in real-time Color InfraRed (CIR) video images. The spectral ranges for SILVACAM are green (490-580 nm), red (580-680 nm), and NIR (760-900 nm). The "false color" technique converts the invisible NIR spectral band into a false (red) color, indicating the state of vegetation health (healthy vegetation has a strong reflection in NIR). Applications of SILVACAM in forestry, agriculture, geology, surveying.[556),557),558)]

| | |
|---|---|
| Image pickup device | 1/2 inch interline CCD array (3 spectral bands: green, red, NIR) |
| Effective number of pixels | 683 (horizontal) x 582 (vertical) in PAL standard, or 670 (horizontal) x 492 (vertical) in NSTC standard |
| Electronic shutter speeds | 1/60 (normal), 1/250, 1/500, 1/1000 (switchable) |
| Color system | PAL (r-Y, B-Y method encoder), MII, VHS, S-VHS, RGB |
| Synchronization | Internal (built-in SPG), or external (composite video or black burst signal) |
| Lens mount | 1/2" bayonet |
| Sensitivity | f/4, 2.000 lux (+18 dB) |
| SNR (standard) | 55 dB typical (contour correction off, gamma 1, bandwidth 5 MHz, Matrix off, chroma off) |
| Horizontal resolution | 500 TV lines |
| Video signal output | |
| 26 pin connector | Composite video signal (VBS) 1 Vp-p, and separate Y/C signals (compatible with S-VHS) or component signal (Y/R-Y/B-Y for MII or R/G/B...0.7 V-p 75 Ohm) |
| 7 pin connector | Separate Y/C signals (in Y/C 433 mode only) compatible with super VHS |
| Test output terminal | Composite video signal (VBS) 1 Vp-p (any one of IR, R or G signals can be selected using the internal select switch |
| System Parameters | |
| Voltage | 12 V DC (10.5 - 15 V) |
| Current | < 1.5 A |
| Mass of SILVACAM | 2.7 kg |
| Camera Mount Configurations | a)   Normal down-looking (FOV = 60º) b)   Forward-looking c)   Across-track scanning |

**Table 208:**   **Technical Parameters of the SILVACAM Video Camera**

556)   B. Braam, "CIR Video: an operational tool for environmental monitoring", Proc. International Symposium 'Operationalization of Remote Sensing', Vol. 2, April 19-23 1993, Enschede, The Netherlands, pp. 191-204
557)   Brochures provided by Karelsilva Oy, Finland
558)   M. Rantasuo, B. Braam, "Color Infrared Airborne Video System Development in Finland", Surveying Science in Finland, Vol 10, Nr. 2, 1992, pp. 59-69

The camera is also provided with a GPS receiver for geo-referencing of the video data (the GPS signal along with UTC is superimposed on the video signal). The video data may be stored on different storage media (S-VHS video cassette recorder is normal, MII, etc.). The video data can be integrated in digital or analog format with other data using GIS techniques.

Data: Ampex FR 3030 high-density digital recorder for data playback, and a quicklook facility (Honeywell 1856 A Visicorder with a 1226 film processor) are available.


## B.105   SLAR (Side-Looking Airborne Radar, NLR)

| Parameter | | Specification |
|---|---|---|
| Frequency | | 9.4 GHz (X-Band), HH-polarized |
| Transceiver | transmitted power | 25 kW |
| | pulse length | 50 ns |
| | pulse repetition frequency | 200 Hz |
| Dynamic range (special amplifier) | | 80 dB |
| Digitalization | video bandwidth (after filter) | dc - 10 MHz |
| | sampling frequency | 50 MHz (20 ns) |
| | samples per line | 4096 |
| | bits per sample | 8 |
| Pixel size (final) | | 7.5 x 7.5 m / 15 x 15 m |
| Range | | max 12 km |
| Resolution | range (across track) | 7.5 m |
| | azimuth (along track) | 10 mrad (two-way) |
| | radiometric | 0.3 dB |
| Flight altitude | | 100 - 6000 m |
| Ground speed | | 90 m/s |

**Table 209:    SLAR Specification**

SLAR[559] is used as a research tool in the area of surveying and monitoring of land and sea. The Dutch digital SLAR is a joint development of the Physics and Electronics Laboratory (TNO) of Delft University of Technology, and NLR. The instrument is installed in the NLR Metro II laboratory aircraft.

In order to enable multi-temporal registration of data, a set of algorithms has been implemented on an NLR computer to geometrically correct acquired SLAR data for variations in aircraft attitude and motion through simultaneously recorded inertial reference parameters. Furthermore, radiometric corrections for system parameters are performed to calculate radar backscatter coefficients.


## B.106   SMIFTS (Spatially Modulated Imaging FTS)

SMIFTS is a cryogenically cooled, spatially modulated imaging Fourier transform interferometer spectrometer for spectral measurements in the 1 - 5 µm range. The instrument was designed and built at the Department of Geology and Geophysics of the University of Hawaii, Honolulu  - sponsored by the Office of Naval Research (ONR) and by DARPA. First successful helicopter test flights with a prototype SMIFTS took place in the summer of 1993. - The new technology provided by SMIFTS invites for a more detailed discussion of some imaging issues and of comparisons with other concepts.[560]

The SMIFTS technology encompasses a combination of characteristics that are not available from other spectral measurement technologies (see Table 210). These are as follows:

---

559)  "SLAR - Side-Looking Airborne Radar", a NLR brochure
560)  P. G. Lucey, T. Williams, K Horton, K. Hinck, C. Budney, "SMIFTS: A Cryogenically Cooled, Spatially Modulated Imaging Fourier Transform Spectrometer for Remote Sensing Applications", SPIE Proceedings, Volume 1937, 14-15 April, 1993, Orlando, Fl.

- Broad, detector-limited wavelength range
- Wide field of view (FOV), the spectral characteristics are independent of the size or geometry of the input aperture at the focal plane of the input optics.
- Simultaneous measurement of all spectral channels
- Compact instrument design
- Moderate spectral resolution ($\lambda/\Delta\lambda$ = 100 to 1000)
- One dimension of imaging
- No moving parts of the instrument

| System Technology | Resolution $\lambda/\Delta\lambda$ | Wavelength range | Moving Parts | Simultaneous acquisition of all spectral channels | Throughput | Remarks |
|---|---|---|---|---|---|---|
| SMIFTS | $10^2$ - $10^3$ | Broad (detector-limited) | no | yes | very high | slit & FOV independent |
| Grating | $10^2$ - $10^5$ | Narrow (optics-limited) | no | yes | low | narrow slit required |
| Prism | $10^2$ - $10^3$ | Narrow (optics-limited) | no | yes | low | |
| Michelson Interferometer | 10 - $10^5$ | Broad (detector-limited) | yes | no | very high | sequential acquisition |
| Filter (electronically tunable) | $10^2$ | Narrow (optics-limited) | no | no | very high | sequential acquisition |
| Filter (mechanical) | 10 - $10^3$ | Broad (detector-limited) | yes | no | very high | sequential acquisition |
| Filter (mask) | $10^2$ | Narrow (optics-limited) | no | yes | very high | |
| Filter (mask) | $10^2$ | Broad (detector-limited) | no | no | very high | sequential acquisition |

**Table 210:  Overview of some Hyperspectral Sensor Technology Characteristics**

| Spectral Parameters | | Spatial Parameters | |
|---|---|---|---|
| Spectral range | 1.0 - 5.2 µm | IFOV | 0.66 mrad |
| Spectral bandwidth | | FOV | 9.7° |
| Nr. of bands | 100 | Pixels/line | 256 |
| Electronic Parameters | | Swath width | 330 m, 2000 m |
| Quantization | 12 bit | | |
| SNR | 1000 - 2000 | Platform | Helicopter |
| Data rate | 8 Mbit/s | Altitude | 2000 m |

**Table 211:  Specification of some SMIFTS Parameters**

The SMIFTS instrument contains three major optical subsystems: a Sagnac interferometer, which produces the spatially modulated interferogram; a Fourier transform lens, which frees the spectral properties of dependence on aperture geometry and allows the wide FOV; and a cylindrical lens, which re-images one axis of the input aperture onto the detector array providing the one dimension of imaging.

The Sagnac interferometer is known as a triangle path or common path interferometer, because the two beams emerging from the beamsplitter, one in reflectance and one in transmission, follow the same path in opposite directions. These two beams, observed beyond the interferometer, form two images of an input source. For an interferometer with perfect symmetry, the two images of the sources are coincident and no interference effects are observed, i.e. there exists perfect constructive interference. If either mirror $M_1$ or $M_2$ is displaced, then the two beams traverse different paths, and the two images of the source are displaced orthogonally to the line of sight in opposite directions. As these two images are of the same source, they are mutually coherent and interference occurs. The detector array is used to sample the resultant interference pattern.

**Broad Wavelength range**. - SMIFTS relies upon the Fourier transformation of the interference pattern produced by 2 mutually coherent source pairs to obtain the spectrum of the source. The range of the spectrum is limited principally by the spectral response of the detector array employed, with second order effects due to transmission and reflection characteristics of the optics. An SMIFTS spectrometer can, with a single detector, cover the entire sensitivity range of that detector. Grating spectrometers covering a range of several μm, suffer significant losses due to methods required to overcome the phenomenon of overlapping orders (e.g. multiply blazed gratings), or require multiple detector arrays or spectrometers.

**Wide Field of View**. - The spectral characteristics of SMIFTS are independent of the size or geometry of the input aperture at the focal plane of the input optics. As an interferometer, the instrument is not constrained by the resolution-luminosity product. Hence, continuous monitoring of large FOVs are possible while preserving spectral resolution and range. A grating spectrometer requires a narrow input slit to preserve spectral resolution, so that large FOVs cannot be continuously monitored, but must be scanned. The input aperture of the SMIFTS can be varied arbitrarily if desired, for example to preserve spatial resolution at a variety of ranges or the increase SNRs at the expense of spatial resolution.

**Simultaneous Measurement of all spectral channels**. - This characteristic is shared by imaging spectrometers based on diffraction gratings or prisms. The importance of this characteristic is that it ensures that each spectral measurement and each spectral channel has observed the target over the same time interval. Spectrometers based upon Michelson interferometers or tunable filters or wedge filters observe targets sequentially. If the object/target varies over the timescale of the data acquisition, intrinsically or due to tracking errors, then sequential spectrometers suffer from the effect of "jitter noise". Simultaneous acquisition ensures that the entire spectrum is a true time average of the target over the integration period.

**Moderate spectral resolution.** - The resolution of interferometer spectrometers is proportional to the number of samples taken of the interferogram. SMIFTS uses a CCD detector array to sample the interferogram. Hence, the spectral resolution is a function of the number of detector elements along the interferogram axis of the array. By measuring a single-sided interferogram, the spectral resolution in inverse centimeters ($\Delta\sigma$) can be defined as : $\Delta\sigma = \sigma_c / (0.8 \times N)$,
where $\sigma_c$ is the high frequency cutoff, and N is the number of pixels along the interferogram axis of the array. The '0.8' reflects that a small portion of the interferogram must also be measured on both sides of the centerburst to preserve phase information. Modern IR arrays range in size from roughly 100 to 1000 pixels on the largest axis, so that the spectral resolution attainable by a SMIFTS in terms of $R = \sigma/\Delta\sigma$ is also roughly 100 to 1000. CCDs in the VIS range are available in much larger sizes so that a resolution in the order of 2000 - 4000 is possible in the range of silicon CCDs.

**One Dimension of Imaging**. - On the 2-D detector array utilized by a SMIFTS, one axis contains the spectral information, the other axis contains the spatial information along one axis of the source. If a slit aperture is used with the SMIFTS, then the information present on the array is directly analogous to that collected by a grating imaging spectrometer. Each row of the array contains the spectrum of the corresponding point along the slit aperture. Unlike the grating spectrometer, the slit of the SMIFTS can be made arbitrarily wide, so that the area from which a row derives the spectrum can be roughly square, or extremely elongated, depending on the needs of the application. In wide field applications, each row of the detector array contains the spectrum of a long narrow rectangle, with a width corresponding to the width of the detector element as mapped on the focal plane, and arbitrary length. One spatial axis is then retained even for wide field monitoring.

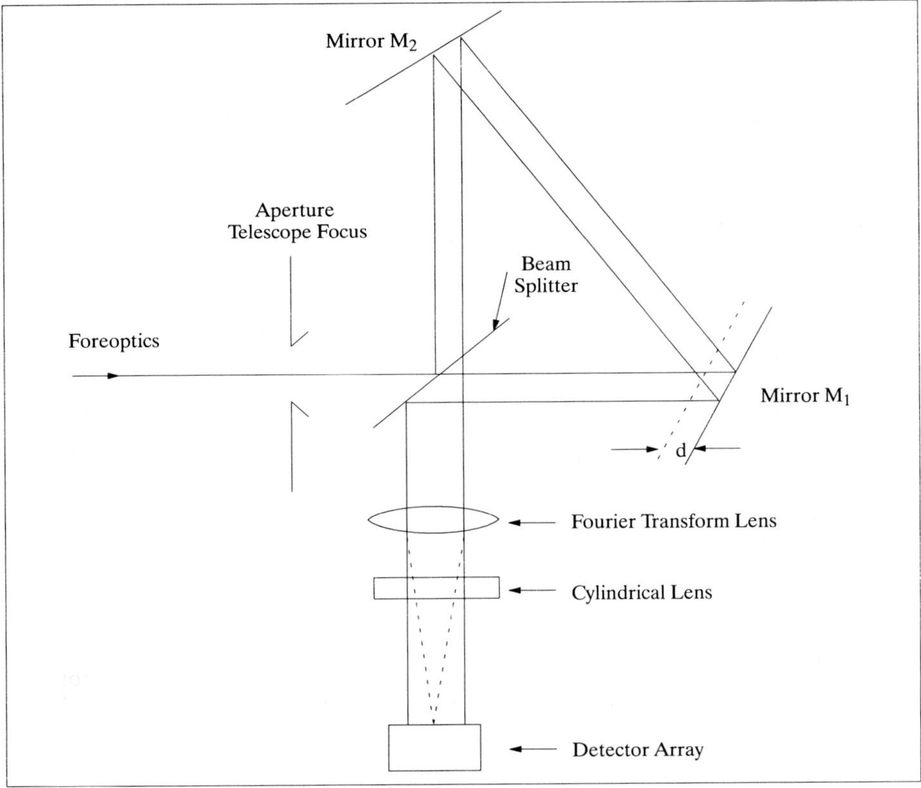


**Figure 137:    Concept of a Spatially Modulated Imaging Interferometer FTS**

The spectrum is recovered from the interferogram via the inverse Fourier transform of the interferograms of such sources. The proven Fourier transform spectroscopy, utilizing a Michelson interferometer, serves as a basis to obtain the sampled interference pattern from the input source. All of the principles which apply to Michelson interferometers, also apply to the spatially-modulated interferometer, as described here; no development in the the basic derivation of the spectrum from the interferogram is required.

Light from a foreoptics is focussed on the aperture. The Sagnac interferometer divides the input image into two images separated laterally with respect to the optic axis. These images are coherent and so produce an interference pattern. The Fourier transform lens removes dependence of the interference pattern on the detector array from any dependence on the position or shape of the input aperture which defines the IFOV. The cylindrical lens re-images the input aperture along the dimension normal to the page onto the detector, which causes each row of the detector element to contain the spectrum (after transformation) of the corresponding point on the input slit. Unlike a dispersive spectrometer, the slit can be made arbitrary in width without affecting the resolution of the spectrum or the wavelength calibration.

SMIFTS is a cryogenically cooled instrument utilizing a 128 x 128 element InSb detector array covering the spectral region from 1 to 5 $\mu$m. This detector configuration provides a 10000 wavenumber (1 $\mu$m) cutoff frequency, and therefore a spectral resolution of about 100 cm$^{-1}$. An upgraded SMIFTS detector version of 256 x 256 elements was built doubling the spectral resolution.

## B.107   SSTR (Sea Surface Temperature Radiometer)

SSTR is a NASA/JPL precision infrared multichannel filter wheel radiometer for sea surface temperature measurements. The radiometer is a basic research instrument, designed to improve understanding of the way, in which water vapor absorbs outgoing thermal infrared radiation from the sea surface, and designed to improve techniques to correct for water vapor absorption effects in satellite-based sea surface temperature retrievals. Since 1986, the instrument has been flown on manned blimps, tethered aerostats, and research turbo-prop aircraft. The radiometer was flown most recently in the TOGA/COARE campaign on the NCAR Electra aircraft, and made extensive observations of the sea surface temperature structure in the tropical Pacific Ocean.[561]

The instrument operates in a nadir pointing position, it has a precision of 0.002 °C at 300 K for an electrical bandwidth of 0.1 Hz.

| Electrical Bandwiths | 0.1, 1.0, 10.0 Hz |
|---|---|
| Chopping frequency | 100 Hz |
| Telescope | Cassegrain |
| Collecting aperture | 20 cm |
| FOV | 1 mrad |
| Optical filters | 6 - contiguous between 825 cm$^{-1}$ and 1005 cm$^{-1}$ (9.95 - 12.12 μm) |
| Detector | HgCdTe (LN$_2$ - cooled) |
| SNR | 0.006 K (@ 1 Hz, 940 cm$^{-1}$, 300 K) |

**Table 212:    Specification of the SSTR Instrument**

## B.108   STAR (<u>S</u>ea-Ice and <u>T</u>errain <u>A</u>ssessment <u>R</u>adar)

Star-1 and Star-2 are X-Band SAR sensors of INTERA Information Technologies Ltd., Calgary, Canada.[562],[563],[564]

Background: Star-1 was developed by ERIM (Environmental Research Institute of Michigan) and INTERA. ERIM designed and built the instrument, INTERA provided the aircraft , the real-time SAR processor, peripheral equipment, the integration of all instruments, and is operating the sensor in the field. The program started in 1982, the sensor became operational for commercial applications at the end of 1983 (some limited amount of research is carried out for INTERA's own in-house activities). The Star imaging SAR system was designed for, and first applied to the monitoring of ice conditions in Arctic waters. The radar imagery provided essential information for the petroleum drilling operations in that region. This first system has since been modified and another one similar to it has been built. As of 1993 INTERA owns and operates 2 airborne radar systems: the Star-1 and Star-2.

Applications:

- Terrain mapping (agricultural, forest cover and land use mapping; geologic exploration, cartographic mapping, etc.)
- Off-shore drilling support in ice-covered waters
- Marine transportation support in ice-covered waters
- Data source for sea ice forecast models
- Oceanographic applications (ocean currents and fronts, wave directional spectrum, shallow water anomalies)

561)  Information provided by D. Hagan of NASA/JPL

562)  M. D. Thompson, et al., "A decade of commercial radar operations: INTERA's Star-1 and Star-2 services", presented at the 1$^{st}$ Thematic International Symposium on Operationalization of Remote Sensing, Enschede, The Netherlands, 19-23 April, 1993

563)  J. B. Mercer, "A new airborne SAR for ice reconnaissance operations", Proceedings of the IGARSS '89 Symposium, Vancouver B. C. 10-14 July, 1989

564)  Handout provided by R. Lowry and L. Lalonde of INTERA

| Parameter | Star-1 Instrument | Star-2 Instrument |
| --- | --- | --- |
| RF center frequency (X-Band) | 9.375 GHz | 9.375 GHz |
| Transmitted pulse length | 30 µs | 30 µs |
| PRF (Pulse Repetition Frequency), scaled to ground speed | 960 Hz (1200 Hz max) | 666 Hz max @ 450 knots (230 m/s) |
| Range pixel size (4096 pixels) <br> Azimuth pixel size | 6 or 12 m <br> 4.2 m | 15 or 25 m <br> 15 or 25 m |
| Swath width | 23 or 46 km | 60 or 100 km |
| Speckle reduction | 7 independent looks | > 20 looks |
| Maximum far range | 70 km | |
| Incidence angle mid-swath | 73 or 78° | 79 or 82° |
| Synthetic aperture length (mid-swath) | 480 or 640 m | 900 or 1200 m |
| Duty factor (percentage of time in which the radar is transmitting the microwave energy) | 4% | 2% |
| Peak transmitted power | 2 kW | 6 kW |
| Receiver noise | 3 dB | 3 dB |
| Antenna gain | 30 dB | 32 dB |
| Scene dynamic range | > 40 dB | > 40 dB |
| Minimum detectable signal | < -30 dB $\sigma^o$ noise equiv. at 70 km max. range | < -30 dB $\sigma^o$ noise equiv. at 120 km max. range |
| Geometric distortion | with GPS < 3 pixels plus terrain height effects | with GPS < 3 pixels plus terrain height effects |
| Slant to ground range conversion | selectable on or off | selectable on or off |
| Peak range side lobe <br> Peak azimuth side lobe | - 20 dB <br> - 25 dB | - 20 dB <br> - 25 dB |
| Range to near side (typical) <br> Range to far side (typical) | 15 to 20 km <br> 40 to 70 km | 20 km <br> 84 to 125 km |
| Image formation time | real-time (10-20 s delay) | real-time (10-20 s delay) |
| System calibration repeatability | 1.2 dB | 1.2 dB |
| System mass | 450 kg | 1130 kg (2 radars) |
| Power requirements | 6 kVA (28V) | 12 kVA (28V), 2 radars |
| Bandwidth of recorded signal | ground speed dependent, max. 250 kByte/s | max. 2 x 250 kByte/s |
| Antenna size | 1.2 x 1.5 x 0.6 m | 5.5 x 1.5 x 1.0 m |
| Antenna gimbal requirements | roll, pitch and yaw synchronization from INS | roll, pitch and yaw (ARINC 429 digital data from LTN 90 INS) |
| Antenna polarization | HH (transmit and receive) | HH (transmit and receive) |
| Aircraft type <br> Observation altitude <br> Along-track velocity | Cessna Conquest <br> 8.84 km <br> 280 knots (144 m/s) | Challenger 600 <br> 10.6 km <br> 400 knots (205 m/s) |
| Note: ARINC 429 is an aircraft data communication bus standard, while LTN 90 is the Litton model number of the Inertial Navigation System (INS) that INTERA uses. | | |

**Table 213:     Typical System Parameters of the Star-1 and Star-2 SAR Instruments**

The Star-1 system is mounted on a Cessna Conquest Turbo prop aircraft. It operates in 2 resolution modes: 6 m pixel mode imaging a 23 km wide swath, and a 12 m pixel mode imaging a 46 km swath. INTERA has used the system to fly surveys in over 20 countries for varies applications such as ice monitoring, geological mapping and base maps for logistical aspects of petroleum and mining programs, classification of vegetation cover and forest management, cartographic mapping and various environmental applications related to geomorphology and natural resources.

The Star-2 (built in 1986) is mounted on a Challenger jet and operates also in two resolution modes: 15 m pixel mode imaging a swath 60 km wide, and a 25 m pixel resolution mode for a swath 100 km wide. The Star-2 is equipped with a dual-sided radar so that it can image up to 2 swaths 100 km on either side of the aircraft simultaneously. This aircraft is used as of Fall 1993 for ice reconnaissance in the Arctic waters and along the Canadian Eastern Seaboard to assist with navigation activities in those regions. The Challenger aircraft with Star-2 is ideally suited for regional repetitive coverage required for surveillance applications.

INTERA has developed all the processing software for generating radar imagery collected by its SAR systems. In particular, it has made significant advances in radar processing software integrating GPS and other navigational information with the radar data to provide accurate geographic control to the imagery, including the capability to generate topographic products with very limited ground control. INTERA combines also radar data with other types of remotely-sensed data, working with the imagery in a GIS environment and deriving interpretation products for numerous applications.

## B.109   STRATO 2C (Piloted High-Altitude Research Aircraft)

A German program initiated by the Ministry for Research and Technology (BMFT) in February 1992 for the development of a high-altitude (24 km) and long-range (18.000 km) composite aircraft for stratospheric research. DLR has been commissioned with the project management, the plane is being developed and built by Grob GmbH in Mattsies, Germany. Plans call for an availability of the aircraft by the end of 1995 (to be operated by DLR), and for a scientific utilization period starting in 1996.[565]

| Parameter | Value |
|---|---|
| Empty mass of aircraft | 8250 kg |
| Maximum Take-off mass | 13500 kg |
| Payload mass | 800 kg (for high-altitude flights of 24 km over 8 hours for a range (max) of 7000 km, cruising speed = 560 km/h) |
| Payload mass | 1000 kg (for long duration flights of 48 hours (max) at 18 km altitude of 18.000 km range, cruising speed = 300 - 400 km/h) |
| Wing span (overall) | 56.5 m |
| Overall length | 24 m |
| Volume of unpressurized payload compartment (behind cabin) | 4 m$^3$ |
| Installed power | two liquid-cooled engines (Teledyne Continental Motors) with 300 kW each; and a turbocharging system PW127, Pratt &Whitney |
| Payload power | 8.4 kW (300 Ampere at 28 V) for all instruments |
| Inner cabin diameter/length | 1.85 m/5.5 m (effective length) |
| Pressurized cabin | for 2 pilots, 2 mission scientists, and for payload provision |

**Table 214:    Performance Parameters of the STRATO 2C Research Aircraft**

A scientific program for the Strato 2C missions is being considered for the following research fields or applications:[566]

- Dynamics and chemistry of the atmosphere [ozone research, in-situ measurements, concentration profiles of tracer gases, existence and propagation of Polar Stratospheric Clouds (PSCs), heterogeneous chemistry on droplets, etc.].
- Investigations of clouds and their influence on climatic processes [exchange processes between the troposphere and the stratosphere, in particular in regions like Europe and in the tropics].
- Exchange processes between the biosphere and the atmosphere.
- Air traffic pollution and its influence on tracer gases.
- Test and calibration of Earth observation sensors and facilities to be flown on satellites.
- Observation of land/ocean/polar ice-caps from high altitudes

A corresponding sensor program for the STRATO 2C missions is being developed as of 1993.

---

565)  "Preliminary Technical Description STRATO 2C", DLR paper, Feb. 22, 1993
566)  U. Schumann, et al., "Scientific-Technical Concept for the Application of the Strato 2C High-Altitude Research Aircraft", DLR paper, Feb. 1993

# B.110   Sun Photometer

A NASA-sponsored instrument developed and operated at ARC. The multi-wavelength airborne tracking sun photometer is used for measuring solar radiation. The measurements provide a record of atmospheric optical depth and transmissivity data and permit the atmospheric correction of remotely-sensed data. The instrument is mounted on the exterior of the aircraft (e.g. DC-8, C-130, C-131, Twin Otter or CV 990) it automatically tracks the sun while simultaneously measuring six wavelengths (channels) of incoming solar radiation.[567]

Each channel consists of a doubly baffled entrance tube, inference filter, photodiode detector, and integral amplifier. The entrance baffles define a detector FOV with a measured half angle of 2.2°. The six filter/detector/preamp sets are mounted in a common heat sink maintained at 45 ± 1°C. Filters are currently (1993) centered at 382, 451, 526, 861, 940, and 1060 nm. Filter full widths at half-maximum (FWHM) are 6 to 15 nm.

Solar tracking is achieved by azimuth and elevation motors driven by error signals derived from a differential-shadowing sun sensor. Data are digitized and recorded every 2 to 10 seconds. The science data set includes the six detector signals, detector temperature, sun tracker azimuth and elevation angles, tracking errors, and time.

| Wavelength (µm) | FWHM (Full width half maximum) Bandwidth (µm) |
|---|---|
| 0.382 | 0.0122 |
| 0.451 | 0.0062 |
| 0.526 | 0.0091 |
| 0.861 | 0.0130 |
| 0.940 | 0.0127 |
| 1.060 | 0.0152 |

**Table 215:**   **Detector Wavelengths and FWHM Wavelengths of the Sun Photometer**

There are a number of applications/objectives for the sun photometer.

- Mapping of optical depth spectra of aerosols, jet fuels, forest fires, etc.
- Assessment of the impact of the Pinatubo cloud on atmospheric radiation and climate, mapping the optical depth spectra of the volcanic cloud in the northern and southern hemisphere (May-June 1993 campaign).[568][569],[570]
- Provision of validation measurements for the spaceborne instrument SAGE II (Stratospheric Aerosol and Gas Experiment II) on ERBS (see chapter A.24).

Optical depth spectra are inverted for the derivation of the stratospheric particle size distributions and ratios relating mass, area, extinction, and lidar backscatter.

## B.110.1  HIRAASS (High Resolution Airborne Autotracking Sun Spectrometer)

A proposed project to DOE by Ames Research Center. HIRAASS is to be flown on airborne missions (also on Perseus A aircraft), with the objective to measure optical depths and vertical extinction profiles of aerosols, clouds, water vapor, ozone, and nitrogen dioxide. The instrument design draws on the experience of the ARC 'Sun Photometer'. The instrument achieves fine spectral resolution (1 to 2 nm) by combining a grating with a linear array detector spanning the spectral range from 350 nm to 1020 nm. An additional photodiode channel extents the spectral range to 1550 nm for better characterization of large aerosol particles.


567) T. Matsumoto, P. B. Russell, C. Mina, W. Van Ark, "Airborne Tracking Sunphotometer", Journal of Atmospheric and Oceanographic Technology, Vol. 4, 1987, pp. 336-339

568) P. B. Russell et al., "Pinatubo and pre-Pinatubo optical depth spectra: Mauna Loa measurements, comparisons, inferred particle size distributions, radiative effects, and relationship to lidar data", J. Geophysical Research, in press, 1993a

569) R. F. Pueschel, J. M. Livingston, "Aerosol Spectral Optical Depths: Jet Fuel and Forest Fire Smokes", Journal of Geophysical Research, Vol. 95, Nr. 22, 1990, pp. 417-422

570) M. A. Spanner R. C. Wrigley, R. F. Pueschel, J. M. Livingston, D. S. Colburn, "Determination of Atmospheric Properties During the first Land Surface Climatology Project Field Experiment", Journal of Spacecraft and Rockets, Vol. 27, Nr. 4, July-August 1990, pp. 373-379

## B.111   TRWIS (TRW Imaging Spectrometer)

TRW of Redondo Beach Ca. has developed and built a set of hyperspectral imaging spectrometers operational since 1991. The instruments provide a real-time measurement display with features for vegetation or mineral identification, or quantification (crop health, biomass algae content, etc.). As of 1/1994 three TRWIS-B instruments and one TRWIS-II instrument are operational.[571]

The TRWIS-II instrument was designed to operate with two spectral ranges: 900-1800 nm and 1500-2500 nm. The two bands are selected by rotating the motorized grating. Note: only 3/4 of the 128 samples (=97) correspond to actual data; the spectral coverage is therefore 1160 nm from 1.46 - 2.62 μm (nominally from 1.5 - 2.5 μm).

TRWIS-B employs a Si CCD camera with 768 spectral and 493 spatial pixels. TRWIS-II employs an InSb direct injection readout 256 x 256 focal plane array. Simultaneous spectral readout for both instruments. Data are calibrated using a spectrally flat and spatially uniform calibration standards.

| Parameter | TRWIS-B | TRWIS-II |
|---|---|---|
| Spectral range | 450 - 880 nm | 900 - 1800 nm or 1500 - 2500 nm |
| Bandwidth | 4.8 nm | 12 nm |
| Nr. of Bands | 90 | 128 |
| IFOV | 0.4 - 2 mrad | 0.45 mrad |
| FOV | 5 - 25° | 6° |
| Pixel/line | 240 | 240 |
| SNR | ≥ 70:1 | 50:1 |
| Data recording | all bands continuous 14000 spectra/s | all bands continuous 14000 spectra/s |
| Data processing | real-time | real-time |
| Platforms | aircraft or helicopter (flight altitudes range from 100 m to 10 km). GPS receiver (GPS coordinates are used to identify targets and determine flight paths) | |

**Table 216:    Specification Parameters of TRWIS Instruments**

| Natural resource management | Environmental Protection |
|---|---|
| Vegetation identification<br>Vegetation condition and biomass<br>Fire regrowth<br>Fishing<br>Phytoplankton survey<br>Ice/snow inventory | Oil spills<br>Toxic material spills/dumps<br>Sewage spills<br>Sediment<br>Nutrient transport |
| Natural resource exploration | Others |
| Petroleum<br>Minerals<br>Water | Bathymetry<br>Land use planning<br>Search and rescue, mapping/charting |

**Table 217:    Typical Applications of TRWIS Instruments**

Data: The video data correspond to: (8 bit/sample) (128 samples/line) (240 lines/frame) (60 frames/s) = 14.75 Mbit/s. The data is recorded onto VHS video recorders.

TRWIS-II has the ability to record data at 12 bit resolution over the 256 x 256 focal plane array corresponding to a digital data rate of 47.2 Mbit/s.

## B.112   TSCC (Translinear Scanning CCD Camera)

The TSCC is a high-definition electro-optical camera (EOC) of NASA/ARC and GSFC (the instrument is also referred to as 'EOC'). The camera is nose-mounted in the aircraft with forward or aft tilt up to 53°. Objectives: bi-directional reflectance and polarization

---

571)  Information provided by R. B. Herrick of TRW, Redondo Beach Ca.

studies. The system captures high-resolution digitized images and stores the imagery on magnetic tape. The camera is a commercially available product of Kodak (KAF-1400) with a silicon array imager. Wavelengths are selected by a rotating filter wheel. There are plans for upgrades (2 k x 2 k camera array and selectable filters). Wavelength channels can be either polarized or non-polarized. The instrument has flown on ER-2 aircraft for the 1991 and 1992 FIRE projects and on the 1993 TOGA/COARE campaign for cloud studies. The tilting scan rate is programmable in order to track and multiple-image cloud scenes at differing altitudes.

| Spectral coverage | 400 - 950 nm |
|---|---|
| Filtration | 6 position filter wheel (6 spectral filters) polarizing filter |
| Nominal spectral filters | 420, 480, 530, 680, 875, 950 nm (with 10 nm bandwidth) |
| Lens (interchangeable) | 28 mm |
| Aperture | f/2.8 |
| IFOV | 0.2 mrad |
| Spatial resolution (ground) | 4.8 m at aircraft altitude of 20000 m |
| Frame size | 1280 pixels x 1025 pixels (recorded) |
| Image size (ground) | 6.2 km x 4.9 km |

**Table 218:** **Characteristics of the TSCC**

Data collection parameters of TSCC:
Frame rate        4 images every 6 seconds
Tilt angle        Forward tilt up to 42°, aft tilt up to 53°
FOV (nominal)  15°
Data storage     Tape cassette
Capacity          5 GByte


# B.113   TU-134A (Tupolev Flying Laboratory)

The Tu-134A is a converted Russian passenger aircraft (owned and operated by NPO Vega) dedicated for airborne observation campaigns as well as for instrument development of future systems. The aircraft is equipped with a number of sensors to suit the requirements of a particular campaign. The aircraft/laboratory is used on a contractual basis. Some of the instruments are: IMARC (Radar), SIR (Scanning Infrared Radiometer), AFA (Aerial Foto Apparatus), and navigation instruments along with their support equipment.[572]

Applications: geology, oceanology, agriculture, ecology, cartography, etc. . IMARC is also used for research (i.e. testing of other SAR instruments, and for algorithm development).


## B.113.1  IMARC (Imaging Multifrequency Airborne Radar Complex)

IMARC is multifrequency polarimetric SAR with the objective of Earth surface sensing. IMARC was developed and built NPO Vega, the instrument went through the following development phases:

- IMARC is operational since 1983 with a 1-frequency SAR, (4 cm wavelength)
- IMARC is operational since 1990 with a 2-frequency SAR, (4 cm and 254 cm)
- IMARC is operational since 1993 with a 3-frequency SAR, (4 cm, 68 cm and 254 cm)
- IMARC will be available with a 4-frequency SAR in 1994, (4, 23, 68, and 254 cm)

The different frequency bands of IMARC are independent modules and may operate individually or simultaneously dependent on the user requirements. During 1993 the IMARC on-board data recording is being changed from holographic photofilm recording to an all-digital recording system, in 1994 the ground system is also changed to an all-digital system (see Figure 138).

---

572) Information provided by Y. Krilov of NPO Vega, Moscow

| Parameter | Waveband | | | |
|---|---|---|---|---|
| Waveband (cm) | 4 | 23 | 68 | 254 |
| Waveband frequency (GHz) | 7.7<br>X-Band | 1.28<br>L-Band | 0.44<br>L-Band | 0.118<br>(VHF) |
| Swath (km) | 6 and 12 km (15 km in 1994) | | | |
| Direction of observation | right side or left side of the aircraft | | | |
| Spatial resolution (m) | 4-6 | 8-10 | 15-20 | 15-25 |
| Radiated power (kW) | 40<br>(impulse) | 0.5 (linear<br>freq. mod) | 0.8 (linear<br>freq. mod) | 0.6<br>(impulse) |
| Type of signals | pulse | chirp | chirp | pulse |
| Polarization | HH, VV, (HV, VH) | | | |
| Viewing angles from nadir (°) | 60 - 83 | | | |
| Antenna type | slotted<br>waveguide | slotted | | |
| Antenna gain (dB) | 30 | 14-17 | 14-17 | 9-11 |
| Radiation @ 3-dB level (°)　in azimuth<br>in elevation | 1.7<br>24 | 24<br>24 | 24<br>24 | 40<br>60 |
| Image scale | 1:200000 | | | |
| Dynamic range (dB)　optical processing | 25 | | 15-20 | |
| Dynamic range (dB)　digital processing | 32 | | 36 | |
| Number of samples (after 1. upgrade '93) | 1024 | | | |
| Number of samples (after 2. upgrade '94) | 3000 | 3000 | 1500 | 750 |
| Number of pixels (m x m) (after 1. upgrade '93) | 15 x 15 | | | |
| Number of pixels (m x m) (after 2. upgrade '94) | 4 x 4 | 4 x 4 | 8 x 8 | 16 x 16 |
| Number of incoherent looks | 4 | | | |
| Quantization level　input (ADC output)<br>output (Image Proc.) | 2 x 6<br>8 | | | |
| Total power consumption (kVA)　110V, 400MHz<br>+27 V DC | 15 kVA<br>4 kVA | | | |
| Mass of instrument (kg) | 1500 | | | |
| Flight altitude (m) | 500 - 3000 | | | |
| Flight speed (km/h) | 500 - 600 | | | |

**Table 219:　Technical Specifications of IMARC**

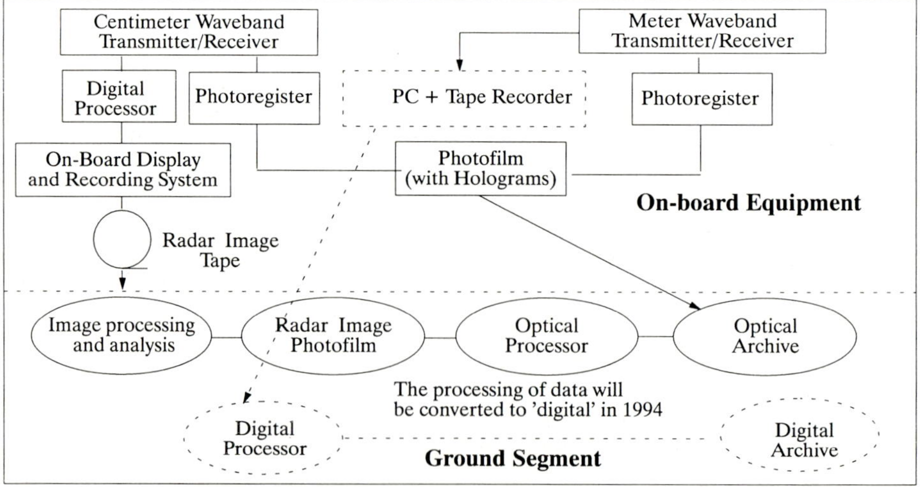


**Figure 138:　Block Diagram of IMARC**

## B.113.2  SIR (Scanning Infrared Radiometer)

| Spectral range | 7 - 14 µm |
|---|---|
| Swath width | 120° |
| V/H ratio (V = aircraft velocity; H = height) | 0.16 - 1.2 s$^{-1}$ |
| Radiometric resolution at 20°C with SNR = 1 | 0.3 °C |
| Line frequency | 500 Hz |
| Image scale on film with 80 mm width | $1 : M_x = 1 : H/ (33.42 \times 10^{-3})$ |
| Observation heights | 200 - 1000 m |

**Table 220:**  **Specification of some SIR Sensor Parameters**

## B.113.3  AFA-41/10 (Aerial Foto Apparatus)

Camera type      AFA 41/10
Focal length      100 mm
Film width      180 mm

## B.114  UV-DIAL (Ultraviolet Differential Absorption Lidar)

The UV-DIAL[573] sensor is a cooperative development of USEPA (US Environmental Protection Agency), 'Environmental Monitoring Systems Laboratory' (EMSL), Las Vegas, with the University of Nevada's 'Desert Research Institute', and the Harry Reid Center for Environmental Studies.

Applications: research tool for the measurement of ozone concentrations, concentrations of other trace gases and pollutants in the atmosphere, determination of air quality, etc.

DIAL is a multi-wavelength lidar concept that uses the wavelength-dependent absorption of atmospheric constituents to measure their range-resolved concentration. Measurements of absorption of laser beams are obtained at "on-wavelengths" 9for which the gaseous constituent is strongly absorbing) and "off-wavelengths" (for which the gaseous constituent is not strongly absorbing) to provide solutions to the lidar equation in ratio form in terms of these wavelengths.

The UV-DIAL instrument is based on a KrF (Krypton Fluoride) excimer laser, which generates 700 mJ, 20 ns pulses at 248 nm with a maximum pulse repetition rate of 20 Hz. The laser beam is split up - 1/3 of the power being focused into a hydrogen ($H_2$) Raman cell, and 2/3 into a deuterium ($D_2$) Raman cell. These Raman cells produce transmitted laser beams at frequencies shifted from the KrF fundamental by integral multiples of the vibrational frequency of hydrogen and deuterium.

The receiver section consists of a downlooking telescope for collection of the backscattered light and a spectrograph/detector system for detection of the individual laser lines. The telescope is a classical Newtonian design with a 0.5 m diameter, f/2.5 primary mirror, arranged in a non-coaxial configuration with the laser transmitter. The telescope images the received light into the entrance slit of a custom Czerny-Turner spectrograph, which in turn images each of five different wavelengths [276.9 nm ($S_1$ of $H_2$), 291.6 nm ($S_2$ of $D_2$), 312.9 nm ($S_2$ of $H_2$), 319.4 nm ($S_3$ of $D_2$), and 359.4 nm ($S_3$ of $H_2$)] onto a separate photomultiplier tube (PMT). The PMTs are mounted directly in the focal plane of the spectrograph. The PMT are gated on for the appropriate measurement range. The PMT signals are digitized by 12 bit A/D converters with a maximum spatial resolution of 15 m. The data of the UV-DIAL

573) H. Moosmüller, R. J. Alvarez, R. M. Jorgensen, C. M. Edmonds, D. H. Bundy, D. Diebel, M. P. Bistrow, J. L. McElroy, "An Airborne Lidar System for Tropospheric Ozone Measurement", presented at the 86th Annual Meeting of AWMA, Denver, Co., June 13-18, 1993

are stored on magnetic tape and simultaneously displayed on a monitor in real-time to allow for operational control of the system. In addition, information of the viewing scene for the DIAL is provided by a downward-looking video camera.

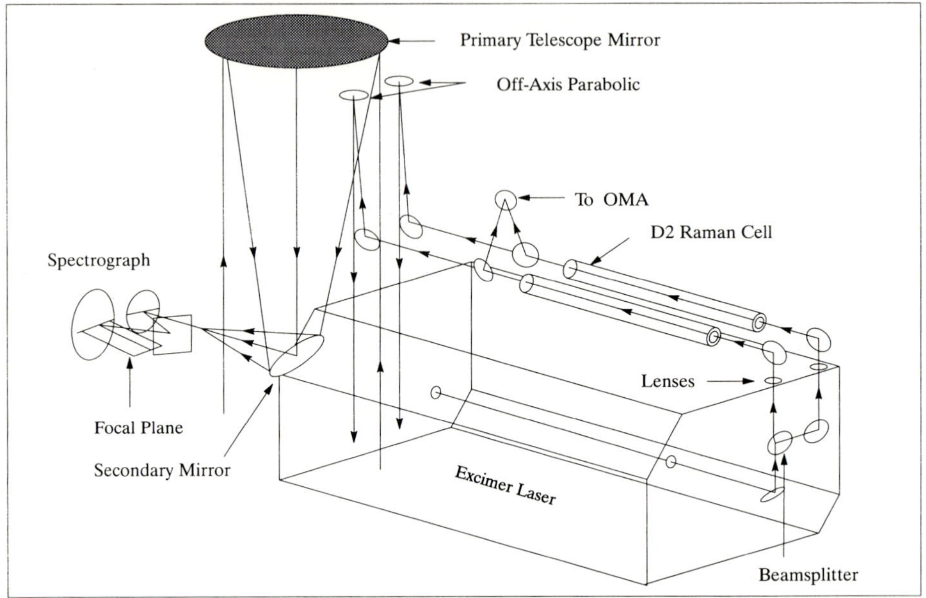


**Figure 139:    Optical Setup of the UV-DIAL System**

Test campaign in Southeastern Michigan:[574] The system has been installed in a mid-sized aircraft (DeHavilland C-7 Caribou) for initial airborne testing in the Great Lakes area in May 1992. During these test flights several two-dimensional ozone distributions were acquired. The laser was operated at a repetition rate of 20 Hz, data from 300 consecutive shots were averaged for each vertical ozone profile (horizontal resolution of 1.17 km at an average flight speed of 280 km/h). The vertical resolution of about 150 m is determined by a sampling interval of 30 m and smoothing over five data points. Ozone concentrations have been calculated from 90 m above ground to 150 m below the aircraft flight altitude.

Results: Range-resolved data on gaseous concentrations and on relative density (in terms of lidar backscatter) of aerosol distributions can be obtained in the lowest two to three kilometers of the atmosphere in cells beneath the aircraft in the vertical plane along the flight path (sounding profile). Cell resolution is a joint function of laser firing rate, data digitization, data averaging and smoothing rate, and horizontal aircraft speed.


## B.115   VIRL (Visible and near Infrared Lidar)

VIRL is a NASA/GSFC airborne lidar instrument with the objective to study the source and characteristics of atmospheric aerosol particles (measurement of aerosol backscatter cross-section at the fundamental and doubled Nd:YAG laser wavelengths of 1.064 µm and 0.532 µm, and in addition, at the wavelength of 1.54 µm). The instrument is operational since 1989 and flown on NASA ER-2 and DC-8 aircraft; it participated in the GLOBE (Global Back-

574) J. L. McElroy, et al., "Airborne UV-DIAL Measurements of Ozone Distributions in Southeastern Michigan", presented at 86th Annual Meeting of AWMA, Denver, Co., June 13-18, 1993

scatter Experiment) flights in November 1989 and in May-June 1990 (to support laser wind sounder development), as well in TOGA/COARE. campaign in 1993 (to explore the maintenance of the West-Pacific warm pool).[575]

The instrument permits simultaneous measurements of aerosol and cloud backscatter at multiple wavelengths. The advantage of a 1.5 µm lidar is eye safe operation and increased sensitivity to aerosol characteristics. An important factor is accurate calibration. The system offers a new calibration technique based on hard target laser measurements.

| Transmitter: | |
|---|---|
| Laser | Nd:YAG I, II |
| Raman cell | Light Age one meter mixing cell, 300 psi methane |
| Wavelengths | 1.064,   0.532,   1.54 µm |
| Power (output) | 150,   25,   20 mJ |
| PRF (Pulse Repetition Frequency) | 50 Hz |
| Divergence | 0.8 mrad |
| Receiver: | |
| Telescope | Nadir or zenith pointing capability, 40 cm diameter |
| FOV | 1.66 mrad |
| Wavelengths | 1.064,   0.532,   1.54 µm |
| Filter bandwidth | 2.0,   0.8,   10.0 nm |
| Detectors | Si:APD Si:APD InGaAs:PD |
| A/D sample rate | 2 MHz |
| A/D resolution | 12 bits |

**Table 221:    VIRL Instrument Characteristics**

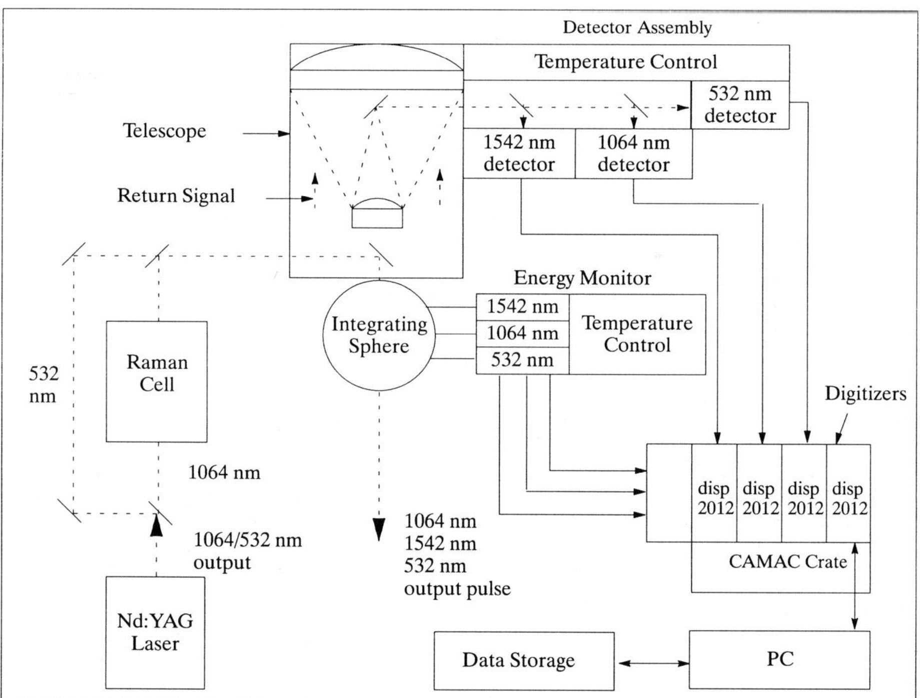


**Figure 140:    Schematic Overview of the VIRL Instrument**

575) J. D. Spinhirne, S. Chudamani, J. F. Cavanaugh, "Visible and Near IR Lidar Backscatter Observations on the GLOBE Pacific Survey Missions", Seventh Symposium on Meteorological Observations and Instrumentation and Special Sessions on Laser Atmospheric Studies, Jan. 14-18, 1991, New Orleans, La.

## B.116   VIS (Video Imaging System)

VIS is a NASA-sponsored instrument at ARC. It consists of a video cam-corder pointed downward, and a video cassette using super VHS tape. VIS records terrain and atmospheric conditions below the aircraft, such as cloud cover or desert overpasses. The data collected are used to justify or explain anomalous results obtained by other sensors flown in conjunction with VIS.[576]

The VIS instrument became operational in 1992, it is typically flown as a tracking camera for cloud-top mapping missions only, as the spatial resolution is too course to be very useful from ER-2 altitudes for terrestrial viewing. The VIS camera has a single objective, one of the listed lenses is installed prior to a flight.

- Hitachi KP-C551 color camera, color system: NTSC
- Resolution: 430 x 350 lines (H x V)
- FOV        4 mm lens       $71^\circ$
             8 mm lens       $35^\circ$
             12 mm lens      $23^\circ$
             16 mm lens      $18^\circ$
             25 mm lens      $12^\circ$
- Minimum illumination: 5 lux
- VCR Panasonic AG-6750A (Time lapse recorder)

| Record modes | Record Intervals (frames) |
|---|---|
| 2 hr | 1/60 s |
| 6 hr | 1/60 s |
| 24 hr | 0.2 s |
| 48 hr | 0.4 s |
| 72 hr | 0.6 s |
| 120 hr | 1.0 s |
| 180 hr | 1.5 s |
| 240 hr | 2.0 s |
| 480 hr | 4.0 s |

- Resolution (S-VHS): 400 lines
- FIGI configuration 24 hour mode, 12 mm lens, 1/2000 @ f/5.6 [note: FIGI refers to the island of Fiji in the South Pacific, the campaign CEPEX (Central Equatorial Pacific Experiment) was conducted in March/April 1993; this was an atmospheric water vapor/ Greenhouse study].

## B.117   WHiRL (Wide-angle High-Resolution Line-imager)

WHiRL is a single-channel prototype CCD imager of CCRS, Ottawa, Canada. The instrument is a research tool with the goal in mind to obtain high-quality digital imagery in remote sensing applications such as forestry and topographic mapping. The sensor offers a wide angular field of view (FOV = $70^\circ$ and 6000 pixel swath width) and a high spatial resolution. WHiRL has been flown in test flights in 1991 providing resolutions (pixel sizes) down to 0.25 m (1070 m altitude). At a flight altitude of 12 km the resolution is 2.8 m and the swath width 12.8 km. The goal of the WHiRL program is the eventual development of a multispectral pushbroom imager for economical monitoring applications.[577]

---

576) Information provided by D. Dokken of NASA-HQ and by J. Myers of ARC
577) R. A. Neville, R. Marois, J. W. Schwarz, S. M. Till, "Wide-angle high-resolution line-imager prototype flight test results", Applied Optics, Vol. 31, Nr. 18, 1992, pp. 3463-3472

| Scanner type | Pushbroom CCD scanner, 1 spectral channel |
|---|---|
| Spectral response | 595 nm |
| Bandwidth (FWHM) | 20 nm |
| FOV (IFOV) | 70° (o.23 mrad.) |
| Detector type | Loral Fairchild CCD 191, buried channel 6000-element linear array |
| Lens type | Custom design f/2.8 retrofocus, focal length = 43 mm |
| Line sample rates | 56.6 to 320 lines/s in steps of $2^{1/2}$ |
| Gains | 1 to 4 x $2^{1/2}$ in steps of $2^{1/2}$ |
| Exposure time | 3.125 to 17.678 ms in steps of $2^{1/2}$ - ( 3.125 ms  corresponds to a flight speed of 80 m/s) |
| Quantization | 12 bit |
| Data format | 12 tracks of HDDT in aircraft transcribed onto CCT in standard (LGSOWG) format post-flight |
| Signal/data processing | none at present (1993) |
| Interfaces | Alice real-time display (quicklook capability for subswath) High-density digital tape recorder (time code generator onto HDDT) Inertial navigation system recorded on a separate tape, can be merged with image data post-flight |

**Table 222:    WHiRL Sensor Characteristics**


# B.118   WILD RC30 (Aerial Camera System)

WILD RC30 is a commercially available aerial survey camera series of Leica AG, Heer-brugg, Switzerland. Background: The instrument series started with the WILD RC10 cam-era in 1968 - '80, several hundred of this series were sold worldwide; this was followed by the WILD RC10A camera model in the period from 1981- '86 (several hundred sold). The RC20 camera has been built during the time frame 1987 - '92 (about 200 were sold in of the series, introduction of digital electronics in aerial cameras). The current model is the RC30 camera introduced in 1993. Each series represents a new generation in camera technology for a wide range of photogrammetric applications.[578]

The RC30 camera system provides such features as:

- PEM (Automatic real-time Exposure Control). The PEM sensor measures a spectral range of 0.4 to 1.0 μm with a maximum sensitivity around 0.7 μm, and a field of view (FOV) of 80°.
- EDI (External Data Interface), link between the camera and the GPS receiver, the air-craft navigation system, or the mission computer.
- FMC (Forward Motion Compensation), the FMC counteracts the forward image mo-tion, range between 1 mm/s to 64 mm/s.
- Filters (for all exposure conditions; absorption of unwanted spectral ranges, suppres-sing stray light, optimizing light distribution in the image plane, correcting color-bal-ance of film emulsion).
- V/H range of the camera is between 6 mrad/s to 300 mrad/s (for 88 mm and 153 mm lenses); the shutter speed is between 1/100 s to 1/1000 s (continuously adjustable); the aperture is between f/4 to f/22 continuously adjustable.

- ASCOT (Aerial Survey Control Tool), an integrated system - hardware and software - to support survey flight navigation. ASCOT provides the following functions:
  - Guidance for the camera operator/navigator and pilot during flight, to accomplish the task according to a predefined plan.
  - Automatic exposure release at predefined locations (pin-point photography).
  - Automatic in-flight data annotation on imagery (freely programmable).
  - Use of output from a directional gyro or other sensors to transmit drift correction to the camera mount.

---

578) Information provided by Leica AG, Heerbrugg, Switzerland

- Data logging and processing for mission evaluation and reporting
- Link with geodetic software packages to compute the exact position of projection centers, and include them in aero-triangulation programs.

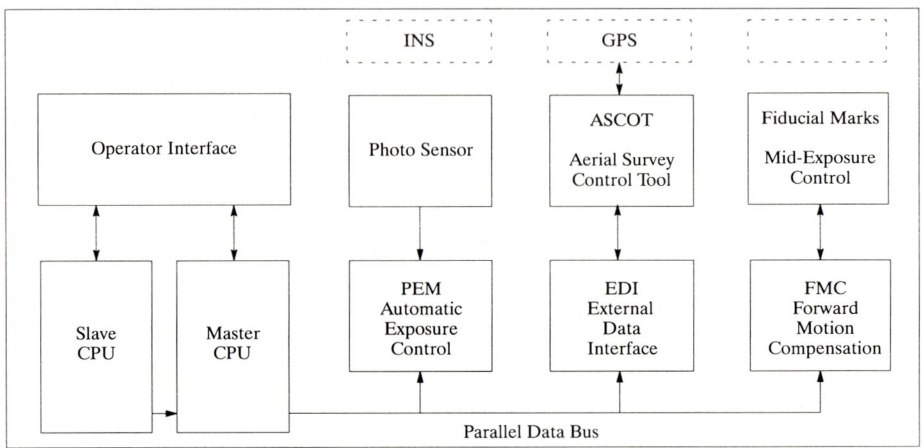


**Figure 141:    Flight and Data Management Elements of the WILD RC30 Camera**


## B.119   WIS (Wedge Imaging Spectrometer)

The WIS is a hyperspectral instrument developed by Hughes Santa Barbara Research Center (SBRC) on Corporate Independent R&D (IR&D). Flight demonstrations of the instrument were performed in 1992, sponsored in part by ARPA and other US government agencies.[579]

The WIS sensor offers a novel spectral separation technique in which the spectral separation filters are mated to the detector array to achieve two-dimensional sampling of the combined spatial/spectral information passed by the filter. The technique obviates the need for a complex aft-optics assembly. The filter mask technique utilized in WIS compares to the general performance characteristics of the last row in Table 210.

The linear spectral wedge filter is a thin-film optical device that transmits radiation at a center wavelength that depends on the spatial position of illumination on the filter. On one side of the substrate, the device has thin-film depositions that are tapered linearly in thickness along one direction, resulting in a wedge shape. Since the bandpass center wavelength for a thin-film stack depends fundamentally on the stack layer thickness, the center wavelength of the passed radiation varies linearly with position parallel to the tapered edge of the filter stack. In the direction parallel to the untapered edge, the center wavelength is constant. Filter blocking stacks, required to control out-of-band leakage in the wedge filter, are on the opposite side of the substrate. Because the spectral bandwidth of a thin-film stack depends primarily on the structure of the stack (i.e. the relative thickness of the layers and their compositions), it is a constant percentage of the center wavelength ($\Delta\lambda/\lambda$). SBRC has developed designs with bandwidths of 3%, 2%, 1%, and 0.5%.

When an array of detectors is placed behind the wedge filter device, each detector in the "spectral" dimension will receive radiation from the scene at a different center wavelength - the array output is, in essence, the sampled spectrum of the scene. By using an area array of

---

579) Information provided by E. S. Putnam of Hughes SBRC

detectors, the detected scene information will vary spatially in one direction and spectrally in the other. Scanning the filter/array assembly along the spectral dimension will then build a spatial image (2-D) in each of the detected spectral bands, thus creating an imaging spectrometer.

The WIS sensor concept does not provide simultaneous spectral sampling of all spatial location within the sensor field of view. Instead, the sensor samples each ground point in all spectral bands over a short period of time (in the order of 1 s) by using the forward motion of the aircraft to "pushbroom" the ground image across the WIS detector array. The data are then registered in postprocessing to superimpose the spectral bands.

In the WIS flight demonstration unit (FDU), each of the 64 spectral bands comprises 128 spatial elements. The corresponding nominal ground coverage and spatial resolution is a function of the principal data collection parameters: altitude, velocity, detector integration time, and optics focal length. In addition, since the desired data collection parameters may require the use of a ground velocity that is too fast for a given aircraft, the WIS data acquisition system allows a selected number of data frames to be skipped for every frame that is recorded.

Two new Focal Plane Arrays (FPAs) are under development (end of 1993), one for the VNIR- and one for the SWIR region. They will be configured as illustrated in Figure 142 and summarized in Table 223, allowing a total deposition to occur on one side of a single substrate, as opposed to the three substrates required for the FDU, when a single filter covered the entire array. In addition, the spectral bandpass of each filter can be tailored to the requirements of the phenomena to be sensed in these spectral regions.

Filter 1 of the new WIS models is optimized for sensing vegetation, plant litter, and iron oxide materials; filter 2 for monitoring biological events near the "red edge"; filter 3 for discrimination of rocks, soils, and contained water; and filter 4 for direct identification of surface materials based on absorption features. The WIS-VNIR is expected to ready in mid-1994 for demonstration and use, the WIS-SWIR in late 1994.

| Parameter | Existing VNIR (WIS-FDU) | New VNIR Instrument (WIS-VNIR) | | New SWIR Instrument (WIS-SWIR) | |
|---|---|---|---|---|---|
| Filter | 1 | 1 | 2 | 3 | 4 |
| Spectral range (μm) | 0.40 - 1.03 | 0.40 - 0.60 | 0.60 - 0.95 | 0.95 -1.80 | 1.80 - 2.50 |
| Band center spacing (nm) | 10.3 | 12.5 | 6.0 | 30 | 23 |
| Number of bands | 64 | 17 | 67 | 27 | 45 |
| Spectral resolution ($\Delta\lambda/\lambda$, %) | 1.8 | 2.4 | 0.9 | 2.1 | 1.0 |
| Spectral resolution (nm) | 7.2 - 18.5 | 9.6 - 14.4 | 5.4 - 8.6 | 20.0 - 37.8 | 18.0 - 25.0 |
| Detector material and type | Si CCD | Si CCD | | InSb | |
| Detector size (μm) | (128 x 64) | 18 (512 x 512) | | 40 (320 x 210) | |
| Telescope focal length (mm) | 55 and 108 | 27.4 | | 38.1 | |
| FOV (swath width) (degree) | 10 and 15 | 19.1 | | 19.1 | |
| IFOV (mrad) | 1.36 | 0.66 | | 1.05 | |
| Spatial resolution (m) | 0.5 - 5 | 1 - 5 | | 1.6 - 8 | |
| Ground Sample Distance (m) at observation altitude of 1.5 km | 2 | 1 | | 1.6 | |
| Data quantization (bit) | 12 | 12 | | 12 | |
| Data rate (Mbit/s) | 22.5 @ 6 ms dwell | 69 @ 10 ms dwell | | 23 @ 16 ms dwell | |

**Table 223: Specification of the WIS Flight Demonstration Unit and future Models**

Image data processing of WIS imagery requires some rectification processes (as with all other image data as well). For instance, for hyperspectral processing it is necessary to insure that the spectra from a point in the image truly corresponds to a single point (finite small region) in the scene. The applied processing involves re-registration of the band images.

The WIS-FDU was flown on a Saberliner 40 aircraft for proof-of-concept demonstrations. The aircraft was equipped with a GPS receiver and other navigational instruments.

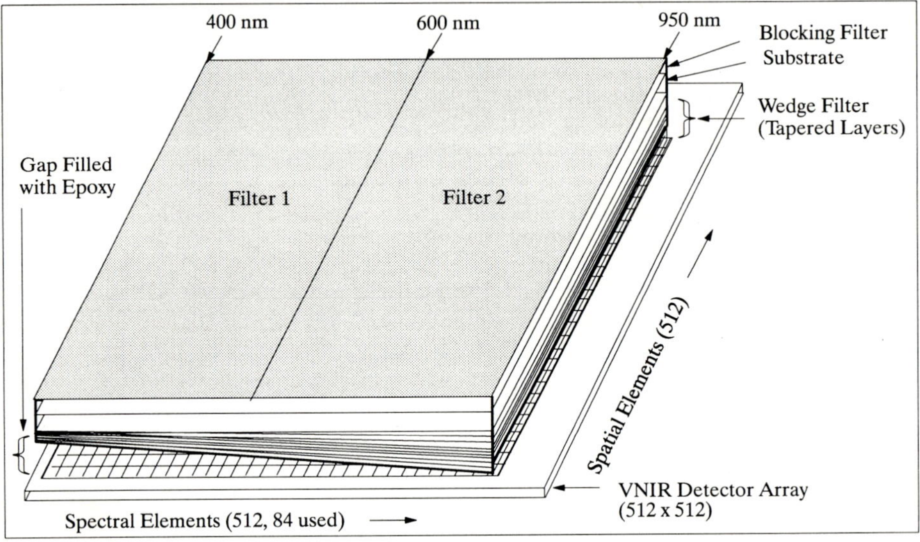


**Figure 142:**    **The WIS-VNIR FPA Scheme of the Linear Spectral Wedge Concept**

| Sensor (Agency/Company) | Nr. of Bands | Spectral Range (nm) | Band Width at FWHM (nm) | IFOV (mrad) | FOV (°) | Data Product | Period of Operation |
|---|---|---|---|---|---|---|---|
| AIS-1 | 128 | 990-2100 1200-2400 | 9.3 | 1.91 | 3.7 | Image cube | 1982-85 |
| AIS-2 (NASA/JPL) | 128 | 800-1600 1200-2400 | 10.6 | 2.05 | 7.3 | Image cube | 1986-87 |
| AISA (Karelsilva Oy) | 1-286 | 450-900 | 1.56-9.36 | 1 | 21 | Image cube | since 1993 |
| AMSS (GEOSCAN) | 32 8 6 | 490-1090 2.020-2370 8500-12000 | 170-240 430-440 550-590 | 2.1 x 3.0 | 92 | Image cube | since 1985 |
| ASAS upgraded ASAS (NASA/GSFC) | 29 62 | 455-873 400-1060 | 15 11.5 | 0.80 0.80 | 25 25 | Image cube Image cube | 1987-91 since 1992 |
| ASTER Simulator (DAIS-2815) (GER) | 1 3 20 | 760-850 3000-5000 8000-12000 | 90 600-700 200 | 1.0, 2.5, or 5.0 | 28.8, 65, or 104 | Image cube | since 1991 |
| AVIRIS (JPL) | 224 | 380-2500 | 9.7-12.0 | 1 | 30 | Image cube | since 1987 |
| CASI (Itres Research) | ≥ 288 | 430-870 | 1.8 | 0.3-2.4 | 35 | Profile image | since 1990 |
| CIS (China) | 64 24 1 2 | 400-1040 2000-2480 3530-3940 10500-12500 | 10 20 410 1000 | 1.2 x 3.6 1.2 x 1.8 1.2 x 1.2 1.2 x 1.2 | 80° | Image cube | since 1993 |
| CHRISS (SAIC) | 40 | 430-860 | 11 | 0.05 | 10 | Image cube | under construction |
| DAIS-7915 (GER/DLR) | 32 8 32 1 6 | 400-1010 1500-1788 1970-2450 3000-5000 8700-12700 | 10-16 36 36 2000 600 | 3.3, 2.5, or 5.0 | 64-78 | Image cube | Fall 1993 |
| DAIS-16115 (GER) | 76 32 32 6 12 2 | 400-1000 1000-1800 2000-2500 3000-5000 8000-12000 400-1000 | 8 25 16 333 333 stereo | 3 | 78 | image cube | Spring 1994 |
| FLI/PMI (Moniteq) | ≥ 288 | 430-805 | 2.5 | 1.3 | 70 | Profile image | 1984-90 |
| GER-63 Channel Scanner (GER) | 24 4 29 6 | 400-1000 1500-2000 2000-2500 8000-12500 | 25 125 17.2 750 | 2.5, 3.3, or 4.5 | 90 | Image cube | since 1986 |
| HYDICE (NRL/Hughes) | 206 | 400-2500 | 7.6 - 14.9 | 0.5 | 8.94 | Image cube | under construction |
| ISM (DESPA/IAS/OPS) | 64 64 | 800-1600 1600-3200 | 12.5 25.0 | 3.3 x 11.7 | 40 (selectable) | Image cube | since 1991 |
| MAS (Daedalus, NASA/AMES) | 50 | 547-14521 | 31-517 | 2.5 | 85.92 | Image cube | since 1992 |
| MIVIS (Daedalus) | 20 8 64 10 | 433-833 1150-1550 2000-2500 8200-12700 | 20 50 8 400-500 | 2.0 | 70 | Image cube | 1993 |
| MUSIC Lockheed | 90 90 | 2500-7000 6000-14500 | 25-70 60-1400 | 0.5 | 1.3 | Image cube | 1989 |
| ROSIS (MBB/DLR/GKSS) | 84 | 430-830 (≤ 2400) | 4-12 | 0.56 | 16 (32) | Image cube | since 1993 (1994) |
| SFSI (CCRS) | 122 | 1200-2400 | 10 | 0.4 | 11.7 | Image cube | Test phase start of 1994 |
| SMIFTS U. of Hawaii | 100 | 1000-5200 | | 0.66 | 9.7 | Image cube | |
| TRWIS-B TRWIS-II (TRW) | 90 128 | 450-880 900-1800 or 1500-2500 | 4.8 12 | 0.4 - 2 0.45 | 5-25 6 | Image cube | 1991 |
| WIS-FDU WIS-VNIR WIS-SWIR (Hughes SBRC) | 64 17+67 27+45 | 400-1030 400-950 950-2500 | 10.3 12.5 & 6 30 & 23 | 1.36 0.66 1.05 | 10 & 15 19.1 19.1 | Image cube Image cube Image cube | Test 1992 1994 1995 |

**Table 224:    Summary of Hyperspectral Airborne Imaging Spectrometers[580)]**

---

580)  Courtesy of K. Staenz of CCRS, "A Decade of Imaging Spectrometry in Canada", Canadian Journal of Remote Sensing, Vol. 18 Nr. 4, October 1992, pp. 187-197

| Instrument - Agency | Frequencies/Band (GHz) | Polarizations | Comment |
|---|---|---|---|
| AIRSAR - NASA/JPL | 0.45/(P), 1.26/(L), 5.31/(C) | H, V, full polarimetric | |
| TOPSAR - NASA/JPL | 5.288/(C) | VV | AIRSAR configuration |
| CARABAS - FOA Sweden | 20-90 MHz/(VHF) | H | Foliage/ground penetration |
| CAS-SAR - CAS/ISRA China | 9.375/(X) | HH, VV, HV, VH, | |
| C/X-SAR - CCRS Canada | 5.3/(C) 9.25/(X) | VV, HH, VH, HV full polarimetric | C-Band interferometry |
| DO-SAR (Dornier, Germany) | 5.3/(C) 9.6/(X) 35/(Ka) | VV, VH, HH, HV VV, VH, HH, HV VV | polarimetric |
| DRA-SAR - DRA/UK | 5.7/(C) 9.65/(X) | full HH | |
| EMISAR - EMI/ TUD Denmark | 5.3/(C) 1.4/(L) | full, polarimetric full, polarimetric | 1993 end of 1994 |
| E-SAR - DLR, Germany | 1.3/(L) 5.3/(C) 9.6/(X) 0.450/(P) | VV or HH, VV or HH VV or HH VV or HH | will be added in 1994 |
| FOLPEN - SRI, Ca. | 100-500 MHz/(VHF/UHF) | HH | Foliage/ground penetration |
| MMW-SAR - MIT | 33.56/(Ka) | V, H, full polarimetric | |
| NEC-SAR - NEC Jap. | 9.53/(X) | HH | Interferometric mode |
| P-3/SAR - ERIM,Navy | 0.350/(UWB) 1.25/(L) 5.30/(C) 9.35/(X) | VV, VH, HH, HV VV, VH, HH, HV VV, VH, HH, HV VV, VH, HH, HV | UBW added in 1994 existing, polarimetric |
| PHARUS - TNO-FEL Delft, The Netherlands | 5.25/(C) | full, polarimetric | |
| SAR Travers - Vega, Moscow | 3.0/(S) | VV or HH | |
| RAMSES - ONERA France | 1.6/(L), 3.0/(S), 6.2/(C) 9.5/(X), 14.5/(Ku), 35/(Ka) 95/(W) | VV, HH, Vh, HV full polarimetric | S-Band and Ka-Band capability will be added by the end of 1994 |
| STAR-1,2 - INTERA | 9.375/(X) | HH | |
| IMARC - Vega Moscow | 118 MHz/(VHF) 0.44/(L), 1.28/(L), 7.7/(X) | HH, VV, HV, VH full polarimetric | |

**Table 225:   Overview of Airborne SAR Systems**

| Aircraft/ Manufacturer | Model | Altitude ft | Range (nm) | Endurance (hr) | Payload mass (kg) | Payload power (kW) | Agency |
|---|---|---|---|---|---|---|---|
| Convair Gen. Dynamics | 580 | 30,000 (7.5 km) | 2,100 | 5.5 | 2,300 | | NRC |
| Twin Otter DeHavilland | DHC-6 -200 | 21,000 | 600 | 5.0 | 1,200 | | |
| Falcon-20 Dassault | | 42,000 | 1,200 | 3.3 | 1,500 | | |
| T-33 (2) Canadair | | 42,000 | 1,200 | 3.0 | 140 | | |
| Convair Gen. Dynamics | 580 | 25,000 | 1,400 | 4.5 | 1,850 | | CCRS |
| Falcon-20C Dassault | | 42,000 | 1,200 | 3.2 | 950 | | Innotech Aviation Ltd. |
| Cessna | T310R | 27,500 | 950 | 5.0 | | | Geodesy |
| King Air Beechcraft | 200 | 35,000 | 1,500 | 6.0 | 1450 | | Kenn Borek Air Ltd. |
| Piper Chieftain | PA31-350 | 24,000 | 850 | 5.0 | 362 | | OMNR AFFMB |
| DC-3 | | 12,000 | 1000 | 8.0 | 1000 | | Innotech Aviation Ltd. |

**Table 226:   Performance Parameters of Canadian Research Aircraft[581],[582]**

---

581)  Source: Airborne Geoscience Newsletter, March 1991, p. 3
582)  Information provided by K Staenz of CCRS

| Aircraft/ Manufacturer | Model | Altitude ft | Range (nm) | Endurance (hr) | Payload mass (kg) | Payload power (kW) | Agency |
|---|---|---|---|---|---|---|---|
| ER-2 (2) Lockheed | | 70,000 (21 km) | 3,200 | 7.0 | 2,700 | 25 | NASA/ARC Moffett Field Ca. |
| C-130 Lockheed | NC-130B | 24,000 (7.3 km) | 2,000 | 7.0 | 20,000 | 26 | |
| DC-8 MACDAC | 72 | 40,000 (12 km) | 6,000 | 12.0 | 30,000 | 80 | |
| Electra Lockheed | L-188 | 25,000 (7.5 km) | 2,000 | 7.5 | 19,000 | 40 | NASA/WFF Wallops, Va. |
| P-3 Orion Lockheed | P-3B | 28,000 (8.5 km) | 4,150 | 12.5 | 13,900 | 33 | |
| Skyvan Short Brothers | SC-7 | 15,000 (4.5 km) | 650 | 4.0 | 5,000 | 2.8 | |
| Sabreliner Rockwell | T-39 | 41,000 (12.3 km) | 1,400 | 3.2 | 1,500 | 3.0 | |
| Helicopter Bell | UH-1B | 10,000 (3 km) | 450 | 4.5 | 2,000 | 3.5 | |
| Lear Jet Gates | 23-049 | 41,000 (12.3 km) | 1,000 | 3.0 | 750 | 4.0 | NASA/ SSC |
| WB-57F Gen. Dynamics | | 65,000 (19.8 km) | 2,500 | 7.0 | 4,000 | 20 | NASA/ JSC, Houston |
| Aero Com. Gulfstream | 500S | 25,000 (7.5 km) | 800 | 5.0 | 1,200 | | NOAA |
| Citation Cessna | 550 | 43,000 (13 km) | 1,900 | 3.5-4.5 | 2,000 | 5.0 | |
| Helicopter Bell | UH-1H (2) | 15,000 (4.5 km) | 500 | 5.5 | 1,400 | 1.0 | |
| Helicopter Bell | BH-12 (2) | 21,000 (6.4 km) | 500 | 5.5 | 2,500 | 1.0 | |
| King Air Beechcraft | C-90 | 25,000 (7.5 km) | 1,300 | 5.0 | 1,250 | 4.2 | |
| Turbo-Com Gulfstream | 690A | 30,000 (9 km) | | 4.0 | 1,000 | | |
| Orion (2) Lockheed | WP-3D | 25,000 (7.5 km) | 3,300 | 10.0 | 12,000 | 14.3 | |
| P3 Lockheed | RP3A | 25,000 (7.5 km) | 3,000 | 15.0 | 7,000 | 50 | NRL |
| P3 Lockheed | EP3A | 25,000 (7.5 km) | 3,000 | 15.0 | 7,000 | 50 | |
| P3 Lockheed | UP3A | 25,000 (7.5 km) | 3,000 | 15.0 | 7,000 | 50 | |
| P3 Lockheed | EP3B | 25,000 (7.5 km) | 3,000 | 15.0 | 13,000 | 50 | |
| Electra Lockheed | L-188C | 30,000 (9 km) | 2,600 | 8.0 | 23,000 | 70 | NCAR Boulder Co. |
| King Air Beechcraft | B200-T | 35,000 (10.6 km) | 1,900 | 7.0 | 1,700 | 10.6 | |
| Hercules Lockheed | C-130 | 32,000 (10 km) | 3,000 | 11 | 29,000 | | |
| WB-57F Gen. Dynamics | | 65,000 (19.8 km) | 2,500 | 7.0 | 4,000 | 20 | |
| NKC-135 Boeing | Tanker | 40,000 (12 km) | 4,200 | 10.0 | 11,000 | 8.8 | AGFL |
| Beech Baron Beechcraft | B58-TC | 25,000 (7.5 km) | 800 | 4.5 | 1,100 | 5.6 | USAF |
| Hercules Lockheed | C-130 | 24,000 (7.3 km) | | 8.0 | | | USCG |
| Citation Cessna | 550 | 43,000 (13 km) | 1,800 | 3.5-4.5 | 2,000 | 11.2 | U. of North Dakota |
| King Air Beechcraft | 200-T | 35,000 (10.6 km) | 1,800 | 5.5 | 2,600 | 11.7 | U. of Wyoming |
| Samaritan Convair | C-131A | 25,000 (7.5 km) | 1,200 | 7.5 | 9,000 | 20 | U. of Washington |

**Table 227:    Performance Parameters of US Research Aircraft[583]**

---

583)  Source: Airborne Geoscience Newsletter, April 1990

| Aircraft/ Manufacturer | Model | Altitude ft | Range (nm) | Endurance (hr) | Payload mass (kg) | Payload power (kW) | Agency |
|---|---|---|---|---|---|---|---|
| Caravelle Aerospatiale | | 38,000 | 1,700 | 4.5 | 13,200 | | CEV France |
| Falcon (2) Dassault | 20 | 42,000 | 1,200 | 3.3 | 1,500 | | IGN France |
| F 27 Fokker | | 25,000 | 1,000 | 4.5 | 3,150 | | |
| King Air (2) Beechcraft | 200 | 35,000 | 2,000 | | 1,200 | | |
| Aero-Com Rockwell | 600 | 27,500 | 1,500 | 8.0 | | | |
| Merlin Swearingen | | 31,000 | 1,500 | 5.5 | 1,370 | | CAM (Centre d'Aviation Météorologique) France |
| Aztek Piper | | 20,000 | 1,000 | 6.0 | 510 | | |
| Hercules Lockheed | C-130 | 32,000 | 3,000 | 11 | 29,000 | | MRF UK |
| Jetstream Scot. Aviation | | 26,000 | 1,200 | 5.0 | | | NPTEC (CERL) UK |
| Metro Swearingen | | 31,000 | 1,500 | 5.5 | 1,370 | | NLR Netherlands |
| Be 80 Beechcraft | | 26,800 | 1,300 | 8.0 | | | |
| Chieftain Piper | | 27,200 | 940 | 4.75 | | | Geosens Netherlands |
| C 406 Cessna | | | | | | | Rijkswaterstaat Netherlands |
| DO 228 Dornier | | 24,000 | 765 | 5.5 | 900 | | Belfotop Belgium |
| Falcon E Dassault | | 42,000 | 1,200 | 3.3 | 1,500 | 3.4 | DLR Germany |
| DO 228 Dornier | 212 | 27,000 | 1,350 | 8.0 | 1,850 | 8.4 | |
| DO 228 (2) Dornier | 101 | 27,000 | 1,600 | 9.0 | 2,000 | 6.3 | |
| C 207 A Cessna | | 13,300 | 600 | 5.0 | 450 | 1.7 | |
| CT 207 A Cessna | | 30,000 | 600 | 5.0 | 450 | | |
| ASK 16 (3) Schleicher | | 12,100 | 380 | 6.5 | 100 | | |
| BO 105 C MBB | Helicopter | 17,000 | 240 | 2.5 | 550 | | |
| HS 125 Hawker Siddley | | 41,000 | 2,000 | 6.0 | 900 | | EPC (Conti-Flug), Germany |
| DO 228 (2) Dornier | | 27,000 | 1,600 | 9.0 | 2,000 | | AWI Germany |
| DO 228 Dornier | | 27,000 | 1,600 | 9.0 | 2,000 | | Topogramm Germany |
| King Air Beechcraft | 200 | 35,000 | 2,000 | | 1,200 | | AZBS Germany |
| DO 128 Dornier | | 20,000 | 780 | 5.5 | | | U. of Braunschweig |
| DO 28 Dornier | | 24,000 | 765 | 5.5 | 900 | | |
| Victor Partenavia | | 20,000 | 850 | 5.0 | 500 | | U. of Berlin |
| C 207 Cessna | | 23,300 | 600 | 5.0 | 450 | | U. of Berlin WIB |
| Cherokee Piper | PA-31 | 16,200 | 700 | 5.0 | 400 | | Umweltdata Germany |
| D 500 Grob | | 50,000 | 1,600 | 16.0 | 1,000 | civil co-utilization | BMVg Germany |
| G 109 Grob | | 15,000 | 900 | 12.0 | 120 | | U. of Munich |

**Table 228:　Performance Parameters of European Research Aircraft**[584]

---

584) Courtesy of V. Harbers, DLR

# Part C    Reference Data and Definitions

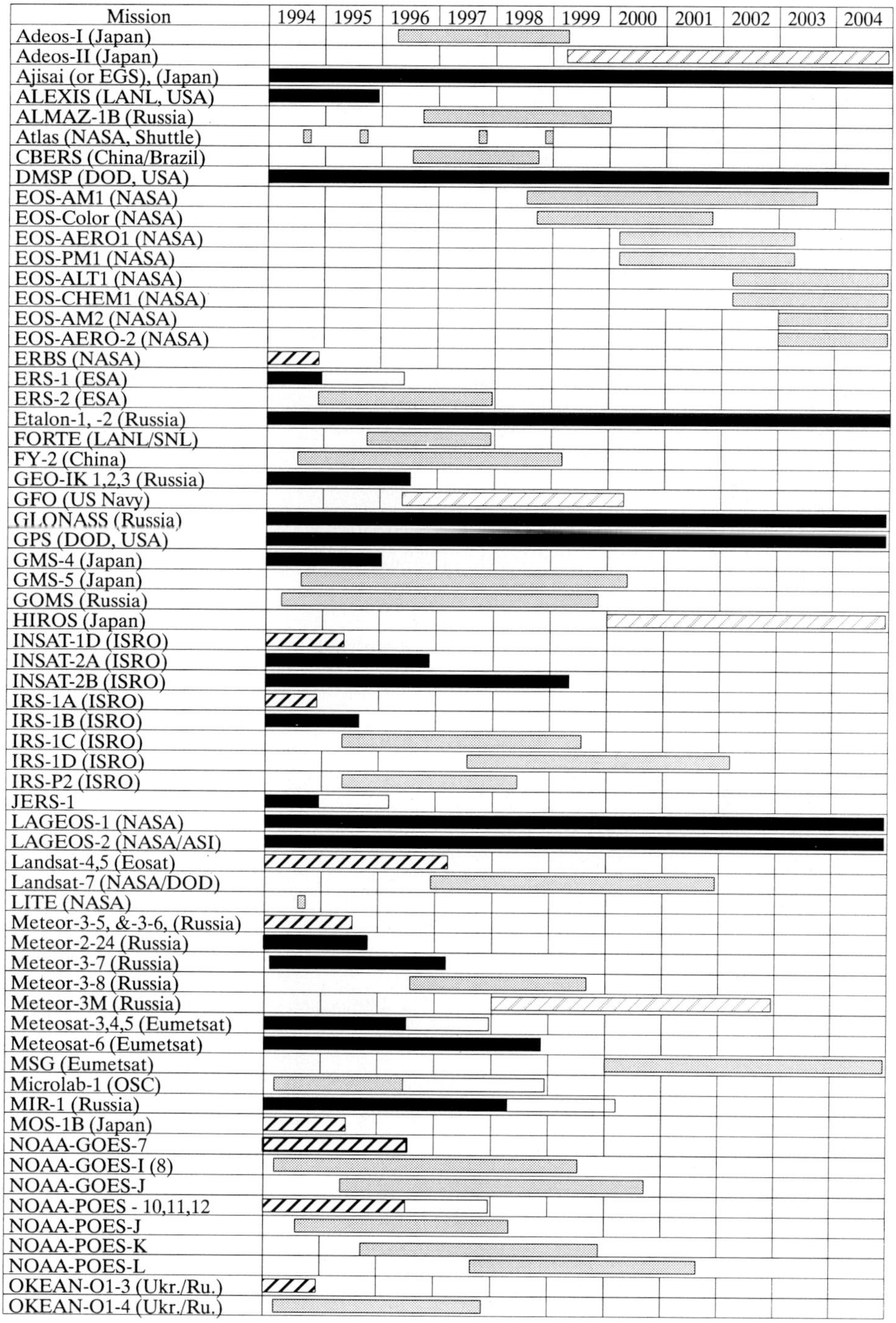

| Mission | 1994 | 1995 | 1996 | 1997 | 1998 | 1999 | 2000 | 2001 | 2002 | 2003 | 2004 |
|---|---|---|---|---|---|---|---|---|---|---|---|
| Adeos-I (Japan) | | | ▓ | ▓ | ▓ | ▓ | | | | | |
| Adeos-II (Japan) | | | | | | ▨ | ▨ | ▨ | ▨ | ▨ | ▨ |
| Ajisai (or EGS), (Japan) | █ | █ | █ | █ | █ | █ | █ | █ | █ | █ | █ |
| ALEXIS (LANL, USA) | █ | █ | | | | | | | | | |
| ALMAZ-1B (Russia) | | | ▓ | ▓ | ▓ | | | | | | |
| Atlas (NASA, Shuttle) | ▪ | ▪ | | ▪ | ▪ | | | | | | |
| CBERS (China/Brazil) | | | ▓ | ▓ | ▓ | | | | | | |
| DMSP (DOD, USA) | █ | █ | █ | █ | █ | █ | █ | █ | █ | █ | █ |
| EOS-AM1 (NASA) | | | | | ▓ | ▓ | ▓ | ▓ | ▓ | | |
| EOS-Color (NASA) | | | | | ▓ | ▓ | ▓ | ▓ | | | |
| EOS-AERO1 (NASA) | | | | | | | ▓ | ▓ | ▓ | ▓ | |
| EOS-PM1 (NASA) | | | | | | | ▓ | ▓ | ▓ | ▓ | |
| EOS-ALT1 (NASA) | | | | | | | | | ▓ | ▓ | ▓ |
| EOS-CHEM1 (NASA) | | | | | | | | | ▓ | ▓ | ▓ |
| EOS-AM2 (NASA) | | | | | | | | | | ▓ | ▓ |
| EOS-AERO-2 (NASA) | | | | | | | | | | ▓ | ▓ |
| ERBS (NASA) | ▨ | | | | | | | | | | |
| ERS-1 (ESA) | █ | | | | | | | | | | |
| ERS-2 (ESA) | | ▓ | ▓ | ▓ | | | | | | | |
| Etalon-1, -2 (Russia) | █ | █ | █ | █ | █ | █ | █ | █ | █ | █ | █ |
| FORTE (LANL/SNL) | | ▓ | ▓ | ▓ | | | | | | | |
| FY-2 (China) | ▓ | ▓ | ▓ | ▓ | ▓ | | | | | | |
| GEO-IK 1,2,3 (Russia) | █ | █ | █ | | | | | | | | |
| GFO (US Navy) | | | ▨ | ▨ | ▨ | ▨ | | | | | |
| GLONASS (Russia) | █ | █ | █ | █ | █ | █ | █ | █ | █ | █ | █ |
| GPS (DOD, USA) | █ | █ | █ | █ | █ | █ | █ | █ | █ | █ | █ |
| GMS-4 (Japan) | █ | █ | | | | | | | | | |
| GMS-5 (Japan) | ▓ | ▓ | ▓ | ▓ | ▓ | ▓ | | | | | |
| GOMS (Russia) | ▓ | ▓ | ▓ | ▓ | ▓ | ▓ | | | | | |
| HIROS (Japan) | | | | | | | ▨ | ▨ | ▨ | ▨ | ▨ |
| INSAT-1D (ISRO) | ▨ | | | | | | | | | | |
| INSAT-2A (ISRO) | █ | █ | █ | | | | | | | | |
| INSAT-2B (ISRO) | █ | █ | █ | █ | █ | █ | | | | | |
| IRS-1A (ISRO) | ▨ | | | | | | | | | | |
| IRS-1B (ISRO) | █ | █ | | | | | | | | | |
| IRS-1C (ISRO) | | ▓ | ▓ | ▓ | ▓ | ▓ | | | | | |
| IRS-1D (ISRO) | | | | ▓ | ▓ | ▓ | ▓ | ▓ | | | |
| IRS-P2 (ISRO) | | ▓ | ▓ | ▓ | ▓ | | | | | | |
| JERS-1 | █ | █ | | | | | | | | | |
| LAGEOS-1 (NASA) | █ | █ | █ | █ | █ | █ | █ | █ | █ | █ | █ |
| LAGEOS-2 (NASA/ASI) | █ | █ | █ | █ | █ | █ | █ | █ | █ | █ | █ |
| Landsat-4,5 (Eosat) | ▨ | ▨ | ▨ | ▨ | | | | | | | |
| Landsat-7 (NASA/DOD) | | | ▓ | ▓ | ▓ | ▓ | ▓ | ▓ | | | |
| LITE (NASA) | ▪ | | | | | | | | | | |
| Meteor-3-5, &-3-6, (Russia) | ▨ | | | | | | | | | | |
| Meteor-2-24 (Russia) | █ | █ | | | | | | | | | |
| Meteor-3-7 (Russia) | █ | █ | █ | █ | | | | | | | |
| Meteor-3-8 (Russia) | | | ▓ | ▓ | ▓ | ▓ | | | | | |
| Meteor-3M (Russia) | | | | | ▨ | ▨ | ▨ | ▨ | ▨ | | |
| Meteosat-3,4,5 (Eumetsat) | █ | █ | █ | | | | | | | | |
| Meteosat-6 (Eumetsat) | █ | █ | █ | █ | █ | | | | | | |
| MSG (Eumetsat) | | | | | | | ▓ | ▓ | ▓ | ▓ | ▓ |
| Microlab-1 (OSC) | ▓ | ▓ | ▓ | ▓ | | | | | | | |
| MIR-1 (Russia) | █ | █ | █ | █ | █ | | | | | | |
| MOS-1B (Japan) | ▨ | ▨ | | | | | | | | | |
| NOAA-GOES-7 | ▨ | ▨ | | | | | | | | | |
| NOAA-GOES-I (8) | ▓ | ▓ | ▓ | ▓ | ▓ | ▓ | | | | | |
| NOAA-GOES-J | | ▓ | ▓ | ▓ | ▓ | ▓ | ▓ | | | | |
| NOAA-POES - 10,11,12 | ▨ | ▨ | ▨ | | | | | | | | |
| NOAA-POES-J | ▓ | ▓ | ▓ | ▓ | ▓ | ▓ | | | | | |
| NOAA-POES-K | | ▓ | ▓ | ▓ | ▓ | ▓ | ▓ | ▓ | | | |
| NOAA-POES-L | | | | | ▓ | ▓ | ▓ | ▓ | | | |
| OKEAN-O1-3 (Ukr./Ru.) | ▨ | | | | | | | | | | |
| OKEAN-O1-4 (Ukr./Ru.) | ▓ | ▓ | ▓ | ▓ | | | | | | | |

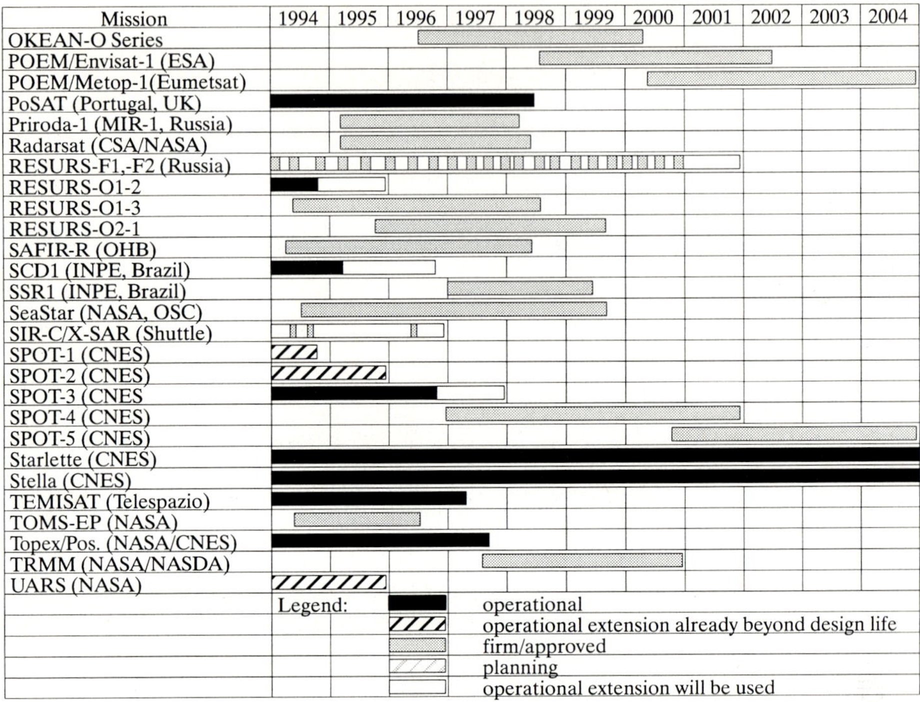

**Table 229:** **Operational and Planned Earth Observation Satellite Programs**

| Mission/Satellite | Orbit | Applications | Sensor Complement |
|---|---|---|---|
| ADEOS (NASDA, Japan), L:1996 | SS-polar, A: 800 km, | Global: Land, Sea, Atmosphere | OCTS, AVNIR, NSCAT, TOMS, POLDER, IMG, ILAS,RIS |
| Ajisai ( or EGS), (Japan), L:12.8.1986 | Circular, A:1500 km, I:50° | Laser ranging, Geodynamics | passive satellite |
| Alexis (LANL), L:25.4.1993 | A:844 km, P:749 km I:70° | Astrophysics, lightning detect. | Alexis, Blackbeard |
| ALMAZ-1 (Russia) L: 31.3.91, End: 17.10.92 | Polar, A:300-360 km, I: 72.7° | Oceanography, Geology, Geophysics | SAR (60 Scenes/Day) |
| ALMAZ-1B, L: end of 1996 | Polar, A:400 km, I: 73° | Oceanography, Geology, Geophysics | SLR-3, SAR-3, SAR-10, SAR70, MSU-SK, MSU-E, SROM, OSSI, Balkan-2 |
| ATLAS, NASA (Shuttle) | -1: 24.3.-2.4.92 -2: 8.4.-16.4.1993 -3: Aug. 1994 -4: Aug. 1995 | Earth remote sensing | ATMOS, MAS, SUSIM, SOL-CON, ACR, SOLSPEC, SSBUV, ALAE, GRILLE, ISO, AEPI, SE-PAC, FAUST, CRISTA, |
| CBERS (China/Brazil ERS), L:1995/6 | SS-polar, A:778 km, I:98.5° | Resource | WFI, CCD, IR-MSS |
| DMSP, DOD (Data from NOAA), ops | SS-polar, A:850 km, I:99° | Oceanography, Climatology | OLS, SSM/I |
| EOS AM-1 (NASA) L: 1998 | SS-polar, A: 705 km, | Energy cycle, Earth surface, Aerosols | CERES, MODIS, MISR, AS-TER, MOPITT |
| EOS Color-1 (Seastar-2) L: 2.. half 1998 | SS-polar, A: 700 km | Phytoplankton, Chlorophyll, Biomass | SeaWiFS II |
| EOS Aerosol-1,L: 2000 | I: 57° | Aerosols in atmosphere | SAGE III, ESOP, |

| Mission/Satellite | Orbit | Applications | Sensor Complement |
|---|---|---|---|
| EOS PM-1 (NASA) L:2000 | SS-polar, A: 705 km | Energy cycle, Snow/Ice, Temperature | CERES, MODIS-N, AIRS, AMSU-A, MHS, MIMR |
| EOS ALT-1 (NASA), L:2002, | | Ocean circulation | ALT, GGI, GLRS-A |
| EOS CHEM-1 (NASA) L:2002 | SS-polar | Atmospheric. Chemist. | HiRDLS, TES, SAGE III, STIK-SCAT |
| ERBS (NASA) L:5.10.84, | A: 610, I: 57°, | Energy cycle | ERBE, SAGE II, |
| ERS-1 (ESA) L: 17.7.91, ops | SS-polar, A:785 km, I: 98.5° | Sea, Land, Atmosphere | AMI-Wave, AMI-SCAT, RA-1, ATSR, LRR, |
| ERS-2 (ESA) L: end of 1994 | SS-polar, A:824 km | Sea, Land, Atmosphere | AMI, AATSR, RA, PRARE, LRR, GOME |
| Etalon-1, -2 (Russia), L:1.10.1989, L:31.5.1989 | Circular, I:64.8° A:19130 km | Laser Ranging, Geodynamics | passive satellite |
| Fengyun-2 (China), L:1994 | GEO, 105° E | Meteorology, climate | Scanning Radiometer |
| FORTE (LANL) L. 11/1995 | A:800 km, I:65° | Ionosphere | RF System, Optical System |
| GEO-IK (Russia), L:30.5.88 | Circular, A:1500 km, I:74° | Laser Ranging, Geodynamics | passive satellite |
| GOMS (Russia), L: 4/1994 | GEO, 76° East | Clouds,Land surface temp. prof. | STR, RMS, |
| GMS (JMA, Japan) L: 1989 | GEO, 140° East | Meteorology | VISSR, SEM, DCS |
| GPS/NAVSTAR (DOD, USA) | | actual Position- & velocity vector | |
| GLONASS (Russia) | | actual Position- & velocity vector | |
| INSAT-2A (ISRO),L:9.7.92 INSAT-2B (ISRO), L:22.7.93 | GEO, 74° East | Meteorology, Communication | VHRR, DCS |
| IRS-1A (ISRO), India L:17.3.88 | SS-polar, A:904 km, I: 99.5°, | Land surface | LISS-1, LISS-2 |
| IRS-1B (ISRO), India L:29.8.91 | SS-polar, A:900 km, I:99.5° | Land surface | LISS-1, LISS-2 |
| IRS-1C (ISRO), India L: Begin 1994 (Vostok) | SS-polar, A:900 km | Land surface | PAN, LISS-3, WiFS |
| IRS-P2 (ISRO), India, L: 1995 | SS-polar,A:817km | Land ocean surface, atmosphere | LISS-II, MOS (DLR) |
| JERS-1 (NASDA), Japan L: 11.2.92 | SS-polar, A:568 km, I:97.7° | Geology, land use | SAR, OPS |
| LAGEOS-1 (NASA), L:4.5.79 | Circular,A: 5950 km, I:110° | Laser Ranging, Earth Crust, Geodynamics | passive Satellite |
| LAGEOS-2 (ASI,NASA), L:22.10.92 | Circular, A:5950 km, I:52° | Laser Ranging, Earth Crust, Geodynamics | passive Satellite |
| Landsat-5 (Eosat) L:1.3.84 | SS-polar, A:705 km, I:98.2° | Land surface | MSS, TM,RBV, |
| Meteor-3-5, 3-6 and Meteor 2-24 in Orbit (Planeta, Russia), | polar, A:1200 km, I:82.5 | Meteorology, Climatology | MR-2000, MR-900B, Klimat, SM, RMK-2,  IR, TOMS |
| Meteor-3-7 (Russia), L: 25.1.1994 | polar, A:1201 km, I:82.5 | Meteorology, Climatology | MR-2000, MR-900B, Klimat, SM, RMK-2, SCARAB, PRARE |
| Meteosat-5 and 6 (EUMET-SAT) | GEO | Met., Climatology | VISSR (MSR), DCS |
| MIR-1 (Russia, NPO Energia) L:20.2.86 | A: 350-410 km, I:51.6° | Land surface | Photo-Complex, TV-Complex |
| MOMS-02 on SL D-2 (Germany), L: April 26, 93 | A: 296 km, I:28.5° | Stereographic Earth imaging | MOMS-02 |
| MOS-1B (NASDA, Japan), L:7.2.90 | SS-polar, A:908 km, I:99.1° | Sea surface(Color), Vegetation | MESSR, VTIR, MSR |
| NOAA-GOES-7 (NOAA) L:26.2.87 | GEO | Meteorology, Climatology | VISSR (VAS), SEM, DCS, |
| NOAA-11 (POES, NOAA) L:24.9.88 | SS-polar, A:800 km, I:98° | Met., Climate, Atmosphere, Energy | AVHRR, HIRS-2,SSU, MSU, SEM, SBUV, ERBE, S&R |

| Mission/Satellite | Orbit | Applications | Sensor Complement |
|---|---|---|---|
| NOAA-12 (POES, NOAA) L:14.5.91 | SS-polar, A:800 km, I:98° | Meteorology, Climate, Atmos. | AVHRR, HIRS, MSU, SEM |
| Okean-O1-3 (Russia), L: June 91, End: Jan. 92 Okean-O1-4 (Russia),L: 2/94 | SS-polar, A:660 km, I:82.5° | Land surface | RLSBO, RM-08, MSU-M, MSU-SK, MSU-S |
| POEM ENVISAT-1 (ESA), L:1998 | SS-polar | Atmosphere, Environment | MERIS, MIPAS, RA-2, MWR, ASAR, GOMOS, SCIAMACHY, AATSR, DORIS, SCARAB, VIRSR, MTS, MHS, IASI,ARGOS, ASCAT, MIMR, S&R, SEM |
| METOP-1 (EUMETSAT), L:2000 | SS-polar | Meteorology, Climate | |
| PoSAT (Portugal, UK), L: Sept. 25, 1993 | SS polar, A:820 km, I:98° | Land surface images | EIS, Star Sensor, CRE, TDE, DSPE, GPS |
| PRIRODA-1 MIR-Module (Russia) L: early 95 | A: 350-410 km, I:51.6° | Land surface, Atmosphere, Sea surface | MW Sensors, Travers (SAR), Greben (R. Alt), Optic Sensors, MOS, MOMS-02 |
| RADARSAT (CSA, Canada), L:Jan. 1995 | SS-polar, A:800 km, I:98.6° | Arctic Ice Topography | SAR |
| RESURS-F (F-1, F-2), Russia | polar, A:240-275 km, I. 82.3° | Land surface images | KFA-200, KFA-1000, MK-4 |
| RESURS-O1-2, Russia L:20.4.88 | SS-polar, A:660 km, I:98°, | Resources | 2 x MSU-E, 2 x MSU-SK, Travers (SAR) |
| RESURS-O1-3, Russia L: 2/1994 | SS-polar, A:660 km, I:98°, | Resources | 2 x MSU-E, 2 x MSU-SK, Travers (SAR) |
| RESURS-O2 Series, Russia, L: 1995 | SS-polar, A:660 km, I:98° | Resources | 2 x MSU-E, 2 x MSU-SK, SAR |
| SAFIR-R (OHB) L: 2/1994 on Resurs-O1-3 | SS polar, A:660 km, I:98° | Data Collection | |
| SCD1 (INPE, Brazil) L: 2/1993 | A:750 km, I:25° | Data Collection | |
| SSR1 (INPE, Brazil) L:1997 | SS, A:640 km | Land surface imaging | WFI |
| SeaStar (NASA, OSC), L: July 94 | SS-polar, A:705 km, I:98.2° | Ocean color, Biology, Phytoplankton | SeaWiFS (commercial and science data use) |
| SIR-C/X-SAR (Shuttle) L1:4/94; L2:8/1994; L3:1996 | A:225 km, I:57° | Land, Ocean, Geology, Topography | L/C-Band SAR (JPL), X-SAR (NASA/JPL, DARA/DLR, ASI) |
| SPOT-2 (CNES), L:22.1.90 | SS-polar, A:832 km, I:98.7° | Land, Ocean, Geology, Topogr. | HRV(MS), HRV(PAN), DORIS |
| SPOT-3 (CNES), L:26.9.93 | SS-polar, A:832 km, I:98.7° | Land, Ocean, Geology, Topogr. | HRV(MS), HRV(PAN), DORIS, POAM-II |
| SPOT-4 (CNES), L:1997 | SS-polar, A:832 km, I:98.7° | Land, Ocean, Geology, Top. | HRVIR(MS), HRVIR(PAN), VEGETATION, PASTEL |
| Starlette (CNES) L:2.6.1975 | A:1100 km, I:50° | Laser ranging | passive satellite |
| Stella (CNES), L: 25.9.1993 | A: 800 km, I:98° | Laser ranging | passive satellite |
| TEMISAT (Telespazio, Italy) | polar, A. 950 km, I. 82.5° | Data collection | |
| TOMS-EP (NASA), L:7/94 | SS-polar, A:955 km, I:99.3° | Atmosphere, Ozone | TOMS |
| TOPEX/POSEIDON (NASA/CNES), L:10.8.92 | Circular, A:1330 km, I:63.5° | Altimetry Mission | ALT (NASA), TMR, GPSDR, LRA, SSALT & DORIS (CNES) |
| TRMM (NASDA/NASA) L: 8/1997 | Circular, A:370 km, I:35° | Atmospheric Circulation, Rain | PR (NASDA), VIRS (NASA), TMI (NASA), CERES (NASA), LIS (NASA) |
| UARS (NASA), L:13.9.91 | A:600 km, I:57° | Energy Cycle, Atmospheric Chemistry | CLAES, ISAMS, HALOE,MLS, SOLSTICE, SUSIM, PEM, HRDI, WINDII, ACRIM2 |

Legend: L = Launch; A = Altitude; SS-polar = Sun-synchronous-polar; I = Inclination
Shaded fields indicate operational missions.

**Table 230:    Survey of EO Missions and Sensors**

# C.1     Definitions, Concepts, Summaries

### Remote Sensing

Remote sensing, in the broadest sense, is the measurement or acquisition of information of some property of an object or phenomenon, that is not in physical contact with the object or phenomenon under study. Remote sensing instruments, such as optical imagers or radar sensors, acquire information about an object of interest by detecting and measuring the changes that the object imposes on the surrounding electromagnetic field. For instance, in the case of visible and near-infrared (NIR) sensors the reflected sunlight is used to acquire information about the chemical composition and physical structure of the object being observed. In the case of thermal infrared and passive microwave sensors, the emitted field is used to acquire information about the thermal properties as well as the composition of the object. In the active microwave region, information about the object's physical structure and electrical properties is acquired by analyzing the reflected field when the sensor illuminates the object with a well-defined generated field of electromagnetic waves.

## C.1.1     Remote Sensing across the Electromagnetic Spectrum

Electromagnetic waves interact with matter by a variety of mechanisms which are related to the composition and structure of the object. These mechanisms modify the waves characteristics (spectral, content, polarization intensity, direction, etc.) in such a way that the object properties can be identified. The emission or absorption spectral lines of a gas in the infrared spectrum will uniquely identify the nature and concentration of its constituents as well as its temperature. The reflectance spectral lines of a solid in the visible and near-infrared could allow the identification of its chemical composition, and, to some extent, its crystalline structure.

The depolarization and spectral reflectivity of the radar echo from a surface provides information about the roughness structure, geometric structure, morphology, and dielectric constant of the surface and immediate subsurface. The time of arrival and spectral contents of the radar echo allow to locate the position of the reflecting object and its speed to extremely high accuracy (few cm in location and few cm/s in speed).

The intensity of the black body radiation field in the thermal and microwave regions allows the derivation of the thermal properties of the object.

## C.1.2     Types and Classes of Remote Sensors and Sensing Data

The survey in Part A represents a great variety of sensor types that need an ordering scheme for a better understanding and overview by the reader. There are many ways of sensor classifications (some of them are trivial or simply problem specific), for instance by:

- the type of application (field of research or operation such as: meteorology, atmospheric research, solid Earth, climatology, biosphere, etc.)
- the type of primary observation targets (oceans, land, coastal regions, ice sheets, or atmosphere.
- the observed frequency range (UV, optical, infrared, microwave, etc.),
- the type of sensing mechanism (passive or active source),
- the type of instrument (imager, altimeter, sounder, spectrometer, radiometer, etc.)
- the type of measurement precision (resolutions, )
- etc.

Another classification scheme can be derived from the type of information being sought (Ch. Elachi, see Ref. 585)). If we consider for instance the three fundamental information parameters as being: **spatial**, **spectral**, and **intensity** information, each located diametrically at the extremities of a triangle, as illustrated in Figure 143, then a number of basic relationships and commonalities are self-evident or may be directly extracted.

The primary sensor types are allocated in the triangle to the primary information parameters. The measurement of two prime parameters result in an instrument type (like an imaging spectrometer) which is located somewhere between the two extremities and is also designated accordingly. There are already attempts to combine all three fundamental information parameters into a single instrument (such as MISR (in EOS)= Multi-Angle Imaging Spectro-Radiometer). Not everything can be characterized and explained by this method, but the scheme illustrates the fundamental principles involved, even beyond the fancy naming conventions of some sensors.

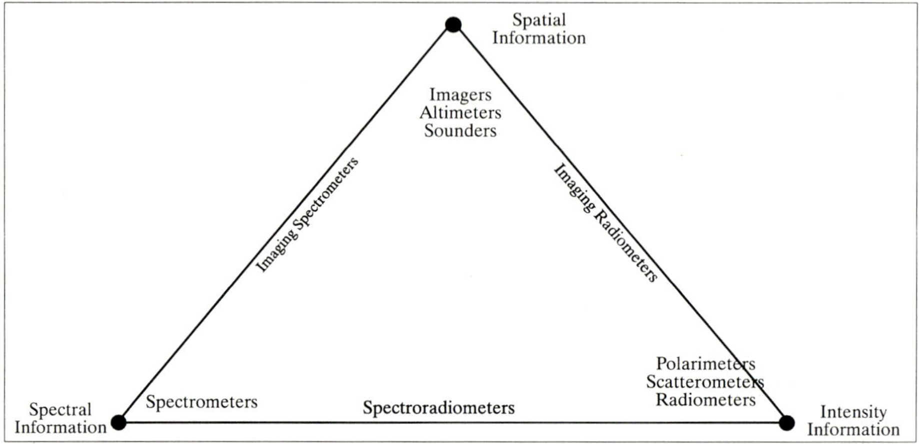


**Figure 143:    Overview of Sensor Class Relationships**[585)]

## 1.    Spatial Information

Spatial information throughout the spectrum (2-D or 3-D) is acquired with an imager (this may be a camera, a mapper, a scanner, etc. or combinations thereof). The major application is in topographic mapping.

a.    Optical imaging techniques[586)] (optical systems are limited by cloud cover, etc.)
-    Stereophotogrammetry - photographic methods have the advantage of very good geometric resolutions, the disadvantages are in the handling of the film material, the data require geometric corrections, detection of reference points and conversion of brightness into digital information
-    Electromechanical line scanner - 2-D images are acquired line by line by scanning a mirror across track. Where the swaths of two adjacent subsatellite tracks overlap, stereo pairs are obtained. For polar orbiting satellites overlappings are normally small near the equatorial regions and large over the polar regions. Examples: MSS and TM (of Landsat)
-    Electronic line scanner - in this method an array of CCDs (Charge-Coupled Devices) is used instead of a single detector and scanning mirror. The advantages are: absence of moving elements, longer integration times per pixel are possible, and better geometric performance. The disadvantages: the number of detectors is limited which limits the resolution. Examples: HRV (Spot), MSU-E (CIS), etc.
-    Stereo line scanner - Three lines of semiconductors are arranged in the focal plane of the pushbroom scanner perpendicular to the direction of flight. The 3 lines scan


585)   Charles Elachi, "Introduction to the Physics and Techniques of Remote Sensing", John Wiley & Sons, 1987, Chapters 1-1 and 8-8.

586)   see chapter 5 in 'The Interdisciplinary Role of Space Geodesy', Springer Verlag, 1987, pp. 164-165

simultaneously in forward-, down-, and aft-direction. Three images of the same scene are acquired from the same swath. The stereo images are taken simultaneously. Example: MOMS-02.

b.   Sounding techniques - Sounders are used for field measurements to map subsurface structures, or to map atmospheric profiles of temperature, composition and pressure as a function of altitude. The nadir sounders scan across the satellite suborbital track to provide good horizontal resolution (in the suborbital plane), but they usually have poor vertical resolution. Limb sounders look at the horizon (the limb) and scan vertically, producing good vertical resolution but poor horizontal resolution (see Figure 146).

The effect of the electromagnetic wave interactions on the Earth's atmosphere are relatively complex due to the multiplicity of mechanisms: scattering, absorption, emission and refraction. The atmospheric sounding technique is based on the fact that carbon dioxide ($CO_2$) is the same percentage ratio of the total composition of the atmosphere at all altitudes - this is called the constant mixing ratio. The atmosphere has an emissivity that varies with wavelength from about 0% at 10 µm to about 100% at 16 µm. An infrared sounder with about 5 channels in the region from 11-16 µm probes the radiation of the atmosphere to different depths. Since the emissivity is high at 16 µm, only the top of the atmosphere is sensed by the 16 µm channel. The 11 µm channel, however, senses the radiation from all altitudes. The other channels are somewhere in between.

All sounders examine either the **emission** or **backscattered** radiation spectrum within a narrow field of view.

In the case of emission, the radiation source is the atmosphere itself. The sensor measures the spectral characteristics and intensity of the emitted radiation. If atmospheric parameters are measured a down-looking geometry (nadir) is used for the spacecraft sounder.

In the scattering technique, the approach is to measure the characteristics of the scattered wave field in the direction (or directions) away from the incident wave direction. The source can be the sun or a star.

Some sounding techniques/applications are:
-   Temperature sounding - measurement of radiance variation as a function of frequency near a spectral line. For instance, the temperature is measured by examining a small part of the $CO_2$ spectral band at 15 µm or microwave emission from $O_2$.
-   Composition sounding - identification of atmospheric constituents (by spectral techniques)
-   Pressure sounding - technique: columnar absorption measurements of a homogeneously mixed gas constituent
-   Density sounding - technique: refraction profile measurements
-   Sounding of atmospheric motion (wind fields) - technique: measurement of the spectral line shift by the Doppler effect.

In the occultation techniques, the approach is to measure the changes that the atmosphere impinges on a signal of known characteristics as the signal propagates through the atmosphere. The signal source can be the sun, the moon, a star, or a man-made source. Usually a limb-looking geometry is used to sound the upper atmosphere by occultation or by emission.

Another sounding technique is called LIDAR (see chapter C.1.10.3). This is an active technique, similar to radar. A laser pulse is generated and transmitted outward. The light is scattered back from the atmosphere. A time-gated detector senses the return. Since the sensor is not active immediately after the generation of the pulse, the detector is not 'blinded' by the backscatter from the mirror. In fact, since it is only active for a very brief period of time (the pulse length), it senses the return from only a small portion of the atmosphere at a specific range. Light travels about 0.33 m in 1 ns; thus, segments of the atmosphere in the order of 1 m in depth can be probed this way. The use of several wavelengths permits a limited amount of constituent identification.[587]

c.   Altimetry

In a number of applications, the information required is strongly related to accurate distance measurements or to the 3-D spatial characteristics of a surface of an object. This is the wide area of topographic measurements with the use of altimeters.

-   Radar altimeters (non-imaging radar sensors) use the ranging capability of radar sensors to measure the surface topographic profile (parallel to the subsatellite track). There are beam-limited altimeter types as well as pulse-limited altimeters.
-   Scanning Laser Altimetry (see lidar ranging techniques under C.1.10.3). In the across-track scanning technique the 3-D topography can be acquired by a series of neighboring profiles. The beam scanning can be achieved electronically by using a phased array antenna or mechanically by scanning the antenna or by spinning the spacecraft.
-   All spaceborne radar altimeters to date have been wide-beam systems limited in accuracy by their pulse duration[588]. Such altimeters are useful for smooth surfaces (oceans), but are ineffective over relatively high relief continental terrain. - A fundamental problem in narrow-beam spaceborne radar altimetry is the physical constraint of antenna size. Large antennas are required for small radar footprints.

d.   SAR (Synthetic Aperture Radar) Mapping
SARs generate microwave images of a surface. By using the overlapping images from adjacent tracks, one can again produce stereo images. However, observation of the two overlapping tracks must occur from the same side. SARs have the disadvantages of high data reduction cost and of poor vertical resolutions.

e.   Radar Interferometry
The radar interferometer applies to two radar receiver units and a baseline across track. The phase differences as a function of location are used for direct measurement of the elevation of the backscattering surface.


**2.   Spectral Information**

Spectral information is acquired with a **spectrometer** (examples: Fourier Transform Spectrometers, grating instruments, Fabry-Perot instruments, or gas correlation spectrometers). Spectrometers are used to detect, measure and analyze the spectral content of the incident electromagnetic field. This information plays a key role in identifying the chemical composition of the object being sensed (for instance atmospheric constituents). In case of atmospheric studies, the spatial aspect is less critical than the spectral aspect due to slow spectral variation in the chemical composition. In the case of surface studies, both spatial and spectral information are essential, leading to the need of **imaging spectrometers** (see chapter C.1.8).

The selection of the number of spectral bands, the bandwidth of each band, the imaging spatial resolution, and the instantaneous field of view leads to trade-offs based on the object being sensed, the sensor data-handling capability, and the detector technological capabilities.

587)  Encyclopedia of Physical Science and Technology, Academic Press, 1987
588)  NASA 'Topographic Science Report', 1988, p. 46

### 3.   Intensity Information

In a number of applications the information needed is mainly contained in the accurate measurement of the intensity of the spectrum over a wide region.

- Radiometers - measurement of radiative intensities (flux, polarization, thermal properties, temperature profiles of the atmosphere and of sea surfaces, etc. ). Imaging radiometers are used to spatially map the variation of these parameters.

- Scatterometers - are used in active remote sensing to accurately measure the backscattered field when the surface is illuminated by a signal with a narrow spectral bandwidth. Measurement of surface roughness.

- Polarimeters - these are a special type of radiometer. They are used for applications in which the key radiative information is embedded in the polarization state of the transmitted, reflected, or scattered wave. The polarization characteristic of reflected or scattered sunlight provides information about the physical properties of planetary atmospheres.

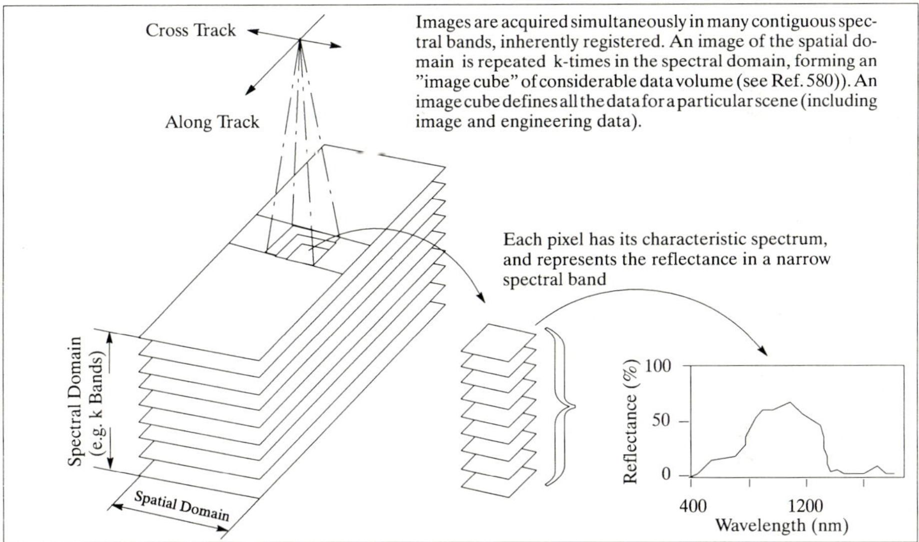


**Figure 144:   The Imaging Spectrometry Concept**

## C.1.3   Some Aspects of Radiometric Instrument Calibration

In remote sensing, radiometry or spectroscopy is the measurement of radiation that is incident upon a sensor's receiver/detector after having passed from a target source through an intervening medium to the sensor. The objective of such a measurement is to characterize the source, in terms of its size, shape, location, composition, temperature, and maybe other parameters of interest. Target attributes can be inferred from the instrument response based upon adequate calibration.[589]

For the limited case of noncoherent and noninterfering radiation, the target can be characterized in terms of three nearly independent measurement domains:

- spatial (geometric extent of measurement)

---

589)  C. L. Wyatt, "Radiometry", in Encyclopedia of Physical Science and Technology, Vol. 11, 1987, pp. 738-749

- spectral (distribution of energy flux as a function of wavelength or frequency)
- polarization (orientation of the electromagnetic vector)

The sensor calibration must reflect the sensor response in these measurement domains. **The calibration objective is to determine a functional relationship between the target source flux and the sensor output signal.** This relationship is also referred to as sensor "responsivity". Inherent in this sensor responsivity is the complete characterization of the instrument (such as IFOV, spectral bandpass , time constant, polarization, the sensor subsystems, the signal conditioning, the throughput of the instrument, etc.).

The quality of the measurement is the most difficult problem of calibration. Optically pure measurements are only approximated with practical instruments, which are always non-ideal. Consequently, the interpretation of field data is subject to some uncertainty and is often dependent upon assumptions that must be made about the target. Some of the measurement problems/objectives in the measurement domains are as follows:

- The sensor is bombarded by unwanted flux that originates from outside the instrument field of view (such as radiation contributions from the Earth's atmosphere, or from the sun, or from other background sources). Objective: The sensor output for a 'spatially pure measurement' is a function of the radiant flux originating from the target within the sensor IFOV and is completely independent of any flux arriving from outside this region (see Figure 145 for a general scenario).
- The sensor is also bombarded with unwanted flux that is out-of-band, that is, outside of the spectral region of interest. Objective: The sensor output for a 'spectrally pure measurement' is a function of the radiant flux originating from within the sensor spectral bandpass and is completely independent of any flux arriving at the sensor from any spectral region outside the bandpass.
- The flux originating from the target may be polarized. The sensor may be sensitive to the polarization of incident flux. Objective: The polarization characteristics may be considered as target attributes for discrimination against the background, or they may be related physical properties of the target.

A major (ideal) objective of the calibration of a sensor type (radiometer, etc.) is to make the measurements independent of the instrument. This means that when a particular physical entity is to be measured at different times and places or with different instruments, the results should always be the same. It also means that attributes inferred by the measurements are target attributes and not instrument attributes.

Unfortunately, the complexity of the physical conditions of the instrument and source parameters existing at the time of the measurement make this ideal of calibration somewhat difficult, if not impossible, to obtain. A basic rule of good performance is that the calibration should be conducted under conditions that **reproduce**, as completely as possible, the conditions under which the measurements are to be made.

A common practice for spectral-response comparison is the use of a calibration source. Obviously, a sensor in orbit cannot be calibrated as extensively as the same sensor in the laboratory, but high-quality sensors have generally built-in standards (along with well-defined calibration procedures), which permit a periodic comparison at least for the primary measured values. Depending on the quantity to be calibrated, such a 'standard' may turn out to be an external source of known reference (star, moon, sun, etc.) or an internal source, a blackbody or a calibration lamp (e.g., a high-temperature gray-body source).

Often temperature variations around the orbit may affect the spectral stability of the instrument. In such cases, spectral calibration needs to be repeated from time to time by either looking onto a built-in source with well known spectral features or by evaluating well known features in the spectrum itself. Calibration sources can be small semiconductor laser-diodes or atomic line-lamps such as low-pressure Hg or PtCrNe lamps. Optical instruments working in the visible or ultraviolet range of the spectrum often use the well known Fraunhofer-structures in the solar spectrum for spectral calibration.

If absolute radiometric accuracy is important then spaceborne measurements may also be checked against underflying airborne measurements that are made with the same sensor type.

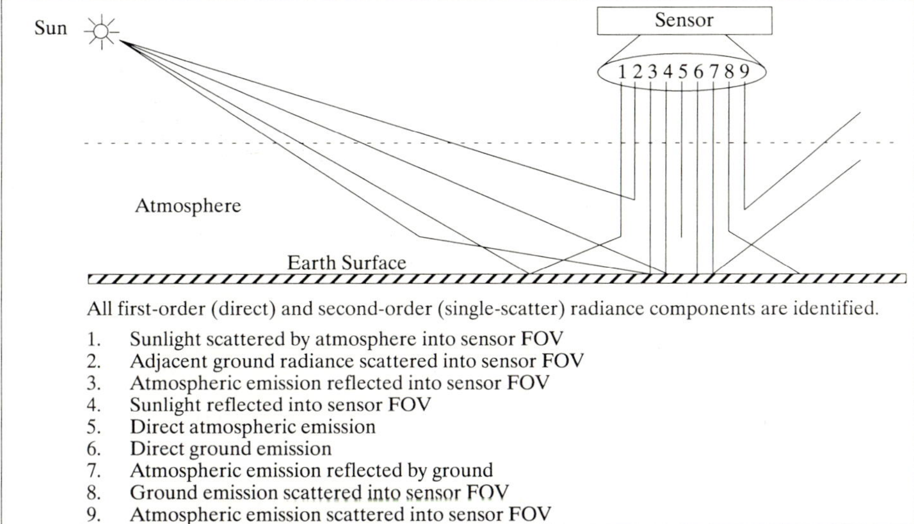


All first-order (direct) and second-order (single-scatter) radiance components are identified.
1. Sunlight scattered by atmosphere into sensor FOV
2. Adjacent ground radiance scattered into sensor FOV
3. Atmospheric emission reflected into sensor FOV
4. Sunlight reflected into sensor FOV
5. Direct atmospheric emission
6. Direct ground emission
7. Atmospheric emission reflected by ground
8. Ground emission scattered into sensor FOV
9. Atmospheric emission scattered into sensor FOV

**Figure 145: Generalized Scenario of Radiative Contributions in Remote Sensing**[590)]

## C.1.4   Correction/Calibration methods on sensor data

Once the functional relationships between the target source flux and the sensor output signals are known then this knowledge can be applied for data correction in the ground processing algorithms.

- **Atmospheric correction**. Applied to sensor data in order to separate and eliminate the atmospheric path radiance contribution from the ground-reflected signal (e. g., surface contribution also referred to as ground radiance). Path radiance is energy, that may originate from the surface outside the 'field of view', it is scattered into the beam by the atmosphere. In the visible region and at shorter wavelengths, Rayleigh scattering, caused by density fluctuations in atmospheric gases, is the dominant contributor to path radiance. Light scattered by particles or aerosols creates both path radiance as well as extinction.

  The algorithms employed are based on model assumptions on the atmospheric state and supplementary measurements. As of 1993 only approximate methods are available. Much effort is given to the development of strict algorithms. The application of atmospheric correction (the atmospheric path radiance contribution of the signal is being eliminated by 'masking') to sensor data permits the experienced user to compare different data sets of the same sensor (and/or of different sensors), measured at different times and different locations (mosaicking).

- **Geometric correction**. The removal of sensor, platform, or scene-induced geometric measurement errors in such a way, that the data conform to a required projection. This

590) Courtesy of J. M. Palmer, University of Arizona, Tucson; published in: J. M. Palmer, "Calibration of satellite sensors in the thermal infrared", 108 / SPIE Vol. 1762 Infrared Technology XVIII, 1992

process involves the creation of a new digital image by resampling the input digital data. - Geometric correction is performed by giving each pixel of the image the correct place on Earth. This process is known as **geocoding**, i.e. matching pixel coordinates with the proper Earth coordinates in the required projection.

- **Radiometric correction**. Processing techniques on sensor data to calibrate and correct the particular radiometric measurement errors of the sensor detectors. A typical error that may effect the actual radiation measurement of a sensor is the brightness intensity variation across the image plane. For sensor data in the optical range, the far-side data is usually brighter than near-side data, thereby reducing the contrast of the image; for SAR data, however, the opposite is true, e. g. the near-side data is brighter than the far-side data. Another error source may result from the fact that an output signal of a particular detector cell is not exactly proportional to the brightness intensity of the input signal. - All radiometric corrections are performed on the digital gray values of the pixels of an image resulting in 'calibrated gray values'.

| Remote Sensor Types | Part of Electromagnetic Spectrum utilized | Remarks/Use<br>Note: 1 Å = $10^{-10}$ m; 1 μm = $10^{-6}$ m |
|---|---|---|
| **Scintillation Counters**<br>Gamma-ray spectrometers<br>Geiger counters | <0.03 - 0.100 Å | Measurement of emitted natural radiation by gamma-ray detectors or Geiger counters |
| **Scanners with photomultipliers**<br>Image orthicons and cameras with filtered infrared film > 2900 Å | 100 Å - 0.4 μm | Records incident natural radiation. Imaged ultraviolet spectroscopy available. |
| Cameras<br>Using conventional B&W and color film | 4,000 - 7,000 Å<br>(0.4 - 0.7 μm) | B&W film for high spatial detail |
| Using infrared film (B&W and IR color) | 6,000 - 9,000 Å<br>(0.6 - 0.9 μm) | Improved spatial detail through contrast. Greater reflectance gradients useful for vegetation surveys |
| Multispectral units | 3,000 - 9,000 Å<br>(0.3 - 0.9 μm) | Individual narrow band scenes available with multi-camera systems |
| **Lidar**<br>Laser radar | 4,000 - 11,000 Å<br>(0.4 - 1.1 μm) | Monochromatic active systems for measuring backscattered radiation from the atmosphere (particles) |
| Radiometers | usually IR and microwave bands | Generally measures total radiation in a wide band in the infrared or microwave regions. Imagery obtained by scanning techniques. |
| Photometers | 4,000 - 7,000 Å<br>(0.4 - 0.7 μm) | Measures luminous flux in various bands of the optical region for distribution, color, etc. |
| Spectrometers | in any spectral region | Narrow-band data available sequentially - Electromagnetic radiation amplitude vs frequency |
| Solid State Detectors<br>Single detectors<br>Line arrays<br>Matrices | 1 μm - 1 mm | Single detecting element used in scanners, radiometers. 1-D and 2-D arrays for sequential data gathering. |
| Radars | 1 mm - 0.8 m | Narrow band active systems. Both analog and imagery available. |
| **Radiometers** (microwave) | 1 mm - 0.8 m | Passive systems. Both analog and imagery available. |

**Table 231:    Survey of Remote Sensor Types and their Applications**[591)]

---

591) Manual of Remote Sensing, Second Edition, American Society of Photogrammetry, 1983, p. 41

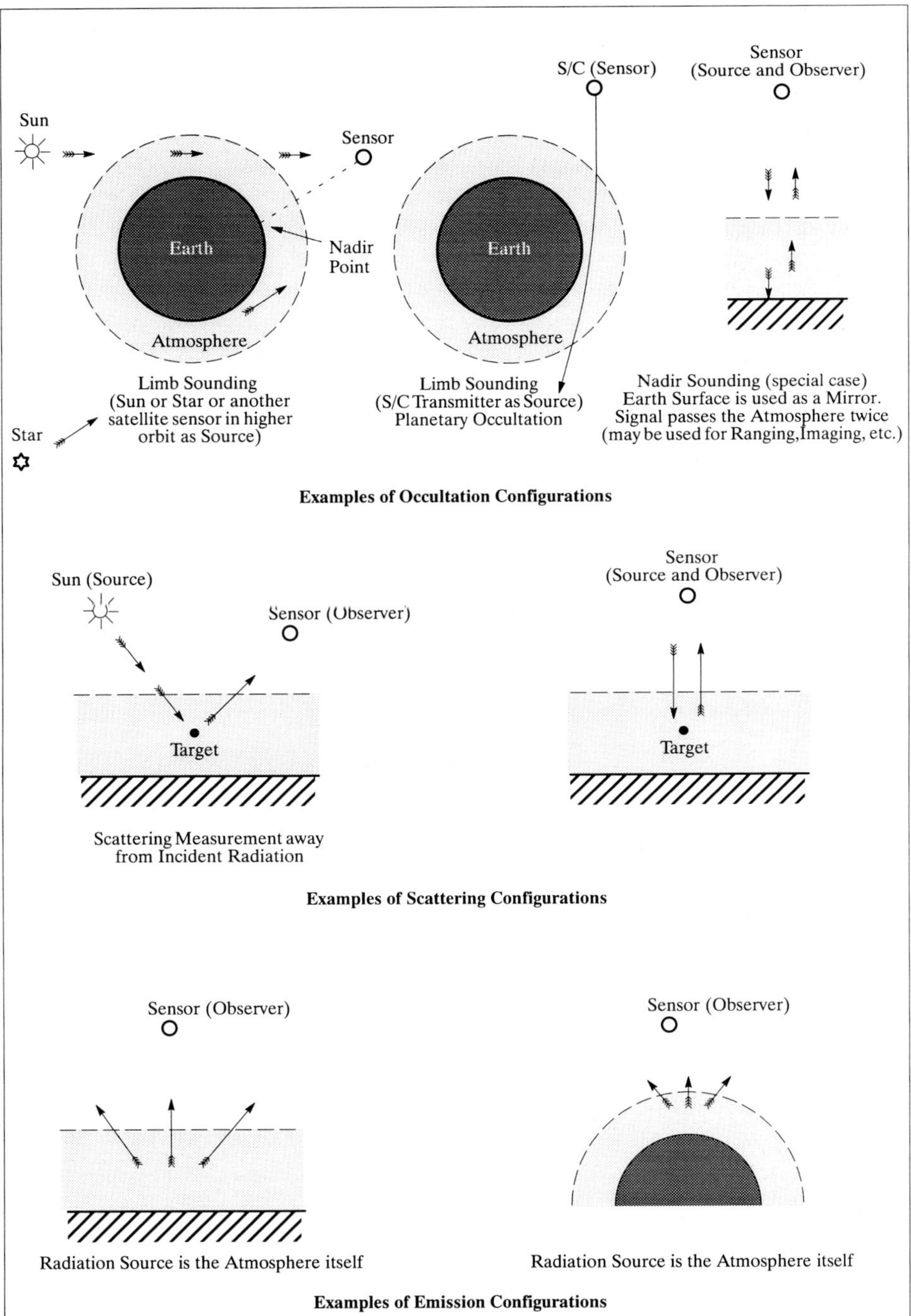



Figure 146: Different Configurations for Atmospheric Sounding[592]

592) Courtesy of Ch. Elachi, JPL

## C.1.5 Imaging Systems in the Visible and IR Spectra

Optical/Infrared imaging is picture-taking in the visible/infrared ranges of the electromagnetic spectrum. There are a number of technologies and several types of sensors measuring in the visible/infrared spectra.

### Photographic Imaging (cameras)

- Mapping cameras[593] - also referred to as metric or cartographic cameras. They feature usually a high spatial resolution and a wide coverage, there is also a high degree of distortion correction. Mapping cameras are usually 'panchromatic cameras', i.e. they contain the entire visible spectrum with a single channel. Panchromatic film for instance has a spectral range from 0.2 - 0.8 µm.
  [Examples: RMK series of Carl Zeiss, KC-series of Fairchild, S190B Earth Terrain Camera of Skylab (Itek), LFC of OSTA-3 shuttle payload,   etc.]

- Multispectral cameras - A camera records a number of images of a scene through several spectral filters. In other instances the various images are scanned and digitized. The spectroradiometric accuracy of the imagery, which is important for feature discrimination and identification, depends on the spectroradiometric response of the particular instrument. Multispectral cameras or mappers feature limited spectral resolution with high spatial resolution. MS cameras may have optionally a panchromatic channel (in parallel to the other discrete channels).
  [Examples: MKF-6, Zeiss Jena, S190A on Skylab, KFA-1000, KFA-200, and MK-4 series of Resurs-F, MKS of MIR, etc. ]

- Panoramic cameras - feature a small instantaneous field of view (IFOV) and high resolving power (which is constant over a large angular field of the camera). All forms of panoramic camera produce an image that is distorted. The following configurations are common according to the relative movements of the lens, slit, and film:
  1. the lens and slit can rotate while the film remains stationary during the exposure (direct scanning)
  2. a prism can rotate in front of the stationary lens at half the rate that the film moves past the stationary slit (rotating prism).
  3. the relative motion is divided between that of the film and the lens-slit combination (referred to as: split-scan panoramic camera by Fairchild, and optical-bar panoramic camera by Itek)

### Electro-Optical Imaging Sensors

- Frame sensors. (of TV heritage, such as Return-Beam Vidicon technique) - an electron beam scans the image on a photosensitive surface (raster scan of a complete image), multispectral imaging capability. - The image is focused onto a photoconductor which causes the intensity of the electron beam, discharged from the electron gun, to vary with the intensity of the light. The values of the intensity are converted into digital information.
  [RBV (Return-Beam Vidicon, Landsat) TV optical instruments of Meteor series,]

- Electromechanical scanning systems - These are mostly multispectral scanners with medium to high spatial resolutions and medium spectral resolutions. There are single-line scanners in use as well as multi-line scanner arrangements (both are called: whiskbroom). The available time to  dwell on each ground cell is very brief (each scan line consists of n ground cells which need to be measured sequentially) which in turn requires a very high bandwidth.
  Scanning methods: scanners may use on-axis optics or telescopes with a flat scan mirror located in front of the collecting aperture (see C.1.7 for more details).
  [Examples: TM and MSS on Landsat, MSU-SK on Resurs and Meteor series, Sea-WiFS, CZCS, etc. ]

---

593)  P. N. Slater, Remote Sensing, Optics and Optical Systems, Addison-Wesley, 1980, pp. 342-351

- Electronic scanning systems [Linear-array systems or CCDs (Charge-Coupled Devices, which scan in a pushbroom fashion, line by line, the total width of the swath) are the type mostly used; matrix sensors may be another type] - These are multispectral devices with high spatial resolutions and good spectral resolutions. A CCD array consists of an arrangement of small, square photosensitive cells onto which the incident radiation is focused. Each CCD cell creates an electrical charge which varies with the intensity of the incident radiation. The entire array forms an area of pixels carrying digital data which can be processed by computer.

The CCD sensor uses a wide-angle optics system **in which the total FOV is imaged on a detector array at once**. Successive lines are imaged and sampled by the multiplexer as the platform moves along the orbital path. The time that elapses between the imaging of two successive lines can be as long as the platform takes to move the distance in the scene. This is referred to as dwell time. Advantages: A long dwell time leads to low noise in the received signals. There are no moving parts on a pushbroom sensor. Disadvantages: a very large number of detectors are needed for high resolution images; the radiometric calibration of all detectors is difficult. In addition the pushbroom scheme requires a wide field-of view optics to obtain the same same swath as for a corresponding scanner.

The useful measurement range of CCDs is normally between 0.3 - 1.2 µm (VIS, NIR), there are also extensions up tp 2 µm. There are also multi-line CCDs in use.
[Examples: MOMS-01,-02 (on Shuttle and Priroda), HRV (Spot), PAN, LISS-III, WiFS (all of IRS-1C), MSU-E on Resurs and Priroda, MOS-Obsor on Priroda, etc.]

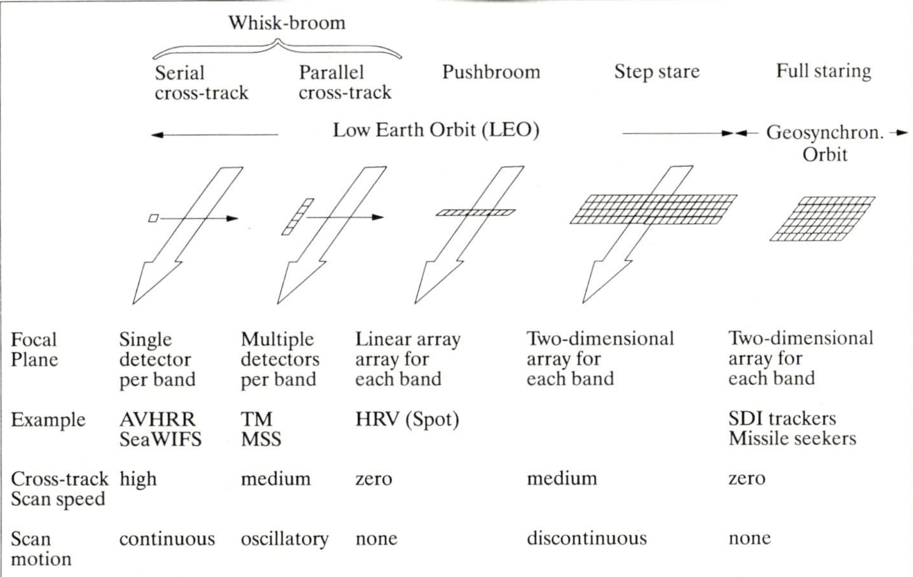


| Focal Plane | Single detector per band | Multiple detectors per band | Linear array array for each band | Two-dimensional array for each band | Two-dimensional array for each band |
|---|---|---|---|---|---|
| Example | AVHRR SeaWIFS | TM MSS | HRV (Spot) | | SDI trackers Missile seekers |
| Cross-track Scan speed | high | medium | zero | medium | zero |
| Scan motion | continuous | oscillatory | none | discontinuous | none |

A 'step-stare' imaging scheme is one which uses motion compensation to allow the array to stare for a certain period at the ground target as the platform is moving forward. The array is then stepped to a new location and held again until the cross-track imaging process is completed. The advantages of this scheme are an increased dwell time, only moderate field-of-view optics are needed. The disadvantages: a larger and more complex focal plane than the pushbroom scheme.

**Figure 147:    Some Optical Remote Sensing Arrangements and Techniques**[594]

---

594) "The Future of Remote Sensing from Space: Civilian Satellite Systems and Applications", Office of Technology Assessment, U.S. Congress, OTA-ISC-558, ISBN 0-16-041884-4, July 1993, pp. 142-143, original source: SBRC

## C.1.6    Resolution (for visible and infrared Imagery)

The amount of data/information contained within an image depends in large part on the image resolution. Definitions of spatial, spectral and radiometric resolution are not unique. There are interdependencies between spatial, spectral, and radiometric resolutions for each remote sensor affecting the various compromises and trade-offs for particular applications[595].

**Spatial (or geometric) Resolution**

Spatial resolution is the smallest unit of distance (usually a side length of a square area in an image) that can be discriminated by a sensor measurement of the target. Spatial resolution is a function of geometry (scale) between sensor and target for the instant of measurement. The image in a mechanical scanner setup is not formed at once over the swath width (as it would be the case in a camera setup), rather it is built up by a sequence of individual small images across the track (see scanners below). Each small image taken by the scanner represents an instant of measurement, hence it is the unit area of observation (actually a 'point') which is also referred to as 'the instantaneous field of view' (**IFOV;** note IFOV is a function of altitude, detector size, and focal length of the optical system). The output value of the sensor detector provides the value for one 'point' (smallest square area in an image), which is referred to as **picture element** or simply as **pixel**, in the overall image. Successive pixels in the scan line are generated by data from successive positions. The advancement to the next scan line is achieved by the orbital motion of the satellite (or by a mirror tilt).

IFOV, the unit area of measurement of a scanner, may be given as an angle (in mradians or in degrees) or directly as a length (edge length of the spatial resolution). The IFOV is chosen depending on the type of application. For instance: observation of small-scale variations with relatively small-area coverage (in this case regional patterns may be difficult to characterize, because generalization is required of the high frequency information in such images), or large-scale variations with frequent large-area coverage (regional patterns may be readily observed, however, details may be averaged within a pixel, implying a loss of information). Sensor arrangements onboard Earth observation satellites are possible that provide specifically high and low resolution images. The high resolution images are used for regional coverage and detail, while the low resolution images offer large-area coverage as well as high temporal coverage for quick synoptic views of an area. The price of spatial resolution is being paid with correspondingly high data rates, having direct consequences on the dimension of all follow-up processes (communications in the space and ground segments, processing, storage, and user access costs).

Naturally, there is a tendency and requirement for ever improving spatial resolutions. The Landsat-5 MSS sensor has a spatial resolution of 80 m, this is adequate for many land-based applications. The Landsat-5 TM sensor has a spatial resolution of 30 m. The Spot HRV sensor has a spatial resolution of 20 m (10 m panchromatic). The trend goes toward a 5 m spatial resolution.

**Spectral Resolution**

The spectral resolution of a remote sensor is determined by the bandwidth of the channels used. High spectral resolution is achieved by narrow bandwidths which, collectively, are likely to provide a more accurate spectral signature for discrete objects than are broad bandwidths. However, narrow-band instruments tend to acquire data with a low signal-to-noise ratio, lowering the system's radiometric resolution. This problem may be somewhat reduced if relatively long look times are used during imaging. In contrast, broad-band sensors usually have good spatial and radiometric resolution.

---

595)  Manual of Remote Sensing, Second Edition, American Society of Photogrammetry, 1983, pp. 20-26

**Radiometric Resolution**

Radiometric resolution refers to the resolving power of a system in wavelength and energy. The limiting factor for a radiometric measurement is the signal/noise ratio (SNR) of the instrument receiver (see chapter C.1.9). Considering the effects of varying illumination, the radiometric dynamic range of a sensor is determined by the maximum radiance value that the sensor can experience for a given band.

- A sensor has not the capability to obtain measurements of reflectivity as a continuous function of wavelength for all pixels in an image obtained with a multi-spectral scanner. All that is obtainable are integrated (i.e. averaged) reflectivities over the wavelength bands used in the receiver.

On the output side of a sensor, the measurement is converted into a number of discrete digital levels, also referred to as 'quantization'. This secondary measurement effect of quantization (the primary effect is SNR) is sometimes referred to as 'radiometric resolution'. In this terminology, a 10 bit radiometric resolution signifies, that a detector's converted digital output signal may range from 0 to 1023 (or 1024 quantization levels). - Some applications do not require a very high radiometric resolution (for instance sea surface temperature measurements), while other observations (in particular for vegetation, chlorophyll, etc.) require as many spectral channels as possible for good interpretation of the data.

# C.1.7   Multispectral Scanners[596]

In remote sensing multispectral scanners are non-photographic instruments which are used for a wide wave spectrum (visible, infrared, ) of radiation measurements. A scanner provides the features of 'scanning' [thereby increasing considerably the width of its observation path; this is different from 'sounding', which considers only a line or surface (plane) observation of the immediate orbital path], and of splitting the received radiation into a number of spectral ranges or 'bands'. An image is not formed all at once, as it is in a camera, but is built up by the process of scanning. The scanner records within its IFOV (instantaneous field of view) a very small area (ground cell), and builds up a line of recorded signals of these small areas referred to as pixels (picture elements). The next scan line depends on the forward movement of the satellite. The scanning process may be accomplished by different techniques such as a 'rotating mirror' or 'pushbroom' or simply by a spinning satellite.

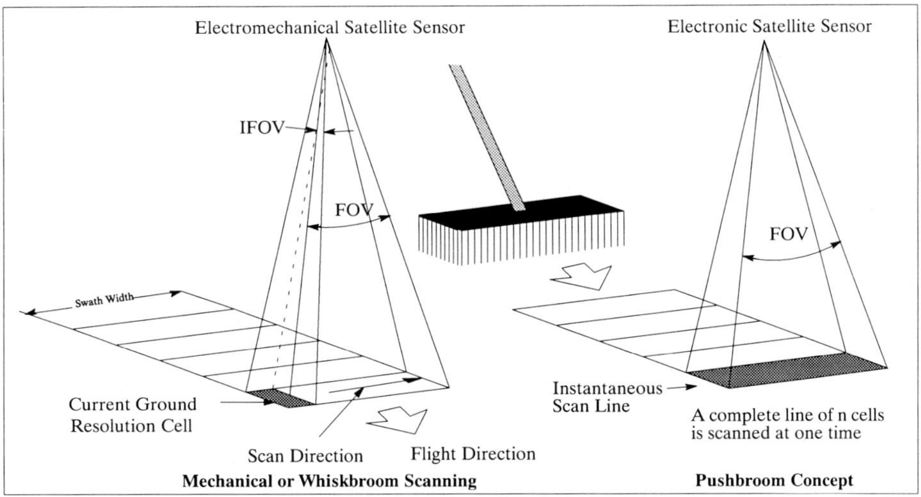


**Figure 148:   Schematic Scanning/Imaging Geometries on a Ground Surface**

596) A. Cracknell, A. Hayes, 'Introduction to Remote Sensing', Taylor & Francis, 1991, pp. 29-30

Note: Conventional imagery (VIS and IR image geometry) is acquired and arranged in a so-called angle-angle coordinate system, the sensor measures angular speed, an image is built up from a great number of contiguous pixels in a raster pattern consisting in turn of many scan lines.

A multispectral scanner consists of a telescope and various other optical and electronic components. At a given instant the telescope receives radiation from a given unit area, the instantaneous field of view (IFOV), on the surface of the Earth in the line of sight of the telescope. The radiation is reflected by a mirror and separated into different spectral bands. The intensity of the radiation is measured by a detector. The output value from the detector gives the intensity for one point (picture element or pixel) in the image. Successive pixels form a scan line. A multispectral scanner produces in parallel several coregistered images, one corresponding to each of the spectral bands. The detector interface of any sensor is of critical importance. It is generally at this detector interface where the overall system sensitivity and dynamic range performance are established.

## C.1.8  Imaging Spectrometers

The imaging spectrometer concept can be applied to the following spectral regions: VIS, NIR, SWIR and TIR (i.e. from about 0.4 μm to about 12 - 13 μm). However, different optics components and detectors have to be used for the different regions. The collecting optics and an imaging slit are used to image one line (one pixel wide) across the swath on an intermediate focal plane. A dispersive element (grating or prism) is then used to spectrally disperse the 'line' signal into a series of lines, each corresponding to a spectral band (see Figure 149). A 2-D detector array is then used to measure the dispersed light. Thus, each array column will give the spectrum corresponding to one pixel. The whole process is repeated as the platform moves by the equivalent of one spatial resolution element along the track.[597)]

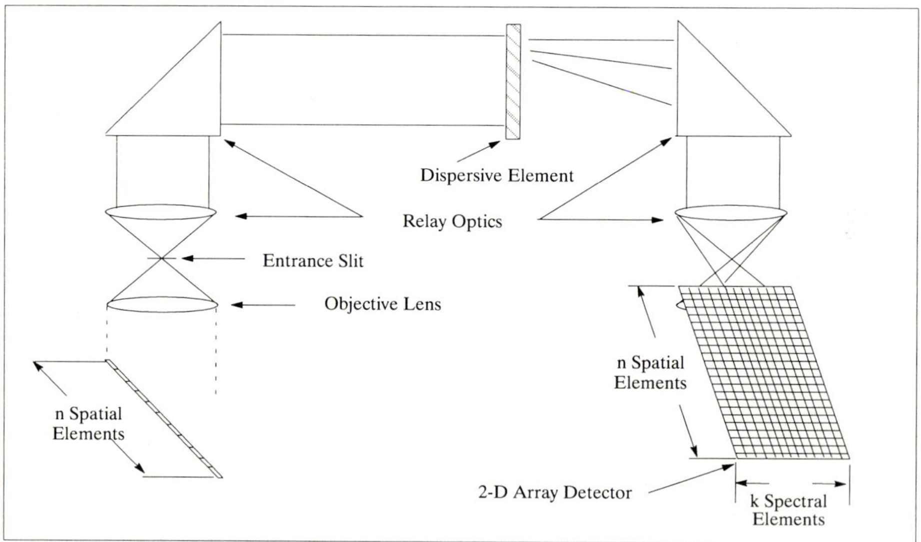


**Figure 149:**  **Collecting Optics of an Imaging Spectrometer**

---

597) Ch. Elachi, "Earth Surface Sensing in the '90s", Progress in Imaging Sensors, Proc. ISPRS Symposium, Stuttgart, 1-5 September 1986, ESA SP-252, November 1986, pp. 1-9

Imaging spectrometers allow the simultaneous acquisition of images in many contiguous, narrow bands. Spaceborne sensors have relatively few and broad bands, while airborne sensors may have hundreds of narrow bands, the high spectral dimensionality of these airborne data enables the construction of laboratory-like spectra for each pixel in the scene (see Figure 144).

Two approaches for the design of imaging spectrometers are in use. The wiskbroom approach includes an optomechanical fore-optics from which light is passed into a spectrometer, where it is dispersed and focused onto a linear detector array. This concept provides as many contiguous spectral bands as there are detector elements in the linear array. It is also possible to use 2-D (area) detectors instead of linear arrays in this configuration.

An alternate solution is the pushbroom approach using 2-D detector arrays at the focal plane of the spectrometer (see Figure 188), thereby eliminating the need for the mechanical scanning device. In this arrangement, one dimension of the array corresponds to the cross-track pixel length while the spectral information is co-registered in the other direction.

The collecting optics of imaging spectrometers (2-D airborne devices), usually designed with many contiguous spectral and spatial bands, generates very high data volumes, which require in turn very high-rate recording devices (sometimes in the order of 50-100 Mbit/s) for contiguous strip imagery acquisition. These recorders are usually very expensive, or not available at all for the intended application. In some cases a solution is provided with an interval acquisition technique by the sensor detector. The sensor abandons the concept of contiguous strip imagery in favor of discrete area or frame imagery. The data rate constraint at the recorder interface is circumvented in this way, because the next set of data frames is provided to the recorder at the instant, when the previous set of frames was stored away.

## C.1.9   Passive Radiometry

Radiometry in general is the measurement of electromagnetic radiation. A passive system is restricted to measuring the incoming radiation of a wave spectrum in question. Hence, a passive system is restrained to that portion of radiation that is emitted with a reasonable intensity from the observed objects (Earth surface or atmospheric volume or whatever).

In general passive microwave radiometry is limited by its poor spatial resolution which depends on range, on the wavelength of the radiation used, on the aperture of the antenna and on the signal/noise ratio. The signal/noise ratio (SNR) is in turn influenced by the strength of the signal produced by the target and by the temperature and sensitivity of the receiver.

- Passive sensors measuring in the visible (VIS) and infrared (IR) spectral ranges are restricted to sun illumination. Passive sensors in the VIS range are not able to see through clouds. If it is cloudy they produce images of the top of the clouds and not from the surface of the Earth.

- Passive sensors measuring radiation in the **microwave spectrum** have the advantage of collecting data during day and night cycles (measurement of emitted radiation rather than reflected solar radiation), the presence of cloud cover does not significantly degrade the measurement accuracy. The spatial resolution of passive microwave sensors is fairly poor compared to VIS and IR sensors (longer wavelengths, by a factor of 100, and much lower intensity of microwave radiation results in poorer resolution), the advantage is the global coverage capability. Passive microwave radiometry involves the detection of thermally generated microwave radiation. The characteristics of the received natural radiation, in terms of the variation in intensity, polarization, frequency and observation angle, is dependent on the nature of the surface being observed and on its emissivity. The natural   The spectrum of passive microwave radiometry is usually from 200 GHz - 1 GHz [or from 0.15 cm to 30 cm in wavelength; the newer planned sensors go to frequencies in the order of 600 GHz (0.5 mm wavelength)].

A passive microwave radiometer is different from traditional radar receivers (i. e. active microwave receivers) in two important respects. First, the input signal processed by a radar receiver may be coherent and nearly monochromatic while secondly, the natural radiation emitted by a surface (target or source) is phase-incoherent and extents over the entire electromagnetic spectrum. This means, it is noiselike in character and similar to the noise power generated by the receiver components.

The second difference relates to SNR (Signal/Noise Ratio) at the receiver output. With the condition: SNR »1, there is good confidence that the information can be extracted from the noise. However, the radiometric signal to be measured in a passive microwave receiver is usually much smaller than the receiver noise power. Passive radiometers are highly sensitive receivers that are configured to measure very small input levels with a high degree of precision. Several different configurations have been developed. Most of them use synchronous modulation at the input and demodulation at the output, and integrate the desired signal over as long a period of time as possible (the integration is set equal to the dwell time of the IFOV).

By inspection, optical wavelengths are about five orders of magnitude ($10^5$) shorter than microwaves; this implies that the resolution capability of an optical radiometer is inherently superior to the resolution of a passive microwave radiometer by about the same factor. For instance, the IFOV (spatial resolution) of an optical radiometer is of the order tens of meters compared to tens of kilometers for a microwave radiometer (for the same-size aperture and same orbital parameters). This means that the multiple-detector configurations employed in optical scanners are not feasible at microwave frequencies unless very large antenna structures are used.

Radiometric imaging of a scene of interest is accomplished by scanning the main beam of the antenna. For a moving platform, scanning in the cross-track dimension is sufficient to produce an image. Both mechanical and electronic (beam-steering) scanning techniques are used in microwave radiometry. In mechanical scanning, the direction of the antenna beam is changed by mechanical rotation or angular movement of the radiating aperture of the antenna system. Alternatively, phased array antennas can be used to steer the direction of the antenna beam electronically (no mechanical motion in the scanning process).

Two generic types of passive radiometers have evolved, profiling instruments (or sounders, also referred to as broadband radiometers), and surface imaging instruments (multi-channel radiometers).

1.  **Atmospheric sounders** provide information about vertical **profiles** of temperature and molecular constituent concentrations in the atmosphere by making measurements near the molecular resonance frequencies (resonance method with nadir pointing). Another approach to atmospheric constituent monitoring is the limb sounding technique. This method can provide substantially improved sensitivities and vertical resolutions (profiles) over those of the nadir-looking instruments.

2.  **Surface imaging sensors**, on the other hand, operate primarily at window frequencies, where atmospheric absorption is low and surface features can be imaged or measured quantitatively. For most of these surface features high spatial resolution is desirable (see also Ref. 598)).
    The nadir technique is employed for surface imaging. The radiometric measurements are affected to some extent by the water vapor, clouds and rainfall. Hence, most surface sensing radiometers include frequency channels sensitive to atmospheric water vapor and liquid water, to measure global distributions of these parameters and to correct for their effects on the measurement of the surface parameters.

A radiometer system for microwave remote sensing typically consists of the following subsystems: An antenna and scan subsystem, which receives incoming radiation from specified beam-pointing directions; a radiometer receiver (detectors plus the electronics subsystem), which detects and amplifies the received radiation within the assigned frequency band; and a data and control subsystem, which provides timing and sequencing signals for the antenna and radiometer subsystems, and performs digitizing and various other functions for the output of a digital data stream.

Many applications have become well established in the field of passive microwave radiometry such as temperature sounding in operational meteorology, sea-ice mapping for navigation in the polar regions, measurements of ocean surface temperatures and surface wind speeds.

Important microwave imaging instruments of passive devices are the multi-channel (multispectral, or multi-frequency) microwave radiometers. The vast majority of observations that have been done at optical wavelengths (mostly photographic cameras and multispectral scanners) involve passive radiometry. All the work done at infrared wavelengths involves passive radiometry, while at microwave wavelengths both passive and active techniques are important.

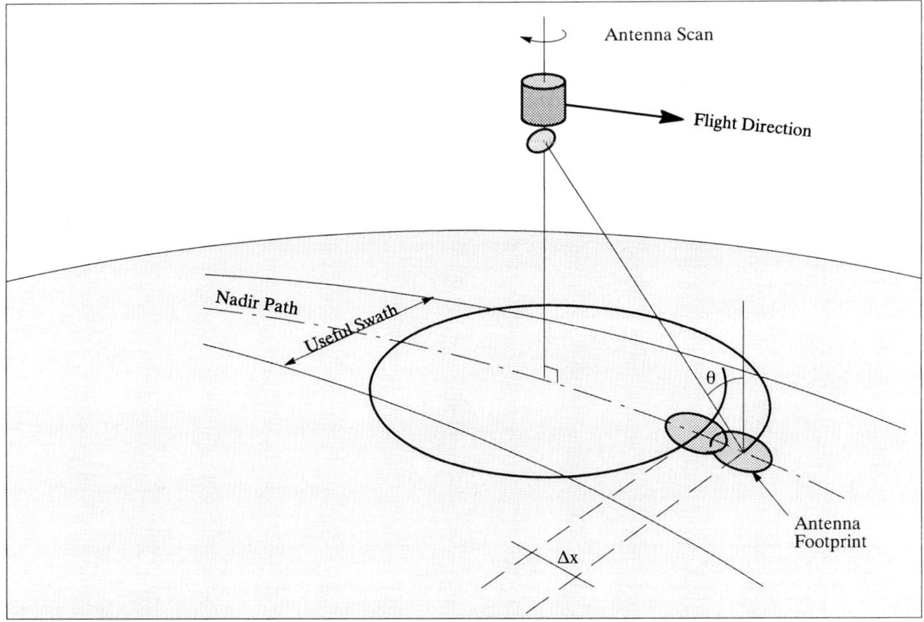


**Figure 150:    Typical Geometries of a Conical Scanning Passive MW Radiometer**

| Launch Year | Spacecraft | Sensor | Frequencies (GHz) | Antenna Type | Swath Width (km) | Spatial Resolution (km) | Prime parameters measured or inferred (i.e. objectives) |
|---|---|---|---|---|---|---|---|
| 1962 | Mariner 2 (Venus fly-by) | | 15.8, 22.2 | Mech. scanned parabola | | 1300 | Limb darkening of planetary emission, temperature |
| 1968<br><br>1970 | Cosmos 243<br><br>Cosmos 384 | | 3.5, 8.8, 22.2, 37,<br>same | Nadir-viewing parabola<br>same | | 13 | Atm: water vapor, liquid water content; Sfc: temp., sea ice concentration |
| 1972 | Nimbus-5 | ESMR<br><br>NEMS | 19.3<br><br>22.2,31.4, 53.6,54.9,58.8 | electrically scanned array<br>5 lens-loaded horns, nadir | 3000 | 25<br><br>200 | Atm: rain rate; Sfc: sea temp., ice concentration<br>Atm: temp. profile, water vapor; Sfc: ice classification, snow cover |
| 1973 | Skylab | S 193<br><br>S 194 | 13.9<br><br>1.4 | Mech. scanned parabola<br>Nadir-viewing | 180 | 16<br><br>115 | Surface: winds, precipitation<br><br>Surface: soil moisture |
| 1974 | Meteor | | 37 | Dual-polariza. 35° from nadir | | | Atm: Liquid water content |
| 1975 | Nimbus-6 | ESMR<br><br>SCAMS | 37<br><br>22.2,31.6,52.8 53.8, 55.4 | elec. scanned array, dual pol.<br>3 rot. hyperb. mirrors | 1300<br><br>2700 | 20 x 43<br><br>150 | Same as Nimbus-5 ESMR<br><br>Same as Nimbus-5 NEMS |
| 1978 | DMSP | SSM/T | 50.5,53.2,54.3 54.9,58.4,58.8 59.4 | Single rotating mirror | 1500 | 174 | Atm: Temp. profile |
| 1978 | Tiros-N | MSU | 50.3,53.7,55.0 57.9, | Dual rotating mirrors | 2300 | 110 | Atm: Temp. profile |
| 1978<br><br>1978 | Seasat,<br><br>Nimbus-7 | SMMR<br><br>same | 6.6,10.7,18, 21, 37 | Offset-fed osc. parabola, dual polariza. | 600<br><br>800 | 18 x 23<br><br>22 x 35 | Atm: water vapor, rain rate, Sfc: sea temp. wind speed, ice concentration, snow cover, |
| 1979 | Cosmos 1076, 1151, | Device ν<br><br>Device π | 3.53, 9.37, 22.2, 37.5<br>9.37 | Nadir-viewing parabola (no scanning) | Swath=18 km at 0.8 cm (37.5)<br>Swath = 80 km at 8.5 cm (3.53 GHz) | | Sea surface: temp, wind speed, vapor in clouds, etc. |
| 1979<br><br>1981 | Bhaskara-I (ISRO)<br>Bhask.-II (ISRO) | SAMIR<br><br>SAMIR | 19.1,19.6,22.2<br><br>19.35,22.235, 31.4 | Fixed antenna spinning S/C<br>same | 150 & 1000<br>same | 150<br><br>125 | Atm: water vapor , liquid water content<br>Atm: same, Sfc: rain, wind |
| 1987 | DMSP | SMM/I | 19.35,22.235, 37,85.5 | Offset-fed rot. parabola, dual-polariz. | 1400 | 16 x 24 | Atm: water vapor, liq. content; Sfc. wind, sea ice, snow cover, soil moisture |
| 1987 | MOS-1 | MSR | 23.8, 31.4 | Offset casse-grain, mech. scan, dual-p. | 317 | 23 | Atm: water vapor content, Sfc: ice, snow, |
| 1988 | Okean-O1 | RU-08 | 37.5 | Scanning | 550 | 15 x 20 | Atm. water content |
| 1991 | ERS-1 | ATSR-MWR | 23.8, 36.5 | Offset antenna | | 20 | Atm. water content (vapor and liquid), MW emission |
| 1991 | UARS | MLS | 63,183,205 | 3 mirror, scan-ning at 5.5 sec/ scan | N/A | 3 km | Atm:limb sounding, con-centration of ClO, $H_2O$, $O_3$, and atm. pressure |
| 1992 | ATLAS-1 | MAS | 61,62,63,183, 184, 204 | | | 10 (vert.) 4 for higher freq. | Atm. limb sounding, pres-sure, temp., concentration of ozone, water, etc. |
| 1992 | Topex/Po-seidon | TMR | 18, 21, 37, | Offset oscillat-ing parabola, dual oscillator | N/A | 30 km | Total water content, in nadir to correct the altimeter path for water vapor |

**Table 232:**     **Survey of flown/operational Passive Microwave Radiometers[598], [599]**

598) The table of E. G. Njoku has been updated: "Passive Microwave Remote Sensing of the Earth from Space - A Review", in Proceedings of the IEEE, Vol. 70 Nr. 7, July 1882, pp. 728-750
599) "The Multi-Frequency Imaging Microwave Radiometer" - Instrument Panel Report, esa SP-1138, Aug. 1990

| Launch Year | Spacecraft | Sensor | Frequencies (GHz) | Antenna Type | Swath Width (km) | Spatial Resolution (km) | Objectives |
|---|---|---|---|---|---|---|---|
| 1995 | MIR-Priroda | IKAR-N | 5, 13.76, 22.22, 37.5, 100, | Horn antenna, nadir pointing | 60 | 60 | MW emissions of atm. /sea surface system at nadir |
| | MIR-Priroda | IKAR-D | 22.22, 37.5, 100 | Parabolic antenna, mech. scan | 400 | 5 - 15 | Scanning radiometer |
| | MIR-Priroda | IKAR-P | 5.0, 13.76 | Fan Horn antenna (3 in parallel at 13.76 GHz, and 6 at 5 GHz) | 700 | 50 - 75 | Ocean surface temperature |
| | MIR-Priroda | R-400 | 7.5 | Parabolic conical scan antenna | 400 | 50 | Earth surface radiation |
| 1995 - 2000 | NOAA K, L, M, N<br><br>EOS-PM-1 | AMSU-A (MTS)<br>AMSU-B (MHS) | 23.8-31.4 (15 cha.)<br>50.3-89 (12 cha.)<br>89, 157, 183 (5 cha each) | Parabolic off-center reflectors<br><br>same | 2200<br><br>2200 | 40<br><br>15.4 | Atm. temp. measurements from 0 - 40 km<br>Atm. humidity, precipitation, |
| 1997/8 | TRMM | TMI | 10 - 91 range | Dual polarization | | 45 - 4.4 | Data will be related to rainfall rates |
| 1996 | ADEOS | NSCAT | 13.995 | | 1200 | 50 | Ocean surface winds |
| 1996 | ODIN | SMR | 119,122,188 553, 575 | Offset Cas-segrain | N/A | | Atm. limb emissions of $H_2O$, Oxygen atoms, ClO radicals, etc. |
| 1998<br><br>2000 | Metop-1<br><br>EOS-PM-1 | MIMR<br><br>MIMR | 6.8, 10.65, 18.7, 23.8, 36.5, 90, | Conical scan, mech. rotat. antenna | 1400 | 60 x 40 (6.8)<br>4.8 x 3.1 (90) | Precip. rate, cloud water, water vapor, temp. profiles, sea surface roughness and temp., soil moisture, snow cover |
| 2002 | Envisat-2 | AMAS | 62, 63, 115, 118, 183, 184, 204, 206 | Offset cas-segrain, elliptical pol. reflector | | 300 5-10 vertical | Atm. limb sounding, pressure (62,63), CO (115), temp. (118), $H_2O$ (183), $O_3$ (184), ClO (204), $O_3$ (206) |
| 2002 | EOS-CHEM-1 | MLS | 63, 215, 440, 535, 640 | Offset cas-segrain scanning | N/A | 3 x 250 horz. 1.4 vertical | MW/thermal emissions from atm. limb, |
| 1999 | ADEOS II | AMSR | 6.6, 10.65, 18.7, 23.8, 36.5, 55, 89 | | 1700 | 5-60 | Ocean vapor profiles, surface temp., wind speed; MW emission from atmosphere |

**Table 233:    Survey of planned Passive Microwave Radiometer Missions**


## C.1.10  Active Radiometry

An active instrument is restricted to wavelength ranges in which reasonable intensities of radiation must be generated by the remote sensing device in order to obtain a measurable return echo radiation (signal attenuation is minimized by choosing a suitable atmospheric "window", see Figure 164).

An active microwave system can improve the 'poor' spatial resolution associated with a passive microwave system. An active microwave system measures not only the intensity of radiation, but also:

-   The time taken for the emitted pulse of radiation to travel from the device (satellite) to the ground and back to the device.
-   The Doppler shift in the frequency of the radiation echo as a result of relative motion of the satellite on the ground

- The polarization of the radiation (note: polarization can also be measured by passive devices)

## C.1.10.1  Types of Radar Sensors

There are three general categories of radar sensors (i.e. active microwave instruments) that are flown on satellites, depending on the type of geophysical measurement, namely:[600])

**Altimeters**

Altimeters are used for surface height measurement (distance) along the satellite track. They are used for land topographic mapping and for very high resolution (few cm) ocean topographic mapping (ocean circulation studies). This is achieved by very accurate measurement of the time delay for a radar pulse to propagate from the sensor to the surface and back. Scanning altimeters can provide topographic measurements across a wide swath thus providing "altimetric images".

An accurate distance measurement (usually vertically down to the sea surface, i.e. nadir pointing) requires in turn a very accurate knowledge of the satellite orbit. The size of the "footprint" of a radar altimeter pulse on the water surface is the area of the surface of the sea that contributes to the return pulse (echo) received by the altimeter.
Accurate measurements of the ocean surface mean level contribute to the detection and measurement of ocean currents, tides and storm surges, and to the accurate mapping of underwater features. Altimeters can also measure the surface wave height.

**Scatterometers/Spectrometers**

Scatterometers/spectrometers are used to measure very accurately the surface reflectivity (back scatter) as a function of the frequency, polarization, and illumination direction of the sensing signal. They are used to characterize quantitatively the surface roughness. Spatial resolution is not of specific importance and is usually sacrificed at the expense of amplitude measurement accuracy. Scatterometry can be combined with high resolution imaging capability within the same sensor, leading to 'imaging scatterometry'.

Scatterometers use a complicated arrangement of radar beams (usually 4) which enable the determination of direction as well as of speed of the wind.

**Imaging SARs** (Synthetic Aperture Radars)

Imaging radars are used to acquire high resolution (few meter to cm range) large scale images of the surface. They are employed in the study of surface features such as geologic structures, ocean surface waves, polar ice cover, land use patterns. SAR systems are characterized by very high data rates, large antennas, and large power requirements.

In a SAR instrument, reflected signals are received from successive positions of the antenna as the satellite moves in its orbit. In this way an image is built up that is similar to the image one would obtain from a real antenna of several hundred meters or even a few kilometers in length. The reconstruction of an image from a SAR device is however complex. The use of a SAR device involves a sophisticated technology for radar system design and signal-processing techniques. The advantages of a SAR device are: operations in all weather conditions, and operations during the day and night cycles of an orbit.

Hybrid radar sensors

Advanced imaging radar sensors are planned and used which will simultaneously operate at a number of frequencies, in all polarization states. They will be calibrated combining the imaging, scatterometry, and spectrometry functions. Stereo imagers and scanning radar al-

---

600)  Charles Elachi, "Spaceborne Radar Remote Sensing: Applications and Techniques", IEEE Press, 1988

timeters will be used to provide topographic maps over large swaths, thus combining the altimetry and imaging functions. Rain mapping radars will use the imaging, altimetry (sounding), and scatterometry functions to provide a calibrated three-dimensional volumetric image of rain regions, but at lower resolution than required with surface imagers or altimeters.

## C.1.10.2  SAR Terminology and Definitions

An imaging radar system is generally arranged in a side-looking configuration.[601] The radar antenna illuminates a surface strip (footprint) to one side of the nadir track. The side-looking configuration is necessary to eliminate right-left ambiguities from two symmetric equidistant points and to get better range resolution on the ground. As the platform moves in its orbit, a continuous strip of swath width is mapped in the along-track direction of the satellite. The look angle $\theta$ is the angle from the vertical to the center line of the radar beam direction (normal to the flight direction)

Actual imaging SARs use a series of pulses instead of a continuous signal. Each transmitted wave front hits the target surface at near range and sweeps across the swath to far range. The pulse bandwidth B determines the cross-track or range resolution. The echos are sensed coherently.

### SAR Images

Synthetic aperture radar images are acquired in an **azimuth-range coordinate system,** the sensor measures the Doppler history (time delay) of sequential wave fronts in the illuminated footprint. (Note: this is fundamentally different from VIS and IR imagery which is acquired in an angle-angle coordinate system). The generation (reconstruction) of SAR imagery involves therefore an extensive amount of computation (with corrections applied for a number of effects) for the simple reason that the formation of each 'pixel' involves the combination of data from many thousands of echoes (i.e. synthetic array elements).

Most of the data handling in a SAR processor can be thought of as a correlation of the received signal with a two-dimensional reference function. The major processing step involves the accurate modeling of the SAR response to a point target and a continuous field of targets. Pixels as such exist only after the reconstruction of an image (representing the smallest distance unit that can be discriminated by computational methods) in their conventional raster pattern.

**Range Resolution** (across-track direction). The range resolution $(X_r)$ corresponds to the minimum distance between two points on the surface which are separable. (B= bandwidth of signal (Hz); c = speed of Light; $\theta$ = look angle)

$X_r = c/(2B \sin \theta)$ (Ground range resolution for SAR)

**Azimuth Resolution** (along-track direction).. The azimuth resolution $(X_a)$ for a real aperture radar (non-SAR) corresponds to the two nearest separable points along the azimuth line, i.e. on a constant delay line. This is identical to the azimuth antenna footprint extension because the echos returned from all the points along the line spanning that width are returned at the same time. (h = height above ground; $\lambda$ = wavelength; L = antenna length)

---

601)  Charles Elachi, "Spaceborne Radar Remote Sensing: Applications and Techniques", IEEE Press, 1988, pp. 63-64

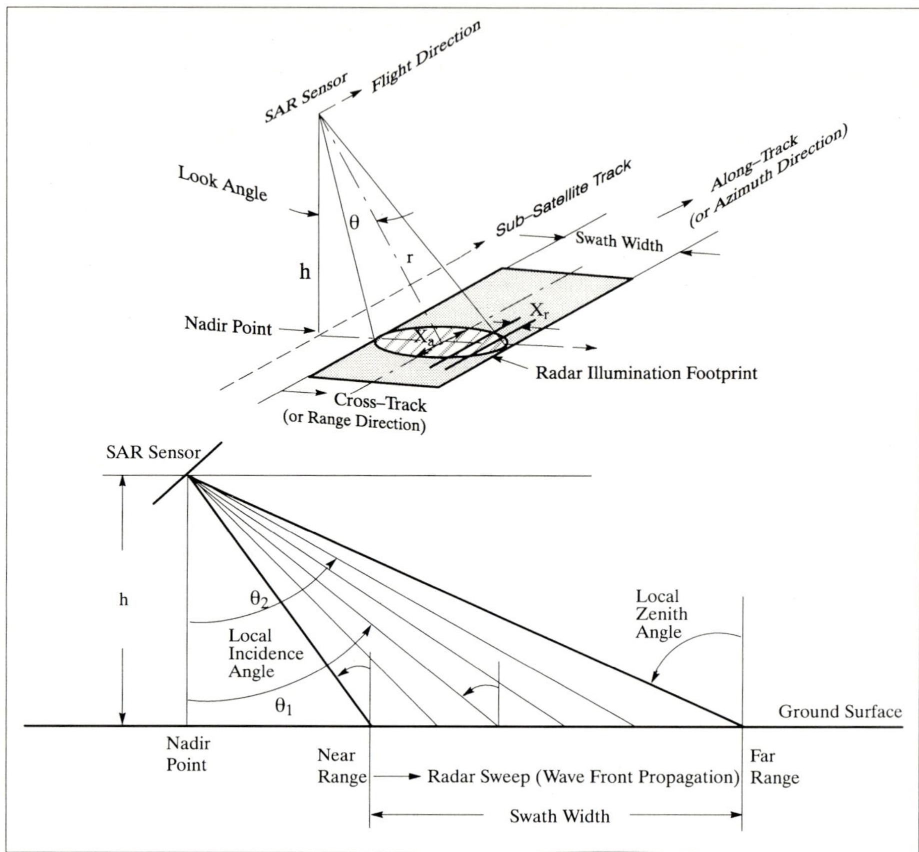


**Figure 151:   Characteristic Geometry Definitions for a Side-looking Imaging Radar**

$X_a = h\lambda/(L \cos \theta)$; (for a real aperture radar)

For a focussed SAR, however, $X_a$ is independent of range.

**$X_a = L/2$ ( for a Synthetic Aperture Radar)** one look highest azimuth resolution

Hence, the spatial azimuth resolution of the SAR is not dependent on the altitude of the sensor. This can be explained by the fact that the imaging mechanism uses the Doppler shifts in the echo and the differential time delays between surface points, neither of which is a function of the distance between the sensor and the surface. Note: the altitude still plays a factor in the power requirements to acquire a detectable echo.

### Looks, Speckles and Radiometric Resolution of SAR Images

The number of neighboring pixels averaged in the processing of a SAR image is referred to as '**the number of looks N**'. This technique is used to improve the noisy appearance of a radar image. A noisy appearance results from the backscatter contributions of the many scattering points of the surface area. Two neighboring areas with the same backscatter cross section may be different in fine detail of the two surfaces, but otherwise homogeneous - the returned signals from the two areas will be different. There will be brightness variations in the resulting image from one resolution element to the next. This is referred to as **speckle**.

The radiometric resolution in SAR images is mainly governed by the speckle phenomenon (coherent scattering) and not so much by the number of possible signal quantization steps. One way of describing radiometric resolution (Q) of a SAR image is by the signal-to-noise ratio. The radiometric resolution represents the logarithmic difference of two discernable backscatter coefficients which can be expressed in dB (neglecting thermal noise).

$$Q = 10 \, \log\left[1 + \frac{1}{\sqrt{N}}\right]$$

Small values of Q define a good radiometric resolution. This is the case for higher values of N (looks). For N=1 $\rightarrow$ Q = 3 dB; N=16 $\rightarrow$ Q = 1 dB (approx.)

| Parameter | Launch Date | Spatial Resolution (km) | Swath Width (km) | Frequency (GHz) | Polarization | Data Rate (kb/s) | Peak Power (W) | Orbit Altitude/ Inclination (km/°) | Remarks/Sensor |
|---|---|---|---|---|---|---|---|---|---|
| Seasat | 27.6.78 | 50 | 500 | 14.6 | VV,HH | | 110 | 800/108 | |
| ERS-1 | 17.7.91 | 50 | 500 | 5.3 | LV | | 4800 | 785/98.5 | AMI in Scatterometer mode |
| ERS-2 | 1994 | 50 | 500 | 5.3 | LV | | 4800 | 785/98.5 | AMI in Scatterometer mode |
| ADEOS | 1996 | 25,50 | 600/1200 | 13.995 | VV,HH | 2.9 | 275 | 797/98.6° | NSCAT |
| EOS | 1998 | 15-30 | 2x200 | | | | | | |

**Table 234:**   **Characteristics of Spaceborne Radar Scatterometers**

| Parameter Mission | Launch Date | Spatial Resolution (m) | Swath Width (km) | Frequency (GHz) | Polarization | Look Angle (°) | Bandwidth (MHz) | Data Rate (Mb/s) | Peak Power (W) | Altitude/ Inclina. (km/°) | Remarks |
|---|---|---|---|---|---|---|---|---|---|---|---|
| Seasat | 27.6.78 | 25 | 100 | 1.28 | HH | 20 | 19 | 110 | 1000 | 800/108 | |
| SIR-A | 12.11.81 | 40 | 50 | 1.28 | HH | 47 | 6 | N/A | 1000 | 260/38 | Shuttle |
| SIR-B | 5.10.84 | 25 | 30 | 1.28 | HH | 15-60 | 12 | 34 | 1000 | 225/57 | Shuttle |
| Kosmos 1870 | 25.7.87 | 25-30 | 20-35 | 3.125 | HH | 25-60 | 10 | - | 250 kW | 270/72 | |
| Almaz-1 | 31.3.91 | 13-20 | 2 x 172 | 3.125 | HH | 25-60 | 10 | 100 | 190 kW | 350/72 | pulse time 0.1 s |
| ERS-1 (AMI) | 17.7.91 | 30 | 100 | 5.3 | VV | 23 | 15.55 | 105 | 4800 | 785/ 98.5 | 3 modes |
| JERS-1 | 11.2.92 | 18 | 75 | 1.275 | HH | 35.21 | 15 | 60 | 744 | 568/97.7 | |
| Priroda | 1993 | 50 | 50 | 1.28, 3.28 | HH, VV | 35 | 5 | 16 | 1000, 300 | 400/52 | SAR Travers |
| SIR-C L,C-Band X-Band | 93,94,96 | 15-25 | 30-100 | 1.28, 5.3 | Quad | 20-55 | 10,20 | 46/Ch. | 3600/5500 | 225/57 | Shuttle |
| Radarsat | 1994 | 25-100 | 100-170 | 5.3 | HH | 20-60 | 11,17 | 110 | 1200 | 800/ | |
| ERS-2 (AMI) | 1994 | 30 | 100 | 5.3 | VV | 23 | 15.35 | 105 | 4800 | 785/98.5 | 3 modes |
| TRMM | 1997 | | 220 | 13.796 13.802 | | | | .0935 | 600 | 370/35° | |
| EOS-SAR | 2000 | 20-500 | 30-500 | 1.28, 5.3, 9.6 | Quad. | 20-55 | 10,20 | 180 | 5800 | 800/ | |

**Table 235:**   **Characteristics of Spaceborne Imaging Radars**[602],[603]

---

602)  Portions of the table are extracted from: Charles Elachi, "Spaceborne Radar Remote Sensing: Applications and Techniques", IEEE Press, 1988, pp. 234-236
603)  Table 235 is being retired (Status of 1. Book). The reason: I am being told by experts, that the peak power column doesn't make sense for certain systems, there should be other criteria instead.

| Parameter / Missions | Launch Date | Range Precision (cm) | Frequency (GHz) | Uncompressed Pulse Width (µs) | Compressed Pulse Width (µs) | Peak Power (W) | Beam Limited Footprint (km) | Pulse Limited Footprint (km) | Orbit Altitude/ Inclina. (km/°) |
|---|---|---|---|---|---|---|---|---|---|
| GEOS-3 | 1975 | 50 | 13.9 | 1 | 12.5 | 2000 | 38 | 3.5 | 843/115 |
| Seasat | 27.6.78 | 10 | 13.5 | 3.2 | 3.1 | 2000 | 22 | 1.7 | 800/108 |
| Geosat | 1985 | 5 | 13.5 | 102.4 | 3.1 | 20 | 29 | 1.7 | 800/108 |
| ERS-1 (RA-1) | 17.7.91 | 5-7 | 13.8 | 20 | 3 | 55 | 18 | 1.7 | 785/98.5 |
| Topex/Pos | 10.8.92 | 2.4 | 13.6/5.3 | 102.4 | 3.1 | 20 | 26/65 | 2.2 | 1334/63.5 |
| Priroda (Greben) | 1993 | 10 | 13.76 | 1.7 | 12.5 | 40 | 13 | 2.3 | 400/52 |
| ERS-1 (RA-1) | 1994 | 5-7 | 13.8 | 20 | 3 | 55 | 18 | 1.7 | 785/98.5 |
| GFO | 1995 | | | | | | | | |
| EOS | 1998 | | | | | | | | |
| Envisat-1 (RA-2) | 1998 | | 13.575 3.2 | 20 | | | | | |

**Table 236:     Characteristics of Spaceborne Radar Altimeters (Ocean Surface Mapping)**

## C.1.10.3  Lidars (Laser-Based Remote Sensing)

Lidar = **L**ight **D**etection **and R**anging. Lidar refers to laser-based remote sensing, i.e. radar principle applied in the **optical (and NIR) regions** of the electromagnetic spectrum. This wave spectrum (shorter wavelengths than the microwave spectrum) implies and promises to be the next level of observation technology and of information interpretation (in the number of phenomena as well as in detail and accuracy), due to the use of a much finer scale of measurement. Lidars are based on the principle of a laser-light pulse being sent into the atmosphere to probe the distance, physical state, or chemical composition of the backscattering layers. Lidar is also a generic term for a variety of sensors operating on different concepts, like:[604],[605]

Backscatter Lidar
In this technique information on a remote target is derived from the manner in which the transmitted energy is backscattered, reflected or reradiated by the target. The technique offers data on the scattering and extinction coefficients of the various atmospheric layers such as extent, height distribution, and optical thickness of aerosol and cloud layers. The prime candidate for the backscatter lidar is the Neodymium doped Yttrium Aluminum Garnet laser (or simply Nd:YAG laser).

Differential Absorption Lidar (DIAL)
In this path absorption technique information is derived relating to the path along which energy is transmitted or received by comparing the radar echoes in a tuneable multi-wavelength laser system (measurement of the differential ratios of the radar returns with those of the transmitted ratios by tuning the laser wavelength to the specific absorption features of atmospheric trace constituents). The technique offers the capability to determine the densities of specific atmospheric constituents as well as water vapor and temperature profiles at better accuracies than passive sounders.

Doppler Wind Lidar
This technique is based on the principle of measuring the Doppler shift of the light backscattered from aerosol particles transported by the wind. The technique offers the capability to measure global wind fields. Since the process inherently measures the line-of-sight velocity component, scan techniques and processing algorithms must be employed to obtain the desired wind field. - Most wind lidars are based on the carbon-dioxide ($CO_2$) gas laser. Such an

---

604)  H.Lutz, E. Armandillo, "Laser-Based Remote Sensing from Space", esa bulletin 66, may 1991, pp. 73
605)  R.T.H. Collis, P.B. Russell "Laser Applications in Remote Sensing", chapter 4 of 'Remote Sensing for Environmental Sciences', Springer Verlag, 1976, pp. 110-146

instrument generally needs a highly coherent and powerful laser beam , combined with a complex receiver and Doppler processor.

Ranging and Altimeter Lidar
Laser ranging and altimetry can provide accurate measurements of the distance from a reference height (i.e. the satellite orbital height) to precise locations on the Earth's surface. It can improve our understanding and knowledge of many processes and phenomena in the solid Earth Sciences such as geodesy, geodynamics, ice dynamics, land topography, and Earth resources. These observational requirements are covered by types of instruments in spaceborne applications:

-   A laser ranging system for accurate point positioning in geodynamics
    This laser ranging concept from space could enhance the capabilities of the present ground-based satellite laser ranging systems (SLR's) to such high-altitude satellites as LAGEOS or Starlette.  A single-wavelength nanosecond pulse laser is sufficient for centimeter-level accuracy. Subcentimeter accuracies require a dual-wavelength picosecond laser to compensate for atmospheric propagation effects (beam refraction).
-   A laser altimeter for altimetry measurements over land, oceans, and ice.[606]
    Laser altimetry is the only way to obtain elevation data with centimeter-level vertical precision and horizontal resolution.
    The laser altimeter, primarily for land and ice applications, is an extension of the backscattering lidar instrument. The main characteristics of a laser altimeter are: the small diffraction-limited footprint (proportional to the laser wavelength), which permits topographic mapping of ice sheets, terrain, forestry or waters with high spatial resolution; the insensitivity to speckles, since direct detection can be successfully employed in a visible-wavelength laser; high single pulse measurement accuracy, since in principle no averaging is required; and the insensitivity of beam propagation to atmospheric water vapor.

The lidar sensor technology is as of 1993 still in an experimental state[607]. So far no atmospheric lidar sensors have been flown on Earth-orbiting spacecraft. On the US side there are programs for the development of several lidar sensors, among them are:

- LITE      = Lidar in-space technology experiment of NASA/LaRC (a shuttle flight is planned for 1994)

- LAWS     = Lidar atmospheric wind sounder

- GLRS      = Geoscience laser ranging system

In Europe the ESA sensors ATLID (Atmospheric Lidar) and ALADIN (Atmospheric Laser and Doppler Instrument) are considered prime candidates as a backscatter lidar and as a Doppler wind lidar respectively for the POEM missions. ALISSA (France) is planned to be flown on PRIRODA.

**Lidar Principle**

In the principal radar-type or "lidar" applications of laser energy, the following basic elements make up the system:

- Laser Transmitter
- Collector (receiving telescope)  and detector of backscattered energy
- Electronics and data processing resources

The laser transmitter, which is the essence of the active lidar technique, provides all the energy available for the determination of the remote target's characteristics (or distance). By far the most effective approach in all forms of active lidar systems is the use of laser energy in

---

606) 'Topographic Science Working Group Report' to the Land Processes Branch, Earth Science and Applications Division, NASA Headquarters, 1988, p. 46
607) 'Lidar in Space', in Optical Remote Sensing of the Atmosphere, 1990 Technical Digest Series of the Optical Society of America, Volume 4, pp. 67-70

high intensity pulsed form, although range information can also be derived from continuous wave (CW) transmissions. At present two somewhat different laser ranging methods are used:[608)

1.  the phase comparison technique using high-frequency modulation of one of the laser light's parameters (amplitude, polarization, wavelength, etc.)

2.  the short pulse direct travel time measurement technique (delays are measured either by the 'single photon detection method', or by the 'multi-photoelectron detection method')

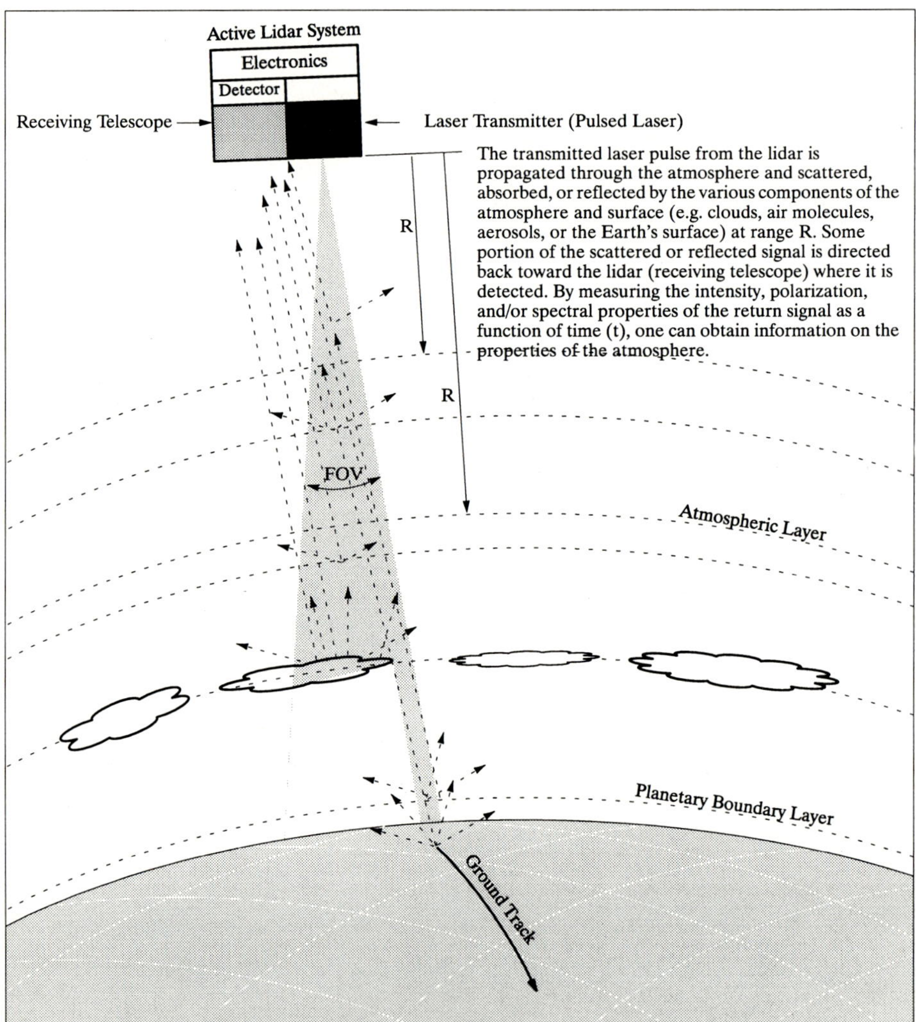


**Figure 152:    Spaceborne Lidar and Principle of Backscattering Lidar Measurements**

The emitted laser beam interacts with the atmospheric constituents, causing alterations in the intensity, polarization and wavelength of the backscattered signal. The distance to the

<hr>

608)  J. Gaignebet, G. Lund, "The Potential of two Color Laser Ranging", in 'Radars and Lidars in Earth and Planetary Sciences', International Symposium 2-4 Sept. 1991, esa SP-328, pp. 123-127

scattering medium can be deduced with a high degree of accuracy from the time delay of the return signal.

The high degree of spatial coherence of laser energy makes it possible to achieve a good measure of directionality, and very narrow transmitted beams (typically less than a milliradian in angle) are used to achieve maximum concentration of energy and high resolution. Returned energy collected in the optical system of the receiver is converted to electrical signals by a photo detector. Due to the monochromaticity of laser energy, light of other wavelengths can readily be excluded from the detector by filters, thus reducing extraneous noise in the received signal.

**Lidar Applications**

- Lidar applications in atmospheric research:
  [meteorology, climatology, boundary layer physics, pollution, visibility, radiative budget, atmospheric dynamics and chemistry (composition)]

- Lidar applications in solid-Earth research:
  [crustal movement, gravity field, Earth kinematics and terrain mapping, laser ranging will provide advances in precision orbit determination]

**Lidar Observation Capabilities**

Spaceborne lidars have the potential to observe directly or indirectly a number of atmospheric parameters, such as:

- altitude (layer) measurements of features (such as cloud tops, planetary boundary layers, temperature inversions, subvisible cirrus clouds and aerosol particles, backscattering profiles and extinction coefficients, ) → domain of backscatter lidars

- horizontal extent and density distribution of scattering layers (e.g. clouds and aerosols)

- vertical humidity, temperature and pressure profiles throughout the depth of the atmosphere (depth-resolved measurements) → domain of DIAL technique

- wind fields at many altitude layers → domain of Doppler Wind Lidars

Conceptually lidar measurements provide good vertical resolutions for altitude/height assignments of atmospheric features (the data is considered to be superior to that of sounding or imaging techniques). Lidars are active instruments, as such they can operate during the day and night cycle of an orbit. A disadvantage: lidars cannot penetrate optically thick layers (e.g. clouds).

Lidar concepts were proven and operational for extended periods throughout the past decade in ground-based and airborne equipment.[609),610)]

## C.1.11   Summary of Microwave Tracking Systems

There are four radiometric tracking systems in use (or in the development phase) addressing a variety of navigation requirements, among them precision orbit determination. These are: GPS, GLONASS, DORIS and PRARE.[611)]
While these systems differ in the source of the radio signals, they share common characteristics: they operate by determining the signal's "carrier" phase, its time derivative (i.e. Doppler shift) and/or its derivative with frequency (i.e. group delay and range).


609) H. Lutz, E. Armandillo, 'Laser Sounding from Space', Report of the ESA Technology Working Group on Space Laser Sounding and Ranging, esa SP-1108, Jan. 1989
610) P. Betout, D. Burridge, Ch. Werner, 'Doppler Lidar Working Group Report, esa SP-1112, June 1989
611) See "orbital Analysis" (Chapter 6.4, pp. 205- 212) in 'The Interdisciplinary Role of Space Geodesy', Springer Verlag, 1989,

**GPS and GLONASS** (one-way broadcast systems)

GPS and GLONASS are very similar general purpose positioning systems consisting of constellations of 21-24 satellites in roughly 12-hour orbits, continuously broadcasting dual frequency navigation signals. The carriers are modulated with pseudonoise ranging codes to permit both Doppler and one-way range measurements (also known as "pseudorange" measurements). The signals from each satellite blanket the Earth, extending roughly 3000 km beyond the Earth limb. At an altitude below 3000 km, any user of GPS and/or GLO-NASS will have four or more satellites (transmitters) in view. The combination of dual data types (phase and pseudorange) and continuous three-dimensional coverage will enable a quasi-geometric positioning technique.

**PRARE and DORIS**

PRARE and DORIS are tracking systems requiring a host satellite (for the space segment instrument) and global ground-based tracking networks for precise orbit determination applications. Both systems (PRARE and DORIS) require an orbit determination process by conventional dynamic techniques (with physical models) whose accuracies depend on the quality of the models.

DORIS is a one-way Doppler system which broadcasts continuously at two frequencies: 401 and 2036 MHz. The DORIS instrument on the host satellite observes individual ground beacons in sequence and measures the Doppler frequency of the received signals.

PRARE is a two-way tracking system, broadcasting dual-frequency signals (2200 and 8500 MHz) from a transmitter onboard the host satellite to receiver-transponders on the ground. The signals are modulated by pseudonoise ranging codes to permit both range and range-rate measurements. The 8500 MHz signal is coherently transponded back to the host satellite (at 7200 MHz) for onboard range and range rate extraction. The 2200 MHz signal is received and tracked on the ground to provide an ionospheric correction.

## C.1.12 Definitions of Orbital Terminology in Remote Sensing

**Inclination**

Inclination (I) is defined as the angle between the Earth's equatorial plane and the satellite's orbital plane (which remains constant, the different orbit periods of the Earth and the satellite provide longitudinal precession for successive orbits). An inclination angle of 30° means for instance that the observation area of the satellite is bounded by ± 30° latitude (tropical region). Polar orbits have an inclination near 90°.

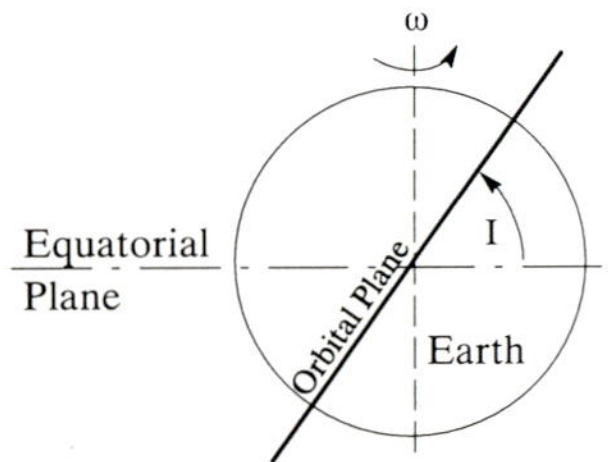


Orientation of the Orbit in Space

**Sun-synchronous Orbit**

An orbit is said to be sun-synchronous, if the orbit precession of the satellite compensates for the Earth's rotation around the sun. For near-Earth satellite orbits (altitude < 1000 km) sun-synchronous orbits are polar orbits at an orbital inclination and altitude such that a satellite passes over a given site always at the same local time. The orbital nodes maintain a near constant solar time. The satellite always crosses the equatorial plane at the same local time. This means that the sun position (lighting conditions) is always the same (within certain limits) for a repeat observation of any given site. There are advantages and disadvantages to the measurement at constant sun-angle depending on the application. Varying sun angles may be of importance to geologists to reveal subtle structural details not visible at higher sun angles.

**Repeat (cycle) Coverage or Temporal Resolution**
Time may be regarded as the fourth variable of resolution. A dynamic Earth environment changes with time, this applies in particular for meteorological observations and vegetation (crops) observations, etc.. Repeat coverage (or temporal resolution) designates the period of time that elapses until the next observation of the same geographic area can take place. Total coverage of the globe can only be provided by polar-orbiting satellites. Their orbit and swath width are usually chosen in such a way that Earth rotation accounts for a continuous and contiguous observation swath (with some overlapping) of the globe. In general each sensor aboard a satellite has a different swath width.

There is a fairly simple trade-off between spatial resolution (or IFOV) and frequency of coverage. Constraints are imposed by the sensor design, the onboard electronics, and the amount of source data that can be handled. Thus the smaller the IFOV, the more data there is to be handled for any given area on the ground and the less frequently data will be available for a given area.

Examples of repeat coverage requirements for some remote sensing applications:

- Operational meteorology (weather forecast) requires a high repeat coverage of large areas at relatively low spatial resolutions (1-5 km), the emphasis is on the measurement of progressive daily cycles. About 20 - 30 min/image of the same area for meteorological geosynchronous satellites (such as METEOSAT, GMS, GOES and INSAT series). Geosynchronous satellites permit the tracking of continuous dynamic processes with the desired viewing frequency (for example: viewing of Earth surface targets at any sun illumination angle, and viewing each point on the ground at exactly the same zenith and azimuth angles over time).

- Monitoring of renewable resources (vegetation, agricultural crops) requires a medium repeat coverage (in the order of several days and up to 2 weeks) but relatively high spectral and spatial resolutions (30 m) on a seasonal basis. Polar orbiting satellites (Landsat, POES, SPOT, Resurs, etc. ) realize these requirements.
  Note: There are also a number of transient phenomena that cannot be adequately seen from polar orbiting satellites, such as the diurnal variability and dynamical behavior of:
  - precipitation and evaporation
  - atmospheric water vapor and wind
  - vegetation color
  - terrestrial ecosystems-land processes
  - land-ocean-atmosphere energy fluxes
  - tropospheric pollutant generation and transport
  - ocean color-productivity
  - soil moisture and net radiation balance
  - etc.
- Cartographic applications (maps) of the Earth's surface require a relatively low repeat coverage but high resolutions (spatial and or spectral). Imagery for urban growth patterns is appropriate in time intervals of a year or so.


**Some Orbit Selection Requirements**
There are a number of factors which may influence the particular orbit selection:
- Global coverage requirement → implies polar or near-polar orbit
- Constant illumination requirement → implies a sun-synchronous orbit
- Continuous observation requirement → implies a geosynchronous orbit
- Measurement of gravity anomalies → implies a low orbit
- Requirement to minimize the atmospheric drag → preference of a high orbit
- Requirement to minimize a radar sensor's power → preference of a low orbit

## Lagrangian Points

Lagrangian[612] points (also referred to as 'libration points') in space are defined as such points, where the gravitational forces due to two or more bodies (the Sun, or planets) balance, and thus where a spacecraft can be positioned in equilibrium at zero velocity.

Lagrangian points are particular solutions of the equations of motion applied to the "Problem of Three Bodies" in an orbital plane of two massive bodies in circular orbits around a common center of gravity and a third body of negligible mass. The equations of motions yield always five points at which the third body can remain at equilibrium. Three of the points are on the line passing through the centers of mass of the two massive bodies - $L_3$ beyond the most massive body, $L_2$ beyond the less massive body, and $L_1$ (the point through which the mass transfer occurs) between the two bodies. The other two points, $L_4$ and $L_5$, are located at the two points in the orbit of the less massive component, which are equidistant (equilateral triangle) from the two main components as illustrated in Figure 153.

A satellite positioned at $L_4$ or $L_5$ is regarded as stable, i.e. it will remain at the point and move along in the specific reference system (here either in the Sun-Earth system or in the Earth-Moon system), while a satellite located at $L_1$ or $L_2$ or $L_3$ is unstable, i.e. it will wander off if slightly disturbed. However, this can be compensated for by small orbit corrections of a thruster engine.[613],[614]

Within the Sun-Jupiter system the $L_4$ and $L_5$ points are known to contain a number of asteroids referred to as the 'Trojans'. The first Trojan (Achilles) was discovered in 1906. By now more than 50 Trojans are known.

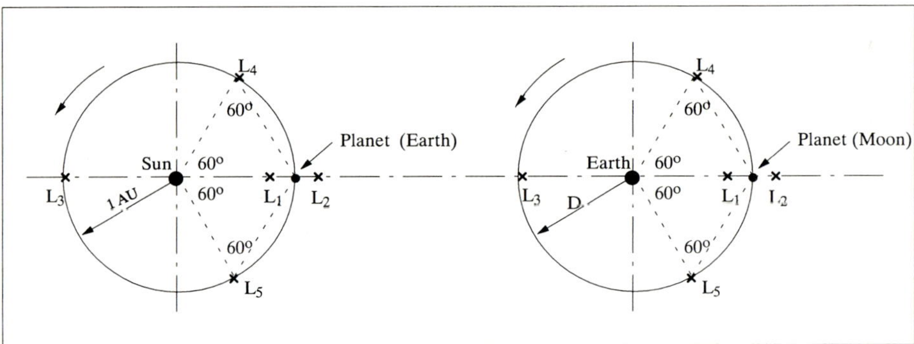


**Figure 153:    Lagrangian Points the Sun-Earth and Earth-Moon Systems**

$L_1$ in the Sun-Earth system is located about 1.5 million km from the Earth towards the Sun ($L_1$ is a possible position from which to study the Sun and the solar wind).


612)  Joseph Louis Lagrange (1736 - 1813), French Mathematician
613)  Fred P. J. Rimrott, "Introductory Orbit Dynamics", Vieweg, Braunschweig/Wiesbaden, 1989, pp. 156-158
614)  Karl Stumpff, "Himmelsmechanik", Band II, VEB Deutscher Verlag der Wissenschaften, Berlin 1965

| Lagrangian Points | Sun - Earth System | | Earth - Moon System | |
|---|---|---|---|---|
| | Distance from Sun | Distance from Earth | Distance from Earth | Distance from Moon |
| $L_1$ | 0.99 AU<br>$= 1.48 \times 10^8$ km | 0.00997 AU<br>$= 1.49 \times 10^6$ km | 0.85 D<br>$= 326740$ km | 0.15 D<br>$= 57660$ km |
| $L_2$ | 1.01 AU<br>$= 1.51 \times 10^8$ km | 0.010 AU<br>$= 1.50 \times 10^6$ km | 1.17 D<br>$= 449748$ km | 0.17 D<br>$= 65348$ km |
| $L_3$ | 0.999 AU<br>$= 1.496 \times 10^8$ km | 1.999 AU<br>$= 2.992 \times 10^8$ km | 0.99 D<br>$= 380556$ km | 1.99 D<br>$= 764956$ km |
| $L_4$ | 1.00 AU | 1.00 AU | 1.00 D | 1.00 D |
| $L_5$ | 1.00 AU | 1.00 AU | 1.00 D | 1.00 D |
| | $d_{Sun} = 1\,392\,000$ km | $d_{Earth} = 12\,740$ km (average diameter) | | $d_{Moon} = 3476$ km |
| Sun-Earth distance (center-center) = 1 AU (Astronomical Unit) = $1.496 \times 10^8$ km | | | Earth-Moon distance D = 384 400 km (center-center) | |

**Table 237:     Values of the Lagrangian Points in the Sun-Earth and Earth-Moon Systems**[615]


## C.2     Summary of World Data Centers (WDCs)

The ICSU (International Council of Scientific Unions) Panel on World Data Centers (Geophysical, Solar & Environment) was established at it's 12th General Assembly in 1968. The Panel oversees the operation of about 40 WDCs, which are maintained by their host countries and are responsible for collecting, archiving, exchanging, and distributing a wide range of data. These data provide baseline information for research in many ICSU discipline, especially for monitoring changes in the geosphere and biosphere.[616]

The WDC system is funded by the national host countries and grouped as follows:

- WDC-A in the USA
- WDC-B in Russia and CIS
- WDC-C1 in Europe
- WDC-C2 in Japan
- WDC-D in China

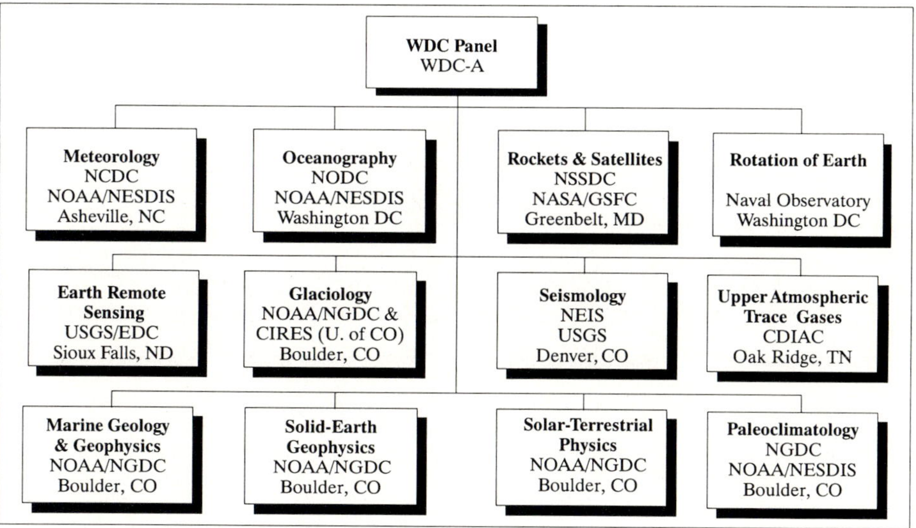


**Figure 154:     Structure of World Data Center A (USA)**

---


615)  Table values: courtesy of F. Jochim of DLR-GSOC
616)  Brochure of WDC Publications Office, J. H. Allen, NOAA/NGDC, E/GCT2, Boulder, CO 80303, USA

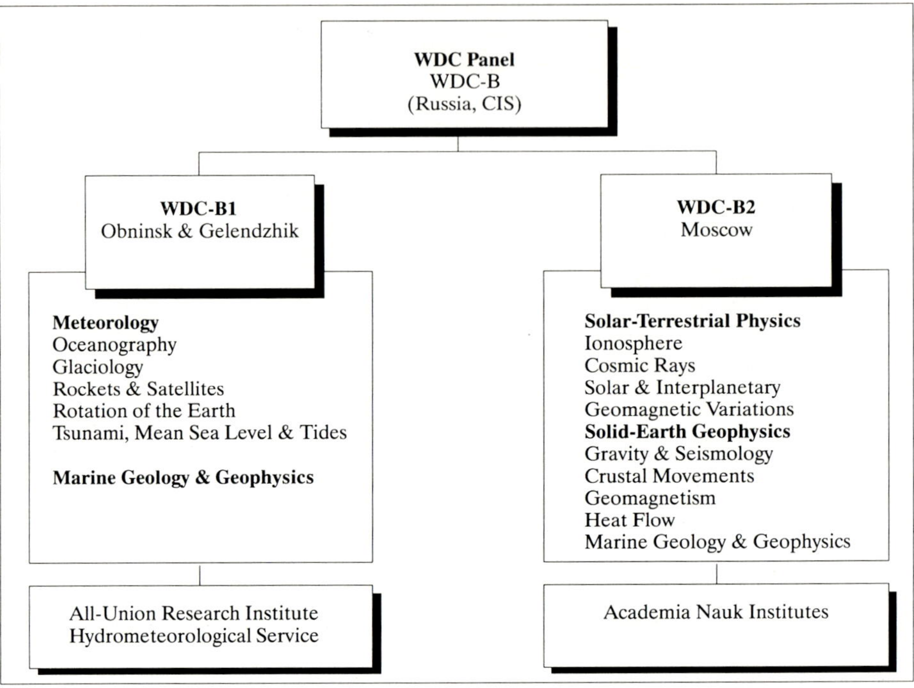


Figure 155:    Structure of the World Data Center B (Russia, CIS)

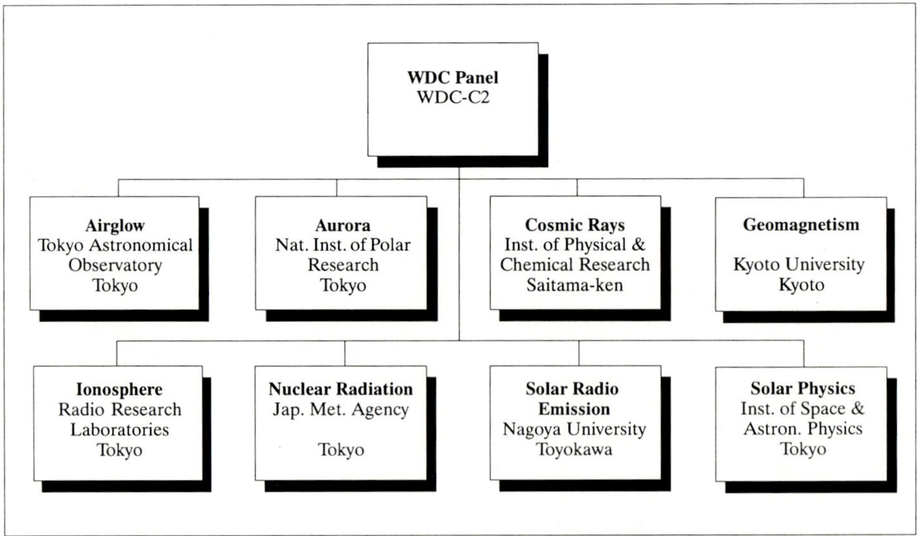


Figure 156:    Structure of World Data Center C2 (Japan)

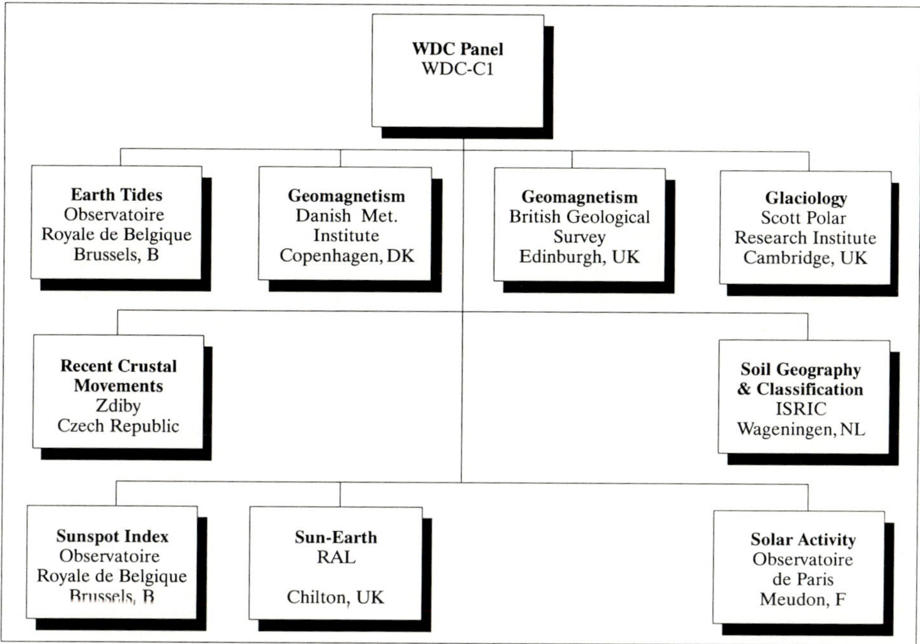


**Figure 157:    Structure of World Data Center C1 (Europe)**

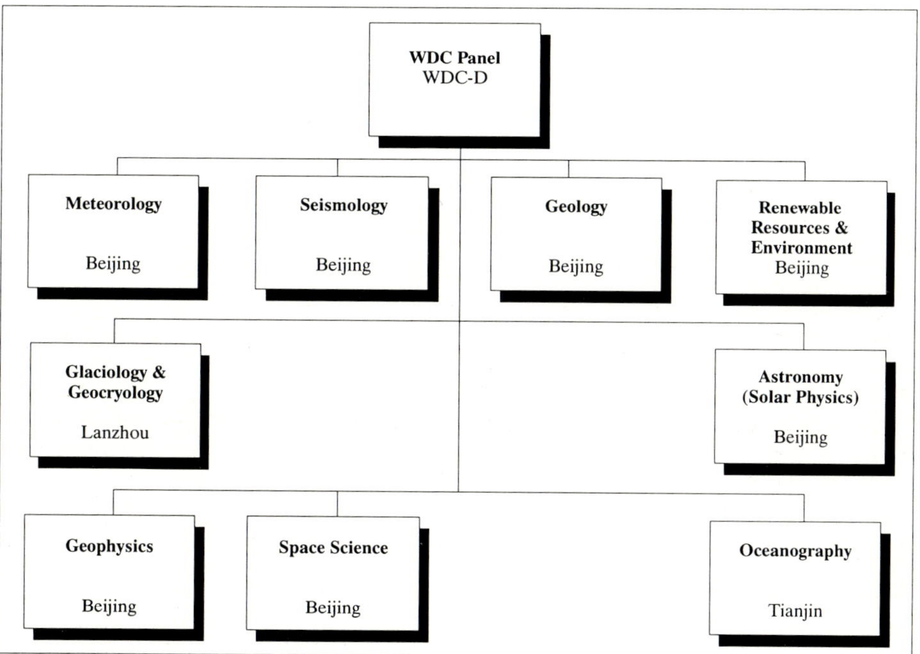


**Figure 158:    Structure of World Data Center D (China)**

## C.3    Committee on Earth Observation Satellites - CEOS

CEOS was created in 1984 as a result of the international Economic Summit of Industrialized Nations (G-7 Versailles Summit of 1982) and serves as the focal point for international coordination of space-related, Earth observation activities.[617]

Policy and technical issues of common interest related to the whole spectrum of Earth observation satellite missions and data received from such are addressed. CEOS encourages complementarity and compatibility among spaceborne Earth observing systems through coordination in mission planning, promotion of full and nondiscriminatory data access, setting of data product standards, and development of compatible products, services, and applications. The user community benefits directly from this international coordination.

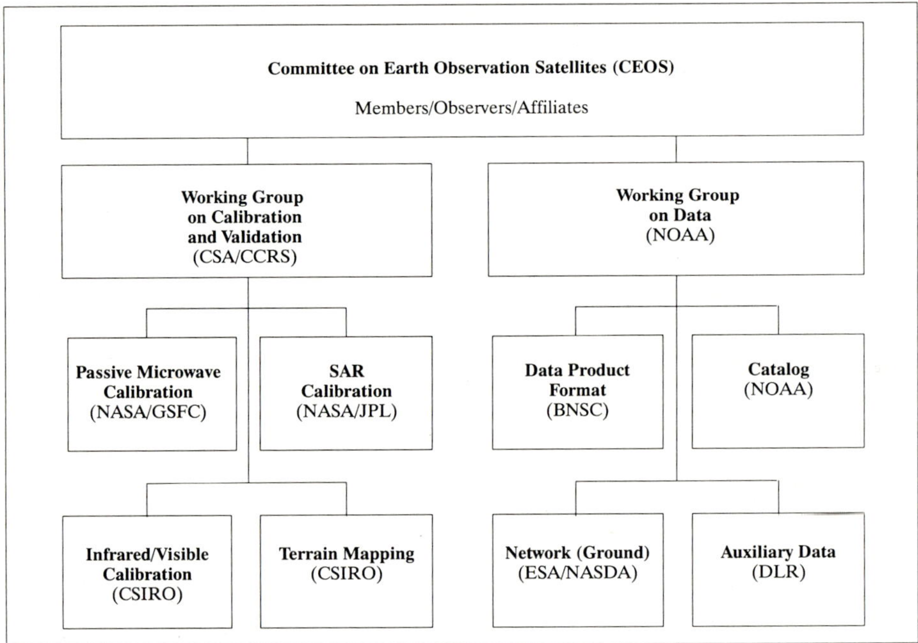


**Figure 159:    CEOS Working Group Structure**

Members are those national and international government agencies with funding and program responsibilities for a satellite Earth observation program currently operating or in the later stages of system development. CEOS members include: CSIRO (Australia), INPE (Brazil), CSA (Canada), CAST (China), ESA and EUMETSAT (Europe), CNES (France), DARA (Germany), ISRO (India) ASI (Italy), STA (Japan), RKA and ROSGIDROMET (Russia), SNSB (Sweden), BNSC (United Kingdom), NASA and NOAA (United States).

Governmental entities with a space-based Earth observation program in the early stages of development or with significant ground segment activity that supports CEOS member agency programs may be invited to participate as an observer. Current (1993) observers are: BOST (Belgium), CCRS (Canada), NRSC (China), CEC (Commission of the European Communities), CRI (New Zealand), and NSC (Norway).

Affiliate members to CEOS are: International Council of Scientific Unions (ICSU), France; International Geosphere-Biosphere Program (IGBP), Sweden; Intergovernmen-

617) L. Moodie, ”Committee on Earth Observation Satellites (CEOS)”, CEOS Newsletter Nr. 1 , Summer 1993

tal Oceanic Commission (IOC), France; World Climate Research Program (WCRP), Switzerland; World Meteorological Organization (WMO), Switzerland; United Nations Environment Program (UNEP), Kenya; Global Climate Observing System (GCOS), Switzerland); Global Ocean Observing System (GOOS), France.

The CEOS plenary meets once a year. There are two working groups, each with four active subgroups as illustrated in Figure 159. The chairmanship of the working groups and subgroups are indicated.

## C.4    Bits and Pieces

| Name of Launch Site | Resp. Organization (Country) | Location (Latitude, Long.) | Remarks |
|---|---|---|---|
| Baikonur Cosmodrome (Tyuratam Leninsk) | CIS (Kazakhstan) | 45°55'N, 63°20'E | since 1957, launch of manned S/C, interplanetary S/C and geostationary S/C |
| Plesetsk (northern Cosmodrome) | CIS, Russia | 62°52'N, 40°42'E | since 1966, launch of polar orbiting satellites |
| Kapustin Yar (Volgograd Cosmodrome) | CIS | 48°30'N, 45°46'E | since 1962, launch of small satellites and interplanetary space probes |
| Cape Canaveral AFB, Fla | USAF/NASA (USA) | 28°30'N, 80°33'W | since 1958, launch of manned S/C, interplanetary S/C, geostationary satellites, etc. |
| Kennedy Space Center, Fla | NASA (USA) | 28°37'N, 80°36'W | since 1967, Complex 39 (launch of Saturn V vehicle, adapted to launch the Shuttle |
| Vandenberg AFB, Ca. | USAF/NASA (USA) | 34°42'N, 120°33'W | since 1959, launches of S/C into polar orbit |
| Wallops Island, Va. | GSFC (USA) | 37°52'N, 75°28'W | since 1961, launch of sounding rockets and Scout launcher for small satellites |
| Kagoshima Space Center | ISAS, Uni. of Tokyo, (Japan) | 31°15'N, 131°05'E | since 1970, launch of scientific satellites |
| Tanegashima Space Center | NASDA (Japan) | 30°24'N, 130°58'E | since 1975, launch sites for scientific and commercial satellites |
| SHAR Sriharikota | ISRO (India) | 13°47'N, 80°15'E | since 1980, launch of some ISRO satellites |
| San Marco Platform (off the coast of Kenya) | CRA, Uni. of Rome (Italy) | 2°56'S, 40°12'E | since 1967, launch of Scout rockets for NASA/ASI |
| Guiana Space Centre Kourou (Fr. Guiana) | CNES/ESA | 5°14'N, 52°46'W | since 1970, launch of scientific and commercial satellites |
| Woomera Range | WRE (Australia) | 31°35'S, 136°50'E | since 1967, launch of sounding rockets |
| Jiuquan SLC | China | 40°50'N, 100°02'E | since 1970, launch of experimental satellites |
| Xichang SLC | China | 27°58'N, 102°13'E | since 1984, launch of satellites into geost. and polar orbits |
| Taiyuan | China | 37°30'N, 112°36'E | since 1988, launches of CZ-4A into polar orbits for remote sensing |
| Esrange, Kiruna | SSC=Swedish Space Corporation (Sweden) | 67°53'N, 21°04'E | launch of sounding rockets |
| Alcântara Launch Center | CTA/INPE (Brazil) | 2°24'S, 44°24'W | in preparation for satellite launches |

**Table 238:    Coordinates of Satellite Launch Sites around the World**[618]

---

618) 'The Cambridge Encyclopedia of Space', Cambridge University Press 1990, p. 121

| Mission Incl., Alt.(km) | Launch/ Landing Date | Orbiter | Mission Highlights / Payload Complement |
|---|---|---|---|
| STS-1 I:40.3°, A:307 | 12. April 1981 14. April 1981 | Columbia | Major systems tested, orbiter sustained tile damage |
| STS-2 I:38°, A:290 | 12. Nov. 1981 14. Nov. 1981 | Columbia | OSTS-1 payload (including MAPS) |
| STS-3 I:38°, A:272 | 22. Mar. 1982 30. Mar. 1982 | Columbia | Get Away Special, spacelab pallet mounted, OSS-1 payload, MLR, EEVT, HBT, SSIP |
| STS-4 I:28.5°, A:364 | 27. June 1982 4. July 1982 | Columbia | DOD Payload, Get Away Specials, CFES, MLR, IECM, SSIP |
| STS-5 I:28.5°, A:340 | 11. Nov. 1982 16. Nov. 1982 | Columbia | ANIK C-3 and SBS C satellites, Get Away Special, SSIP |
| STS-6 I:28.5°, A:330 | 4. April 1983 9. April 1983 | Challenger | 1st EVA, TDRS-1, CFES, MLR, RME, NOSL, Get Away Specials |
| STS-7 I:28.5°, A:360 km | 18. June 1983 24. June 1983 | Challenger | ANIK 2C and PALAPA-B1 satellites, Get Away Specials, 10 experiments mounted on SPAS-01, OSTA-2 payload, CFES, MLR, SSIP |
| STS-8 I:28.5°, A:353 | 30. Aug. 1983 5. Sept. 1983 | Challenger | INSAT-1B satellite, CFES, SSIP, ICAT, RME, Get Away Specials, |
| STS-9 I:57°, A:287 | 28. Nov. 1983 8. Dec. 1983 | Columbia | Spacelab-1 mission (ESA), see A.102 |
| STS-10 41-B I:28.5°, A:350 km | 3. Feb. 1984 11. Feb. 1984 | Challenger | WESTAR-VI and PALAPA-B2 satellites, SPAS-2, Get Away Specials, Cinema-360 camera, ACES, MLR, RME, IEF, MOMS-01 |
| STS-11 41-C I:28.5°, A:580 | 6. April 1984 13. April 1984 | Challenger | SMM satellite repair in orbit, LDEF deployment (57 experiments), IMAX camera, RME, Cinema-360, SSIP |
| STS-12 41-D I:28.5°, A:340 km | 30. Aug. 1984 5. Sept. 1984 | Discovery | 3 satellites (SBS-D, SYNCOM IV-2, TELSTAR), OAST-1, CFES III, RME, SSIP, IMAX camera, CLOUDS |
| STS-13 41-G I:57°, A:411 | 5. Oct. 1984 13. Oct. 1984 | Challenger | ERBS satellite,OSTA-3, LFC, IMAX camera, CANEX, APE, RME, TLD, Get Away Specials |
| STS-14 51-A I:28.5°, A:342 | 8. Nov. 1984 16. Nov. 1984 | Discovery | TELESAT-H (ANIK) and SYNCOM IV-1, retrieval of PALAPA-B-2 and WESTAR-VI, DMOS, RME, |
| STS-15 51-C I:28.5°, A:407 | 24. Jan. 1985 27. Jan. 1985 | Discovery | DOD payload, |
| STS-16 51-D I:28.5°, A:527 | 12. April 1985 19. April 1985 | Discovery | TELSAT-I (ANIK C-1) and SYNCOM IV-3 satellites, CFES III, SSIP, AFE, Get Away Specials, PPE, |
| STS-17 51-B I:57° A:411 km | 29. April 1985 6. May 1985 | Challenger | Spacelab-3 mission ESA,see A.103 [material sciences, life sciences, fluid mechanics, atmospheric physics (AT-MOS), and astronomy], Get Away Specials, |
| STS-18 51-G I:28.5°, A:405 | 17. June 1985 24. June 1985 | Discovery | MORELOS-A, ARABSAT-A and TELSTAR-3D satellites, SPARTAN-1, Get Away Specials, HPTE, ADSF |
| STS-19 51-F I:49.5° A:383 km | 29. July 1985 6. Aug. 1985 | Challenger | Spacelab-2 mission, [life sciences, plasma physics (PDP), astronomy (XRT, IRT), high-energy astro-physics, solar physics (SUSIM, HRTS, CHASE, SOUP), atmospheric physics, and technology research] |
| STS-20 51-I I:28.5°, A:514 | 27. Aug. 1985 3. Sept. 1985 | Discovery | ASC-1, AUSSAT-1, and SYNCOM IV-4 satellites, PVTOS |
| STS-21 52-J I:28.5°, A:590 | 3. Oct. 1985 7. Oct. 1985 | Atlantis | DOD mission |
| STS-22 61-A I:57°, A:383 | 30. Oct. 1985 6. Nov. 1985 | Challenger | Spacelab D1 (German mission), GLOMR satellite deployed from Get Away Special canister |
| STS-23 61-B I:28.5°, A:416 km | 26. Nov. 1985 3. Dec. 1985 | Atlantis | MORELOS-B, AUSSAT-2, and SATCOM satellites, EASE, ACCESS, CFES, DMOS, MPSE, OEX, Get Away Special, IMAX cargo bay camera (ICBC) |
| STS-24 61-C I;28.5°, A:392 | 12. Jan. 1986 18. Jan. 1986 | Columbia | SATCOM KU-1 satellite, CHAMP, MSL-2, Hitchhiker G-1, IR-IE, IBSE, HPCG, SSIP, Get Away Specials, |
| STS-25 51-L | 28. Jan. 1986 | Challenger | accident !!!! |

| Mission Incl., Alt.(km) | Launch/ Landing Date | Orbiter | Mission Highlights / Payload Complement |
|---|---|---|---|
| STS-26 I:28.5°, A:376 | 29. Sept. 1988 3. Oct. 1988 | Discovery | TDRS-3 satellite, PVTOS, PCG, IRCFE, ARC, IFE, MLE, PPE, ELRAD, ADSF, SSIP, OASIS-1 |
| STS-27 I:57°, A: | 2. Dec. 1988 6. Dec. 1988 | Atlantis | DOD payload |
| STS-29 I:28.5°, A:340 | 13. Mar. 1989 18. Mar. 1989 | Discovery | TDRS-4 satellite, OASIS-1, SHARE, PCG, CHRO-MEX, SSIP, AMOS, IMAX camera |
| STS-30 I:28.8°, A:340 | 4. May 1989 8. May 1989 | Atlantis | Magellan/Venus radar mapper S/C, MLE, FEA, AMOS |
| STS-28 I:57°, A: | 8. Aug. 1989 13. Aug. 1989 | Columbia | DOD payload |
| STS-34 I:34.3°, A:342 | 18. Oct. 1989 23. Oct. 1989 | Atlantis | Galileo S/C to Jupiter, SSBUV, GHCD, PM, STEX, MLE, IMAX camera, AMOS, SSIP |
| STS-33 I:28.5°, A: | 22. Nov. 1989 27. Nov. 1989 | Discovery | DOD payload |
| STS-32 I:28.5°, A:389 | 9. Jan. 1990 20. Jan. 1990 | Columbia | SYNCOM IV-F5 satellite, retrieval of LDEF, CNCR, PCG, FEA, AFE, L3, MLE, IMAX, AMOS |
| STS-36 I:62°, A: | 28. Feb. 1990 4. Mar. 1990 | Atlantis | DOD payload |
| STS-31 I:28.5°, A:703 | 24. April 1990 29. April 1990 | Discovery | HST satellite, IMAX (ICBC), APM, PCG, RME III, IPMP, SSIP, AMOS |
| STS-41 I:28.5°, A:340 | 6. Oct. 1990 10. Oct. 1990 | Discovery | Ulysses spacecraft (ESA), SSBUV, ISAC, CHROMEX, VCS, SSCE, IPMP, PSE, RME III, SSIP, AMOS |
| STS-38 | 15. Nov. 1990 20. Nov. 1990 | Atlantis | DOD payload |
| STS-35 I:28.5°, A:520 | 2. Dec. 1990 10. Dec. 1990 | Columbia | ASTRO-1 observatory (consisting of HUT, WUPPE, UIT, and BBXRT), SAREX-2, AMOS |
| STS-37 I:28.5°, A:518 | 5. April 1991 11. April 1991 | Atlantis | GRO (Gamma Ray Observatory), CETA, APM, SAREX II, PCG, BIMDA, RME III, AMOS |
| STS-39 I:57°, A:298 | 28. April 1991 6. May 1991 | Discovery | DOD mission, including unclassified payload: AFP-675, IBSS, CIV, CRO, SPAS-II, STP-1, RME III, CLOUDS-1 |
| STS-40 I:28.5°, A:340 | 5. June 1991 14. June 1991 | Columbia | SLS-1 (Spacelab Life Sciences-1), Get Away Specials, MODE, OEX |
| STS-43 I:28.5° A:340 km | 2. Aug. 1991 11. Aug. 1991 | Atlantis | TDRS-5 satellite, SHARE II, SSBUV, TPCE, OCTW, APE-B, PCG III, BIMDA, IPMP, SAMS, SSCE, UVPI, AMOS |
| STS-48 I:57°, A:657 | 12. Sept. 1991 18. Sept. 1991 | Discovery | UARS satellite, APM, MODE, SAM, CREAM, PARE, PCG II-2, IPMP, AMOS |
| STS-44 I:28.5° A:416 km | 24. Nov. 1991 1. Dec. 1991 | Atlantis | DOD mission, DSP (Defense Support Satellite), plus unclassified payloads: IOCM, AMOS, CREAM, SAM, RME III, VFT-1, UVPI, |
| STS-42 I:57° A:348 km | 22. Jan. 1992 30. Jan. 1992 | Discovery | IML-1 (International Microgravity Laboratory-1), Get Away Specials, GOSAMR-1, IMAX, IPMP, RME III, SSIP |
| STS-45 I:57°, A:340 | 24. Mar. 1992 2. April 1992 | Atlantis | ATLAS-1 payload, SSBUV/A, Get Away Special, IPMP, STL-01, RME III, VFT-2, CLOUDS, SAREX II |
| STS-49 I:28.4°, A:390 | 7. May 1992 16. May 1992 | Endeavour | EVA, Intelsat-VI capture and redeployment, ASEM, CPCG, UVPI, AMOS |
| STS-50 I:28.5°, A:350 | 25. June 1992 9. July 1992 | Columbia | USML-1 (US Microgravity Lab-1), IPMP, SAREX II, UVPI |
| STS-46 I:28.5° A:502 km | 31. July 1992 8. Aug. 1992 | Atlantis | EURECA-1 satellite, TSS (Tethered Satellite System), EOIM-III/TEMP, CONCAP II and III, ICBC, LDCE, AMOS, PHCF, UVPI |
| STS-47 I:57°, A:346 | 12. Sept. 1992 20. Sept. 1992 | Endeavour | Spacelab-J (Japan), Get Away Specials, ISAIAH, SSCE, SAREX II, AMOS, UVPI |
| STS-52 I:28.5°, A:302 km | 22. Oct. 1992 1. Nov. 1992 | Columbia | LAGEOS-II, USMP-1, CANEX-2, SVS, MELEO, QUELD, PARLIQ, OGLOW-2, SATO, CTA, CMIX, CPCG, CVTE, HPP, PSE, SPIE, UVPI |

| Mission<br>Incl., Alt.(km) | Launch/<br>Landing Date | Orbiter | · | Mission Highlights / Payload Complement |
|---|---|---|---|---|
| STS-53<br>I:57°,<br>A:370 km | 2. Dec.  1992<br>9. Dec.  1992 | Discovery | | DOD payload, Get Away Specials, CRYOHP, ODER-ACS, BLAST, CLOUDS, CREAM, FARE, HERCULES, MIS-1, RME III, STL, VFT-2 |
| STS-54<br>I:28.5°, A:296 | 13. Jan.  1993<br>19. Jan.  1993 | Endeavour | | TDRS-6 satellite, DXS, CGBA, CHROMEX, PARE, SAMS, SSCE, |
| STS-56<br>I:57°, A:296 | 8. April  1993<br>17. April 1993 | Discovery | | ATLAS-2 payload, SPARTAN-201, SUVE, CMIX, PARE, STL-1, CREAM, HERCULES, RME III, AMOS |
| STS-55<br>I:28.5°, A:296 | 26. April 1993<br>6. May    1993 | Columbia | | Spacelab D2 (German mission MEDEA,WL, HOLOP, USS, MAUS, AOET, AR, BB, BA, ROTEX),  SAREX, |
| STS-57<br>I:28.5°,<br>A:474 km | 21. June  1993<br>1. July    1993 | Endeavour | | SPACEHAB-1 (22 experiments), EURECA-1 retrieval, Get Away Specials, CONCAP IV, SHOOT, FARE, SA-REX-II, AMOS |
| STS-51<br>I:28.5°<br>A:296 km | 12. Sept.  1993<br>22. Sept.  1993 | Discovery | | ACTS satellite, ASTRO-SPAS deployment of ORFEUS, IMAX, LDCE, CPCG, CHROMEX-04, HRSGS-A, APE-B, IPMP, RME III, AMOS |
| STS-58<br>I:39°, A:283 | 18. Oct.  1993<br>1. Nov.    1993 | Columbia | | SLS-2 (Spacelab Life Sciences-2), DEEFD, OARE, SA-REX-II, PILOT |
| STS-61<br>I:28.5°, A:594 | 2. Dec.  1993<br>13. Dec. 1993 | Endeavour | | Hubble Space Telescope (HST) repair mission (optical correction), IMAX |
| STS-60<br>I:57°, A:404 | 3. Feb.    1994<br>11. Feb.  1994 | Discovery | | WSF-01, Spacehab-2 |
| Legend: | I = Inclination (°) | | A = Nominal Altitude (km) | |
| Projected Flights | | | | |
| STS-62, I:39° | March 94 | Columbia | | USMP-2, OAST-2 payload |
| STS-59, I:57° | April 94 | Endeavour | | SRL-01 (Space Radar Laboratory, also referred to as SIR-C/X-SAR, see A.96), |
| STS-65,I:28.5° | July 94 | Columbia | | IML-2 |
| STS-68, I:57° | Aug. 94 | Atlantis | | SRL-02 |
| STS-64, I:57° | Sept. 94 | Discovery | | Lite-1 (see A.58), SPTN-201 |
| STS-66, I:57° | Oct. 94 | Endeavour | | ATLAS-03, CRISTA, SPAS-01 |
| STS-67,I:28.5° | Dec. 94 | Columbia | | ASTRO-02 |
| STS-63,I:51.6° | Jan.  95 | Discovery | | Spacehab-3, SPTN-204 |
| STS-69,I:51.6° | Feb. 95 | | | SPAS-III, OAST-Flyer |
| STS-71,I:51.6° | May 95 | | | SL-M Spacelab - MIR |

**Table 239:    Space Shuttle Mission Chronology**[619),620),621)]


| Agency | Mission/Program |
|---|---|
| NASDA/Japan | ADEOS series (Advanced Orbiting Observing Satellite)<br>Launch: ADEOS in Feb. 1986; ADEOS-II in 1999, |
| NASA/NASDA | TRMM series (Tropical Rainfall Measuring Mission)<br>Launch: Aug. 1997 |
| ESA | POEM-ENVISAT series; (Environmental Satellite)<br>Launch of ENVISAT-1 in 1988 |
| NOAA | POES series (Polar Orbiting Environmental Satellite) for POES-N and<br>follow-on missions, Launch of POES-N in 1999 |
| EUMETSAT | POEM-METOP series (Meteorological Operational Satellite)<br>Launch of METOP-1 in 2000 |

**Table 240:    Overview of IEOS Mission/Program Series**

---

619)  Information taken from NASA publication: "Space Shuttle Mission Chronology 1981-1992"
620)  The acronyms of the Shuttle payload complements are listed in Appendix B "Definition of Acronyms"
621)  Shuttle orbits provided by J. Gass of NASA/GSFC

| Program | GMS | GOES | METEOSAT |
|---|---|---|---|
| Spin Rate (RPM) | 100 | 100 | 100 |
| Line Scan Direction | W - E | W - E | E - W |
| Telescope Step Direction | N - S | N - S | S - N |
| Number of Scan Lines for full Disk IR | 2500 | 1750 | 2500 |
| Resolution of Sub-Satellite Point (km): <br> VIS <br> IR <br> WV | 1.25 <br> 5.0 <br> - | 0.9 <br> 7-14 <br> - | 2.5 <br> 5.0 <br> 5.0 |
| Full Disk Scan Time (minutes) | 25 | 17.5 | 25 |
| Sensor Type: <br> VIS <br> IR <br> WV | Photomultipliers <br> HgCdTe <br> - | | Si Photodiode <br> HgCdTe <br> HgCdTe |
| Spectral Response (μm): <br> VIS <br> IR <br> WV | 0.55 - <br> 0.75 10.5 <br> - 12.5 | 0.55 - 0.72 <br> 3.9 - 15 <br> - | 0.5 - 0.9 <br> 10.5 - 12.5 <br> 5.7 - 7.1 |

**Table 241:** **Summary of International Geostationary Radiometer Characteristics**[622]

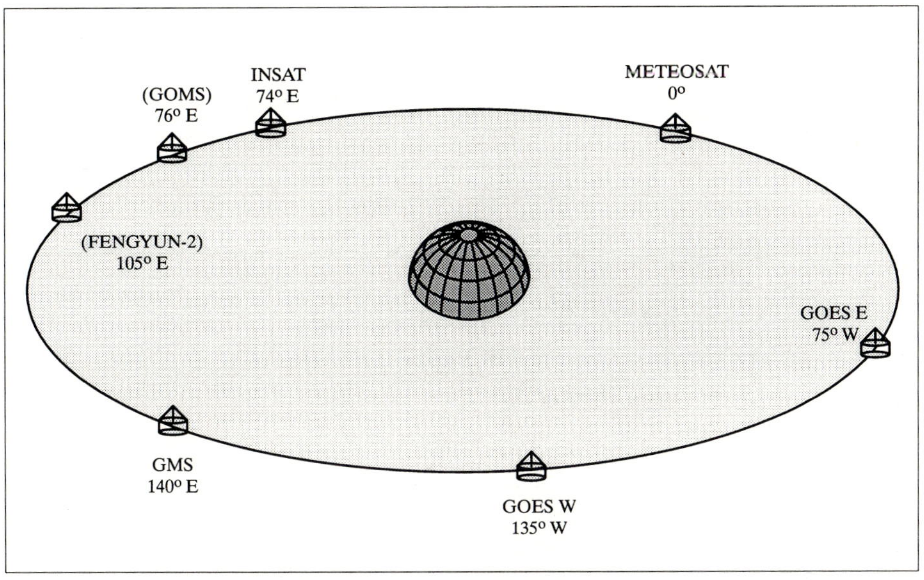


**Figure 160:** **Summary of the Geostationary Meteorological Satellite Families**

622) Introduction to the METEOSAT Operational System, esa BR-32 ISSn 250-1589, Sept. 1987, p. 39
Note: As an alternative to normal operation GOES satellites include the possibility of an Atmospheric Sounding mode using a multispectral radiometer. In addition the GOES and GMS satellites carry the SEM sensor.

| International Designation | Name of Frequency Band | Frequency Band | Wavelength ($\lambda$) | Remarks/Use |
|---|---|---|---|---|
| ELF | Extremely Low Frequency (Megametric waves) | 30 - 3000 Hz | 10000 - 100 km | Alternating current (50 - 60 Hz) |
| VF | Voice Frequency | 300 - 3000 Hz | 1000 - 100 km | Voice |
| VLF | Very-low Frequency (Myriametric waves) | 10 - 30 kHz | 30 - 10 km | Radiotelegraphy |
| LF | Low Frequency (Kilometric waves) | 30 - 300 kHz | 10 - 1 km | Radiotelegraphy, Radio |
| MF | Medium Frequency (Hectometric waves) | 300 - 3000 kHz | 1000 - 100 m | Radio, (AM Radio) |
| HF | High Frequency (Decametric waves) | 3 - 30 MHz | 100 - 10 m | Radiotelephony, Radio Navigation, amateur radio |
| VHF | Very-High Frequency (Metric waves) | 30 - 300 MHz | 10 - 1 m | Radio, TV, Radio Navigation, radiobeacon, (FM radio) |
| UHF | Ultra-High Frequency (Decimetric waves) | 300 - 3000 MHz | 1 m - 0.1 m | TV, radiobeacon, Satellite control, radiolocation, radiorange |
| SHF | Super-High Frequency (Centimetric waves) | 3 - 30 GHz | 10 - 1 cm | Radar, radiobeacon, Satellite broadcast, Maser, MW Heating |
| EHF | Extremely-High Frequency (Millimetric waves) | 30 - 300 GHz | 10 - 1 mm | |
| - | (Decimillimetric waves) | 300 - 3000 GHz | 1 - 0.1 mm | |

▢ ◀—— ITU Radio Spectrum          ▨ ◀—— Microwave Spectrum

| International Designation | Name of Frequency Band | Frequency Band | Wavelength ($\lambda$) | Remarks/Use |
|---|---|---|---|---|
| IR | Infrared | $3 \times 10^{11}$ Hz - $3.8 \times 10^{14}$ Hz | 1 mm - 0.78 $\mu$m | Heat Sounding, Laser Communication |
| VIS | Visible Light | $3.8 \times 10^{14}$ Hz - $7.9 \times 10^{14}$ Hz | 0.78 - 0.38 $\mu$m | Light Telephony, Laser, Electro-optical distance measurements |
| UV | Ultraviolet | $7.9 \times 10^{14}$ Hz - $3 \times 10^{16}$ Hz | 0.38 - 0.01 $\mu$m | Light Telephony, Laser, Electro-optical distance measurements |
| X-Ray | X-Ray (Soft band) | $5 \times 10^{15}$ Hz - $3 \times 10^{19}$ Hz | 60 - 0.1 nm | X-Ray diagnostics, -therapy |
| X-Ray | X-Ray (Medium band) | $3 \times 10^{19}$ Hz - $3 \times 10^{20}$ Hz | $10^{-2}$ - $10^{-3}$ nm | Material Testing |
| X-Ray | X-Ray (Hard band) | $3 \times 10^{20}$ Hz - $2 \times 10^{25}$ Hz | $10^{-3}$ - $10^{-8}$ nm | Atomic particle transitions |
| $\gamma$-Ray | Gamma rays | $8 \times 10^{17}$ Hz - $4.7 \times 10^{21}$ Hz | 0.4 - $10^{-4}$ Hz | Radiology, Material Testing |

**Table 242:     ITU Frequency Band Allocation of the Electromagnetic Spectrum[623]**


## Radio Wave Propagation

Electromagnetic waves propagate in space at the speed of light (c), approximately 300000 km/s. The wave amplitude varies periodically in time with frequency (f) measured in periods per second (Hz). The distance from crest to crest is the wavelength ($\lambda$), measured in meters. The relationship between these three parameters is $f = c/\lambda$.

Radio waves with frequencies of the order of 1 megahertz (MHz) can be reflected by the ionized atmosphere, or refracted when entering or coming out of these ionospheric layers, or even attenuated when crossing them. These effects which influence the ray path, depend on the wavelength (or frequency), and the angle of incidence. Thus, radio messages can be sent to considerable distances, over the horizon and to the antipodes. Yet, some frequency limitations remain; the highest frequencies are not reflected at all. This last property makes space telecommunication possible. However, at the higher frequencies, the radiation is affected by the atmospheric particles and molecules encountered, such as ice, rain drops, water vapor, and oxygen (see also Figure 162).

---

623) 'Reference Data for Radio Engineers', ITT, Sixth Edition 1982

| Letter Designation | Subbands | Frequency Band (GHz) | Wavelength $\lambda$ (cm) | Remarks/Use |
|---|---|---|---|---|
| P-Band | 2 (Nr. of subbands) | 0.225 - 0.390 | 133.3 - 76.9 | |
| L -Band | 10 (p,c,l,y,t,s,x,k,f,z) | 0.390 - 1.550 | 76.9 - 19.3 | |
| S-Band | 13 (e,f,t,c,q,y,g,s,a,w,h,z,d) | 1.55 - 5.20 | 19.3 - 5.77 | |
| C-Band | 5 ($S_z$, $S_d$, $X_a$, $X_q$, $X_y$) | 3.90 - 6.20 | 7.69 - 4.84 | C-Band is a portion of the S-Band and a portion of the X-Band |
| X-Band | 12 (a,q,y,d,b,r,c,l,s,x,f,k) | 5.20 - 10.90 | 5.77 - 2.75 | |
| K-Band | 12 (p,s,e,c,u,t,q,r,m,n,l,a) | 10.90 - 36.00 | 2.75 - 0.834 | |
| $K_u$-Band | 1 ($K_u$) | 15.35 - 17.25 | 1.95 - 1.74 | K-Subband |
| $K_1$-Band | 3 ($K_u$ through $K_q$) | 15.35 - 24.50 | 1.74 - 1.22 | K-Subband |
| $K_a$-Band | 1 ($K_a$) | 33.00 - 36.00 | 0.909 - 0.834 | K-Subband |
| Q-Band | 5 (a,b,c,d,e) | 36.0 - 46.0 | 0.834 - 0.652 | |
| V-Band | 5 (a,b,c,d,e) | 46.0 - 56.0 | 0.652 - 0.536 | |
| W-Band | 2 | 56.0 - 100.0 | 0.536 - 0.300 | |

**Table 243:    International Frequency Range Allocations of Microwave Bands**

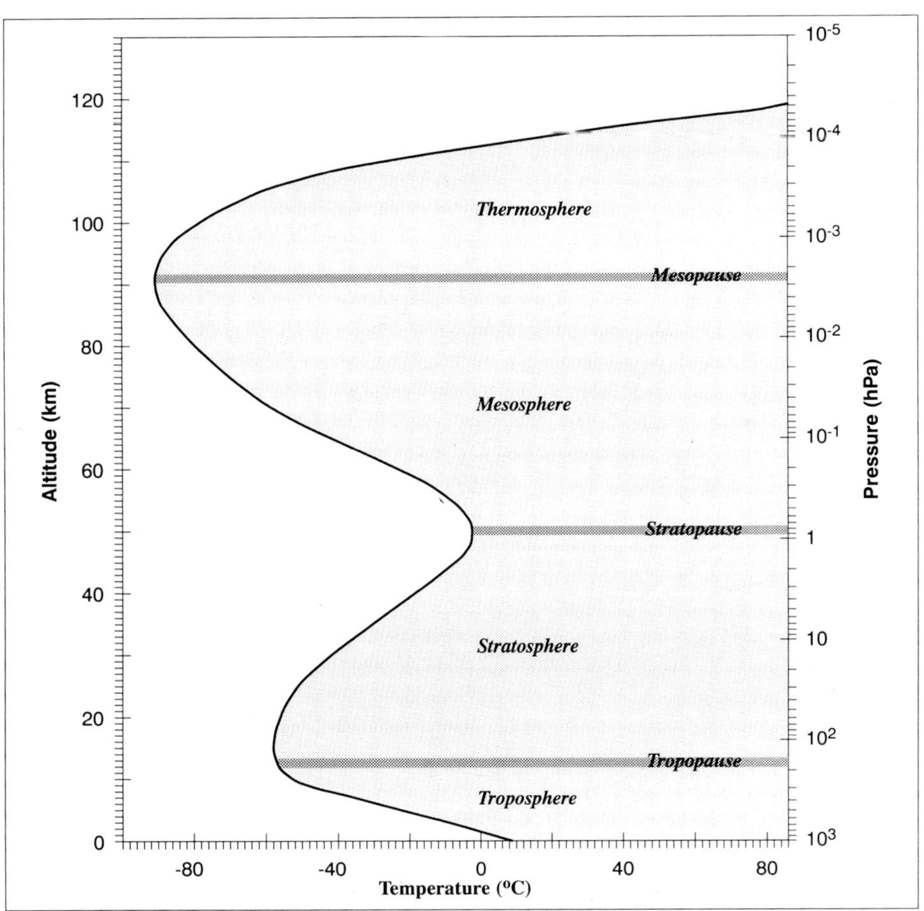


**Figure 161:    Temperature Profile of the Atmosphere in its vertical Layers**

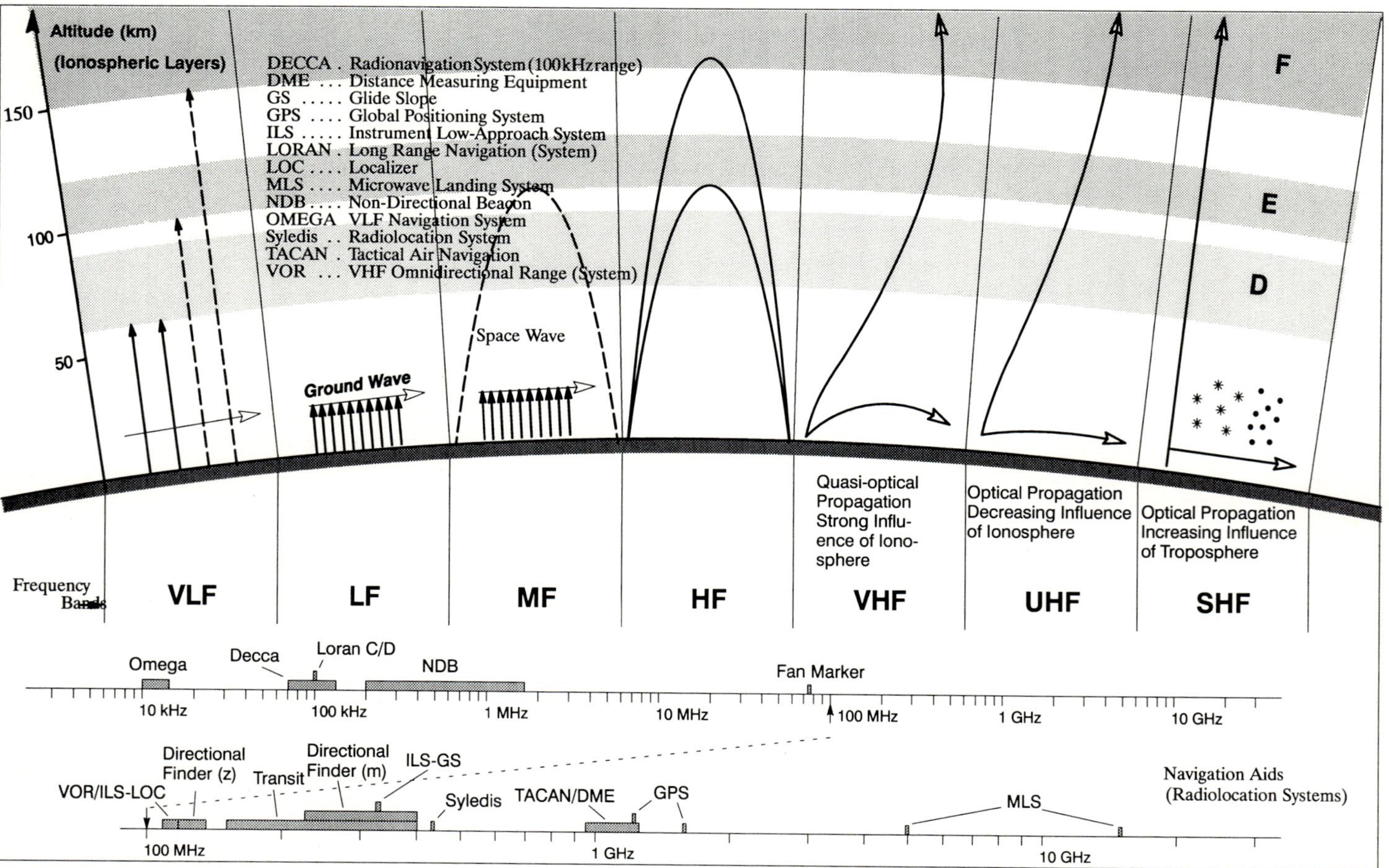


**Figure 162:    Radiolocation Systems/Radio Wave Propagation Behavior in their Bands**

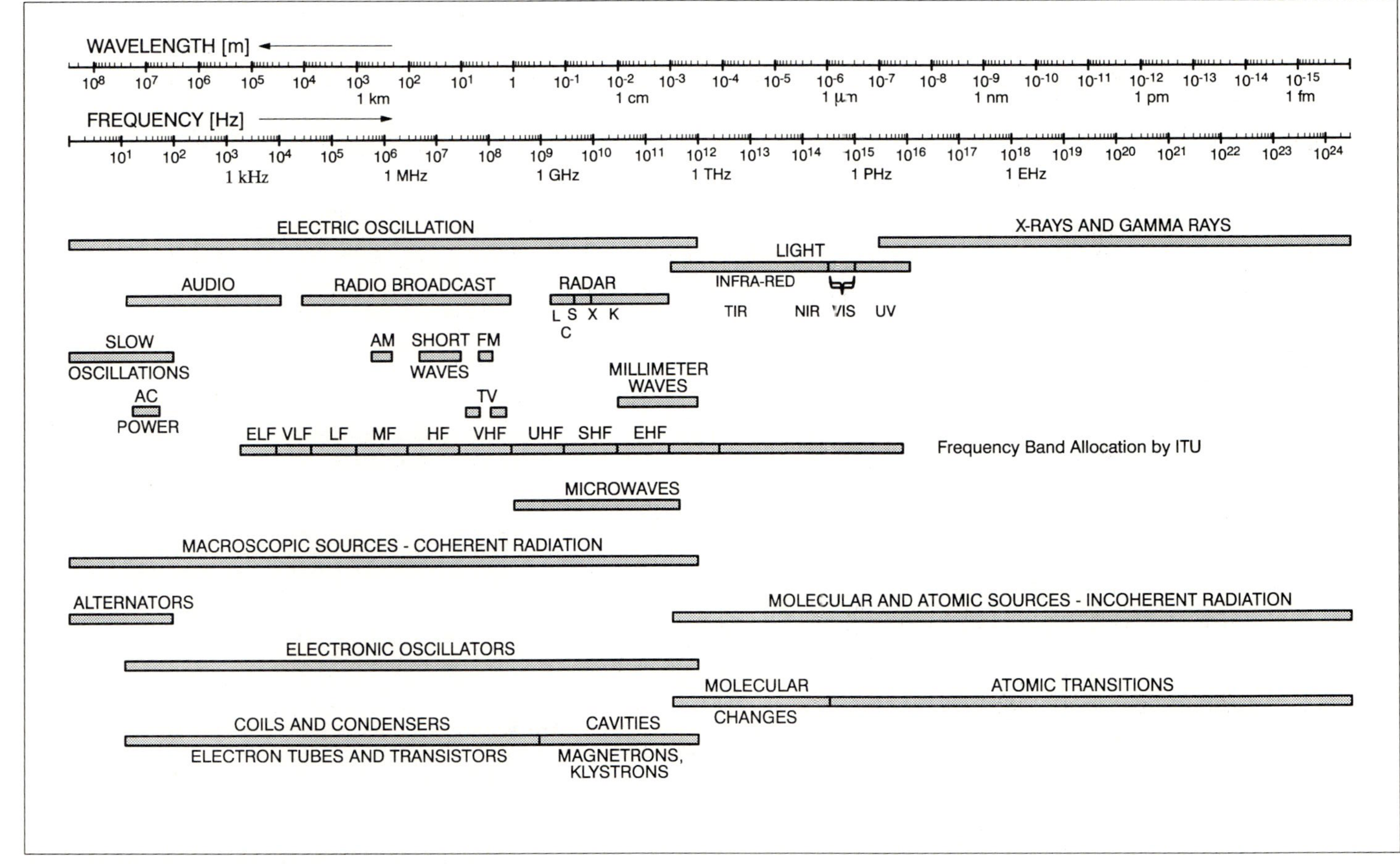


**Figure 163:** Electromagnetic Spectrum with characteristic Sources/Frequency Bands

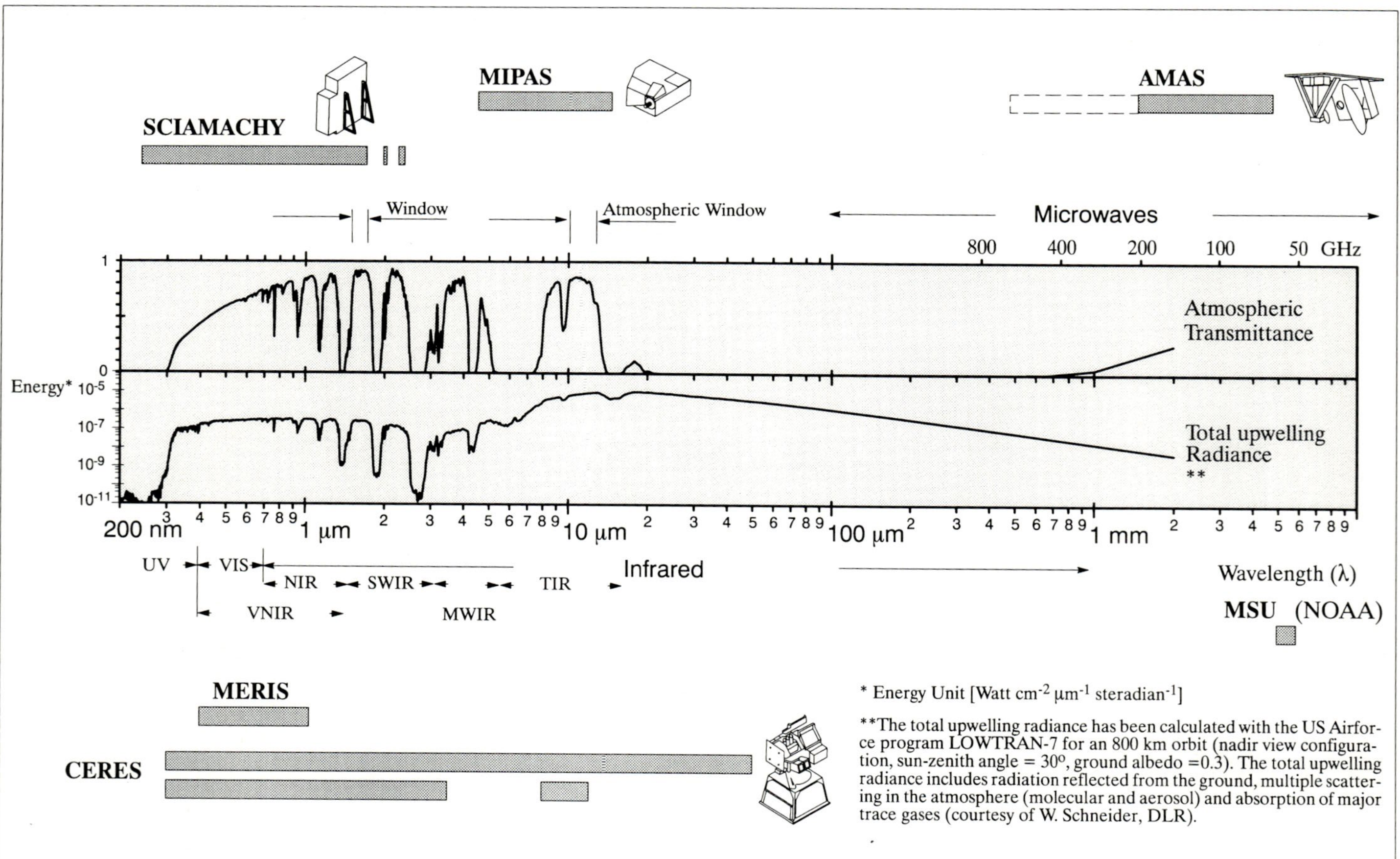


**Figure 164:**    Atmospheric Parameters and Spectral Ranges of some Sensors

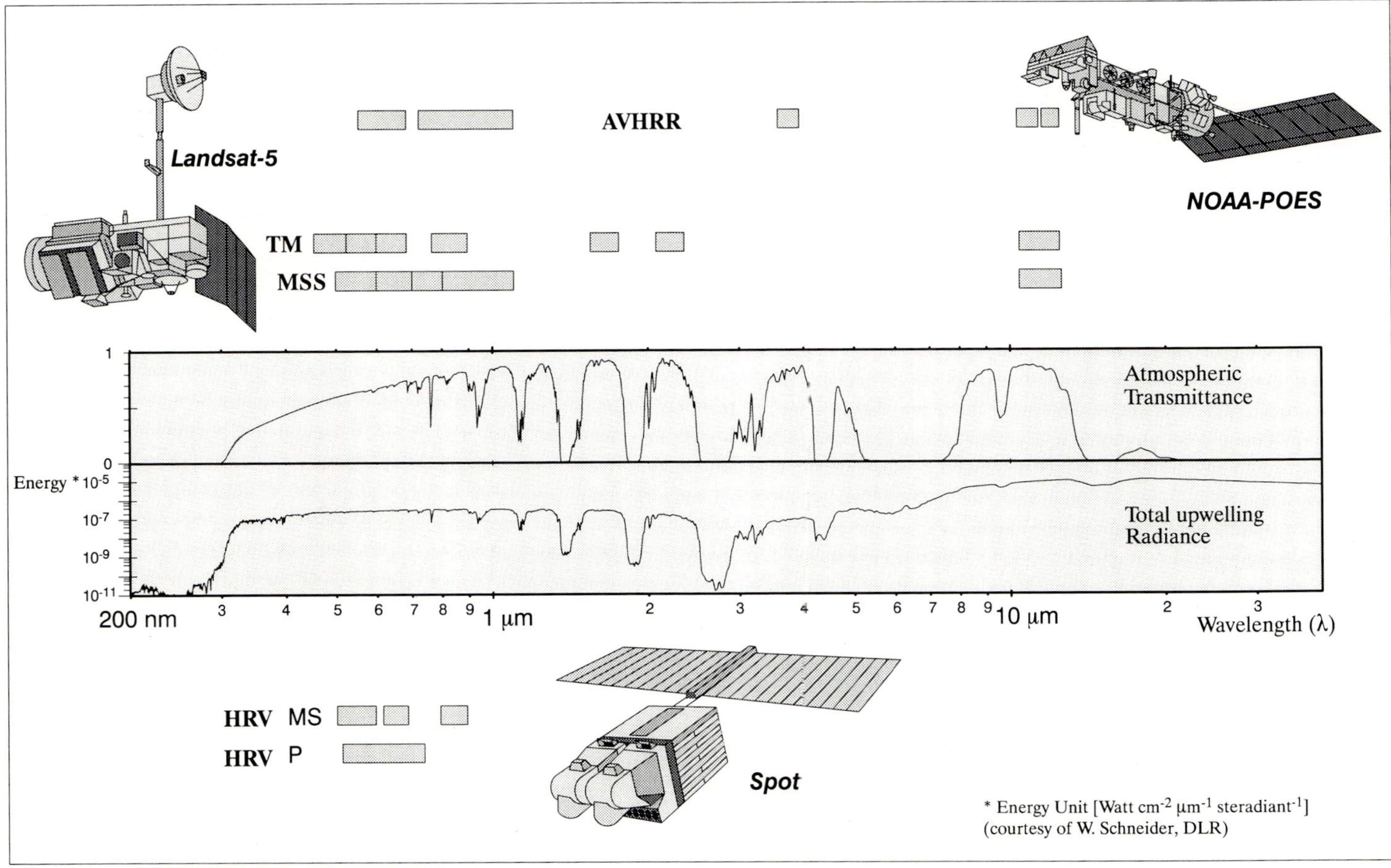


**Figure 165:** Atmospheric Parameters and Spectral Ranges of some Missions/Sensors

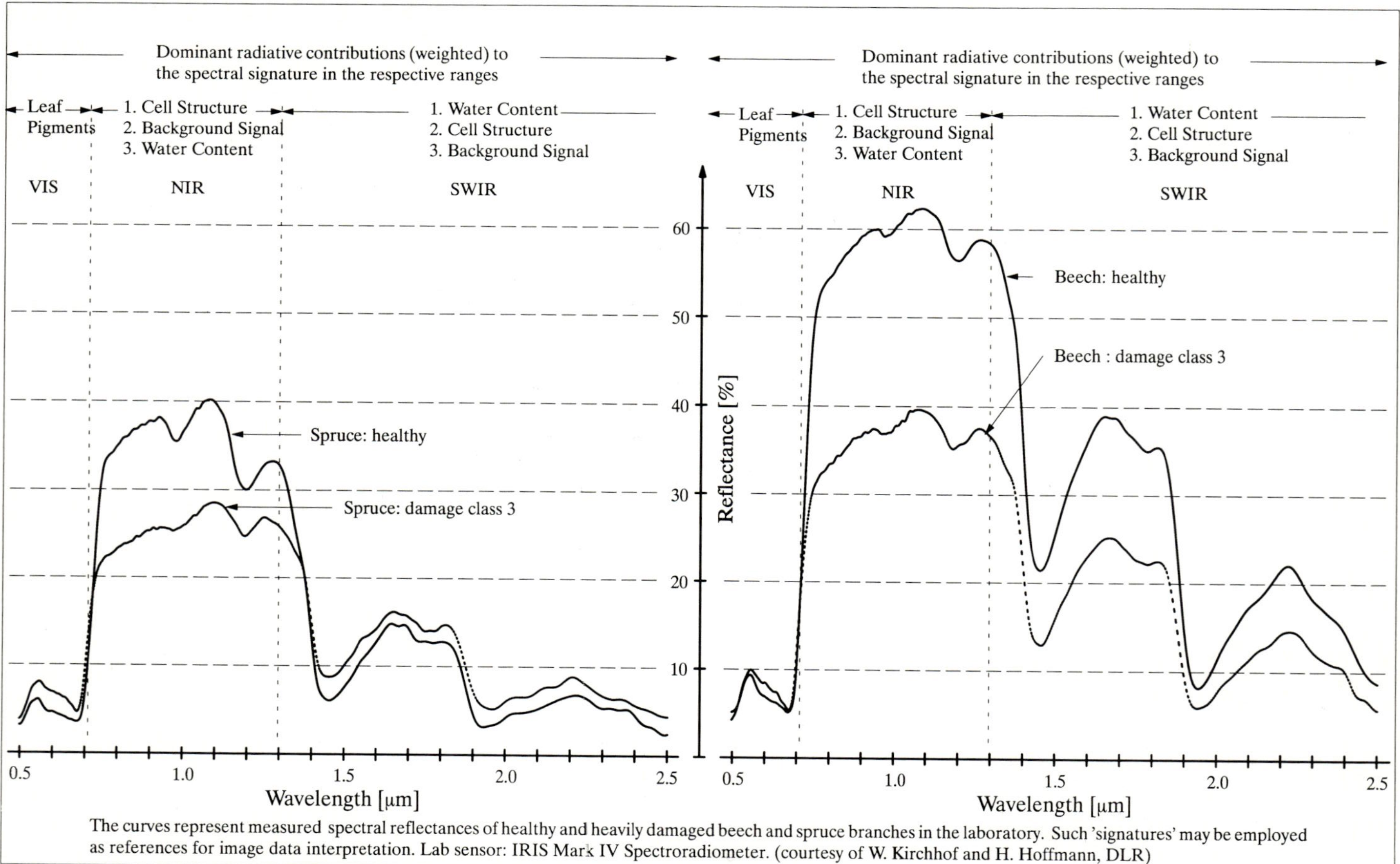


The curves represent measured spectral reflectances of healthy and heavily damaged beech and spruce branches in the laboratory. Such 'signatures' may be employed as references for image data interpretation. Lab sensor: IRIS Mark IV Spectroradiometer. (courtesy of W. Kirchhof and H. Hoffmann, DLR)

**Figure 166:　Spectral Signatures of Vegetation in the Electromagnetic Spectrum**

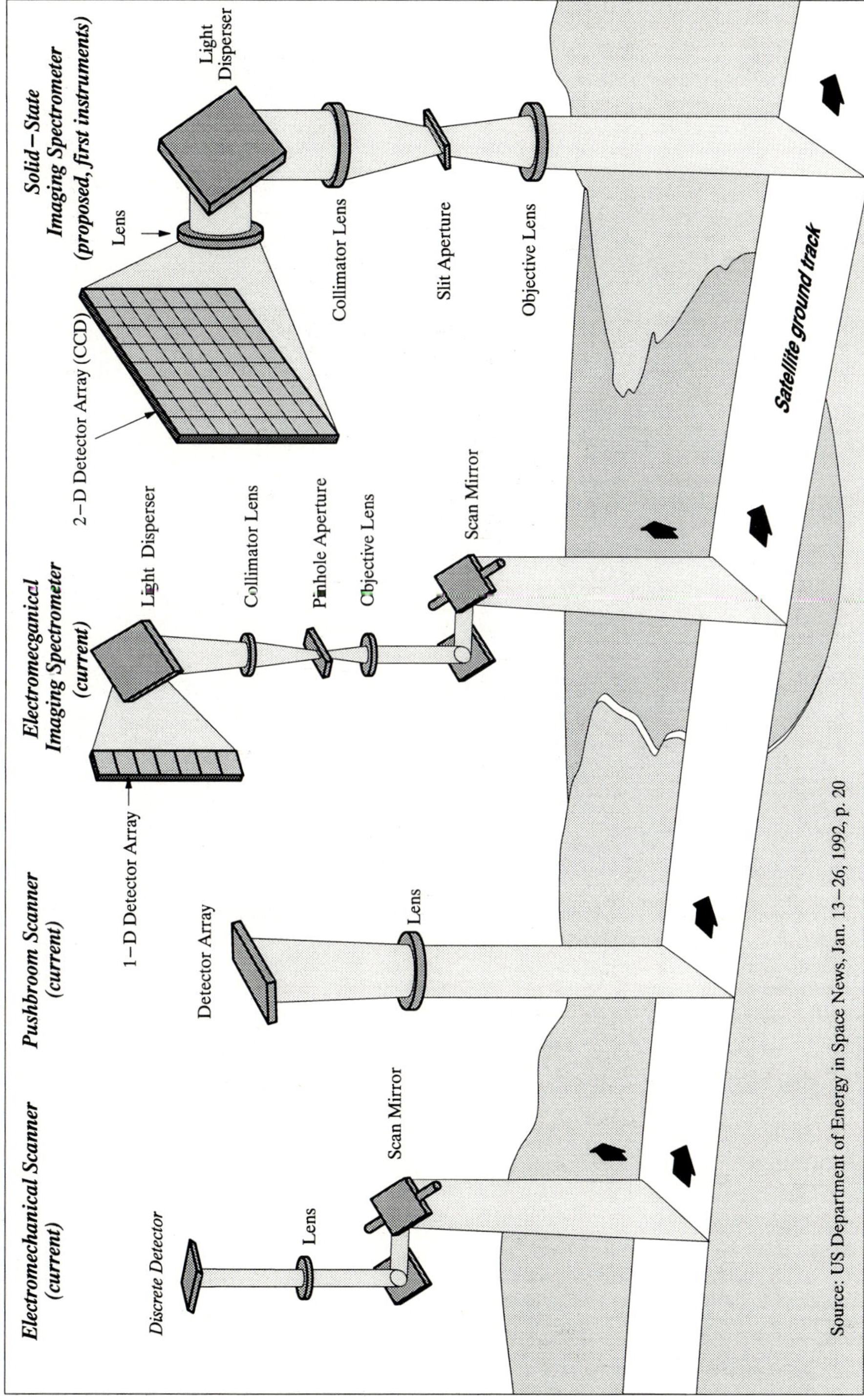


Source: US Department of Energy in Space News, Jan. 13–26, 1992, p. 20

**Figure 167:** Evolution of Imaging Scanner/Spectrometer Concepts

## C.5    Solar Wind and the Magnetosphere - an Introduction

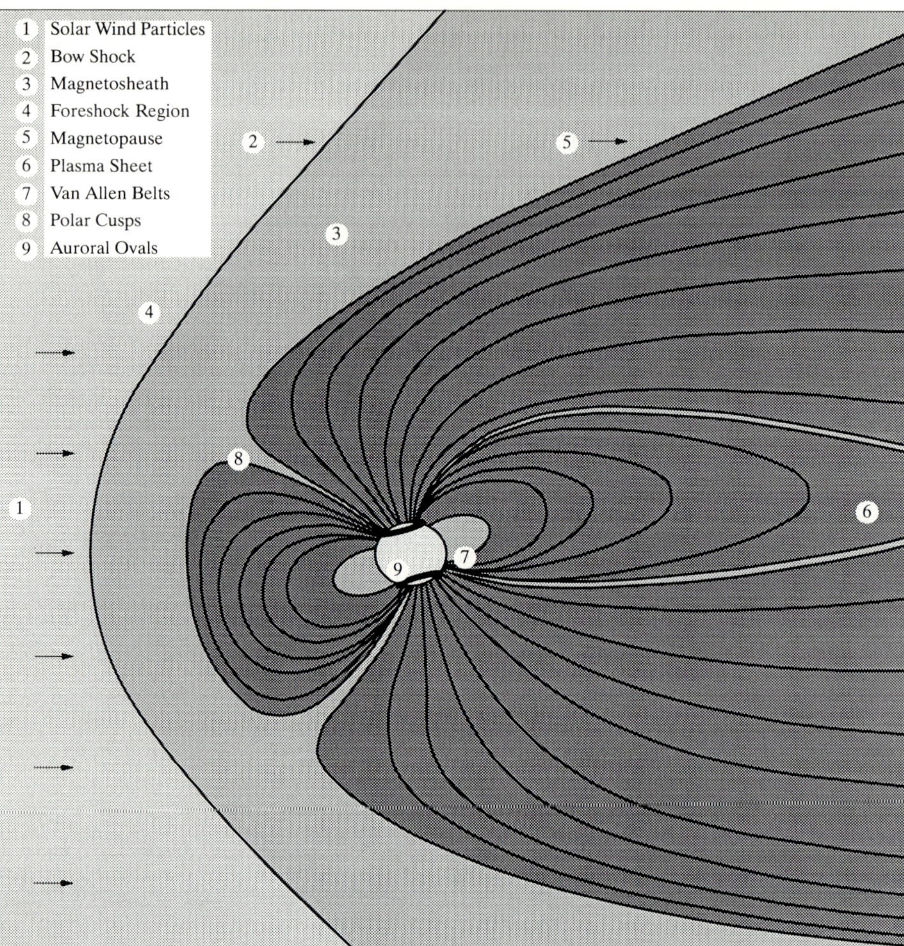


**Figure 168:    Characteristic Model of the Earth's Magnetosphere**

Throughout history mankind has always known that life on Earth is dependent on light and heat from the Sun, but beyond our senses, electromagnetic forces also link the Sun and the Earth in a dynamic interplay that generates and characterizes our protective near-Earth environment and sparks the displays called the Northern and Southern Lights. The space missions and space science of modern times have greatly improved our knowledge of these electromagnetic forces and the overall energy transport in the Sun-Earth system.[624]

The regions of space defined by the electromagnetic link include the Sun and its sphere of influence - the heliosphere - and the Earth and its much smaller sphere of influence - geospace. Geospace includes the near-Earth space and reaches toward the Sun, where the Sun's heliosphere is disturbed by Earth's magnetic field.

---

624)  Taken from the introduction of the NASA/GSFC brochure: "ISTP Global GEOSPACE Science - Energy Transfer in Geospace (ESA/NASA/ISAS)", 1992, Courtesy of GSFC, Greenbelt, Md.

Energy streams out from the Sun toward the Earth in the form of a solar wind of electrified particles (1, see Figure 168). This hot, ionized gas, called plasma, streams toward Earth at about 1.5 million km per hour, carrying particles and magnetic fields from the Sun outward past the planets (with the Sun-Earth distance of 150 million km, the plasma needs 100 hours of travelling time). Earth is shielded from the full blast of these particles by its magnetosphere, the region around the Earth dominated by the Earth's magnetic field.

As the solar wind approaches the Earth's magnetic field, a highly supersonic shock wave is created sunward of the Earth, similar in shape to the shock wave created when a jet plane breaks the sound barrier, but much stronger. This shock wave is called the bow shock (2). Most of the solar wind particles are heated and slowed down at the bow shock and detour around the Earth through a volume of space called the magnetosheath (3). Some particles are actually reflected back from the bow shock into the solar wind stream in a region of turbulence called the foreshock (4).

As the solar wind flows around the Earth, it stretches the Earth's magnetosphere out into a long tail, the magnetotail (not shown in Figure 168). Some of the particles being carried past the Earth leak through the barrier at the boundary of the Earth's magnetic field, called magnetopause (5), and are trapped inside the magnetosphere and stored in the plasma sheet (6) and Van Allen radiation belts (7). Some particles rush through funnel-like openings at the poles, called polar cusps (8). Some energetic particles come down along magnetic field lines and enter into the Earth's upper atmosphere. Particles accelerated in the magnetotail excite atoms and molecules in the Earth's atmosphere. These atoms and molecules then emit light known as the Northern and Southern Lights (or auroras) in the auroral ovals (9), giving a visible signature of this energy transfer from the Sun to the Earth.

The Sun is an active star whose variability affects the flow of the solar wind. For example, solar-flare explosions, associated with sunspots, can cause strong gusts of solar wind. Alterations in the Earth's environment caused by these solar phenomena happen on different time scales from less than a minute to over a century. The Sun's variations (for example, solar x-ray bursts) can affect specific regions on Earth within the time required for light to travel from the Sun to the Earth (8 minutes). Longer timescale, global solar variations may affect long-term climatic changes.

The best known terrestrial effects of solar activity are the geomagnetic storms and auroras that occur within a few days following major solar flares. In turn, the auroras contribute to the heating and ionizing of the upper atmosphere that generate the ionosphere, located at ~150-250 km above the Earth, where the neutral atmosphere gives way to ionized plasma.

Above our atmosphere, ions and charged particles bounce along and spiral around magnetic field lines, deflected from direct impact on the atmosphere and the people below. Thus, the geomagnetic field forms a mantle protecting us from harmful cosmic radiation.

Events on the Sun can trigger changes in the electrical and chemical properties of the atmosphere, the ionosphere, the magnetosphere, the ozone layer, and high-altitude temperatures and wind patterns. These changes cause magnetic storms, communications static, power blackouts, and navigation problems for ships and airplanes with magnetic compasses. Also, satellites and spacecraft can be damaged or can reenter Earth's atmosphere prematurely because of solar storms.

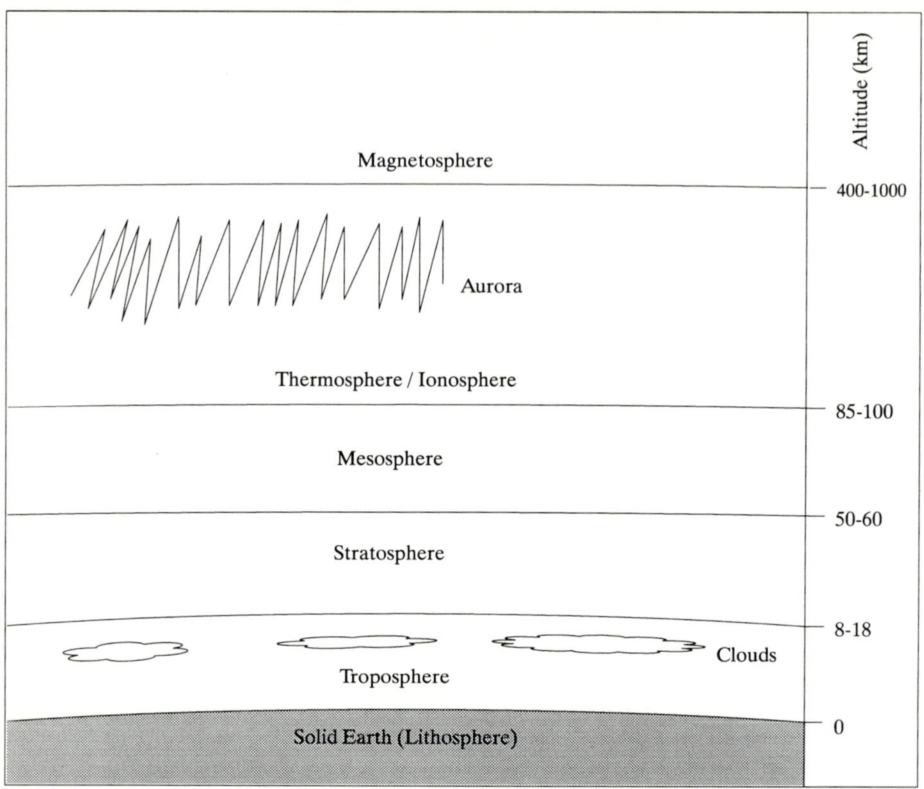



**Figure 169:   Schematic of the Layers of the Earth's Atmosphere**

| Mission | 1994 | 1995 | 1996 | 1997 | 1998 | 1999 | 2000 | 2001 | 2002 | 2003 | 2004 |
|---|---|---|---|---|---|---|---|---|---|---|---|
| ACE (NASA) | | | | firm/approved | | | | | | | |
| APEX (Russia) | planning | | | | | | | | | | |
| CLUSTER (ESA/NASA) | | | firm/approved | | | | | | | | |
| CORONAS (Russia) | | firm/approved | | | | | | | | | |
| Equator-S (MPE, Germany) | | | firm/approved | | | | | | | | |
| FAST (NASA) | | firm/approved | | | | | | | | | |
| FREJA (Sweden/Germany) | operational | | | | | | | | | | |
| GEOTAIL (ISAS/NASA) | operational | | | | | | | | | | |
| Interball (Russia) | | firm/approved | | | | | | | | | |
| ODIN (Sweden) | | | | firm/approved | | | | | | | |
| Ørsted (Denmark) | | | firm/approved | | | | | | | | |
| POLAR (NASA) | | firm/approved | | | | | | | | | |
| SAMPEX (NASA) | operational | | | | | | | | | | |
| SOHO (ESA/NASA) | | firm/approved | | | | | | | | | |
| WIND (NASA) | | firm/approved | | | | | | | | | |

Legend:
- operational
- extension beyond design life
- firm/approved
- planning

**Table 244:   Operational/Planned Solar-Terrestrial Energy Transport Missions**

# Appendix A    Glossary

Only brief definitions are given to provide a quicklook reference for the reader with regard to terminology.[625],[626]

**Aberration**. Geometrical errors in imagery whereby a perfect image is not formed. Typical aberrations include spherical aberration, astigmatism, coma, and chromatic aberrations. Lens bendings, locations, powers, materials, and the number of lenses and position of the aperture stop are all used to minimize aberrations.

**Accuracy.** A measure of the absolute truth of a measurement requiring absolute (traceable) standards . Accuracy is a quality that characterizes the ability of a measuring instrument to give indications equivalent to the 'true value' of the quantity measured. The quantitative expression of accuracy may also be given in terms of uncertainty. The actual or 'true value' of a quantity cannot be determined; it can only be said to exist within tolerance limits of a measured value. The measurement error is the algebraic difference between the measured (or indicated) value and the true value. Hence, accuracy is by it's very nature only an estimation of the true value taking into account all aspects of measurement.

**Active sensor**. A sensor having its own source of EMR (Electromagnetic Radiation) by transmitting a series of signals and detecting the echo. A SAR instrument, a lidar, a radar altimeter, etc., are examples of such active sensors.

**Aerosol**. A gaseous suspension of fine particles such as dust, smoke particles, water droplets, etc. The smallest aerosols are the atoms of the various atmospheric gases.

**Albedo.** The fraction of the total solar radiation incident on a body that is reflected by it (see also 'planetary albedo').

**Aliasing**. A term in data processing referring to two or more distinctly different signals having identical sample values.

**Altimetry**. Altimetry can be described as vertical tracking from a satellite to the geoid. The spaceborne radar altimeter maps the sea surface which is complex, but which largely conforms to the equipotential surface known as the geoid. The ocean surface departs significantly (at the meter level) from the geoid - this departure is of great interest to the science community (oceanographers, meteorologists, geophysicists, etc.) for it reflects the global circulation patterns and also, in the form of tides and changes in sea level, fundamental climatic and solid Earth phenomena.
Furthermore, altimetry provides approximate estimates of the geoid itself, in the form of a mean surface obtained by averaging measurements taken over many satellite passes over the same locations.

**Aperture Stop**. Location within a lens system where the principle ray passes through and crosses the optical axis. The presence of a mechanical limiting aperture typically creates a limiting size.

**Atmosphere**. The envelope of gases surrounding the Earth and bound to it by the Earth's gravitational attraction. Studies of the chemical and radiative properties, dynamic motions, and physical processes of the system constitute the field of meteorology.

**Atmospheric boundary layer** (also referred to as **Planetary boundary layer, PBL**). This boundary layer includes the bottom part of the atmosphere within which the energy exchange processes (between the Earth's surface and the atmosphere) occur mainly through

---

625)  Portions of this glossary are taken from: "Glossary and list of Acronyms/Abbreviations", Earth Observation System (EOS), July 1992, Courtesy of EOS Project Science Office (V. V. Salomonson), GSFC, Greenbelt, Md.

626)  Jeanne Hopkins, "Glossary of Astronomy and Astrophysics", The University of Chicago Press, Second Edition, 1985

vertical transport mechanisms of momentum and heat (production of wind shear turbulence plus convective heat turbulence).  The whole PBL is heated by convection. Temperature gradients are strongest near the surface, because there convective 'eddies' are relatively small and inefficient in carrying the heat upward. The thickness of the daytime PBL is typically in the order of a kilometer, however, it may be tree times as high with strong heating. The PBL may also be limited by a top inversion. On a clear day the whole PBL is turbulent, it increases rapidly in the morning and only very slowly after the time of maximum heating, until it decreases to a minimum during the night (sensitivity to diurnal cycle).

**Atmospheric Refraction.** As a signal (electromagnetic radiation) traverses the atmosphere, it experiences a propagation delay and bending due to the variable characteristics of the medium through which it is passing. The Earth's ionosphere introduces significant systematic perturbations on all microwave tracking data. At lower frequencies (150 MHz) daytime ionospheric biases can easily reach several kilometers in range, several meters per second in range rate, and up to two or three milliradians in position. Since most of the effects decrease as the inverse square of frequency, modern radiometric systems track at dual and well separated higher frequencies.
As a signal travers the troposphere, it sees a varying refractive index resulting primarily from the spatial variations in atmospheric pressure, temperature and humidity. For radiometric technologies, variations in the water vapor content is of chief concern. Optical signals have a much weaker dependence on water vapor. The varying refractive index influences the propagating signal in several ways. For optical signals, the most important effect is the varying group velocity where the pulse speeds up as it travels from the ground station to low pressure regions at higher altitudes. This change is a consequence of Snell's law of refraction which predicts the bending and speed of a light ray as it moves through atmospheric layers with differing refractive indices.[627])

**Aurora.** Light radiated by ions in the Earth's upper atmosphere, mainly near the geomagnetic poles, stimulated by bombardment of energetic charged particles of the solar wind. Aurorae appear about two days after a solar flare and reach their peak about two years after a sunspot maximum. The northern aurora is also referred to as 'aurora borealis' while the southern aurora is also called 'aurora australis'.

**Azimuth Plane (direction).** Observation of an instrument in the along-track direction, i. e. in the direction of the sub-satellite track.

**Backscatter.** Scattering of radiation (or particles) through angles greater than 90° with respect to the original direction of motion.

**Band.** A specification of a spectral range (say, from 0.4 - 0.5 µm) that is used for radiative measurements. The term 'channel' is also in common use with the same meaning as 'band'. In the ITU convention (see Table 243) for the electromagnetic spectrum, the term 'band' refers to a specific frequency range, designated as L-Band, S-Band, X-Band, etc.

**Band-to-Band Registration.** Refers to multispectral image resolution, i.e. how well the same scene is recorded in different spectral bands.

**Bandwidth.** Range of frequencies over which an instrument can be used. It is usually specified in terms of 3 dB points, that is, frequencies at which the response has fallen by 3 dB or 30% from the mid-frequency response.

**Bathymetry.** Measurement of water body depths, in particular of ocean floor surveys. General bathymetry surveys cannot be directly performed from a satellite. However, there are some areas of satellite applications: - bathymetry surveys in coastal regions (with a SAR instrument) of shallow waters, the other method is the interpretation and correlation of radar altimeter data (the technique relies on the assumption that the relationship between the gravity field and bathymetry is uniform over relatively small areas ($\approx 200 \times 200$ km).

---

627) Ivan I. Mueller, S. Zerbini, "The Interdisciplinary Role of Space Geodesy", Lecture Notes in Earth Sciences,
       Springer Verlag, 1989, p. 187

**Biological Productivity.** The amount of organic matter, carbon, or energy content that is accumulated during a given time period.

**Biomass.** The total dry organic matter or stored energy content of living organisms that is present at a specific time in a defined unit (community, ecosystem, crop, etc.) of the Earth's surface.

**Biosphere.** The portion of the Earth and its atmosphere that can support life. The part (reservoir) of the global carbon cycle that includes living organisms (plants, animals,) and life-derived organic matter (litter, detritus). The terrestrial biosphere includes the living biota (plants and animals) and the litter and soil organic matter on land; the marine biosphere includes the biota and detrius in the oceans.

**Blackbody**. Idealized body that absorbs all the radiation incident upon it, and reradiates according the Planck's law.

**Blaze wavelength**. The wavelength of the highest efficiency for the ruled diffraction grating, the "blaze" being the controlled shape of the rulings on the grating.

**Brightness Temperature**. Refers to the equivalent blackbody temperature for a given frequency according to Planck's law.

**Calibration.** Characterization of a sensor (radiometer, spectrometer, etc., see also C.1.3) in the spatial, spectral, temporal and polarization responsive domains. The quality of the measurement is the most difficult aspect of calibration.
1. The activities involved in adjusting an instrument to be intrinsically accurate, either before or after launch (i.e. "instrument calibration").
2. The process of collecting instrument characterization information (scale, offset, nonlinearity, operational, and environmental effects), using either laboratory standards, field standards, or modeling, which is used to interpret instrument measurements (i.e. "data calibration").

**Carbon Cycle.** All parts (reservoirs) and fluxes of carbon; usually thought of as a series of the four main reservoirs of carbon interconnected by pathways of exchange. The four reservoirs - regions of the Earth in which carbon behaves in a systematic manner - are the atmosphere, terrestrial biosphere (usually includes freshwater systems), oceans, and sediments (includes fossil fuels). Each of these global reservoirs may be subdivided into smaller pools ranging in size from individual communities or ecosystems to the total of all living organisms (biota). Carbon exchanges from reservoir to reservoir by various chemical, physical, geological, and biological processes.

**Cassegrain telescope**. A reflective telescope in which a small hyperboloidal primary mirror reflects the convergent beam from the paraboloidal primary mirror through a hole in the primary mirror to an eyepiece in back of the primary mirror.

**Charge-Coupled Device (CCD)**. A type of analog shift register (detector) built into an integrated circuit form. Isolated charge packets are transported by manipulating potential wells (place of minimum potential) within a substrate. - A CCD consists of a linear array of solid-state capacitors. Charge is transferred along the CCD in isolated packets by manipulating the voltages on the solid-state gates. A CCD imager employs solid-state photodetectors (each photodetector element is used as a well) and CCD readout techniques.

**Charge Injection Device (CID)**. An image sensor in which charge packets are typically measured by injecting them into a substrate or by shifting charge packets under an electrode to induce a voltage on the capacitance formed by the electrode and the substrate.

**Chlorofluorocarbons (CFC's)**. A family of inert, nontoxic, and easily liquefied chemicals used in refrigeration, air conditioning, packaging, and insulation, or as solvents or aerosol propellants. Because they are not destroyed in the lower atmosphere, they drift into the upper atmosphere, where - given suitable conditions - their chlorine components destroy ozone.

**Climate.** The statistical collection and representation of the weather conditions for a specified area during a specified time interval, usually decades, together with a description of the state of the external system or boundary conditions. The properties that characterize the climate are thermal (temperatures of the surface air, water, land, and ice), kinetic (wind and ocean currents, together with associated vertical motions and the motions of air masses, humidity, cloudiness and cloud water content, groundwater, lake winds, and water content of snow on land and sea ice), and static (pressure and density of the atmosphere and ocean, composition of the dry air, salinity of the oceans, and the geometric boundaries and physical constants of the system). These properties are interconnected by various physical processes such as precipitation, evaporation, infrared radiation, convection, advection, and turbulence.

**Climate Change.** The long-term fluctuations in temperature, precipitation, wind, and all other aspects of the Earth's climate. External processes, such as solar-irradiance variations, variations of the Earth's orbital parameters (eccentricity, precession, and inclination), lithosphere motions, and volcanic activity, are factors in climatic variation. Internal variations of the climate, e.g., changes in the abundance of greenhouse gases, also may produce fluctuations of sufficient magnitude and variability to explain observed climate change through the feedback processes interrelating the components of the climate system.

**Correction/Calibration methods on sensor data** (see chapters C.1.3 and C.1.4)

**Cloud.** A visible mass of condensed water vapor particles or ice suspended above the Earth's surface. Clouds may be classified by their visual appearance, height, or form.

**Cloud Albedo.** Reflectivity that varies from less than 10 to more than 90 percent of the insolation and depends on drop sizes, liquid water content, water vapor content, thickness of the cloud, and the sun's zenith angle. The smaller the drops and the greater the liquid water content, the greater the cloud albedo, if all factors are the same.

**Cloud Feedback.** The coupling between cloudiness and surface air temperature in which a change in surface temperature could lead to a change in clouds, which could then amplify or diminish the initial temperature perturbation. For example, an increase in surface temperature could increase the evaporation; this in turn might increase the extent of cloud cover. Increased cloud cover would reduce the solar radiation reaching the Earth's surface, thereby lowering the surface temperature. This is an example of negative feedback and does not include the effects of longwave radiation or the advection in the oceans and the atmosphere, which must also be considered in the overall relationship of the climate system.

**Cloud Microphysics.** Study of cloud and precipitation particles (individual or populations) and their interactions with the environment. Of key importance are mass exchange processes such as nucleation, growth, and fallout leading to the broad characteristics of clouds and precipitation.

**Coherence.** A fixed relationship between the phase of waves in a beam of radiation of a single frequency. Two beams of light are coherent when the phase difference between their waves are constant; they are noncoherent if there is a random phase relationship. - In active measurement systems like radars, coherence refers to the availability of phase and amplitude measurements of the radar cross-section of the recovered signals.

**Coma.** An off-axis aberration whereby the outer periphery of a lens system has a higher (or lower) magnification then the central portion of the lens. The image typically is comet shaped.

**Crossover** (difference). A crossover is defined as the intersection of the satellite ground track with itself. At this location, the two crossing passes (one ascending and one descending) provide independent sub-satellite ground track measurements at the same location but at different times. In altimetry crossover differences contain information about uncertainties in the satellite ephemeris and therefore enable correction of radial orbit error.

**Convolution**. Mathematical process, appearing in linear or circular form, that models the input-output filtering process.

**Cryosphere**. The Earth's cryosphere consists of four main elements: sea ice, seasonal snow on land, land ice (including glaciers, ice sheets, and ice shelves), and permafrost. The time scales on which these elements impact human activity range from daily to seasonal for sea ice and snow, and to annual and decadal for ice sheets and glaciers.

**Deforestation.** The removal of forest stands by cutting and burning to provide land for agricultural purposes, residential or industrial building sites, roads, etc., or by harvesting the trees for building materials or fuel. Oxidation of organic matter releases $CO_2$ to the atmosphere, and regional and global impacts may result.

**Depolarization ratio**. The ratio of intensities of light scattered perpendicular and parallel to the E-vector of the incident radiation.

**Detector**. A device that detects and linearly transduces radiative power into an electrical signal. A detector typically takes an average value of the signal over a length of time (from about 1 ns to about 1 ms). Thus it is not possible to measure any signal variations at the rate of either the optical or the infrared frequency spectrum. The electrical signal is usually proportional to the average power or photon rate or the time integral of them - over the integration time. Direct detectors may be categorized as photon detectors (an electrical signal is produced due to free charges on the detector surface by the incident photons) or thermal detectors (an electrical signal is produced due to the temperature change). Detectors may also be classified according to their arrangement: single detectors, array detectors.
Thermal detectors require usually cooling (active or passive). Infrared radiation is 'thermal' by nature, hence the detector is affected by the medium that is measured.
Principal detection methods used in spectroscopy are:
- photographic (plates and films - simultaneous recording of the entire spectrum)
- photoemissive ($\rightarrow$photodiodes, photomultipliers, image tubes)
- radiometric (bolometer, thermocouple, Golat cell)
- semiconductor (photoconductive cell, photovoltaic cells, etc.)

**Digital Filter**. A digital device that is capable of altering the magnitude, frequency or phase response of a digitally encoded input signal.

**Dipole.** An electric dipole is a system composed of two equal charges of opposite sign, separated by a finite distance; e.g. the nucleus and orbital electron of a hydrogen atom. An ordinary bar magnet is a magnetic dipole.

**Dipole Antenna.** A type of array consisting of a system of dipoles. A dipole antenna differs from a dish antenna in that it consists of many separate antennas that collect energy by feeding all their weak individual signals into one common receiving set.

**Discrete Fourier Transform**. A mathematical method of transforming a time series into a set of harmonics in the frequency domain and vice versa.

**Dispersion**. The separation of light into its component colors by its passage through a diffraction grating or by refraction such as that provided by a prism.

**Doppler effect**. The alteration in frequency of a wave radiation caused by a relative motion between the observer and the source of radiation.- Acoustic Doppler effect applies to the propagation of source waves, - optical Doppler effect depends on the relative velocity of the light source and the observer, thermal Doppler effect causes a widening of the spectral lines.

**Doppler shift.** Displacement of spectral lines (frequency) in the radiation received from a source due to its relative motion in the line of sight. Sources approaching (-) the observer are shifted toward the blue; those receding (+), toward the red. Determination of radial distance.

**Duty Cycle.** Fraction of orbital period in which a sensor (or a sensor mode) is actually operational due to overall power limitations of the payload. The concept of a duty cycle applies in particular to sensors with large power requirements such as active sensors, in particular SAR instruments.

**Dwell Time.** The short period of time in which a detector is collecting radiation from a target area or volume. - A very short dwell time results usually in a low (i.e. poor) signal-to-noise (SNR) ratio with all its problems of proper signal recognition and discrimination. A small dwell time also implies 'fast' detectors and electronics.

**Electron volt (eV).** A unit of energy used in atomic and nuclear physics; the kinetic energy acquired by one electron in passing through a potential difference of 1 Volt in vacuum. $1\,\text{eV} = 1.602 \times 10^{-12}$ erg (or $= 1.602 \times 10^{-19}$ joules). An electron with an energy of 1 eV has a velocity of about 580 km/s. The wavelength associated with 1 eV is 12,398 Å.

**Eötvös experiment.** An experiment performed in 1909 by the Hungarian physicist Eötvös to establish that the gravitational acceleration of a body does not depend on its composition - i.e. that the inertial mass and gravitational mass are exactly equal. Today, the linear gradient of gravity is defined in units of Eötvös, where 1 Eötvös $= 10^{-9}\,\text{s}^{-2}$; i.e. difference of $10^{-9}\,\text{ms}^{-2}$ acceleration per meter. The vertical gradient of gravity at the Earth's surface is about 3100 Eötvös.

**Ephemeris.** A tabular statement of the spatial coordinates of a celestial body or a spacecraft as a function of time.

**Erlang.** A measure of communication (telephone) traffic load expressed in units of hundred calls seconds per hour (CCS). One Erlang is defined as the traffic load sufficient to keep one trunk busy on the average and is equivalent to 36 CCS. The measure is also used for DCS (Data Collection Satellite) access capabilities.

**Errors.** Any measurement is affected by random errors and by systematic errors.

**Equinox.** Either of two points on the celestial sphere where the celestial equator intersects the ecliptic (see also 'vernal equinox').

**Equivalent Isotropic Radiated Power (EIRP).** A measure of power radiated by an antenna in the direction of a receiver expressed as the equivalent power that would have to be radiated uniformly in all directions.

**Fabry-Perot Interferometer.** A type of interferometer wherein the beam of light is passed through a series of pairs of partly reflecting surfaces set at various angles to it and spaced at certain pre-chosen numbers of the wavelength to be examined. It differs from a Michelson interferometer in that it has only one 'arm'.

**Fast Fourier Transform.** An algorithm that is often used to implement Discrete Fourier Transforms.

**Feedback Mechanisms.** A sequence of interactions in which the final interaction influences the original one. <u>Negative Feedback</u>: An interaction that reduces or dampens the response of the system in which it is incorporated. <u>Positive Feedback</u>: An interaction that increases or amplifies the response of the system in which it is incorporated.

**Fluorescence.** The absorption of a photon of one wavelength and re-emission of one or more photons at longer wavelengths, especially the transformation of ultraviolet radiation into visible light.

**Focal Length**. Distance measured along the optical axis from the image to the plane, where the axial imaging cone of light intersects the input light bundle.

**Footprint**. Refers to the projection of the instantaneous area of coverage of a sensor onto the Earth's surface. A footprint may also be the instantaneous area of visibility of a data collection platform on a satellite.

**Frequency Division Multiple Access (FDMA)**. A process that shares a spectrum of frequencies among many users by assigning to each a subset of frequencies in which to transmit signals.

**Geoid.** The gravitational equipotential surface (an ellipsoid) near the mean sea level which is used as a datum for gravity surveys. The geoid serves also as a reference surface for topographic heights, for example, as they are shown on maps. The geoid itself represents the surface of zero height.

**Gradiometry.** Study of the spatial gradient of the Earth's gravitational field.

**Gravity Gradient Boom**. A deployable extension of a spacecraft intended to give the spacecraft elongated mass properties to contribute to gravity gradient stability. An elongated "dumbbell"- shaped spacecraft is the most gravity gradient stable configuration with the long axis oriented vertically in orbit, i.e. pointing toward the center of the Earth.

**Greenhouse Gases.** Those gases, such as water vapor, carbon dioxide, methane, and CFC's, that are largely transparent to solar radiation but opaque to outgoing longwave radiation. Their action is similar to that of glass in a greenhouse. Some of the longwave (infrared) radiation is absorbed and remitted by the greenhouse gases. The effect of this is to warm the surface and the lower atmosphere of the Earth.

**Hyperspectral**. In remote sensing the term implies a spectral signature of narrow, continuous and contiguous spectral bands per pixel, e.g. a fine spectral resolution ($\Delta\lambda/\lambda = 1 - 5\%$). A hyperspectral sensor has a minimum of 15 spectral bands (normally 30 - 200 bands). On the other hand, "multispectral" usually implies fewer, spectrally broader bands, which may be non-contiguously spaced color bands per pixel. "**Hyperspectral imaging**" is then a hyperspectral signature for each pixel of the image. The detailed spectral information captured in a hyperspectral image, allows for a detailed examination of the scene.

**Image.**  A remotely sensed image (by CCD camera or radar) is a two-dimensional grid of data; each of its elements is a pixel (picture element) whose coordinates are known and whose light intensity has a DN (Digital Number) value. The coordinates of the pixels and their DN values describe the image as rows, called **lines**, and columns, called **samples**. An 8-bit pixel provides up to 256 brightness levels (level 0 is set to black, while level 255 is set to white), the brightness levels are also referred to as 'grey levels'.
In **false color image processing**, those pixels which have the same DN value are given an arbitrary color. This technique is used, for example, to differentiate between varies types of terrain or species of vegetation - to show changes, which are otherwise not perceptible to the human eye.

**Imaging Array**. A solid-state imaging array consists of a 1-D or 2-D set of photodetectors onto which an image can be focused together with an integral electronic readout scheme. The devices are classified according to the particular readout scheme and detector type. A linear array is an imaging array consisting of a single line of detectors.

**Imaging Spectrometry**. The simultaneous acquisition of images in many contiguous spectral bands is known as imaging spectrometry.

**Infrared Radiation.** Electromagnetic radiation lying in the wavelength interval from 0.7 µm to 1000 µm (or roughly between 1 µm and 1 mm wavelength). Its lower limit is bounded by visible radiation, and its upper limit by microwave radiation. Most of the energy emitted by the Earth and its atmosphere is at infrared wavelengths. Infrared radiation is generated almost entirely by large-scale intermolecular processes. The tri-atomic gases, such as water vapor, carbon dioxide, and ozone, absorb infrared radiation and play important roles in the propagation of infrared radiation in the atmosphere.

**In-situ Soundings.** An observation method by aircraft and balloon systems with the objective to measure parameters in the immediate environment of the observation platform.

**Instantaneous Field of View (IFOV).** A term denoting the smallest spatial ground projection (i. e. a cell) whose radiation can be detected by a sensor (scanner) in a single measurement period (instant). The IFOV may be expressed either as a small angle (in mrad or μrad, in this case the value is independent of the orbital altitude of the sensor), or as a unit area (e. g. 6 m x 6 m), or simply as the pixel size. Hence, the IFOV represents actually the spatial resolution of a measurement.

**Integration time.** Refers to the short time period allocated for the radiative measurement of the instantaneous area of observation by the detector of a sensor (see also dwell time). Depending on sensor type the integration time may be very short (as is the case with electromechanical scanning systems, this instrument measures each individual cell (IFOV) across the swath sequentially), while the entire swath width (FOV) is measured by a CCD detector array in a single measurement.

**Interference.** Signals that arise from sources extraneous to the measurement system and result in errors in the measured value.

**Interferometer.** An instrument class for dispersing spectra. The technique determines the relative phase of two (or more) wave fronts as a function of spatial location by observing interference fringes. Radiation is split into two or more beams which traverse different path lengths. The beams are reflected by mirrors and recombined for interference analysis. - Interferometers are used to measure wavelengths, or to measure the angular width of sources, or to determine the angular position of sources, etc.

- **Michelson interferometer.** Incident radiation from source S is partly reflected and partly transmitted by a beam splitter to two mirrors. Both mirrors reflect the radiation back to the beam splitter where interference occurs. One of the mirrors is movable in Fourier spectroscopy.

- **Mach-Zehnder interferometer.** Variation of Michelson-type - incident radiation moves in only one direction with the use of two beam splitters. A test object is only once traversed.

- **Fabry-Perot interferometer.** A multi-beam interferometer consisting of two parallel glass plates that are coated with highly reflecting and partially transparent films on their inner surfaces. A radiation source S from a spectrometer generates multiple reflections between the two surfaces, producing concentric interference rings.

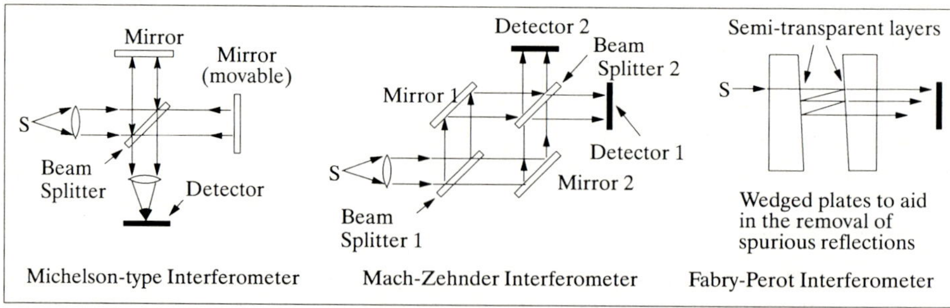


**Figure 170:    Illustration of some Interferometer Concepts**

**Ionosphere.** Rarefied, ionized region of the Earth's atmosphere, between approximately 60 and 400 km.

**Irradiance.** Received radiative power per unit area.

**Langmuir probe.** An instrument employed for the measurement of the current-voltage characteristics of a plasma (single and double probes are in use). Determination of the plasma density.

**Limb/Occultation Sounding.** A horizon-looking observation technique that uses a distant object [(for occultation sounding) sun, star, or a sensor on another satellite in a different orbit, (see Figure 146)] as a source to observe the signal on its path through the atmosphere that is essentially tangential to the Earth's surface. Very long atmospheric paths (up to 400 km) with high sensitivities or dynamic ranges can obtained in this 'limb-sounding' configuration to detect for instance trace gases or to analyze the signal attenuation/retardation.

**Lithosphere.** Outer rigid crust of the Earth.

**Look angle.** The direction in which the antenna is pointing when transmitting and receiving from a particular cell (for an active instrument). Refers to the instrument pointing direction from nadir.

**Layman-alpha radiation.** The radiation emitted by hydrogen at 1,216 Å, first observed in the solar spectrum by rocket-borne spectrographs.

**Magnetometer.** An instrument for measuring changes in the Earth's magnetic field.

**Magnetopause.** The surface defining an interface between the magnetic field of a star and matter in the disk; a surface where the average magnetic pressure of the magnetic field is in pressure balance with the plasma pressure. Earth's magnetopause is the region in the ionosphere where the magnetosphere meets the solar wind.

**Magnetosphere.** The region of space surrounding a rotating, magnetized sphere. Specifically, the outer region of the Earth's ionosphere, starting at about 1000 km above Earth's surface and extending to about 60000 km (or considerably farther such as 100 $R_E$ on the side away from the Sun).

**Maser.** An amplifier utilizing the principle of 'microwave amplification by stimulated emission of radiation'.

**Measurement Mode - Duty Cycle.** The fraction of available time during which an instrument is actively performing Earth measurements and producing meaningful data, including incidental calibration and overhead (such as scan retrace). High data rate, high power consumption, and steerable instruments may have small duty cycles. Daylight-only instruments may have measurement mode duty cycles averaging 50 percent.

**Mesosphere.** Region of the atmosphere between approximately 50 and 85 km.

**Meteoroid.** A small particle in space (sources are comets, collisions between asteroids, and interstellar dust). A meteoroid that survives the passage through the Earth's atmosphere and reaches the Earth's surface is known as a **meteorite.**

**Microwave radiation.** Electromagnetic radiation generally considered in the wavelength range from approximately 1 mm to 1 m (3 orders of magnitude). In microwave radiometry polarization is often used as a discriminant parameter, because microwave antennas are easily built with a single polarization direction. - On the other hand, most optical sensors are relatively independent of polarization, a special effort must be made to polarize the signal prior to detection. - The largest portion of the microwave region is also the 'radar' region, hence both terms are used interchangeably (this applies also to the instruments, i. e. microwave instruments and radar instruments), see also C.1.9 and C.1.10.

**Modeling.** An investigative technique that uses a mathematical or physical representation of a system or theory that accounts for all or some of its known properties. Models are often used to test the effects of changes of system components on the overall performance of the system.

**Mosaic.** An assemblage of overlapping airborne or spaceborne photographs or images whose edges have been matched to form a continuous pictorial representation of a portion of the Earth's surface.

**Nadir.** Direction toward the center of the Earth. Opposite of zenith.

**Non-imaging sensors.** These instruments measure directly such quantities as radiant flux, irradiance, and radiance, which describe the intensity of a radiation field or the optical properties of a surface or a region of space. The sensors are non-imaging in the sense that they do not produce an image (a picture), but rather, integrate over time, space, and wavelength to produce a spectral curve, or a set of numbers that characterize the electromagnetic radiation. Typical measurement products are profiles, such as flux profiles, temperature profiles, moisture profiles, etc. Typical non-imaging instruments are: radar altimeters, sounders, scatterometers, radiometers (there are also imaging radiometers obtained by scanning techniques), etc.

**Nowcasting.** The term is used in meteorology, it is concerned with the development of atmospheric features on time scales between 0 and 3 hours over regional and local areas. Nowcasting is closely linked to very short-range-forecasting which covers developments over the time scale of 3 to 12 hours over regional areas.[628]
A typical example of the application of nowcasting is to incidences of severe weather where small-scale features undergo rapid development. Nowcasting requires the (almost) continuous monitoring of the area(s) of interest. Furthermore, as many of the features to be observed are fairly small, quite high spatial resolutions are needed.

**Nyquist sample rate.**[629] A sample rate above which a band-limited signal can be reconstructed from its sample value. - - If a signal s(t) contains no frequency components at or above $f_N$ Hz [s(t) is then said to be band-limited to $f_N$ Hz], then s(t) can be completely reconstructed from its sample values provided the samples are taken at a rate equal to or in excess of $f_S = 2 f_N$ samples/s. This condition is known as the Nyquist sampling theorem, $f_S$ is referred to as the Nyquist sample frequency, and $f_N$ is sometimes called the Nyquist frequency.

**Occultation.** Distortion or interruption of a direct observation path between the observer (sensor) and a target by an intervening medium (such as an atmosphere or a celestial body).

**Ocean Mixing.** Processes that involve rates of advection, upwelling/downwelling, and eddy diffusion and that determine, for example, how rapidly excess atmospheric carbon dioxide can be taken by the oceans.

**Optical radiation.** Electromagnetic radiation of frequency (or wavelength) that can be focused, dispersed, and detected using optical components such as lenses, mirrors, and gratings.

**Ozone.** A molecule made up of three atoms of oxygen. In the stratosphere, it occurs naturally and provides a protective layer shielding the Earth from ultraviolet radiation and subsequent harmful health effects on humans and the environment. In the troposphere, it is a chemical oxidant and major component of photochemical smog. Ozone is an effective greenhouse gas especially in the middle and upper troposphere and lower stratosphere.

628)  "Meteosat Second Generation Programme Proposal", ESA/PB-EO (92), 57, 9. November, 1992, p. 14
629)  F. J. Taylor, "Digital Signal Processing", Encyclopedia of Physical Science and Technology, Academic Press,
         1987, Vol. 12. p. 600

**Paleoclimatology.**[630] Science which deals with past climate periods of the Earth in very large space-time scales. Long-term baselines of past climate changes are studied and reconstructed to understand climate processes and predict future climate change (climate models). Paleoclimate data are derived from ice cores, tree rings, marine and lake sediments, fossil pollen, plant macro fossils, paleovegetation, past sea surface and lake level data, terrestrial ice sheet height and extent, land surface properties, etc.

**Peak Antenna Gain.** Maximum radiated power intensity of an antenna expressed as a ratio to the radiation power intensity of a hypothetical isotropic antenna fed with the same transmitting power.

**Perigee.** The point in an orbit at which the spacecraft is nearest to the Earth.

**Panchromatic Channel.** A channel of a sensor detector system covering usually the entire width of a spectral range, in particular the visible range (VIS).

**Passive system.** A sensing system that detects or measures radiation emitted by the target. A multispectral scanner is an example of a passive sensor.

**Phase modulation (PM).** Angle modulation in which the phase of a sinewave carrier is caused to depart from the carrier phase by an amount proportional to the instantaneous value of the modulating wave.

**Plasma.** Electrically neutral gas consisting of charged particles such as ions, electrons, and neutrals. A plasma has a high temperature but its density is very low.

**Pixel** (Picture Element). The smallest area unit of an image which is generated by a single digital measurement.

**Photodiode principle.** Photons incident on the photodiode form electron-hole pairs in the detector material (silicon for instance). Electron-hole pairs formed in or near the depletion region are swept apart by the electric field and will ultimately combine with ionized atoms in the depletion region, reducing the effective depletion width. This reduction corresponds to a reduction in charge stored on the p-n junction capacitance.

**Photoelectric cell.** A transducer which converts electromagnetic radiation from the UV, VIS and IR regions of the spectrum into electrical quantities such as voltage, current, or resistance.

**Photomultiplier.** A photoemissive detector in which amplification is obtained by secondary emission.

**Photosynthesis.** Conversion of inorganic matter into organic matter. The manufacture by plants of carbohydrates and oxygen from carbon dioxide and water in the presence of chlorophyll with sunlight as the energy source. Oxygen and water vapor are released in the process. Photosynthesis is dependent on favorable temperature and moisture conditions as well as on the atmospheric carbon dioxide concentration. Increased levels of carbon dioxide can increase net photosynthesis in many plants.

**Phytoplankton.** That portion of the plankton community in a body of water made up of tiny plants (e.g., algae and diatoms).

**Planetary Albedo.** The fraction of incident solar radiation that is reflected by a planet and returned to space. The planetary albedo of the Earth-atmosphere system is approximately 30%, most of which is due to backscatter from clouds in the atmosphere.

**Plasma.** A completely ionized gas, the so-called fourth state of matter (besides solid, liquid, and gas) in which the temperature is too high for atoms as such to exist and which consists of free electrons and free atomic nuclei.

---

630)  D. M. Anderson, R. S. Webb, J. T. Overpeck, B. A. Bauer, "The NOAA Paleoclimatology Program", NOAA Earth System Monitor, Vol. 3 , Nr. 3, March 1993, pp. 6-8

**Plasmoid.** Bubble of plasma. Refers to the merging of magnetic lines in the magnetotail which is thought to produce a bubble of plasma, called a plasmoid, that flows down the tail during the active periods.

**Polarization.** Defines the orientation of the electric (and magnetic) fields of an electromagnetic wave. Horizontal (H) / vertical (V) polarization refers to the electric field (magnetic field) vector being parallel to the surface of the medium that the wave is incident upon.

- Like polarization: HH or VV (one component for the transmit and one for the receive signal, as is the case for active sensors)
- Cross polarization: HV or VH
- Alternating dual polarization: alternate transmit and/or receive polarization so that two polarization combinations are measured - e.g. HH and HV or HH and VV

**Precision.** The precision of a measurement is a measure of the reproducibility or consistency of measurements made with the same sensor. [The effect of random errors can be reduced by repeated measurement or by averaging, which increases the precision of a measurement].

**Preprocessing.** Commonly used to describe corrections and preprocessing done to sensor data prior to information extraction. Preprocessing includes geometric and radiometric correction, mosaicking, resampling, reformatting, etc.

**Primary Productivity.** The rate of carbon fixation by marine photosynthetic organisms (phytoplankton). Primary Productivity results in the reduction of dissolved inorganic carbon to form organic carbon, with concomitant release of oxygen.

**Pulse Code Modulation (PCM).** Any modulation that involves a code made up of pulses to represent binary information. This is a generic term, additional specification is required for a specific purpose.

**Quantization.** The process of converting from continuous values of information to a finite number of discrete values. A 10 bit quantization means that the measured signal can be represented by a total of 1024 digital values, say from 0 to 1023.

**Radar (Radio Detection and Ranging).** A method, a system, or a technique for using beam-, reflected-, and timed electromagnetic radiation to detect, locate, and track objects, to measure distance (altitude), and to acquire terrain imagery (see also chapter C.1.10.2). The term 'radar' in remote sensing terminology refers to active microwave systems (from about 1 GHz - 100 GHz; the majority of current instruments operate below 10 GHz).

Radar systems may be classified by the signal measurement technique employed - there is the pulsed radar class and the continuous-wave radar class. The pulsed radar is the most widely used type of radar system. It is so called because the transmitter sends out pulses of microwave energy with relatively long intervals between pulses. The receiver picks up the echoes of the returned signals - the elapsed time is a measure of the distance travelled. The continuous-wave radar transmits continuous microwave energy - the resultant continuous echo cannot be associated with a specific part of the transmitted signal (hence, range information cannot be obtained). However, the system can determine the speed of a target by measuring the Doppler shift (change in frequency). A more sophisticated continuous-wave instrument, known as 'frequency-modulated radar', is also able to measure range. This is done by tagging each part of the transmitted microwave signal (by changing the frequency continuously), rendering it recognizable upon reception. With the rate of frequency change known, the difference in frequency can be interpreted as a range measurement.

Radar instruments consist of the following elements: transmitter, receiver, antenna, data processing system, and recorder. Radar instruments are built for a specific transmission frequency in the microwave spectrum such as P-Band, L-Band, S-Band, C-Band, X-Band, Ka-Band, etc.; some very advanced instruments offer observation in multiple frequencies.

- **RAR** (Real-Aperture Radar). The term RAR is used because the along-track resolution of a surface image is determined by the actual length of the antenna aperture (the along-track resolution is given by the width of the antenna sweep, the across-track resolution is determined by the range-resolving capability of the instrument). RAR systems are usually much simpler than SAR systems in design and data processing. The RAR pulsed signals are not required to be coherent (only the signal amplitude information is recovered and processed), representing and displaying backscatter characteristics from the surface sweep, that are recorded on film or on magnetic tape. Microwave energy reflected from the surface terrain (target) is converted by the RAR instrument into electrical signals and recorded as a function of distance (along-track and across-track direction). The radar returns from the different positions in the sweep (and at the different ranges) are separated in time by the radar receiver (the across-track range measurement is a function of signal return time). A new pulse is transmitted after reception and recording of the previous pulse for a new radar sweep. - The density of the image varies with the surface properties (roughness, moisture content, etc.). The image can be interpreted in terms of the topographical features of the terrain.

- **SAR** (Synthetic Aperture Radar), see also C.1.10.2. This radar type permits high-resolution imagery at long ranges (a SLAR device from a satellite orbit is of limited use due to the poor resolution obtained by the angular geometry constraints of the radar beam). A SAR instrument is also referred to as a 'coherent SLAR' - SAR is a concept that uses complex signal processing techniques to recover an image by the coherent processing of all return signals of all targets in a sweep. The across-track range of all targets is determined by the signal return time. The pulse bandwidth determines the cross-track or range resolution. Coherence in this SAR context refers to the fact that the phase as well as amplitude information of the radar cross-section (complex radar cross-section) is measured for all recovered signals.
A disadvantage of the SAR observation/processing technique is the generation of very high data rates (between 20 and 100 Mbit/s and more), this implies high communication rates and large storage volumes. On-board recorders are strained to their very limits to handle SAR data. First attempts are being made of real-time on-board preprocessing for the purpose of data reduction.

- **SLAR** (Side-Looking Airborne Radar - an active sensor with RAR technology).

- **Radar Altimeter**. An active device observing the vertical distance between the instrument and the ground by the measurement of elapsed time between the emitted and returned signals of electromagnetic pulses. Determination (mapping) of the height profile of the surface (topographic applications).

- **Scatterometer**. A scatterometer is a non-imaging radar, it is distinguished from other radars by its ability to measure radiation amplitude. A radar scatterometer is an active device measuring the backscattering coefficient of the illuminated cell (area or volume under observation) at a specified configuration of incidence angles, wavelengths, and wave polarization orientations. The backscattering coefficient describes the target backscattering characteristics and varies as a function of surface roughness, moisture content, and dielectric properties. Another application: the surface backscattering coefficient may be used to derive the surface wind vector.

- **Lidar** (Light Detection and Ranging), see C.1.10.3. The lidar instrument is also referred to as the 'optical radar' since it utilizes the optical (and NIR) portion of the electromagnetic spectrum (0.4 - 1.5 µm wavelength range, or a frequency of about 1000 - 100 THz). A very narrow beam (pulse) of laser light is emitted, the echo is analyzed.

**Radar Meteorology**. A discipline that uses backscattered electromagnetic radiation within the microwave band to gain information about the state of the atmosphere, especially with respect to clouds and precipitation. The return signal allows only the interpretation of four fundamental properties of the spectrum: amplitude, phase, frequency, and polarization.

**Radiometer**. An instrument for the quantitative measurement of the intensity of electromagnetic radiation in some band of wavelengths in the spectrum. Usually a radiometer is prefixed such as IR-Radiometer, or Microwave-Radiometer to indicate the spectrum to be measured.

**Raman spectroscopy**[631]. A technique that uses scattered light resulting from photon-molecule collisions to investigate molecular properties. When a monochromatic light beam is incident on systems such as transparent gases, liquids, or solids, most of it is transmitted without change. However, a very small portion of the incident light is scattered. Although most of the scattered light has the same wavelength as the incident radiation, a small part of it occurs at different wavelengths. The scattering of light at different wavelengths is called **Raman scattering** (Indian scientist Sir C. V. Raman, who, with K. S. Krishnan, first reported the phenomenon in 1928). The physical origin of Raman scattering lies in inelastic collisions between the molecules composing the system (e. g. the liquid) and photons, the particles composing the light beam. 'Inelastic collisions' means, that there is an exchange of energy between the photon molecule with a consequent change in energy, and hence wavelength, of the photon.

**Range Direction**. Observation of an instrument in the cross-track direction (normal to the sub-satellite track).

**Rayleigh scattering**. Scattering by particles small in size compared with the wavelengths being scattered, e. g., scattering of blue light by the atmosphere.

**Refractive index**. Ratio of the velocity of light at a given wavelength in air to that in a refractive medium (water).

**Relative aperture**. For a photographic or telescopic lens, the ratio of the equivalent focal length to the diameter of the entrance slit. It is expressed as f/45 or f/5.6, also called 'f number' or speed of lens.

**Resolution**. Minimum (spatial) separation between two measurements in order for a sensor to be able to discriminate between them. Spectral and radiometric resolutions refer to the resolving power of a system in wavelength and energy, respectively.

**Satellite Charging**. All bodies which are placed in a plasma in thermal equilibrium acquire a negative electric charge. The negative potential depends on the plasma temperature. At altitudes of 300 to 500 km, the average kinetic energy of the plasma is low ($<$ 1 eV), hence, satellites become only weakly charged. At high altitudes ( geostationary orbit and further out) the kinetic energy of the plasma is considerably larger (the plasma is referred to as 'hot'), hence, satellites acquire a high potential with respect to it (in the order of several keV).

**Satellite Laser Ranging (SLR)**. Very precise range measurements from ground reference stations to geodynamic satellites (like Lageos, Starlette, Stella, Geo-IK, Etalon, EGS, etc.). The SLR technique employs short pulse lasers from the ground to reflectors on satellites. The quantity of interest is time-of-light (round trip) corrected for ranging system internal delay (calibration), atmospheric refraction (delay), retroreflector offset to the S/C center-of-mass, and network epoch synchronization. SLR techniques are a strong contributor to advances in precision orbit determination.

**Scanning**. The sweep of a mirror, prism, antenna, or other element across a track (normal to the direction of flight); the footprint may be a straight line, a circle or any other shape.

**Scanner**. Any device that scans and by this means produces an image.

**Scattering**. Light absorbed and subsequently re-emitted in all directions at about the same frequency.

---

631)  Encyclopedia of Physical Science and Technology, Academic Press, 1987

**Scintillation**. Variations in the brightness of starlight (i.e. 'twinkling') caused by turbulent strata very high in the Earth's atmosphere.

**Scintillation Counter.** A device that uses a photomultiplier tube to detect or count charged particles (which produce scintillations of radiation when they impact upon phosphor) or $\gamma$-rays.

**Sensor**. An instrument, usually consisting of optics, detectors, and electronics, that collects radiation and converts it to some other form. The form may be a certain pattern (an image, a profile, etc.), a warning, a control signal, or some other signal.

**Sensor Characteristics**. The ability of a sensor to detect and to resolve the incoming radiation. For imaging sensors a very prominent characteristic is the 'ground resolution', its ability to distinguish objects on the Earth's surface. Other sensor characteristics are: scene size, spectral range, spectral resolution, radiometric resolution, pointing accuracy (location knowledge), the timeliness (with which the images are returned to the user, the frequency, with which a given target can be revisited, the fraction of time that the sensor uses for taking an image).

**SNR (Signal-To-Noise Ratio)**. The ratio of the level of the information-bearing signal power to the level of the noise power. The maximum SNR of a device is called the 'dynamic range'.

**Solar Absorption Technique**. A method for the measurement of atmospheric constituents. As sunlight passes through the Earth's atmosphere, certain wavelengths are selectively absorbed by gaseous constituents. In the infrared region, nearly all gases have characteristic, discrete absorptions, whose positions and relative strengths are known from laboratory measurements of pure gas samples. This permits gaseous atmospheric constituents between the sun and an observer to be identified and quantified from high resolution solar spectra.

**Solar Cell**. A device that converts the radiant energy of sunlight directly into electrical power.

**Solar Cycle.** The 11 year period between maxima (or minima) of solar activity. About every 11 years the magnetic field of the Sun reverses polarity; hence the more basic period may be 22 years. It is generally accepted that the solar cycle is maintained by a dynamo driven by the differential rotation of the Sun's envelope.

**Solar Wind.** A radial outflow of plasma from the solar corona, carrying mass and angular momentum away from the Sun (see chapter C.5).

**Sounder**. A remote sensing instrument that measures profiles of certain parameters (like temperature, pressure, moisture, etc.) in a particular plane of observation. Two basic configurations are in use:

- In the nadir-viewing configuration the observation plane is the orbit plane of the platform (along the suborbital track). The scan technique provides good horizontal resolution of the measurements, but usually poor vertical resolutions.

- Limb sounders look at the horizon (the limb) and scan vertically, producing good vertical resolution but poor horizontal resolution.

**Spacecraft Stabilization**. Techniques that control the orientation of the spacecraft in orbit with respect to certain known references. Several methods are in use:

- Single-spin stabilization. The whole S/C body rotates about the axis of the principal moment of inertia. These satellites cannot have oriented antennas, a severe drawback for certain applications.

- Dual-spin stabilization. A configuration in which the S/C consists of two parts: the platform, which is oriented toward the Earth, and the rotor, which rotates about the principal axis of the S/C thereby providing gyroscopic stiffness (example: Meteosat).

- Three-axis stabilization. A configuration in which the entire S/C is oriented toward Earth. The control torques for attitude control are provided by a combination of momentum wheels, reaction wheels, and thrusters.

- Gravity gradient stabilization. A S/C consisting of two masses (main mass and small mass) that are connected by a rod or a boom. This two-mass arrangement produces a gravity gradient along the boom axis and an associated small torque which is employed for S/C orientation. This technique is usually implemented along with magnetic torquing (yet another stabilization method) for better attitude control of microsatellites.

- etc.

**Spectra (methods of dispersing spectra)**. see also Table 210

- Refraction →prisms are used to break up or disperse electromagnetic radiation into its component colors. The path of the radiation bends (refracts) when it passes from one medium into another.
- Diffraction →a light wave breaks up into waves travelling in all directions as it strikes a surface. Diffraction gratings are composed of closely spaced transmitting slits on a flat surface (transmission gratings), or alternate reflecting and non-reflecting grooves on a surface (reflecting gratings).
- Interference →see interferometer.
- Filter (electronically tunable filters)
- Filter (mechanical)
- Filter (mask)

**Spectral band**. An interval in the electromagnetic spectrum defined by two wavelengths, two frequencies, or two wave numbers.

**Spectral and spatial purity**. An evaluation of the quality of radiometric measurements in the spectral and spatial domains.

**Spectral signature**. Quantitative measurement of the spectral properties of an object at one or several wavelength intervals.

**Spectrometer**. A device to detect, measure and analyze the spectral content of the incident electromagnetic radiation. Conventional imaging spectrometers use gratings or prisms (along with collimating and re-imaging optical and mechanical components) to disperse light for spectral discrimination. In a wedge spectrometer spectral discrimination occurs in a focused beam.

**Spread spectrum technology**. A transmission technique that allows multiple senders and receivers to share the same portion of the spectrum by having each sender encode its transmission in a unique way decipherable by only its intended receiver. Part of this technique is used in GPS and GLONASS communication. Spread spectrum technology will also be used for PCS (Personal Communication Services) via satellite on such systems as 'Iridium' and 'Globalstar'.

**Stereoscopy**. Three-dimensional or "stereo" images essential for map-making and for many other applications (flight simulators, etc.). A remotely sensed stereo image combines images taken from cameras at slightly different angles (locations) for better surface relief mapping capabilities.

**Stratopause**. Stratosphere-mesosphere boundary (about 50-55 km) where a relative temperature maxima is found (see Figure 161).

**Stratosphere.** Region of the atmosphere between the troposphere and mesosphere, having a lower boundary of approximately 8 km at the poles to 18 km at the equator and an upper boundary of approximately 50 km. Depending upon latitude and season, the temperature in the lower stratosphere can increase, be isothermal, or even decrease with altitude, but the temperature in the upper stratosphere generally increases with height due to absorption of solar radiation by ozone.

**Sunspot.** A temporary disturbed area in the solar photosphere that appears dark because it is cooler than the surrounding areas. Sunspots are concentrations of strong magnetic flux (2000 - 3000 gauss), with diameters less than about 50 000 km and lifetimes of a few weeks.

**Sun-synchronous orbit** (see chapter C.1.12).

**Telemetry.** A space-to-ground data stream of measured values (normally including instrument science data, instrument engineering data, and spacecraft engineering data) that does not include commands, tracking, computer memory transfer, audio, or video signals.

**Thermoluminescence.** A property of certain minerals that, after electrons are excited into traps by ionizing radiation, they emit light when moderately heated.

**Thermosphere.** Outermost layer of the atmosphere, above the mesosphere.

**Time Division Multiple Access (TDMA).** A process that shares the time domain of a single carrier among many users by assigning to each time intervals in which to transmit signal bursts.

**Trace Gas.** A minor constituent of the atmosphere. The most important trace gases contributing to the greenhouse effect are water vapor, carbon dioxide, ozone, methane, nitrous oxide, and chlorofluorocarbons. Other trace gases include ammonia, nitric oxide, ethylene, sulfur dioxide, methyl chloride, carbon monoxide, and carbon tetrachloride.

**Transducer.** A device chancing one form of signal energy into another, such as a microphone, a thermocouple, a photocell, etc.

**Transponder.** A combined receiver and transmitter whose function is to transmit signals automatically when triggered by an interrogating signal.

**Tropical Year.** The interval of time between two successive vernal equinoxes. It is equal to 365.242 mean solar days.

**Tropopause.** Boundary between the upper troposphere and the lower stratosphere that varies in altitude between approximately 8 km at the poles to 18 km at the equator. The temperature gradient of the Tropopause goes to zero (a relative temperature minima exists).

**Troposphere.** Lowest atmospheric layer, between the surface and the tropopause (lowest 10-20 km of the atmosphere). The troposphere is characterized by decreasing temperature with height, large vertical motion, and large water vapor content. This is the region where most of the 'weather' occurs.

**Upwelling.** The vertical motion of water in the ocean by which subsurface water of lower temperature and greater density moves toward the surface of the ocean. Upwelling occurs most commonly along the western coastlines of continents, but may occur anywhere in the ocean. Upwelling results when winds blowing nearly parallel to a continental coastline transport the light surface water away from the coast. Subsurface water of greater density and lower temperature replaces the surface water, and exerts a considerable influence on the weather of coastal regions. Carbon dioxide is transferred to the atmosphere in regions of upwelling. This is especially important in the Pacific equatorial regions, where 1 to 2 gigatons of carbon per year may be released to the atmosphere. Upwelling also results in increased ocean productivity by transporting nutrient-rich waters to the surface layer of the ocean.

**Van Allen Belt.** Regions or belts in the Earth's magnetosphere (at about 1.4-1.5 $R_E$ and 4.5-6 $R_E$) where many energetic charged particles from the solar wind are trapped in the Earth's magnetic field.

**Vernal Equinox.** The point of intersection between the ecliptic and the celestial equator, where the Sun crosses from the south to the north (Spring Point).

**Very Long Baseline Interferometry (VLBI).** In radio astronomy, a system of two or more antennas placed several hundred or even several thousand kilometers apart, which are operated together as an interferometer. VLBI techniques are also considered in the field of Solid Earth Physics (geodynamics) for the determination of plate motions with accuracies of better than 1 cm/year. VLBI systems offer a superb tie to an inertial celestial reference frame based on extragalactic radio sources, which in turn is used to maintain the terrestrial reference frame.

**Video.** In general used to mean television, or used in the communication of the television image. Specifically, pertains to the bandwidth and spectrum position of the signal which results from television scanning and which is used to reproduce a picture.

**Video imagery.** Although video data does not have the detailed resolution of film, it offers the advantage of immediate processing capability. This is particularly important in time-sensitive applications.

**Wind shear.** The rate of change of a wind component with distance.

**Zodiacal Light.** A faint glow that extends away from the Sun in the ecliptic plane of the sky, visible to the naked eye in the western sky shortly after sunset or in the eastern sky shortly before sunrise. Its spectrum indicates it to be sunlight scattered by interplanetary dust. The zodiacal light contributes about a third of the total light in the sky on a moonless night.

# Appendix B     Definition of the Acronyms

AAOE . . . . . . . . . Airborne Antarctic Ozone Experiment (1987)

AASE . . . . . . . . . Airborne Arctic Stratospheric Expedition (AASE-II took place during the Winter of 1991-1992)

ACCESS . . . . . . . Assembly Concept for Construction of Erectable Space Structure (Shuttle)

ACE . . . . . . . . . . Advanced Composition Explorer (NASA, APL, etc., see A.1)

ACES . . . . . . . . . Acoustic Containerless Experiment System (Shuttle)

ACR . . . . . . . . . . Active Cavity Radiometer (ATLAS Sensor)

ACRES . . . . . . . Australian Centre for Remote Sensing (Belconnen Australia)

ACRIM . . . . . . . Active Cavity Radiometer Irradiance Monitor (EOS Program Sensor)

ACSE . . . . . . . . . Association Control Service Element (Application Layer)

ACTS . . . . . . . . . Advanced Communications Technology Satellite

A/D . . . . . . . . . . Analog/Digital converter

ADEOS . . . . . . . Advanced Earth Observation Satellite (NASDA, A.3, A.4)

ADSF . . . . . . . . . Automated Directional Solidification Furnace (Shuttle payload)

AEM-1 . . . . . . . . Applications Explorers Mission-1 (see HCMM A.46)

AEM-2 . . . . . . . . Applications Explorers Mission-2 (A.5)

AEPI . . . . . . . . . Atmospheric Emissions Photometric Imaging Experiment (ATLAS Sensor)

AES . . . . . . . . . . . Atmospheric Environment Service (Canada)

AESA . . . . . . . . Atmospheric Effects of Stratospheric Aircraft (NASA)

AF . . . . . . . . . . . US Air Force

AFB . . . . . . . . . . Air Force Base (US Air Force)

AFC . . . . . . . . . . Affiliated Data Center (these are institutional facilities that are affiliated with EOSDIS, in particular NOAA facilities are AFCs)

AFE . . . . . . . . . . American Flight Echocardiograph (Shuttle payload)

AFGL . . . . . . . . Air Force Geophysics Lab (USA)

AFP-675 . . . . . . . Air Force Program 675 (Shuttle payload)

AFSCN . . . . . . . Air Force Satellite Control Network (USA)

ADEOS . . . . . . . Advanced Earth Observing Satellite (NASDA), A.3

AGARD . . . . . . . Advisory Group for Aerospace Research and Development

AGASP . . . . . . . Arctic Gas and Aerosols Sampling Project (airborne campaign)

AGC . . . . . . . . . Antenna Gain Control

AGU . . . . . . . . . American Geophysical Union

AIAA . . . . . . . . . American Institute of Astronautics and Aeronautics (New York)

AIRS . . . . . . . . . Atmospheric Infrared Sounder (EOS Sensor)

AIP . . . . . . . . . . Astrophysikalisches Institut Potsdam, Germany

AKR . . . . . . . . . Auroral Kilometric Radiation

ALADIN . . . . . . . Atmospheric Laser and Doppler Instrument (ESA sensor for POEM)

ALEXIS . . . . . . Array of Low Energy X-Ray Imaging Sensors (LANL, A.6)

ALISSA . . . . . . . l'Atmoshere par LIdar Sur SAliout (the French sensor was at first proposed by CNES for a Salyut flight)

ALMAZ . . . . . . . ALMAZ = 'rough Diamond' (Earth observation series, Russia), A.7

ALOHA . . . . . . . One of several communication access methods

AM . . . . . . . . . . . Amplitude Modulation (Type of modulation of the main carrier)

AMAS . . . . . . . . Advanced Millimeter-Wave Atmospheric Sounder (ATMOS Sensor)

AMI . . . . . . . . . . Active Microwave Instrument (ERS-1 Sensor)

AMOS . . . . . . . . Air Force experiment using orbiter as calibration target for ground-based experiment for Air Force Maui Optical Site (Shuttle experiment)

AMPTE . . . . . . . Active Magnetosphere Particle Tracer Explorers (cooperative mission of US/ NASA, Germany and UK, A.8)

AMR .......... Active Microwave Radar (TRMM Sensor)
AMS .......... American Meteorological Society
AO ........... Announcement of Opportunity (usually for sensor on a particular mission)
AOCS ........ Attitude and Orbit Control System
AOET ........ Atomic Oxygen Exposure Tray (Shuttle D2 mission)
APD .......... Avalanche Photo Diode (detector type)
APE .......... Auroral Photography Experiment (Shuttle payload)
APEX ........ Active Plasma Experiment (Intercosmos, A.9)
APL .......... Applied Physics Laboratory (a facility of Johns Hopkins University in Baltimore)
APM ......... Ascent Particle Monitor (Shuttle experiment)
APT .......... Automatic Picture Transmission (one type of NOAA downlink transmission; APT transmits data from two channels of the AVHRR at a reduced resolution of 4 Km in the VHF frequency band (at 137.50 and 137.62 MHz)).
APV .......... Autonomously Piloted Vehicle (Condor)
AR ........... Anthrorack (Shuttle D2 mission)
ARAT ........ Avion de Recherche Atmosphérique et de Télédétection (Atmospheric Research and Remote Sensing Aircraft), ARAT is operated by IGN (Institut Geographique National) for various French organizations including CNES and CNRS
ARC .......... AMES Research Center (NASA facility at Moffett Field Ca.)
ARC .......... Aggregation of Red Blood Cells (Shuttle experiment)
Archimedes I, II . Coordinated European airborne campaigns in the North Sea region (start in 1983, Archimedes IIa took place in April 1988)
ARGOS ....... Argos (CNES System) is a data collection and location system with a space segment and a ground segment. ARGOS is operational on NOAA polar-orbiting S/C. A.74.2, A.10
ARM ......... Atmospheric Radiation Measurement (program of DOE)
ARISTOTELES . Applications and Research Involving Space Techniques Observing The Earth's Field from Low Earth Orbiting Satellite (planned ESA Mission), A.11
ARPA ........ Advanced Research Project Agency (US)
ASAP ........ Airborne Science and Application Program (USGS, NASA)
ASEM ........ Assembly of Station by EVA Methods (Shuttle demonstration)
ASF .......... Alaska SAR Facility in Fairbanks, Alaska (DAAC of NASA EOS Program). ASF will in effect be a US-PAF for ERS-1 data as well as for JERS-1 and RADARSAT data.
ASHOE ....... Airborne Southern Hemisphere Ozone Experiment (1994)
ASI .......... Agenzia Spaziale Italiana (formerly PSN)
ASPRS ....... American Society of Photogrammetry and Remote Sensing
ASTER ....... Advanced Spaceborne Thermal Emission and Reflection Radiometer (Japanese Sensor on EOS missions)
ASTEX ....... Atlantic Stratocumulus Transition Experiment (airborne campaign at the Azores in 1992)
ASTRO-SPAS ... Astronomy Platform - Shuttle Pallet Satellite
ATI .......... Along-Track Interferometry
ATLAS ....... Atmospheric Laboratory for Application and Science (NASA Program, Payload series on Shuttle), A.12
ATLID ....... Atmospheric Lidar (Sensor), an ESA Backscatter Lidar
ATN .......... Advanced TIROS-N Series (NOAA, launched from 1983 on)
ATS .......... Application Technology Satellite (ESSA Satellite)
ATSR ........ Along-Track Scanning Radiometer and Microwave Sounder (ERS-1 Sensor)

ATMOS ......... Atmospheric Trace Molecule Spectroscopy (ATLAS Sensor)
ATMOS ......... Atmospheric and Oceanic Sensors (portion of German Earth Observation program), A.13
AU ............ Astronomical Unit, Sun-Earth distance $= 1.496 \times 10^8$ km (average)
AVHRR ....... Advanced Very-High Resolution Radiometer (NOAA Sensor, AVHRR/3 on NOAA-K,L,M,N is to be renamed in VIRSR for NOAA-O,P,Q)
AVISO ......... Archivage Validation and Interprétation des données des Satellites Océanographiques (CNES oceanographic data center in build-up phase for Topex/Poseidon)
AVNIR ........ Advanced Visible and Near-Infrared Radiometer (NASDA Sensor on ADEOS)
AWI ........... Alfred Wegener Institut, Bremerhaven, Germany
AZBS .......... Avionik Zentrum Braunschweig, Germany
BA ............ Baroreflex (Shuttle D2 mission)
BB ............ Biolabor (Shuttle D2 mission)
BBXRT ........ Broad Band X-Ray Telescope (part of ASTRO-1 observatory)
BCRS .......... Netherlands Remote Sensing Board
BEST .......... Bilan Energétique du Système Tropical (Tropical System Energy Budget), a CNES proposed mission
BIMDA ........ Bioserve/Instrumentation Technology Associates Materials Dispersion Apparatus (Shuttle payload)
BIRA .......... Belgisch Instituut voor Ruimte Aeronomie, Brussels
BLAST ........ Battlefield Laser Acquisition Sensor Test (Shuttle experiment)
BMFT ......... Bundesministerium für Forschung und Technologie (German Ministry for Research and Technology)
BMV .......... Bundesministerium für Verkehr (German Ministry of Transportation)
BMVg ........ Bundesministerium für Verteidigung (German Ministry of Defense)
BNSC .......... British National Space Centre, London
BOST ......... Belgian Office of Science and Technology
BPDF ......... Bidirectional Polarization Distribution Function
BPSK ......... Binary Phase Shift Keying (Modulation Type)
BRDF ......... Bidirectional Reflectance and Distribution Function
CAM .......... Centre d'Aviation Météorologique , France
CAMEX ....... Convection and Atmospheric Moisture Experiment (airborne campaign conducted at NASA Wallops Flight Facility)
CANEX ........ Canadian Experiments (Shuttle payload)
CAO .......... Central Aerological Observatory (Moscow)
CAPE ......... Convection and Precipitation Electrification Experiment
CAS ........... Chinese Academy of Sciences, Beijing
CAST ......... Chinese Academy of Space Technology (Beijing)
CBERS ........ China/Brazil - Earth Resources Satellite, A.14
CCD .......... Charge-Coupled Device (solid-state detector type)
CCE ........... Charge Composition Explorer (S/C of AMPTE mission, A.8.3)
CCITT ........ Comité Consultatif International Téléphonique et Télégraphique (one of three bodies for the definition of OSI, CCITT is a permanent organ of ITU)
CCRS ......... Canada Center for Remote Sensing (Ottawa, Ontario; established in 1972 , 'Department of Energy, Mines and Resources', Canada)
CCSDS ........ Consultative Committee for Space Data Systems
CDA ........... Command and Data Acquisition (NOAA Antenna, downlink concept)
CDDIS ........ Crustal Dynamics Data Information System (database at GSFC)
CDMA ......... Code Division Multiple Access
CEC ........... Commission of European Communities (Brussels)

CEES . . . . . . . . . .   Committee on Earth and Environmental Sciences (US interagency committee)
CEOS . . . . . . . . .   Committee on Earth Observation Satellites
CEPEX . . . . . . . .   Central Equatorial Pacific Experiment (campaign)
CERES . . . . . . . .   Clouds and Earth's Radiant Energy System (EOS Sensor)
CESR . . . . . . . . . .   Centre d'Etude Spatiale des Rayonnements, Toulouse, France (Part of CNRS)
CETA . . . . . . . . . .   Crew and Equipment Translation Aids (Shuttle experiment)
CEV . . . . . . . . . . .   Centre d'Essais en Vol (French Test Flight Center)
CFCs . . . . . . . . . .   Chlorofluorocarbons
CFES . . . . . . . . . .   Continuous Flow Electrophoresis System (Shuttle)
CFRP . . . . . . . . . .   Carbon Fiber Reinforced Plastic
CGBA . . . . . . . . .   Commercial General Bioprocessing Apparatus (Shuttle experiment)
CHAMP . . . . . . .   Comet Halley Active Monitoring Program (Shuttle experiment)
CHASE . . . . . . . .   Coronal Helium Abundance Spacelab Experiment (Spacelab-2)
CHROMEX . . . .   Chromosomes and Plant Cell Division (Shuttle experiment)
CIESIN . . . . . . . .   Consortium for International Earth Science Information Network. (The University of Michigan is a member of CIESIN. CIESIN provides EOS data to non-research users, such as government or the general public).
CIGNET . . . . . . .   Cooperative International GPS Network of IAG (International Association of Geodesy), A.43.4
CIMSS . . . . . . . . .   Cooperative Institute for Meteorological Satellite Studies (University of Wisconsin, Madison
CIR . . . . . . . . . . .   Color Infrared (video images)
CIRES . . . . . . . . .   Cooperative Institute for Research in Environmental Sciences (Boulder, Co.)
CIS . . . . . . . . . . . .   Commonwealth of Independent States (former Soviet Union or USSR)
CIT . . . . . . . . . . . .   California Institute of Technology, Pasadena
CIV . . . . . . . . . . . .   Critical Ionization Velocity (Shuttle experiment)
CLAES . . . . . . . .   Cryogenic Limb Array Etalon Spectrometer (UARS Sensor)
CLOUDS . . . . . .   Cloud Logic to Optimize Use of Defense Systems (Shuttle payload)
CLUSTER . . . . .   ESA/NASA Solar-Terrestrial Mission (A.15)
CMIX . . . . . . . . .   Commercial Materials Dispersion Apparatus Instrument Technology Associates Experiments (Shuttle experiment)
CMS . . . . . . . . . . .   Centre de Météorologie Spatiale, (in Lannion, France)
CNCR . . . . . . . . .   Characterization of Neurospora Circadian Rhythms (Shuttle payload)
CNES . . . . . . . . . .   Centre National D'Etudes Spatiales (Space Agency of France, Toulouse)
CNET . . . . . . . . .   Centre de Recherche en Physique de l'Environnement (Issy-les-Moulineaux, France)
CNIE . . . . . . . . . .   Comision Nacional de Investigaciones Espaciales (Space Agency of Argentina)
CNR/IFSI . . . . . .   Consiglio Nazionale delle Ricerche / Instituto de Fisica dello Spazio Interplanetario, Frascati, Italy
CNR/PSN . . . . . .   Consiglio Nazionale delle Ricerche / Piano Spaziale Nationale (Italy)
CNRS . . . . . . . . .   Centre National de la Recherche Scientifique (Service D'Aeronomie), in Verrieres Le Buisson, CESR in Toulouse, France
COARE . . . . . . . .   Coupled Ocean Atmosphere Response Experiment
COHMEX . . . . .   Cooperative Huntsville Meteorological Experiment (aircraft campaign in 1986)
CONCAP . . . . . .   Consortium for Materials Development in Space Complex Autonomous Payload (Shuttle experiment)
CORONAS . . . . .   Complex of Orbital Observations of the Activity of the Sun (Satellite of the Russian Space Agency, A.16)

COSMOS ...... The term 'Cosmos' or 'Kosmos' is used in Russia to designate any of a series of unmanned satellites that were launched starting in 1962 with Cosmos-1 (the counting in 1988 was around 1800, in 1993 it is around Cosmos-2200). The Cosmos satellite series has been used for a wide variety of purposes, including scientific research, Earth observation, experimental/technological payloads, preoperational meteorological satellites, navigation satellites, etc. There are also many satellites with military payloads under the Cosmos designation.

COSPAR ....... Committee on Space Research (UN sponsored)

COSPAS ....... Space System for the Search of Distressed Vessels (Russia's equipment flown on polar-orbiting S/C.

COTES ........ Conventional Terrestrial Reference System (an IERS program for the specifications of positions on or near the Earth's surface)[632]

CPCG ......... Commercial Protein Crystal Growth (Shuttle experiment)

CRA .......... Centro Ricerche Aerospaziali (University of Rome/Italy)

CRC .......... Communication Research Center (a facility of the Canadian Department of Communications)

CREAM ....... Cosmic Ray Effects and Activation Monitor (Shuttle payload)

CRI .......... Crown Research Institute (New Zealand)

CRISTA ....... Cryogenic Infrared Spectrometers and Telescopes for the Atmosphere (ATLAS Sensor)

CRO .......... Chemical Release Observation (Shuttle experiment)

CRPE ........ Centre de Recherches en Physique de l'Environnement Terrestre et Planetaire, Saint-Maur-des-Fosses (also: Issy-les Moulineaux), France (Lab is part of CNRS and of CNET)

CRRES ........ Combine Release and Radiation Effects Satellite (A.17)

CRT .......... Cathode Ray Tube

CRYOHP ...... Cryogenic Heat Pipe Experiment (Shuttle)

CSA .......... Canadian Space Agency

CSIRO ........ Commonwealth Science and Industrial Research Organization (Australia)

CSMA/CD ..... Carrier Sense Multiple Access / Collision Detection (commercially known under Ethernet)

CSR .......... Centro de Sensores Remote (Italy)

CSTG ......... Commission on International Coordination of Space Techniques for Geodesy and Geodynamics (since 1979), (Commission VIII of the International Association of Geodesy)

CTA .......... Canadian Target Assembly (Shuttle payload)

CTA .......... Centro Tecnico Aerospacial (Saò José dos Campos, S.P., Brazil)

CTP .......... Cloud Top Pressure

CVTE ........ Chemical Vapor Transport Experiment (Shuttle)

CZCS ......... Coastal Zone Color Scanner (NIMBUS-7 Sensor)

DAAC ........ Distributed Active Archive Center (NASA EOS Program; a total of 7 DAACs are planned, each with a different focus on science)

DARA ........ Deutsche Agentur für Raumfahrtangelegenheiten, Bonn (German space agency since 1989)

DARPA ....... Defense Advanced Research Projects Agency (now ARPA, US, DOD agency)

DASA ........ Deutsche Aerospace AG (Daimler holding of MBB, Dornier, etc.)

DBMS ........ Database Management System

DBS .......... Direct Broadcasting Satellite

DCRS ........ Digital Cassette Recorder System

---

632) See : "The International Earth Rotation Service", in 'The Interdisciplinary Role of Space Geodesy', Springer Verlag, 1989, pp. 229-232

DCP . . . . . . . . . . Data Collection Platform (ground segment platform for environmental data measurement, Meteosat, GOES, GMS)
DCPI . . . . . . . . . Data Collection Platform Interrogation (GOES)
DCS . . . . . . . . . . Data Collection System (NOAA- GOES series, Meteosat series, GMS series, geostationary satellites).
DCW . . . . . . . . . Digital Chart of the World (a vector map database by DMA, Fairfax)
DEM . . . . . . . . . Digital Elevation Model
DESPA . . . . . . . . Départment de Recherche Spatiale de L'Oservatoire de Paris/Meudon (France)
DFD . . . . . . . . . . Deutsches Fernerkundungsdatenzentrum (German Remote Sensing Data Center, DLR in Oberpfaffenhofen)
DGFI . . . . . . . . . Deutsches Geodätisches Forschungsinstitut (München)
DGGTN . . . . . . . Direction General de Geografica del Territorio Nacional, Mexico
DGON . . . . . . . . Deutsche Gesellschaft für Ortung und Navigation (Düsseldorf)
DGPF . . . . . . . . Deutsche Gesellschaft für Photogrammetrie und Fernerkundung
DGPS . . . . . . . . . Differential GPS, A.43.5
DFN . . . . . . . . . . Deutsches Forschungsnetz
DIAL . . . . . . . . . Differential Absorption Lidar (Lidar technique)
DISCOS . . . . . . . Database and Information System Characterizing Objects in Space (ESA/ESOC database for space debris and meteoroids, since 1990)
DLR . . . . . . . . . . Deutsche Forschungsanstalt für Luft- und Raumfahrt (German Aerospace Research Establishment)
DMA . . . . . . . . . Defense Mapping Agency, Fairfax, Va. (mapping, charting & geodetic products & services to the military)
DMOS . . . . . . . . Diffusive Mixing of Organic Solutions (Shuttle payload)
DMSP . . . . . . . . Defense Meteorological Satellite Program (USA), A.19
DOAS . . . . . . . . . Differential Optical Absorption Spectroscopy
DOC . . . . . . . . . . Department of Commerce (USA)
DOD . . . . . . . . . Department of Defense (USA)
DOE . . . . . . . . . Department of Energy (USA)
DOM . . . . . . . . . Dissolved Organic Matter
DORIS . . . . . . . . Determination Orbite Radiopositionncment Integres Satellite (CNES one-way tracking system for the measurement of precision orbits); another name convention is: Doppler Orbitography and Radiopositioning Integrated by Satellite, A.110.1
DRA . . . . . . . . . . Defense Research Agency, Malvern, UK
DRS . . . . . . . . . . Data Relay Satellite (ESA system to relay information from the European space plane)
DSN . . . . . . . . . . Deep Space Network (NASA/JPL)
DSRI . . . . . . . . . Danish Space Research Institute
DUT . . . . . . . . . Delft University of Technology, the Netherlands
EARSEC . . . . . . . European Airborne Remote Sensing Capabilities [program since 1990 between CEC (JRC in Ispra, Italy) and ESA]
EARSeL . . . . . . . European Association of Remote Sensing Laboratories (Paris)
EARTHNET . . . . ESA Program since 1977. Earthnet refers to an ESA organization responsible for the ground segment of Earth Observation. Functions: acquisition, archiving .... and distribution of Earth science data.
EASE . . . . . . . . . Experimental Assembly of Structures in Extravehicular Activity (Shuttle)
ECMWF . . . . . . . European Centre for Medium-Range Weather Forecasts (Reading, UK)
ECS . . . . . . . . . . EOSDIS Core System
EDC . . . . . . . . . . EROS Data Center of the US Geological Survey in Sioux Falls, S. D. (DAAC of NASA EOS Program for Land Processes)

EDI . . . . . . . . . . Electronic Data Interchange, (Format Specification according to ANSI Standard X.12; (an existing but non-ISO Protocol)
EDIFACT . . . . . . Electronic Data Interchange for Administration, Commerce, and Transport
EECF . . . . . . . . . Earthnet ERS-1 Central Facility (ESA facility at ESRIN, Italy)
EEVT . . . . . . . . Electrophoresis Equipment Verification Test (Shuttle)
EGS . . . . . . . . . . Experimental Geodetic Satellite of NASDA, (Ajisai, A.20)
EHF . . . . . . . . . . Extremely High Frequency (30 - 300 GHz band)
EHIC . . . . . . . . . Energetic Heavy Ion Composition Experiment
EMAC . . . . . . . . European Multi-Sensor Airborne Campaign (in the framework of ESA/JRC collaboration)
EIRP . . . . . . . . . Effective Isotropic Radiated Power
EISAC . . . . . . . . European Imaging Spectrometry Aircraft Campaign (1989-90)
EISCAT . . . . . . . European Incoherent Scatter Radar
ELAC . . . . . . . . European Lidar Airborne Campaign (1990 to demonstrate the usefulness of a spaceborne lidar)
ELF . . . . . . . . . . Extremely Low Frequency (30 - 3000 Hz)
ELRAD . . . . . . . Earth-Limb Radiance Experiment (Shuttle payload)
ELT-121.5 . . . . . . Emergency Locator Transmitter (see S&RSAT and COSPAS)
E-Mail . . . . . . . . Electronic Mail
EMSL . . . . . . . . Environmental Monitoring Systems Laboratory (Las Vegas, EPA facility)
ENACEOS . . . . . Energetic Neutral Atom Camera for EOS (EOS Sensor)
ENAP . . . . . . . . Energetic Neutral Atom Precipitation (ATLAS Sensor)
ENVISAT . . . . . . Environmental Satellite (ESA, see A.78.1)
EO . . . . . . . . . . . Earth Observation
EOPP . . . . . . . . . Earth Observation Preparatory Programme (ESA)
EOS . . . . . . . . . . Earth Observing System (NASA), A.21 and A.22
EOSAT . . . . . . . Earth Observation Satellite Company (Commercial distributor of Landsat science data, NOAA contractor, in Lanham, Md)
EOSDIS . . . . . . . EOS Data and Information System
EOSP . . . . . . . . . Earth Observing Scanning Polarimeter (EOS Program Sensor)
EOS-SAR . . . . . . EOS Synthetic Aperture Radar (EOS Sensor)
EPA . . . . . . . . . . Environmental Protection Agency (USA)
EPIRB-406 . . . . . Emergency Position Indicating Radio Beacon (on COSPAS and S&RSAT)
EPOCS . . . . . . . . Equatorial Pacific Ocean Climate Studies
EPOP . . . . . . . . . European Polar Platform (old name, now POEM)
Equator-S . . . . . . Solar Terrestrial Mission (A.23)
ERB . . . . . . . . . . Earth Radiation Budget
ERBE . . . . . . . . . Earth Radiation Budget Experiment (NOAA Sensor on NOAA-9 and -10)
ERBS . . . . . . . . . Earth Radiation Budget Satellite (NASA), A.24
ERIM . . . . . . . . Environmental Research Institute of Michigan (USA)
EROS . . . . . . . . Earth Resources Observation System (Data Center of USGS in Sioux Falls, S. D., Archive of Landsat Data)
ERS-1,2 . . . . . . . . European Remote Sensing Satellite (ESA Program), A.25 and A.26
ERS . . . . . . . . . . Earth Resource Satellite
ERTS-1 . . . . . . . . Earth Resources Technology Satellite (NASA satellite, was later renamed to Landsat-1) see Landsat A.55
ESA . . . . . . . . . . European Space Agency
ESA-IRS . . . . . . . ESA - Information Retrieval Service (online database at ESRIN)
ESA/PB-EO . . . . ESA/Programme Board - Earth Observation
ESDIS . . . . . . . . Earth Science Data and Information System (NASA/GSFC)
ESIS . . . . . . . . . . European Space Information System (ESA data system)

ESOC .......... European Space Operation Centre (ESA facility in Darmstadt, Germany)
ESRIN ......... European Space Research Institute (ESA facility in Frascati, Italy)
ESSA .......... Environmental Science and Services Administration (this was a predecessor organization of NOAA)
ESTEC ........ European Space Research and Technology Centre (ESA facility in Noordwijk, Holland)
ETALON ...... Russian passive satellite series for geodetic measurements, A.27
ETM .......... Enhanced Thematic Mapper (Landsat-6 sensor)
EU ............ European Union (formerly EC = European Community)
EUCREX ...... European Cloud and Radiation Experiment (planned for 1994)
EUMETSAT .... European Organization for the Exploitation of Meteorological Satellites (Darmstadt)
EURIMAGE ... European Consortium for Satellite Image Dissemination (Rome)
EURECA ...... European Retrievable Carrier (platform deployed / retrieved on Shuttle, A.28
EURISY ....... European Association for ISY (one of two ISY organizers in Europe, ▶ SAFISY)
EUV .......... Extreme Ultra Violet (spectral range) ▶XUV
EXOS ......... Exospheric Observations, ISAS program (A.29)
FAA .......... Federal Aviation Administration (USA)
FAGS ......... Federation of Astronomical and Geophysical Services
FAO .......... Food and Agriculture Organization (of the UN)
FARE ......... Fluid Acquisition and Resupply Experiment (Shuttle)
FAST ......... Fast Auroral Snapshot Explorer (GSFC mission, A.30)
FCC .......... Federal Communications Commission (USA)
FDDI ......... Fiber Distributed Data Interface
FDMA ........ Frequency Division Multiple Access
FEA .......... Fluid Experiment Apparatus (Shuttle)
Fengyun (FY) ... Chinese meteorological satellite series, A.31 and A.32
FFT .......... Fast Fourier Transform
FIAN ......... Lebedev Physical Institute of the Russian Academy of Sciences, Moscow
FIFE ......... First ISLSCP Field Experiment
FILE ......... Feature Identification and Location Experiment (part of OSTA-1 payload on shuttle STS-2 in Nov. 1981)
FIRE ......... First ISCCP Regional Experiment (1987)
FLINN ........ Fiducial Laboratories for an International Network (a global network supporting Crustal Dynamics Test Sites)
FM ........... Frequency Modulation (Type of modulation of the main carrier)
FMC .......... Forward Motion Compensation
FM-CW ....... Frequency Modulation Continuous Wave
FOA .......... Försvarets Forskningsanstalt (National Defense Research Establishment, Department of Information Technology) Linköpping, Sweden
FORTE ....... Fast On-Orbit Recording of Transient Events (LANL, A.33)
FOV .......... Field of View
FPA .......... Focal Plane Array
FREJA ........ Swedish Solar-Terrestrial Mission (A.34)
FSK .......... Frequency Shift Keying (Modulation type)
FTAM ........ File Transfer Access and Management (OSI File Transfer Method)
FTIR ......... Fourier Transform Infrared Radiometer
FTS .......... Fourier Transform Spectrometer
FWHM ....... Full-Width-Half-Maximum (of distribution curve)
GaAs ......... Gallium Arsenide (a material used for solar panels and for fast chips)
GAC .......... Global Area Coverage

GAF . . . . . . . . . . .  Gesellschaft für Angewandte Fernerkundung (German commercial distributor of Earth observation data, such as Resurs data, Landsat data (via EOSAT), etc.

gal  . . . . . . . . . . .  (from Galileo) unit of acceleration of free fall: 1 gal $= 10^{-2}$ m s$^{-2}$ $= 1$ cm s$^{-2}$; 1 mgal $= 10^{-5}$ m s$^{-2}$

GALE  . . . . . . . .  Genesis of Atlantic Lows Experiment (aircraft campaign in 1986)

GAMES  . . . . . . .  Gravity and Magnetic Earth Surveyor (a GSFC mission in the early planning phase as of 1993)

GARP  . . . . . . . .  Global Atmospheric Research Programme

GAUSS  . . . . . . . .  Galaktische Ultraweitwinkel Schmidt System, Shuttle payload (Galactic super wide angle Schmidt system)

GCMD  . . . . . . .  Global Change Master Directory (NASA)

GCOS  . . . . . . . .  Global Climate Observing System (Program)

GCTE  . . . . . . . .  Global Change and Terrestrial Ecosystems (Program)

Ge  . . . . . . . . . . .  Germanium (detector material)

GEMINI  . . . . . . .  NASA program of the sixties, A.35

GENIUS  . . . . . . .  Global Environmental Network Information and User System (European Study)

GEO-IK  . . . . . . .  Russian S/C for solid Earth research, A.36

GEOS-3 . . . . . . . .  Geodynamics Experimental Ocean Satellite, A.37, GEOS-3 is the first radar altimeter mission, end of mission in 1978)

GEOS  . . . . . . . .  Geodetic Earth Orbiting Satellite

GEOSAT . . . . . . .  US Navy satellite (altimeter mission), A.39

GEOTAIL . . . . . .  Japanese (ISAS) mission to study the structure and dynamics of the geomagnetic tail (part of ISTP), A.41

GER  . . . . . . . . . .  Geophysical & Environmental Research Corp. (Millbrook, New York)

GEWEX  . . . . . . .  Global Energy and Water Experiment (Program)

GFO . . . . . . . . . .  Geosat Follow-On (Satellite), A.40

GFU . . . . . . . . . .  Geophysical Institute of the Academy of Sciences of the Czech Republic, Prague

GFZ . . . . . . . . . . .  GeoForschungsZentrum, Potsdam, Germany

GGI . . . . . . . . . . .  GPS Geoscience Instrument (EOS Sensor)

GGS . . . . . . . . . . .  Global Geospace Science (US program within ISTP with two spacecraft: **Wind** and **Polar**)

GHCD . . . . . . . . .  Growth Hormone Crystal Distribution (Shuttle experiment)

GIPME  . . . . . . . .  Global Investigation of Pollution in the Marine Environment

GIS . . . . . . . . . . . .  Geographic Information System (an archive in particular for forestry data)

GKSS . . . . . . . . . .  Gesellschaft für Kernergieverwertung in Schiffbau und Schiffahrt (Geesthacht, Germany)

GLAS . . . . . . . . . .  Geoscience Laser Altimeter System (previously GLRS)

GLIS  . . . . . . . . . .  Global Land Information System (an online land data directory guide, a public information system operated by USGS at EROS Data Center)

Glavkosmos . . . . .  Russian space organization agency with the objective to develop the commercial side of space activities (created in 1985)

GLOBSAT  . . . . .  Proposed Earth Observation Satellite by the French Earth Science Community.

GLONASS . . . . .  Global Orbiting and Navigation Satellite System (USSR), A.42

GLRS  . . . . . . . . .  Geoscience Laser Ranging System (EOS Sensor), renamed in 1992 to GLAS = Geoscience Laser Altimeter System (The Earth Observer, May/June 1992, p. 16)

GMSK . . . . . . . . .  Gaussian Minimum Shift Keying

GMS  . . . . . . . . . .  Geostationary Meteorological Satellite, Operational Program of JMA (Japan Meteorological Agency), A.44

GNSS . . . . . . . . . .  Global Navigation Satellite System (EC-sponsored study for a civil satellite navigation system)
GOES  . . . . . . . . .  Geostationary Operational Environmental Satellite (NOAA Series), A.73
GOFS  . . . . . . . . .  Global Ocean Flux Study (Program)
GOME  . . . . . . . .  Global Ozone Monitoring Experiment (Sensor on ERS-2)
GOMOS . . . . . . .  Global Ozone Monitoring By Occultation of Stars (French Sensor)
GOMS . . . . . . . . .  Geostationary Operational Environmental Satellite (Russian geostationary meteorological satellite series (at longitude 76 deg. East), A.45
GORC . . . . . . . . .  Global Ocean Carbon Research Program
GOS . . . . . . . . . . .  Geomagnetic Observing System (EOS Sensor)
GOSAMR-1 . . . .  Gelatin of Sols: Applied Microgravity Research-1 (Shuttle experiment)
GOSIP . . . . . . . .  Government Open System Interconnection Profile ( US Government Standard,  GOSIP is a subset of OSI)
GPS . . . . . . . . . . .  Global Positioning System, A.43
GPCP . . . . . . . . . .  Global Precipitation Climatology Project (by ICSU and WMO)
GRGS . . . . . . . . .  Groupe de Recherches de Géodésie Spatiale (Grasse and Toulouse, France)
GRID  . . . . . . . . .  (UNEP) Global Resources Information Database (at EDC)
GSFC . . . . . . . . . .  Goddard Space Flight Center in Greenbelt, Md. (DAAC of NASA EOS Program)
GSOC  . . . . . . . .  German Space Operations Center (DLR facility in Oberpfaffenhofen)
GTC . . . . . . . . . . .  Global Tropospheric Chemistry Program
GTS . . . . . . . . . . .  Global Telecommunications System (of the World Meteorological Organization)
HALE  . . . . . . . .  High Altitude Long Endurance (aircraft)
HALOE . . . . . . . .  Halogen Experiment (UARS Sensor)
HAPEX-II-Sahel  Hydrological and Atmosphere Pilot Experiment (international airborne campaign in the Sahel region of West-Africa, August-October 1992, land surface-atmosphere observation program)
HBT . . . . . . . . . . .  Heflex Bioengineering Test (Shuttle)
HCMM  . . . . . . . .  Heat Capacity Mapping Mission (NASA Sensor), A.46
HCMR . . . . . . . . .  Heat Capacity Mapping Radiometer (sensor of HCMM)
HDDT . . . . . . . . .  High Density Digital Tape
HDT  . . . . . . . . . .  High Density Tape
HDLC . . . . . . . . .  High-Level Data Link Control (Bit-oriented Protocol)
HERCULES . . . .  Hand-held, Earth-oriented, Real-time, Cooperative, User-friendly, Location-targeting and Environmental System (Shuttle experiment)
HF . . . . . . . . . . . .  High Frequency (3 - 30 MHz band)
HgCdTe . . . . . . . .  Mercury Cadmium Telluride (mercadtelluride, a detector material)
HH . . . . . . . . . . . .  Horizontal transmit - Horizontal receive Polarization
HIMSS . . . . . . . . .  High-Resolution Microwave Spectrometer Sounder (EOS Sensor)
HiRDLS  . . . . . . .  High-Resolution Dynamics Limb Sounder (EOS Sensor)
HIRIS . . . . . . . . .  High-Resolution Imaging Spectrometer (EOS Sensor)
HIROS  . . . . . . . .  High Resolution Observing Satellite, A.47
HMS  . . . . . . . . . .  Heterodyne Millimeter-wave Spectrometer (MOSES Sensor)
HOLOP . . . . . . . .  Holographic Optics Laboratory (Shuttle D2 mission)
hPa . . . . . . . . . . . .  hecto Pascal (international standard of pressure, 1 hPa = 1 mbar)
HPCG  . . . . . . . .  Hand-held Protein Crystal Growth (Shuttle Payload)
HPTE  . . . . . . . .  High Precision Tracking Experiment (Shuttle payload)
HRIR . . . . . . . . . .  High-Resolution Infrared Radiometer (NOAA Sensor)
HRPT  . . . . . . . .  High Resolution Picture Transmission (NOAA Broadcast Technique in S-Band at frequencies of 1698.0 and 1707.0 MHz; data from all AVHRR channels (plus TOVS and SEM) is provided at full 1.1 km resolution)

HRSGS-A ...... High Resolution Shuttle Glow Spectroscopy (Shuttle payload)
HRTS ......... High Resolution Telescope and Spectrograph (Shuttle, Spacelab-2 sensor, a 30 cm, f/15 Gregorian telescope, spectrograph in UV range 1170-1700 Å, and a spectroheliograph observing at 1550 Å)
HRV .......... Haute Resolution Visible (High-Resolution Visible Sensor of Spot)
HST .......... Hubble Space Telescope (Shuttle launch)
HSRP ......... High-Speed Research Program (NASA)
HUT .......... Helsinki University of Technology (Helsinki, Finland)
HUT .......... Hopkins Ultraviolet Telescope (part of ASTRO-1 observatory)
HV ........... Horizontal transmit - Vertical receive Polarization
HYDROMET .. Committee for Hydrometeorology (USSR/CIS agency in the field of Meteorology)
Hz ........... Hertz
IAF .......... International Astronautical Federation (Paris)
IAG .......... International Association of Geodesy
IAHS ......... International Association of Hydrological Sciences
IAMAP ....... International Association of Meteorology and Atmospheric Physics
IAS .......... Institut d'Astrophysique Spatiale (Verrières-le-Buisson, France, Lab is part of CNRS)
IASC ......... International Arctic Science Committee (Arctic Centre, University of Lapland, Finland)
IBSE ......... Initial Blood Storage Experiment (Shuttle payload)
ICAO ......... International Civil Aviation Organization
ICAT ......... Incubator-Cell Attachment Test (Shuttle)
ICBC ......... IMAX Cargo Bay Camera (Shuttle)
ICC .......... Instrument Control Center (EOSDIS Facility)
ICE .......... International Cometary Explorer (renamed ISEE-3 mission), A.52.2
ICET ......... International Center for Earth Tides
ICSU ......... International Council of Scientific Unions
ICWG-EO ..... International Coordination Working Group for Earth Observation
IDHT ......... Instrument Data Handling and Transmission (ERS-1 S-Band Antenna)
IDN .......... International Directory Network (CEOS-defined for databases, former designation PID →)
IECM ......... Induced Environment Contamination Monitor (Shuttle)
IEE .......... Institution of Electrical Engineers
IEEE ......... Institute of Electrical and Electronics Engineers (USA)
IEOS ......... International Earth Observing System (Committee dealing with the policies and principles of data exchange, etc. ; partner agencies are: ESA, EUMETSAT, NASA, NOAA, STA (Japan), NASDA, MITI, JMA, and CSA (Canada))
IERS ......... International Earth Rotation Service (Geodetic research)
IERS ......... International Earth Reference System
IFE .......... Isoelectric Focusing Experiment (Shuttle)
IFEOS ........ International Forum on Earth Observations Using Space Station Elements (since 1986)
IFOV ......... Instantaneous Field of View
IGAC ......... International Global Atmosphere Chemistry (IGBP core program)
IGBP ......... International Geosphere/Biosphere Programme of ICSU
IGARSS ....... International Geoscience and Remote Sensing Society
IGFOV ........ Instantaneous Geometric Field of View
IGN .......... Institut Geographique National (French National Geographic Institute)
IGRF ......... International Geomagnetic Reference Field
IGS .......... International GPS Service for Geodynamics
IGU .......... International Geographical Union

IGY ........... International Geophysical Year
IHO ........... International Hydrographic Organization
IKF ........... Institut für Kosmosforschung, Berlin. (Note: as of Jan. 1992 the IKF
was renamed "Institute of Space Sensors", it is part of DLR)
IKI ........... Space Research Institute (of the Russian Academy of Sciences, Mo-
scow; extraterrestrial physics and remote sensing)
IKI-BAN ....... Space Research Institute, Bulgarian Academy of Sciences, Sofia
ILAS .......... Improved Limb Atmospheric Spectrometer (Sensor on ADEOS)
IMAX ......... Image Maximum (a large screen motion picture camera/format used by
the NASA/Smithsonian project to document significant space activi-
ties)
IMF ........... Interplanetary Magnetic Field
IMG .......... Interferometric Monitor for Greenhouse Gases (Sensor on ADEOS)
IML .......... International Microgravity Laboratory (Shuttle payload)
IMO .......... International Maritime Organization
IMP .......... International Monitoring Platform, A.48
IMS .......... Information Management System at GSFC (The top-level function of
EOS DAACs)
IMU .......... Inertial Measurement Unit
INCA ......... Indian National Cartographic Association
INMARSAT .... International Maritime Satellite Organization (London, UK)
INPE ......... Instituto de Pesquisas Espaciais (Civil Aerospace Agency of Brazil, Saò
José dos Campos, S.P.)
INR .......... Image Navigation and Registration (GOES Second Generation S/C)
INRA ......... Institut National de la Recherche Agronomique (France)
INS ........... Inertial Navigation System (for aircraft navigation)
InSb .......... Indium Antimonide (detector type material)
INSAT ........ Indian National Satellite (series, employed for meteorology and com-
munication), A.49
Intercosmos .... USSR/CIS space program for collaborative science projects among its
nine members and with other nations. Intercosmos was created in 1967
inviting the former Soviet-affiliated countries (like, East-Germany,
Hungary, Bulgaria, Poland, etc.) to participate in the Soviet space program with their own national contributions (one area of participation
was in remote sensing, building sensors for specific missions, dissemination and scientific interpretation of data, etc. ). Activities in international manned space flight missions were also under the label of
Intercosmos. There are also satellites in the Intercosmos program
named 'Intercosmos-n', etc.
INTA ......... Instituto National de Tecnica Aeroespacial (space agency of Spain)
INTELSAT ..... International Telecommunications Satellite Organization (Washing-
ton, D. C.)
INTERBALL ... IKI mission program (solar-terrestrial interaction) within ISTP, A.50
IOC ........... Intergovernmental Oceanographic Commission (of UNESCO)
IOCM ........ Interim Operational Contamination Monitor (Shuttle payload)
IPCC ......... Inter-Governmental Panel for Climate Change (set up by WMO)
IPEI .......... Ionospheric Plasma and Electrodynamics Instrument (EOS Sensor)
IPG .......... Institute of Applied Geophysics, Moscow
IPMP ......... Investigations into Polymer Membrane Processing (Shuttle experi-
ment)
IPOMS ........ International Polar-Orbiting Meteorological Satellite (Group)
IPS ........... Instrument Pointing System (Spacelab-2, built by ESA, structure for
mounting telescopes)
IRCFE ........ Infrared Communications Flight Experiment (Shuttle)
IR&D ......... Independent Research & Development (company internal funding)

IRE .......... Institute of Radioengineering and Electronics (of the Russian Academy of Sciences in Moscow; IRE is involved in remote sensing, etc., also providing general management services)
IRF .......... Swedish Institute of Space Physics, Kiruna, Sweden
IR-IE ......... Infrared Imaging Experiment (Shuttle payload)
IRIS .......... International Radio Interferometric Surveying (Subcommittee of the International Association of Geodesy)
IRLS .......... Interrogation, Recording and Location Subsystem (NOAA)
IRM .......... Ion Release Module (S/C of the AMPTE mission, A.8.1)
IRMB ........ Institut Royal de Météorologie (Royal Meteorological Institute of Belgium), Brussels
IRF-U ........ Swedish Institute of Space Physics, Uppsala
IRS .......... Information Retrieval System (ESA data system)
IRS .......... Indian Remote Sensing Satellite (ISRO), A.51 (IRS-1A, 1B, 1C, 1D, 1E)
IRS .......... Inertial Reference System
IRSA ......... Institute for Remote Sensing Applications (of JRC, Ispra Italy)
IRSA ......... Institute for Remote Sensing Applications (Beijing, Chinese Academy of Sciences)
IRT .......... Infrared Telescope (Spacelab-2 instrument, a 15 cm f/4 Herschelian telescope)
ISA .......... Institute of Space Aeronomy (Brussels)
ISA .......... Israel Space Agency
ISAC ......... ISRO Satellite Center, Bangalore, India
ISAC ......... Intelsat Solar Array Coupon (Shuttle experiment)
ISAIAH ....... Israeli Space Agency Investigation about Hornets (Shuttle experiment)
ISAL ......... Investigation of STS Atmospheric Luminosities (Shuttle)
ISAMS ........ Improved Stratospheric and Mesospheric Sounder (UARS Sensor)
ISAS ......... Institute for Space and Astronomical Science (University of Tokyo, Japan)
ISCCP ........ International Satellite Cloud Climatology Project (by ICSU and WMO)
ISDE ......... Institute of Space Device Engineering, Moscow
ISDN ......... Integrated Services Digital Network
ISEE ......... International Sun Earth Explorer (3 S/C mission), A.52
ISIS .......... Intelligentes Satellitenbild Informationssystem (a DFD archival system and service in the operational test phase)
ISLR ......... Integrated Side Lobe Ratio
ISLSCP ....... International Satellite Land-Surface Climatology Programme
ISO .......... International Standards Organization (one of three bodies responsible for the definition of OSI)
IEF .......... Isoelectric Focusing (Shuttle payload)
ISOPS ........ International Space Conference of Pacific-Basin Societies
ISRO ......... Indian Space Research Organization
ISPRS ........ International Society of Photogrammetry and Remote Sensing
IST .......... Instrument Support Terminal (EOSDIS Facility)
ISTP ......... International Solar-Terrestrial Physics Program [involves a total of 12 satellites provided by ESA (SOHO, CLUSTER), NASA (GGS), IKI (CIS/Russia, Interball, ECOS-A), ISAS (Geotail)]
ISTRAC ....... ISRO Telemetry & Command Network, Bangalore, India
ISY .......... International Space Year (1992)
ITIR[633] ....... Intermediate Thermal Infrared Radiation (EOS Sensor); ITIR was renamed in 1990 in ASTER = Advanced Spaceborne Thermal Emission and Reflection Radiometer
ITOS ......... Improved TIROS Operational System (NOAA)

ITRF .......... International Terrestrial Reference Frame
ITT ........... International Telephone and Telegraph Co.
ITU .......... International Telecommunication Union (body for frequency alloca-
              tions, Geneva)
IUGG ........ International Union of Geodesy and Geophysics
IVHS ......... Intelligent Vehicle/Highway Systems
IZMIRAN ...... Institute of Terrestrial Magnetism, Ionosphere and Propagation of Ra-
              dio Waves, Troitsk, Moscow region
JAROS ........ Japan Resources Observation System Organization
JEA .......... Japan Environmental Agency
JEM .......... Japanese Experiment Module (Japan's pressurized module directly at-
              tached to the Space Station Freedom, JEM-1 (1998) and JEM-2, STS
              launch)
JEOS ......... Japanese Earth Observation System
JERS ......... Japanese Earth Resources Satellite, A.53
JGOFS ........ Joint Global Ocean Flux Study (Program)
JGPSC ........ Japan GPS Council (over 80 manufacturers, major users, research insti-
              tutes, etc.)
JHU .......... Johns Hopkins University (Baltimore, Md)
JMA .......... Japan Meteorological Agency
JPL .......... Jet Propulsion Laboratory, Pasadena, Ca. (DAAC of NASA EOS Pro-
              gram). JPL is the only NASA center that is managed by a university,
              namely the California Institute of Technology.
JPOP ......... Japanese Polar Platform ( launch planned for 1998), A.4
JRC .......... Joint Research Centre (umbrella agency of CEC coordinating 9 re-
              search institutes at four sites (Geel, Belgium; Karlsruhe, Germany; Pet-
              ten, Netherlands; and ISPRA, Italy; IRSA at Ispra is an Institute for Re-
              mote Sensing Applications)
JSC .......... Johnson Space Center (Houston)
JWGA ........ Joint Working Group ATMOS
K ............ degree Kelvin
keV .......... Kilo Electron Volt
KNMI ........ Royal Netherlands Meteorological Institute
KSC .......... Kennedy Space Center (NASA facility at Cape Canaveral, Florida)
L3 ........... Latitude/Longitude Locator (Shuttle experiment)
LAGEOS-I,II ... Laser Geodynamics Satellite (NASA/ASI), A.54
LAN .......... Local Area Network
LANDSAT ..... Land (Remote Sensing) Satellite, US EO program, A.55
LANL ........ Los Alamos National Laboratory (Los Alamos NM, DOE facility, op-
              erated by the University of California)
LAPAN ....... Lembaga Penerbangan dan Antariksa Nasional (Indonesian National
              Institute of Aeronautics and Space, Jakarta)
LAP-B ........ Link Access Protocol (for B Channels)
LaRC ......... Langley Research Center, Hampton Va. (DAAC of NASA EOS Pro-
              gram)
LASA ......... Lidar Atmospheric Sounder and Altimeter (EOS sensor)
LASER ....... Light Amplification by Stimulated Emission of Radiation
LASSO ....... Laser Synchronization from (Geo)Stationary Orbit (ESA, Meteosat)
LAWS ........ Laser Atmospheric Wind Sounder (EOS Sensor)
LDCE ........ Limited Duration Space Environment Candidate Materials Exposure
              (Shuttle experiment)
LDEF ........ Long Duration Exposure Facility, NASA S/C, A.56
LDR .......... Linear Depolarization Ratio
LeRC ......... NASA Lewis Research Center, Cleveland, Ohio

---

633) "The Earth Observer", Vol. 2 No. 10, Dec. 1990, pp. 3

LEDA . . . . . . . . .  Landsat On-Line Earthnet Data Availability (ESA Database File)
LEO . . . . . . . . . .  Low Earth Orbit (usually for all satellite orbits up to 1000 or 1200 km altitude; this is to be seen in contrast to geostationary orbits at altitudes of 36000 km)
LERTS . . . . . . . .  Laboratoire d'Etudes et de Recherches en Télédétection Spatiale, Toulouse, France (belongs to CNES/CNRS)
LEWEX . . . . . . .  Labrador Extreme Wave Experiment (airborne campaign in 1987)
LF . . . . . . . . . . .  Low Frequency (30 - 300 kHz band)
LFC . . . . . . . . . .  Large Format Camera, A.57
LFTI . . . . . . . . .  Physical Technical Institute (St. Petersburg)
LIDAR . . . . . . . .  Light Detection and Ranging
LIMS . . . . . . . . .  Limb Infrared Monitor of the Stratosphere (NIMBUS-7 Sensor)
LIS . . . . . . . . . . .  Lighting Imaging Sensor (EOS Sensor)
LISS . . . . . . . . . .  Linear Imaging Self-Scanning Sensor (ISRO sensor series)
LITE . . . . . . . . .  Lidar In-space Technology Experiment, shuttle mission, A.58
LLNL . . . . . . . . .  Lawrence Livermore National Laboratory (Livermore, Ca., a DOE Lab at the University of California)
LMD . . . . . . . . .  Laboratoire de Météorologie Dynamique, Palaiseau (Lab of CNRS)
LOA . . . . . . . . . .  Laboratoire D'Optique Atmosphérique, at the University of Sciences and Technology, Lille, France
LPCE . . . . . . . . .  Laboratoire de Physique de l'Environnement, Orleans, France
LRPT . . . . . . . . .  Low Resolution Picture Transmission (NOAA Downlink Technique in S-Band)
LRR . . . . . . . . . .  Laser Retro Reflector (ERS-1 Sensor)
LS . . . . . . . . . . . .  Landsat Satellite Series of NOAA
LUT . . . . . . . . . .  Local User Terminal (NOAA concept for S&R reception)
MAESA . . . . . . . .  Measurement for Assessing the Effects of Stratospheric Aircraft, 1994
MAESTRO . . . . .  Multiple Airborne Experiments Towards Radar Observations
Magnolia/MFE . .  (MFE = Magnetic Field Experiment) A joint French/US program (proposal status) for long-term (>5 years) monitoring of the Earth's magnetic field and its temporal variations (objectives: main field model, secular variations, core motion determination, electrical conductivity of the mantle)
MAPS . . . . . . . . .  Measurement of Air Pollution from Space Radiometer (shuttle OSTA-1 experiment of STS-2 in Nov. 1981), A.60
MAS . . . . . . . . . . .  Millimeterwave Absorption Sounder (ATLAS Sensor)
MASER . . . . . . . .  Microwave Amplification by Stimulated Emission of Radiation
MAUS . . . . . . . . .  Material Science Autonomous Payload (Shuttle D2 mission)
mb (mbar) . . . . . .  millibar
MBB . . . . . . . . . .  Messerschmitt Bölkow & Blohm (Munich)
MCC . . . . . . . . . .  Mission Control Center
MCHIP/s . . . . . . .  CHIP stands for Yes/No sequences in data transmissions.. One MCHIP/s = 1 Million information sequences/s
MCP . . . . . . . . . .  Meteorological Communications Package. MCP permits a direct data access to the operational meteorological instruments in full resolution during a pass. MCP allows in addition the transmission of global data sets for central ground stations.
MCP . . . . . . . . . .  Micro Channel Plate (detector)
MCT . . . . . . . . . .  Mercury Cadmium Telluride (detector type material)
MDL . . . . . . . . . .  Multi-use Data Link (GOES Second Generation S/C)
MDT . . . . . . . . . .  Mean Down Time
MEDEA . . . . . . .  Material Science Experiment Double Rack for Experiment Modules and Apparatus (Shuttle experiment)
MEI . . . . . . . . . . .  Moscow Power Engineering Institute
MELEO . . . . . . .  Materials Exposure in Low Earth Orbit (Shuttle experiment)

MERIT ........ Measure Earth Rotation and Intercompare the Techniques (an International Earth Rotation Service Program)
MERIS ........ Medium Resolution Imaging Spectrometer (ESA Sensor)
MESSR ....... Multispectral Electronic Self-Scanning Radiometer (MOS Sensor)
METEOR ...... Russian meteorological satellite family, A.61 - A.65
METEOSAT .... European meteorological satellite series of EUMETSAT, A.66
METOP ....... EUMETSAT Meteorological Operational satellite series, A.78.2
MeV .......... Mega Electron Volt
MF ........... Medium Frequency (300 - 3000 kHz band)
MFLOPS ...... Million Floating Point Operations per Second (a measure of computer power)
MHS ......... Message Handling System (MOTIS is the ISO definition of MHS)
MHz .......... Megahertz ($10^6$ Hertz)
Microlab ...... OSC satellite series, A.67
MIFI ......... Moscow Engineering Physical Institute
MIMR ........ Multi-frequency Imaging Microwave Radiometer (ESA sensor)
MIPAS ........ Michelson Interferometer for Passive Atmospheric Sounding (ATMOS Sensor)
MIR .......... Russian Space Station, A.68
MIRAS ....... MIR Infrared Spectrometer (note: this is a modified GRILLE sensor by ISA on the shuttle ATLAS-1 mission)
MIRP ......... Manipulated Information Rate Processor (NOAA S/C subsystem)
MIS-1 ........ Microcapsules in Space-1 (Shuttle experiment)
MISR ......... Multi-Angle Imaging Spectro-Radiometer (EOS Sensor)
MISU ......... Meteorological Institute of Stockholm University, Sweden
MIT .......... Massachusetts Institute of Technology, Cambridge, Ma.
MITI ......... Ministry of International Trade and Industry (Japan)
mJ ........... milli Joule ($10^{-3}$ J)
MLE ......... Mesoscale Lightning Experiment (Shuttle Payload)
MLR ......... Monodisperse Latex Reactor (Shuttle experiemnt)
MLS .......... Microwave Limb Sounder (UARS and EOS Sensor)
MLTI ......... Mesosphere and Lower-Thermosphere/Ionosphere
MODE ........ Middeck O-Gravity Dynamics Experiment (Shuttlc payload)
MO Disk ...... Magneto-Optical Disk
MODIS ....... Moderate-Resolution Imaging Spectrometer (EOS Sensor)
MOMS ....... Modular Optoelectronic Multispectral Scanner ( Shuttle payload of 1983 and 84), A.69 and A.70
MOP ......... Meteosat Operational Programme (European series of weather satellites from  EUMETSAT)
MOPITT ...... Measurement of Pollution in the Troposphere (EOS Sensor)
MOS .......... Marine Observation Satellite (NASDA Satellite, MOS-1 Launch: 1987, MOS-1b launch: Feb. 1989), A.71
MOS .......... Modular Optoelectronic Scanner (IKF Sensor on PRIRODA)
MOSES ....... Molecules in Outer Space and Earth Stratosphere (Swedish Mission, renamed to ODIN), A.75
MPE .......... Max Planck Institut für Extraterrestrik, Garching (Germany)
MPI .......... Max Planck Institut, Germany
MPSE ........ Morelos Payload Specialist Experiments (Shuttle)
MRF ......... Meteorological Research Flight (UK)
MSFC ........ Marshall Space Flight Center, Huntsville, Ala.  (DAAC of NASA EOS Program)
MSK .......... Minimum Shift Keying
MSL-2 ........ Material Science Laboratory-2 (Shuttle payload)
MSR ......... Microwave Scanning Radiometer (MOS Sensor)
MSS .......... Multi Spectral Scanner (Landsat Sensor)

MSU-E ........ Multispectral Scanner - Electronic Scanning
MSU-K ........ Multispectral Scanner - Circular Scanning
MSU-M ........ Multispectral Scanner - Low Resolution
MSU-S ........ Multispectral Scanner-Moderate Resolution
MSU-SK ....... Multispectral Scanner-Moderate Resolution,Conical Scanning
MTBF ......... Mean Time Between Failure
MTF .......... Modulation Transfer Function
MUVIS ....... MOSES UVIsible Spectrometer (MOSES Sensor)
MW ........... Microwave
MWIR ........ Mid-Wavelength Infrared ($\sim 3 - 5\ \mu m$)
MWR ......... Microwave Radiometer
N/A .......... Not Applicable
NADC ......... Naval Air Development Center, Warminster Penna., USA
NASA ........ National Aeronautics and Space Administration (USA)
NASDA ....... National Space Development Agency (Japan)
NAVSTAR-GPS . Navigation System with Time and Ranging - Global Positioning System
              (Precision real-time position determination system of the US Airforce)
NCAR ......... National Center for Atmospheric Research (Boulder Col., NCAR is
              operated by the University Corporation for Atmospheric Research
              (UCAR) under the sponsorship of the National Science Foundation
              (NSF))
NCC ........... National Climatic Center (USA)
NCDC ......... National Climatic Data Center (of NOAA/NESDIS, Asheville, NC)
NCDS ......... NASA Climate Data Center (at GSFC, Science data archive for atmo-
              spheric chemistry and climate (ERBE, etc.))
NDOC ......... National Oceanographic Data Center (USA)
NDVI ......... Normalized Difference Vegetation Index
NEIS ......... National Earthquake Information Service (USGS, Denver Co.)
NEDRES       National Environmental Data Referential Service (NOAA service)
NEΔR ......... Noise Equivalent Delta (or Differential) Radiance (system sensitivity)
NEΔT ......... Noise Equivalent Delta (or Differential) Temperature (system sensitiv-
              ity)
NER .......... Noise Equivalent Radiance
NESDIS ....... National Environmental Satellite Data and Information Service (a
              NOAA database service program, Suitland Md.)
NEWS ......... NOAA Earth Watch Service (information system)
NGDC ......... National Geophysical Data Center (NOAA facility at Boulder Co.)
NIMBUS ....... NASA EO missions series, A.72
NIR .......... Near Infrared (spectrum, from 0.75 to about 1.3 $\mu m$)
NIST ......... National Institute of Standards and Technology (US, formerly National
              Bureau of Standards)
NIVR ......... Netherlands Agency for Aerospace Programs (Delft)
NLR .......... Nationaal Lucht- en Ruimtevaartlaboratorium (National Aerospace
              Laboratory, Amsterdam, The Netherlands)
nm ........... Nanometer ($10^{-9}$ m)
nm ........... Nautical Mile (international, 1,852 m)
nT ........... Nano Tesla ($10^{-9}$ T)
NOAA ......... National Oceanic and Atmospheric Administration (NOAA is an insti-
              tution of the  US Commerce Department)
NOAADIR ..... NOAA Earth System Data Directory (since 1988, still in development),
              also known under NESDD
NODC ......... National Oceanographic Data Center (NOAA/NESDIS, Washington
              DC)
NODS ......... NASA Ocean Data System (located at JPL; Measurements in the ar-
              chive are related to altimetry, scatterometry, and microwave radiome-

|  |  |
|---|---|
| | try. NODS will archive and distribute data products for TOPEX/PO-SEIDON) |
| NOSL | Night/Day Optical Survey of Lightning (Shuttle experiment) |
| NPO | Naulshno Proizwodstwennoje Objedijenie (Scientific/Research Production Association) |
| NPO Energia | Russian space industry consortium (Kaliningrad, Moscow region); responsibility of all Russian manned space projects; builders of S/C (i.e. MIR), payloads, sensors, etc. |
| NPO Machinostroyenia | Russian company, Reutov Moscow Region, builder/integrator of S/C (ALMAZ series), builder of large launch vehicles (Proton, etc.) |
| NPO Planeta | Russian space research institute (Moscow), operators of satellites (Meteor, Okean, Resurs, GOMS series) along with corresponding ground segments, providers of services to the user community in the areas of meteorology/climate, oceanography, Earth resources, and ecological monitoring. From an organizational point of view, NPO Planeta is an agency positioned under ROSGIDROMET, the "Committee for Hydrometeorology and Environmental Monitoring" |
| NPO PM | Research and Production Association of Applied Mechanics, Krasnoyarsk (a closed city until 1991), since 1977, builder/integrator of communication satellites (Gorizont, Express, Gonet, Arkos, Mayak, etc.), navigation satellites, and geodetical satellites) |
| NPO Vega | Russian space/defense industry consortium, Moscow, designers and builders of SAR instruments, etc., operators of airborne instruments |
| NPOP | NASA Polar Platform |
| NRC | National Research Council, Ottawa, Canada |
| NRL | Naval Research Laboratory, Washington D. C. |
| NROSS | Navy Remote Ocean Sensing System (US satellite) |
| NRSA | National Remote Sensing Agency (of India, Balangar Hyderabad) |
| NRSC | National Remote Sensing Centre (UK, this agency was privatized in 1989, commercial sale of remote sensing data) |
| NRSC | National Remote Sensing Center, Beijing |
| NRZ | Non Return to Zero (communication signal parameter) |
| NSC | Norwegian Space Centre |
| NSCAT | NASA Scatterometer (NASA Sensor) |
| NSF | National Science Foundation (USA) |
| NSI | NASA Science Internet - an international dual protocol (TCP/IP and DECnet) network (successor to SPAN) |
| NSIDC | National Snow and Ice Data Center, Boulder, Co. (NOAA Facility at University of Colorado, established in 1982). NSIDC is co-located with WDC-A (World Data Center A for Glaciology). NSIDC is also a DAAC site of the EOS Program. NSIDC has extensive holdings of cryospheric and polar ocean surface-flux data, routine production of sea ice maps from SSM/I sensor). |
| NSSDC | NASA Space Science Data Center (at GSFC) |
| NTSC | National Television System Committee (US TV display standard) |
| NWS | National Weather Service (USA) |
| OACT | Office of Advanced Concepts and Technology (NASA, formerly OAST) |
| OARE | Orbital Acceleration Research Experiment (Shuttle) |
| OASIS-1 | Orbiter Autonomous Supporting Instrumentation System (Shuttle payload) |
| OASIS | On-Line Data Access and Service Information System (Catalog system at NOAA-NCDC) |
| OAST | Office of Application and Space Technology (NASA, Shuttle payloads are also designated by this name - OAST-1, OAST-2, etc.) |
| OBS | Observatoire Paris-Mendon (France) |

OCE . . . . . . . . . . .  Ocean Color Experiment (part of OSTA-1 payload on shuttle)
OCEAN  . . . . . . .  Ocean Color Environment Archive Network (ESA Program)
OCTS . . . . . . . . .  Ocean Color Temperature Scanner (Sensor on ADEOS)
OCTW . . . . . . . .  Optical Communications Through Windows (Shuttle experiment)
ODA . . . . . . . . . .  Office Document Architecture
ODIN  . . . . . . . . .  Proposed Swedish astronomy and aeronomy mission (A.75, in Norse mythology Odin (also called Woden or Wotan) is one of the principal gods)
OEX  . . . . . . . . . .  Orbiter Experiments (Shuttle)
OGLOW . . . . . . .  Sun Orbiter Glow (Shuttle experiment)
OKEAN  . . . . . . .  Ukrainian/Russian Satellite Series, A.76
ONERA  . . . . . . .  Office National d'Etudes et Recherches Aérospatiales (French Aerospace Research Establishment, Chatillon)
ONR  . . . . . . . . . .  Office of Naval Research, Arlington Va.
ORFEUS  . . . . . .  Orbiting Retrievable Far and Extreme Ultraviolet Spectrograph (German Shuttle payload)
Ørsted  . . . . . . . .  Danish research satellite, A.77
OSDPD . . . . . . . .  Office of Satellite Data Processing and Distribution (of NOAA)
OSC . . . . . . . . . . .  Orbital Sciences Corporation (Fairfax Va., owner/operator of commercial  launch services for small payloads, Pegasus vehicle)
OSI . . . . . . . . . . . .  Open System Interconnect (a Standard for open communication)
OSS  . . . . . . . . . . .  NASA's Office of Space Science (Shuttle payload)
OSTA . . . . . . . . . .  Office of Space and Terrestrial Applications (a designation that was also given to the early shuttle payloads)
PAF  . . . . . . . . . . .  Processing and Archiving Facility (a specific DFD facility configuration for the ERS-1 mission)
PAL  . . . . . . . . . . .  Phase Alternation Line (German TV display standard)
PAN . . . . . . . . . . .  Panchromatic Camera (ISRO sensor)
PARE . . . . . . . . .  Physiological and Anatomical Rodent Experiment (Shuttle experiment)
PARLIQ  . . . . . . .  Phase Partitioning in Liquids (Shuttle experiment)
PBL . . . . . . . . . . .  Planetary Boundary Layer
PbS . . . . . . . . . . .  Lead Sulfide (detector type material)
PbSi . . . . . . . . . .  Lead Silicon (detector type material)
PCG . . . . . . . . . .  Protein Crystal Growth (Shuttle experiment)
PCM  . . . . . . . . . .  Pulse Code Modulation
PD . . . . . . . . . . . .  Photo Diode (detector)
PDP . . . . . . . . . . .  Plasma Diagnostics Package (Spacelab-2 sensor, studies of the interaction between the Earth's magnetic field and charged particles in the ionosphere)
PHCF . . . . . . . . . .  Pituitary-Growth Hormone Cell Function (Shuttle experiment)
PI . . . . . . . . . . . . .  Principal Investigator
PID . . . . . . . . . . .  Prototype International Directory (CEOS-defined Directory Interchange Format (DIF)); CEOS members operating an archive with PID capability are: CCRS, DLR-DFD, ESA-ESRIN, NASA, NASDA, NOAA,  RAE, etc. . Hence, standardized archival access is possible (see → IDN).
PILOT . . . . . . . . .  Portable Inflight Landing Operations Trainer (Shuttle experiment)
PIPOR . . . . . . . . .  Program for International Polar Ocean Research
PIXEL . . . . . . . . .  Picture Element
PM . . . . . . . . . . . .  Phase Modulation (Type of modulation of the main carrier)
PM . . . . . . . . . . . .  Polymer Morphology (Shuttle experiment)
PMOD/WRC . . .  Physikalisch-Meteorologisches Observatorium Davos, World Radiation Center, Switzerland
PMS . . . . . . . . . . .  Particle Measuring Systems Inc. (of Boulder Co.)

PMT . . . . . . . . . . . Photomultiplier Tube
PN  . . . . . . . . . . . . Pseudo Noise (code)
PNR . . . . . . . . . . . Pseudo Noise Number (a GPS series designation)
PRN . . . . . . . . . . . Pseudo Random Noise
POCC  . . . . . . . . . Payload Operations and Control Center
POEM-1 . . . . . . . Polar-Orbit Earth-Observation Mission (planned ESA Series)  A.78
POEMS . . . . . . . . Positron Electron Magnet Spectrometer (EOS Sensor)
POES . . . . . . . . . . Polar Orbiting Operational Environmental Satellites (NOAA series of
                           operational polar orbiting satellites), A.74
POGO  . . . . . . . . Polar-Orbiting Geophysical Observatory
POLAR . . . . . . . . NASA/GSFC Solar-Terrestrial Mission (A.79)
POLDER  . . . . . . Polarization and Directionality of Earth's Reflectances (CNES Sensor)
PoSAT . . . . . . . . Portuguese Satellite, A.80
PPE  . . . . . . . . . . Phase Positioning Experiments (Shuttle payload)
PPF  . . . . . . . . . . Polar Platform (ESA Columbus program, PPF is utilized for POEM
                           payloads)
ppb . . . . . . . . . . . Parts per Billion ($10^{-9}$)
ppbv . . . . . . . . . . Parts per Billion, volume
ppm  . . . . . . . . . . Parts per Million ($10^{-6}$)
ppmv . . . . . . . . . Parts per Million, volume
ppt . . . . . . . . . . . Parts per Trillion ($10^{-12}$)
PPS  . . . . . . . . . . Precise Positioning Service (GPS)
pps . . . . . . . . . . . Pulses per second
PRARE . . . . . . . . Precision Rate and Range Rate Equipment, A.81
PRF . . . . . . . . . . Pulse Repetition Frequency
PRIRODA . . . . . Research module of the Space Station MIR (A.82)
PRN . . . . . . . . . . Pseudo Random Noise
PSC . . . . . . . . . . Polar Stratospheric Clouds
PSE . . . . . . . . . . Physiological Systems Experiment (Shuttle)
PSK . . . . . . . . . . Phase Shift Keying (a modulation technique)
PSLR . . . . . . . . . Peak Side Lobe Ratio
PSLV . . . . . . . . . Polar Satellite Launch Vehicle (of ISRO)
PSN  . . . . . . . . . . Piano Spaziale Nationale (previous name of Space agency of Italy, now
                           → ASI)
PSVL . . . . . . . . . Polar Space Launch Vehicle (ISRO)
PtSi . . . . . . . . . . . Platinum Silicon (detector type material)
PTT  . . . . . . . . . . Platform Transmitter Terminal (data collection platform for ARGOS
                           system)
PTT  . . . . . . . . . . Public (Postal) Telephone and Telegraph (utility company)
PVTOS  . . . . . . . Physical Vapor Transport of Organic Solids (Shuttle experiment)
QPSK . . . . . . . . . Quadra-Phase Shift Keying
RADAR  . . . . . . . Radio Detection and Ranging
RADARSAT . . . . A Canadian (CSA/CCRS) EO mission with a SAR instrument (A.83)
RADFET  . . . . . . Radiation-sensitive Field Effect Transistor
RAE . . . . . . . . . . Royal Aerospace Establishment (Farnborough, UK)
RAL . . . . . . . . . . Rutherford Appleton Laboratory (Chilton, Oxon, UK)
RBDS  . . . . . . . . Radio Broadcast Data System
RBV . . . . . . . . . . Return Beam Vidicon Camera (Landsat Sensor)
RCVR . . . . . . . . Receiver
R&D . . . . . . . . . . Research & Development
Resurs . . . . . . . . Russian satellite series for resource monitoring, A.84, A.85
RF (R/F) . . . . . . . Radio Frequency (of active sensors, also of data transmission link, etc.)
RFI  . . . . . . . . . . Radio Frequency Interference
RGB  . . . . . . . . . Red, Blue, Green
RIS . . . . . . . . . . . Retroreflector in Space (Sensor on ADEOS)

RIT ........... Royal Institute of Technology, Sweden
RKA ........... Russian Space Agency, Moscow
RLSBO ........ Radiolokazionnaja Sistema Bokowo Obzora (Side view radar system)
RME .......... Radiation Monitoring Experiment (Shuttle)
rms ........... root mean square
ROSGIDROMET Committee for Hydrometeorology and Environmental Monitoring (Russian Government Agency, similar in functions and services to EU-METSAT and NOAA)
ROSIS ........ Reflective Optics Imaging Spectrometer (airborne sensor)
ROTEX ........ Robotic Experiment (Shuttle experiment)
rpm ........... revolutions per minute
RSRE ......... Royal Signals and Radar Establishment, Great Malvern, Worcestershire, UK
SAFIR ........ Satellite for Information Relay, A.86
SAFIRE ....... Spectroscopy of the Atmosphere Far Infrared Emission (EOS Sensor)
SAFISY ....... Space Agency Forum for the International Space Year in Europe (in 1992)
SAGE ......... Stratospheric Aerosol and Gas Experiment (NASA mission, A.5)
SAGE III ...... Stratospheric Aerosol and Gas Experiment III (EOS Sensor)
SAM .......... Shuttle Activation Monitor (Shuttle experiment)
SAMIR ........ Satellite Microwave Radiometer (ISRO sensor on Bhaskara S/C)
SAMPEX ...... Solar Anomalous and Magnetospheric Explorer (GSFC mission, A.88)
SAMS ......... Space Acceleration Measurement Systems (Shuttle experiment)
SAMS ......... Stratospheric and Mesospheric Sounder (NIMBUS-7 Sensor)
SAN MARCO .. Cooperative Italian/NASA mission (A.89)
SAO .......... Smithsonian Astrophysical Laboratory, USA
SAR .......... Synthetic Aperture Radar (a high-rate imaging technique)
SAREX-2 ...... Shuttle Amateur Radio Experiment (Shuttle payload)
S&R .......... Search and Rescue (Emergency System on NOAA S/C)
S&RSAT ....... Search and Rescue Satellite Aided Tracking System (Canada/France/NOAA). A.74.3
SAS&R ........ Satellite Aided Search & Rescue (INSAT-2 system)
SATO ......... Space Adaptation Tests and Observations (Shuttle experiment)
Sb ............ Antimonide (detector type material)
SBUV ......... Solar Backscatter Ultraviolet (NOAA Sensor on NIMBUS-7, TIROS, POES)
S/C ........... Spacecraft
SCARAB ....... Scanner for Radiation Budget (Sensor, French, USSR, German),
SCATT ........ (Wind) Scatterometer (ESA)
SCD1 ......... Satélite de Coleta de Dados (Data Collection Satellite of Brazil), A.90
SCIAMACHY .. Scanning Imaging Absorption Spectrometer for Atmospheric Cartography (ATMOS Sensor)
SCSI .......... Small Computer Systems Interface
Seasat ......... NASA/JPL EO mission (A.91)
SeaStar ....... A NASA/GSFC mission with the sensor SeaWiFS (A.92)
SeaWiFS ....... Sea Wide Field Sensor (this sensor is considered the CZCS successor)
SECAM ........ Sequential Color and Memory (French TV display standard)
SECDED ...... Single Error Correction - Double Error Detection
SEDIS ........ SeaWiFS European Data Information System (ESA/ESRIN)
SEE .......... Société des Electriciens et des Electroniciens
SEL .......... Space Environment Laboratory (NOAA, Boulder Co., real-time processing of all SEM package data, space environment forecasts)
SEM .......... Space Environment Monitor (NOAA Sensor package on GOES and POES series; Note: the GOES series SEM package arrangement differs considerably from the POES series SEM package)

SERC . . . . . . . . . .  Science and Engineering Research Council (UK, the Mullard Space
                         Science Laboratory of SERC)
SERON . . . . . . . .  South Eastern (US) Regional Oxidant Network (field program to study
                         atmospheric chemistry, July-August 1991)
SFDU . . . . . . . . .  Standard Format Data Unit (a CCSDS format concept)
SGGM . . . . . . . . .  Superconducting Gravity Gradiometer Mission, NASA (planned to fly
                         after the year 2000)
SGS 85 . . . . . . . .  Soviet Geodetic System 1985
SHARE . . . . . . . .  Space-Station Heat Pipe Advanced Radiator Experiment (Shuttle)
SHF . . . . . . . . . .  Super High Frequency (3 - 30 GHz band)
SHOOT . . . . . . .  Super Fluid Helium On Orbit Transfer (Shuttle experiment)
Si . . . . . . . . . . . .  Silicon (detector)
SIR . . . . . . . . . . .  Shuttle Imaging Radar (SIR-A with Payload A; SIR-B with Payload B,
                         etc.), see A.94 - A.96
SITP . . . . . . . . . .  Shanghai Institute of Technical Physics (of the Academy of Sciences of
                         China)
SKYLAB . . . . . . .  Sky Laboratory, NASA Space Station of the seventies (A.97)
SL . . . . . . . . . . . .  Spacelab
SLAR . . . . . . . . .  Side-Looking Airborne Radar (an active sensor with Real Aperture
                         Radar technology)
SLC . . . . . . . . . .  Satellite Launch Center (Complex)
SLR . . . . . . . . . .  Satellite Laser Ranging
SLS . . . . . . . . . . .  Space Life Sciences (Shuttle payload)
SME . . . . . . . . . .  Solar Mesosphere Explorer (NASA, A.98)
SMEX . . . . . . . .  Small Explorer Program (GSFC program since 1988 supporting disci-
                         plines in astrophysics, space physics and upper atmospheric science;
                         SMEX missions are SAMPEX, FAST, SWAS, TOMS, etc.)
SMM . . . . . . . . . .  Solar Maximum Mission (NASA,A.99)
SMMR . . . . . . . .  Scanning Multichannel Microwave Radiometer (NIMBUS-7 and Sea-
                         sat Sensor)
SMS . . . . . . . . . .  Synchronous Meteorological Satellite (designation of the first US
                         weather satellites (1974); this series was later renamed to GOES,
                         (NOAA) )
SNL . . . . . . . . . .  Sandia National Laboratories, Albuquerque, New Mexico, Livermore
                         Ca., USA; SNL is part of DOE and operated by AT&T)
SNR . . . . . . . . . .  Signal-to-Noise Ratio
SNSB . . . . . . . . .  Swedish National Space Board (RYDSTYRELSEN), Solna Sweden
SOCC . . . . . . . .  Satellite Operations and Control Center (NOAA)
SOHO . . . . . . . .  Solar and Heliospheric Observatory (see A.100)
SOLAR-A . . . . . .  ISAS Solar-Terrestrial Mission, A.101
SOLSTICE . . . . .  Solar Stellar Irradiance Comparison Experiment (NASA Sensor)
SOUP . . . . . . . .  Solar Optical Universal Polarimeter (Spacelab-2 sensor)
SPACEHAB . . . .  A concept for commercially sponsored and procured payloads and ser-
                         vices on Shuttle. SPACEHAB Inc., of Arlington Va. has a NASA con-
                         tract leasing Shuttle space on a commercial basis in the so-called 'Com-
                         mercial Middeck Augmentation Module' (CMAM), a pressurized re-
                         search lab owned by SPACEHAB® (an extension of the Shuttle orbiter
                         middeck in the Shuttle cargo bay). SPACEHAB in turn sells its services,
                         providing the needed support for commercial development of space
                         payloads as well as physical and operational integration, and all services
                         (training, etc.) for these payloads. Once in flight, SPACEHAB payloads
                         are crew-tended on request. The SPACEHAB contract was awarded in
                         Nov. 1990, the first SPACEHAB flight took place on STS-57 in June
                         1993. - SPACEHAB-1, -2 identifies also a series of Shuttle payloads.
SPACELAB . . . .  Space Laboratory on NASA Shuttle missions (A.102 - A.103)

SPADE .......... Stratospheric Photochemistry, Aerosols and Dynamics Experiment (NASA, airborne campaign)

SPAN .......... Space Physics Analysis Network (based on the DECnet protocol). [The US - SPAN (NASA) Service was discontinued at the end of 1990; the E-SPAN (ESA) Service will be continued]. SPAN permits user access to data archives. The successor of SPAN in the US is NSI (NASA Science Internet), a dual protocol (TCP/IP and DECnet) network.

SPARTAN ...... Shuttle Pointed Autonomous Research Tool for Astronomy (Shuttle)

SPAS .......... Shuttle Pallet Satellite (a Shuttle retrievable free-flyer platform for payloads, SPAS was built by MBB), SPAS-1 on STS-7 in 1983, ASTRO-SPAS is a direct successor of SPAS, ASTRO-SPAS-1 on STS-51 in Sept. 1993

SPIE .......... Society of Photo-Optical Instrumentation Engineering (international)

SPIE .......... Shuttle Plume Impingement Experiment

SPOT .......... Système Probatoire d'Observation de la Terre (French Earth Observing Satellite), (A.104)

SPOT Image .... Spot image data distributor (Toulouse and Reston Va.)

SPS .......... Standard Positioning Service (GPS)

sr .......... Steradian - a unit of measurement of a solid angle (a steradian is the solid angle that cuts out on a sphere a surface described around the vertex of the angle of area equal to the square of the radius of the sphere. A complete sphere forms a solid angle equal to $4\pi$.

SRI .......... Stanford Research Institute (original designation, now SRI International, the institute separated from the University for legal reasons)

SRL .......... Space Radar Laboratory (shuttle missions of SIR-C/X-SAR payloads)

SSBUV ........ Shuttle Solar Backscatter Ultraviolet (Shuttle Experiment)

SSC .......... Swedish Space Corporation of Solna, Sweden (a government-owned limited corporation under the Ministry of Industry)

SSC .......... Stennis Space Center, Mississippi (NASA)

SSCE .......... Solid Surface Combustion Experiment (Shuttle)

SSEOP ........ Space Shuttle Earth Observation Project

SSIP .......... Shuttle Student Involvement Program

SSM/I .......... Special Sensor Microwave/Image (US Department of Defense, US Air Force Sensor)

SSP .......... Sub-Satellite Point

SST .......... Sea Surface Temperature

SSU .......... Stratospheric Sounding Unit (UK sensor on NOAA S/C)

STA .......... Science and Technology Agency (of Japan)

STALO ........ Stable Local Oscillator

Starlette ........ CNES 'Solid Earth' mission, a passive satellite for geodetic studies with SLR observations (A.105)

STC .......... Sensitive Time Control (SAR antenna parameter)

STC .......... Star Tracker Camera

STDN ........ Standard Tracking and Data Network (NASA)

Stella .......... CNES experiment onboard Spot-3 for gravity field studies of the Earth (A.106)

STEP .......... Solar-Terrestrial Energy Program (International Program)

STEP .......... Stratosphere Troposphere Exchange Project (airborne campaign)

STEX .......... Sensor Technology Experiment (Shuttle)

STIB .......... Stratosphere Troposphere Interactions and the Biosphere (Program)

STIKSCAT ..... Stick Scatterometer (EOS sensor)

STL-1 .......... Space Tissue Loss-1 (Shuttle experiment)

STORM-FEST .. STORM - Fronts Experiment Systems Test (airborne campaign)

STP-1 .......... Space Test Payload-1 (Shuttle)

STS .......... Space (Shuttle) Transport System

SUSIM . . . . . . . . .  Stratosphere Ultraviolet Spectral Irradiance Monitor (ATLAS Sensor)
SUVE  . . . . . . . . .  Solar Ultraviolet Experiment (Shuttle experiment)
SVHS . . . . . . . . . .  Super Video Home System (a tape recorder system)
SVI . . . . . . . . . . .  Spectral Vegetation Index
SVN . . . . . . . . . .  Satellite Vehicle NAVSTAR (a GPS series numbering system)
SVS  . . . . . . . . . .  Space Vision System (Shuttle experiment)
SWADE . . . . . . . .  Surface Wave Dynamics Experiment (airborne in 1991)
SWAS . . . . . . . . .  Submillimeter Wave Astronomy Satellite (NASA/GSFC)
SWIR . . . . . . . . . .  Short Wave Infrared (spectrum, from about 1.3 $\mu$m to 3 $\mu$m)
SWIRLS  . . . . . . .  Stratospheric Wind Infrared Limb Sounder (EOS Sensor)
TB  . . . . . . . . . . .  Terabyte ($10^{12}$ Byte)
TBD . . . . . . . . . . .  To be defined (or to be determined)
TCP/IP . . . . . . . .  Transmission Control Protocol/Interchange Protocol
T&DR . . . . . . . . .  Tracking and Data Relay (NOAA)
TDL . . . . . . . . . . .  Tunable Diode Laser (spectrometer)
TDMA . . . . . . . .  Time Division Multiple Access
TDRS  . . . . . . . . .  Tracking and Data Relay Satellite (NASA)
TEMISAT . . . . . .  Telespazio Micro Satellite (see A.107)
TERS . . . . . . . . . .  Tropical Earth Resources Satellite [a joint program conceived by the
                        Netherlands (NIVR) and Indonesia (LAPAN) in 1985, the program got
                        stalled after phase A because of a lack of funds]
TES  . . . . . . . . . .  Tropospheric Emission Spectrometer (EOS Sensor)
Tesla . . . . . . . . . .  Standard international unit of magnetic flux density. 1 T = $10^4$ gauss.
TFOV  . . . . . . . . .  Total Field of View
THIR . . . . . . . . . .  Temperature Humidity Infrared Sounder (NIMBUS-7 Sensor)
THz . . . . . . . . . . .  Tera Hertz ($10^{12}$ Hz)
TIMED . . . . . . . .  Thermosphere, Ionosphere, Mesosphere Energetics and Dynamics
                        (NASA/GSFC mission), A.108
TIP . . . . . . . . . . .  TIROS (or Telemetry) Information Processor (onboard of POES S/C,
                        also a downlink data stream of NOAA S/C)
TIR  . . . . . . . . . .  Thermal Infrared (spectrum, from about 6 $\mu$m to about 15 - 20 $\mu$m)
TIROS . . . . . . . . .  Television and Infrared Observation Satellite (US Environmental/Me-
                        teorological Remote Sensing Program; TIROS 1-10 = 1. generation,
                        ESSA 1-9 = 2. generation, ITOS (TIROS-M) = 3. generation,)
TIROS-N  . . . . . .  TIROS-NOAA (fourth-generation TIROS satellite series)
TLD . . . . . . . . . . .  Thermoluminescent Dosimeter (Shuttle payload)
TLM . . . . . . . . . .  Telemetry
TM . . . . . . . . . . .  Thematic Mapper (Landsat Sensor)
TNO . . . . . . . . . .  Delft University of Technology, Delft, The Netherlands
TOF . . . . . . . . . . .  Time-of-Flight (measurement)
TOGA . . . . . . . . .  Tropical Oceans and Global Atmosphere Experiment (Program)
TOGA/COARE .  Tropical Oceans and Global Atmosphere Experiment / Coupled Ocean
                        Atmosphere Response Experiment
TOMS  . . . . . . . .  NASA missions (A.109)
Topex/Poseidon . .  Topography Experiment for Ocean Circulation (NASA/CNES EO Mis-
                        sion), A.110
TOS . . . . . . . . . . .  TIROS Operational System (NOAA)
TOVS . . . . . . . . .  TIROS Operational Vertical Sounder (NOAA, a three instrument sys-
                        tem consisting of : HIRS-2; SSU; and the MSU)
TPCE . . . . . . . . . .  Tank Pressure Control Equipment (Shuttle payload)
TRACER  . . . . . .  Tropospheric Radiometer for Atmospheric Chemistry and Environ-
                        ment Research (EOS Sensor)
TRAMAR . . . . . .  Tropical Rain Mapping Radar (EOS Sensor)
TREE . . . . . . . . .  Tropical Rain-Forest Ecology Experiment (NASA airborne cam-
                        paigns)

TREES ........ Tropical Ecosystem Environment Observation by Satellites (Joint CEC (Commission of the European Community) and ESA Program for long-term monitoring of global forest cover and the rate of deforestation. The program will use data from the following satellites: NOAA-11, NOAA-12, Landsat-5,6 , Spot, ERS-1,2, etc.)[634]

TRMM ........ Tropical Rainfall Measuring Mission (NASA-NASDA Mission), A.111

TRW (Inc.) ..... Thompson, Ramo and Wooldridge (Remote Sensing Center of TRW is at Redondo Beach Ca.)

TT&C ........ Telemetry, Tracking & Command (Data for S/C Operations)

TUD .......... Technical University of Denmark, Lyngby, Denmark

TWERLE ...... Tropical Wind Energy Conversion Reference Level Equipment (NOAA sensor on NIMBUS-6)

$\mu$ or $\mu$m ....... micron, $= 10^{-6}$ m

UARP ......... Upper Atmospheric Research Program (NASA)

UARS ........ Upper Atmosphere Research Satellite (NASA Satellite, launch: Sept. 1991) A.112

UAV .......... Unmanned Aerospace Vehicle (PERSEUS, CONDOR, etc.)

UCAR ......... University Corporation for Atmospheric Research (Boulder Co., UCAR is sponsored by NSF - there are over 60 member institutions in UCAR)

UCB .......... University of California, Berkeley

UCLA ......... University of California, Los Angeles

UHB .......... User Home Base

UHF .......... Ultra High Frequency (300   3000 MHz band)

UIT .......... Ultraviolet Imaging Telescope (part of ASTRO-1 payload on Shuttle)

UKS .......... United Kingdom Subsatellite (S/C of the AMPTE mission, A.8.2)

UNAVCO ...... University Navstar Consortium (a US Earth sciences community initiative to foster GPS applications in particular in the area of surveying)

UNCED ....... United Nations Conference on Environment & Development

UNEP ......... United Nations Environmental Program

UNEP/GRID ... UNEP Global Resource Information Database

UoSAT ........ University of Surrey Satellite (UK)

USAF ......... US Air Force

USCG ......... US Coast Guard

USDA ........ US Department of Agriculture

USEPA ....... US Environmental Protection Agency

USGCRP ...... US Global Change Research Program

USGS ......... United States Geological Survey (an agency of the Department of the Interior, DOI)

USMP ........ US Microgravity Payload (Shuttle payload)

USS .......... Unique Support Structure (Shuttle)

USSR ......... Union of Soviet Socialist Republics (former)

UTC .......... Universal Time Coordinated

UV ........... Ultra Violet (spectral range from 0.01 - 0.38 $\mu$m)

UVPI ......... Ultraviolet Plume Instrument (Shuttle experiment)

UWB ......... Ultra Wideband

VCR .......... Video Cassette Recorder

VCS .......... Voice Command System (Shuttle)

VDA ......... VHF Collection System Antenna (NOAA)

VFT-1 ........ Visual Function Tester-1 (Shuttle experiment)

VH ........... Vertical transmit - Horizontal receive Polarization

VHF .......... Very High Frequency (30 - 300 MHz band)

VHRR ........ Very High-Resolution Radiometer (Insat sensor)

---

634)  'European Initiative to Assess Global Forest Damage', Space News June 10-16, 1991, p. 24

VHS ........... Video Home System
VI ............. Vegetation Index
Viking ........ Swedish satellite mission for the study of the Earth's magnetosphere, A.113
VIR ........... Visible Infrared (spectrum)
VIS ........... Visible (spectrum 0.4 - 0.7 μm)
VLBI .......... Very Long Baseline Interferometry
VLF ........... Very Low Frequency (frequency band of 10 - 30 kHz)
VNIIEM ....... The All-Russian Research Institute of Electromechanics (S/C builder/ integrator, Meteor series, Okean series, Resurs series, GOMS, etc.)
VNIR .......... Visible Near Infrared (spectral range)
VOXEL ........ Volume Element
VRA .......... VHF Realtime Antenna (NOAA)
VRAM ........ Video RAM
VSAT ......... Very Small Aperture Terminal (small ground antenna for satellite communication)
VSWR ......... Voltage Standing Wave Ratio
VT ............ Virtual Terminal
VTIR ......... Visible and Thermal Infrared Radiometer (MOS Sensor)
VV ............ Vertical transmit - Vertical receive Polarization
WARC ........ World Administrative Radio Conference (of the ITU)
WCRP ........ World Climate Research Program (of WMO)
WDC .......... World Data Center
WEFAX ....... Weather Facsimile (NOAA broadcast service of GOES S/C; transmission of environmental data in WEFAX format to ground stations)
WFF ........... Wallops Flight Facility (of GSFC)
WGS 84 ....... World Geodetic System (DOD reference system for GPS)
WiFS ......... Wide Field Sensor (ISRO sensor)
WIND ......... NASA/GSFC Solar-Terrestrial Mission (A.114)
WL ........... Werkstofflabor (Shuttle D2 mission)
WMO ......... World Meteorological Organization (Geneva, Switzerland)
WMSCC ....... World Meteorological Service Computing Center
WOCE ......... World Ocean Circulation Experiment (Program)
WSF .......... Wake Shield Facility (Shuttle payload, small retrievable satellite)
WUPPE ........ Wisconsin Ultraviolet Photo Polarimeter Experiment (part of ASTRO-1 payload on Shuttle)
WV ........... Water Vapor (in the 5.7 - 7.1 μm water vapor absorption band)
WWW ........ World Weather Watch (Program)
XIE .......... X-Ray Imaging Radar (EOS Sensor)
XRT .......... X-Ray Telescope (Spacelab-2 sensor, energy detection 2.5-25 keV)
XTI .......... Cross-Track Interferometry
XTR .......... Transmitter
XUV ......... Extreme Ultra Violet (same as EUV, i.e. < 130 nm spectral range)
YAG .......... Yttrium Aluminum Garnet (a type of solid state crystal laser)
YLF .......... Yttrium Lithium Fluoride

# Appendix C    Index Table of Sensors

## Numbers

3D-Electron Analyzer, 33, 83

3D-Ion Analyzer, 33

## A

AA3500  = Airborne Bispectral Scanner (ABS), 348

AADS1220 = Dual Channel Terrain Surveillance Scanner, 346

AADS1221 = Maritime Surveillance Scanner, 347

AADS1230 = Dual Channel Quantitative Infrared Scanner, 347

AADS1250 = Eleven Channel Analog Multispectral System, 348

AADS1260 = Daedalus Multispectral Scanner (predecessor of ATM), 345

AADS1268 = ATM (Airborne Thematic Mapper), 346

AADS1278 = ATMX (ATM extended), 345

AADS1280 = Daedalus Multispectral Scanner, 346

AATSR = Advanced Along Track Scanning Radiometer, 91, 214

AC = Actinometric instrument (radiation budget sensor), 164

AC = Radiation Budget Sensor, 162

ACR = Active Cavity Radiometer, 41

ACRIM = Active Cavity Radiometer Irradiance Monitor, 69, 79, 81, 266, 297

ADALT = Advanced Radar Altimeter, 19

ADS = Analyzer of Dynamic Spectra, 138

AEPI = Atmospheric Emissions Photometric Imaging Experiment, 42

AES = Airborne Emission Spectrometer, 305

AFA-41/10 (Aerial Foto Apparatus), 453

AFA-41/20 = Aerial Foto Apparatus, 433

AIMR = Airborne Imaging Microwave Radiometer, 306

AIR-93 = Airborne Imaging 93 GHz Radiometer, 306

Airglow Instrument (SME), 264

AIRS = Atmospheric Infrared Sounder, 69, 79, 81, 200

AIRSAR = Airborne SAR, 307, 462

AIS = Airborne Imaging Spectrometer, 310, 461

AISA = Airborne Imaging Spectrometer for different Applications, 311, 461

AKR-2 = Analyzer of Kilometric Radiation, 138

ALADIN = Atmospheric Laser and Doppler Instrument, 220, 493

ALAE = Atmospheric Lyman-Alpha Emissions, 42

ALAS = Airborne Laser Altimeter System, 312

ALEX = Aerosol Lidar Experiment, 354

ALEXIS = Array of Low Energy X-Ray Imaging Sensors, 20

ALF = Airborne Laser Fluorosensor, 314

ALIAS = Aircraft Laser Infrared Absorption Spectrometer, 315

ALIAS-I on ER-2 Aircraft, 316

ALIAS-II on Perseus Aircraft, 316, 425

ALISSA= l'Atmosphere par LIdar Sur SAliout, 178, 236, 493

ALPHA-3 = Ion Trap Experiment, 137

ALPHA-3 = Trapped Ion Experiment, 136

ALPS = Airborne Laser Polarization Sensor, 317

ALT = Altimeter (active microwave sensor), 69, 81

Altimeter, further considerations, 89, 113, 214, 288

AMAS = Advanced Millimeter-Wave Atmospheric Sounder, 45, 487, 512

AMEI-2 = Energy Mass Analyzer, 137

AMI = Active Microwave Instrument, 86 further considerations, 91

AMI-SCAT = Active MW Instrument - Scatterometer (ERS-1), 86

AMIE = Airglow Measurement of Infrared Emitters, 42

# K

Printing: Mercedesdruck, Berlin
Binding: Buchbinderei Lüderitz & Bauer, Berlin